# Statistical Inference

Second Edition

## George Casella
*University of Florida*

## Roger L. Berger
*North Carolina State University*

**DUXBURY**

**THOMSON LEARNING**

Australia • Canada • Mexico • Singapore • Spain • United Kingdom • United States

**DUXBURY**

**THOMSON LEARNING**

Sponsoring Editor: *Carolyn Crockett*
Marketing Representative: *Tom Ziolkowski*
Editorial Assistant: *Jennifer Jenkins*
Production Editor: *Tom Novack*
Assistant Editor: *Ann Day*
Manuscript Editor: *Carol Reitz*
Permissions Editor: *Sue Ewing*

Cover Design: *Jennifer Mackres*
Interior Illustration: *Lori Heckelman*
Print Buyer: *Vena Dyer*
Typesetting: *Integre Technical Publishing Co.*
Cover Printing: *Phoenix Color Corp*

**Printing and Binding: RR Donnelley**

*For more information about this or any other Duxbury products, contact:*
DUXBURY
511 Forest Lodge Road
Pacific Grove, CA 93950 USA
www.duxbury.com
1-800-423-0563 (Thomson Learning Academic Resource Center)

Printed in United States of America

10 9

**Library of Congress Cataloging-in-Publication Data**

Casella, George.
    Statistical inference / George Casella, Roger L. Berger.—2nd ed.
        p. cm.
    Includes bibliographical references and indexes.
    ISBN-10: 0-534-24312-6
    ISBN-13: 978-0-534-24312-8
    1. Mathematical statistics. 2. Probabilities.   I. Berger, Roger L.
II. Title.
QA276.C37 2001
519.5—dc21

                                                    2001025794

*To Anne and Vicki*

*Duxbury titles of related interest*

# Preface to the Second Edition

Although Sir Arthur Conan Doyle is responsible for most of the quotes in this book, perhaps the best description of the life of this book can be attributed to the Grateful Dead sentiment, "What a long, strange trip it's been."

Plans for the second edition started about six years ago, and for a long time we struggled with questions about what to add and what to delete. Thankfully, as time passed, the answers became clearer as the flow of the discipline of statistics became clearer. We see the trend moving away from elegant proofs of special cases to algorithmic solutions of more complex and practical cases. This does not undermine the importance of mathematics and rigor; indeed, we have found that these have become more important. But the manner in which they are applied is changing.

For those familiar with the first edition, we can summarize the changes succinctly as follows. Discussion of asymptotic methods has been greatly expanded into its own chapter. There is more emphasis on computing and simulation (see Section 5.5 and the computer algebra Appendix); coverage of the more applicable techniques has been expanded or added (for example, bootstrapping, the EM algorithm, p-values, logistic and robust regression); and there are many new Miscellanea and Exercises. We have de-emphasized the more specialized theoretical topics, such as equivariance and decision theory, and have restructured some material in Chapters 3–11 for clarity.

There are two things that we want to note. First, with respect to computer algebra programs, although we believe that they are becoming increasingly valuable tools, we did not want to force them on the instructor who does not share that belief. Thus, the treatment is "unobtrusive" in that it appears only in an appendix, with some hints throughout the book where it may be useful. Second, we have changed the numbering system to one that facilitates finding things. Now theorems, lemmas, examples, and definitions are numbered together; for example, Definition 7.2.4 is followed by Example 7.2.5 and Theorem 10.1.3 precedes Example 10.1.4.

The first four chapters have received only minor changes. We reordered some material (in particular, the inequalities and identities have been split), added some new examples and exercises, and did some general updating. Chapter 5 has also been reordered, with the convergence section being moved further back, and a new section on generating random variables added. The previous coverage of invariance, which was in Chapters 7–9 of the first edition, has been greatly reduced and incorporated into Chapter 6, which otherwise has received only minor editing (mostly the addition of new exercises). Chapter 7 has been expanded and updated, and includes a new section on the EM algorithm. Chapter 8 has also received minor editing and updating, and has a new section on p-values. In Chapter 9 we now put more emphasis on pivoting (having realized that "guaranteeing an interval" was merely "pivoting the cdf"). Also, the material that was in Chapter 10 of the first edition (decision theory) has been reduced, and small sections on loss function optimality of point estimation, hypothesis testing, and interval estimation have been added to the appropriate chapters.

Chapter 10 is entirely new and attempts to lay out the fundamentals of large sample inference, including the delta method, consistency and asymptotic normality, bootstrapping, robust estimators, score tests, etc. Chapter 11 is classic oneway ANOVA and linear regression (which was covered in two different chapters in the first edition). Unfortunately, coverage of randomized block designs has been eliminated for space reasons. Chapter 12 covers regression with errors-in-variables and contains new material on robust and logistic regression.

After teaching from the first edition for a number of years, we know (approximately) what can be covered in a one-year course. From the second edition, it should be possible to cover the following in one year:

| | | | |
|---|---|---|---|
| Chapter 1: | Sections 1–7 | Chapter 6: | Sections 1–3 |
| Chapter 2: | Sections 1–3 | Chapter 7: | Sections 1–3 |
| Chapter 3: | Sections 1–6 | Chapter 8: | Sections 1–3 |
| Chapter 4: | Sections 1–7 | Chapter 9: | Sections 1–3 |
| Chapter 5: | Sections 1–6 | Chapter 10: | Sections 1, 3, 4 |

Classes that begin the course with some probability background can cover more material from the later chapters.

Finally, it is almost impossible to thank all of the people who have contributed in some way to making the second edition a reality (and help us correct the mistakes in the first edition). To all of our students, friends, and colleagues who took the time to send us a note or an e-mail, we thank you. A number of people made key suggestions that led to substantial changes in presentation. Sometimes these suggestions were just short notes or comments, and some were longer reviews. Some were so long ago that their authors may have forgotten, but we haven't. So thanks to Arthur Cohen, Sir David Cox, Steve Samuels, Rob Strawderman and Tom Wehrly. We also owe much to Jay Beder, who has sent us numerous comments and suggestions over the years and possibly knows the first edition better than we do, and to Michael Perlman and his class, who are sending comments and corrections even as we write this.

This book has seen a number of editors. We thank Alex Kugashev, who in the mid-1990s first suggested doing a second edition, and our editor, Carolyn Crockett, who constantly encouraged us. Perhaps the one person (other than us) who is most responsible for this book is our first editor, John Kimmel, who encouraged, published, and marketed the first edition. Thanks, John.

*George Casella*
*Roger L. Berger*

# Preface to the First Edition

When someone discovers that you are writing a textbook, one (or both) of two questions will be asked. The first is "Why are you writing a book?" and the second is "How is your book different from what's out there?" The first question is fairly easy to answer. You are writing a book because you are not entirely satisfied with the available texts. The second question is harder to answer. The answer can't be put in a few sentences so, in order not to bore your audience (who may be asking the question only out of politeness), you try to say something quick and witty. It usually doesn't work.

The purpose of this book is to build theoretical statistics (as different from mathematical statistics) from the first principles of probability theory. Logical development, proofs, ideas, themes, etc., evolve through statistical arguments. Thus, starting from the basics of probability, we develop the theory of statistical inference using techniques, definitions, and concepts that are statistical and are natural extensions and consequences of previous concepts. When this endeavor was started, we were not sure how well it would work. The final judgment of our success is, of course, left to the reader.

The book is intended for first-year graduate students majoring in statistics or in a field where a statistics concentration is desirable. The prerequisite is one year of calculus. (Some familiarity with matrix manipulations would be useful, but is not essential.) The book can be used for a two-semester, or three-quarter, introductory course in statistics.

The first four chapters cover basics of probability theory and introduce many fundamentals that are later necessary. Chapters 5 and 6 are the first statistical chapters. Chapter 5 is transitional (between probability and statistics) and can be the starting point for a course in statistical theory for students with some probability background. Chapter 6 is somewhat unique, detailing three statistical principles (sufficiency, likelihood, and invariance) and showing how these principles are important in modeling data. Not all instructors will cover this chapter in detail, although we strongly recommend spending some time here. In particular, the likelihood and invariance principles are treated in detail. Along with the sufficiency principle, these principles, and the thinking behind them, are fundamental to total statistical understanding.

Chapters 7–9 represent the central core of statistical inference, estimation (point and interval) and hypothesis testing. A major feature of these chapters is the division into methods of *finding* appropriate statistical techniques and methods of *evaluating* these techniques. Finding and evaluating are of interest to both the theorist and the practitioner, but we feel that it is important to separate these endeavors. Different concerns are important, and different rules are invoked. Of further interest may be the sections of these chapters titled Other Considerations. Here, we indicate how the rules of statistical inference may be relaxed (as is done every day) and still produce meaningful inferences. Many of the techniques covered in these sections are ones that are used in consulting and are helpful in analyzing and inferring from actual problems.

The final three chapters can be thought of as special topics, although we feel that some familiarity with the material is important in anyone's statistical education. Chapter 10 is a thorough introduction to decision theory and contains the most modern material we could include. Chapter 11 deals with the analysis of variance (oneway and randomized block), building the theory of the complete analysis from the more simple theory of treatment contrasts. Our experience has been that experimenters are most interested in inferences from contrasts, and using principles developed earlier, most tests and intervals can be derived from contrasts. Finally, Chapter 12 treats the theory of regression, dealing first with simple linear regression and then covering regression with "errors in variables." This latter topic is quite important, not only to show its own usefulness and inherent difficulties, but also to illustrate the limitations of inferences from ordinary regression.

As more concrete guidelines for basing a one-year course on this book, we offer the following suggestions. There can be two distinct types of courses taught from this book. One kind we might label "more mathematical," being a course appropriate for students majoring in statistics and having a solid mathematics background (at least $1\frac{1}{2}$ years of calculus, some matrix algebra, and perhaps a real analysis course). For such students we recommend covering Chapters 1–9 in their entirety (which should take approximately 22 weeks) and spend the remaining time customizing the course with selected topics from Chapters 10–12. Once the first nine chapters are covered, the material in each of the last three chapters is self-contained, and can be covered in any order.

Another type of course is "more practical." Such a course may also be a first course for mathematically sophisticated students, but is aimed at students with one year of calculus who may not be majoring in statistics. It stresses the more practical uses of statistical theory, being more concerned with understanding basic statistical concepts and deriving reasonable statistical procedures for a variety of situations, and less concerned with formal optimality investigations. Such a course will necessarily omit a certain amount of material, but the following list of sections can be covered in a one-year course:

| Chapter | Sections |
|---|---|
| 1 | All |
| 2 | 2.1, 2.2, 2.3 |
| 3 | 3.1, 3.2 |
| 4 | 4.1, 4.2, 4.3, 4.5 |
| 5 | 5.1, 5.2, 5.3.1, 5.4 |
| 6 | 6.1.1, 6.2.1 |
| 7 | 7.1, 7.2.1, 7.2.2, 7.2.3, 7.3.1, 7.3.3, 7.4 |
| 8 | 8.1, 8.2.1, 8.2.3, 8.2.4, 8.3.1, 8.3.2, 8.4 |

| 9  | 9.1, 9.2.1, 9.2.2, 9.2.4, 9.3.1, 9.4 |
|----|---------------------------------------|
| 11 | 11.1, 11.2                            |
| 12 | 12.1, 12.2                            |

If time permits, there can be some discussion (with little emphasis on details) of the material in Sections 4.4, 5.5, and 6.1.2, 6.1.3, 6.1.4. The material in Sections 11.3 and 12.3 may also be considered.

The exercises have been gathered from many sources and are quite plentiful. We feel that, perhaps, the only way to master this material is through practice, and thus we have included much opportunity to do so. The exercises are as varied as we could make them, and many of them illustrate points that are either new or complementary to the material in the text. Some exercises are even taken from research papers. (It makes you feel old when you can include exercises based on papers that were new research during your own student days!) Although the exercises are not subdivided like the chapters, their ordering roughly follows that of the chapter. (Subdivisions often give too many hints.) Furthermore, the exercises become (again, roughly) more challenging as their numbers become higher.

As this is an introductory book with a relatively broad scope, the topics are not covered in great depth. However, we felt some obligation to guide the reader one step further in the topics that may be of interest. Thus, we have included many references, pointing to the path to deeper understanding of any particular topic. (The *Encyclopedia of Statistical Sciences*, edited by Kotz, Johnson, and Read, provides a fine introduction to many topics.)

To write this book, we have drawn on both our past teachings and current work. We have also drawn on many people, to whom we are extremely grateful. We thank our colleagues at Cornell, North Carolina State, and Purdue—in particular, Jim Berger, Larry Brown, Sir David Cox, Ziding Feng, Janet Johnson, Leon Gleser, Costas Goutis, Dave Lansky, George McCabe, Chuck McCulloch, Myra Samuels, Steve Schwager, and Shayle Searle, who have given their time and expertise in reading parts of this manuscript, offered assistance, and taken part in many conversations leading to constructive suggestions. We also thank Shanti Gupta for his hospitality, and the library at Purdue, which was essential. We are grateful for the detailed reading and helpful suggestions of Shayle Searle and of our reviewers, both anonymous and non-anonymous (Jim Albert, Dan Coster, and Tom Wehrly). We also thank David Moore and George McCabe for allowing us to use their tables, and Steve Hirdt for supplying us with data. Since this book was written by two people who, for most of the time, were at least 600 miles apart, we lastly thank Bitnet for making this entire thing possible.

*George Casella*
*Roger L. Berger*

*"We have got to the deductions and the inferences," said Lestrade, winking at me. "I find it hard enough to tackle facts, Holmes, without flying away after theories and fancies."*

**Inspector Lestrade to Sherlock Holmes**

*The Boscombe Valley Mystery*

# Contents

# List of Tables

# List of Figures

# List of Examples

# Probability Theory

*"You can, for example, never foretell what any one man will do, but you can say with precision what an average number will be up to. Individuals vary, but percentages remain constant. So says the statistician."*

**Sherlock Holmes**
*The Sign of Four*

The subject of probability theory is the foundation upon which all of statistics is built, providing a means for modeling populations, experiments, or almost anything else that could be considered a random phenomenon. Through these models, statisticians are able to draw inferences about populations, inferences based on examination of only a part of the whole.

The theory of probability has a long and rich history, dating back at least to the seventeenth century when, at the request of their friend, the Chevalier de Meré, Pascal and Fermat developed a mathematical formulation of gambling odds.

The aim of this chapter is not to give a thorough introduction to probability theory; such an attempt would be foolhardy in so short a space. Rather, we attempt to outline some of the basic ideas of probability theory that are fundamental to the study of statistics.

Just as statistics builds upon the foundation of probability theory, probability theory in turn builds upon set theory, which is where we begin.

## 1.1 Set Theory

One of the main objectives of a statistician is to draw conclusions about a population of objects by conducting an experiment. The first step in this endeavor is to identify the possible outcomes or, in statistical terminology, the sample space.

**Definition 1.1.1**  The set, $S$, of all possible outcomes of a particular experiment is called the *sample space* for the experiment.

If the experiment consists of tossing a coin, the sample space contains two outcomes, heads and tails; thus,

$$S = \{\mathrm{H}, \mathrm{T}\}.$$

If, on the other hand, the experiment consists of observing the reported SAT scores of randomly selected students at a certain university, the sample space would be the set of positive integers between 200 and 800 that are multiples of ten—that is, $S = \{200, 210, 220, \ldots, 780, 790, 800\}$. Finally, consider an experiment where the observation is reaction time to a certain stimulus. Here, the sample space would consist of all positive numbers, that is, $S = (0, \infty)$.

We can classify sample spaces into two types according to the number of elements they contain. Sample spaces can be either countable or uncountable; if the elements of a sample space can be put into 1–1 correspondence with a subset of the integers, the sample space is countable. Of course, if the sample space contains only a finite number of elements, it is countable. Thus, the coin-toss and SAT score sample spaces are both countable (in fact, finite), whereas the reaction time sample space is uncountable, since the positive real numbers cannot be put into 1–1 correspondence with the integers. If, however, we measured reaction time to the nearest second, then the sample space would be (in seconds) $S = \{0, 1, 2, 3, \ldots\}$, which is then countable.

This distinction between countable and uncountable sample spaces is important only in that it dictates the way in which probabilities can be assigned. For the most part, this causes no problems, although the mathematical treatment of the situations is different. On a philosophical level, it might be argued that there can only be countable sample spaces, since measurements cannot be made with infinite accuracy. (A sample space consisting of, say, all ten-digit numbers is a countable sample space.) While in practice this is true, probabilistic and statistical methods associated with uncountable sample spaces are, in general, less cumbersome than those for countable sample spaces, and provide a close approximation to the true (countable) situation.

Once the sample space has been defined, we are in a position to consider collections of possible outcomes of an experiment.

**Definition 1.1.2**   An *event* is any collection of possible outcomes of an experiment, that is, any subset of $S$ (including $S$ itself).

Let $A$ be an event, a subset of $S$. We say the event $A$ occurs if the outcome of the experiment is in the set $A$. When speaking of probabilities, we generally speak of the probability of an event, rather than a set. But we may use the terms interchangeably.

We first need to define formally the following two relationships, which allow us to order and equate sets:

$$A \subset B \Leftrightarrow x \in A \Rightarrow x \in B; \qquad \text{(containment)}$$

$$A = B \Leftrightarrow A \subset B \text{ and } B \subset A. \qquad \text{(equality)}$$

Given any two events (or sets) $A$ and $B$, we have the following elementary set operations:

**Union:** The union of $A$ and $B$, written $A \cup B$, is the set of elements that belong to either $A$ or $B$ or both:

$$A \cup B = \{x : x \in A \text{ or } x \in B\}.$$

**Intersection:** The intersection of $A$ and $B$, written $A \cap B$, is the set of elements that belong to both $A$ and $B$:

$$A \cap B = \{x : x \in A \text{ and } x \in B\}.$$

**Complementation:** The complement of $A$, written $A^c$, is the set of all elements that are not in $A$:

$$A^c = \{x \colon x \notin A\}.$$

**Example 1.1.3 (Event operations)** Consider the experiment of selecting a card at random from a standard deck and noting its suit: clubs (C), diamonds (D), hearts (H), or spades (S). The sample space is

$$S = \{C, D, H, S\},$$

and some possible events are

$$A = \{C, D\} \quad \text{and} \quad B = \{D, H, S\}.$$

From these events we can form

$$A \cup B = \{C, D, H, S\}, \quad A \cap B = \{D\}, \quad \text{and} \quad A^c = \{H, S\}.$$

Furthermore, notice that $A \cup B = S$ (the event $S$) and $(A \cup B)^c = \emptyset$, where $\emptyset$ denotes the *empty set* (the set consisting of no elements). ‖

The elementary set operations can be combined, somewhat akin to the way addition and multiplication can be combined. As long as we are careful, we can treat sets as if they were numbers. We can now state the following useful properties of set operations.

**Theorem 1.1.4** *For any three events, $A$, $B$, and $C$, defined on a sample space $S$,*

| | | |
|---|---|---|
| **a.** Commutativity | $A \cup B = B \cup A,$ | |
| | $A \cap B = B \cap A;$ | |
| **b.** Associativity | $A \cup (B \cup C) = (A \cup B) \cup C,$ | |
| | $A \cap (B \cap C) = (A \cap B) \cap C;$ | |
| **c.** Distributive Laws | $A \cap (B \cup C) = (A \cap B) \cup (A \cap C),$ | |
| | $A \cup (B \cap C) = (A \cup B) \cap (A \cup C);$ | |
| **d.** DeMorgan's Laws | $(A \cup B)^c = A^c \cap B^c,$ | |
| | $(A \cap B)^c = A^c \cup B^c.$ | |

**Proof:** The proof of much of this theorem is left as Exercise 1.3. Also, Exercises 1.9 and 1.10 generalize the theorem. To illustrate the technique, however, we will prove the Distributive Law:

$$A \cap (B \cup C) = (A \cap B) \cup (A \cap C).$$

(You might be familiar with the use of Venn diagrams to "prove" theorems in set theory. We caution that although Venn diagrams are sometimes helpful in visualizing a situation, they do not constitute a formal proof.) To prove that two sets are equal, it must be demonstrated that each set contains the other. Formally, then

$$A \cap (B \cup C) = \{x \in S \colon x \in A \text{ and } x \in (B \cup C)\};$$

$$(A \cap B) \cup (A \cap C) = \{x \in S \colon x \in (A \cap B) \text{ or } x \in (A \cap C)\}.$$

We first show that $A \cap (B \cup C) \subset (A \cap B) \cup (A \cap C)$. Let $x \in (A \cap (B \cup C))$. By the definition of intersection, it must be that $x \in (B \cup C)$, that is, either $x \in B$ or $x \in C$. Since $x$ also must be in $A$, we have that either $x \in (A \cap B)$ or $x \in (A \cap C)$; therefore,

$$x \in ((A \cap B) \cup (A \cap C)),$$

and the containment is established.

Now assume $x \in ((A \cap B) \cup (A \cap C))$. This implies that $x \in (A \cap B)$ or $x \in (A \cap C)$. If $x \in (A \cap B)$, then $x$ is in both $A$ and $B$. Since $x \in B, x \in (B \cup C)$ and thus $x \in (A \cap (B \cup C))$. If, on the other hand, $x \in (A \cap C)$, the argument is similar, and we again conclude that $x \in (A \cap (B \cup C))$. Thus, we have established $(A \cap B) \cup (A \cap C) \subset A \cap (B \cup C)$, showing containment in the other direction and, hence, proving the Distributive Law.                                                                                    □

The operations of union and intersection can be extended to infinite collections of sets as well. If $A_1, A_2, A_3, \ldots$ is a collection of sets, all defined on a sample space $S$, then

$$\bigcup_{i=1}^{\infty} A_i = \{x \in S : x \in A_i \text{ for some } i\},$$

$$\bigcap_{i=1}^{\infty} A_i = \{x \in S : x \in A_i \text{ for all } i\}.$$

For example, let $S = (0, 1]$ and define $A_i = [(1/i), 1]$. Then

$$\bigcup_{i=1}^{\infty} A_i = \bigcup_{i=1}^{\infty} [(1/i), 1] = \{x \in (0,1] : x \in [(1/i), 1] \text{ for some } i\}$$

$$= \{x \in (0, 1]\} = (0, 1];$$

$$\bigcap_{i=1}^{\infty} A_i = \bigcap_{i=1}^{\infty} [(1/i), 1] = \{x \in (0,1] : x \in [(1/i), 1] \text{ for all } i\}$$

$$= \{x \in (0, 1] : x \in [1, 1]\} = \{1\}. \qquad \text{(the point 1)}$$

It is also possible to define unions and intersections over uncountable collections of sets. If $\Gamma$ is an index set (a set of elements to be used as indices), then

$$\bigcup_{a \in \Gamma} A_a = \{x \in S : x \in A_a \text{ for some } a\},$$

$$\bigcap_{a \in \Gamma} A_a = \{x \in S : x \in A_a \text{ for all } a\}.$$

If, for example, we take $\Gamma = \{\text{all positive real numbers}\}$ and $A_a = (0, a]$, then $\cup_{a \in \Gamma} A_a = (0, \infty)$ is an uncountable union. While uncountable unions and intersections do not play a major role in statistics, they sometimes provide a useful mechanism for obtaining an answer (see Section 8.2.3).

Finally, we discuss the idea of a partition of the sample space.

**Definition 1.1.5** Two events $A$ and $B$ are *disjoint* (or *mutually exclusive*) if $A \cap B = \emptyset$. The events $A_1, A_2, \ldots$ are *pairwise disjoint* (or *mutually exclusive*) if $A_i \cap A_j = \emptyset$ for all $i \neq j$.

Disjoint sets are sets with no points in common. If we draw a Venn diagram for two disjoint sets, the sets do not overlap. The collection

$$A_i = [i, i+1), \quad i = 0, 1, 2, \ldots,$$

consists of pairwise disjoint sets. Note further that $\cup_{i=0}^{\infty} A_i = [0, \infty)$.

**Definition 1.1.6** If $A_1, A_2, \ldots$ are pairwise disjoint and $\cup_{i=1}^{\infty} A_i = S$, then the collection $A_1, A_2, \ldots$ forms a *partition* of $S$.

The sets $A_i = [i, i+1)$ form a partition of $[0, \infty)$. In general, partitions are very useful, allowing us to divide the sample space into small, nonoverlapping pieces.

## 1.2 Basics of Probability Theory

When an experiment is performed, the realization of the experiment is an outcome in the sample space. If the experiment is performed a number of times, different outcomes may occur each time or some outcomes may repeat. This "frequency of occurrence" of an outcome can be thought of as a probability. More probable outcomes occur more frequently. If the outcomes of an experiment can be described probabilistically, we are on our way to analyzing the experiment statistically.

In this section we describe some of the basics of probability theory. We do not define probabilities in terms of frequencies but instead take the mathematically simpler axiomatic approach. As will be seen, the axiomatic approach is not concerned with the interpretations of probabilities, but is concerned only that the probabilities are defined by a function satisfying the axioms. Interpretations of the probabilities are quite another matter. The "frequency of occurrence" of an event is one example of a particular interpretation of probability. Another possible interpretation is a subjective one, where rather than thinking of probability as frequency, we can think of it as a belief in the chance of an event occurring.

### 1.2.1 Axiomatic Foundations

For each event $A$ in the sample space $S$ we want to associate with $A$ a number between zero and one that will be called the probability of $A$, denoted by $P(A)$. It would seem natural to define the domain of $P$ (the set where the arguments of the function $P(\cdot)$ are defined) as all subsets of $S$; that is, for each $A \subset S$ we define $P(A)$ as the probability that $A$ occurs. Unfortunately, matters are not that simple. There are some technical difficulties to overcome. We will not dwell on these technicalities; although they are of importance, they are usually of more interest to probabilists than to statisticians. However, a firm understanding of statistics requires at least a passing familiarity with the following.

**Definition 1.2.1**    A collection of subsets of $S$ is called a *sigma algebra* (or *Borel field*), denoted by $\mathcal{B}$, if it satisfies the following three properties:

**a.** $\emptyset \in \mathcal{B}$ (the empty set is an element of $\mathcal{B}$).

**b.** If $A \in \mathcal{B}$, then $A^c \in \mathcal{B}$ ($\mathcal{B}$ is closed under complementation).

**c.** If $A_1, A_2, \ldots \in \mathcal{B}$, then $\cup_{i=1}^{\infty} A_i \in \mathcal{B}$ ($\mathcal{B}$ is closed under countable unions).

The empty set $\emptyset$ is a subset of any set. Thus, $\emptyset \subset S$. Property (a) states that this subset is always in a sigma algebra. Since $S = \emptyset^c$, properties (a) and (b) imply that $S$ is always in $\mathcal{B}$ also. In addition, from DeMorgan's Laws it follows that $\mathcal{B}$ is closed under countable intersections. If $A_1, A_2, \ldots \in \mathcal{B}$, then $A_1^c, A_2^c, \ldots \in \mathcal{B}$ by property (b), and therefore $\cup_{i=1}^{\infty} A_i^c \in \mathcal{B}$. However, using DeMorgan's Law (as in Exercise 1.9), we have

$$(1.2.1) \qquad \left( \bigcup_{i=1}^{\infty} A_i^c \right)^c = \bigcap_{i=1}^{\infty} A_i.$$

Thus, again by property (b), $\cap_{i=1}^{\infty} A_i \in \mathcal{B}$.

Associated with sample space $S$ we can have many different sigma algebras. For example, the collection of the two sets $\{\emptyset, S\}$ is a sigma algebra, usually called the trivial sigma algebra. The only sigma algebra we will be concerned with is the smallest one that contains all of the open sets in a given sample space $S$.

**Example 1.2.2 (Sigma algebra–I)**    If $S$ is finite or countable, then these technicalities really do not arise, for we define for a given sample space $S$,

$$\mathcal{B} = \{\text{all subsets of } S, \text{ including } S \text{ itself}\}.$$

If $S$ has $n$ elements, there are $2^n$ sets in $\mathcal{B}$ (see Exercise 1.14). For example, if $S = \{1, 2, 3\}$, then $\mathcal{B}$ is the following collection of $2^3 = 8$ sets:

$$\begin{array}{ccc} \{1\} & \{1, 2\} & \{1, 2, 3\} \\ \{2\} & \{1, 3\} & \emptyset \\ \{3\} & \{2, 3\} & \end{array} \qquad \|$$

In general, if $S$ is uncountable, it is not an easy task to describe $\mathcal{B}$. However, $\mathcal{B}$ is chosen to contain any set of interest.

**Example 1.2.3 (Sigma algebra–II)**    Let $S = (-\infty, \infty)$, the real line. Then $\mathcal{B}$ is chosen to contain all sets of the form

$$[a, b], \quad (a, b], \quad (a, b), \quad \text{and} \quad [a, b)$$

for all real numbers $a$ and $b$. Also, from the properties of $\mathcal{B}$, it follows that $\mathcal{B}$ contains all sets that can be formed by taking (possibly countably infinite) unions and intersections of sets of the above varieties.                    $\|$

We are now in a position to define a probability function.

**Definition 1.2.4**    Given a sample space $S$ and an associated sigma algebra $\mathcal{B}$, a *probability function* is a function $P$ with domain $\mathcal{B}$ that satisfies

1. $P(A) \geq 0$ for all $A \in \mathcal{B}$.
2. $P(S) = 1$.
3. If $A_1, A_2, \ldots \in \mathcal{B}$ are pairwise disjoint, then $P(\cup_{i=1}^{\infty} A_i) = \sum_{i=1}^{\infty} P(A_i)$.

The three properties given in Definition 1.2.4 are usually referred to as the Axioms of Probability (or the Kolmogorov Axioms, after A. Kolmogorov, one of the fathers of probability theory). Any function $P$ that satisfies the Axioms of Probability is called a probability function. The axiomatic definition makes no attempt to tell what particular function $P$ to choose; it merely requires $P$ to satisfy the axioms. For any sample space many different probability functions can be defined. Which one(s) reflects what is likely to be observed in a particular experiment is still to be discussed.

**Example 1.2.5 (Defining probabilities–I)**    Consider the simple experiment of tossing a fair coin, so $S = \{H, T\}$. By a "fair" coin we mean a balanced coin that is equally as likely to land heads up as tails up, and hence the reasonable probability function is the one that assigns equal probabilities to heads and tails, that is,

$$(1.2.2) \qquad\qquad P(\{H\}) = P(\{T\}).$$

Note that (1.2.2) does not follow from the Axioms of Probability but rather is outside of the axioms. We have used a symmetry interpretation of probability (or just intuition) to impose the requirement that heads and tails be equally probable. Since $S = \{H\} \cup \{T\}$, we have, from Axiom 2, $P(\{H\} \cup \{T\}) = 1$. Also, $\{H\}$ and $\{T\}$ are disjoint, so $P(\{H\} \cup \{T\}) = P(\{H\}) + P(\{T\})$ and

$$(1.2.3) \qquad\qquad P(\{H\}) + P(\{T\}) = 1.$$

Simultaneously solving (1.2.2) and (1.2.3) shows that $P(\{H\}) = P(\{T\}) = \frac{1}{2}$.

Since (1.2.2) is based on our knowledge of the particular experiment, not the axioms, any nonnegative values for $P(\{H\})$ and $P(\{T\})$ that satisfy (1.2.3) define a legitimate probability function. For example, we might choose $P(\{H\}) = \frac{1}{9}$ and $P(\{T\}) = \frac{8}{9}$. ‖

We need general methods of defining probability functions that we know will always satisfy Kolmogorov's Axioms. We do not want to have to check the Axioms for each new probability function, like we did in Example 1.2.5. The following gives a common method of defining a legitimate probability function.

**Theorem 1.2.6**    *Let $S = \{s_1, \ldots, s_n\}$ be a finite set. Let $\mathcal{B}$ be any sigma algebra of subsets of $S$. Let $p_1, \ldots, p_n$ be nonnegative numbers that sum to 1. For any $A \in \mathcal{B}$, define $P(A)$ by*

$$P(A) = \sum_{\{i : s_i \in A\}} p_i.$$

*(The sum over an empty set is defined to be 0.) Then $P$ is a probability function on $\mathcal{B}$. This remains true if $S = \{s_1, s_2, \ldots\}$ is a countable set.*

**Proof:** We will give the proof for finite $S$. For any $A \in \mathcal{B}$, $P(A) = \sum_{\{i:s_i \in A\}} p_i \geq 0$, because every $p_i \geq 0$. Thus, Axiom 1 is true. Now,

$$P(S) = \sum_{\{i:s_i \in S\}} p_i = \sum_{i=1}^n p_i = 1.$$

Thus, Axiom 2 is true. Let $A_1, \ldots, A_k$ denote pairwise disjoint events. ($\mathcal{B}$ contains only a finite number of sets, so we need consider only finite disjoint unions.) Then,

$$P\left(\bigcup_{i=1}^k A_i\right) = \sum_{\{j:s_j \in \cup_{i=1}^k A_i\}} p_j = \sum_{i=1}^k \sum_{\{j:s_j \in A_i\}} p_j = \sum_{i=1}^k P(A_i).$$

The first and third equalities are true by the definition of $P(A)$. The disjointedness of the $A_i$s ensures that the second equality is true, because the same $p_j$s appear exactly once on each side of the equality. Thus, Axiom 3 is true and Kolmogorov's Axioms are satisfied.  □

The physical reality of the experiment might dictate the probability assignment, as the next example illustrates.

**Example 1.2.7 (Defining probabilities–II)**    The game of darts is played by throwing a dart at a board and receiving a score corresponding to the number assigned to the region in which the dart lands. For a novice player, it seems reasonable to assume that the probability of the dart hitting a particular region is proportional to the area of the region. Thus, a bigger region has a higher probability of being hit.

Referring to Figure 1.2.1, we see that the dart board has radius $r$ and the distance between rings is $r/5$. If we make the assumption that the board is always hit (see Exercise 1.7 for a variation on this), then we have

$$P(\text{scoring } i \text{ points}) = \frac{\text{Area of region } i}{\text{Area of dart board}}.$$

For example

$$P(\text{scoring } 1 \text{ point}) = \frac{\pi r^2 - \pi(4r/5)^2}{\pi r^2} = 1 - \left(\frac{4}{5}\right)^2.$$

It is easy to derive the general formula, and we find that

$$P(\text{scoring } i \text{ points}) = \frac{(6-i)^2 - (5-i)^2}{5^2}, \quad i = 1, \ldots, 5,$$

independent of $\pi$ and $r$. The sum of the areas of the disjoint regions equals the area of the dart board. Thus, the probabilities that have been assigned to the five outcomes sum to 1, and, by Theorem 1.2.6, this is a probability function (see Exercise 1.8).  ‖

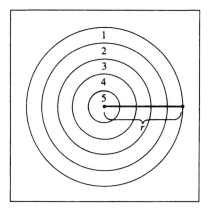

Figure 1.2.1. *Dart board for Example 1.2.7*

Before we leave the axiomatic development of probability, there is one further point to consider. Axiom 3 of Definition 1.2.4, which is commonly known as the Axiom of Countable Additivity, is not universally accepted among statisticians. Indeed, it can be argued that axioms should be simple, self-evident statements. Comparing Axiom 3 to the other axioms, which are simple and self-evident, may lead us to doubt whether it is reasonable to assume the truth of Axiom 3.

The Axiom of Countable Additivity is rejected by a school of statisticians led by deFinetti (1972), who chooses to replace this axiom with the Axiom of Finite Additivity.

*Axiom of Finite Additivity:*   If $A \in \mathcal{B}$ and $B \in \mathcal{B}$ are disjoint, then

$$P(A \cup B) = P(A) + P(B).$$

While this axiom may not be entirely self-evident, it is certainly simpler than the Axiom of Countable Additivity (and is implied by it – see Exercise 1.12).

Assuming only finite additivity, while perhaps more plausible, can lead to unexpected complications in statistical theory – complications that, at this level, do not necessarily enhance understanding of the subject. We therefore proceed under the assumption that the Axiom of Countable Additivity holds.

### 1.2.2 The Calculus of Probabilities

From the Axioms of Probability we can build up many properties of the probability function, properties that are quite helpful in the calculation of more complicated probabilities. Some of these manipulations will be discussed in detail in this section; others will be left as exercises.

We start with some (fairly self-evident) properties of the probability function when applied to a single event.

**Theorem 1.2.8**  *If $P$ is a probability function and $A$ is any set in $\mathcal{B}$, then*

**a.** $P(\emptyset) = 0$, *where $\emptyset$ is the empty set;*

**b.** $P(A) \leq 1$;

**c.** $P(A^c) = 1 - P(A)$.

**Proof:** It is easiest to prove (c) first. The sets $A$ and $A^c$ form a partition of the sample space, that is, $S = A \cup A^c$. Therefore,

$$(1.2.4) \qquad\qquad P(A \cup A^c) = P(S) = 1$$

by the second axiom. Also, $A$ and $A^c$ are disjoint, so by the third axiom,

$$(1.2.5) \qquad\qquad P(A \cup A^c) = P(A) + P(A^c).$$

Combining (1.2.4) and (1.2.5) gives (c).

Since $P(A^c) \geq 0$, (b) is immediately implied by (c). To prove (a), we use a similar argument on $S = S \cup \emptyset$. (Recall that both $S$ and $\emptyset$ are always in $\mathcal{B}$.) Since $S$ and $\emptyset$ are disjoint, we have

$$1 = P(S) = P(S \cup \emptyset) = P(S) + P(\emptyset),$$

and thus $P(\emptyset) = 0$. $\qquad\qquad\qquad\qquad\qquad\qquad\qquad\qquad\qquad\qquad\square$

Theorem 1.2.8 contains properties that are so basic that they also have the flavor of axioms, although we have formally proved them using only the original three Kolmogorov Axioms. The next theorem, which is similar in spirit to Theorem 1.2.8, contains statements that are not so self-evident.

**Theorem 1.2.9**  *If $P$ is a probability function and $A$ and $B$ are any sets in $\mathcal{B}$, then*

**a.** $P(B \cap A^c) = P(B) - P(A \cap B)$;

**b.** $P(A \cup B) = P(A) + P(B) - P(A \cap B)$;

**c.** *If $A \subset B$, then $P(A) \leq P(B)$.*

**Proof:** To establish (a) note that for any sets $A$ and $B$ we have

$$B = \{B \cap A\} \cup \{B \cap A^c\},$$

and therefore

$$(1.2.6) \qquad P(B) = P(\{B \cap A\} \cup \{B \cap A^c\}) = P(B \cap A) + P(B \cap A^c),$$

where the last equality in (1.2.6) follows from the fact that $B \cap A$ and $B \cap A^c$ are disjoint. Rearranging (1.2.6) gives (a).

To establish (b), we use the identity

$$(1.2.7) \qquad\qquad A \cup B = A \cup \{B \cap A^c\}.$$

A Venn diagram will show why (1.2.7) holds, although a formal proof is not difficult (see Exercise 1.2). Using (1.2.7) and the fact that $A$ and $B \cap A^c$ are disjoint (since $A$ and $A^c$ are), we have

(1.2.8) $\qquad P(A \cup B) = P(A) + P(B \cap A^c) = P(A) + P(B) - P(A \cap B)$

from (a).

If $A \subset B$, then $A \cap B = A$. Therefore, using (a) we have

$$0 \leq P(B \cap A^c) = P(B) - P(A),$$

establishing (c). $\qquad\qquad\qquad\qquad\qquad\qquad\qquad\qquad\qquad\qquad\qquad\qquad$ □

Formula (b) of Theorem 1.2.9 gives a useful inequality for the probability of an intersection. Since $P(A \cup B) \leq 1$, we have from (1.2.8), after some rearranging,

(1.2.9) $\qquad\qquad\qquad\qquad P(A \cap B) \geq P(A) + P(B) - 1.$

This inequality is a special case of what is known as *Bonferroni's Inequality* (Miller 1981 is a good reference). Bonferroni's Inequality allows us to bound the probability of a simultaneous event (the intersection) in terms of the probabilities of the individual events.

**Example 1.2.10 (Bonferroni's Inequality)**     Bonferroni's Inequality is particularly useful when it is difficult (or even impossible) to calculate the intersection probability, but some idea of the size of this probability is desired. Suppose $A$ and $B$ are two events and each has probability .95. Then the probability that both will occur is bounded below by

$$P(A \cap B) \geq P(A) + P(B) - 1 = .95 + .95 - 1 = .90.$$

Note that unless the probabilities of the individual events are sufficiently large, the Bonferroni bound is a useless (but correct!) negative number. $\qquad\qquad$ ‖

We close this section with a theorem that gives some useful results for dealing with a collection of sets.

**Theorem 1.2.11**    *If $P$ is a probability function, then*
**a.** $P(A) = \sum_{i=1}^{\infty} P(A \cap C_i)$ *for any partition $C_1, C_2, \ldots$;*
**b.** $P(\cup_{i=1}^{\infty} A_i) \leq \sum_{i=1}^{\infty} P(A_i)$ *for any sets $A_1, A_2, \ldots$.*        (*Boole's Inequality*)

**Proof:** Since $C_1, C_2, \ldots$ form a partition, we have that $C_i \cap C_j = \emptyset$ for all $i \neq j$, and $S = \cup_{i=1}^{\infty} C_i$. Hence,

$$A = A \cap S = A \cap \left( \bigcup_{i=1}^{\infty} C_i \right) = \bigcup_{i=1}^{\infty} (A \cap C_i),$$

where the last equality follows from the Distributive Law (Theorem 1.1.4). We therefore have

$$P(A) = P\left(\bigcup_{i=1}^{\infty}(A \cap C_i)\right).$$

Now, since the $C_i$ are disjoint, the sets $A \cap C_i$ are also disjoint, and from the properties of a probability function we have

$$P\left(\bigcup_{i=1}^{\infty}(A \cap C_i)\right) = \sum_{i=1}^{\infty} P(A \cap C_i),$$

establishing (a).

To establish (b) we first construct a disjoint collection $A_1^*, A_2^*, \ldots$, with the property that $\cup_{i=1}^{\infty} A_i^* = \cup_{i=1}^{\infty} A_i$. We define $A_i^*$ by

$$A_1^* = A_1, \quad A_i^* = A_i \backslash \left(\bigcup_{j=1}^{i-1} A_j\right), \quad i = 2, 3, \ldots,$$

where the notation $A \backslash B$ denotes the part of $A$ that does not intersect with $B$. In more familiar symbols, $A \backslash B = A \cap B^c$. It should be easy to see that $\cup_{i=1}^{\infty} A_i^* = \cup_{i=1}^{\infty} A_i$, and we therefore have

$$P\left(\bigcup_{i=1}^{\infty} A_i\right) = P\left(\bigcup_{i=1}^{\infty} A_i^*\right) = \sum_{i=1}^{\infty} P(A_i^*),$$

where the last equality follows since the $A_i^*$ are disjoint. To see this, we write

$$A_i^* \cap A_k^* = \left\{A_i \backslash \left(\bigcup_{j=1}^{i-1} A_j\right)\right\} \cap \left\{A_k \backslash \left(\bigcup_{j=1}^{k-1} A_j\right)\right\} \quad \text{(definition of } A_i^*)$$

$$= \left\{A_i \cap \left(\bigcup_{j=1}^{i-1} A_j\right)^c\right\} \cap \left\{A_k \cap \left(\bigcup_{j=1}^{k-1} A_j\right)^c\right\} \quad \text{(definition of ``\backslash'')}$$

$$= \left\{A_i \cap \bigcap_{j=1}^{i-1} A_j^c\right\} \cap \left\{A_k \cap \bigcap_{j=1}^{k-1} A_j^c\right\} \quad \text{(DeMorgan's Laws)}$$

Now if $i > k$, the first intersection above will be contained in the set $A_k^c$, which will have an empty intersection with $A_k$. If $k > i$, the argument is similar. Further, by construction $A_i^* \subset A_i$, so $P(A_i^*) \le P(A_i)$ and we have

$$\sum_{i=1}^{\infty} P(A_i^*) \le \sum_{i=1}^{\infty} P(A_i),$$

establishing (b). □

There is a similarity between Boole's Inequality and Bonferroni's Inequality. In fact, they are essentially the same thing. We could have used Boole's Inequality to derive (1.2.9). If we apply Boole's Inequality to $A^c$, we have

$$P\left(\bigcup_{i=1}^{n} A_i^c\right) \leq \sum_{i=1}^{n} P(A_i^c),$$

and using the facts that $\cup A_i^c = (\cap A_i)^c$ and $P(A_i^c) = 1 - P(A_i)$, we obtain

$$1 - P\left(\bigcap_{i=1}^{n} A_i\right) \leq n - \sum_{i=1}^{n} P(A_i).$$

This becomes, on rearranging terms,

$$(1.2.10) \qquad P\left(\bigcap_{i=1}^{n} A_i\right) \geq \sum_{i=1}^{n} P(A_i) - (n - 1),$$

which is a more general version of the Bonferroni Inequality of (1.2.9).

### 1.2.3 Counting

The elementary process of counting can become quite sophisticated when placed in the hands of a statistician. Most often, methods of counting are used in order to construct probability assignments on finite sample spaces, although they can be used to answer other questions also.

**Example 1.2.12 (Lottery–I)**   For a number of years the New York state lottery operated according to the following scheme. From the numbers 1, 2, ..., 44, a person may pick any six for her ticket. The winning number is then decided by randomly selecting six numbers from the forty-four. To be able to calculate the probability of winning we first must count how many different groups of six numbers can be chosen from the forty-four.                                                                                    ‖

**Example 1.2.13 (Tournament)**   In a single-elimination tournament, such as the U.S. Open tennis tournament, players advance only if they win (in contrast to double-elimination or round-robin tournaments). If we have 16 entrants, we might be interested in the number of paths a particular player can take to victory, where a path is taken to mean a sequence of opponents.                                                                      ‖

Counting problems, in general, sound complicated, and often we must do our counting subject to many restrictions. The way to solve such problems is to break them down into a series of simple tasks that are easy to count, and employ known rules of combining tasks. The following theorem is a first step in such a process and is sometimes known as the Fundamental Theorem of Counting.

**Theorem 1.2.14**   *If a job consists of $k$ separate tasks, the $i$th of which can be done in $n_i$ ways, $i = 1, \ldots, k$, then the entire job can be done in $n_1 \times n_2 \times \cdots \times n_k$ ways.*

**Proof:** It suffices to prove the theorem for $k = 2$ (see Exercise 1.15). The proof is just a matter of careful counting. The first task can be done in $n_1$ ways, and for each of these ways we have $n_2$ choices for the second task. Thus, we can do the job in

$$\underbrace{(1 \times n_2) + (1 \times n_2) + \cdots + (1 \times n_2)}_{n_1 \text{ terms}} = n_1 \times n_2$$

ways, establishing the theorem for $k = 2$.                                              □

**Example 1.2.15 (Lottery–II)**  Although the Fundamental Theorem of Counting is a reasonable place to start, in applications there are usually more aspects of a problem to consider. For example, in the New York state lottery the first number can be chosen in 44 ways, and the second number in 43 ways, making a total of $44 \times 43 = 1{,}892$ ways of choosing the first two numbers. However, if a person is allowed to choose the same number twice, then the first two numbers can be chosen in $44 \times 44 = 1{,}936$ ways.                                              ‖

The distinction being made in Example 1.2.15 is between counting *with replacement* and counting *without replacement*. There is a second crucial element in any counting problem, whether or not the ordering of the tasks is important. To illustrate with the lottery example, suppose the winning numbers are selected in the order 12, 37, 35, 9, 13, 22. Does a person who selected 9, 12, 13, 22, 35, 37 qualify as a winner? In other words, does the order in which the task is performed actually matter? Taking all of these considerations into account, we can construct a $2 \times 2$ table of possibilities:

*Possible methods of counting*

|  | Without replacement | With replacement |
|---|---|---|
| Ordered |  |  |
| Unordered |  |  |

Before we begin to count, the following definition gives us some extremely helpful notation.

**Definition 1.2.16**  For a positive integer $n$, $n!$ (read $n$ factorial) is the product of all of the positive integers less than or equal to $n$. That is,

$$n! = n \times (n - 1) \times (n - 2) \times \cdots \times 3 \times 2 \times 1.$$

Furthermore, we define $0! = 1$.

Let us now consider counting all of the possible lottery tickets under each of these four cases.

1. *Ordered, without replacement*  From the Fundamental Theorem of Counting, the first number can be selected in 44 ways, the second in 43 ways, etc. So there are

$$44 \times 43 \times 42 \times 41 \times 40 \times 39 = \frac{44!}{38!} = 5{,}082{,}517{,}440$$

possible tickets.

2. *Ordered, with replacement*  Since each number can now be selected in 44 ways (because the chosen number is replaced), there are

$$44 \times 44 \times 44 \times 44 \times 44 \times 44 = 44^6 = 7{,}256{,}313{,}856$$

possible tickets.

3. *Unordered, without replacement*  We know the number of possible tickets when the ordering must be accounted for, so what we must do is divide out the redundant orderings. Again from the Fundamental Theorem, six numbers can be arranged in $6 \times 5 \times 4 \times 3 \times 2 \times 1$ ways, so the total number of unordered tickets is

$$\frac{44 \times 43 \times 42 \times 41 \times 40 \times 39}{6 \times 5 \times 4 \times 3 \times 2 \times 1} = \frac{44!}{6! \, 38!} = 7{,}059{,}052.$$

This form of counting plays a central role in much of statistics—so much, in fact, that it has earned its own notation.

**Definition 1.2.17**    For nonnegative integers $n$ and $r$, where $n \geq r$, we define the symbol $\binom{n}{r}$, read *n choose r*, as

$$\binom{n}{r} = \frac{n!}{r! \, (n-r)!}.$$

In our lottery example, the number of possible tickets (unordered, without replacement) is $\binom{44}{6}$. These numbers are also referred to as *binomial coefficients*, for reasons that will become clear in Chapter 3.

4. *Unordered, with replacement*  This is the most difficult case to count. You might first guess that the answer is $44^6/(6 \times 5 \times 4 \times 3 \times 2 \times 1)$, but this is not correct (it is too small).

To count in this case, it is easiest to think of placing 6 markers on the 44 numbers. In fact, we can think of the 44 numbers defining bins in which we can place the six markers, M, as shown, for example, in this figure.

| M |   |   | MM | M |   |   |   | M | M |   |   |   |
|---|---|---|----|---|---|---|---|---|---|---|---|---|
| 1 | 2 | 3 | 4 | 5 | $\cdots$ | 41 | 42 | 43 | 44 | | | |

The number of possible tickets is then equal to the number of ways that we can put the 6 markers into the 44 bins. But this can be further reduced by noting that all we need to keep track of is the arrangement of the markers and the walls of the bins. Note further that the two outermost walls play no part. Thus, we have to count all of the arrangements of 43 walls (44 bins yield 45 walls, but we disregard the two end walls) and 6 markers. We therefore have $43 + 6 = 49$ objects, which can be arranged in $49!$ ways. However, to eliminate the redundant orderings we must divide by both $6!$ and $43!$, so the total number of arrangements is

$$\frac{49!}{6! \, 43!} = 13{,}983{,}816.$$

Although all of the preceding derivations were done in terms of an example, it should be easy to see that they hold in general. For completeness, we can summarize these situations in Table 1.2.1.

Table 1.2.1. *Number of possible arrangements of size r from n objects*

|  | Without replacement | With replacement |
|---|---|---|
| Ordered | $\dfrac{n!}{(n-r)!}$ | $n^r$ |
| Unordered | $\dbinom{n}{r}$ | $\dbinom{n+r-1}{r}$ |

*1.2.4 Enumerating Outcomes*

The counting techniques of the previous section are useful when the sample space $S$ is a finite set and all the outcomes in $S$ are equally likely. Then probabilities of events can be calculated by simply counting the number of outcomes in the event. To see this, suppose that $S = \{s_1, \ldots, s_N\}$ is a finite sample space. Saying that all the outcomes are equally likely means that $P(\{s_i\}) = 1/N$ for every outcome $s_i$. Then, using Axiom 3 from Definition 1.2.4, we have, for any event $A$,

$$P(A) = \sum_{s_i \in A} P(\{s_i\}) = \sum_{s_i \in A} \frac{1}{N} = \frac{\text{\# of elements in } A}{\text{\# of elements in } S}.$$

For large sample spaces, the counting techniques might be used to calculate both the numerator and denominator of this expression.

**Example 1.2.18 (Poker)**   Consider choosing a five-card poker hand from a standard deck of 52 playing cards. Obviously, we are sampling without replacement from the deck. But to specify the possible outcomes (possible hands), we must decide whether we think of the hand as being dealt sequentially (ordered) or all at once (unordered). If we wish to calculate probabilities for events that depend on the order, such as the probability of an ace in the first two cards, then we must use the ordered outcomes. But if our events do not depend on the order, we can use the unordered outcomes. For this example we use the unordered outcomes, so the sample space consists of all the five-card hands that can be chosen from the 52-card deck. There are $\binom{52}{5} = 2{,}598{,}960$ possible hands. If the deck is well shuffled and the cards are randomly dealt, it is reasonable to assign probability $1/2{,}598{,}960$ to each possible hand.

We now calculate some probabilities by counting outcomes in events. What is the probability of having four aces? How many different hands are there with four aces? If we specify that four of the cards are aces, then there are 48 different ways of specifying the fifth card. Thus,

$$P(\text{four aces}) = \frac{48}{2{,}598{,}960},$$

less than 1 chance in 50,000. Only slightly more complicated counting, using Theorem 1.2.14, allows us to calculate the probability of having four of a kind. There are 13 ways to specify which denomination there will be four of. After we specify these four cards, there are 48 ways of specifying the fifth. Thus, the total number of hands with four of a kind is $(13)(48)$ and

$$P(\text{four of a kind}) = \frac{(13)(48)}{2{,}598{,}960} = \frac{624}{2{,}598{,}960}.$$

To calculate the probability of exactly one pair (not two pairs, no three of a kind, etc.) we combine some of the counting techniques. The number of hands with exactly one pair is

$$(1.2.11) \qquad\qquad 13 \binom{4}{2} \binom{12}{3} 4^3 = 1{,}098{,}240.$$

Expression (1.2.11) comes from Theorem 1.2.14 because

$13 = \#$ of ways to specify the denomination for the pair,

$\binom{4}{2} = \#$ of ways to specify the two cards from that denomination,

$\binom{12}{3} = \#$ of ways of specifying the other three denominations,

$4^3 = \#$ of ways of specifying the other three cards from those denominations.

Thus,

$$P(\text{exactly one pair}) = \frac{1{,}098{,}240}{2{,}598{,}960}. \qquad\qquad \|$$

When sampling without replacement, as in Example 1.2.18, if we want to calculate the probability of an event that does not depend on the order, we can use either the ordered or unordered sample space. Each outcome in the unordered sample space corresponds to $r!$ outcomes in the ordered sample space. So, when counting outcomes in the ordered sample space, we use a factor of $r!$ in both the numerator and denominator that will cancel to give the same probability as if we counted in the unordered sample space.

The situation is different if we sample with replacement. Each outcome in the unordered sample space corresponds to some outcomes in the ordered sample space, but the number of outcomes differs.

**Example 1.2.19 (Sampling with replacement)**  Consider sampling $r = 2$ items from $n = 3$ items, with replacement. The outcomes in the ordered and unordered sample spaces are these.

| Unordered | $\{1,1\}$ | $\{2,2\}$ | $\{3,3\}$ | $\{1,2\}$ | $\{1,3\}$ | $\{2,3\}$ |
|---|---|---|---|---|---|---|
| Ordered | $(1,1)$ | $(2,2)$ | $(3,3)$ | $(1,2),(2,1)$ | $(1,3),(3,1)$ | $(2,3),(3,2)$ |
| Probability | $1/9$ | $1/9$ | $1/9$ | $2/9$ | $2/9$ | $2/9$ |

The probabilities come from considering the nine outcomes in the ordered sample space to be equally likely. This corresponds to the common interpretation of "sampling with replacement"; namely, one of the three items is chosen, each with probability $1/3$; the item is noted and replaced; the items are mixed and again one of the three items is chosen, each with probability $1/3$. It is seen that the six outcomes in the unordered sample space are not equally likely under this kind of sampling. The formula for the number of outcomes in the unordered sample space is useful for enumerating the outcomes, but ordered outcomes must be counted to correctly calculate probabilities.

$\parallel$

Some authors argue that it is appropriate to assign equal probabilities to the unordered outcomes when "randomly distributing $r$ indistinguishable balls into $n$ distinguishable urns." That is, an urn is chosen at random and a ball placed in it, and this is repeated $r$ times. The order in which the balls are placed is not recorded so, in the end, an outcome such as $\{1, 3\}$ means one ball is in urn 1 and one ball is in urn 3.

But here is the problem with this interpretation. Suppose two people observe this process, and Observer 1 records the order in which the balls are placed but Observer 2 does not. Observer 1 will assign probability $2/9$ to the event $\{1, 3\}$. Observer 2, who is observing exactly the same process, should also assign probability $2/9$ to this event. But if the six unordered outcomes are written on identical pieces of paper and one is randomly chosen to determine the placement of the balls, then the unordered outcomes each have probability $1/6$. So Observer 2 will assign probability $1/6$ to the event $\{1, 3\}$.

The confusion arises because the phrase "with replacement" will typically be interpreted with the sequential kind of sampling we described above, leading to assigning a probability $2/9$ to the event $\{1, 3\}$. This is the correct way to proceed, as probabilities should be determined by the sampling mechanism, not whether the balls are distinguishable or indistinguishable.

**Example 1.2.20 (Calculating an average)**   As an illustration of the distinguishable/indistinguishable approach, suppose that we are going to calculate all possible averages of four numbers selected from

$$2, 4, 9, 12$$

where we draw the numbers with replacement. For example, possible draws are $\{2, 4, 4, 9\}$ with average 4.75 and $\{4, 4, 9, 9\}$ with average 6.5. If we are only interested in the average of the sampled numbers, the ordering is unimportant, and thus the total number of distinct samples is obtained by counting according to unordered, with-replacement sampling.

The total number of distinct samples is $\binom{n+n-1}{n}$. But now, to calculate the probability distribution of the sampled averages, we must count the different ways that a particular average can occur.

The value 4.75 can occur only if the sample contains one 2, two 4s, and one 9. The number of possible samples that have this configuration is given in the following table:

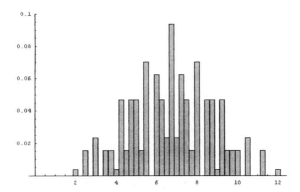

Figure 1.2.2. *Histogram of averages of samples with replacement from the four numbers* $\{2, 4, 9, 12\}$

| Unordered | Ordered |
|---|---|
| $\{2, 4, 4, 9\}$ | $(2, 4, 4, 9), (2, 4, 9, 4), (2, 9, 4, 4), (4, 2, 4, 9),$ $(4, 2, 9, 4), (4, 4, 2, 9), (4, 4, 9, 2), (4, 9, 2, 4),$ $(4, 9, 4, 2), (9, 2, 4, 4), (9, 4, 2, 4), (9, 4, 4, 2)$ |

The total number of ordered samples is $n^n = 4^4 = 256$, so the probability of drawing the unordered sample $\{2, 4, 4, 9\}$ is $12/256$. Compare this to the probability that we would have obtained if we regarded the unordered samples as equally likely – we would have assigned probability $1/\binom{n+n-1}{n} = 1/\binom{7}{4} = 1/35$ to $\{2, 4, 4, 9\}$ and to every other unordered sample.

To count the number of ordered samples that would result in $\{2, 4, 4, 9\}$, we argue as follows. We need to enumerate the possible orders of the four numbers $\{2, 4, 4, 9\}$, so we are essentially using counting method 1 of Section 1.2.3. We can order the sample in $4 \times 3 \times 2 \times 1 = 24$ ways. But there is a bit of double counting here, since we cannot count distinct arrangements of the two 4s. For example, the 24 ways would count $\{9, 4, 2, 4\}$ twice (which would be OK if the 4s were different). To correct this, we divide by 2! (there are 2! ways to arrange the two 4s) and obtain $24/2 = 12$ ordered samples. In general, if there are $k$ places and we have $m$ different numbers repeated $k_1, k_2, \ldots, k_m$ times, then the number of ordered samples is $\dfrac{k!}{k_1! k_2! \cdots k_m!}$. This type of counting is related to the *multinomial distribution*, which we will see in Section 4.6. Figure 1.2.2 is a histogram of the probability distribution of the sample averages, reflecting the multinomial counting of the samples.

There is also one further refinement that is reflected in Figure 1.2.2. It is possible that two different unordered samples will result in the same mean. For example, the unordered samples $\{4, 4, 12, 12\}$ and $\{2, 9, 9, 12\}$ both result in an average value of 8. The first sample has probability $3/128$ and the second has probability $3/64$, giving the value 8 a probability of $9/128 = .07$. See Example A.0.1 in Appendix A for details on constructing such a histogram. The calculation that we have done in this example is an elementary version of a very important statistical technique known as the *bootstrap* (Efron and Tibshirani 1993). We will return to the bootstrap in Section 10.1.4.    ‖

## 1.3 Conditional Probability and Independence

All of the probabilities that we have dealt with thus far have been unconditional probabilities. A sample space was defined and all probabilities were calculated with respect to that sample space. In many instances, however, we are in a position to update the sample space based on new information. In such cases, we want to be able to update probability calculations or to calculate *conditional probabilities.*

**Example 1.3.1 (Four aces)**   Four cards are dealt from the top of a well-shuffled deck. What is the probability that they are the four aces? We can calculate this probability by the methods of the previous section. The number of distinct groups of four cards is

$$\binom{52}{4} = 270{,}725.$$

Only one of these groups consists of the four aces and every group is equally likely, so the probability of being dealt all four aces is $1/270{,}725$.

We can also calculate this probability by an "updating" argument, as follows. The probability that the first card is an ace is $4/52$. *Given that the first card is an ace,* the probability that the second card is an ace is $3/51$ (there are 3 aces and 51 cards left). Continuing this argument, we get the desired probability as

$$\frac{4}{52} \times \frac{3}{51} \times \frac{2}{50} \times \frac{1}{49} = \frac{1}{270{,}725}. \qquad\qquad \|$$

In our second method of solving the problem, we updated the sample space after each draw of a card; we calculated conditional probabilities.

**Definition 1.3.2**   If $A$ and $B$ are events in $S$, and $P(B) > 0$, then the *conditional probability of $A$ given $B$*, written $P(A|B)$, is

$$(1.3.1) \qquad\qquad P(A|B) = \frac{P(A \cap B)}{P(B)}.$$

Note that what happens in the conditional probability calculation is that $B$ becomes the sample space: $P(B|B) = 1$. The intuition is that our original sample space, $S$, has been updated to $B$. All further occurrences are then calibrated with respect to their relation to $B$. In particular, note what happens to conditional probabilities of disjoint sets. Suppose $A$ and $B$ are disjoint, so $P(A \cap B) = 0$. It then follows that $P(A|B) = P(B|A) = 0$.

**Example 1.3.3 (Continuation of Example 1.3.1)**   Although the probability of getting all four aces is quite small, let us see how the conditional probabilities change given that some aces have already been drawn. Four cards will again be dealt from a well-shuffled deck, and we now calculate

$$P(4 \text{ aces in 4 cards} \,|\, i \text{ aces in } i \text{ cards}), \quad i = 1, 2, 3.$$

The event {4 aces in 4 cards} is a subset of the event {$i$ aces in $i$ cards}. Thus, from the definition of conditional probability, (1.3.1), we know that

$$P(4 \text{ aces in 4 cards} \,|\, i \text{ aces in } i \text{ cards})$$

$$= \frac{P(\{4 \text{ aces in 4 cards}\} \cap \{i \text{ aces in } i \text{ cards}\})}{P(i \text{ aces in } i \text{ cards})}$$

$$= \frac{P(4 \text{ aces in 4 cards})}{P(i \text{ aces in } i \text{ cards})}.$$

The numerator has already been calculated, and the denominator can be calculated with a similar argument. The number of distinct groups of $i$ cards is $\binom{52}{i}$, and

$$P(i \text{ aces in } i \text{ cards}) = \frac{\binom{4}{i}}{\binom{52}{i}}.$$

Therefore, the conditional probability is given by

$$P(4 \text{ aces in 4 cards} \,|\, i \text{ aces in } i \text{ cards}) = \frac{\binom{52}{i}}{\binom{52}{4}\binom{4}{i}} = \frac{(4-i)!48!}{(52-i)!} = \frac{1}{\binom{52-i}{4-i}}.$$

For $i = 1$, 2, and 3, the conditional probabilities are .00005, .00082, and .02041, respectively.                                                                                        ‖

For any $B$ for which $P(B) > 0$, it is straightforward to verify that the probability function $P(\cdot|B)$ satisfies Kolmogorov's Axioms (see Exercise 1.35). You may suspect that requiring $P(B) > 0$ is redundant. Who would want to condition on an event of probability 0? Interestingly, sometimes this is a particularly useful way of thinking of things. However, we will defer these considerations until Chapter 4.

Conditional probabilities can be particularly slippery entities and sometimes require careful thought. Consider the following often-told tale.

**Example 1.3.4 (Three prisoners)**   Three prisoners, A, B, and C, are on death row. The governor decides to pardon one of the three and chooses at random the prisoner to pardon. He informs the warden of his choice but requests that the name be kept secret for a few days.

The next day, A tries to get the warden to tell him who had been pardoned. The warden refuses. A then asks which of B or C will be executed. The warden thinks for a while, then tells A that B is to be executed.

**Warden's reasoning:** Each prisoner has a $\frac{1}{3}$ chance of being pardoned. Clearly, either B or C must be executed, so I have given A no information about whether A will be pardoned.

**A's reasoning:** Given that B will be executed, then either A or C will be pardoned. My chance of being pardoned has risen to $\frac{1}{2}$.

It should be clear that the warden's reasoning is correct, but let us see why. Let $A, B$, and $C$ denote the events that A, B, or C is pardoned, respectively. We know that $P(A) = P(B) = P(C) = \frac{1}{3}$. Let $\mathcal{W}$ denote the event that the warden says B will die. Using (1.3.1), A can update his probability of being pardoned to

$$P(A|\mathcal{W}) = \frac{P(A \cap \mathcal{W})}{P(\mathcal{W})}.$$

What is happening can be summarized in this table:

| Prisoner pardoned | Warden tells A | |
| --- | --- | --- |
| A | B dies | each with equal |
| A | C dies | probability |
| B | C dies | |
| C | B dies | |

Using this table, we can calculate

$$
\begin{aligned}
P(\mathcal{W}) &= P(\text{warden says B dies}) \\
&= P(\text{warden says B dies and A pardoned}) \\
&\quad + P(\text{warden says B dies and C pardoned}) \\
&\quad + P(\text{warden says B dies and B pardoned}) \\
&= \frac{1}{6} + \frac{1}{3} + 0 \;=\; \frac{1}{2}.
\end{aligned}
$$

Thus, using the warden's reasoning, we have

$$P(A|\mathcal{W}) = \frac{P(A \cap \mathcal{W})}{P(\mathcal{W})}$$

(1.3.2)
$$= \frac{P(\text{warden says B dies and A pardoned})}{P(\text{warden says B dies})} = \frac{1/6}{1/2} = \frac{1}{3}.$$

However, A falsely interprets the event $\mathcal{W}$ as equal to the event $B^c$ and calculates

$$P(A|B^c) = \frac{P(A \cap B^c)}{P(B^c)} = \frac{1/3}{2/3} = \frac{1}{2}.$$

We see that conditional probabilities can be quite slippery and require careful interpretation. For some other variations of this problem, see Exercise 1.37.     ‖

Re-expressing (1.3.1) gives a useful form for calculating intersection probabilities,

(1.3.3)                    $$P(A \cap B) = P(A|B)P(B),$$

which is essentially the formula that was used in Example 1.3.1. We can take advantage of the symmetry of (1.3.3) and also write

(1.3.4)                    $$P(A \cap B) = P(B|A)P(A).$$

When faced with seemingly difficult calculations, we can break up our calculations according to (1.3.3) or (1.3.4), whichever is easier. Furthermore, we can equate the right-hand sides of these equations to obtain (after rearrangement)

$$(1.3.5) \qquad P(A|B) = P(B|A)\frac{P(A)}{P(B)},$$

which gives us a formula for "turning around" conditional probabilities. Equation (1.3.5) is often called Bayes' Rule for its discoverer, Sir Thomas Bayes (although see Stigler 1983).

Bayes' Rule has a more general form than (1.3.5), one that applies to partitions of a sample space. We therefore take the following as the definition of Bayes' Rule.

**Theorem 1.3.5 (Bayes' Rule)**  *Let $A_1, A_2, \ldots$ be a partition of the sample space, and let $B$ be any set. Then, for each $i = 1, 2, \ldots$,*

$$P(A_i|B) = \frac{P(B|A_i)P(A_i)}{\sum_{j=1}^{\infty} P(B|A_j)P(A_j)}.$$

**Example 1.3.6 (Coding)**    When coded messages are sent, there are sometimes errors in transmission. In particular, Morse code uses "dots" and "dashes," which are known to occur in the proportion of 3:4. This means that for any given symbol,

$$P(\text{dot sent}) = \frac{3}{7} \quad \text{and} \quad P(\text{dash sent}) = \frac{4}{7}.$$

Suppose there is interference on the transmission line, and with probability $\frac{1}{8}$ a dot is mistakenly received as a dash, and vice versa. If we receive a dot, can we be sure that a dot was sent? Using Bayes' Rule, we can write

$$P(\text{dot sent} \,|\, \text{dot received}) = P(\text{dot received} \,|\, \text{dot sent})\frac{P(\text{dot sent})}{P(\text{dot received})}.$$

Now, from the information given, we know that $P(\text{dot sent}) = \frac{3}{7}$ and $P(\text{dot received}|\text{dot sent}) = \frac{7}{8}$. Furthermore, we can also write

$$P(\text{dot received}) = P(\text{dot received} \cap \text{dot sent}) + P(\text{dot received} \cap \text{dash sent})$$
$$= P(\text{dot received} \,|\, \text{dot sent})P(\text{dot sent})$$
$$+ P(\text{dot received} \,|\, \text{dash sent})P(\text{dash sent})$$
$$= \frac{7}{8} \times \frac{3}{7} + \frac{1}{8} \times \frac{4}{7} = \frac{25}{56}.$$

Combining these results, we have that the probability of correctly receiving a dot is

$$P(\text{dot sent} \,|\, \text{dot received}) = \frac{(7/8) \times (3/7)}{25/56} = \frac{21}{25}. \qquad \|$$

In some cases it may happen that the occurrence of a particular event, $B$, has no effect on the probability of another event, $A$. Symbolically, we are saying that

$$(1.3.6) \qquad P(A|B) = P(A).$$

If this holds, then by Bayes' Rule (1.3.5) and using (1.3.6) we have

$$(1.3.7) \qquad P(B|A) = P(A|B)\frac{P(B)}{P(A)} = P(A)\frac{P(B)}{P(A)} = P(B),$$

so the occurrence of $A$ has no effect on $B$. Moreover, since $P(B|A)P(A) = P(A \cap B)$, it then follows that

$$P(A \cap B) = P(A)P(B),$$

which we take as the definition of statistical independence.

**Definition 1.3.7** Two events, $A$ and $B$, are *statistically independent* if

$$(1.3.8) \qquad P(A \cap B) = P(A)P(B).$$

Note that independence could have been equivalently defined by either (1.3.6) or (1.3.7) (as long as either $P(A) > 0$ or $P(B) > 0$). The advantage of (1.3.8) is that it treats the events symmetrically and will be easier to generalize to more than two events.

Many gambling games provide models of independent events. The spins of a roulette wheel and the tosses of a pair of dice are both series of independent events.

**Example 1.3.8 (Chevalier de Meré)** The gambler introduced at the start of the chapter, the Chevalier de Meré, was particularly interested in the event that he could throw at least 1 six in 4 rolls of a die. We have

$$P(\text{at least 1 six in 4 rolls}) = 1 - P(\text{no six in 4 rolls})$$

$$= 1 - \prod_{i=1}^{4} P(\text{no six on roll } i),$$

where the last equality follows by independence of the rolls. On any roll, the probability of *not* rolling a six is $\frac{5}{6}$, so

$$P(\text{at least 1 six in 4 rolls}) = 1 - \left(\frac{5}{6}\right)^4 = .518. \qquad \|$$

Independence of $A$ and $B$ implies independence of the complements also. In fact, we have the following theorem.

**Theorem 1.3.9**     *If A and B are independent events, then the following pairs are also independent:*

**a.** *A and $B^c$,*

**b.** *$A^c$ and B,*

**c.** *$A^c$ and $B^c$.*

**Proof:** We will prove only (a), leaving the rest as Exercise 1.40. To prove (a) we must show that $P(A \cap B^c) = P(A)P(B^c)$. From Theorem 1.2.9a we have

$$
\begin{aligned}
P(A \cap B^c) &= P(A) - P(A \cap B) \\
&= P(A) - P(A)P(B) \qquad (A \text{ and } B \text{ are independent}) \\
&= P(A)(1 - P(B)) \\
&= P(A)P(B^c). \qquad\qquad\qquad \square
\end{aligned}
$$

Independence of more than two events can be defined in a manner similar to (1.3.8), but we must be careful. For example, we might think that we could say $A, B$, and $C$ are independent if $P(A \cap B \cap C) = P(A)P(B)P(C)$. However, this is not the correct condition.

**Example 1.3.10 (Tossing two dice)**     Let an experiment consist of tossing two dice. For this experiment the sample space is

$$S = \{(1,1), (1,2), \ldots, (1,6), (2,1), \ldots, (2,6), \ldots, (6,1), \ldots, (6,6)\};$$

that is, $S$ consists of the 36 ordered pairs formed from the numbers 1 to 6. Define the following events:

$$
\begin{aligned}
A &= \{\text{doubles appear}\} &= \{(1,1), (2,2), (3,3), (4,4), (5,5), (6,6)\}, \\
B &= \{\text{the sum is between 7 and 10}\}, \\
C &= \{\text{the sum is 2 or 7 or 8}\}.
\end{aligned}
$$

The probabilities can be calculated by counting among the 36 possible outcomes. We have

$$P(A) = \frac{1}{6}, \quad P(B) = \frac{1}{2}, \quad \text{and} \quad P(C) = \frac{1}{3}.$$

Furthermore,

$$
\begin{aligned}
P(A \cap B \cap C) &= P(\text{the sum is 8, composed of double 4s}) \\
&= \frac{1}{36} \\
&= \frac{1}{6} \times \frac{1}{2} \times \frac{1}{3} \\
&= P(A)P(B)P(C).
\end{aligned}
$$

However,

$$P(B \cap C) = P(\text{sum equals 7 or 8}) = \frac{11}{36} \neq P(B)P(C).$$

Similarly, it can be shown that $P(A \cap B) \neq P(A)P(B)$; therefore, the requirement $P(A \cap B \cap C) = P(A)P(B)P(C)$ is not a strong enough condition to guarantee pairwise independence. ‖

A second attempt at a general definition of independence, in light of the previous example, might be to define $A, B,$ and $C$ to be independent if all the pairs are independent. Alas, this condition also fails.

**Example 1.3.11 (Letters)** Let the sample space $S$ consist of the 3! permutations of the letters a, b, and c along with the three triples of each letter. Thus,

$$S = \left\{ \begin{array}{ccc} \text{aaa} & \text{bbb} & \text{ccc} \\ \text{abc} & \text{bca} & \text{cba} \\ \text{acb} & \text{bac} & \text{cab} \end{array} \right\}.$$

Furthermore, let each element of $S$ have probability $\frac{1}{9}$. Define

$$A_i = \{i\text{th place in the triple is occupied by a}\}.$$

It is then easy to count that

$$P(A_i) = \frac{1}{3}, \quad i = 1, 2, 3,$$

and

$$P(A_1 \cap A_2) = P(A_1 \cap A_3) = P(A_2 \cap A_3) = \frac{1}{9},$$

so the $A_i$s are pairwise independent. But

$$P(A_1 \cap A_2 \cap A_3) = \frac{1}{9} \neq P(A_1)P(A_2)P(A_3),$$

so the $A_i$s do not satisfy the probability requirement. ‖

The preceding two examples show that simultaneous (or mutual) independence of a collection of events requires an extremely strong definition. The following definition works.

**Definition 1.3.12** A collection of events $A_1, \ldots, A_n$ are *mutually independent* if for any subcollection $A_{i_1}, \ldots, A_{i_k}$, we have

$$P\left( \bigcap_{j=1}^{k} A_{i_j} \right) = \prod_{j=1}^{k} P\left( A_{i_j} \right).$$

**Example 1.3.13 (Three coin tosses–I)**    Consider the experiment of tossing a coin three times. A sample point for this experiment must indicate the result of each toss. For example, HHT could indicate that two heads and then a tail were observed. The sample space for this experiment has eight points, namely,

$$\{\text{HHH}, \text{HHT}, \text{HTH}, \text{THH}, \text{TTH}, \text{THT}, \text{HTT}, \text{TTT}\}.$$

Let $H_i$, $i = 1, 2, 3$, denote the event that the $i$th toss is a head. For example,

(1.3.9)                        $H_1 = \{\text{HHH}, \text{HHT}, \text{HTH}, \text{HTT}\}.$

If we assign probability $\frac{1}{8}$ to each sample point, then using enumerations such as (1.3.9), we see that $P(H_1) = P(H_2) = P(H_3) = \frac{1}{2}$. This says that the coin is fair and has an equal probability of landing heads or tails on each toss.

Under this probability model, the events $H_1$, $H_2$, and $H_3$ are also mutually independent. To verify this we note that

$$P(H_1 \cap H_2 \cap H_3) = P(\{\text{HHH}\}) = \frac{1}{8} = \frac{1}{2} \cdot \frac{1}{2} \cdot \frac{1}{2} = P(H_1)P(H_2)P(H_3).$$

To verify the condition in Definition 1.3.12, we also must check each pair. For example,

$$P(H_1 \cap H_2) = P(\{\text{HHH}, \text{HHT}\}) = \frac{2}{8} = \frac{1}{2} \cdot \frac{1}{2} = P(H_1)P(H_2).$$

The equality is also true for the other two pairs. Thus, $H_1$, $H_2$, and $H_3$ are mutually independent. That is, the occurrence of a head on any toss has no effect on any of the other tosses.

It can be verified that the assignment of probability $\frac{1}{8}$ to each sample point is the only probability model that has $P(H_1) = P(H_2) = P(H_3) = \frac{1}{2}$ and $H_1, H_2$, and $H_3$ mutually independent.                                                                                    ‖

## 1.4 Random Variables

In many experiments it is easier to deal with a summary variable than with the original probability structure. For example, in an opinion poll, we might decide to ask 50 people whether they agree or disagree with a certain issue. If we record a "1" for agree and "0" for disagree, the sample space for this experiment has $2^{50}$ elements, each an ordered string of 1s and 0s of length 50. We should be able to reduce this to a reasonable size! It may be that the only quantity of interest is the number of people who agree (equivalently, disagree) out of 50 and, if we define a variable $X =$ number of 1s recorded out of 50, we have captured the essence of the problem. Note that the sample space for $X$ is the set of integers $\{0, 1, 2, \ldots, 50\}$ and is much easier to deal with than the original sample space.

In defining the quantity $X$, we have defined a mapping (a function) from the original sample space to a new sample space, usually a set of real numbers. In general, we have the following definition.

**Definition 1.4.1**    A *random variable* is a function from a sample space $S$ into the real numbers.

**Example 1.4.2 (Random variables)** In some experiments random variables are implicitly used; some examples are these.

*Examples of random variables*

| Experiment | Random variable |
|---|---|
| Toss two dice | $X$ = sum of the numbers |
| Toss a coin 25 times | $X$ = number of heads in 25 tosses |
| Apply different amounts of fertilizer to corn plants | $X$ = yield/acre |

‖

In defining a random variable, we have also defined a new sample space (the range of the random variable). We must now check formally that our probability function, which is defined on the original sample space, can be used for the random variable.

Suppose we have a sample space

$$S = \{s_1, \ldots, s_n\}$$

with a probability function $P$ and we define a random variable $X$ with range $\mathcal{X} = \{x_1, \ldots, x_m\}$. We can define a probability function $P_X$ on $\mathcal{X}$ in the following way. We will observe $X = x_i$ if and only if the outcome of the random experiment is an $s_j \in S$ such that $X(s_j) = x_i$. Thus,

$$(1.4.1) \qquad P_X(X = x_i) = P(\{s_j \in S : X(s_j) = x_i\}).$$

Note that the left-hand side of (1.4.1), the function $P_X$, is an *induced* probability function on $\mathcal{X}$, defined in terms of the original function $P$. Equation (1.4.1) formally defines a probability function, $P_X$, for the random variable $X$. Of course, we have to verify that $P_X$ satisfies the Kolmogorov Axioms, but that is not a very difficult job (see Exercise 1.45). Because of the equivalence in (1.4.1), we will simply write $P(X = x_i)$ rather than $P_X(X = x_i)$.

*A note on notation*: Random variables will always be denoted with uppercase letters and the realized values of the variable (or its range) will be denoted by the corresponding lowercase letters. Thus, the random variable $X$ can take the value $x$.

**Example 1.4.3 (Three coin tosses–II)** Consider again the experiment of tossing a fair coin three times from Example 1.3.13. Define the random variable $X$ to be the number of heads obtained in the three tosses. A complete enumeration of the value of $X$ for each point in the sample space is

| $s$ | HHH | HHT | HTH | THH | TTH | THT | HTT | TTT |
|---|---|---|---|---|---|---|---|---|
| $X(s)$ | 3 | 2 | 2 | 2 | 1 | 1 | 1 | 0 |

The range for the random variable $X$ is $\mathcal{X} = \{0, 1, 2, 3\}$. Assuming that all eight points in $S$ have probability $\frac{1}{8}$, by simply counting in the above display we see that the induced probability function on $\mathcal{X}$ is given by

| $x$ | 0 | 1 | 2 | 3 |
|---|---|---|---|---|
| $P_X(X = x)$ | $\frac{1}{8}$ | $\frac{3}{8}$ | $\frac{3}{8}$ | $\frac{1}{8}$ |

For example, $P_X(X = 1) = P(\{\text{HTT}, \text{THT}, \text{TTH}\}) = \frac{3}{8}$.                    ‖

**Example 1.4.4 (Distribution of a random variable)**    It may be possible to determine $P_X$ even if a complete listing, as in Example 1.4.3, is not possible. Let $S$ be the $2^{50}$ strings of 50 0s and 1s, $X$ = number of 1s, and $\mathcal{X} = \{0, 1, 2, \ldots, 50\}$, as mentioned at the beginning of this section. Suppose that each of the $2^{50}$ strings is equally likely. The probability that $X = 27$ can be obtained by counting all of the strings with 27 1s in the original sample space. Since each string is equally likely, it follows that

$$P_X(X = 27) = \frac{\# \text{ strings with 27 1s}}{\# \text{ strings}} = \frac{\binom{50}{27}}{2^{50}}.$$

In general, for any $i \in \mathcal{X}$,

$$P_X(X = i) = \frac{\binom{50}{i}}{2^{50}}.$$                    ‖

The previous illustrations had both a finite $S$ and finite $\mathcal{X}$, and the definition of $P_X$ was straightforward. Such is also the case if $\mathcal{X}$ is countable. If $\mathcal{X}$ is uncountable, we define the induced probability function, $P_X$, in a manner similar to (1.4.1). For any set $A \subset \mathcal{X}$,

(1.4.2)                    $P_X(X \in A) = P(\{s \in S : X(s) \in A\})\,.$

This does define a legitimate probability function for which the Kolmogorov Axioms can be verified. (To be precise, we use (1.4.2) to define probabilities only for a certain sigma algebra of subsets of $\mathcal{X}$. But we will not concern ourselves with these technicalities.)

## 1.5 Distribution Functions

With every random variable $X$, we associate a function called the cumulative distribution function of $X$.

**Definition 1.5.1**    The *cumulative distribution function* or *cdf* of a random variable $X$, denoted by $F_X(x)$, is defined by

$$F_X(x) = P_X(X \leq x), \quad \text{for all } x.$$

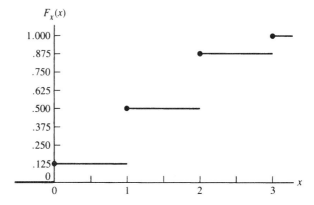

Figure 1.5.1. *Cdf of Example 1.5.2*

**Example 1.5.2 (Tossing three coins)**   Consider the experiment of tossing three fair coins, and let $X$ = number of heads observed. The cdf of $X$ is

(1.5.1)
$$F_X(x) = \begin{cases} 0 & \text{if } -\infty < x < 0 \\ \frac{1}{8} & \text{if } 0 \leq x < 1 \\ \frac{1}{2} & \text{if } 1 \leq x < 2 \\ \frac{7}{8} & \text{if } 2 \leq x < 3 \\ 1 & \text{if } 3 \leq x < \infty. \end{cases}$$

The step function $F_X(x)$ is graphed in Figure 1.5.1. There are several points to note from Figure 1.5.1. $F_X$ is defined for all values of $x$, not just those in $\mathcal{X} = \{0, 1, 2, 3\}$. Thus, for example,

$$F_X(2.5) = P(X \leq 2.5) = P(X = 0, 1, \text{or } 2) = \frac{7}{8}.$$

Note that $F_X$ has jumps at the values of $x_i \in \mathcal{X}$ and the size of the jump at $x_i$ is equal to $P(X = x_i)$. Also, $F_X(x) = 0$ for $x < 0$ since $X$ cannot be negative, and $F_X(x) = 1$ for $x \geq 3$ since $x$ is certain to be less than or equal to such a value.   ‖

As is apparent from Figure 1.5.1, $F_X$ can be discontinuous, with jumps at certain values of $x$. By the way in which $F_X$ is defined, however, at the jump points $F_X$ takes the value at the top of the jump. (Note the different inequalities in (1.5.1).) This is known as *right-continuity*—the function is continuous when a point is approached from the right. The property of right-continuity is a consequence of the definition of the cdf. In contrast, if we had defined $F_X(x) = P_X(X < x)$ (note strict inequality), $F_X$ would then be *left-continuous*. The size of the jump at any point $x$ is equal to $P(X = x)$.

Every cdf satisfies certain properties, some of which are obvious when we think of the definition of $F_X(x)$ in terms of probabilities.

**Theorem 1.5.3**    *The function $F(x)$ is a cdf if and only if the following three conditions hold:*

**a.** $\lim_{x \to -\infty} F(x) = 0$ *and* $\lim_{x \to \infty} F(x) = 1$.

**b.** $F(x)$ *is a nondecreasing function of $x$.*

**c.** $F(x)$ *is right-continuous; that is, for every number $x_0$, $\lim_{x \downarrow x_0} F(x) = F(x_0)$.*

**Outline of proof:**  To prove necessity, the three properties can be verified by writing $F$ in terms of the probability function (see Exercise 1.48). To prove sufficiency, that if a function $F$ satisfies the three conditions of the theorem then it is a cdf for some random variable, is much harder. It must be established that there exists a sample space $S$, a probability function $P$ on $S$, and a random variable $X$ defined on $S$ such that $F$ is the cdf of $X$.                                                     □

**Example 1.5.4 (Tossing for a head)**    Suppose we do an experiment that consists of tossing a coin until a head appears. Let $p =$ probability of a head on any given toss, and define a random variable $X =$ number of tosses required to get a head. Then, for any $x = 1, 2, \ldots$,

$$(1.5.2) \qquad\qquad P(X = x) = (1 - p)^{x-1} p,$$

since we must get $x - 1$ tails followed by a head for the event to occur and all trials are independent. From (1.5.2) we calculate, for any positive integer $x$,

$$(1.5.3) \qquad\qquad P(X \leq x) = \sum_{i=1}^{x} P(X = i) = \sum_{i=1}^{x} (1 - p)^{i-1} p.$$

The partial sum of the geometric series is

$$(1.5.4) \qquad\qquad \sum_{k=1}^{n} t^{k-1} = \frac{1 - t^n}{1 - t}, \qquad t \neq 1,$$

a fact that can be established by induction (see Exercise 1.50). Applying (1.5.4) to our probability, we find that the cdf of the random variable $X$ is

$$F_X(x) = P(X \leq x)$$
$$= \frac{1 - (1 - p)^x}{1 - (1 - p)} p$$
$$= 1 - (1 - p)^x, \qquad x = 1, 2, \ldots .$$

The cdf $F_X(x)$ is flat between the nonnegative integers, as in Example 1.5.2.

It is easy to show that if $0 < p < 1$, then $F_X(x)$ satisfies the conditions of Theorem 1.5.3. First,

$$\lim_{x \to -\infty} F_X(x) = 0$$

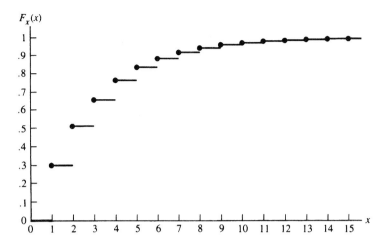

Figure 1.5.2. *Geometric cdf, p = .3*

since $F_X(x) = 0$ for all $x < 0$, and

$$\lim_{x \to \infty} F_X(x) = \lim_{x \to \infty} 1 - (1-p)^x = 1,$$

where $x$ goes through only integer values when this limit is taken. To verify property (b), we simply note that the sum in (1.5.3) contains more *positive* terms as $x$ increases. Finally, to verify (c), note that, for any $x$, $F_X(x + \epsilon) = F_X(x)$ if $\epsilon > 0$ is sufficiently small. Hence,

$$\lim_{\epsilon \downarrow 0} F_X(x + \epsilon) = F_X(x),$$

so $F_X(x)$ is right-continuous. $F_X(x)$ is the cdf of a distribution called the *geometric distribution* (after the series) and is pictured in Figure 1.5.2.                            ‖

**Example 1.5.5 (Continuous cdf)**   An example of a continuous cdf is the function

(1.5.5)                            $$F_X(x) = \frac{1}{1 + e^{-x}},$$

which satisfies the conditions of Theorem 1.5.3. For example,

$$\lim_{x \to -\infty} F_X(x) = 0 \quad \text{since} \quad \lim_{x \to -\infty} e^{-x} = \infty$$

and

$$\lim_{x \to \infty} F_X(x) = 1 \quad \text{since} \quad \lim_{x \to \infty} e^{-x} = 0.$$

Differentiating $F_X(x)$ gives

$$\frac{d}{dx} F_X(x) = \frac{e^{-x}}{(1 + e^{-x})^2} > 0,$$

showing that $F_X(x)$ is increasing. $F_X$ is not only right-continuous, but also continuous. This is a special case of the *logistic distribution*. ‖

**Example 1.5.6 (Cdf with jumps)** If $F_X$ is not a continuous function of $x$, it is possible for it to be a mixture of continuous pieces and jumps. For example, if we modify $F_X(x)$ of (1.5.5) to be, for some $\epsilon, 1 > \epsilon > 0$,

(1.5.6)
$$F_Y(y) = \begin{cases} \dfrac{1-\epsilon}{1+e^{-y}} & \text{if } y < 0 \\[2mm] \epsilon + \dfrac{(1-\epsilon)}{1+e^{-y}} & \text{if } y \geq 0, \end{cases}$$

then $F_Y(y)$ is the cdf of a random variable $Y$ (see Exercise 1.47). The function $F_Y$ has a jump of height $\epsilon$ at $y = 0$ and otherwise is continuous. This model might be appropriate if we were observing the reading from a gauge, a reading that could (theoretically) be anywhere between $-\infty$ and $\infty$. This particular gauge, however, sometimes sticks at 0. We could then model our observations with $F_Y$, where $\epsilon$ is the probability that the gauge sticks. ‖

Whether a cdf is continuous or has jumps corresponds to the associated random variable being continuous or not. In fact, the association is such that it is convenient to define continuous random variables in this way.

**Definition 1.5.7** A random variable $X$ is *continuous* if $F_X(x)$ is a continuous function of $x$. A random variable $X$ is *discrete* if $F_X(x)$ is a step function of $x$.

We close this section with a theorem formally stating that $F_X$ completely determines the probability distribution of a random variable $X$. This is true if $P(X \in A)$ is defined only for events $A$ in $\mathcal{B}^1$, the smallest sigma algebra containing all the intervals of real numbers of the form $(a, b)$, $[a, b)$, $(a, b]$, and $[a, b]$. If probabilities are defined for a larger class of events, it is possible for two random variables to have the same distribution function but not the same probability for every event (see Chung 1974, page 27). In this book, as in most statistical applications, we are concerned only with events that are intervals, countable unions or intersections of intervals, etc. So we do not consider such pathological cases. We first need the notion of two random variables being identically distributed.

**Definition 1.5.8** The random variables $X$ and $Y$ are *identically distributed* if, for every set $A \in \mathcal{B}^1$, $P(X \in A) = P(Y \in A)$.

Note that two random variables that are identically distributed are not necessarily equal. That is, Definition 1.5.8 does not say that $X = Y$.

**Example 1.5.9 (Identically distributed random variables)** Consider the experiment of tossing a fair coin three times as in Example 1.4.3. Define the random variables $X$ and $Y$ by

$$X = \text{number of heads observed} \quad \text{and} \quad Y = \text{number of tails observed}.$$

The distribution of $X$ is given in Example 1.4.3, and it is easily verified that the distribution of $Y$ is exactly the same. That is, for each $k = 0, 1, 2, 3$, we have $P(X = k) = P(Y = k)$. So $X$ and $Y$ are identically distributed. However, for no sample points do we have $X(s) = Y(s)$.       ‖

**Theorem 1.5.10**    *The following two statements are equivalent:*

**a.** *The random variables $X$ and $Y$ are identically distributed.*

**b.** $F_X(x) = F_Y(x)$ *for every $x$.*

**Proof:** To show equivalence we must show that each statement implies the other. We first show that (a) $\Rightarrow$ (b).

Because $X$ and $Y$ are identically distributed, for any set $A \in \mathcal{B}^1$, $P(X \in A) = P(Y \in A)$. In particular, for every $x$, the set $(-\infty, x]$ is in $\mathcal{B}^1$, and

$$F_X(x) = P(X \in (-\infty, x]) = P(Y \in (-\infty, x]) = F_Y(x).$$

The converse implication, that (b) $\Rightarrow$ (a), is much more difficult to prove. The above argument showed that if the $X$ and $Y$ probabilities agreed on all sets, then they agreed on intervals. We now must prove the opposite; that is, if the $X$ and $Y$ probabilities agree on all intervals, then they agree on all sets. To show this requires heavy use of sigma algebras; we will not go into these details here. Suffice it to say that it is necessary to prove only that the two probability functions agree on all intervals (Chung 1974, Section 2.2).     □

## 1.6 Density and Mass Functions

Associated with a random variable $X$ and its cdf $F_X$ is another function, called either the probability density function (pdf) or probability mass function (pmf). The terms pdf and pmf refer, respectively, to the continuous and discrete cases. Both pdfs and pmfs are concerned with "point probabilities" of random variables.

**Definition 1.6.1**    The *probability mass function (pmf)* of a discrete random variable $X$ is given by

$$f_X(x) = P(X = x) \quad \text{for all } x.$$

**Example 1.6.2 (Geometric probabilities)**    For the geometric distribution of Example 1.5.4, we have the pmf

$$f_X(x) = P(X = x) = \begin{cases} (1-p)^{x-1}p & \text{for } x = 1, 2, \ldots \\ 0 & \text{otherwise.} \end{cases}$$

Recall that $P(X = x)$ or, equivalently, $f_X(x)$ is the size of the jump in the cdf at $x$. We can use the pmf to calculate probabilities. Since we can now measure the probability of a single point, we need only sum over all of the points in the appropriate event. Hence, for positive integers $a$ and $b$, with $a \le b$, we have

$$P(a \le X \le b) = \sum_{k=a}^{b} f_X(k) = \sum_{k=a}^{b} (1-p)^{k-1}p.$$

As a special case of this we get

$$(1.6.1) \qquad P(X \le b) = \sum_{k=1}^{b} f_X(k) = F_X(b). \qquad\qquad \|$$

A widely accepted convention, which we will adopt, is to use an uppercase letter for the cdf and the corresponding lowercase letter for the pmf or pdf.

We must be a little more careful in our definition of a pdf in the continuous case. If we naively try to calculate $P(X = x)$ for a continuous random variable, we get the following. Since $\{X = x\} \subset \{x - \epsilon < X \le x\}$ for any $\epsilon > 0$, we have from Theorem 1.2.9(c) that

$$P(X = x) \le P(x - \epsilon < X \le x) = F_X(x) - F_X(x - \epsilon)$$

for any $\epsilon > 0$. Therefore,

$$0 \le P(X = x) \le \lim_{\epsilon \downarrow 0} [F_X(x) - F_X(x - \epsilon)] = 0$$

by the continuity of $F_X$. However, if we understand the purpose of the pdf, its definition will become clear.

From Example 1.6.2, we see that a pmf gives us "point probabilities." In the discrete case, we can sum over values of the pmf to get the cdf (as in (1.6.1)). The analogous procedure in the continuous case is to substitute integrals for sums, and we get

$$P(X \le x) = F_X(x) = \int_{-\infty}^{x} f_X(t)\, dt.$$

Using the Fundamental Theorem of Calculus, if $f_X(x)$ is continuous, we have the further relationship

$$(1.6.2) \qquad \frac{d}{dx} F_X(x) = f_X(x).$$

Note that the analogy with the discrete case is almost exact. We "add up" the "point probabilities" $f_X(x)$ to obtain interval probabilities.

**Definition 1.6.3**    The *probability density function* or *pdf*, $f_X(x)$, of a continuous random variable $X$ is the function that satisfies

$$(1.6.3) \qquad F_X(x) = \int_{-\infty}^{x} f_X(t)\, dt \quad \text{for all } x.$$

*A note on notation*: The expression "$X$ has a distribution given by $F_X(x)$" is abbreviated symbolically by "$X \sim F_X(x)$," where we read the symbol "$\sim$" as "is distributed as." We can similarly write $X \sim f_X(x)$ or, if $X$ and $Y$ have the same distribution, $X \sim Y$.

In the continuous case we can be somewhat cavalier about the specification of interval probabilities. Since $P(X = x) = 0$ if $X$ is a continuous random variable,

$$P(a < X < b) = P(a < X \le b) = P(a \le X < b) = P(a \le X \le b).$$

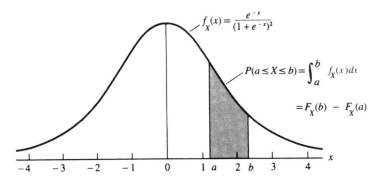

Figure 1.6.1. *Area under logistic curve*

It should be clear that the pdf (or pmf) contains the same information as the cdf. This being the case, we can use either one to solve problems and should try to choose the simpler one.

**Example 1.6.4 (Logistic probabilities)** For the logistic distribution of Example 1.5.5 we have

$$F_X(x) = \frac{1}{1 + e^{-x}}$$

and, hence,

$$f_X(x) = \frac{d}{dx} F_X(x) = \frac{e^{-x}}{(1 + e^{-x})^2}.$$

The area under the curve $f_X(x)$ gives us interval probabilities (see Figure 1.6.1):

$$
\begin{aligned}
P(a < X < b) \;&=\; F_X(b) - F_X(a) \\
&=\; \int_{-\infty}^{b} f_X(x)\,dx - \int_{-\infty}^{a} f_X(x)\,dx \\
&=\; \int_{a}^{b} f_X(x)\,dx. \qquad\qquad \|
\end{aligned}
$$

There are really only two requirements for a pdf (or pmf), both of which are immediate consequences of the definition.

**Theorem 1.6.5**  *A function $f_X(x)$ is a pdf (or pmf) of a random variable $X$ if and only if*

**a.** $f_X(x) \geq 0$ *for all* $x$.

**b.** $\sum_x f_X(x) = 1$ *(pmf)*   or   $\int_{-\infty}^{\infty} f_X(x)\,dx = 1$ *(pdf)*.

**Proof:** If $f_X(x)$ is a pdf (or pmf), then the two properties are immediate from the definitions. In particular, for a pdf, using (1.6.3) and Theorem 1.5.3, we have that

$$1 = \lim_{x \to \infty} F_X(x) = \int_{-\infty}^{\infty} f_X(t)\, dt.$$

The converse implication is equally easy to prove. Once we have $f_X(x)$, we can define $F_X(x)$ and appeal to Theorem 1.5.3. □

From a purely mathematical viewpoint, any nonnegative function with a finite positive integral (or sum) can be turned into a pdf or pmf. For example, if $h(x)$ is any nonnegative function that is positive on a set $A$, 0 elsewhere, and

$$\int_{\{x \in A\}} h(x)\, dx = K < \infty$$

for some constant $K > 0$, then the function $f_X(x) = h(x)/K$ is a pdf of a random variable $X$ taking values in $A$.

Actually, the relationship (1.6.3) does not always hold because $F_X(x)$ may be continuous but not differentiable. In fact, there exist continuous random variables for which the integral relationship does not exist for *any* $f_X(x)$. These cases are rather pathological and we will ignore them. Thus, in this text, we will assume that (1.6.3) holds for any continuous random variable. In more advanced texts (for example, Billingsley 1995, Section 31) a random variable is called *absolutely continuous* if (1.6.3) holds.

## 1.7 Exercises

**1.1** For each of the following experiments, describe the sample space.

   (a) Toss a coin four times.

   (b) Count the number of insect-damaged leaves on a plant.

   (c) Measure the lifetime (in hours) of a particular brand of light bulb.

   (d) Record the weights of 10-day-old rats.

   (e) Observe the proportion of defectives in a shipment of electronic components.

**1.2** Verify the following identities.

   (a) $A \backslash B = A \backslash (A \cap B) = A \cap B^c$

   (b) $B = (B \cap A) \cup (B \cap A^c)$

   (c) $B \backslash A = B \cap A^c$

   (d) $A \cup B = A \cup (B \cap A^c)$

**1.3** Finish the proof of Theorem 1.1.4. For any events $A$, $B$, and $C$ defined on a sample space $S$, show that

(a) $A \cup B = B \cup A$ and $A \cap B = B \cap A$.                 (commutativity)
(b) $A \cup (B \cup C) = (A \cup B) \cup C$ and $A \cap (B \cap C) = (A \cap B) \cap C$.     (associativity)
(c) $(A \cup B)^c = A^c \cap B^c$ and $(A \cap B)^c = A^c \cup B^c$.        (DeMorgan's Laws)

**1.4** For events $A$ and $B$, find formulas for the probabilities of the following events in terms of the quantities $P(A)$, $P(B)$, and $P(A \cap B)$.

(a) either $A$ or $B$ or both
(b) either $A$ or $B$ but not both
(c) at least one of $A$ or $B$
(d) at most one of $A$ or $B$

**1.5** Approximately one-third of all human twins are identical (one-egg) and two-thirds are fraternal (two-egg) twins. Identical twins are necessarily the same sex, with male and female being equally likely. Among fraternal twins, approximately one-fourth are both female, one-fourth are both male, and half are one male and one female. Finally, among all U.S. births, approximately 1 in 90 is a twin birth. Define the following events:

$$A = \{\text{a U.S. birth results in twin females}\}$$

$$B = \{\text{a U.S. birth results in identical twins}\}$$

$$C = \{\text{a U.S. birth results in twins}\}$$

(a) State, in words, the event $A \cap B \cap C$.
(b) Find $P(A \cap B \cap C)$.

**1.6** Two pennies, one with $P(\text{head}) = u$ and one with $P(\text{head}) = w$, are to be tossed together independently. Define

$$p_0 = P(0 \text{ heads occur}),$$

$$p_1 = P(1 \text{ head occurs}),$$

$$p_2 = P(2 \text{ heads occur}).$$

Can $u$ and $w$ be chosen such that $p_0 = p_1 = p_2$? Prove your answer.

**1.7** Refer to the dart game of Example 1.2.7. Suppose we do not assume that the probability of hitting the dart board is 1, but rather is proportional to the area of the dart board. Assume that the dart board is mounted on a wall that is hit with probability 1, and the wall has area $A$.

(a) Using the fact that the probability of hitting a region is proportional to area, construct a probability function for $P(\text{scoring } i \text{ points})$, $i = 0, \ldots, 5$. (No points are scored if the dart board is not hit.)
(b) Show that the conditional probability distribution $P(\text{scoring } i \text{ points}|\text{board is hit})$ is exactly the probability distribution of Example 1.2.7.

**1.8** Again refer to the game of darts explained in Example 1.2.7.

(a) Derive the general formula for the probability of scoring $i$ points.
(b) Show that $P(\text{scoring } i \text{ points})$ is a decreasing function of $i$, that is, as the points increase, the probability of scoring them decreases.
(c) Show that $P(\text{scoring } i \text{ points})$ is a probability function according to the Kolmogorov Axioms.

**1.9** Prove the general version of DeMorgan's Laws. Let $\{A_\alpha: \alpha \in \Gamma\}$ be a (possibly uncountable) collection of sets. Prove that

(a) $\left(\cup_\alpha A_\alpha\right)^c = \cap_\alpha A_\alpha^c$.                      (b) $\left(\cap_\alpha A_\alpha\right)^c = \cup_\alpha A_\alpha^c$.

**1.10** Formulate and prove a version of DeMorgan's Laws that applies to a finite collection of sets $A_1, \ldots, A_n$.

**1.11** Let $S$ be a sample space.

(a) Show that the collection $\mathcal{B} = \{\emptyset, S\}$ is a sigma algebra.

(b) Let $\mathcal{B} = \{$all subsets of $S$, including $S$ itself$\}$. Show that $\mathcal{B}$ is a sigma algebra.

(c) Show that the intersection of two sigma algebras is a sigma algebra.

**1.12** It was noted in Section 1.2.1 that statisticians who follow the deFinetti school do not accept the Axiom of Countable Additivity, instead adhering to the Axiom of Finite Additivity.

(a) Show that the Axiom of Countable Additivity implies Finite Additivity.

(b) Although, by itself, the Axiom of Finite Additivity does not imply Countable Additivity, suppose we supplement it with the following. Let $A_1 \supset A_2 \supset \cdots \supset A_n \supset \cdots$ be an infinite sequence of nested sets whose limit is the empty set, which we denote by $A_n \downarrow \emptyset$. Consider the following:

**Axiom of Continuity:**     If $A_n \downarrow \emptyset$, then $P(A_n) \to 0$.

Prove that the Axiom of Continuity and the Axiom of Finite Additivity imply Countable Additivity.

**1.13** If $P(A) = \frac{1}{3}$ and $P(B^c) = \frac{1}{4}$, can $A$ and $B$ be disjoint? Explain.

**1.14** Suppose that a sample space $S$ has $n$ elements. Prove that the number of subsets that can be formed from the elements of $S$ is $2^n$.

**1.15** Finish the proof of Theorem 1.2.14. Use the result established for $k = 2$ as the basis of an induction argument.

**1.16** How many different sets of initials can be formed if every person has one surname and

(a) exactly two given names?               (b) either one or two given names?

(c) either one or two or three given names?

(Answers: (a) $26^3$ (b) $26^3 + 26^2$ (c) $26^4 + 26^3 + 26^2$)

**1.17** In the game of dominoes, each piece is marked with two numbers. The pieces are symmetrical so that the number pair is not ordered (so, for example, $(2,6) = (6,2)$). How many different pieces can be formed using the numbers $1, 2, \ldots, n$?

(Answer: $n(n+1)/2$)

**1.18** If $n$ balls are placed at random into $n$ cells, find the probability that exactly one cell remains empty.

(Answer: $\binom{n}{2} n! / n^n$)

**1.19** If a multivariate function has continuous partial derivatives, the order in which the derivatives are calculated does not matter. Thus, for example, the function $f(x, y)$ of two variables has equal third partials

$$\frac{\partial^3}{\partial x^2 \partial y} f(x, y) = \frac{\partial^3}{\partial y \partial x^2} f(x, y).$$

(a) How many fourth partial derivatives does a function of three variables have?

(b) Prove that a function of $n$ variables has $\binom{n+r-1}{r}$ $r$th partial derivatives.

**1.20** My telephone rings 12 times each week, the calls being randomly distributed among the 7 days. What is the probability that I get at least one call each day?

(Answer: .2285)

**1.21** A closet contains $n$ pairs of shoes. If $2r$ shoes are chosen at random $(2r < n)$, what is the probability that there will be no matching pair in the sample? (Answer: $\binom{n}{2r}2^{2r}/\binom{2n}{2r}$)

**1.22** (a) In a draft lottery containing the 366 days of the year (including February 29), what is the probability that the first 180 days drawn (without replacement) are evenly distributed among the 12 months?

(b) What is the probability that the first 30 days drawn contain none from September? (Answers: (a) $.167 \times 10^{-8}$ (b) $\binom{336}{30}/\binom{366}{30}$)

**1.23** Two people each toss a fair coin $n$ times. Find the probability that they will toss the same number of heads. (Answer: $\left(\frac{1}{4}\right)^n \binom{2n}{n}$)

**1.24** Two players, A and B, alternately and independently flip a coin and the first player to obtain a head wins. Assume player A flips first.

(a) If the coin is fair, what is the probability that A wins?

(b) Suppose that $P(\text{head}) = p$, not necessarily $\frac{1}{2}$. What is the probability that A wins?

(c) Show that for all $p, 0 < p < 1, P(\text{A wins}) > \frac{1}{2}$. (*Hint:* Try to write $P(\text{A wins})$ in terms of the events $E_1, E_2, \ldots$, where $E_i = \{\text{head first appears on } i\text{th toss}\}$.)

(Answers: (a) 2/3 (b) $\frac{p}{1-(1-p)^2}$)

**1.25** The Smiths have two children. At least one of them is a boy. What is the probability that both children are boys? (See Gardner 1961 for a complete discussion of this problem.)

**1.26** A fair die is cast until a 6 appears. What is the probability that it must be cast more than five times?

**1.27** Verify the following identities for $n \geq 2$.

(a) $\sum_{k=0}^{n}(-1)^k \binom{n}{k} = 0$  
(b) $\sum_{k=1}^{n} k \binom{n}{k} = n2^{n-1}$  
(c) $\sum_{k=1}^{n}(-1)^{k+1}k \binom{n}{k} = 0$

**1.28** A way of approximating large factorials is through the use of *Stirling's Formula*:

$$n! \approx \sqrt{2\pi}n^{n+(1/2)}e^{-n},$$

a complete derivation of which is difficult. Instead, prove the easier fact,

$$\lim_{n\to\infty} \frac{n!}{n^{n+(1/2)}e^{-n}} = \text{a constant.}$$

(*Hint:* Feller 1968 proceeds by using the monotonicity of the logarithm to establish that

$$\int_{k-1}^{k} \log x\, dx < \log k < \int_{k}^{k+1} \log x\, dx, \quad k = 1, \ldots, n,$$

and hence

$$\int_{0}^{n} \log x\, dx < \log n! < \int_{1}^{n+1} \log x\, dx.$$

Now compare $\log n!$ to the average of the two integrals. See Exercise 5.35 for another derivation.)

**1.29** (a) For the situation of Example 1.2.20, enumerate the ordered samples that make up the unordered samples $\{4, 4, 12, 12\}$ and $\{2, 9, 9, 12\}$.

(b) Suppose that we had a collection of six numbers, $\{1, 2, 7, 8, 14, 20\}$. What is the probability of drawing, with replacement, the unordered sample $\{2, 7, 7, 8, 14, 14\}$?

(c) Verify that an unordered sample of size $k$, from $m$ different numbers repeated $k_1, k_2, \ldots, k_m$ times, has $\dfrac{k!}{k_1! k_2! \cdots k_m!}$ ordered components, where $k_1 + k_2 + \cdots + k_m = k$.

(d) Establish that the number of multinomial coefficients, and hence the number of distinct bootstrap samples, is $\binom{k+m-1}{k}$. In other words,

$$\sum_{k_1, k_2, \ldots, k_m} I_{\{k_1 + k_2 + \cdots + k_m = k\}} = \binom{k + m - 1}{k}.$$

**1.30** For the collection of six numbers, $\{1, 2, 7, 8, 14, 20\}$, draw a histogram of the distribution of all possible sample averages calculated from samples drawn with replacement.

**1.31** For the situation of Example 1.2.20, the average of the original set of numbers $\{2, 4, 9, 12\}$ is $\frac{27}{4}$, which has the highest probability.

(a) Prove that, in general, if we sample with replacement from the set $\{x_1, x_2, \ldots, x_n\}$, the outcome with average $(x_1 + x_2 + \cdots + x_n)/n$ is the most likely, having probability $\frac{n!}{n^n}$.

(b) Use Stirling's Formula (Exercise 1.28) to show that $n!/n^n \approx \sqrt{2n\pi}/e^n$ (Hall 1992, Appendix I).

(c) Show that the probability that a particular $x_i$ is missing from an outcome is $(1 - \frac{1}{n})^n \to e^{-1}$ as $n \to \infty$.

**1.32** An employer is about to hire one new employee from a group of $N$ candidates, whose future potential can be rated on a scale from 1 to $N$. The employer proceeds according to the following rules:

(a) Each candidate is seen in succession (in random order) and a decision is made whether to hire the candidate.

(b) Having rejected $m-1$ candidates $(m > 1)$, the employer can hire the $m$th candidate only if the $m$th candidate is better than the previous $m - 1$.

Suppose a candidate is hired on the $i$th trial. What is the probability that the best candidate was hired?

**1.33** Suppose that 5% of men and .25% of women are color-blind. A person is chosen at random and that person is color-blind. What is the probability that the person is male? (Assume males and females to be in equal numbers.)

**1.34** Two litters of a particular rodent species have been born, one with two brown-haired and one gray-haired (litter 1), and the other with three brown-haired and two gray-haired (litter 2). We select a litter at random and then select an offspring at random from the selected litter.

(a) What is the probability that the animal chosen is brown-haired?

(b) Given that a brown-haired offspring was selected, what is the probability that the sampling was from litter 1?

**1.35** Prove that if $P(\cdot)$ is a legitimate probability function and $B$ is a set with $P(B) > 0$, then $P(\cdot|B)$ also satisfies Kolmogorov's Axioms.

**1.36** If the probability of hitting a target is $\frac{1}{5}$, and ten shots are fired independently, what is the probability of the target being hit at least twice? What is the conditional probability that the target is hit at least twice, given that it is hit at least once?

**1.37** Here we look at some variations of Example 1.3.4.

(a) In the warden's calculation of Example 1.3.4 it was assumed that if A were to be pardoned, then with equal probability the warden would tell A that either B or C would die. However, this need not be the case. The warden can assign probabilities $\gamma$ and $1 - \gamma$ to these events, as shown here:

| Prisoner pardoned | Warden tells A | |
|---|---|---|
| A | B dies | with probability $\gamma$ |
| A | C dies | with probability $1 - \gamma$ |
| B | C dies | |
| C | B dies | |

Calculate $P(A|W)$ as a function of $\gamma$. For what values of $\gamma$ is $P(A|W)$ less than, equal to, or greater than $\frac{1}{3}$?

(b) Suppose again that $\gamma = \frac{1}{2}$, as in the example. After the warden tells A that B will die, A thinks for a while and realizes that his original calculation was false. However, A then gets a bright idea. A asks the warden if he can swap fates with C. The warden, thinking that no information has been passed, agrees to this. Prove that A's reasoning is now correct and that his probability of survival has jumped to $\frac{2}{3}$!

A similar, but somewhat more complicated, problem, the "Monte Hall problem" is discussed by Selvin (1975). The problem in this guise gained a fair amount of notoriety when it appeared in a Sunday magazine (vos Savant 1990) along with a correct answer but with questionable explanation. The ensuing debate was even reported on the front page of the Sunday *New York Times* (Tierney 1991). A complete and somewhat amusing treatment is given by Morgan *et al.* (1991) [see also the response by vos Savant 1991]. Chun (1999) pretty much exhausts the problem with a very thorough analysis.

**1.38** Prove each of the following statements. (Assume that any conditioning event has positive probability.)

(a) If $P(B) = 1$, then $P(A|B) = P(A)$ for any $A$.

(b) If $A \subset B$, then $P(B|A) = 1$ and $P(A|B) = P(A)/P(B)$.

(c) If $A$ and $B$ are mutually exclusive, then

$$P(A|A \cup B) = \frac{P(A)}{P(A) + P(B)}.$$

(d) $P(A \cap B \cap C) = P(A|B \cap C)P(B|C)P(C)$.

**1.39** A pair of events $A$ and $B$ cannot be simultaneously *mutually exclusive* and *independent*. Prove that if $P(A) > 0$ and $P(B) > 0$, then:

(a) If $A$ and $B$ are mutually exclusive, they cannot be independent.

(b) If $A$ and $B$ are independent, they cannot be mutually exclusive.

**1.40** Finish the proof of Theorem 1.3.9 by proving parts (b) and (c).

**1.41** As in Example 1.3.6, consider telegraph signals "dot" and "dash" sent in the proportion 3:4, where erratic transmissions cause a dot to become a dash with probability $\frac{1}{4}$ and a dash to become a dot with probability $\frac{1}{3}$.

(a) If a dash is received, what is the probability that a dash has been sent?

(b) Assuming independence between signals, if the message dot–dot was received, what is the probability distribution of the four possible messages that could have been sent?

**1.42** The *inclusion-exclusion identity* of Miscellanea 1.8.1 gets it name from the fact that it is proved by the method of inclusion and exclusion (Feller 1968, Section IV.1). Here we go into the details. The probability $P(\cup_{i=1}^{n} A_i)$ is the sum of the probabilities of all the sample points that are contained in at least one of the $A_i$s. The method of inclusion and exclusion is a recipe for counting these points.

(a) Let $E_k$ denote the set of all sample points that are contained in exactly $k$ of the events $A_1, A_2, \ldots, A_n$. Show that $P(\cup_{i=1}^{n} A_i) = \sum_{i=1}^{n} P(E_i)$.

(b) Without loss of generality, assume that $E_k$ is contained in $A_1, A_2, \ldots, A_k$. Show that $P(E_k)$ appears $k$ times in the sum $P_1$, $\binom{k}{2}$ times in the sum $P_2$, $\binom{k}{3}$ times in the sum $P_3$, etc.

(c) Show that

$$k - \binom{k}{2} + \binom{k}{3} - \cdots \pm \binom{k}{k} = 1.$$

(See Exercise 1.27.)

(d) Show that parts $(a) - (c)$ imply $\sum_{i=1}^{n} P(E_i) = P_1 - P_2 = \cdots \pm P_n$, establishing the inclusion-exclusion identity.

**1.43** For the *inclusion-exclusion identity* of Miscellanea 1.8.1:

(a) Derive both Boole's and Bonferroni's Inequality from the inclusion-exclusion identity.

(b) Show that the $P_i$ satisfy $P_i \geq P_j$ if $i \leq j$ and that the sequence of bounds in Miscellanea 1.8.1 improves as the number of terms increases.

(c) Typically as the number of terms in the bound increases, the bound becomes more useful. However, Schwager (1984) cautions that there are some cases where there is not much improvement, in particular if the $A_i$s are highly correlated. Examine what happens to the sequence of bounds in the extreme case when $A_i = A$ for every $i$. (See Worsley 1982 and the correspondence of Worsley 1985 and Schwager 1985.)

**1.44** Standardized tests provide an interesting application of probability theory. Suppose first that a test consists of 20 multiple-choice questions, each with 4 possible answers. If the student guesses on each question, then the taking of the exam can be modeled as a sequence of 20 independent events. Find the probability that the student gets at least 10 questions correct, given that he is guessing.

**1.45** Show that the induced probability function defined in (1.4.1) defines a legitimate probability function in that it satisfies the Kolmogorov Axioms.

**1.46** Seven balls are distributed randomly into seven cells. Let $X_i =$ the number of cells containing exactly $i$ balls. What is the probability distribution of $X_3$? (That is, find $P(X_3 = x)$ for every possible $x$.)

**1.47** Prove that the following functions are cdfs.

(a) $\frac{1}{2} + \frac{1}{\pi}\tan^{-1}(x)$, $x \in (-\infty, \infty)$     (b) $(1 + e^{-x})^{-1}$, $x \in (-\infty, \infty)$

(c) $e^{-e^{-x}}$, $x \in (-\infty, \infty)$                          (d) $1 - e^{-x}$, $x \in (0, \infty)$

(e) the function defined in (1.5.6)

**1.48** Prove the necessity part of Theorem 1.5.3.

**1.49** A cdf $F_X$ is *stochastically greater* than a cdf $F_Y$ if $F_X(t) \le F_Y(t)$ for all $t$ and $F_X(t) < F_Y(t)$ for some $t$. Prove that if $X \sim F_X$ and $Y \sim F_Y$, then

$$P(X > t) \ge P(Y > t) \quad \text{for every } t$$

and

$$P(X > t) > P(Y > t) \quad \text{for some } t,$$

that is, $X$ tends to be bigger than $Y$.

**1.50** Verify formula (1.5.4), the formula for the partial sum of the geometric series.

**1.51** An appliance store receives a shipment of 30 microwave ovens, 5 of which are (unknown to the manager) defective. The store manager selects 4 ovens at random, without replacement, and tests to see if they are defective. Let $X =$ number of defectives found. Calculate the pmf and cdf of $X$ and plot the cdf.

**1.52** Let $X$ be a continuous random variable with pdf $f(x)$ and cdf $F(x)$. For a fixed number $x_0$, define the function

$$g(x) = \begin{cases} f(x)/[1 - F(x_0)] & x \ge x_0 \\ 0 & x < x_0. \end{cases}$$

Prove that $g(x)$ is a pdf. (Assume that $F(x_0) < 1$.)

**1.53** A certain river floods every year. Suppose that the low-water mark is set at 1 and the high-water mark $Y$ has distribution function

$$F_Y(y) = P(Y \le y) = 1 - \frac{1}{y^2}, \quad 1 \le y < \infty.$$

(a) Verify that $F_Y(y)$ is a cdf.

(b) Find $f_Y(y)$, the pdf of $Y$.

(c) If the low-water mark is reset at 0 and we use a unit of measurement that is $\frac{1}{10}$ of that given previously, the high-water mark becomes $Z = 10(Y - 1)$. Find $F_Z(z)$.

**1.54** For each of the following, determine the value of $c$ that makes $f(x)$ a pdf.

(a) $f(x) = c \sin x, \ 0 < x < \pi/2$      (b) $f(x) = c e^{-|x|}, \ -\infty < x < \infty$

**1.55** An electronic device has lifetime denoted by $T$. The device has value $V = 5$ if it fails before time $t = 3$; otherwise, it has value $V = 2T$. Find the cdf of $V$, if $T$ has pdf

$$f_T(t) = \frac{1}{1.5} e^{-t/(1.5)}, \quad t > 0.$$

## 1.8 Miscellanea

### 1.8.1 Bonferroni and Beyond

The Bonferroni bound of (1.2.10), or Boole's Inequality (Theorem 1.2.11), provides simple bounds on the probability of an intersection or union. These bounds can be made more and more precise with the following expansion.

For sets $A_1, A_2, \ldots A_n$, we create a new set of nested intersections as follows. Let

$$P_1 = \sum_{i=1}^{n} P(A_i)$$

$$P_2 = \sum_{1 \le i < j \le n}^{n} P(A_i \cap A_j)$$

$$P_3 = \sum_{1 \le i < j < k \le n}^{n} P(A_i \cap A_j \cap A_k)$$

$$\vdots$$

$$P_n = P(A_1 \cap A_2 \cap \cdots \cap A_n).$$

Then the *inclusion-exclusion identity* says that

$$P(A_1 \cup A_2 \cup \cdots \cup A_n) = P_1 - P_2 + P_3 - P_4 + \cdots \pm P_n.$$

Moreover, the $P_i$ are ordered in that $P_i \ge P_j$ if $i \le j$, and we have the sequence of upper and lower bounds

$$P_1 \ge P(\cup_{i=1}^n A_i) \ge P_1 - P_2$$
$$P_1 - P_2 + P_3 \ge P(\cup_{i=1}^n A_i) \ge P_1 - P_2 + P_3 - P_4$$

$$\vdots$$

See Exercises 1.42 and 1.43 for details.

These bounds become increasingly tighter as the number of terms increases, and they provide a refinement of the original Bonferroni bounds. Applications of these bounds include approximating probabilities of runs (Karlin and Ost 1988) and multiple comparisons procedures (Naiman and Wynn 1992).

Chapter 2

# Transformations and Expectations

*"We want something more than mere theory and preaching now, though."*
**Sherlock Holmes**
*A Study in Scarlet*

Often, if we are able to model a phenomenon in terms of a random variable $X$ with cdf $F_X(x)$, we will also be concerned with the behavior of functions of $X$. In this chapter we study techniques that allow us to gain information about functions of $X$ that may be of interest, information that can range from very complete (the distributions of these functions) to more vague (the average behavior).

## 2.1 Distributions of Functions of a Random Variable

If $X$ is a random variable with cdf $F_X(x)$, then any function of $X$, say $g(X)$, is also a random variable. Often $g(X)$ is of interest itself and we write $Y = g(X)$ to denote the new random variable $g(X)$. Since $Y$ is a function of $X$, we can describe the probabilistic behavior of $Y$ in terms of that of $X$. That is, for any set $A$,

$$P(Y \in A) = P(g(X) \in A),$$

showing that the distribution of $Y$ depends on the functions $F_X$ and $g$. Depending on the choice of $g$, it is sometimes possible to obtain a tractable expression for this probability.

Formally, if we write $y = g(x)$, the function $g(x)$ defines a mapping from the original sample space of $X$, $\mathcal{X}$, to a new sample space, $\mathcal{Y}$, the sample space of the random variable $Y$. That is,

$$g(x)\colon \mathcal{X} \to \mathcal{Y}.$$

We associate with $g$ an inverse mapping, denoted by $g^{-1}$, which is a mapping from subsets of $\mathcal{Y}$ to subsets of $\mathcal{X}$, and is defined by

(2.1.1) $$g^{-1}(A) = \{x \in \mathcal{X}\colon g(x) \in A\}.$$

Note that the mapping $g^{-1}$ takes sets into sets; that is, $g^{-1}(A)$ is the set of points in $\mathcal{X}$ that $g(x)$ takes into the set $A$. It is possible for $A$ to be a point set, say $A = \{y\}$. Then

$$g^{-1}(\{y\}) = \{x \in \mathcal{X}\colon g(x) = y\}.$$

In this case we often write $g^{-1}(y)$ instead of $g^{-1}(\{y\})$. The quantity $g^{-1}(y)$ can still be a set, however, if there is more than one $x$ for which $g(x) = y$. If there is only one $x$ for which $g(x) = y$, then $g^{-1}(y)$ is the point set $\{x\}$, and we will write $g^{-1}(y) = x$. If the random variable $Y$ is now defined by $Y = g(X)$, we can write for any set $A \subset \mathcal{Y}$,

$$P(Y \in A) = P(g(X) \in A)$$

(2.1.2)
$$= P(\{x \in \mathcal{X}\colon g(x) \in A\})$$

$$= P\left(X \in g^{-1}(A)\right).$$

This defines the probability distribution of $Y$. It is straightforward to show that this probability distribution satisfies the Kolmogorov Axioms.

If $X$ is a discrete random variable, then $\mathcal{X}$ is countable. The sample space for $Y = g(X)$ is $\mathcal{Y} = \{y\colon y = g(x),\ x \in \mathcal{X}\}$, which is also a countable set. Thus, $Y$ is also a discrete random variable. From (2.1.2), the pmf for $Y$ is

$$f_Y(y) = P(Y = y) = \sum_{x \in g^{-1}(y)} P(X = x) = \sum_{x \in g^{-1}(y)} f_X(x), \quad \text{for } y \in \mathcal{Y},$$

and $f_Y(y) = 0$ for $y \notin \mathcal{Y}$. In this case, finding the pmf of $Y$ involves simply identifying $g^{-1}(y)$, for each $y \in \mathcal{Y}$, and summing the appropriate probabilities.

**Example 2.1.1 (Binomial transformation)**   A discrete random variable $X$ has a *binomial distribution* if its pmf is of the form

(2.1.3)      $$f_X(x) = P(X = x) = \binom{n}{x} p^x (1 - p)^{n-x}, \quad x = 0, 1, \ldots, n,$$

where $n$ is a positive integer and $0 \le p \le 1$. Values such as $n$ and $p$ that can be set to different values, producing different probability distributions, are called *parameters*. Consider the random variable $Y = g(X)$, where $g(x) = n - x$; that is, $Y = n - X$. Here $\mathcal{X} = \{0, 1, \ldots, n\}$ and $\mathcal{Y} = \{y\colon y = g(x),\ x \in \mathcal{X}\} = \{0, 1, \ldots, n\}$. For any $y \in \mathcal{Y}$, $n - x = g(x) = y$ if and only if $x = n - y$. Thus, $g^{-1}(y)$ is the single point $x = n - y$, and

$$
\begin{aligned}
f_Y(y) &= \sum_{x \in g^{-1}(y)} f_X(x) \\
&= f_X(n - y) \\
&= \binom{n}{n - y} p^{n-y} (1 - p)^{n-(n-y)} \\
&= \binom{n}{y} (1 - p)^y p^{n-y}. \qquad \left( \begin{array}{l} \text{Definition 1.2.17} \\ \text{implies } \binom{n}{y} = \binom{n}{n-y} \end{array} \right)
\end{aligned}
$$

Thus, we see that $Y$ also has a binomial distribution, but with parameters $n$ and $1 - p$.                                                                                          ‖

If $X$ and $Y$ are continuous random variables, then in some cases it is possible to find simple formulas for the cdf and pdf of $Y$ in terms of the cdf and pdf of $X$ and the function $g$. In the remainder of this section, we consider some of these cases.

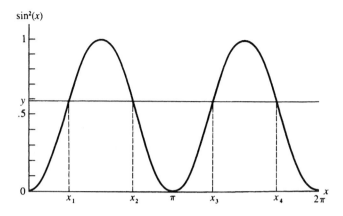

Figure 2.1.1. *Graph of the transformation* $y = \sin^2(x)$ *of Example 2.1.2*

The cdf of $Y = g(X)$ is

$$
\begin{aligned}
F_Y(y) &= P(Y \le y) \\
&= P(g(X) \le y) \\
&= P(\{x \in \mathcal{X}: g(x) \le y\}) \\
&= \int_{\{x \in \mathcal{X}: \, g(x) \le y\}} f_X(x)\,dx.
\end{aligned}
$$

(2.1.4)

Sometimes there may be difficulty in identifying $\{x \in \mathcal{X}: g(x) \le y\}$ and carrying out the integration of $f_X(x)$ over this region, as the next example shows.

**Example 2.1.2 (Uniform transformation)**   Suppose $X$ has a uniform distribution on the interval $(0, 2\pi)$, that is,

$$
f_X(x) = \begin{cases} 1/(2\pi) & 0 < x < 2\pi \\ 0 & \text{otherwise.} \end{cases}
$$

Consider $Y = \sin^2(X)$. Then (see Figure 2.1.1)

(2.1.5)      $P(Y \le y) = P(X \le x_1) + P(x_2 \le X \le x_3) + P(X \ge x_4).$

From the symmetry of the function $\sin^2(x)$ and the fact that $X$ has a uniform distribution, we have

$$P(X \le x_1) = P(X \ge x_4) \quad \text{and} \quad P(x_2 \le X \le x_3) = 2P(x_2 \le X \le \pi),$$

so

(2.1.6)          $P(Y \le y) = 2P(X \le x_1) + 2P(x_2 \le X \le \pi),$

where $x_1$ and $x_2$ are the two solutions to

$$\sin^2(x) = y, \qquad 0 < x < \pi.$$

Thus, even though this example dealt with a seemingly simple situation, the resulting expression for the cdf of $Y$ was not simple.                                             $\|$

When transformations are made, it is important to keep track of the sample spaces of the random variables; otherwise, much confusion can arise. When the transformation is from $X$ to $Y = g(X)$, it is most convenient to use

(2.1.7)  $\mathcal{X} = \{x \colon f_X(x) > 0\}$      and      $\mathcal{Y} = \{y \colon y = g(x) \text{ for some } x \in \mathcal{X}\}.$

The pdf of the random variable $X$ is positive only on the set $\mathcal{X}$ and is 0 elsewhere. Such a set is called the *support set* of a distribution or, more informally, the *support* of a distribution. This terminology can also apply to a pmf or, in general, to any nonnegative function.

It is easiest to deal with functions $g(x)$ that are *monotone*, that is, those that satisfy either

$$u > v \Rightarrow g(u) > g(v) \ \text{(increasing)} \quad \text{or} \quad u < v \Rightarrow g(u) > g(v) \ \text{(decreasing)}.$$

If the transformation $x \to g(x)$ is monotone, then it is *one-to-one and onto* from $\mathcal{X} \to \mathcal{Y}$. That is, each $x$ goes to only one $y$ and each $y$ comes from at most one $x$ (one-to-one). Also, for $\mathcal{Y}$ defined as in (2.1.7), for each $y \in \mathcal{Y}$ there is an $x \in \mathcal{X}$ such that $g(x) = y$ (onto). Thus, the transformation $g$ uniquely pairs $x$s and $y$s. If $g$ is monotone, then $g^{-1}$ is single-valued; that is, $g^{-1}(y) = x$ if and only if $y = g(x)$. If $g$ is increasing, this implies that

$$\{x \in \mathcal{X} \colon g(x) \le y\} = \{x \in \mathcal{X} \colon g^{-1}(g(x)) \le g^{-1}(y)\}$$
(2.1.8)
$$= \{x \in \mathcal{X} \colon x \le g^{-1}(y)\}.$$

If $g$ is decreasing, this implies that

$$\{x \in \mathcal{X} \colon g(x) \le y\} = \{x \in \mathcal{X} \colon g^{-1}(g(x)) \ge g^{-1}(y)\}$$
(2.1.9)
$$= \{x \in \mathcal{X} \colon x \ge g^{-1}(y)\}.$$

(A graph will illustrate why the inequality reverses in the decreasing case.) If $g(x)$ is an increasing function, then using (2.1.4), we can write

$$F_Y(y) = \int_{\{x \in \mathcal{X} \colon x \le g^{-1}(y)\}} f_X(x)\,dx = \int_{-\infty}^{g^{-1}(y)} f_X(x)\,dx = F_X\left(g^{-1}(y)\right).$$

If $g(x)$ is decreasing, we have

$$F_Y(y) = \int_{g^{-1}(y)}^{\infty} f_X(x)\,dx = 1 - F_X\left(g^{-1}(y)\right).$$

The continuity of $X$ is used to obtain the second equality. We summarize these results in the following theorem.

**Theorem 2.1.3**    *Let $X$ have cdf $F_X(x)$, let $Y = g(X)$, and let $\mathcal{X}$ and $\mathcal{Y}$ be defined as in (2.1.7).*

**a.** *If $g$ is an increasing function on $\mathcal{X}$, $F_Y(y) = F_X\left(g^{-1}(y)\right)$ for $y \in \mathcal{Y}$.*

**b.** *If $g$ is a decreasing function on $\mathcal{X}$ and $X$ is a continuous random variable, $F_Y(y) = 1 - F_X\left(g^{-1}(y)\right)$ for $y \in \mathcal{Y}$.*

**Example 2.1.4 (Uniform-exponential relationship–I)**    Suppose $X \sim f_X(x) = 1$ if $0 < x < 1$ and $0$ otherwise, the *uniform*$(0, 1)$ *distribution*. It is straightforward to check that $F_X(x) = x$, $0 < x < 1$. We now make the transformation $Y = g(X) = -\log X$. Since

$$\frac{d}{dx}g(x) = \frac{d}{dx}(-\log x) = \frac{-1}{x} < 0, \qquad \text{for} \quad 0 < x < 1,$$

$g(x)$ is a decreasing function. As $X$ ranges between $0$ and $1$, $-\log x$ ranges between $0$ and $\infty$, that is, $\mathcal{Y} = (0, \infty)$. For $y > 0$, $y = -\log x$ implies $x = e^{-y}$, so $g^{-1}(y) = e^{-y}$. Therefore, for $y > 0$,

$$F_Y(y) = 1 - F_X\left(g^{-1}(y)\right) = 1 - F_X(e^{-y}) = 1 - e^{-y}. \qquad (F_X(x) = x)$$

Of course, $F_Y(y) = 0$ for $y \leq 0$. Note that it was necessary only to verify that $g(x) = -\log x$ is monotone on $(0, 1)$, the support of $X$.                    ‖

If the pdf of $Y$ is continuous, it can be obtained by differentiating the cdf. The resulting expression is given in the following theorem.

**Theorem 2.1.5**    *Let $X$ have pdf $f_X(x)$ and let $Y = g(X)$, where $g$ is a monotone function. Let $\mathcal{X}$ and $\mathcal{Y}$ be defined by (2.1.7). Suppose that $f_X(x)$ is continuous on $\mathcal{X}$ and that $g^{-1}(y)$ has a continuous derivative on $\mathcal{Y}$. Then the pdf of $Y$ is given by*

$$(2.1.10) \qquad f_Y(y) = \begin{cases} f_X(g^{-1}(y))\left|\dfrac{d}{dy}g^{-1}(y)\right| & y \in \mathcal{Y} \\ 0 & \text{otherwise.} \end{cases}$$

**Proof:** From Theorem 2.1.3 we have, by the chain rule,

$$f_Y(y) = \frac{d}{dy}F_Y(y) = \begin{cases} f_X(g^{-1}(y))\dfrac{d}{dy}g^{-1}(y) & \text{if } g \text{ is increasing} \\ -f_X(g^{-1}(y))\dfrac{d}{dy}g^{-1}(y) & \text{if } g \text{ is decreasing,} \end{cases}$$

which can be expressed concisely as (2.1.10).                    □

**Example 2.1.6 (Inverted gamma pdf)**    Let $f_X(x)$ be the *gamma pdf*

$$f(x) = \frac{1}{(n-1)!\beta^n}x^{n-1}e^{-x/\beta}, \qquad 0 < x < \infty,$$

where $\beta$ is a positive constant and $n$ is a positive integer. Suppose we want to find the pdf of $g(X) = 1/X$. Note that here the support sets $\mathcal{X}$ and $\mathcal{Y}$ are both the interval

$(0, \infty)$. If we let $y = g(x)$, then $g^{-1}(y) = 1/y$ and $\frac{d}{dy}g^{-1}(y) = -1/y^2$. Applying the above theorem, for $y \in (0, \infty)$, we get

$$f_Y(y) = f_X\left(g^{-1}(y)\right)\left|\frac{d}{dy}g^{-1}(y)\right|$$

$$= \frac{1}{(n-1)!\beta^n}\left(\frac{1}{y}\right)^{n-1}e^{-1/(\beta y)}\frac{1}{y^2}$$

$$= \frac{1}{(n-1)!\beta^n}\left(\frac{1}{y}\right)^{n+1}e^{-1/(\beta y)},$$

a special case of a pdf known as the *inverted gamma pdf*.                                     ‖

In many applications, the function $g$ may be neither increasing nor decreasing; hence the above results will not apply. However, it is often the case that $g$ will be monotone over certain intervals and that allows us to get an expression for $Y = g(X)$. (If $g$ is not monotone over certain intervals, then we are in *deep* trouble.)

**Example 2.1.7 (Square transformation)**    Suppose $X$ is a continuous random variable. For $y > 0$, the cdf of $Y = X^2$ is

$$F_Y(y) = P(Y \le y) = P(X^2 \le y) = P(-\sqrt{y} \le X \le \sqrt{y}).$$

Because $x$ is continuous, we can drop the equality from the left endpoint and obtain

$$F_Y(y) = P(-\sqrt{y} < X \le \sqrt{y})$$
$$= P(X \le \sqrt{y}) - P(X \le -\sqrt{y}) \quad = \quad F_X(\sqrt{y}) - F_X(-\sqrt{y}).$$

The pdf of $Y$ can now be obtained from the cdf by differentiation:

$$f_Y(y) = \frac{d}{dy}F_Y(y)$$

$$= \frac{d}{dy}[F_X(\sqrt{y}) - F_X(-\sqrt{y})]$$

$$= \frac{1}{2\sqrt{y}}f_X(\sqrt{y}) + \frac{1}{2\sqrt{y}}f_X(-\sqrt{y}),$$

where we use the chain rule to differentiate $F_X(\sqrt{y})$ and $F_X(-\sqrt{y})$. Therefore, the pdf is

(2.1.11)                    $$f_Y(y) = \frac{1}{2\sqrt{y}}\left(f_X(\sqrt{y}) + f_X(-\sqrt{y})\right).$$

Notice that the pdf of $Y$ in (2.1.11) is expressed as the sum of two pieces, pieces that represent the intervals where $g(x) = x^2$ is monotone. In general, this will be the case.                                     ‖

**Theorem 2.1.8** *Let $X$ have pdf $f_X(x)$, let $Y = g(X)$, and define the sample space $\mathcal{X}$ as in (2.1.7). Suppose there exists a partition, $A_0, A_1, \ldots, A_k$, of $\mathcal{X}$ such that $P(X \in A_0) = 0$ and $f_X(x)$ is continuous on each $A_i$. Further, suppose there exist functions $g_1(x), \ldots, g_k(x)$, defined on $A_1, \ldots, A_k$, respectively, satisfying*

i. *$g(x) = g_i(x)$, for $x \in A_i$,*

ii. *$g_i(x)$ is monotone on $A_i$,*

iii. *the set $\mathcal{Y} = \{y\colon y = g_i(x) \text{ for some } x \in A_i\}$ is the same for each $i = 1, \ldots, k$,*

*and*

iv. *$g_i^{-1}(y)$ has a continuous derivative on $\mathcal{Y}$, for each $i = 1, \ldots, k$.*

*Then*

$$
f_Y(y) = \begin{cases} \sum_{i=1}^{k} f_X\left(g_i^{-1}(y)\right) \left|\dfrac{d}{dy} g_i^{-1}(y)\right| & y \in \mathcal{Y} \\ 0 & \text{otherwise.} \end{cases}
$$

The important point in Theorem 2.1.8 is that $\mathcal{X}$ can be divided into sets $A_1, \ldots, A_k$ such that $g(x)$ is monotone on each $A_i$. We can ignore the "exceptional set" $A_0$ since $P(X \in A_0) = 0$. It is a technical device that is used, for example, to handle endpoints of intervals. It is important to note that each $g_i(x)$ is a one-to-one transformation from $A_i$ onto $\mathcal{Y}$. Furthermore, $g_i^{-1}(y)$ is a one-to-one function from $\mathcal{Y}$ onto $A_i$ such that, for $y \in \mathcal{Y}$, $g_i^{-1}(y)$ gives the unique $x = g_i^{-1}(y) \in A_i$ for which $g_i(x) = y$. (See Exercise 2.7 for an extension.)

**Example 2.1.9 (Normal-chi squared relationship)**   Let $X$ have the *standard normal distribution*,

$$
f_X(x) = \frac{1}{\sqrt{2\pi}} e^{-x^2/2}, \qquad -\infty < x < \infty.
$$

Consider $Y = X^2$. The function $g(x) = x^2$ is monotone on $(-\infty, 0)$ and on $(0, \infty)$. The set $\mathcal{Y} = (0, \infty)$. Applying Theorem 2.1.8, we take

$$
\begin{aligned}
A_0 &= \{0\}; \\
A_1 &= (-\infty, 0), & g_1(x) &= x^2, & g_1^{-1}(y) &= -\sqrt{y}; \\
A_2 &= (0, \infty), & g_2(x) &= x^2, & g_2^{-1}(y) &= \sqrt{y}.
\end{aligned}
$$

The pdf of $Y$ is

$$
\begin{aligned}
f_Y(y) &= \frac{1}{\sqrt{2\pi}} e^{-(-\sqrt{y})^2/2} \left|-\frac{1}{2\sqrt{y}}\right| + \frac{1}{\sqrt{2\pi}} e^{-(\sqrt{y})^2/2} \left|\frac{1}{2\sqrt{y}}\right| \\
&= \frac{1}{\sqrt{2\pi}} \frac{1}{\sqrt{y}} e^{-y/2}, \qquad 0 < y < \infty.
\end{aligned}
$$

The pdf of $Y$ is one that we will often encounter, that of a *chi squared random variable* with 1 degree of freedom. $\quad\|$

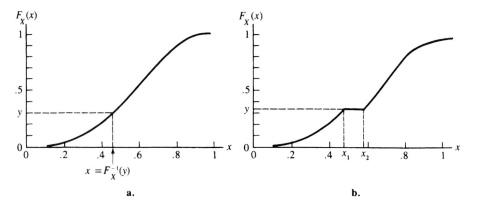

Figure 2.1.2. **(a)** $F(x)$ *strictly increasing;* **(b)** $F(x)$ *nondecreasing*

We close this section with a special and very useful transformation.

**Theorem 2.1.10 (Probability integral transformation)** *Let $X$ have continuous cdf $F_X(x)$ and define the random variable $Y$ as $Y = F_X(X)$. Then $Y$ is uniformly distributed on $(0, 1)$, that is, $P(Y \le y) = y, 0 < y < 1$.*

Before we prove this theorem, we will digress for a moment and look at $F_X^{-1}$, the inverse of the cdf $F_X$, in some detail. If $F_X$ is strictly increasing, then $F_X^{-1}$ is well defined by

$$(2.1.12) \qquad F_X^{-1}(y) = x \quad \Leftrightarrow \quad F_X(x) = y.$$

However, if $F_X$ is constant on some interval, then $F_X^{-1}$ is not well defined by (2.1.12), as Figure 2.1.2 illustrates. Any $x$ satisfying $x_1 \le x \le x_2$ satisfies $F_X(x) = y$.

This problem is avoided by defining $F_X^{-1}(y)$ for $0 < y < 1$ by

$$(2.1.13) \qquad F_X^{-1}(y) = \inf \{x\colon F_X(x) \ge y\},$$

a definition that agrees with (2.1.12) when $F_X$ is nonconstant and provides an $F_X^{-1}$ that is single-valued even when $F_X$ is not strictly increasing. Using this definition, in Figure 2.1.2b, we have $F_X^{-1}(y) = x_1$. At the endpoints of the range of $y$, $F_X^{-1}(y)$ can also be defined. $F_X^{-1}(1) = \infty$ if $F_X(x) < 1$ for all $x$ and, for any $F_X$, $F_X^{-1}(0) = -\infty$.

**Proof of Theorem 2.1.10:** For $Y = F_X(X)$ we have, for $0 < y < 1$,

$$
\begin{aligned}
P(Y \le y) &= P(F_X(X) \le y) \\
&= P(F_X^{-1}[F_X(X)] \le F_X^{-1}(y)) \qquad (F_X^{-1} \text{ is increasing}) \\
&= P(X \le F_X^{-1}(y)) \qquad\qquad\qquad \text{(see paragraph below)} \\
&= F_X\left(F_X^{-1}(y)\right) \qquad\qquad\qquad\quad \text{(definition of } F_X) \\
&= y. \qquad\qquad\qquad\qquad\qquad\quad\ \text{(continuity of } F_X)
\end{aligned}
$$

At the endpoints we have $P(Y \leq y) = 1$ for $y \geq 1$ and $P(Y \leq y) = 0$ for $y \leq 0$, showing that $Y$ has a uniform distribution.

The reasoning behind the equality

$$P\left(F_X^{-1}(F_X(X)) \leq F_X^{-1}(y)\right) = P(X \leq F_X^{-1}(y))$$

is somewhat subtle and deserves additional attention. If $F_X$ is strictly increasing, then it is true that $F_X^{-1}(F_X(x)) = x$. (Refer to Figure 2.1.2a.) However, if $F_X$ is flat, it may be that $F_X^{-1}(F_X(x)) \neq x$. Suppose $F_X$ is as in Figure 2.1.2b and let $x \in [x_1, x_2]$. Then $F_X^{-1}(F_X(x)) = x_1$ for any $x$ in this interval. Even in this case, though, the probability equality holds, since $P(X \leq x) = P(X \leq x_1)$ for any $x \in [x_1, x_2]$. The flat cdf denotes a region of 0 probability ($P(x_1 < X \leq x) = F_X(x) - F_X(x_1) = 0$). $\qquad\square$

One application of Theorem 2.1.10 is in the generation of random samples from a particular distribution. If it is required to generate an observation $X$ from a population with cdf $F_X$, we need only generate a uniform random number $U$, between 0 and 1, and solve for $x$ in the equation $F_X(x) = u$. (For many distributions there are other methods of generating observations that take less computer time, but this method is still useful because of its general applicability.)

## 2.2 Expected Values

The expected value, or expectation, of a random variable is merely its average value, where we speak of "average" value as one that is weighted according to the probability distribution. The expected value of a distribution can be thought of as a measure of center, as we think of averages as being middle values. By weighting the values of the random variable according to the probability distribution, we hope to obtain a number that summarizes a typical or expected value of an observation of the random variable.

**Definition 2.2.1**   The *expected value* or *mean* of a random variable $g(X)$, denoted by $\mathrm{E}\, g(X)$, is

$$\mathrm{E}\, g(X) = \begin{cases} \int_{-\infty}^{\infty} g(x) f_X(x)\, dx & \text{if } X \text{ is continuous} \\ \sum_{x \in \mathcal{X}} g(x) f_X(x) = \sum_{x \in \mathcal{X}} g(x) P(X = x) & \text{if } X \text{ is discrete,} \end{cases}$$

provided that the integral or sum exists. If $\mathrm{E}|g(X)| = \infty$, we say that $\mathrm{E}\, g(X)$ does not exist. (Ross 1988 refers to this as the "law of the unconscious statistician." We do not find this amusing.)

**Example 2.2.2 (Exponential mean)**   Suppose $X$ has an *exponential* ($\lambda$) *distribution*, that is, it has pdf given by

$$f_X(x) = \frac{1}{\lambda} e^{-x/\lambda}, \qquad 0 \leq x < \infty, \quad \lambda > 0.$$

Then $\mathrm{E}\,X$ is given by

$$
\begin{aligned}
\mathrm{E}\,X &= \int_0^\infty \frac{1}{\lambda} x e^{-x/\lambda}\, dx \\
&= -x e^{-x/\lambda}\Big|_0^\infty + \int_0^\infty e^{-x/\lambda}\, dx \qquad \text{(integration by parts)} \\
&= \int_0^\infty e^{-x/\lambda}\, dx \;=\; \lambda.
\end{aligned}
$$
∥

**Example 2.2.3 (Binomial mean)**   If $X$ has a *binomial distribution*, its pmf is given by

$$
P(X = x) = \binom{n}{x} p^x (1-p)^{n-x}, \qquad x = 0, 1, \ldots, n,
$$

where $n$ is a positive integer, $0 \le p \le 1$, and for every fixed pair $n$ and $p$ the pmf sums to 1. The expected value of a binomial random variable is given by

$$
\mathrm{E}\,X = \sum_{x=0}^n x \binom{n}{x} p^x (1-p)^{n-x} = \sum_{x=1}^n x \binom{n}{x} p^x (1-p)^{n-x}
$$

(the $x = 0$ term is 0). Using the identity $x \binom{n}{x} = n \binom{n-1}{x-1}$, we have

$$
\begin{aligned}
\mathrm{E}\,X &= \sum_{x=1}^n n \binom{n-1}{x-1} p^x (1-p)^{n-x} \\
&= \sum_{y=0}^{n-1} n \binom{n-1}{y} p^{y+1} (1-p)^{n-(y+1)} \qquad \text{(substitute } y = x - 1\text{)} \\
&= np \sum_{y=0}^{n-1} \binom{n-1}{y} p^y (1-p)^{n-1-y} \\
&= np,
\end{aligned}
$$

since the last summation must be 1, being the sum over all possible values of a binomial($n - 1$, $p$) pmf.
∥

**Example 2.2.4 (Cauchy mean)**   A classic example of a random variable whose expected value does not exist is a *Cauchy random variable*, that is, one with pdf

$$
f_X(x) = \frac{1}{\pi} \frac{1}{1 + x^2}, \qquad -\infty < x < \infty.
$$

It is straightforward to check that $\int_{-\infty}^\infty f_X(x)\, dx = 1$, but $\mathrm{E}|X| = \infty$. Write

$$
\mathrm{E}|X| = \int_{-\infty}^\infty \frac{|x|}{\pi} \frac{1}{1 + x^2}\, dx = \frac{2}{\pi} \int_0^\infty \frac{x}{1 + x^2}\, dx.
$$

For any positive number $M$,

$$\int_0^M \frac{x}{1+x^2}\, dx = \frac{\log(1+x^2)}{2}\Big|_0^M = \frac{\log(1+M^2)}{2}.$$

Thus,

$$E|X| = \lim_{M\to\infty} \frac{2}{\pi}\int_0^M \frac{x}{1+x^2}\, dx = \frac{1}{\pi}\lim_{M\to\infty}\log(1+M^2) = \infty$$

and $E\,X$ does not exist.                                                      ∥

The process of taking expectations is a linear operation, which means that the expectation of a linear function of $X$ can be easily evaluated by noting that for any constants $a$ and $b$,

(2.2.1)                          $E(aX + b) = aE\,X + b.$

For example, if $X$ is binomial$(n, p)$, so that $E\,X = np$, then

$$E(X - np) = E\,X - np = np - np = 0.$$

The expectation operator, in fact, has many properties that can help ease calculational effort. Most of these properties follow from the properties of the integral or sum, and are summarized in the following theorem.

**Theorem 2.2.5**   *Let $X$ be a random variable and let $a, b$, and $c$ be constants. Then for any functions $g_1(x)$ and $g_2(x)$ whose expectations exist,*
**a.** $E(ag_1(X) + bg_2(X) + c) = aE\,g_1(X) + bE\,g_2(X) + c.$
**b.** *If $g_1(x) \geq 0$ for all $x$, then $E\,g_1(X) \geq 0$.*
**c.** *If $g_1(x) \geq g_2(x)$ for all $x$, then $E\,g_1(X) \geq E\,g_2(X)$.*
**d.** *If $a \leq g_1(x) \leq b$ for all $x$, then $a \leq E\,g_1(X) \leq b$.*

**Proof:** We will give details for only the continuous case, the discrete case being similar. By definition,

$$E(ag_1(X) + bg_2(X) + c)$$

$$= \int_{-\infty}^{\infty} (ag_1(x) + bg_2(x) + c)f_X(x)\, dx$$

$$= \int_{-\infty}^{\infty} ag_1(x)f_X(x)\, dx + \int_{-\infty}^{\infty} bg_2(x)f_X(x)\, dx + \int_{-\infty}^{\infty} cf_X(x)\, dx$$

by the additivity of the integral. Since $a$, $b$, and $c$ are constants, they factor out of their respective integrals and we have

$$E(ag_1(X) + bg_2(X) + c)$$

$$= a\int_{-\infty}^{\infty} g_1(x)f_X(x)\, dx + b\int_{-\infty}^{\infty} g_2(x)f_X(x)\, dx + c\int_{-\infty}^{\infty} f_X(x)\, dx$$

$$= aE\,g_1(X) + bE\,g_2(x) + c,$$

establishing (a). The other three properties are proved in a similar manner.          □

**Example 2.2.6 (Minimizing distance)**   The expected value of a random variable has another property, one that we can think of as relating to the interpretation of $EX$ as a good guess at a value of $X$.

Suppose we measure the distance between a random variable $X$ and a constant $b$ by $(X - b)^2$. The closer $b$ is to $X$, the smaller this quantity is. We can now determine the value of $b$ that minimizes $E(X - b)^2$ and, hence, will provide us with a good predictor of $X$. (Note that it does no good to look for a value of $b$ that minimizes $(X - b)^2$, since the answer would depend on $X$, making it a useless predictor of $X$.)

We could proceed with the minimization of $E(X - b)^2$ by using calculus, but there is a simpler method. (See Exercise 2.19 for a calculus-based proof.) Using the belief that there is something special about $EX$, we write

$$\begin{aligned}
E(X - b)^2 &= E(X - EX + EX - b)^2 & \left(\begin{array}{c}\text{add } \pm EX, \text{ which} \\ \text{changes nothing}\end{array}\right) \\
&= E\left((X - EX) + (EX - b)\right)^2 & \text{(group terms)} \\
&= E(X - EX)^2 + (EX - b)^2 + 2E\left((X - EX)(EX - b)\right),
\end{aligned}$$

where we have expanded the square. Now, note that

$$E\left((X - EX)(EX - b)\right) = (EX - b)E(X - EX) = 0,$$

since $(EX - b)$ is constant and comes out of the expectation, and $E(X - EX) = EX - EX = 0$. This means that

$$(2.2.2) \qquad\qquad E(X - b)^2 = E(X - EX)^2 + (EX - b)^2.$$

We have no control over the first term on the right-hand side of (2.2.2), and the second term, which is always greater than or equal to 0, can be made equal to 0 by choosing $b = EX$. Hence,

$$(2.2.3) \qquad\qquad \min_b E(X - b)^2 = E(X - EX)^2.$$

See Exercise 2.18 for a similar result about the median.                            ‖

When evaluating expectations of nonlinear functions of $X$, we can proceed in one of two ways. From the definition of $E\,g(X)$, we could directly calculate

$$(2.2.4) \qquad\qquad E\,g(X) = \int_{-\infty}^{\infty} g(x) f_X(x)\,dx.$$

But we could also find the pdf $f_Y(y)$ of $Y = g(X)$ and we would have

$$(2.2.5) \qquad\qquad E\,g(X) = EY = \int_{-\infty}^{\infty} y f_Y(y)\,dy.$$

**Example 2.2.7 (Uniform–exponential relationship–II)**   Let $X$ have a uniform$(0, 1)$ distribution, that is, the pdf of $X$ is given by

$$f_X(x) = \begin{cases} 1 & \text{if } 0 \le x \le 1 \\ 0 & \text{otherwise,} \end{cases}$$

and define a new random variable $g(X) = -\log X$. Then

$$\mathrm{E}\, g(X) = \mathrm{E}(-\log X) = \int_0^1 -\log x \, dx = x - x \log x \big|_0^1 = 1.$$

But we also saw in Example 2.1.4 that $Y = -\log X$ has cdf $1 - e^{-y}$ and, hence, pdf $f_Y(y) = \frac{d}{dy}(1 - e^{-y}) = e^{-y}$, $0 < y < \infty$, which is a special case of the exponential pdf with $\lambda = 1$. Thus, by Example 2.2.2, $\mathrm{E}\, Y = 1$.                                   ||

## 2.3 Moments and Moment Generating Functions

The various moments of a distribution are an important class of expectations.

**Definition 2.3.1**    For each integer $n$, the $n$th *moment of* $X$ (or $F_X(x)$), $\mu'_n$, is

$$\mu'_n = \mathrm{E}\, X^n.$$

The $n$th *central moment of* $X$, $\mu_n$, is

$$\mu_n = \mathrm{E}(X - \mu)^n,$$

where $\mu = \mu'_1 = \mathrm{E}\, X$.

Aside from the mean, $\mathrm{E}\, X$, of a random variable, perhaps the most important moment is the second central moment, more commonly known as the variance.

**Definition 2.3.2**    The *variance* of a random variable $X$ is its second central moment, $\mathrm{Var}\, X = \mathrm{E}(X - \mathrm{E}\, X)^2$. The positive square root of $\mathrm{Var}\, X$ is the *standard deviation* of $X$.

The variance gives a measure of the degree of spread of a distribution around its mean. We saw earlier in Example 2.2.6 that the quantity $\mathrm{E}(X - b)^2$ is minimized by choosing $b = \mathrm{E}\, X$. Now we consider the absolute size of this minimum. The interpretation attached to the variance is that larger values mean $X$ is more variable. At the extreme, if $\mathrm{Var}\, X = \mathrm{E}(X - \mathrm{E}\, X)^2 = 0$, then $X$ is equal to $\mathrm{E}\, X$ with probability 1, and there is no variation in $X$. The standard deviation has the same qualitative interpretation: Small values mean $X$ is very likely to be close to $\mathrm{E}\, X$, and large values mean $X$ is very variable. The standard deviation is easier to interpret in that the measurement unit on the standard deviation is the same as that for the original variable $X$. The measurement unit on the variance is the square of the original unit.

**Example 2.3.3 (Exponential variance)**    Let $X$ have the exponential($\lambda$) distribution, defined in Example 2.2.2. There we calculated $\mathrm{E}X = \lambda$, and we can now calculate the variance by

$$\mathrm{Var}\, X = \mathrm{E}(X - \lambda)^2 = \int_0^\infty (x - \lambda)^2 \frac{1}{\lambda} e^{-x/\lambda} \, dx$$

$$= \int_0^\infty (x^2 - 2x\lambda + \lambda^2) \frac{1}{\lambda} e^{-x/\lambda} \, dx.$$

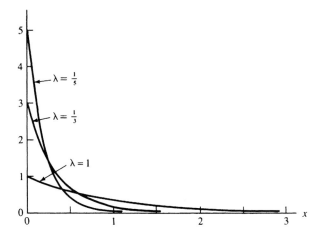

Figure 2.3.1. *Exponential densities for* $\lambda = 1, \frac{1}{3}, \frac{1}{5}$

To complete the integration, we can integrate each of the terms separately, using integration by parts on the terms involving $x$ and $x^2$. Upon doing this, we find that $\mathrm{Var}\, X = \lambda^2$.                                                                    ‖

We see that the variance of an exponential distribution is directly related to the parameter $\lambda$. Figure 2.3.1 shows several exponential distributions corresponding to different values of $\lambda$. Notice how the distribution is more concentrated about its mean for smaller values of $\lambda$. The behavior of the variance of an exponential, as a function of $\lambda$, is a special case of the variance behavior summarized in the following theorem.

**Theorem 2.3.4**   *If $X$ is a random variable with finite variance, then for any constants $a$ and $b$,*

$$\mathrm{Var}(aX + b) = a^2 \, \mathrm{Var}\, X.$$

**Proof:** From the definition, we have

$$
\begin{aligned}
\mathrm{Var}(aX + b) &= \mathrm{E}\left((aX + b) - \mathrm{E}(aX + b)\right)^2 \\
&= \mathrm{E}(aX - a\mathrm{E}\, X)^2 \qquad\qquad (\mathrm{E}(aX + b) = a\mathrm{E}\, X + b) \\
&= a^2 \mathrm{E}(X - \mathrm{E}\, X)^2 \\
&= a^2 \, \mathrm{Var}\, X. \qquad\qquad\qquad\qquad\qquad\qquad\quad \square
\end{aligned}
$$

It is sometimes easier to use an alternative formula for the variance, given by

(2.3.1)                          $\mathrm{Var}\, X = \mathrm{E}\, X^2 - (\mathrm{E}\, X)^2,$

which is easily established by noting

$$\begin{aligned}
\operatorname{Var} X = \operatorname{E}(X - \operatorname{E} X)^2 \;&=\; \operatorname{E}[X^2 - 2X\operatorname{E} X + (\operatorname{E} X)^2] \\
&= \operatorname{E} X^2 - 2(\operatorname{E} X)^2 + (\operatorname{E} X)^2 \\
&= \operatorname{E} X^2 - (\operatorname{E} X)^2,
\end{aligned}$$

where we use the fact that $\operatorname{E}(X\operatorname{E} X) = (\operatorname{E} X)(\operatorname{E} X) = (\operatorname{E} X)^2$, since $\operatorname{E} X$ is a constant. We now illustrate some moment calculations with a discrete distribution.

**Example 2.3.5 (Binomial variance)**   Let $X \sim \text{binomial}(n, p)$, that is,

$$P(X = x) = \binom{n}{x} p^x (1 - p)^{n-x}, \qquad x = 0, 1, \dots, n.$$

We have previously seen that $\operatorname{E} X = np$. To calculate $\operatorname{Var} X$ we first calculate $\operatorname{E} X^2$. We have

$$(2.3.2) \qquad\qquad \operatorname{E} X^2 = \sum_{x=0}^{n} x^2 \binom{n}{x} p^x (1 - p)^{n-x}.$$

In order to sum this series, we must first manipulate the binomial coefficient in a manner similar to that used for $\operatorname{E} X$ (Example 2.2.3). We write

$$(2.3.3) \qquad\qquad x^2 \binom{n}{x} = x\frac{n!}{(x-1)!(n-x)!} = xn\binom{n-1}{x-1}.$$

The summand in (2.3.2) corresponding to $x = 0$ is 0, and using (2.3.3), we have

$$\begin{aligned}
\operatorname{E} X^2 \;&=\; n\sum_{x=1}^{n} x\binom{n-1}{x-1} p^x (1 - p)^{n-x} \\
&=\; n\sum_{y=0}^{n-1} (y+1)\binom{n-1}{y} p^{y+1}(1 - p)^{n-1-y} \qquad\qquad (\text{setting } y = x - 1)\\
&=\; np\sum_{y=0}^{n-1} y\binom{n-1}{y} p^y (1 - p)^{n-1-y} + np\sum_{y=0}^{n-1} \binom{n-1}{y} p^y (1 - p)^{n-1-y}.
\end{aligned}$$

Now it is easy to see that the first sum is equal to $(n-1)p$ (since it is the mean of a binomial$(n-1, p)$), while the second sum is equal to 1. Hence,

$$(2.3.4) \qquad\qquad\qquad \operatorname{E} X^2 = n(n-1)p^2 + np.$$

Using (2.3.1), we have

$$\operatorname{Var} X = n(n-1)p^2 + np - (np)^2 = -np^2 + np = np\,(1 - p). \qquad\qquad \|$$

Calculation of higher moments proceeds in an analogous manner, but usually the mathematical manipulations become quite involved. In applications, moments of order 3 or 4 are sometimes of interest, but there is usually little statistical reason for examining higher moments than these.

We now introduce a new function that is associated with a probability distribution, the *moment generating function* (mgf). As its name suggests, the mgf can be used to generate moments. In practice, it is easier in many cases to calculate moments directly than to use the mgf. However, the main use of the mgf is not to generate moments, but to help in characterizing a distribution. This property can lead to some extremely powerful results when used properly.

**Definition 2.3.6**   Let $X$ be a random variable with cdf $F_X$. The *moment generating function (mgf) of* $X$ (or $F_X$), denoted by $M_X(t)$, is

$$M_X(t) = \mathrm{E}\, e^{tX},$$

provided that the expectation exists for $t$ in some neighborhood of 0. That is, there is an $h > 0$ such that, for all $t$ in $-h < t < h$, $\mathrm{E}e^{tX}$ exists. If the expectation does not exist in a neighborhood of 0, we say that the moment generating function does not exist.

More explicitly, we can write the mgf of $X$ as

$$M_X(t) = \int_{-\infty}^{\infty} e^{tx} f_X(x)\, dx \qquad \text{if } X \text{ is continuous,}$$

or

$$M_X(t) = \sum_x e^{tx} P(X = x) \qquad \text{if } X \text{ is discrete.}$$

It is very easy to see how the mgf generates moments. We summarize the result in the following theorem.

**Theorem 2.3.7**   *If $X$ has mgf $M_X(t)$, then*

$$\mathrm{E}\, X^n = M_X^{(n)}(0),$$

*where we define*

$$M_X^{(n)}(0) = \left. \frac{d^n}{dt^n} M_X(t) \right|_{t=0}.$$

*That is, the nth moment is equal to the nth derivative of $M_X(t)$ evaluated at $t = 0$.*

**Proof:** Assuming that we can differentiate under the integral sign (see the next section), we have

$$\frac{d}{dt} M_X(t) = \frac{d}{dt} \int_{-\infty}^{\infty} e^{tx} f_X(x)\, dx$$

$$= \int_{-\infty}^{\infty} \left( \frac{d}{dt} e^{tx} \right) f_X(x)\, dx$$

$$= \int_{-\infty}^{\infty} (x e^{tx}) f_X(x)\, dx$$

$$= \mathrm{E}\, X e^{tX}.$$

Thus,

$$\frac{d}{dt}M_X(t)\Big|_{t=0} = \mathrm{E}\,Xe^{tX}\big|_{t=0} = \mathrm{E}\,X.$$

Proceeding in an analogous manner, we can establish that

$$\frac{d^n}{dt^n}M_X(t)\Big|_{t=0} = \mathrm{E}\,X^n e^{tX}\big|_{t=0} = \mathrm{E}\,X^n. \qquad \square$$

**Example 2.3.8 (Gamma mgf)**   In Example 2.1.6 we encountered a special case of the gamma pdf,

$$f(x) = \frac{1}{\Gamma(\alpha)\beta^\alpha}x^{\alpha-1}e^{-x/\beta}, \quad 0 < x < \infty, \quad \alpha > 0, \quad \beta > 0,$$

where $\Gamma(\alpha)$ denotes the gamma function, some of whose properties are given in Section 3.3. The mgf is given by

$$
\begin{aligned}
M_X(t) &= \frac{1}{\Gamma(\alpha)\beta^\alpha}\int_0^\infty e^{tx}x^{\alpha-1}e^{-x/\beta}\,dx \\
(2.3.5) \qquad &= \frac{1}{\Gamma(\alpha)\beta^\alpha}\int_0^\infty x^{\alpha-1}e^{-(\frac{1}{\beta}-t)x}\,dx \\
&= \frac{1}{\Gamma(\alpha)\beta^\alpha}\int_0^\infty x^{\alpha-1}e^{-x/(\frac{\beta}{1-\beta t})}\,dx.
\end{aligned}
$$

We now recognize the integrand in (2.3.5) as the *kernel* of another gamma pdf. (The *kernel* of a function is the main part of the function, the part that remains when constants are disregarded.) Using the fact that, for any positive constants $a$ and $b$,

$$f(x) = \frac{1}{\Gamma(a)b^a}x^{a-1}e^{-x/b}$$

is a pdf, we have that

$$\int_0^\infty \frac{1}{\Gamma(a)b^a}x^{a-1}e^{-x/b}\,dx = 1$$

and, hence,

$$(2.3.6) \qquad \int_0^\infty x^{a-1}e^{-x/b}\,dx = \Gamma(a)b^a.$$

Applying (2.3.6) to (2.3.5), we have

$$M_X(t) = \frac{1}{\Gamma(\alpha)\beta^\alpha}\Gamma(\alpha)\left(\frac{\beta}{1-\beta t}\right)^\alpha = \left(\frac{1}{1-\beta t}\right)^\alpha \qquad \text{if } t < \frac{1}{\beta}.$$

If $t \geq 1/\beta$, then the quantity $(1/\beta) - t$, in the integrand of (2.3.5), is nonpositive and the integral in (2.3.6) is infinite. Thus, the mgf of the gamma distribution exists only if $t < 1/\beta$. (In Section 3.3 we will explore the gamma function in more detail.)

The mean of the gamma distribution is given by

$$\mathrm{E}\,X = \frac{d}{dt} M_X(t)\bigg|_{t=0} = \frac{\alpha\beta}{(1 - \beta t)^{\alpha+1}}\bigg|_{t=0} = \alpha\beta.$$

Other moments can be calculated in a similar manner.        $\|$

**Example 2.3.9 (Binomial mgf)**   For a second illustration of calculating a moment generating function, we consider a discrete distribution, the binomial distribution. The binomial$(n, p)$ pmf is given in (2.1.3). So

$$M_X(t) = \sum_{x=0}^{n} e^{tx} \binom{n}{x} p^x (1 - p)^{n-x} = \sum_{x=0}^{n} \binom{n}{x} (pe^t)^x (1 - p)^{n-x}.$$

The binomial formula (see Theorem 3.2.2) gives

(2.3.7) $$\sum_{x=0}^{n} \binom{n}{x} u^x v^{n-x} = (u + v)^n.$$

Hence, letting $u = pe^t$ and $v = 1 - p$, we have

$$M_X(t) = [pe^t + (1 - p)]^n.$$        $\|$

As previously mentioned, the major usefulness of the moment generating function is not in its ability to generate moments. Rather, its usefulness stems from the fact that, in many cases, the moment generating function can characterize a distribution. There are, however, some technical difficulties associated with using moments to characterize a distribution, which we will now investigate.

If the mgf exists, it characterizes an infinite set of moments. The natural question is whether characterizing the infinite set of moments uniquely determines a distribution function. The answer to this question, unfortunately, is no. Characterizing the set of moments is not enough to determine a distribution uniquely because there may be two distinct random variables having the same moments.

**Example 2.3.10 (Nonunique moments)**   Consider the two pdfs given by

$$f_1(x) = \frac{1}{\sqrt{2\pi}x} e^{-(\log x)^2/2}, \qquad 0 \leq x < \infty,$$

$$f_2(x) = f_1(x)[1 + \sin(2\pi \log x)], \qquad 0 \leq x < \infty.$$

(The pdf $f_1$ is a special case of a *lognormal pdf*.)

It can be shown that if $X_1 \sim f_1(x)$, then

$$\mathrm{E}\,X_1^r = e^{r^2/2}, \qquad r = 0, 1, \ldots,$$

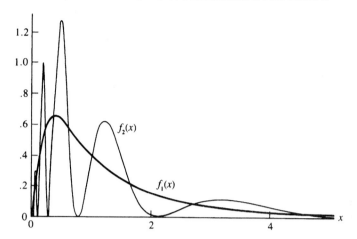

Figure 2.3.2. *Two pdfs with the same moments:* $f_1(x) = \frac{1}{\sqrt{2\pi}x}e^{-(\log x)^2/2}$ *and* $f_2(x) = f_1(x)[1 + \sin(2\pi \log x)]$

so $X_1$ has all of its moments. Now suppose that $X_2 \sim f_2(x)$. We have

$$\mathrm{E}\,X_2^r = \int_0^\infty x^r f_1(x)[1 + \sin(2\pi \log x)]\,dx$$

$$= \mathrm{E}\,X_1^r + \int_0^\infty x^r f_1(x)\sin(2\pi \log x)\,dx.$$

However, the transformation $y = \log x - r$ shows that this last integral is that of an odd function over $(-\infty, \infty)$ and hence is equal to 0 for $r = 0, 1, \ldots$. Thus, even though $X_1$ and $X_2$ have distinct pdfs, they have the same moments for all $r$. The two pdfs are pictured in Figure 2.3.2.

See Exercise 2.35 for details and also Exercises 2.34, 2.36, and 2.37 for more about mgfs and distributions.                                                                    ‖

The problem of uniqueness of moments does not occur if the random variables have bounded support. If that is the case, then the infinite sequence of moments does uniquely determine the distribution (see, for example, Billingsley 1995, Section 30). Furthermore, if the mgf exists in a neighborhood of 0, then the distribution *is* uniquely determined, no matter what its support. Thus, existence of all moments is not equivalent to existence of the moment generating function. The following theorem shows how a distribution can be characterized.

**Theorem 2.3.11**    *Let $F_X(x)$ and $F_Y(y)$ be two cdfs all of whose moments exist.*

**a.** *If $X$ and $Y$ have bounded support, then $F_X(u) = F_Y(u)$ for all $u$ if and only if $\mathrm{E}\,X^r = \mathrm{E}\,Y^r$ for all integers $r = 0, 1, 2, \ldots$.*

**b.** *If the moment generating functions exist and $M_X(t) = M_Y(t)$ for all $t$ in some neighborhood of 0, then $F_X(u) = F_Y(u)$ for all $u$.*

In the next theorem, which deals with a sequence of mgfs that converges, we do not treat the bounded support case separately. Note that the uniqueness assumption is automatically satisfied if the limiting mgf exists in a neighborhood of 0 (see Miscellanea 2.6.1).

**Theorem 2.3.12 (Convergence of mgfs)**   *Suppose* $\{X_i,\, i = 1, 2, \ldots\}$ *is a sequence of random variables, each with mgf* $M_{X_i}(t)$. *Furthermore, suppose that*

$$\lim_{i \to \infty} M_{X_i}(t) = M_X(t), \qquad \text{for all } t \text{ in a neighborhood of } 0,$$

*and* $M_X(t)$ *is an mgf. Then there is a unique cdf* $F_X$ *whose moments are determined by* $M_X(t)$ *and, for all* $x$ *where* $F_X(x)$ *is continuous, we have*

$$\lim_{i \to \infty} F_{X_i}(x) = F_X(x).$$

*That is, convergence, for* $|t| < h$, *of mgfs to an mgf implies convergence of cdfs.*

The proofs of Theorems 2.3.11 and 2.3.12 rely on the theory of *Laplace transforms*. (The classic reference is Widder 1946, but Laplace transforms also get a comprehensive treatment by Feller 1971.) The defining equation for $M_X(t)$, that is,

$$(2.3.8) \qquad\qquad M_X(t) = \int_{-\infty}^{\infty} e^{tx} f_X(x)\, dx,$$

defines a Laplace transform ($M_X(t)$ is the Laplace transform of $f_X(x)$). A key fact about Laplace transforms is their uniqueness. If (2.3.8) is valid for all $t$ such that $|t| < h$, where $h$ is some positive number, then given $M_X(t)$ there is only one function $f_X(x)$ that satisfies (2.3.8). Given this fact, the two previous theorems are quite reasonable. While rigorous proofs of these theorems are not beyond the scope of this book, the proofs are technical in nature and provide no real understanding. We omit them.

The possible nonuniqueness of the moment sequence is an annoyance. If we show that a sequence of moments converges, we will not be able to conclude formally that the random variables converge. To do so, we would have to verify the uniqueness of the moment sequence, a generally horrible job (see Miscellanea 2.6.1). However, if the sequence of mgfs converges in a neighborhood of 0, then the random variables converge. Thus, we can consider the convergence of mgfs as a sufficient, but not necessary, condition for the sequence of random variables to converge.

**Example 2.3.13 (Poisson approximation)**   One approximation that is usually taught in elementary statistics courses is that binomial probabilities (see Example 2.3.5) can be approximated by *Poisson* probabilities, which are generally easier to calculate. The binomial distribution is characterized by two quantities, denoted by $n$ and $p$. It is taught that the Poisson approximation is valid "when $n$ is large and $np$ is small," and rules of thumb are sometimes given.

The Poisson($\lambda$) pmf is given by

$$P(X = x) = \frac{e^{-\lambda}\lambda^x}{x!}, \qquad x = 0, 1, 2, \ldots,$$

where $\lambda$ is a positive constant. The approximation states that if $X \sim$ binomial$(n, p)$ and $Y \sim$ Poisson$(\lambda)$, with $\lambda = np$, then

$$(2.3.9) \qquad\qquad\qquad P(X = x) \approx P(Y = x)$$

for large $n$ and small $np$. We now show that the mgfs converge, lending credence to this approximation. Recall that

$$(2.3.10) \qquad\qquad\qquad M_X(t) = [pe^t + (1 - p)]^n.$$

For the Poisson$(\lambda)$ distribution, we can calculate (see Exercise 2.33)

$$M_Y(t) = e^{\lambda(e^t - 1)},$$

and if we define $p = \lambda/n$, then $M_X(t) \to M_Y(t)$ as $n \to \infty$. The validity of the approximation in (2.3.9) will then follow from Theorem 2.3.12.

We first must digress a bit and mention an important limit result, one that has wide applicability in statistics. The proof of this lemma may be found in many standard calculus texts.

**Lemma 2.3.14**     *Let $a_1, a_2, \ldots$ be a sequence of numbers converging to $a$, that is, $\lim_{n \to \infty} a_n = a$. Then*

$$\lim_{n \to \infty} \left(1 + \frac{a_n}{n}\right)^n = e^a.$$

Returning to the example, we have

$$M_X(t) = [pe^t + (1 - p)]^n = \left[1 + \frac{1}{n}(e^t - 1)(np)\right]^n = \left[1 + \frac{1}{n}(e^t - 1)\lambda\right]^n,$$

because $\lambda = np$. Now set $a_n = a = (e^t - 1)\lambda$, and apply Lemma 2.3.14 to get

$$\lim_{n \to \infty} M_X(t) = e^{\lambda(e^t - 1)} = M_Y(t),$$

the moment generating function of the Poisson.

The Poisson approximation can be quite good even for moderate $p$ and $n$. In Figure 2.3.3 we show a binomial mass function along with its Poisson approximation, with $\lambda = np$. The approximation appears to be satisfactory.                                   $\|$

We close this section with a useful result concerning mgfs.

**Theorem 2.3.15**     *For any constants $a$ and $b$, the mgf of the random variable $aX + b$ is given by*

$$M_{aX+b}(t) = e^{bt} M_X(at).$$

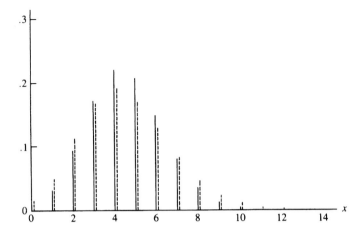

Figure 2.3.3. *Poisson (dotted line) approximation to the binomial (solid line), $n = 15$, $p = .3$*

**Proof:** By definition,

$$
\begin{aligned}
M_{aX+b}(t) &= \mathrm{E}\left(e^{(aX+b)t}\right) \\[2mm]
&= \mathrm{E}\left(e^{(aX)t}e^{bt}\right) && \text{(properties of exponentials)} \\[2mm]
&= e^{bt}\mathrm{E}\left(e^{(at)X}\right) && (e^{bt}\text{ is constant}) \\[2mm]
&= e^{bt}M_X(at), && \text{(definition of mgf)}
\end{aligned}
$$

proving the theorem.       □

## 2.4 Differentiating Under an Integral Sign

In the previous section we encountered an instance in which we desired to interchange the order of integration and differentiation. This situation is encountered frequently in theoretical statistics. The purpose of this section is to characterize conditions under which this operation is legitimate. We will also discuss interchanging the order of differentiation and summation.

Many of these conditions can be established using standard theorems from calculus and detailed proofs can be found in most calculus textbooks. Thus, detailed proofs will not be presented here.

We first want to establish the method of calculating

$$
(2.4.1) \qquad \frac{d}{d\theta}\int_{a(\theta)}^{b(\theta)} f(x,\theta)\,dx,
$$

where $-\infty < a(\theta), b(\theta) < \infty$ for all $\theta$. The rule for differentiating (2.4.1) is called Leibnitz's Rule and is an application of the Fundamental Theorem of Calculus and the chain rule.

**Theorem 2.4.1 (Leibnitz's Rule)** *If $f(x, \theta)$, $a(\theta)$, and $b(\theta)$ are differentiable with respect to $\theta$, then*

$$\frac{d}{d\theta} \int_{a(\theta)}^{b(\theta)} f(x, \theta)\, dx = f(b(\theta), \theta)\frac{d}{d\theta}b(\theta) - f(a(\theta), \theta)\frac{d}{d\theta}a(\theta) + \int_{a(\theta)}^{b(\theta)} \frac{\partial}{\partial\theta} f(x, \theta)\, dx.$$

Notice that if $a(\theta)$ and $b(\theta)$ are constant, we have a special case of Leibnitz's Rule:

$$\frac{d}{d\theta} \int_{a}^{b} f(x, \theta)\, dx = \int_{a}^{b} \frac{\partial}{\partial\theta} f(x, \theta)\, dx.$$

Thus, in general, if we have the integral of a differentiable function over a finite range, differentiation of the integral poses no problem. If the range of integration is infinite, however, problems can arise.

Note that the interchange of derivative and integral in the above equation equates a partial derivative with an ordinary derivative. Formally, this must be the case since the left-hand side is a function of only $\theta$, while the integrand on the right-hand side is a function of both $\theta$ and $x$.

The question of whether interchanging the order of differentiation and integration is justified is really a question of whether limits and integration can be interchanged, since a derivative is a special kind of limit. Recall that if $f(x, \theta)$ is differentiable, then

$$\frac{\partial}{\partial\theta} f(x, \theta) = \lim_{\delta \to 0} \frac{f(x, \theta + \delta) - f(x, \theta)}{\delta},$$

so we have

$$\int_{-\infty}^{\infty} \frac{\partial}{\partial\theta} f(x, \theta)\, dx = \int_{-\infty}^{\infty} \lim_{\delta \to 0} \left[ \frac{f(x, \theta + \delta) - f(x, \theta)}{\delta} \right] dx,$$

while

$$\frac{d}{d\theta} \int_{-\infty}^{\infty} f(x, \theta)\, dx = \lim_{\delta \to 0} \int_{-\infty}^{\infty} \left[ \frac{f(x, \theta + \delta) - f(x, \theta)}{\delta} \right] dx.$$

Therefore, if we can justify the interchanging of the order of limits and integration, differentiation under the integral sign will be justified. Treatment of this problem in full generality will, unfortunately, necessitate the use of measure theory, a topic that will not be covered in this book. However, the statements and conclusions of some important results can be given. The following theorems are all corollaries of Lebesgue's Dominated Convergence Theorem (see, for example, Rudin 1976).

**Theorem 2.4.2** *Suppose the function $h(x, y)$ is continuous at $y_0$ for each $x$, and there exists a function $g(x)$ satisfying*

i. $|h(x, y)| \le g(x)$ *for all $x$ and $y$,*
ii. $\int_{-\infty}^{\infty} g(x)\, dx < \infty$.

   *Then*

$$\lim_{y \to y_0} \int_{-\infty}^{\infty} h(x, y)\, dx = \int_{-\infty}^{\infty} \lim_{y \to y_0} h(x, y)\, dx.$$

The key condition in this theorem is the existence of a dominating function $g(x)$, with a finite integral, which ensures that the integrals cannot be too badly behaved. We can now apply this theorem to the case we are considering by identifying $h(x, y)$ with the difference $(f(x, \theta + \delta) - f(x, \theta))/\delta$.

**Theorem 2.4.3** *Suppose $f(x, \theta)$ is differentiable at $\theta = \theta_0$, that is,*

$$\lim_{\delta \to 0} \frac{f(x, \theta_0 + \delta) - f(x, \theta_0)}{\delta} = \frac{\partial}{\partial \theta} f(x, \theta) \Big|_{\theta = \theta_0}$$

*exists for every $x$, and there exists a function $g(x, \theta_0)$ and a constant $\delta_0 > 0$ such that*

i. $\left| \dfrac{f(x, \theta_0 + \delta) - f(x, \theta_0)}{\delta} \right| \leq g(x, \theta_0)$, *for all $x$ and $|\delta| \leq \delta_0$,*

ii. $\int_{-\infty}^{\infty} g(x, \theta_0) \, dx < \infty.$

*Then*

$$(2.4.2) \qquad \frac{d}{d\theta} \int_{-\infty}^{\infty} f(x, \theta) \, dx \Big|_{\theta = \theta_0} = \int_{-\infty}^{\infty} \left[ \frac{\partial}{\partial \theta} f(x, \theta) \Big|_{\theta = \theta_0} \right] dx.$$

Condition (i) is similar to what is known as a *Lipschitz condition*, a condition that imposes smoothness on a function. Here, condition (i) is effectively bounding the variability in the first derivative; other smoothness constraints might bound this variability by a constant (instead of a function $g$), or place a bound on the variability of the second derivative of $f$.

The conclusion of Theorem 2.4.3 is a little cumbersome, but it is important to realize that although we seem to be treating $\theta$ as a variable, the statement of the theorem is for one value of $\theta$. That is, for each value $\theta_0$ for which $f(x, \theta)$ is differentiable at $\theta_0$ and satisfies conditions (i) and (ii), the order of integration and differentiation can be interchanged. Often the distinction between $\theta$ and $\theta_0$ is not stressed and (2.4.2) is written

$$(2.4.3) \qquad \frac{d}{d\theta} \int_{-\infty}^{\infty} f(x, \theta) \, dx = \int_{-\infty}^{\infty} \frac{\partial}{\partial \theta} f(x, \theta) \, dx.$$

Typically, $f(x, \theta)$ is differentiable at all $\theta$, not at just one value $\theta_0$. In this case, condition (i) of Theorem 2.4.3 can be replaced by another condition that often proves easier to verify. By an application of the mean value theorem, it follows that, for fixed $x$ and $\theta_0$, and $|\delta| \leq \delta_0$,

$$\frac{f(x, \theta_0 + \delta) - f(x, \theta_0)}{\delta} = \frac{\partial}{\partial \theta} f(x, \theta) \Big|_{\theta = \theta_0 + \delta^*(x)}$$

for some number $\delta^*(x)$, where $|\delta^*(x)| \leq \delta_0$. Therefore, condition (i) will be satisfied if we find a $g(x, \theta)$ that satisfies condition (ii) and

$$(2.4.4) \qquad \left| \frac{\partial}{\partial \theta} f(x, \theta) \Big|_{\theta = \theta'} \right| \leq g(x, \theta) \qquad \text{for all } \theta' \text{ such that } |\theta' - \theta| \leq \delta_0.$$

Note that in (2.4.4) $\delta_0$ is implicitly a function of $\theta$, as is the case in Theorem 2.4.3. This is permitted since the theorem is applied to each value of $\theta$ individually. From (2.4.4) we get the following corollary.

**Corollary 2.4.4** *Suppose $f(x, \theta)$ is differentiable in $\theta$ and there exists a function $g(x, \theta)$ such that (2.4.4) is satisfied and $\int_{-\infty}^{\infty} g(x, \theta)\, dx < \infty$. Then (2.4.3) holds.*

Notice that both condition (i) of Theorem 2.4.3 and (2.4.4) impose a uniformity requirement on the functions to be bounded; some type of uniformity is generally needed before derivatives and integrals can be interchanged.

**Example 2.4.5 (Interchanging integration and differentiation–I)** Let $X$ have the exponential($\lambda$) pdf given by $f(x) = (1/\lambda)e^{-x/\lambda}$, $0 < x < \infty$, and suppose we want to calculate

$$(2.4.5) \qquad \frac{d}{d\lambda} \, \mathrm{E}\, X^n = \frac{d}{d\lambda} \int_0^\infty x^n \left(\frac{1}{\lambda}\right) e^{-x/\lambda}\, dx$$

for integer $n > 0$. If we could move the differentiation inside the integral, we would have

$$\frac{d}{d\lambda} \, \mathrm{E}\, X^n = \int_0^\infty \frac{\partial}{\partial \lambda} x^n \left(\frac{1}{\lambda}\right) e^{-x/\lambda}\, dx$$

$$(2.4.6) \qquad = \int_0^\infty \frac{x^n}{\lambda^2} \left(\frac{x}{\lambda} - 1\right) e^{-x/\lambda}\, dx$$

$$= \frac{1}{\lambda^2} \, \mathrm{E}\, X^{n+1} - \frac{1}{\lambda} \, \mathrm{E}\, X^n.$$

To justify the interchange of integration and differentiation, we bound the derivative of $x^n (1/\lambda)e^{-x/\lambda}$. Now

$$\left| \frac{\partial}{\partial \lambda} \left( \frac{x^n e^{-x\lambda}}{\lambda} \right) \right| = \frac{x^n e^{-x/\lambda}}{\lambda^2} \left| \frac{x}{\lambda} - 1 \right| \leq \frac{x^n e^{-x/\lambda}}{\lambda^2} \left( \frac{x}{\lambda} + 1 \right). \qquad \text{(since } \tfrac{x}{\lambda} > 0\text{)}$$

For some constant $\delta_0$ satisfying $0 < \delta_0 < \lambda$, take

$$g(x, \lambda) = \frac{x^n e^{-x/(\lambda+\delta_0)}}{(\lambda - \delta_0)^2} \left( \frac{x}{\lambda - \delta_0} + 1 \right).$$

We then have

$$\left| \frac{\partial}{\partial \lambda} \left( \frac{x^n e^{-x/\lambda}}{\lambda} \right) \right|_{\lambda=\lambda'} \leq g(x, \lambda) \qquad \text{for all } \lambda' \text{ such that } |\lambda' - \lambda| \leq \delta_0.$$

Since the exponential distribution has all of its moments, $\int_{-\infty}^{\infty} g(x, \lambda)\, dx < \infty$ as long as $\lambda - \delta_0 > 0$, so the interchange of integration and differentiation is justified. $\qquad \|$

The property illustrated for the exponential distribution holds for a large class of densities, which will be dealt with in Section 3.4.

Notice that (2.4.6) gives us a recursion relation for the moments of the exponential distribution,

$$(2.4.7) \qquad \mathrm{E}\, X^{n+1} = \lambda \mathrm{E}\, X^n + \lambda^2 \frac{d}{d\lambda}\, \mathrm{E}\, X^n,$$

making the calculation of the $(n+1)$st moment relatively easy. This type of relationship exists for other distributions. In particular, if $X$ has a normal distribution with mean $\mu$ and variance 1, so it has pdf $f(x) = (1/\sqrt{2\pi})e^{-(x-\mu)^2/2}$, then

$$\mathrm{E}\, X^{n+1} = \mu \mathrm{E}\, X^n - \frac{d}{d\mu}\, \mathrm{E}\, X^n.$$

We illustrate one more interchange of differentiation and integration, one involving the moment generating function.

**Example 2.4.6 (Interchanging integration and differentiation–II)**     Again let $X$ have a normal distribution with mean $\mu$ and variance 1, and consider the mgf of $X$,

$$M_X(t) = \mathrm{E}\, e^{tX} = \frac{1}{\sqrt{2\pi}} \int_{-\infty}^{\infty} e^{tx} e^{-(x-\mu)^2/2}\, dx.$$

In Section 2.3 it was stated that we can calculate moments by differentiation of $M_X(t)$ and differentiation under the integral sign was justified:

$$(2.4.8) \qquad \frac{d}{dt} M_X(t) = \frac{d}{dt} \mathrm{E}\, e^{tX} = \mathrm{E}\, \frac{\partial}{\partial t} e^{tX} = \mathrm{E}(X e^{tX}).$$

We can apply the results of this section to justify the operations in (2.4.8). Notice that when applying either Theorem 2.4.3 or Corollary 2.4.4 here, we identify $t$ with the variable $\theta$ in Theorem 2.4.3. The parameter $\mu$ is treated as a constant.

From Corollary 2.4.4, we must find a function $g(x, t)$, with finite integral, that satisfies

$$(2.4.9) \qquad \left. \frac{\partial}{\partial t} e^{tx} e^{-(x-\mu)^2/2} \right|_{t=t'} \le g(x, t) \quad \text{for all } t' \text{ such that } |t' - t| \le \delta_0.$$

Doing the obvious, we have

$$\left| \frac{\partial}{\partial t} e^{tx} e^{-(x-\mu)^2/2} \right| = \left| x e^{tx} e^{-(x-\mu)^2/2} \right| \le |x| e^{tx} e^{-(x-\mu)^2/2}.$$

It is easiest to define our function $g(x, t)$ separately for $x \ge 0$ and $x < 0$. We take

$$g(x, t) = \begin{cases} |x| e^{(t-\delta_0)x} e^{-(x-\mu)^2/2} & \text{if } x < 0 \\ |x| e^{(t+\delta_0)x} e^{-(x-\mu)^2/2} & \text{if } x \ge 0. \end{cases}$$

It is clear that this function satisfies (2.4.9); it remains to check that its integral is finite.

For $x \geq 0$ we have

$$g(x,t) = xe^{-(x^2-2x(\mu+t+\delta_0)+\mu^2)/2}.$$

We now complete the square in the exponent; that is, we write

$$x^2 - 2x(\mu+t+\delta_0) + \mu^2$$

$$= x^2 - 2x(\mu+t+\delta_0) + (\mu+t+\delta_0)^2 - (\mu+t+\delta_0)^2 + \mu^2$$

$$= (x - (\mu+t+\delta_0))^2 + \mu^2 - (\mu+t+\delta_0)^2,$$

and so, for $x \geq 0$,

$$g(x,t) = xe^{-[x-(\mu+t+\delta_0)]^2/2}e^{-[\mu^2-(\mu+t+\delta_0)^2]/2}.$$

Since the last exponential factor in this expression does not depend on $x$, $\int_0^\infty g(x,t)\,dx$ is essentially calculating the mean of a normal distribution with mean $\mu+t+\delta_0$, except that the integration is only over $[0, \infty)$. However, it follows that the integral is finite because the normal distribution has a finite mean (to be shown in Chapter 3). A similar development for $x < 0$ shows that

$$g(x,t) = |x|e^{-[x-(\mu+t-\delta_0)]^2/2}e^{-[\mu^2-(\mu+t-\delta_0)^2]/2}$$

and so $\int_{-\infty}^0 g(x,t)\,dx < \infty$. Therefore, we have found an integrable function satisfying (2.4.9) and the operation in (2.4.8) is justified.                                    ‖

We now turn to the question of when it is possible to interchange differentiation and summation, an operation that plays an important role in discrete distributions. Of course, we are concerned only with infinite sums, since a derivative can always be taken inside a finite sum.

**Example 2.4.7 (Interchanging summation and differentiation)**   Let $X$ be a discrete random variable with the *geometric distribution*

$$P(X = x) = \theta(1-\theta)^x, \quad x = 0, 1, \ldots, \quad 0 < \theta < 1.$$

We have that $\sum_{x=0}^\infty \theta(1-\theta)^x = 1$ and, provided that the operations are justified,

$$\frac{d}{d\theta}\sum_{x=0}^\infty \theta(1-\theta)^x = \sum_{x=0}^\infty \frac{d}{d\theta}\theta(1-\theta)^x$$

$$= \sum_{x=0}^\infty \left[(1-\theta)^x - \theta x(1-\theta)^{x-1}\right]$$

$$= \frac{1}{\theta}\sum_{x=0}^\infty \theta(1-\theta)^x - \frac{1}{1-\theta}\sum_{x=0}^\infty x\theta(1-\theta)^x.$$

Since $\sum_{x=0}^\infty \theta(1-\theta)^x = 1$ for all $0 < \theta < 1$, its derivative is 0. So we have

(2.4.10)          $$\frac{1}{\theta}\sum_{x=0}^\infty \theta(1-\theta)^x - \frac{1}{1-\theta}\sum_{x=0}^\infty x\theta(1-\theta)^x = 0.$$

Now the first sum in (2.4.10) is equal to 1 and the second sum is $\mathrm{E}\,X$; hence (2.4.10) becomes

$$\frac{1}{\theta} - \frac{1}{1-\theta}\,\mathrm{E}\,X = 0,$$

or

$$\mathrm{E}\,X = \frac{1-\theta}{\theta}.$$

We have, in essence, summed the series $\sum_{x=0}^{\infty} x\theta(1-\theta)^x$ by differentiating.    ‖

Justification for taking the derivative inside the summation is more straightforward than the integration case. The following theorem provides the details.

**Theorem 2.4.8**    *Suppose that the series $\sum_{x=0}^{\infty} h(\theta, x)$ converges for all $\theta$ in an interval $(a, b)$ of real numbers and*

i. *$\frac{\partial}{\partial\theta} h(\theta, x)$ is continuous in $\theta$ for each $x$,*

ii. *$\sum_{x=0}^{\infty} \frac{\partial}{\partial\theta} h(\theta, x)$ converges uniformly on every closed bounded subinterval of $(a, b)$.*

*Then*

$$(2.4.11) \qquad \frac{d}{d\theta} \sum_{x=0}^{\infty} h(\theta, x) = \sum_{x=0}^{\infty} \frac{\partial}{\partial\theta} h(\theta, x).$$

The condition of uniform convergence is the key one to verify in order to establish that the differentiation can be taken inside the summation. Recall that a series converges uniformly if its sequence of partial sums converges uniformly, a fact that we use in the following example.

**Example 2.4.9 (Continuation of Example 2.4.7)**    To apply Theorem 2.4.8 we identify

$$h(\theta, x) = \theta(1 - \theta)^x$$

and

$$\frac{\partial}{\partial\theta} h(\theta, x) = (1 - \theta)^x - \theta x(1 - \theta)^{x-1},$$

and verify that $\sum_{x=0}^{\infty} \frac{\partial}{\partial\theta} h(\theta, x)$ converges uniformly. Define $S_n(\theta)$ by

$$S_n(\theta) = \sum_{x=0}^{n} \left[(1 - \theta)^x - \theta x(1 - \theta)^{x-1}\right].$$

The convergence will be uniform on $[c, d] \subset (0, 1)$ if, given $\epsilon > 0$, we can find an $N$ such that

$$n > N \Rightarrow |S_n(\theta) - S_\infty(\theta)| < \epsilon \quad \text{for all } \theta \in [c, d].$$

Recall the partial sum of the geometric series (1.5.3). If $y \neq 1$, then we can write

$$\sum_{k=0}^{n} y^k = \frac{1 - y^{n+1}}{1 - y}.$$

Applying this, we have

$$\sum_{x=0}^{n} (1 - \theta)^x = \frac{1 - (1 - \theta)^{n+1}}{\theta}$$

$$\sum_{x=0}^{n} \theta x (1 - \theta)^{x-1} = \theta \sum_{x=0}^{n} -\frac{\partial}{\partial \theta} (1 - \theta)^x$$

$$= -\theta \frac{d}{d\theta} \sum_{x=0}^{n} (1 - \theta)^x$$

$$= -\theta \frac{d}{d\theta} \left[ \frac{1 - (1 - \theta)^{n+1}}{\theta} \right].$$

Here we (justifiably) pull the derivative through the finite sum. Calculating this derivative gives

$$\sum_{x=0}^{n} \theta x (1 - \theta)^{x-1} = \frac{(1 - (1 - \theta)^{n+1}) - (n+1)\theta(1 - \theta)^n}{\theta},$$

and, hence,

$$S_n(\theta) = \frac{1 - (1 - \theta)^{n+1}}{\theta} - \frac{(1 - (1 - \theta)^{n+1}) - (n+1)\theta(1 - \theta)^n}{\theta}$$

$$= (n+1)(1 - \theta)^n.$$

It is clear that, for $0 < \theta < 1$, $S_\infty = \lim_{n \to \infty} S_n(\theta) = 0$. Since $S_n(\theta)$ is continuous, the convergence is uniform on any closed bounded interval. Therefore, the series of derivatives converges uniformly and the interchange of differentiation and summation is justified.          ‖

We close this section with a theorem that is similar to Theorem 2.4.8, but treats the case of interchanging the order of summation and integration.

**Theorem 2.4.10**  *Suppose the series $\sum_{x=0}^{\infty} h(\theta, x)$ converges uniformly on $[a, b]$ and that, for each $x$, $h(\theta, x)$ is a continuous function of $\theta$. Then*

$$\int_a^b \sum_{x=0}^{\infty} h(\theta, x) \, d\theta = \sum_{x=0}^{\infty} \int_a^b h(\theta, x) \, d\theta.$$

## 2.5 Exercises

**2.1** In each of the following find the pdf of $Y$. Show that the pdf integrates to 1.

(a) $Y = X^3$ and $f_X(x) = 42x^5(1 - x)$, $0 < x < 1$

(b) $Y = 4X + 3$ and $f_X(x) = 7e^{-7x}$, $0 < x < \infty$

(c) $Y = X^2$ and $f_X(x) = 30x^2(1 - x)^2$, $0 < x < 1$

(See Example A.0.2 in Appendix A.)

**2.2** In each of the following find the pdf of $Y$.

(a) $Y = X^2$ and $f_X(x) = 1$, $0 < x < 1$

(b) $Y = -\log X$ and $X$ has pdf

$$f_X(x) = \frac{(n + m + 1)!}{n!\,m!}\,x^n(1 - x)^m, \quad 0 < x < 1, \quad m, n \text{ positive integers}$$

(c) $Y = e^X$ and $X$ has pdf

$$f_X(x) = \frac{1}{\sigma^2}\,xe^{-(x/\sigma)^2/2}, \quad 0 < x < \infty, \quad \sigma^2 \text{ a positive constant}$$

**2.3** Suppose $X$ has the geometric pmf $f_X(x) = \frac{1}{3}\left(\frac{2}{3}\right)^x$, $x = 0, 1, 2, \ldots$. Determine the probability distribution of $Y = X/(X + 1)$. Note that here both $X$ and $Y$ are discrete random variables. To specify the probability distribution of $Y$, specify its pmf.

**2.4** Let $\lambda$ be a fixed positive constant, and define the function $f(x)$ by $f(x) = \frac{1}{2}\lambda e^{-\lambda x}$ if $x \geq 0$ and $f(x) = \frac{1}{2}\lambda e^{\lambda x}$ if $x < 0$.

(a) Verify that $f(x)$ is a pdf.

(b) If $X$ is a random variable with pdf given by $f(x)$, find $P(X < t)$ for all $t$. Evaluate all integrals.

(c) Find $P(|X| < t)$ for all $t$. Evaluate all integrals.

**2.5** Use Theorem 2.1.8 to find the pdf of $Y$ in Example 2.1.2. Show that the same answer is obtained by differentiating the cdf given in (2.1.6).

**2.6** In each of the following find the pdf of $Y$ and show that the pdf integrates to 1.

(a) $f_X(x) = \frac{1}{2}e^{-|x|}$, $-\infty < x < \infty$; $Y = |X|^3$

(b) $f_X(x) = \frac{3}{8}(x + 1)^2$, $-1 < x < 1$; $Y = 1 - X^2$

(c) $f_X(x) = \frac{3}{8}(x + 1)^2$, $-1 < x < 1$; $Y = 1 - X^2$ if $X \leq 0$ and $Y = 1 - X$ if $X > 0$

**2.7** Let $X$ have pdf $f_X(x) = \frac{2}{9}(x + 1)$, $-1 \leq x \leq 2$.

(a) Find the pdf of $Y = X^2$. Note that Theorem 2.1.8 is not directly applicable in this problem.

(b) Show that Theorem 2.1.8 remains valid if the sets $A_0, A_1, \ldots, A_k$ *contain* $\mathcal{X}$, and apply the extension to solve part (a) using $A_0 = \emptyset$, $A_1 = (-1, 1)$, and $A_2 = (1, 2)$.

**2.8** In each of the following show that the given function is a cdf and find $F_X^{-1}(y)$.

(a) $F_X(x) = \begin{cases} 0 & \text{if } x < 0 \\ 1 - e^{-x} & \text{if } x \geq 0 \end{cases}$

(b) $F_X(x) = \begin{cases} e^x/2 & \text{if } x < 0 \\ 1/2 & \text{if } 0 \le x < 1 \\ 1 - (e^{1-x}/2) & \text{if } 1 \le x \end{cases}$

(c) $F_X(x) = \begin{cases} e^x/4 & \text{if } x < 0 \\ 1 - (e^{-x}/4) & \text{if } x \ge 0 \end{cases}$

Note that, in part (c), $F_X(x)$ is discontinuous but (2.1.13) is still the appropriate definition of $F_X^{-1}(y)$.

**2.9** If the random variable $X$ has pdf

$$ f(x) = \begin{cases} \frac{x-1}{2} & 1 < x < 3 \\ 0 & \text{otherwise,} \end{cases} $$

find a monotone function $u(x)$ such that the random variable $Y = u(X)$ has a uniform$(0, 1)$ distribution.

**2.10** In Theorem 2.1.10 the probability integral transform was proved, relating the uniform cdf to any continuous cdf. In this exercise we investigate the relationship between discrete random variables and uniform random variables. Let $X$ be a discrete random variable with cdf $F_X(x)$ and define the random variable $Y$ as $Y = F_X(X)$.

(a) Prove that $Y$ is stochastically greater than a uniform$(0, 1)$; that is, if $U \sim$ uniform $(0, 1)$, then

$$ P(Y > y) \ge P(U > y) = 1 - y, \qquad \text{for all } y, \ 0 < y < 1, $$

$$ P(Y > y) > P(U > y) = 1 - y, \qquad \text{for some } y, \ 0 < y < 1. $$

(Recall that *stochastically greater* was defined in Exercise 1.49.)

(b) Equivalently, show that the cdf of $Y$ satisfies $F_Y(y) \le y$ for all $0 < y < 1$ and $F_Y(y) < y$ for some $0 < y < 1$. (*Hint:* Let $x_0$ be a jump point of $F_X$, and define $y_0 = F_X(x_0)$. Show that $P(Y \le y_0) = y_0$. Now establish the inequality by considering $y = y_0 + \epsilon$. Pictures of the cdfs will help.)

**2.11** Let $X$ have the standard normal pdf, $f_X(x) = (1/\sqrt{2\pi})e^{-x^2/2}$.

(a) Find $EX^2$ directly, and then by using the pdf of $Y = X^2$ from Example 2.1.7 and calculating $EY$.

(b) Find the pdf of $Y = |X|$, and find its mean and variance.

**2.12** A random right triangle can be constructed in the following manner. Let $X$ be a random angle whose distribution is uniform on $(0, \pi/2)$. For each $X$, construct a triangle as pictured below. Here, $Y = $ height of the random triangle. For a fixed constant $d$, find the distribution of $Y$ and $EY$.

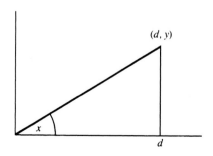

**2.13** Consider a sequence of independent coin flips, each of which has probability $p$ of being heads. Define a random variable $X$ as the length of the run (of either heads or tails) started by the first trial. (For example, $X = 3$ if either TTTH or HHHT is observed.) Find the distribution of $X$, and find $\mathrm{E}\,X$.

**2.14** (a) Let $X$ be a continuous, nonnegative random variable $[f(x) = 0$ for $x < 0]$. Show that

$$\mathrm{E}\,X = \int_0^\infty [1 - F_X(x)]\,dx,$$

where $F_X(x)$ is the cdf of $X$.

(b) Let $X$ be a discrete random variable whose range is the nonnegative integers. Show that

$$\mathrm{E}\,X = \sum_{k=0}^\infty (1 - F_X(k)),$$

where $F_X(k) = P(X \leq k)$. Compare this with part (a).

**2.15** Betteley (1977) provides an interesting addition law for expectations. Let $X$ and $Y$ be any two random variables and define

$$X \wedge Y = \min(X, Y) \quad \text{and} \quad X \vee Y = \max(X, Y).$$

Analogous to the probability law $P(A \cup B) = P(A) + P(B) - P(A \cap B)$, show that

$$\mathrm{E}(X \vee Y) = \mathrm{E}\,X + \mathrm{E}\,Y - \mathrm{E}(X \wedge Y).$$

(*Hint:* Establish that $X + Y = (X \vee Y) + (X \wedge Y)$.)

**2.16** Use the result of Exercise 2.14 to find the mean duration of certain telephone calls, where we assume that the duration, $T$, of a particular call can be described probabilistically by $P(T > t) = ae^{-\lambda t} + (1-a)e^{-\mu t}$, where $a$, $\lambda$, and $\mu$ are constants, $0 < a < 1$, $\lambda > 0$, $\mu > 0$.

**2.17** A *median* of a distribution is a value $m$ such that $P(X \leq m) \geq \frac{1}{2}$ and $P(X \geq m) \geq \frac{1}{2}$. (If $X$ is continuous, $m$ satisfies $\int_{-\infty}^m f(x)\,dx = \int_m^\infty f(x)\,dx = \frac{1}{2}$.) Find the median of the following distributions.

(a) $f(x) = 3x^2, \quad 0 < x < 1$        (b) $f(x) = \dfrac{1}{\pi(1 + x^2)}, \quad -\infty < x < \infty$

**2.18** Show that if $X$ is a continuous random variable, then

$$\min_a \mathrm{E}\,|X - a| = \mathrm{E}\,|X - m|,$$

where $m$ is the median of $X$ (see Exercise 2.17).

**2.19** Prove that

$$\frac{d}{da}\mathrm{E}(X - a)^2 = 0 \Leftrightarrow \mathrm{E}\,X = a$$

by differentiating the integral. Verify, using calculus, that $a = \mathrm{E}\,X$ is indeed a minimum. List the assumptions about $F_X$ and $f_X$ that are needed.

**2.20** A couple decides to continue to have children until a daughter is born. What is the expected number of children of this couple? (*Hint:* See Example 1.5.4.)

**2.21** Prove the "two-way" rule for expectations, equation (2.2.5), which says $\mathrm{E}\,g(X) = \mathrm{E}\,Y$, where $Y = g(X)$. Assume that $g(x)$ is a monotone function.

**2.22** Let $X$ have the pdf

$$f(x) = \frac{4}{\beta^3 \sqrt{\pi}} x^2 e^{-x^2/\beta^2}, \quad 0 < x < \infty, \quad \beta > 0.$$

   (a) Verify that $f(x)$ is a pdf.         (b) Find $EX$ and Var $X$.

**2.23** Let $X$ have the pdf

$$f(x) = \frac{1}{2}(1+x), \quad -1 < x < 1.$$

   (a) Find the pdf of $Y = X^2$.         (b) Find $EY$ and Var $Y$.

**2.24** Compute $EX$ and Var $X$ for each of the following probability distributions.

   (a) $f_X(x) = ax^{a-1}, \, 0 < x < 1, \, a > 0$
   (b) $f_X(x) = \frac{1}{n}, \, x = 1, 2, \ldots, n, \, n > 0$ an integer
   (c) $f_X(x) = \frac{3}{2}(x - 1)^2, \, 0 < x < 2$

**2.25** Suppose the pdf $f_X(x)$ of a random variable $X$ is an *even function*. ($f_X(x)$ is an *even function* if $f_X(x) = f_X(-x)$ for every $x$.) Show that

   (a) $X$ and $-X$ are identically distributed.
   (b) $M_X(t)$ is symmetric about 0.

**2.26** Let $f(x)$ be a pdf and let $a$ be a number such that, for all $\epsilon > 0$, $f(a + \epsilon) = f(a - \epsilon)$. Such a pdf is said to be *symmetric* about the point $a$.

   (a) Give three examples of symmetric pdfs.
   (b) Show that if $X \sim f(x)$, symmetric, then the median of $X$ (see Exercise 2.17) is the number $a$.
   (c) Show that if $X \sim f(x)$, symmetric, and $EX$ exists, then $EX = a$.
   (d) Show that $f(x) = e^{-x}$, $x \geq 0$, is not a symmetric pdf.
   (e) Show that for the pdf in part (d), the median is less than the mean.

**2.27** Let $f(x)$ be a pdf, and let $a$ be a number such that if $a \geq x \geq y$, then $f(a) \geq f(x) \geq f(y)$, and if $a \leq x \leq y$, then $f(a) \geq f(x) \geq f(y)$. Such a pdf is called *unimodal* with a *mode* equal to $a$.

   (a) Give an example of a unimodal pdf for which the mode is unique.
   (b) Give an example of a unimodal pdf for which the mode is not unique.
   (c) Show that if $f(x)$ is both symmetric (see Exercise 2.26) and unimodal, then the point of symmetry is a mode.
   (d) Consider the pdf $f(x) = e^{-x}$, $x \geq 0$. Show that this pdf is unimodal. What is its mode?

**2.28** Let $\mu_n$ denote the $n$th central moment of a random variable $X$. Two quantities of interest, in addition to the mean and variance, are

$$\alpha_3 = \frac{\mu_3}{(\mu_2)^{3/2}} \quad \text{and} \quad \alpha_4 = \frac{\mu_4}{\mu_2^2}.$$

The value $\alpha_3$ is called the *skewness* and $\alpha_4$ is called the *kurtosis*. The skewness measures the lack of symmetry in the pdf (see Exercise 2.26). The kurtosis, although harder to interpret, measures the peakedness or flatness of the pdf.

(a) Show that if a pdf is symmetric about a point $a$, then $\alpha_3 = 0$.

(b) Calculate $\alpha_3$ for $f(x) = e^{-x}$, $x \geq 0$, a pdf that is *skewed to the right*.

(c) Calculate $\alpha_4$ for each of the following pdfs and comment on the peakedness of each.

$$f(x) = \frac{1}{\sqrt{2\pi}} e^{-x^2/2}, \quad -\infty < x < \infty$$

$$f(x) = \frac{1}{2}, \quad -1 < x < 1$$

$$f(x) = \frac{1}{2} e^{-|x|}, \quad -\infty < x < \infty$$

Ruppert (1987) uses *influence functions* (see Miscellanea 10.6.4) to explore further the meaning of kurtosis, and Groeneveld (1991) uses them to explore skewness; see also Balanda and MacGillivray (1988) for more on the interpretation of $\alpha_4$.

**2.29** To calculate moments of discrete distributions, it is often easier to work with the *factorial moments* (see Miscellanea 2.6.2).

(a) Calculate the factorial moment $E[X(X - 1)]$ for the binomial and Poisson distributions.

(b) Use the results of part (a) to calculate the variances of the binomial and Poisson distributions.

(c) A particularly nasty discrete distribution is the *beta-binomial*, with pmf

$$P(y = y) = a \left( \frac{1}{y + a} \right) \frac{\binom{n}{y} \binom{a+b-1}{a}}{\binom{n+a+b-1}{y+a}},$$

where $n$, $a$, and $b$ are integers, and $y = 0, 1, 2, \ldots, n$. Use factorial moments to calculate the variance of the beta-binomial. (See Exercise 4.34 for another approach to this calculation.)

**2.30** Find the moment generating function corresponding to

(a) $f(x) = \frac{1}{c}, \quad 0 < x < c$.

(b) $f(x) = \frac{2x}{c^2}, \quad 0 < x < c$.

(c) $f(x) = \frac{1}{2\beta} e^{-|x-\alpha|/\beta}, \quad -\infty < x < \infty, \quad -\infty < \alpha < \infty, \quad \beta > 0$.

(d) $P(X = x) = \binom{r + x - 1}{x} p^r (1-p)^x, \quad x = 0, 1, \ldots, \quad 0 < p < 1, r > 0$ an integer.

**2.31** Does a distribution exist for which $M_X(t) = t/(1 - t), |t| < 1$? If yes, find it. If no, prove it.

**2.32** Let $M_X(t)$ be the moment generating function of $X$, and define $S(t) = \log(M_X(t))$. Show that

$$\frac{d}{dt} S(t) \Big|_{t=0} = E X \quad \text{and} \quad \frac{d^2}{dt^2} S(t) \Big|_{t=0} = \text{Var} X.$$

**2.33** In each of the following cases verify the expression given for the moment generating function, and in each case use the mgf to calculate $E X$ and $\text{Var} X$.

(a) $P(X = x) = \frac{e^{-\lambda}\lambda^x}{x!}$, $M_X(t) = e^{\lambda(e^t - 1)}$, $x = 0, 1, \ldots$; $\lambda > 0$

(b) $P(X = x) = p(1 - p)^x$, $M_X(t) = \frac{p}{1 - (1-p)e^t}$, $x = 0, 1, \ldots$; $0 < p < 1$

(c) $f_X(x) = \frac{e^{-(x-\mu)^2/(2\sigma^2)}}{\sqrt{2\pi}\sigma}$, $M_X(t) = e^{\mu t + \sigma^2 t^2/2}$, $-\infty < x < \infty$; $-\infty < \mu < \infty$, $\sigma > 0$

**2.34** A distribution cannot be uniquely determined by a finite collection of moments, as this example from Romano and Siegel (1986) shows. Let $X$ have the normal distribution, that is, $X$ has pdf

$$f_X(x) = \frac{1}{\sqrt{2\pi}} e^{-x^2/2}, \quad -\infty < x < \infty.$$

Define a discrete random variable $Y$ by

$$P\left(Y = \sqrt{3}\right) = P\left(Y = -\sqrt{3}\right) = \frac{1}{6}, \qquad P\left(Y = 0\right) = \frac{2}{3}.$$

Show that

$$E\, X^r = E\, Y^r \quad \text{for } r = 1, 2, 3, 4, 5.$$

(Romano and Siegel point out that for any finite $n$ there exists a discrete, and hence nonnormal, random variable whose first $n$ moments are equal to those of $X$.)

**2.35** Fill in the gaps in Example 2.3.10.

(a) Show that if $X_1 \sim f_1(x)$, then

$$E\, X_1^r = e^{r^2/2}, \quad r = 0, 1, \ldots.$$

So $f_1(x)$ has all of its moments, and all of the moments are finite.

(b) Now show that

$$\int_0^\infty x^r f_1(x) \sin(2\pi \log x)\, dx = 0$$

for all positive integers $r$, so $EX_1^r = E\, X_2^r$ for all $r$. (Romano and Siegel 1986 discuss an extreme version of this example, where an entire class of distinct pdfs have the same moments. Also, Berg 1988 has shown that this moment behavior can arise with simpler transforms of the normal distribution such as $X^3$.)

**2.36** The *lognormal distribution*, on which Example 2.3.10 is based, has an interesting property. If we have the pdf

$$f(x) = \frac{1}{\sqrt{2\pi}x} e^{-(\log x)^2/2}, \quad 0 \le x < \infty,$$

then Exercise 2.35 shows that all moments exist and are finite. However, this distribution does not have a moment generating function, that is,

$$M_X(t) = \int_0^\infty \frac{e^{tx}}{\sqrt{2\pi}x} e^{-(\log x)^2/2}\, dx$$

does not exist. Prove this.

**2.37** Referring to the situation described in Miscellanea 2.6.3:

(a) Plot the pdfs $f_1$ and $f_2$ to illustrate their difference.

(b) Plot the cumulant generating functions $K_1$ and $K_2$ to illustrate their similarity.

(c) Calculate the moment generating functions of the pdfs $f_1$ and $f_2$. Are they similar or different?

(d) How do the pdfs $f_1$ and $f_2$ relate to the pdfs described in Example 2.3.10?

**2.38** Let $X$ have the negative binomial distribution with pmf

$$f(x) = \binom{r + x - 1}{x} p^r (1 - p)^x, \quad x = 0, 1, 2, \ldots,$$

where $0 < p < 1$ and $r > 0$ is an integer.

(a) Calculate the mgf of $X$.

(b) Define a new random variable by $Y = 2pX$. Show that, as $p \downarrow 0$, the mgf of $Y$ converges to that of a chi squared random variable with $2r$ degrees of freedom by showing that

$$\lim_{p \to 0} M_Y(t) = \left( \frac{1}{1 - 2t} \right)^r, \quad |t| < \frac{1}{2}.$$

**2.39** In each of the following cases calculate the indicated derivatives, justifying all operations.

(a) $\frac{d}{dx} \int_0^x e^{-\lambda t}\, dt$  (b) $\frac{d}{d\lambda} \int_0^\infty e^{-\lambda t}\, dt$

(c) $\frac{d}{dt} \int_t^1 \frac{1}{x^2}\, dx$  (d) $\frac{d}{dt} \int_1^\infty \frac{1}{(x - t)^2}\, dx$

**2.40** Prove

$$\sum_{k=0}^{x} \binom{n}{k} p^k (1 - p)^{n-k} = (n - x) \binom{n}{x} \int_0^{1-p} t^{n-x-1} (1 - t)^x\, dt.$$

(*Hint:* Integrate by parts or differentiate both sides with respect to $p$.)

## 2.6 Miscellanea

### 2.6.1 Uniqueness of Moment Sequences

A distribution is not necessarily determined by its moments. But if $\sum_{r=1}^{\infty} \mu^r r^k / k!$ has a positive radius of convergence, where $X \sim F_X$ and $E X^r = \mu'_r$, then the moment sequence is unique, and hence the distribution is uniquely determined (Billingsley 1995, Section 30). Convergence of this sum also implies that the moment-generating function exists in an interval, and hence the moment-generating function determines the distribution

A sufficient condition for the moment sequence to be unique is *Carleman's Condition* (Chung 1974). If $X \sim F_X$ and we denote $E X^r = \mu'_r$, then the moment sequence is unique if

$$\sum_{r=1}^{\infty} \frac{1}{(\mu'_{2r})^{1/(2r)}} = +\infty.$$

This condition is, in general, not easy to verify.

Feller (1971) has a very complete development of Laplace transforms, of which mgfs are a special case. In particular, Feller shows (similar to Billingsley) that whenever

$$M_X(t) = \sum_{r=0}^{\infty} \frac{(-1)^r \mu'_r t^r}{r!}$$

converges on an interval $-t_0 \leq t < t_0$, $t_0 > 0$, the distribution $F_X$ is uniquely determined. Thus, when the mgf exists, the moment sequence determines the distribution $F_X$ uniquely.

It should be clear that using the mgf to determine the distribution is a difficult task. A better method is through the use of *characteristic functions*, which are explained below. Although characteristic functions simplify the characterization of a distribution, they necessitate understanding complex analysis. You win some and you lose some.

### 2.6.2 Other Generating Functions

In addition to the moment generating function, there are a number of other generating functions available. In most cases, the characteristic function is the most useful of these. Except for rare circumstances, the other generating functions are less useful, but there are situations where they can ease calculations.

*Cumulant generating function*   For a random variable $X$, the cumulant generating function is the function $\log[M_X(t)]$. This function can be used to generate the *cumulants* of $X$, which are defined (rather circuitously) as the coefficients in the Taylor series of the cumulant generating function (see Exercise 2.32).

*Factorial moment generating function*   The factorial moment-generating function of $X$ is defined as $Et^X$, if the expectation exists. The name comes from the fact that this function satisfies

$$\left. \frac{d^r}{dt^r} E t^X \right|_{t=1} = E\{X(X-1)\cdots(X-r+1)\},$$

where the right-hand side is a *factorial moment*. If $X$ is a discrete random variable, then we can write

$$E t^X = \sum_x t^x P(X = x),$$

and the factorial moment generating function is called the *probability-generating function*, since the coefficients of the power series give the probabilities. That is, to obtain the probability that $X = k$, calculate

$$\left. \frac{1}{k!} \frac{d^k}{dt^k} E t^X \right|_{t=0} = P(X = k).$$

*Characteristic function*   Perhaps the most useful of all of these types of functions is the characteristic function. The characteristic function of $X$ is defined by

$$\phi_X(t) = \mathrm{E}\, e^{itX},$$

where $i$ is the complex number $\sqrt{-1}$, so the above expectation requires complex integration. The characteristic function does much more than the mgf does. When the moments of $F_X$ exist, $\phi_X$ can be used to generate them, much like an mgf. The characteristic function always exists and it completely determines the distribution. That is, every cdf has a unique characteristic function. So we can state a theorem like Theorem 2.3.11, for example, but without qualification.

**Theorem 2.6.1 (Convergence of characteristic functions)**   *Suppose $X_k$, $k = 1, 2, \ldots$, is a sequence of random variables, each with characteristic function $\phi_{X_k}(t)$. Furthermore, suppose that*

$$\lim_{k \to \infty} \phi_{X_k}(t) = \phi_X(t), \text{ for all } t \text{ in a neighborhood of } 0,$$

*and $\phi_X(t)$ is a characteristic function. Then, for all $x$ where $F_X(x)$ is continuous,*

$$\lim_{k \to \infty} F_{X_k}(x) = F_X(x).$$

A full treatment of generating functions is given by Feller (1968). Characteristic functions can be found in almost any advanced probability text; see Billingsley (1995) or Resnick (1999).

*2.6.3 Does the Moment Generating Function Characterize a Distribution?*

In an article with the above title, McCullagh (1994) looks at a pair of densities similar to those in Example 2.3.10 but having pdfs

$$f_1 = n(0, 1) \quad \text{and} \quad f_2 = f_1(x) \left[ 1 + \frac{1}{2} \sin(2\pi x) \right]$$

with cumulant generating functions

$$K_1(t) = t^2/2 \quad \text{and} \quad K_2(t) = K_1(t) + \log \left[ 1 + \frac{1}{2} e^{-2\pi^2} \sin(2\pi t) \right].$$

He notes that although the densities are visibly dissimilar, the cgfs are virtually identical, with maximum difference less than $1.34 \times 10^{-9}$ over the entire range (less than the size of one pixel). So the answer to the question posed in the title is "*yes* for mathematical purposes but a resounding *no* for numerical purposes." In contrast, Waller (1995) illustrates that although the mgfs fail to numerically distinguish the distributions, the *characteristic functions* do a fine job. (Waller *et al.* 1995 and Luceño 1997 further investigate the usefulness of the characteristic function in numerically obtaining the cdfs.) See Exercise 2.37 for details.

Chapter 3

# Common Families of Distributions

*"How do all these unusuals strike you, Watson?"*
*"Their cumulative effect is certainly considerable, and yet each of them is quite possible in itself."*

**Sherlock Holmes and Dr. Watson**
*The Adventure of the Abbey Grange*

## 3.1 Introduction

Statistical distributions are used to model populations; as such, we usually deal with a *family* of distributions rather than a single distribution. This family is indexed by one or more parameters, which allow us to vary certain characteristics of the distribution while staying with one functional form. For example, we may specify that the normal distribution is a reasonable choice to model a particular population, but we cannot precisely specify the mean. Then, we deal with a parametric family, normal distributions with mean $\mu$, where $\mu$ is an unspecified parameter, $-\infty < \mu < \infty$.

In this chapter we catalog many of the more common statistical distributions, some of which we have previously encountered. For each distribution we will give its mean and variance and many other useful or descriptive measures that may aid understanding. We will also indicate some typical applications of these distributions and some interesting and useful interrelationships. Some of these facts are summarized in tables at the end of the book. This chapter is by no means comprehensive in its coverage of statistical distributions. That task has been accomplished by Johnson and Kotz (1969–1972) in their multiple-volume work *Distributions in Statistics* and in the updated volumes by Johnson, Kotz, and Balakrishnan (1994, 1995) and Johnson, Kotz, and Kemp (1992).

## 3.2 Discrete Distributions

A random variable $X$ is said to have a discrete distribution if the range of $X$, the sample space, is countable. In most situations, the random variable has integer-valued outcomes.

*Discrete Uniform Distribution*

A random variable $X$ has a *discrete uniform* $(1, N)$ *distribution* if

(3.2.1) $$P(X = x|N) = \frac{1}{N}, \quad x = 1, 2, \ldots, N,$$

where $N$ is a specified integer. This distribution puts equal mass on each of the outcomes $1, 2, \ldots, N$.

*A note on notation*: When we are dealing with parametric distributions, as will almost always be the case, the distribution is dependent on values of the parameters. In order to emphasize this fact and to keep track of the parameters, we write them in the pmf preceded by a "$|$" (given). This convention will also be used with cdfs, pdfs, expectations, and other places where it might be necessary to keep track of the parameters. When there is no possibility of confusion, the parameters may be omitted in order not to clutter up notation too much.

To calculate the mean and variance of $X$, recall the identities (provable by induction)

$$\sum_{i=1}^{k} i = \frac{k(k+1)}{2} \quad \text{and} \quad \sum_{i=1}^{k} i^2 = \frac{k(k+1)(2k+1)}{6}.$$

We then have

$$\mathrm{E}X = \sum_{x=1}^{N} xP(X = x|N) = \sum_{x=1}^{N} x\frac{1}{N} = \frac{N+1}{2}$$

and

$$\mathrm{E}X^2 = \sum_{x=1}^{N} x^2\frac{1}{N} = \frac{(N+1)(2N+1)}{6},$$

and so

$$\mathrm{Var}\, X = \mathrm{E}X^2 - (\mathrm{E}X)^2$$

$$= \frac{(N+1)(2N+1)}{6} - \left(\frac{N+1}{2}\right)^2$$

$$= \frac{(N+1)(N-1)}{12}.$$

This distribution can be generalized so that the sample space is any range of integers, $N_0, N_0 + 1, \ldots, N_1$, with pmf $P(X = x|N_0, N_1) = 1/(N_1 - N_0 + 1)$.

*Hypergeometric Distribution*

The hypergeometric distribution has many applications in finite population sampling and is best understood through the classic example of the urn model.

Suppose we have a large urn filled with $N$ balls that are identical in every way except that $M$ are red and $N - M$ are green. We reach in, blindfolded, and select $K$ balls at random (the $K$ balls are taken all at once, a case of sampling without replacement). What is the probability that exactly $x$ of the balls are red?

The total number of samples of size $K$ that can be drawn from the $N$ balls is $\binom{N}{K}$, as was discussed in Section 1.2.3. It is required that $x$ of the balls be red, and this can be accomplished in $\binom{M}{x}$ ways, leaving $\binom{N-M}{K-x}$ ways of filling out the sample with $K - x$ green balls. Thus, if we let $X$ denote the number of red balls in a sample of size $K$, then $X$ has a *hypergeometric distribution* given by

$$(3.2.2) \qquad P(X = x | N, M, K) = \frac{\binom{M}{x}\binom{N-M}{K-x}}{\binom{N}{K}}, \qquad x = 0, 1, \ldots, K.$$

Note that there is, implicit in (3.2.2), an additional assumption on the range of $X$. Binomial coefficients of the form $\binom{n}{r}$ have been defined only if $n \geq r$, and so the range of $X$ is additionally restricted by the pair of inequalities

$$M \geq x \quad \text{and} \quad N - M \geq K - x,$$

which can be combined as

$$M - (N - K) \leq x \leq M.$$

In many cases $K$ is small compared to $M$ and $N$, so the range $0 \leq x \leq K$ will be contained in the above range and, hence, will be appropriate. The formula for the hypergeometric probability function is usually quite difficult to deal with. In fact, it is not even trivial to verify that

$$\sum_{x=0}^{K} P(X = x) = \sum_{x=0}^{K} \frac{\binom{M}{x}\binom{N-M}{K-x}}{\binom{N}{K}} = 1.$$

The hypergeometric distribution illustrates the fact that, statistically, dealing with finite populations (finite $N$) is a difficult task.

The mean of the hypergeometric distribution is given by

$$EX = \sum_{x=0}^{K} x \frac{\binom{M}{x}\binom{N-M}{K-x}}{\binom{N}{K}} = \sum_{x=1}^{K} x \frac{\binom{M}{x}\binom{N-M}{K-x}}{\binom{N}{K}}. \qquad \text{(summand is 0 at } x = 0)$$

To evaluate this expression, we use the identities (already encountered in Section 2.3)

$$x \binom{M}{x} = M \binom{M-1}{x-1},$$

$$\binom{N}{K} = \frac{N}{K} \binom{N-1}{K-1},$$

and obtain

$$\mathrm{E}X = \sum_{x=1}^{K} \frac{M \binom{M-1}{x-1}\binom{N-M}{K-x}}{\frac{N}{K}\binom{N-1}{K-1}} = \frac{KM}{N} \sum_{x=1}^{K} \frac{\binom{M-1}{x-1}\binom{N-M}{K-x}}{\binom{N-1}{K-1}}.$$

We now can recognize the second sum above as the sum of the probabilities for another hypergeometric distribution based on parameter values $N-1$, $M-1$, and $K-1$. This can be seen clearly by defining $y = x - 1$ and writing

$$\sum_{x=1}^{K} \frac{\binom{M-1}{x-1}\binom{N-M}{K-x}}{\binom{N-1}{K-1}} = \sum_{y=0}^{K-1} \frac{\binom{M-1}{y}\binom{(N-1)-(M-1)}{K-1-y}}{\binom{N-1}{K-1}}$$

$$= \sum_{y=0}^{K-1} P(Y = y | N - 1, M - 1, K - 1) = 1,$$

where $Y$ is a hypergeometric random variable with parameters $N-1$, $M-1$, and $K-1$. Therefore, for the hypergeometric distribution,

$$\mathrm{E}X = \frac{KM}{N}.$$

A similar, but more lengthy, calculation will establish that

$$\mathrm{Var}\, X = \frac{KM}{N}\left(\frac{(N-M)(N-K)}{N(N-1)}\right).$$

Note the manipulations used here to calculate $\mathrm{E}X$. The sum was transformed to another hypergeometric distribution with different parameter values and, by recognizing this fact, we were able to sum the series.

**Example 3.2.1 (Acceptance sampling)**    The hypergeometric distribution has application in acceptance sampling, as this example will illustrate. Suppose a retailer buys goods in lots and each item can be either acceptable or defective. Let

$$N = \# \text{ of items in a lot,}$$

$$M = \# \text{ of defectives in a lot.}$$

Then we can calculate the probability that a sample of size $K$ contains $x$ defectives. To be specific, suppose that a lot of 25 machine parts is delivered, where a part is considered acceptable only if it passes tolerance. We sample 10 parts and find that none are defective (all are within tolerance). What is the probability of this event if there are 6 defectives in the lot of 25? Applying the hypergeometric distribution with $N = 25, M = 6, K = 10$, we have

$$P(X = 0) = \frac{\binom{6}{0}\binom{19}{10}}{\binom{25}{10}} = .028,$$

showing that our observed event is quite unlikely if there are 6 (or more!) defectives in the lot.                                                                                    ‖

*Binomial Distribution*

The binomial distribution, one of the more useful discrete distributions, is based on the idea of a *Bernoulli trial*. A Bernoulli trial (named for James Bernoulli, one of the founding fathers of probability theory) is an experiment with two, and only two, possible outcomes. A random variable $X$ has a *Bernoulli(p) distribution* if

$$(3.2.3) \qquad X = \begin{cases} 1 & \text{with probability } p \\ 0 & \text{with probability } 1 - p, \end{cases} \qquad 0 \le p \le 1.$$

The value $X = 1$ is often termed a "success" and $p$ is referred to as the success probability. The value $X = 0$ is termed a "failure." The mean and variance of a Bernoulli($p$) random variable are easily seen to be

$$\text{E}X = 1p + 0(1 - p) = p,$$
$$\text{Var}\,X = (1 - p)^2 p + (0 - p)^2 (1 - p) = p(1 - p).$$

Many experiments can be modeled as a sequence of Bernoulli trials, the simplest being the repeated tossing of a coin; $p = $ probability of a head, $X = 1$ if the coin shows heads. Other examples include gambling games (for example, in roulette let $X = 1$ if red occurs, so $p = $ probability of red), election polls ($X = 1$ if candidate A gets a vote), and incidence of a disease ($p = $ probability that a random person gets infected).

If $n$ identical Bernoulli trials are performed, define the events

$$A_i = \{X = 1 \text{ on the } i\text{th trial}\}, \quad i = 1, 2, \ldots, n.$$

If we assume that the events $A_1, \ldots, A_n$ are a collection of independent events (as is the case in coin tossing), it is then easy to derive the distribution of the total number of successes in $n$ trials. Define a random variable $Y$ by

$$Y = \text{ total number of successes in } n \text{ trials.}$$

The event $\{Y = y\}$ will occur only if, out of the events $A_1, \ldots, A_n$, exactly $y$ of them occur, and necessarily $n - y$ of them do not occur. One particular outcome (one particular ordering of occurrences and nonoccurrences) of the $n$ Bernoulli trials might be $A_1 \cap A_2 \cap A_3^c \cap \cdots \cap A_{n-1} \cap A_n^c$. This has probability of occurrence

$$P(A_1 \cap A_2 \cap A_3^c \cap \cdots \cap A_{n-1} \cap A_n^c) = pp(1 - p) \cdots \cdot p(1 - p)$$
$$= p^y (1 - p)^{n-y},$$

where we have used the independence of the $A_i$s in this calculation. Notice that the calculation is not dependent on *which* set of $y$ $A_i$s occurs, only that *some* set of $y$ occurs. Furthermore, the event $\{Y = y\}$ will occur no matter which set of $y$ $A_i$s occurs. Putting this all together, we see that a particular sequence of $n$ trials with exactly $y$ successes has probability $p^y (1 - p)^{n-y}$ of occurring. Since there are $\binom{n}{y}$

such sequences (the number of orderings of $y$ 1s and $n - y$ 0s), we have

$$P(Y = y | n, p) = \binom{n}{y} p^y (1 - p)^{n-y}, \quad y = 0, 1, 2, \ldots, n,$$

and $Y$ is called a *binomial$(n, p)$ random variable.*

The random variable $Y$ can be alternatively, and equivalently, defined in the following way: In a sequence of $n$ identical, independent Bernoulli trials, each with success probability $p$, define the random variables $X_1, \ldots, X_n$ by

$$X_i = \begin{cases} 1 & \text{with probability } p \\ 0 & \text{with probability } 1 - p. \end{cases}$$

The random variable

$$Y = \sum_{i=1}^{n} X_i$$

has the binomial$(n, p)$ distribution.

The fact that $\sum_{y=0}^{n} P(Y = y) = 1$ follows from the following general theorem.

**Theorem 3.2.2 (Binomial Theorem)**   *For any real numbers $x$ and $y$ and integer $n \geq 0$,*

(3.2.4)
$$(x + y)^n = \sum_{i=0}^{n} \binom{n}{i} x^i y^{n-i}.$$

**Proof:** Write

$$(x + y)^n = (x + y)(x + y) \cdots \cdots (x + y),$$

and consider how the right-hand side would be calculated. From each factor $(x + y)$ we choose either an $x$ or $y$, and multiply together the $n$ choices. For each $i = 0, 1, \ldots, n$, the number of such terms in which $x$ appears exactly $i$ times is $\binom{n}{i}$. Therefore, this term is of the form $\binom{n}{i} x^i y^{n-i}$ and the result follows.   $\Box$

If we take $x = p$ and $y = 1 - p$ in (3.2.4), we get

$$1 = (p + (1 - p))^n = \sum_{i=0}^{n} \binom{n}{i} p^i (1 - p)^{n-i},$$

and we see that each term in the sum is a binomial probability. As another special case, take $x = y = 1$ in Theorem 3.2.2 and get the identity

$$2^n = \sum_{i=0}^{n} \binom{n}{i}.$$

The mean and variance of the binomial distribution have already been derived in Examples 2.2.3 and 2.3.5, so we will not repeat the derivations here. For completeness, we state them. If $X \sim \text{binomial}(n, p)$, then

$$\text{E}X = np, \quad \text{Var}\, X = np(1 - p).$$

The mgf of the binomial distribution was calculated in Example 2.3.9. It is

$$M_X(t) = [pe^t + (1 - p)]^n.$$

**Example 3.2.3 (Dice probabilities)**    Suppose we are interested in finding the probability of obtaining at least one 6 in four rolls of a fair die. This experiment can be modeled as a sequence of four Bernoulli trials with success probability $p = \frac{1}{6} = P(\text{die}$ shows 6). Define the random variable $X$ by

$$X = \text{total number of 6s in four rolls.}$$

Then $X \sim \text{binomial}\left(4, \frac{1}{6}\right)$ and

$$P(\text{at least one } 6) = P(X > 0) = 1 - P(X = 0)$$

$$= 1 - \binom{4}{0}\left(\frac{1}{6}\right)^0\left(\frac{5}{6}\right)^4$$

$$= 1 - \left(\frac{5}{6}\right)^4$$

$$= .518.$$

Now we consider another game; throw a pair of dice 24 times and ask for the probability of at least one double 6. This, again, can be modeled by the binomial distribution with success probability $p$, where

$$p = P(\text{roll a double } 6) = \frac{1}{36}.$$

So, if $Y = $ number of double 6s in 24 rolls, $Y \sim \text{binomial}\left(24, \frac{1}{36}\right)$ and

$$P(\text{at least one double } 6) = P(Y > 0)$$

$$= 1 - P(Y = 0)$$

$$= 1 - \binom{24}{0}\left(\frac{1}{36}\right)^0\left(\frac{35}{36}\right)^{24}$$

$$= 1 - \left(\frac{35}{36}\right)^{24}$$

$$= .491.$$

This is the calculation originally done in the eighteenth century by Pascal at the request of the gambler de Meré, who thought both events had the same probability. (He began to believe he was wrong when he started losing money on the second bet.)

‖

*Poisson Distribution*

The Poisson distribution is a widely applied discrete distribution and can serve as a model for a number of different types of experiments. For example, if we are modeling a phenomenon in which we are waiting for an occurrence (such as waiting for a bus, waiting for customers to arrive in a bank), the number of occurrences in a given time interval can sometimes be modeled by the Poisson distribution. One of the basic assumptions on which the Poisson distribution is built is that, for small time intervals, the probability of an arrival is proportional to the length of waiting time. This makes it a reasonable model for situations like those indicated above. For example, it makes sense to assume that the longer we wait, the more likely it is that a customer will enter the bank. See the Miscellanea section for a more formal treatment of this.

Another area of application is in spatial distributions, where, for example, the Poisson may be used to model the distribution of bomb hits in an area or the distribution of fish in a lake.

The Poisson distribution has a single parameter $\lambda$, sometimes called the intensity parameter. A random variable $X$, taking values in the nonnegative integers, has a *Poisson($\lambda$) distribution* if

(3.2.5) $$P(X = x|\lambda) = \frac{e^{-\lambda}\lambda^x}{x!}, \quad x = 0, 1, \ldots.$$

To see that $\sum_{x=0}^{\infty} P(X = x|\lambda) = 1$, recall the Taylor series expansion of $e^y$,

$$e^y = \sum_{i=0}^{\infty} \frac{y^i}{i!}.$$

Thus,

$$\sum_{x=0}^{\infty} P(X = x|\lambda) = e^{-\lambda}\sum_{x=0}^{\infty} \frac{\lambda^x}{x!} = e^{-\lambda}e^{\lambda} = 1.$$

The mean of $X$ is easily seen to be

$$EX = \sum_{x=0}^{\infty} x\frac{e^{-\lambda}\lambda^x}{x!}$$

$$= \sum_{x=1}^{\infty} x\frac{e^{-\lambda}\lambda^x}{x!}$$

$$= \lambda e^{-\lambda}\sum_{x=1}^{\infty} \frac{\lambda^{x-1}}{(x-1)!}$$

$$= \lambda e^{-\lambda}\sum_{y=0}^{\infty} \frac{\lambda^y}{y!} \qquad \text{(substitute } y = x - 1)$$

$$= \lambda$$

A similar calculation will show that

$$\text{Var } X = \lambda,$$

and so the parameter $\lambda$ is both the mean and the variance of the Poisson distribution.

The mgf can also be obtained by a straightforward calculation, again following from the Taylor series of $e^y$. We have

$$M_X(t) = e^{\lambda(e^t - 1)}.$$

(See Exercise 2.33 and Example 2.3.13.)

**Example 3.2.4 (Waiting time)**     As an example of a waiting-for-occurrence application, consider a telephone operator who, on the average, handles five calls every 3 minutes. What is the probability that there will be no calls in the next minute? At least two calls?

If we let $X =$ number of calls in a minute, then $X$ has a Poisson distribution with $EX = \lambda = \frac{5}{3}$. So

$$P(\text{no calls in the next minute}) = P(X = 0)$$

$$= \frac{e^{-5/3} \left(\frac{5}{3}\right)^0}{0!}$$

$$= e^{-5/3} = .189;$$

$$P(\text{at least two calls in the next minute}) = P(X \geq 2)$$

$$= 1 - P(X = 0) - P(X = 1)$$

$$= 1 - .189 - \frac{e^{-5/3} \left(\frac{5}{3}\right)^1}{1!}$$

$$= .496. \qquad \qquad \|$$

Calculation of Poisson probabilities can be done rapidly by noting the following recursion relation:

$$(3.2.6) \qquad\qquad P(X = x) = \frac{\lambda}{x} P(X = x - 1), \quad x = 1, 2, \ldots.$$

This relation is easily proved by writing out the pmf of the Poisson. Similar relations hold for other discrete distributions. For example, if $Y \sim$ binomial$(n, p)$, then

$$(3.2.7) \qquad\qquad P(Y = y) = \frac{(n - y + 1)}{y} \frac{p}{1 - p} P(Y = y - 1).$$

The recursion relations (3.2.6) and (3.2.7) can be used to establish the Poisson approximation to the binomial, which we have already seen in Section 2.3, where the approximation was justified using mgfs. Set $\lambda = np$ and, if $p$ is small, we can write

$$\frac{n - y + 1}{y} \frac{p}{1 - p} = \frac{np - p(y - 1)}{y - py} \approx \frac{\lambda}{y}$$

since, for small $p$, the terms $p(y-1)$ and $py$ can be ignored. Therefore, to this level of approximation, (3.2.7) becomes

(3.2.8)
$$P(Y = y) = \frac{\lambda}{y}P(Y = y - 1),$$

which is the Poisson recursion relation. To complete the approximation, we need only establish that $P(X = 0) \approx P(Y = 0)$, since all other probabilities will follow from (3.2.8). Now

$$P(Y = 0) = (1 - p)^n = \left(1 - \frac{np}{n}\right)^n = \left(1 - \frac{\lambda}{n}\right)^n$$

upon setting $np = \lambda$. Recall from Section 2.3 that for fixed $\lambda$, $\lim_{n\to\infty}(1 - (\lambda/n))^n = e^{-\lambda}$, so for large $n$ we have the approximation

$$P(Y = 0) = \left(1 - \frac{\lambda}{n}\right)^n \approx e^{-\lambda} = P(X = 0),$$

completing the Poisson approximation to the binomial.

The approximation is valid when $n$ is large and $p$ is small, which is exactly when it is most useful, freeing us from calculation of binomial coefficients and powers for large $n$.

**Example 3.2.5 (Poisson approximation)**   A typesetter, on the average, makes one error in every 500 words typeset. A typical page contains 300 words. What is the probability that there will be no more than two errors in five pages?

If we assume that setting a word is a Bernoulli trial with success probability $p = \frac{1}{500}$ (notice that we are labeling an error as a "success") and that the trials are independent, then $X =$ number of errors in five pages (1500 words) is binomial$\left(1500, \frac{1}{500}\right)$. Thus

$$P(\text{no more than two errors}) = P(X \le 2)$$

$$= \sum_{x=0}^{2} \binom{1500}{x} \left(\frac{1}{500}\right)^x \left(\frac{499}{500}\right)^{1500-x}$$

$$= .4230,$$

which is a fairly cumbersome calculation. If we use the Poisson approximation with $\lambda = 1500\left(\frac{1}{500}\right) = 3$, we have

$$P(X \le 2) \approx e^{-3}\left(1 + 3 + \frac{3^2}{2}\right) = .4232. \qquad \|$$

*Negative Binomial Distribution*

The binomial distribution counts the number of successes in a fixed number of Bernoulli trials. Suppose that, instead, we count the number of Bernoulli trials required to get a fixed number of successes. This latter formulation leads to the negative binomial distribution.

In a sequence of independent Bernoulli($p$) trials, let the random variable $X$ denote the trial at which the $r$th success occurs, where $r$ is a fixed integer. Then

$$(3.2.9) \qquad P(X = x | r, p) = \binom{x-1}{r-1} p^r (1-p)^{x-r}, \qquad x = r,\ r+1,\ldots,$$

and we say that $X$ has a *negative binomial($r, p$) distribution*.

The derivation of (3.2.9) follows quickly from the binomial distribution. The event $\{X = x\}$ can occur only if there are exactly $r - 1$ successes in the first $x - 1$ trials, and a success on the $x$th trial. The probability of $r - 1$ successes in $x - 1$ trials is the binomial probability $\binom{x-1}{r-1} p^{r-1}(1-p)^{x-r}$, and with probability $p$ there is a success on the $x$th trial. Multiplying these probabilities gives (3.2.9).

The negative binomial distribution is sometimes defined in terms of the random variable $Y =$ number of failures before the $r$th success. This formulation is statistically equivalent to the one given above in terms of $X =$ trial at which the $r$th success occurs, since $Y = X - r$. Using the relationship between $Y$ and $X$, the alternative form of the negative binomial distribution is

$$(3.2.10) \qquad P(Y = y) = \binom{r+y-1}{y} p^r (1-p)^y, \qquad y = 0, 1, \ldots.$$

Unless otherwise noted, when we refer to the negative binomial($r, p$) distribution we will use this pmf.

The negative binomial distribution gets its name from the relationship

$$\binom{r+y-1}{y} = (-1)^y \binom{-r}{y} = (-1)^y \frac{(-r)(-r-1)(-r-2)\cdots\cdots(-r-y+1)}{(y)(y-1)(y-2)\cdots\cdots(2)(1)},$$

which is, in fact, the defining equation for binomial coefficients with negative integers (see Feller 1968 for a complete treatment). Substituting into (3.2.10) yields

$$P(Y = y) = (-1)^y \binom{-r}{y} p^r (1-p)^y,$$

which bears a striking resemblance to the binomial distribution.

The fact that $\sum_{y=0}^{\infty} P(Y = y) = 1$ is not easy to verify but follows from an extension of the Binomial Theorem, an extension that includes negative exponents. We will not pursue this further here. An excellent exposition on binomial coefficients can be found in Feller (1968).

The mean and variance of $Y$ can be calculated using techniques similar to those used for the binomial distribution:

$$EY = \sum_{y=0}^{\infty} y \binom{r+y-1}{y} p^r (1-p)^y$$

$$= \sum_{y=1}^{\infty} \frac{(r+y-1)!}{(y-1)!(r-1)!} p^r (1-p)^y$$

$$= \sum_{y=1}^{\infty} r \binom{r+y-1}{y-1} p^r (1-p)^y.$$

Now write $z = y - 1$, and the sum becomes

$$EY = \sum_{z=0}^{\infty} r \binom{r+z}{z} p^r (1-p)^{z+1}$$

$$= r \frac{(1-p)}{p} \sum_{z=0}^{\infty} \binom{(r+1)+z-1}{z} p^{r+1} (1-p)^z \qquad \binom{\text{summand is negative}}{\text{binomial pmf}}$$

$$= r \frac{(1-p)}{p}.$$

Since the sum is over all values of a negative binomial$(r+1, p)$ distribution, it equals 1. A similar calculation will show

$$\text{Var}\, Y = \frac{r(1-p)}{p^2}.$$

There is an interesting, and sometimes useful, reparameterization of the negative binomial distribution in terms of its mean. If we define the parameter $\mu = r(1-p)/p$, then $EY = \mu$ and a little algebra will show that

$$\text{Var}\, Y = \mu + \frac{1}{r}\mu^2.$$

The variance is a quadratic function of the mean. This relationship can be useful in both data analysis and theoretical considerations (Morris 1982).

The negative binomial family of distributions includes the Poisson distribution as a limiting case. If $r \to \infty$ and $p \to 1$ such that $r(1-p) \to \lambda, 0 < \lambda < \infty$, then

$$EY = \frac{r(1-p)}{p} \to \lambda,$$

$$\text{Var}\, Y = \frac{r(1-p)}{p^2} \to \lambda,$$

which agree with the Poisson mean and variance. To demonstrate that the negative binomial$(r, p) \to$ Poisson$(\lambda)$, we can show that all of the probabilities converge. The fact that the mgfs converge leads us to expect this (see Exercise 3.15).

**Example 3.2.6 (Inverse binomial sampling)**     A technique known as inverse binomial sampling is useful in sampling biological populations. If the proportion of individuals possessing a certain characteristic is $p$ and we sample until we see $r$ such individuals, then the number of individuals sampled is a negative binomial random variable.

For example, suppose that in a population of fruit flies we are interested in the proportion having vestigial wings and decide to sample until we have found 100 such flies. The probability that we will have to examine at least $N$ flies is (using (3.2.9))

$$P(X \geq N) = \sum_{x=N}^{\infty} \binom{x-1}{99} p^{100}(1-p)^{x-100}$$

$$= 1 - \sum_{x=100}^{N-1} \binom{x-1}{99} p^{100}(1-p)^{x-100}.$$

For given $p$ and $N$, we can evaluate this expression to determine how many fruit flies we are likely to look at. (Although the evaluation is cumbersome, the use of a recursion relation will speed things up.)                                                    ‖

Example 3.2.6 shows that the negative binomial distribution can, like the Poisson, be used to model phenomena in which we are waiting for an occurrence. In the negative binomial case we are waiting for a specified number of successes.

### Geometric Distribution

The geometric distribution is the simplest of the waiting time distributions and is a special case of the negative binomial distribution. If we set $r = 1$ in (3.2.9) we have

$$P(X = x|p) = p(1-p)^{x-1}, \quad x = 1, 2, \ldots,$$

which defines the pmf of a *geometric random variable* $X$ with success probability $p$. $X$ can be interpreted as the trial at which the first success occurs, so we are "waiting for a success." The fact that $\sum_{x=1}^{\infty} P(X = x) = 1$ follows from properties of the geometric series. For any number $a$ with $|a| < 1$,

$$\sum_{x=1}^{\infty} a^{x-1} = \frac{1}{1-a},$$

which we have already encountered in Example 1.5.4.

The mean and variance of $X$ can be calculated by using the negative binomial formulas and by writing $X = Y + 1$ to obtain

$$EX = EY + 1 = \frac{1}{p} \quad \text{and} \quad \text{Var } X = \frac{1-p}{p^2}.$$

The geometric distribution has an interesting property, known as the "memoryless" property. For integers $s > t$, it is the case that

(3.2.11)                        $$P(X > s|X > t) = P(X > s - t);$$

that is, the geometric distribution "forgets" what has occurred. The probability of getting an additional $s - t$ failures, having already observed $t$ failures, is the same as the probability of observing $s - t$ failures at the start of the sequence. In other words, the probability of getting a run of failures depends only on the length of the run, not on its position.

To establish (3.2.11), we first note that for any integer $n$,

$$P(X > n) = P(\text{no successes in } n \text{ trials})$$

(3.2.12)
$$= (1 - p)^n,$$

and hence

$$P(X > s | X > t) = \frac{P(X > s \text{ and } X > t)}{P(X > t)}$$

$$= \frac{P(X > s)}{P(X > t)}$$

$$= (1 - p)^{s-t}$$

$$= P(X > s - t).$$

**Example 3.2.7 (Failure times)**   The geometric distribution is sometimes used to model "lifetimes" or "time until failure" of components. For example, if the probability is .001 that a light bulb will fail on any given day, then the probability that it will last at least 30 days is

$$P(X > 30) = \sum_{x=31}^{\infty} .001(1 - .001)^{x-1} = (.999)^{30} = .970. \qquad \|$$

The memoryless property of the geometric distribution describes a very special "lack of aging" property. It indicates that the geometric distribution is not applicable to modeling lifetimes for which the probability of failure is expected to increase with time. There are other distributions used to model various types of aging; see, for example, Barlow and Proschan (1975).

## 3.3  Continuous Distributions

In this section we will discuss some of the more common families of continuous distributions, those with well-known names. The distributions mentioned here by no means constitute all of the distributions used in statistics. Indeed, as was seen in Section 1.6, any nonnegative, integrable function can be transformed into a pdf.

*Uniform Distribution*

The continuous *uniform distribution* is defined by spreading mass uniformly over an interval $[a, b]$. Its pdf is given by

(3.3.1)
$$f(x|a, b) = \begin{cases} \dfrac{1}{b - a} & \text{if } x \in [a, b] \\ 0 & \text{otherwise.} \end{cases}$$

It is easy to check that $\int_a^b f(x)\,dx = 1$. We also have

$$EX = \int_a^b \frac{x}{b-a}\,dx = \frac{b+a}{2};$$

$$\operatorname{Var} X = \int_a^b \frac{\left(x - \frac{b+a}{2}\right)^2}{b-a}\,dx = \frac{(b-a)^2}{12}.$$

*Gamma Distribution*

The gamma family of distributions is a flexible family of distributions on $[0, \infty)$ and can be derived by the construction discussed in Section 1.6. If $\alpha$ is a positive constant, the integral

$$\int_0^\infty t^{\alpha-1}e^{-t}\,dt$$

is finite. If $\alpha$ is a positive integer, the integral can be expressed in closed form; otherwise, it cannot. In either case its value defines the *gamma function*,

(3.3.2) $$\Gamma(\alpha) = \int_0^\infty t^{\alpha-1}e^{-t}\,dt.$$

The gamma function satisfies many useful relationships, in particular,

(3.3.3) $$\Gamma(\alpha+1) = \alpha\Gamma(\alpha), \quad \alpha > 0,$$

which can be verified through integration by parts. Combining (3.3.3) with the easily verified fact that $\Gamma(1) = 1$, we have for any integer $n > 0$,

(3.3.4) $$\Gamma(n) = (n-1)!.$$

(Another useful special case, which will be seen in (3.3.15), is that $\Gamma(\frac{1}{2}) = \sqrt{\pi}$.)

Expressions (3.3.3) and (3.3.4) give recursion relations that ease the problems of calculating values of the gamma function. The recursion relation allows us to calculate any value of the gamma function from knowing only the values of $\Gamma(c)$, $0 < c \le 1$.

Since the integrand in (3.3.2) is positive, it immediately follows that

(3.3.5) $$f(t) = \frac{t^{\alpha-1}e^{-t}}{\Gamma(\alpha)}, \quad 0 < t < \infty,$$

is a pdf. The full gamma family, however, has two parameters and can be derived by changing variables to get the pdf of the random variable $X = \beta T$ in (3.3.5), where $\beta$ is a positive constant. Upon doing this, we get the *gamma$(\alpha, \beta)$ family*,

(3.3.6) $$f(x|\alpha, \beta) = \frac{1}{\Gamma(\alpha)\beta^\alpha}x^{\alpha-1}e^{-x/\beta}, \quad 0 < x < \infty, \quad \alpha > 0, \quad \beta > 0.$$

The parameter $\alpha$ is known as the shape parameter, since it most influences the peakedness of the distribution, while the parameter $\beta$ is called the scale parameter, since most of its influence is on the spread of the distribution.

The mean of the gamma$(\alpha, \beta)$ distribution is

$$(3.3.7) \qquad EX = \frac{1}{\Gamma(\alpha)\beta^\alpha} \int_0^\infty x x^{\alpha-1} e^{-x/\beta} \, dx.$$

To evaluate (3.3.7), notice that the integrand is the kernel of a gamma$(\alpha+1, \beta)$ pdf. From (3.3.6) we know that, for any $\alpha$, $\beta > 0$,

$$(3.3.8) \qquad \int_0^\infty x^{\alpha-1} e^{-x/\beta} \, dx = \Gamma(\alpha)\beta^\alpha,$$

so we have

$$EX = \frac{1}{\Gamma(\alpha)\beta^\alpha} \int_0^\infty x^\alpha e^{-x/\beta} \, dx$$

$$= \frac{1}{\Gamma(\alpha)\beta^\alpha} \Gamma(\alpha+1)\beta^{\alpha+1}$$

$$= \frac{\alpha\Gamma(\alpha)\beta}{\Gamma(\alpha)} \qquad \text{(from (3.3.3))}$$

$$= \alpha\beta.$$

Note that to evaluate $EX$ we have again used the technique of recognizing the integral as the kernel of another pdf. (We have already used this technique to calculate the gamma mgf in Example 2.3.8 and, in a discrete case, to do binomial calculations in Examples 2.2.3 and 2.3.5.)

The variance of the gamma$(\alpha, \beta)$ distribution is calculated in a manner analogous to that used for the mean. In particular, in calculating $EX^2$ we deal with the kernel of a gamma$(\alpha+2, \beta)$ distribution. The result is

$$\text{Var}\, X = \alpha\beta^2.$$

In Example 2.3.8 we calculated the mgf of a gamma$(\alpha, \beta)$ distribution. It is given by

$$M_X(t) = \left(\frac{1}{1-\beta t}\right)^\alpha, \quad t < \frac{1}{\beta}.$$

**Example 3.3.1 (Gamma-Poisson relationship)**     There is an interesting relationship between the gamma and Poisson distributions. If $X$ is a gamma$(\alpha, \beta)$ random variable, where $\alpha$ is an integer, then for any $x$,

$$(3.3.9) \qquad P(X \le x) = P(Y \ge \alpha),$$

where $Y \sim \text{Poisson}(x/\beta)$. Equation (3.3.9) can be established by successive integrations by parts, as follows. Since $\alpha$ is an integer, we write $\Gamma(\alpha) = (\alpha-1)!$ to get

$$P(X \le x) = \frac{1}{(\alpha-1)!\beta^\alpha} \int_0^x t^{\alpha-1} e^{-t/\beta} \, dt$$

$$= \frac{1}{(\alpha-1)!\beta^\alpha} \left[ -t^{\alpha-1}\beta e^{-t/\beta}\Big|_0^x + \int_0^x (\alpha-1)t^{\alpha-2}\beta e^{-t/\beta} \, dt \right],$$

where we use the integration by parts substitution $u = t^{\alpha-1}$, $dv = e^{-t/\beta} dt$. Continuing our evaluation, we have

$$P(X \le x) = \frac{-1}{(\alpha-1)!\beta^{\alpha-1}} x^{\alpha-1} e^{-x/\beta} + \frac{1}{(\alpha-2)!\beta^{\alpha-1}} \int_0^x t^{\alpha-2} e^{-t/\beta} dt$$

$$= \frac{1}{(\alpha-2)!\beta^{\alpha-1}} \int_0^x t^{\alpha-2} e^{-t/\beta} dt - P(Y = \alpha - 1),$$

where $Y \sim \text{Poisson}(x/\beta)$. Continuing in this manner, we can establish (3.3.9). (See Exercise 3.19.) ‖

There are a number of important special cases of the gamma distribution. If we set $\alpha = p/2$, where $p$ is an integer, and $\beta = 2$, then the gamma pdf becomes

$$(3.3.10) \qquad f(x|p) = \frac{1}{\Gamma(p/2)2^{p/2}} x^{(p/2)-1} e^{-x/2}, \quad 0 < x < \infty,$$

which is the *chi squared pdf with $p$ degrees of freedom*. The mean, variance, and mgf of the chi squared distribution can all be calculated by using the previously derived gamma formulas.

The chi squared distribution plays an important role in statistical inference, especially when sampling from a normal distribution. This topic will be dealt with in detail in Chapter 5.

Another important special case of the gamma distribution is obtained when we set $\alpha = 1$. We then have

$$(3.3.11) \qquad f(x|\beta) = \frac{1}{\beta} e^{-x/\beta}, \quad 0 < x < \infty,$$

the *exponential pdf* with scale parameter $\beta$. Its mean and variance were calculated in Examples 2.2.2 and 2.3.3.

The exponential distribution can be used to model lifetimes, analogous to the use of the geometric distribution in the discrete case. In fact, the exponential distribution shares the "memoryless" property of the geometric. If $X \sim \text{exponential}(\beta)$, that is, with pdf given by (3.3.11), then for $s > t \ge 0$,

$$P(X > s|X > t) = P(X > s - t),$$

since

$$P(X > s|X > t) = \frac{P(X > s, X > t)}{P(X > t)}$$

$$= \frac{P(X > s)}{P(X > t)} \qquad \text{(since } s > t\text{)}$$

$$= \frac{\int_s^\infty \frac{1}{\beta} e^{-x/\beta} dx}{\int_t^\infty \frac{1}{\beta} e^{-x/\beta} dx}$$

$$= \frac{e^{-s/\beta}}{e^{-t/\beta}}$$

$$= e^{-(s-t)/\beta}$$

$$= P(X > s - t).$$

Another distribution related to both the exponential and the gamma families is the *Weibull distribution*. If $X \sim$ exponential$(\beta)$, then $Y = X^{1/\gamma}$ has a Weibull$(\gamma, \beta)$ distribution,

$$(3.3.12) \qquad f_Y(y|\gamma, \beta) = \frac{\gamma}{\beta} y^{\gamma-1} e^{-y^\gamma/\beta}, \quad 0 < y < \infty, \quad \gamma > 0, \quad \beta > 0.$$

Clearly, we could have started with the Weibull and then derived the exponential as a special case ($\gamma = 1$). This is a matter of taste. The Weibull distribution plays an extremely important role in the analysis of failure time data (see Kalbfleisch and Prentice 1980 for a comprehensive treatment of this topic). The Weibull, in particular, is very useful for modeling *hazard functions* (see Exercises 3.25 and 3.26).

*Normal Distribution*

The normal distribution (sometimes called the *Gaussian distribution*) plays a central role in a large body of statistics. There are three main reasons for this. First, the normal distribution and distributions associated with it are very tractable analytically (although this may not seem so at first glance). Second, the normal distribution has the familiar bell shape, whose symmetry makes it an appealing choice for many population models. Although there are many other distributions that are also bell-shaped, most do not possess the analytic tractability of the normal. Third, there is the Central Limit Theorem (see Chapter 5 for details), which shows that, under mild conditions, the normal distribution can be used to approximate a large variety of distributions in large samples.

The normal distribution has two parameters, usually denoted by $\mu$ and $\sigma^2$, which are its mean and variance. The pdf of the *normal distribution* with mean $\mu$ and variance $\sigma^2$ (usually denoted by n$(\mu, \sigma^2)$) is given by

$$(3.3.13) \qquad f(x|\mu, \sigma^2) = \frac{1}{\sqrt{2\pi}\sigma} e^{-(x-\mu)^2/(2\sigma^2)}, \quad -\infty < x < \infty.$$

If $X \sim$ n$(\mu, \sigma^2)$, then the random variable $Z = (X - \mu)/\sigma$ has a n$(0, 1)$ distribution, also known as the *standard normal*. This is easily established by writing

$$P(Z \le z) = P\left(\frac{X - \mu}{\sigma} \le z\right)$$

$$= P(X \le z\sigma + \mu)$$

$$= \frac{1}{\sqrt{2\pi}\sigma} \int_{-\infty}^{z\sigma+\mu} e^{-(x-\mu)^2/(2\sigma^2)} \, dx$$

$$= \frac{1}{\sqrt{2\pi}} \int_{-\infty}^{z} e^{-t^2/2} \, dt, \qquad \left(\text{substitute } t = \frac{x - \mu}{\sigma}\right)$$

showing that $P(Z \le z)$ is the standard normal cdf.

It therefore follows that all normal probabilities can be calculated in terms of the standard normal. Furthermore, calculations of expected values can be simplified by carrying out the details in the $n(0,1)$ case, then transforming the result to the $n(\mu, \sigma^2)$ case. For example, if $Z \sim n(0,1)$,

$$\text{E}Z = \frac{1}{\sqrt{2\pi}} \int_{-\infty}^{\infty} z e^{-z^2/2} dz = -\frac{1}{\sqrt{2\pi}} e^{-z^2/2} \Big|_{-\infty}^{\infty} = 0,$$

and so, if $X \sim n(\mu, \sigma^2)$, it follows from Theorem 2.2.5 that

$$\text{E}X = \text{E}(\mu + \sigma Z) = \mu + \sigma \text{E}Z = \mu.$$

Similarly, we have that Var $Z = 1$ and, from Theorem 2.3.4, Var $X = \sigma^2$.

We have not yet established that (3.3.13) integrates to 1 over the whole real line. By applying the standardizing transformation, we need only to show that

$$\frac{1}{\sqrt{2\pi}} \int_{-\infty}^{\infty} e^{-z^2/2} dz = 1.$$

Notice that the integrand above is symmetric around 0, implying that the integral over $(-\infty, 0)$ is equal to the integral over $(0, \infty)$. Thus, we reduce the problem to showing

(3.3.14) $$\int_{0}^{\infty} e^{-z^2/2} dz = \frac{\sqrt{2\pi}}{2} = \sqrt{\frac{\pi}{2}}.$$

The function $e^{-z^2/2}$ does not have an antiderivative that can be written explicitly in terms of elementary functions (that is, in closed form), so we cannot perform the integration directly. In fact, this is an example of an integration that either you know how to do or else you can spend a very long time going nowhere. Since both sides of (3.3.14) are positive, the equality will hold if we establish that the squares are equal. Square the integral in (3.3.14) to obtain

$$\left( \int_{0}^{\infty} e^{-z^2/2} dz \right)^2 = \left( \int_{0}^{\infty} e^{-t^2/2} dt \right) \left( \int_{0}^{\infty} e^{-u^2/2} du \right)$$

$$= \int_{0}^{\infty} \int_{0}^{\infty} e^{-(t^2+u^2)/2} \, dt \, du.$$

The integration variables are just dummy variables, so changing their names is allowed. Now, we convert to polar coordinates. Define

$$t = r \cos\theta \quad \text{and} \quad u = r \sin\theta.$$

Then $t^2 + u^2 = r^2$ and $dt \, du = r \, d\theta \, dr$ and the limits of integration become $0 < r < \infty$, $0 < \theta < \pi/2$ (the upper limit on $\theta$ is $\pi/2$ because $t$ and $u$ are restricted to be positive). We now have

$$\int_0^\infty \int_0^\infty e^{-(t^2+u^2)/2} \, dt \, du = \int_0^\infty \int_0^{\pi/2} re^{-r^2/2} \, d\theta \, dr$$

$$= \frac{\pi}{2} \int_0^\infty re^{-r^2/2} dr$$

$$= \frac{\pi}{2} \left[ -e^{-r^2/2} \Big|_0^\infty \right]$$

$$= \frac{\pi}{2},$$

which establishes (3.3.14).

This integral is closely related to the gamma function; in fact, by making the substitution $w = \frac{1}{2}z^2$ in (3.3.14), we see that this integral is essentially $\Gamma(\frac{1}{2})$. If we are careful to get the constants correct, we will see that (3.3.14) implies

(3.3.15)
$$\Gamma\left(\frac{1}{2}\right) = \int_0^\infty w^{-1/2} e^{-w} dw = \sqrt{\pi}.$$

The normal distribution is somewhat special in the sense that its two parameters, $\mu$ (the mean) and $\sigma^2$ (the variance), provide us with complete information about the exact shape and location of the distribution. This property, that the distribution is determined by $\mu$ and $\sigma^2$, is not unique to the normal pdf, but is shared by a family of pdfs called location–scale families, to be discussed in Section 3.5.

Straightforward calculus shows that the normal pdf (3.3.13) has its maximum at $x = \mu$ and inflection points (where the curve changes from concave to convex) at $\mu \pm \sigma$. Furthermore, the probability content within 1, 2, or 3 standard deviations of the mean is

$$P(|X - \mu| \leq \sigma) = P(|Z| \leq 1) = .6826,$$

$$P(|X - \mu| \leq 2\sigma) = P(|Z| \leq 2) = .9544,$$

$$P(|X - \mu| \leq 3\sigma) = P(|Z| \leq 3) = .9974,$$

where $X \sim n(\mu, \sigma^2)$, $Z \sim n(0, 1)$, and the numerical values can be obtained from many computer packages or from tables. Often, the two-digit values reported are .68, .95, and .99, respectively. Although these do not represent the rounded values, they are the values commonly used. Figure 3.3.1 shows the normal pdf along with these key features.

Among the many uses of the normal distribution, an important one is its use as an approximation to other distributions (which is partially justified by the Central Limit Theorem). For example, if $X \sim$ binomial$(n, p)$, then $EX = np$ and $\text{Var}\, X = np(1-p)$, and under suitable conditions, the distribution of $X$ can be approximated by that of a normal random variable with mean $\mu = np$ and variance $\sigma^2 = np(1-p)$. The "suitable conditions" are that $n$ should be large and $p$ should not be extreme (near 0 or 1). We want $n$ large so that there are enough (discrete) values of $X$ to make an approximation by a continuous distribution reasonable, and $p$ should be "in the middle" so the binomial is nearly symmetric, as is the normal. As with most approximations there

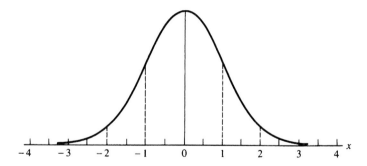

Figure 3.3.1. *Standard normal density*

are no absolute rules, and each application should be checked to decide whether the approximation is good enough for its intended use. A conservative rule to follow is that the approximation will be good if $\min(np, n(1-p)) \geq 5$.

**Example 3.3.2 (Normal approximation)**    Let $X \sim$ binomial$(25, .6)$. We can approximate $X$ with a normal random variable, $Y$, with mean $\mu = 25(.6) = 15$ and standard deviation $\sigma = ((25)(.6)(.4))^{1/2} = 2.45$. Thus

$$P(X \leq 13) \approx P(Y \leq 13) = P\left(Z \leq \frac{13-15}{2.45}\right) = P(Z \leq -.82) = .206,$$

while the exact binomial calculation gives

$$P(X \leq 13) = \sum_{x=0}^{13} \binom{25}{x}(.6)^x(.4)^{25-x} = .267,$$

showing that the normal approximation is good, but not terrific. The approximation can be greatly improved, however, by a "continuity correction." To see how this works, look at Figure 3.3.2, which shows the binomial$(25, .6)$ pmf and the $n(15, (2.45)^2)$ pdf. We have drawn the binomial pmf using bars of width 1, with height equal to the probability. Thus, the areas of the bars give the binomial probabilities. In the approximation, notice how the area of the approximating normal is smaller than the binomial area (the normal area is everything to the left of the line at 13, whereas the binomial area includes the entire bar at 13 up to 13.5). The continuity correction adds this area back by adding $\frac{1}{2}$ to the cutoff point. So instead of approximating $P(X \leq 13)$, we approximate the equivalent expression (because of the discreteness), $P(X \leq 13.5)$ and obtain

$$P(X \leq 13) = P(X \leq 13.5) \approx P(Y \leq 13.5) = P(Z \leq -.61) = .271,$$

a much better approximation. In general, the normal approximation with the continuity correction is far superior to the approximation without the continuity correction.

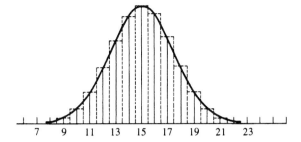

Figure 3.3.2. *Normal*$(15, (2.45)^2)$ *approximation to the binomial*$(25, .6)$

We also make the correction on the lower end. If $X \sim$ binomial$(n, p)$ and $Y \sim$ n$(np, np(1 - p))$, then we approximate

$$P(X \leq x) \approx P(Y \leq x + 1/2),$$
$$P(X \geq x) \approx P(Y \geq x - 1/2). \qquad \|$$

*Beta Distribution*

The beta family of distributions is a continuous family on $(0, 1)$ indexed by two parameters. The *beta*$(\alpha, \beta)$ *pdf* is

(3.3.16) $\quad f(x|\alpha, \beta) = \dfrac{1}{B(\alpha, \beta)} x^{\alpha - 1}(1 - x)^{\beta - 1}, \quad 0 < x < 1, \quad \alpha > 0, \quad \beta > 0,$

where $B(\alpha, \beta)$ denotes the beta function,

$$B(\alpha, \beta) = \int_0^1 x^{\alpha - 1}(1 - x)^{\beta - 1} \, dx.$$

The beta function is related to the gamma function through the following identity:

(3.3.17) $\qquad\qquad\qquad\qquad B(\alpha, \beta) = \dfrac{\Gamma(\alpha)\Gamma(\beta)}{\Gamma(\alpha + \beta)}.$

Equation (3.3.17) is very useful in dealing with the beta function, allowing us to take advantage of the properties of the gamma function. In fact, we will never deal directly with the beta function, but rather will use (3.3.17) for all of our evaluations.

The beta distribution is one of the few common "named" distributions that give probability 1 to a finite interval, here taken to be $(0, 1)$. As such, the beta is often used to model proportions, which naturally lie between 0 and 1. We will see illustrations of this in Chapter 4.

Calculation of moments of the beta distribution is quite easy, due to the particular form of the pdf. For $n > -\alpha$ we have

$$\mathrm{E}X^n = \frac{1}{B(\alpha, \beta)} \int_0^1 x^n x^{\alpha - 1}(1 - x)^{\beta - 1} \, dx$$

$$= \frac{1}{B(\alpha, \beta)} \int_0^1 x^{(\alpha + n) - 1}(1 - x)^{\beta - 1} \, dx.$$

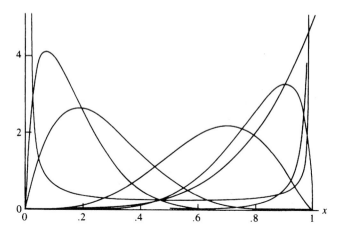

Figure 3.3.3. *Beta densities*

We now recognize the integrand as the kernel of a beta$(\alpha + n, \beta)$ pdf; hence,

$$(3.3.18) \qquad \mathrm{E}X^n = \frac{B(\alpha + n, \beta)}{B(\alpha, \beta)} = \frac{\Gamma(\alpha + n)\Gamma(\alpha + \beta)}{\Gamma(\alpha + \beta + n)\Gamma(\alpha)}.$$

Using (3.3.3) and (3.3.18) with $n = 1$ and $n = 2$, we calculate the mean and variance of the beta$(\alpha, \beta)$ distribution as

$$\mathrm{E}X = \frac{\alpha}{\alpha + \beta} \quad \text{and} \quad \mathrm{Var}\, X = \frac{\alpha\beta}{(\alpha + \beta)^2(\alpha + \beta + 1)}.$$

As the parameters $\alpha$ and $\beta$ vary, the beta distribution takes on many shapes, as shown in Figure 3.3.3. The pdf can be strictly increasing ($\alpha > 1$, $\beta = 1$), strictly decreasing ($\alpha = 1$, $\beta > 1$), U-shaped ($\alpha < 1$, $\beta < 1$), or unimodal ($\alpha > 1$, $\beta > 1$). The case $\alpha = \beta$ yields a pdf symmetric about $\frac{1}{2}$ with mean $\frac{1}{2}$ (necessarily) and variance $(4(2\alpha + 1))^{-1}$. The pdf becomes more concentrated as $\alpha$ increases, but stays symmetric, as shown in Figure 3.3.4. Finally, if $\alpha = \beta = 1$, the beta distribution reduces to the uniform$(0, 1)$, showing that the uniform can be considered to be a member of the beta family. The beta distribution is also related, through a transformation, to the $F$ distribution, a distribution that plays an extremely important role in statistical analysis (see Section 5.3).

*Cauchy Distribution*

The *Cauchy distribution* is a symmetric, bell-shaped distribution on $(-\infty, \infty)$ with pdf

$$(3.3.19) \qquad f(x|\theta) = \frac{1}{\pi} \frac{1}{1 + (x - \theta)^2}, \quad -\infty < x < \infty, \quad -\infty < \theta < \infty.$$

(See Exercise 3.39 for a more general version of the Cauchy pdf.) To the eye, the Cauchy does not appear very different from the normal distribution. However, there

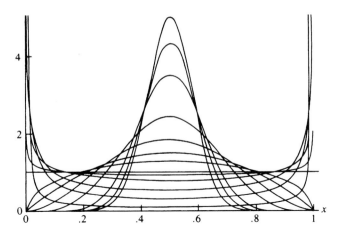

Figure 3.3.4. *Symmetric beta densities*

is a very great difference, indeed. As we have already seen in Chapter 2, the mean of the Cauchy distribution does not exist; that is,

$$(3.3.20) \qquad \mathrm{E}|X| = \int_{-\infty}^{\infty} \frac{1}{\pi} \frac{|x|}{1 + (x - \theta)^2} \, dx = \infty.$$

It is easy to see that (3.3.19) defines a proper pdf for all $\theta$. Recall that $\frac{d}{dt} \arctan(t) = (1 + t^2)^{-1}$; hence,

$$\int_{-\infty}^{\infty} \frac{1}{\pi} \frac{1}{1 + (x - \theta)^2} \, dx = \frac{1}{\pi} \arctan(x - \theta) \Big|_{-\infty}^{\infty} = 1,$$

since $\arctan(\pm\infty) = \pm\pi/2$.

Since $\mathrm{E}|X| = \infty$, it follows that no moments of the Cauchy distribution exist or, in other words, all absolute moments equal $\infty$. In particular, the mgf does not exist.

The parameter $\theta$ in (3.3.19) does measure the center of the distribution; it is the median. If $X$ has a Cauchy distribution with parameter $\theta$, then from Exercise 3.37 it follows that $P(X \geq \theta) = \frac{1}{2}$, showing that $\theta$ is the median of the distribution. Figure 3.3.5 shows a Cauchy(0) distribution together with a n(0, 1), where we see the similarity in shape but the much thicker tails of the Cauchy.

The Cauchy distribution plays a special role in the theory of statistics. It represents an extreme case against which conjectures can be tested. But do not make the mistake of considering the Cauchy distribution to be only a pathological case, for it has a way of turning up when you least expect it. For example, it is common practice for experimenters to calculate ratios of observations, that is, ratios of random variables. (In measures of growth, it is common to combine weight and height into one measurement weight-for-height, that is, weight/height.) A surprising fact is that the ratio of two standard normals has a Cauchy distribution (see Example 4.3.6). Taking ratios can lead to ill-behaved distributions.

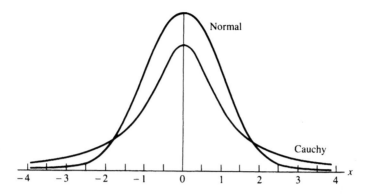

Figure 3.3.5. *Standard normal density and Cauchy density*

## Lognormal Distribution

If $X$ is a random variable whose logarithm is normally distributed (that is, $\log X \sim$ n$(\mu, \sigma^2)$), then $X$ has a lognormal distribution. The pdf of $X$ can be obtained by straightforward transformation of the normal pdf using Theorem 2.1.5, yielding

(3.3.21)

$$f(x|\mu, \sigma^2) = \frac{1}{\sqrt{2\pi}\sigma} \frac{1}{x} e^{-(\log x - \mu)^2/(2\sigma^2)}, \quad 0 < x < \infty, \quad -\infty < \mu < \infty, \quad \sigma > 0,$$

for the *lognormal pdf*. The moments of $X$ can be calculated directly using (3.3.21), or by exploiting the relationship to the normal and writing

$$\begin{aligned} EX &= Ee^{\log X} \\ &= Ee^Y \qquad\qquad (Y = \log X \sim \text{n}(\mu, \sigma^2)) \\ &= e^{\mu + (\sigma^2/2)}. \end{aligned}$$

The last equality is obtained by recognizing the mgf of the normal distribution (set $t = 1$, see Exercise 2.33). We can use a similar technique to calculate $EX^2$ and get

$$\operatorname{Var} X = e^{2(\mu + \sigma^2)} - e^{2\mu + \sigma^2}.$$

The lognormal distribution is similar in appearance to the gamma distribution, as Figure 3.3.6 shows. The distribution is very popular in modeling applications when the variable of interest is skewed to the right. For example, incomes are necessarily skewed to the right, and modeling with a lognormal allows the use of normal-theory statistics on log(income), a very convenient circumstance.

## Double Exponential Distribution

The *double exponential distribution* is formed by reflecting the exponential distribution around its mean. The pdf is given by

(3.3.22)      $$f(x|\mu, \sigma) = \frac{1}{2\sigma} e^{-|x-\mu|/\sigma}, \quad -\infty < x < \infty, \quad -\infty < \mu < \infty, \quad \sigma > 0.$$

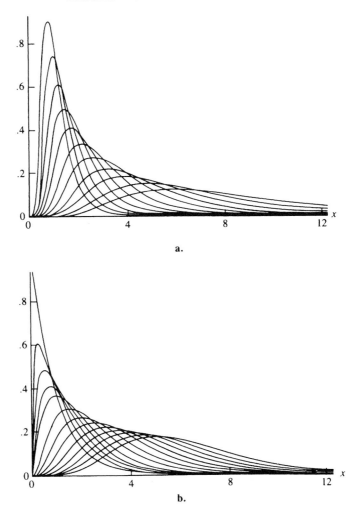

Figure 3.3.6. **(a)** *Some lognormal densities;* **(b)** *some gamma densities*

The double exponential provides a symmetric distribution with "fat" tails (much fatter than the normal) but still retains all of its moments. It is straightforward to calculate

$$EX = \mu \quad \text{and} \quad \text{Var}\, X = 2\sigma^2.$$

The double exponential distribution is not bell-shaped. In fact, it has a peak (or more formally, a point of nondifferentiability) at $x = \mu$. When we deal with this distribution analytically, it is important to remember this point. The absolute value signs can also be troublesome when performing integrations, and it is best to divide the integral into regions around $x = \mu$:

$$\text{EX} = \int_{-\infty}^{\infty} \frac{x}{2\sigma} e^{-|x-\mu|/\sigma} \, dx$$

(3.3.23)
$$= \int_{-\infty}^{\mu} \frac{x}{2\sigma} e^{(x-\mu)/\sigma} \, dx + \int_{\mu}^{\infty} \frac{x}{2\sigma} e^{-(x-\mu)/\sigma} \, dx.$$

Notice that we can remove the absolute value signs over the two regions of integration. (This strategy is useful, in general, in dealing with integrals containing absolute values; divide up the region of integration so the absolute value signs can be removed.) Evaluation of (3.3.23) can be completed by performing integration by parts on each integral.

There are many other continuous distributions that have uses in different statistical applications, many of which will appear throughout the rest of the book. The comprehensive work by Johnson and co-authors, mentioned at the beginning of this chapter, is a valuable reference for most useful statistical distributions.

## 3.4 Exponential Families

A family of pdfs or pmfs is called an *exponential family* if it can be expressed as

(3.4.1)
$$f(x|\boldsymbol{\theta}) = h(x)c(\boldsymbol{\theta}) \exp\left( \sum_{i=1}^{k} w_i(\boldsymbol{\theta}) t_i(x) \right).$$

Here $h(x) \geq 0$ and $t_1(x), \ldots, t_k(x)$ are real-valued functions of the observation $x$ (they cannot depend on $\boldsymbol{\theta}$), and $c(\boldsymbol{\theta}) \geq 0$ and $w_1(\boldsymbol{\theta}), \ldots, w_k(\boldsymbol{\theta})$ are real-valued functions of the possibly vector-valued parameter $\boldsymbol{\theta}$ (they cannot depend on $x$). Many common families introduced in the previous section are exponential families. These include the continuous families—normal, gamma, and beta, and the discrete families—binomial, Poisson, and negative binomial.

To verify that a family of pdfs or pmfs is an exponential family, we must identify the functions $h(x)$, $c(\boldsymbol{\theta})$, $w_i(\boldsymbol{\theta})$, and $t_i(x)$ and show that the family has the form (3.4.1). The next example illustrates this.

**Example 3.4.1 (Binomial exponential family)**　　Let $n$ be a positive integer and consider the binomial$(n, p)$ family with $0 < p < 1$. Then the pmf for this family, for $x = 0, \ldots, n$ and $0 < p < 1$, is

$$f(x|p) = \binom{n}{x} p^x (1 - p)^{n-x}$$

(3.4.2)
$$= \binom{n}{x} (1 - p)^n \left( \frac{p}{1 - p} \right)^x$$

$$= \binom{n}{x} (1 - p)^n \exp\left( \log\left( \frac{p}{1 - p} \right) x \right).$$

Define

$$h(x) = \begin{cases} \binom{n}{x} & x = 0, \ldots, n \\ 0 & \text{otherwise,} \end{cases} \qquad c(p) = (1-p)^n, \quad 0 < p < 1,$$

$$w_1(p) = \log\left(\frac{p}{1-p}\right), \quad 0 < p < 1, \quad \text{and} \quad t_1(x) = x.$$

Then we have

(3.4.3)                    $f(x|p) = h(x)c(p)\exp[w_1(p)t_1(x)],$

which is of the form (3.4.1) with $k = 1$. In particular, note that $h(x) > 0$ only if $x = 0, \ldots, n$ and $c(p)$ is defined only if $0 < p < 1$. This is important, as (3.4.3) must match (3.4.2) for *all* values of $x$ and is an exponential family only if $0 < p < 1$ (so the functions of the parameter are only defined here). Also, the parameter values $p = 0$ and 1 are sometimes included in the binomial model, but we have not included them here because the set of $x$ values for which $f(x|p) > 0$ is different for $p = 0$ and 1 than for other $p$ values.                                                                      ‖

The specific form of (3.4.1) results in exponential families having many nice mathematical properties. But more important for a statistical model, the form of (3.4.1) results in many nice statistical properties. We next illustrate a calculational shortcut for moments of an exponential family.

**Theorem 3.4.2**     *If $X$ is a random variable with pdf or pmf of the form (3.4.1), then*

(3.4.4)     $\mathrm{E}\left(\sum_{i=1}^{k} \frac{\partial w_i(\boldsymbol{\theta})}{\partial \theta_j} t_i(X)\right) = -\frac{\partial}{\partial \theta_j} \log c(\boldsymbol{\theta});$

(3.4.5)   $\mathrm{Var}\left(\sum_{i=1}^{k} \frac{\partial w_i(\boldsymbol{\theta})}{\partial \theta_j} t_i(X)\right) = -\frac{\partial^2}{\partial \theta_j^2} \log c(\boldsymbol{\theta}) - \mathrm{E}\left(\sum_{i=1}^{k} \frac{\partial^2 w_i(\boldsymbol{\theta})}{\partial \theta_j^2} t_i(X)\right).$

Although these equations may look formidable, when applied to specific cases they can work out quite nicely. Their advantage is that we can replace integration or summation by differentiation, which is often more straightforward.

**Example 3.4.3 (Binomial mean and variance)**   From Example 3.4.1 we have

$$\frac{d}{dp} w_1(p) = \frac{d}{dp} \log \frac{p}{1-p} = \frac{1}{p(1-p)}$$

$$\frac{d}{dp} \log c(p) = \frac{d}{dp} n \log(1-p) = \frac{-n}{1-p}$$

and thus from Theorem 3.4.2 we have

$$\mathrm{E}\left(\frac{1}{p(1-p)} X\right) = \frac{n}{1-p}$$

and a bit of rearrangement yields $E(X) = np$. The variance identity works in a similar manner. ||

The proof of Theorem 3.4.2 is a calculus excursion and is relegated to Exercise 3.31. See also Exercise 3.32 for a special case.

We now look at another example and some other features of exponential families.

**Example 3.4.4 (Normal exponential family)**     Let $f(x|\mu, \sigma^2)$ be the $n(\mu, \sigma^2)$ family of pdfs, where $\boldsymbol{\theta} = (\mu, \sigma), -\infty < \mu < \infty, \sigma > 0$. Then

$$f(x|\mu, \sigma^2) = \frac{1}{\sqrt{2\pi}\sigma} \exp\left(-\frac{(x-\mu)^2}{2\sigma^2}\right)$$

(3.4.6)

$$= \frac{1}{\sqrt{2\pi}\sigma} \exp\left(-\frac{\mu^2}{2\sigma^2}\right) \exp\left(-\frac{x^2}{2\sigma^2} + \frac{\mu x}{\sigma^2}\right).$$

Define

$$h(x) = 1 \text{ for all } x;$$

$$c(\boldsymbol{\theta}) = c(\mu, \sigma) = \frac{1}{\sqrt{2\pi}\sigma} \exp\left(\frac{-\mu^2}{2\sigma^2}\right), \quad -\infty < \mu < \infty, \sigma > 0;$$

$$w_1(\mu, \sigma) = \frac{1}{\sigma^2}, \quad \sigma > 0; \qquad w_2(\mu, \sigma) = \frac{\mu}{\sigma^2}, \sigma > 0;$$

$$t_1(x) = -x^2/2; \quad \text{and} \quad t_2(x) = x.$$

Then

$$f(x|\mu, \sigma^2) = h(x)c(\mu, \sigma)\exp[w_1(\mu, \sigma)t_1(x) + w_2(\mu, \sigma)t_2(x)],$$

which is the form (3.4.1) with $k = 2$. Note again that the parameter functions are defined only over the range of the parameter. ||

In general, the set of $x$ values for which $f(x|\boldsymbol{\theta}) > 0$ cannot depend on $\boldsymbol{\theta}$ in an exponential family. The entire definition of the pdf or pmf must be incorporated into the form (3.4.1). This is most easily accomplished by incorporating the range of $x$ into the expression for $f(x|\boldsymbol{\theta})$ through the use of an indicator function.

**Definition 3.4.5**     The *indicator function* of a set $A$, most often denoted by $I_A(x)$, is the function

$$I_A(x) = \begin{cases} 1 & x \in A \\ 0 & x \notin A. \end{cases}$$

An alternative notation is $I(x \in A)$.

Thus, the normal pdf of Example 3.4.4 would be written

$$f(x|\mu, \sigma^2) = h(x)c(\mu, \sigma)\exp[w_1(\mu, \sigma)t_1(x) + w_2(\mu, \sigma)t_2(x)]I_{(-\infty,\infty)}(x).$$

Since the indicator function is a function of only $x$, it can be incorporated into the function $h(x)$, showing that this pdf is of the form (3.4.1).

From (3.4.1), since the factor $\exp(\cdot)$ is always positive, it can be seen that for any $\boldsymbol{\theta} \in \Theta$, that is, for any $\boldsymbol{\theta}$ for which $c(\boldsymbol{\theta}) > 0, \{x : f(x|\boldsymbol{\theta}) > 0\} = \{x : h(x) > 0\}$ and this set does not depend on $\boldsymbol{\theta}$. So, for example, the set of pdfs given by $f(x|\theta) = \theta^{-1}\exp(1 - (x/\theta)), 0 < \theta < x < \infty$, is not an exponential family even though we can write $\theta^{-1}\exp(1 - (x/\theta)) = h(x)c(\theta)\exp(w(\theta)t(x))$, where $h(x) = e^1$, $c(\theta) = \theta^{-1}$, $w(\theta) = \theta^{-1}$, and $t(x) = -x$. Writing the pdf with indicator functions makes this very clear. We have

$$f(x|\theta) = \theta^{-1}\exp\left(1 - \left(\frac{x}{\theta}\right)\right) I_{[\theta,\infty)}(x).$$

The indicator function cannot be incorporated into any of the functions of (3.4.1) since it is not a function of $x$ alone, not a function of $\theta$ alone, and cannot be expressed as an exponential. Thus, this is not an exponential family.

An exponential family is sometimes reparameterized as

(3.4.7)
$$f(x|\eta) = h(x)c^*(\eta)\exp\left(\sum_{i=1}^{k} \eta_i t_i(x)\right).$$

Here the $h(x)$ and $t_i(x)$ functions are the same as in the original parameterization (3.4.1). The set $\mathcal{H} = \left\{\eta = (\eta_1, \ldots, \eta_k) : \int_{-\infty}^{\infty} h(x)\exp\left(\sum_{i=1}^{k} \eta_i t_i(x)\right) dx < \infty\right\}$ is called the *natural parameter space* for the family. (The integral is replaced by a sum over the values of $x$ for which $h(x) > 0$ if $X$ is discrete.) For the values of $\eta \in \mathcal{H}$, we must have $c^*(\eta) = \left[\int_{-\infty}^{\infty} h(x)\exp\left(\sum_{i=1}^{k} \eta_i t_i(x)\right) dx\right]^{-1}$ to ensure that the pdf integrates to 1. Since the original $f(x|\boldsymbol{\theta})$ in (3.4.1) is a pdf or pmf, the set $\{\eta = (w_1(\boldsymbol{\theta}), \ldots, w_k(\boldsymbol{\theta})) : \boldsymbol{\theta} \in \Theta\}$ must be a subset of the natural parameter space. But there may be other values of $\eta \in \mathcal{H}$ also. The natural parameterization and the natural parameter space have many useful mathematical properties. For example, $\mathcal{H}$ is convex.

**Example 3.4.6 (Continuation of Example 3.4.4)**   To determine the natural parameter space for the normal family of distributions, replace $w_i(\mu, \sigma)$ with $\eta_i$ in (3.4.6) to obtain

(3.4.8)
$$f(x|\eta_1, \eta_2) = \frac{\sqrt{\eta_1}}{\sqrt{2\pi}}\exp\left(-\frac{\eta_2^2}{2\eta_1}\right)\exp\left(-\frac{\eta_1 x^2}{2} + \eta_2 x\right).$$

The integral will be finite if and only if the coefficient on $x^2$ is negative. This means $\eta_1$ must be positive. If $\eta_1 > 0$, the integral will be finite regardless of the value of $\eta_2$. Thus the natural parameter space is $\{(\eta_1, \eta_2) : \eta_1 > 0, -\infty < \eta_2 < \infty\}$. Identifying (3.4.8) with (3.4.6), we see that $\eta_2 = \mu/\sigma^2$ and $\eta_1 = 1/\sigma^2$. Although natural parameters provide a convenient mathematical formulation, they sometimes lack simple interpretations like the mean and variance.                    ‖

In the representation (3.4.1) it is often the case that the dimension of the vector $\boldsymbol{\theta}$ is equal to $k$, the number of terms in the sum in the exponent. This need not be so, and it is possible for the dimension of the vector $\boldsymbol{\theta}$ to be equal to $d < k$. Such an exponential family is called a *curved exponential family*.

**Definition 3.4.7** A *curved exponential family* is a family of densities of the form (3.4.1) for which the dimension of the vector $\boldsymbol{\theta}$ is equal to $d < k$. If $d = k$, the family is a *full exponential family*. (See also Miscellanea 3.8.3.)

**Example 3.4.8 (A curved exponential family)** The normal family of Example 3.4.4 is a full exponential family. However, if we assume that $\sigma^2 = \mu^2$, the family becomes curved. (Such a model might be used in the analysis of variance; see Exercises 11.1 and 11.2.) We then have

$$f(x|\mu) = \frac{1}{\sqrt{2\pi\mu^2}} \exp\left(-\frac{(x-\mu)^2}{2\mu^2}\right)$$

(3.4.9)

$$= \frac{1}{\sqrt{2\pi\mu^2}} \exp\left(-\frac{1}{2}\right) \exp\left(-\frac{x^2}{2\mu^2} + \frac{x}{\mu}\right).$$

For the normal family the full exponential family would have parameter space $(\mu, \sigma^2) = \Re \times (0, \infty)$, while the parameter space of the curved family $(\mu, \sigma^2) = (\mu, \mu^2)$ is a parabola. ‖

Curved exponential families are useful in many ways. The next example illustrates a simple use.

**Example 3.4.9 (Normal approximations)** In Chapter 5 we will see that if $X_1, \ldots, X_n$ is a sample from a Poisson($\lambda$) population, then the distribution of $\bar{X} = \Sigma_i X_i / n$ is approximately

$$\bar{X} \sim n(\lambda, \lambda/n),$$

a curved exponential family.

The $n(\lambda, \lambda/n)$ approximation is justified by the Central Limit Theorem (Theorem 5.5.14). In fact, we might realize that most such CLT approximations will result in a curved normal family. We have seen the normal binomial approximation (Example 3.3.2): If $X_1, \ldots, X_n$ are iid Bernoulli($p$), then

$$\bar{X} \sim n(p, p(1-p)/n),$$

approximately. For another illustration, see Example 5.5.16. ‖

Although the fact that the parameter space is a lower-dimensional space has some influence on the properties of the family, we will see that curved families still enjoy many of the properties of full families. In particular, Theorem 3.4.2 applies to curved exponential families. Moreover, full and curved exponential families have other statistical properties, which will be discussed throughout the remainder of the text. For example, suppose we have a large number of data values from a population that has a pdf or pmf of the form (3.4.1). Then only $k$ numbers ($k$ = number of terms in the sum in (3.4.1)) that can be calculated from the data summarize all the information about $\boldsymbol{\theta}$ that is in the data. This "data reduction" property is treated in more detail in Chapter 6 (Theorem 6.2.10), where we discuss sufficient statistics.

For more of an introduction to exponential families, see Lehmann (1986, Section 2.7) or Lehmann and Casella (1998, Section 1.5 and Note 1.10.6). A thorough introduction, at a somewhat more advanced level, is given in the classic monograph by Brown (1986).

## 3.5 Location and Scale Families

In Sections 3.3 and 3.4, we discussed several common families of continuous distributions. In this section we discuss three techniques for constructing families of distributions. The resulting families have ready physical interpretations that make them useful for modeling as well as convenient mathematical properties.

The three types of families are called location families, scale families, and location–scale families. Each of the families is constructed by specifying a single pdf, say $f(x)$, called the *standard pdf* for the family. Then all other pdfs in the family are generated by transforming the standard pdf in a prescribed way. We start with a simple theorem about pdfs.

**Theorem 3.5.1** *Let $f(x)$ be any pdf and let $\mu$ and $\sigma > 0$ be any given constants. Then the function*

$$g(x|\mu,\sigma) = \frac{1}{\sigma} f\left(\frac{x-\mu}{\sigma}\right)$$

*is a pdf.*

**Proof:** To verify that the transformation has produced a legitimate pdf, we need to check that $(1/\sigma)f((x-\mu)/\sigma)$, as a function of $x$, is a pdf for every value of $\mu$ and $\sigma$ we might substitute into the formula. That is, we must check that $(1/\sigma)f((x-\mu)/\sigma)$ is nonnegative and integrates to 1. Since $f(x)$ is a pdf, $f(x) \geq 0$ for all values of $x$. So, $(1/\sigma)f((x-\mu)/\sigma) \geq 0$ for all values of $x, \mu$, and $\sigma$. Next we note that

$$\int_{-\infty}^{\infty} \frac{1}{\sigma} f\left(\frac{x-\mu}{\sigma}\right) dx = \int_{-\infty}^{\infty} f(y)\, dy \qquad \left(\text{substitute } y = \frac{x-\mu}{\sigma}\right)$$

$$= 1, \qquad\qquad (\text{since } f(y) \text{ is a pdf})$$

as was to be verified.                                                                 $\square$

We now turn to the first of our constructions, that of location families.

**Definition 3.5.2** Let $f(x)$ be any pdf. Then the family of pdfs $f(x-\mu)$, indexed by the parameter $\mu$, $-\infty < \mu < \infty$, is called the *location family with standard pdf* $f(x)$ and $\mu$ is called the *location parameter* for the family.

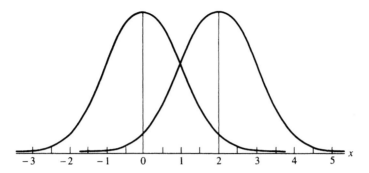

Figure 3.5.1. *Two members of the same location family: means at 0 and 2*

To see the effect of introducing the location parameter $\mu$, consider Figure 3.5.1. At $x = \mu$, $f(x - \mu) = f(0)$; at $x = \mu + 1$, $f(x - \mu) = f(1)$; and, in general, at $x = \mu + a$, $f(x - \mu) = f(a)$. Of course, $f(x - \mu)$ for $\mu = 0$ is just $f(x)$. Thus the location parameter $\mu$ simply shifts the pdf $f(x)$ so that the shape of the graph is unchanged but the point on the graph that was above $x = 0$ for $f(x)$ is above $x = \mu$ for $f(x - \mu)$. It is clear from Figure 3.5.1 that the area under the graph of $f(x)$ between $x = -1$ and $x = 2$ is the same as the area under the graph of $f(x - \mu)$ between $x = \mu - 1$ and $x = \mu + 2$. Thus if $X$ is a random variable with pdf $f(x - \mu)$, we can write

$$P(-1 \le X \le 2|0) = P(\mu - 1 \le X \le \mu + 2|\mu),$$

where the random variable $X$ has pdf $f(x - 0) = f(x)$ on the left of the equality and pdf $f(x - \mu)$ on the right.

Several of the families introduced in Section 3.3 are, or have as subfamilies, location families. For example, if $\sigma > 0$ is a specified, known number and we define

$$f(x) = \frac{1}{\sqrt{2\pi}\sigma}e^{-x^2/(2\sigma^2)}, \quad -\infty < x < \infty,$$

then the location family with standard pdf $f(x)$ is the set of normal distributions with unknown mean $\mu$ and known variance $\sigma^2$. To see this, check that replacing $x$ by $x - \mu$ in the above formula yields pdfs of the form defined in (3.3.13). Similarly, the Cauchy family and the double exponential family, with $\sigma$ a specified value and $\mu$ a parameter, are examples of location families. But the point of Definition 3.5.2 is that we can start with *any* pdf $f(x)$ and generate a family of pdfs by introducing a location parameter.

If $X$ is a random variable with pdf $f(x - \mu)$, then $X$ may be represented as $X = Z + \mu$, where $Z$ is a random variable with pdf $f(z)$. This representation is a consequence of Theorem 3.5.6 (with $\sigma = 1$), which will be proved later. Consideration of this representation indicates when a location family might be an appropriate model for an observed variable $X$. We will describe two such situations.

First, suppose an experiment is designed to measure some physical constant $\mu$, say the temperature of a solution. But there is some measurement error involved in the observation. So the actual observed value $X$ is $Z + \mu$, where $Z$ is the measurement error. $X$ will be greater than $\mu$ if $Z > 0$ for this observation and less than $\mu$ if $Z < 0$. The distribution of the random measurement error might be well known from previous experience in using this measuring device to measure other solutions. If this distribution has pdf $f(z)$, then the pdf of the observed value $X$ is $f(x - \mu)$.

As another example, suppose the distribution of reaction times of drivers on a coordination test is known from previous experimentation. Denote the reaction time for a randomly chosen driver by the random variable $Z$. Let the pdf of $Z$ describing the known distribution be $f(z)$. Now, consider "applying a treatment" to the population. For example, consider what would happen if everyone drank three glasses of beer. We might assume that everyone's reaction time would change by some unknown amount $\mu$. (This very simple model, in which everyone's reaction time changes by the same amount $\mu$, is probably not the best model. For example, it is known that the effect of alcohol is weight-dependent, so heavier people are likely to be less affected by the beers.) Being open-minded scientists, we might even allow the possibility that $\mu < 0$, that is, that the reaction times decrease. Then, if we observe the reaction time of a randomly selected driver after "treatment," the reaction time would be $X = Z + \mu$ and the family of possible distributions for $X$ would be given by $f(x - \mu)$.

If the set of $x$ for which $f(x) > 0$ is not the whole real line, then the set of $x$ for which $f(x - \mu) > 0$ will depend on $\mu$. Example 3.5.3 illustrates this.

**Example 3.5.3 (Exponential location family)**  Let $f(x) = e^{-x}$, $x \geq 0$, and $f(x) = 0$, $x < 0$. To form a location family we replace $x$ with $x - \mu$ to obtain

$$f(x|\mu) = \begin{cases} e^{-(x-\mu)} & x - \mu \geq 0 \\ 0 & x - \mu < 0 \end{cases}$$

$$= \begin{cases} e^{-(x-\mu)} & x \geq \mu \\ 0 & x < \mu. \end{cases}$$

Graphs of $f(x|\mu)$ for various values of $\mu$ are shown in Figure 3.5.2. As in Figure 3.5.1, the graph has been shifted. Now the positive part of the graph starts at $\mu$ rather than at $0$. If $X$ measures time, then $\mu$ might be restricted to be nonnegative so that $X$ will be positive with probability 1 for every value of $\mu$. In this type of model, where $\mu$ denotes a bound on the range of $X$, $\mu$ is sometimes called a *threshold parameter*. ‖

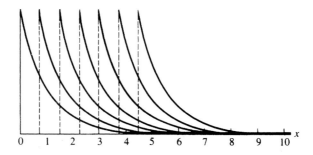

Figure 3.5.2. *Exponential location densities*

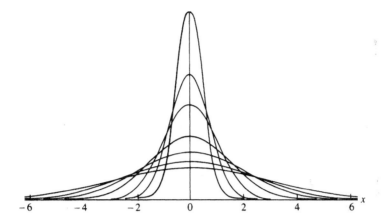

Figure 3.5.3. *Members of the same scale family*

The other two types of families to be discussed in this section are scale families and location–scale families.

**Definition 3.5.4**　　Let $f(x)$ be any pdf. Then for any $\sigma > 0$, the family of pdfs $(1/\sigma)f(x/\sigma)$, indexed by the parameter $\sigma$, is called the *scale family with standard pdf* $f(x)$ and $\sigma$ is called the *scale parameter* of the family.

The effect of introducing the scale parameter $\sigma$ is either to stretch ($\sigma > 1$) or to contract ($\sigma < 1$) the graph of $f(x)$ while still maintaining the same basic shape of the graph. This is illustrated in Figure 3.5.3. Most often when scale parameters are used, $f(x)$ is either symmetric about 0 or positive only for $x > 0$. In these cases the stretching is either symmetric about 0 or only in the positive direction. But, in the definition, any pdf may be used as the standard.

Several of the families introduced in Section 3.3 either are scale families or have scale families as subfamilies. These are the gamma family if $\alpha$ is a fixed value and $\beta$ is the scale parameter, the normal family if $\mu = 0$ and $\sigma$ is the scale parameter, the exponential family, and the double exponential family if $\mu = 0$ and $\sigma$ is the scale parameter. In each case the standard pdf is the pdf obtained by setting the scale parameter equal to 1. Then all other members of the family can be shown to be of the form in Definition 3.5.4.

**Definition 3.5.5**　　Let $f(x)$ be any pdf. Then for any $\mu$, $-\infty < \mu < \infty$, and any $\sigma > 0$, the family of pdfs $(1/\sigma)f((x - \mu)/\sigma)$, indexed by the parameter $(\mu, \sigma)$, is called the *location–scale family with standard pdf* $f(x)$; $\mu$ is called the *location parameter* and $\sigma$ is called the *scale parameter*.

The effect of introducing both the location and scale parameters is to stretch ($\sigma > 1$) or contract ($\sigma < 1$) the graph with the scale parameter and then shift the graph so that the point that was above 0 is now above $\mu$. Figure 3.5.4 illustrates this transformation of $f(x)$. The normal and double exponential families are examples of location–scale families. Exercise 3.39 presents the Cauchy as a location–scale family.

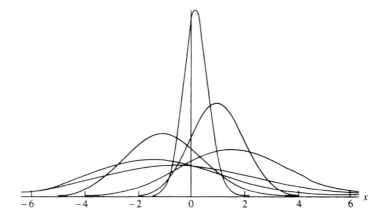

Figure 3.5.4. *Members of the same location–scale family*

The following theorem relates the transformation of the pdf $f(x)$ that defines a location–scale family to the transformation of a random variable $Z$ with pdf $f(z)$. As mentioned earlier in the discussion of location families, the representation in terms of $Z$ is a useful mathematical tool and can help us understand when a location–scale family might be appropriate in a modeling context. Setting $\sigma = 1$ in Theorem 3.5.6 yields a result for location (only) families, and setting $\mu = 0$ yields a result for scale (only) families.

**Theorem 3.5.6**  *Let $f(\cdot)$ be any pdf. Let $\mu$ be any real number, and let $\sigma$ be any positive real number. Then $X$ is a random variable with pdf $(1/\sigma)f((x-\mu)/\sigma)$ if and only if there exists a random variable $Z$ with pdf $f(z)$ and $X = \sigma Z + \mu$.*

**Proof:** To prove the "if" part, define $g(z) = \sigma z + \mu$. Then $X = g(Z)$, $g$ is a monotone function, $g^{-1}(x) = (x - \mu)/\sigma$, and $\left|(d/dx)g^{-1}(x)\right| = 1/\sigma$. Thus by Theorem 2.1.5, the pdf of $X$ is

$$f_X(x) = f_Z\big(g^{-1}(x)\big)\left|\frac{d}{dx}g^{-1}(x)\right| = f\left(\frac{x-\mu}{\sigma}\right)\frac{1}{\sigma}.$$

To prove the "only if" part, define $g(x) = (x - \mu)/\sigma$ and let $Z = g(X)$. Theorem 2.1.5 again applies: $g^{-1}(z) = \sigma z + \mu$, $\left|(d/dz)g^{-1}(z)\right| = \sigma$, and the pdf of $Z$ is

$$f_Z(z) = f_X\big(g^{-1}(z)\big)\left|\frac{d}{dz}g^{-1}(z)\right| = \frac{1}{\sigma}f\left(\frac{(\sigma z + \mu) - \mu}{\sigma}\right)\sigma = f(z).$$

Also,

$$\sigma Z + \mu = \sigma g(X) + \mu = \sigma\left(\frac{X-\mu}{\sigma}\right) + \mu = X. \qquad \square$$

An important fact to extract from Theorem 3.5.6 is that the random variable $Z = (X - \mu)/\sigma$ has pdf

$$f_Z(z) = \frac{1}{1} f\left(\frac{z - 0}{1}\right) = f(z).$$

That is, the distribution of $Z$ is that member of the location–scale family corresponding to $\mu = 0, \sigma = 1$. This was already proved for the special case of the normal family in Section 3.3.

Often, calculations can be carried out for the "standard" random variable $Z$ with pdf $f(z)$ and then the corresponding result for the random variable $X$ with pdf $(1/\sigma)f((x - \mu)/\sigma)$ can be easily derived. An example is given in the following, which is a generalization of a computation done in Section 3.3 for the normal family.

**Theorem 3.5.7** *Let $Z$ be a random variable with pdf $f(z)$. Suppose $EZ$ and $\operatorname{Var} Z$ exist. If $X$ is a random variable with pdf $(1/\sigma)f((x - \mu)/\sigma)$, then*

$$EX = \sigma EZ + \mu \quad \text{and} \quad \operatorname{Var} X = \sigma^2 \operatorname{Var} Z.$$

*In particular, if $EZ = 0$ and $\operatorname{Var} Z = 1$, then $EX = \mu$ and $\operatorname{Var} X = \sigma^2$.*

**Proof:** By Theorem 3.5.6, there is a random variable $Z^*$ with pdf $f(z)$ and $X = \sigma Z^* + \mu$. So $EX = \sigma EZ^* + \mu = \sigma EZ + \mu$ and $\operatorname{Var} X = \sigma^2 \operatorname{Var} Z^* = \sigma^2 \operatorname{Var} Z$. $\square$

For any location–scale family with a finite mean and variance, the standard pdf $f(z)$ can be chosen in such a way that $EZ = 0$ and $\operatorname{Var} Z = 1$. (The proof that this choice can be made is left as Exercise 3.40.) This results in the convenient interpretation of $\mu$ and $\sigma^2$ as the mean and variance of $X$, respectively. This is the case for the usual definition of the normal family as given in Section 3.3. However, this is not the choice for the usual definition of the double exponential family as given in Section 3.3. There, $\operatorname{Var} Z = 2$.

Probabilities for any member of a location–scale family may be computed in terms of the standard variable $Z$ because

$$P(X \le x) = P\left(\frac{X - \mu}{\sigma} \le \frac{x - \mu}{\sigma}\right) = P\left(Z \le \frac{x - \mu}{\sigma}\right).$$

Thus, if $P(Z \le z)$ is tabulated or easily calculable for the standard variable $Z$, then probabilities for $X$ may be obtained. Calculations of normal probabilities using the standard normal table are examples of this.

## 3.6 Inequalities and Identities

Statistical theory is literally brimming with inequalities and identities—so many that entire books are devoted to the topic. The major work by Marshall and Olkin (1979) contains many inequalities using the concept of majorization. The older work by Hardy, Littlewood, and Polya (1952) is a compendium of classic inequalities. In this section and in Section 4.7 we will mix some old and some new, giving some idea of the types of results that exist. This section is devoted to those identities and inequalities that arise from probabilistic concerns, while those in Section 4.7 rely more on basic properties of numbers and functions.

### 3.6.1 Probability Inequalities

The most famous, and perhaps most useful, probability inequality is Chebychev's Inequality. Its usefulness comes from its wide applicability. As with many important results, its proof is almost trivial.

**Theorem 3.6.1 (Chebychev's Inequality)**   *Let $X$ be a random variable and let $g(x)$ be a nonnegative function. Then, for any $r > 0$,*

$$P\left(g(X) \geq r\right) \leq \frac{Eg(X)}{r}.$$

**Proof:**

$$Eg(X) = \int_{-\infty}^{\infty} g(x) f_X(x) \, dx$$

$$\geq \int_{\{x:g(x)\geq r\}} g(x) f_X(x) \, dx \qquad (g \text{ is nonnegative})$$

$$\geq r \int_{\{x:g(x)\geq r\}} f_X(x) \, dx$$

$$= rP\left(g(X) \geq r\right). \qquad (\text{definition})$$

Rearranging now produces the desired inequality.                              □

**Example 3.6.2 (Illustrating Chebychev)**   The most widespread use of Chebychev's Inequality involves means and variances. Let $g(x) = (x-\mu)^2/\sigma^2$, where $\mu = EX$ and $\sigma^2 = \operatorname{Var} X$. For convenience write $r = t^2$. Then

$$P\left(\frac{(X-\mu)^2}{\sigma^2} \geq t^2\right) \leq \frac{1}{t^2} E\frac{(X-\mu)^2}{\sigma^2} = \frac{1}{t^2}.$$

Doing some obvious algebra, we get the inequality

$$P(|X - \mu| \geq t\sigma) \leq \frac{1}{t^2}$$

and its companion

$$P(|X - \mu| < t\sigma) \geq 1 - \frac{1}{t^2},$$

which gives a universal bound on the deviation $|X - \mu|$ in terms of $\sigma$. For example, taking $t = 2$, we get

$$P(|X - \mu| \geq 2\sigma) \leq \frac{1}{2^2} = .25,$$

so there is at least a 75% chance that a random variable will be within $2\sigma$ of its mean (no matter what the distribution of $X$).                                    ‖

While Chebychev's Inequality is widely applicable, it is necessarily conservative. (See, for example, Exercise 3.46 and Miscellanea 3.8.2.) In particular, we can often get tighter bounds for some specific distributions.

**Example 3.6.3 (A normal probability inequality)**     If $Z$ is standard normal, then

$$(3.6.1) \qquad P(|Z| \geq t) \leq \sqrt{\frac{2}{\pi}} \frac{e^{-t^2/2}}{t}, \qquad \text{for all } t > 0.$$

Compare this with Chebychev's Inequality. For $t = 2$, Chebychev gives $P(|Z| \geq t) \leq .25$ but $\sqrt{(2/\pi)}e^{-2}/2 = .054$, a vast improvement.

To prove (3.6.1), write

$$P(Z \geq t) = \frac{1}{\sqrt{2\pi}} \int_t^\infty e^{-x^2/2} \, dx$$

$$\leq \frac{1}{\sqrt{2\pi}} \int_t^\infty \frac{x}{t} e^{-x^2/2} \, dx \qquad \left( \begin{array}{c} \text{since } x/t > 1 \\ \text{for } x > t \end{array} \right)$$

$$= \frac{1}{\sqrt{2\pi}} \frac{e^{-t^2/2}}{t}$$

and use the fact that $P(|Z| \geq t) = 2P(Z \geq t)$. A lower bound on $P(|Z| \geq t)$ can be established in a similar way (see Exercise 3.47).                                    ‖

Many other probability inequalities exist, and almost all of them are similar in spirit to Chebychev's. For example, we will see (Exercise 3.45) that

$$P(X \geq a) \leq e^{-at} M_X(t),$$

but, of course, this inequality requires the existence of the mgf. Other inequalities, tighter than Chebychev but requiring more assumptions, exist (as detailed in Miscellanea 3.8.2).

*3.6.2 Identities*

In this section we present a sampling of various identities that can be useful not only in establishing theorems but also in easing numerical calculations. An entire class of identities can be thought of as "recursion relations," a few of which we have already seen. Recall that if $X$ is Poisson($\lambda$), then

$$(3.6.2) \qquad P(X = x + 1) = \frac{\lambda}{x + 1} P(X = x),$$

allowing us to calculate Poisson probabilities recursively starting from $P(X = 0) = e^{-\lambda}$. Relations like (3.6.2) exist for almost all discrete distributions (see Exercise 3.48). Sometimes they exist in a slightly different form for continuous distributions.

**Theorem 3.6.4** *Let $X_{\alpha,\beta}$ denote a gamma$(\alpha, \beta)$ random variable with pdf $f(x|\alpha, \beta)$, where $\alpha > 1$. Then for any constants $a$ and $b$,*

$$(3.6.3) \quad P(a < X_{\alpha,\beta} < b) = \beta \left( f(a|\alpha, \beta) - f(b|\alpha, \beta) \right) + P(a < X_{\alpha-1,\beta} < b).$$

**Proof:** By definition,

$$P(a < X_{\alpha,\beta} < b) = \frac{1}{\Gamma(\alpha)\beta^\alpha} \int_a^b x^{\alpha-1} e^{-x/\beta} \, dx$$

$$= \frac{1}{\Gamma(\alpha)\beta^\alpha} \left[ -x^{\alpha-1} \beta e^{-x/\beta} \Big|_a^b + \int_a^b (\alpha - 1) x^{\alpha-2} \beta e^{-x/\beta} \, dx \right],$$

where we have done an integration by parts with $u = x^{\alpha-1}$ and $dv = e^{-x/\beta} \, dx$. Continuing, we have

$$P(a < X_{\alpha,\beta} < b) = \beta \left( f(a|\alpha, \beta) - f(b|\alpha, \beta) \right) + \frac{(\alpha - 1)}{\Gamma(\alpha)\beta^{\alpha-1}} \int_a^b x^{\alpha-2} e^{-x/\beta} \, dx.$$

Using the fact that $\Gamma(\alpha) = (\alpha - 1)\Gamma(\alpha - 1)$, we see that the last term is $P(a < X_{\alpha-1,\beta} < b)$. $\qquad\square$

If $\alpha$ is an integer, repeated use of (3.6.3) will eventually lead to an integral that can be evaluated analytically (when $\alpha = 1$, the exponential distribution). Thus, we can easily compute these gamma probabilities.

There is an entire class of identities that rely on integration by parts. The first of these is attributed to Charles Stein, who used it in his work on estimation of multivariate normal means (Stein 1973, 1981).

**Lemma 3.6.5 (Stein's Lemma)** *Let $X \sim n(\theta, \sigma^2)$, and let $g$ be a differentiable function satisfying $\mathrm{E} |g'(X)| < \infty$. Then*

$$\mathrm{E}[g(X)(X - \theta)] = \sigma^2 \mathrm{E} g'(X).$$

**Proof:** The left-hand side is

$$\mathrm{E}[g(X)(X - \theta)] = \frac{1}{\sqrt{2\pi}\sigma} \int_{-\infty}^\infty g(x)(x - \theta) e^{-(x-\theta)^2/(2\sigma^2)} \, dx.$$

Use integration by parts with $u = g(x)$ and $dv = (x - \theta)e^{-(x-\theta)^2/2\sigma^2} \, dx$ to get

$$\mathrm{E}[g(X)(X - \theta)] = \frac{1}{\sqrt{2\pi}\sigma} \left[ -\sigma^2 g(x) e^{-(x-\theta)^2/(2\sigma^2)} \Big|_{-\infty}^\infty + \sigma^2 \int_{-\infty}^\infty g'(x) e^{-(x-\theta)^2/(2\sigma^2)} \, dx \right]$$

The condition on $g'$ is enough to ensure that the first term is 0 and what remains on the right-hand side is $\sigma^2 \mathrm{E} g'(X)$. $\qquad\square$

**Example 3.6.6 (Higher-order normal moments)** Stein's Lemma makes calculation of higher-order moments quite easy. For example, if $X \sim n(\theta, \sigma^2)$, then

$$\begin{aligned} EX^3 &= EX^2(X - \theta + \theta) \\ &= EX^2(X - \theta) + \theta EX^2 \\ &= 2\sigma^2 EX + \theta EX^2 \qquad\qquad (g(x) = x^2,\ g'(x) = 2x) \\ &= 2\sigma^2\theta + \theta(\sigma^2 + \theta^2) \\ &= 3\theta\sigma^2 + \theta^3. \qquad\qquad\qquad\qquad\qquad\qquad \| \end{aligned}$$

Similar integration-by-parts identities exist for many distributions (see Exercise 3.49 and Hudson 1978). One can also get useful identities by exploiting properties of a particular distribution, as the next theorem shows.

**Theorem 3.6.7** *Let $\chi_p^2$ denote a chi squared random variable with $p$ degrees of freedom. For any function $h(x)$,*

(3.6.4)
$$Eh(\chi_p^2) = pE\left(\frac{h\left(\chi_{p+2}^2\right)}{\chi_{p+2}^2}\right)$$

*provided the expectations exist.*

**Proof:** The phrase "provided the expectations exist" is a lazy way of avoiding specification of conditions on $h$. In general, reasonable functions will satisfy (3.6.4). We have

$$\begin{aligned} Eh(\chi_p^2) &= \frac{1}{\Gamma(p/2)2^{p/2}}\int_0^\infty h(x)x^{(p/2)-1}e^{-x/2}\,dx \\ &= \frac{1}{\Gamma(p/2)2^{p/2}}\int_0^\infty \left(\frac{h(x)}{x}\right)x^{((p+2)/2)-1}e^{-x/2}\,dx, \end{aligned}$$

where we have multiplied the integrand by $x/x$. Now write

$$\Gamma\left(\frac{p}{2}\right)2^{p/2} = \frac{\Gamma((p+2)/2)2^{(p+2)/2}}{p},$$

so we have

$$\begin{aligned} Eh(\chi_p^2) &= \frac{p}{\Gamma((p+2)/2)2^{(p+2)/2}}\int_0^\infty\left(\frac{h(x)}{x}\right)x^{((p+2)/2)-1}e^{-x/2}\,dx \\ &= pE\left(\frac{h(\chi_{p+2}^2)}{\chi_{p+2}^2}\right). \qquad\qquad\qquad \Box \end{aligned}$$

Some moment calculations are very easy with (3.6.4). For example, the mean of a $\chi_p^2$ is

$$E\chi_p^2 = pE\left(\frac{\chi_{p+2}^2}{\chi_{p+2}^2}\right) = pE(1) = p,$$

and the second moment is

$$
\mathrm{E}(\chi_p^2)^2 = p\mathrm{E}\left(\frac{\left(\chi_{p+2}^2\right)^2}{\chi_{p+2}^2}\right) = p\mathrm{E}\left(\chi_{p+2}^2\right) = p(p+2).
$$

So $\operatorname{Var} \chi_p^2 = p(p+2) - p^2 = 2p$.

We close our section on identities with some discrete analogs of the previous identities. A general version of the two identities in Theorem 3.6.8 is due to Hwang (1982).

**Theorem 3.6.8 (Hwang)**    *Let $g(x)$ be a function with $-\infty < \mathrm{E}g(X) < \infty$ and $-\infty < g(-1) < \infty$. Then:*

**a.** *If $X \sim Poisson(\lambda)$,*

$$
(3.6.5) \qquad\qquad \mathrm{E}\left(\lambda g(X)\right) = \mathrm{E}\left(Xg(X-1)\right).
$$

**b.** *If $X \sim negative\ binomial(r, p)$,*

$$
(3.6.6) \qquad\qquad \mathrm{E}\left((1-p)g(X)\right) = \mathrm{E}\left(\frac{X}{r+X-1}g(X-1)\right).
$$

**Proof:** We will prove part (a), saving part (b) for Exercise 3.50. We have

$$
\begin{aligned}
\mathrm{E}\left(\lambda g(X)\right) &= \sum_{x=0}^{\infty} \lambda g(x) \frac{e^{-\lambda}\lambda^x}{x!} \\
&= \sum_{x=0}^{\infty} g(x) \frac{e^{-\lambda}\lambda^{x+1}}{x!} \frac{(x+1)}{(x+1)} \\
&= \sum_{x=0}^{\infty} (x+1)g(x) \frac{e^{-\lambda}\lambda^{x+1}}{(x+1)!}.
\end{aligned}
$$

Now transform the summation index, writing $y = x + 1$. As $x$ goes from 0 to $\infty$, $y$ goes from 1 to $\infty$. Thus

$$
\begin{aligned}
\mathrm{E}\left(\lambda g(X)\right) &= \sum_{y=1}^{\infty} yg(y-1) \frac{e^{-\lambda}\lambda^y}{y!} \\
&= \sum_{y=0}^{\infty} yg(y-1) \frac{e^{-\lambda}\lambda^y}{y!} \qquad \text{(added term is 0)} \\
&= \mathrm{E}\left(Xg(X-1)\right),
\end{aligned}
$$

since this last sum is a Poisson($\lambda$) expectation.                    $\square$

Hwang (1982) used his identity in a manner similar to Stein, proving results about multivariate estimators. The identity has other applications, in particular in moment calculations.

**Example 3.6.9 (Higher-order Poisson moments)**    For $X \sim$ Poisson($\lambda$), take $g(x) = x^2$ and use (3.6.5):

$$E(\lambda X^2) = E\left(X(X-1)^2\right) = E(X^3 - 2X^2 + X).$$

Therefore, the third moment of a Poisson($\lambda$) is

$$EX^3 = \lambda EX^2 + 2EX^2 - EX$$
$$= \lambda(\lambda + \lambda^2) + 2(\lambda + \lambda^2) - \lambda$$
$$= \lambda^3 + 3\lambda^2 + \lambda.$$

For the negative binomial, the mean can be calculated by taking $g(x) = r + x$ in (3.6.6):

$$E\left((1-p)(r+X)\right) = E\left(\frac{X}{r+X-1}(r+X-1)\right) = EX,$$

so, rearranging, we get

$$(EX)\left((1-p) - 1\right) = -r(1-p)$$

or

$$EX = \frac{r(1-p)}{p}.$$

Other moments can be calculated similarly.                                      ‖

## 3.7 Exercises

**3.1** Find expressions for $EX$ and Var $X$ if $X$ is a random variable with the general discrete uniform($N_0, N_1$) distribution that puts equal probability on each of the values $N_0, N_0 + 1, \ldots, N_1$. Here $N_0 \leq N_1$ and both are integers.

**3.2** A manufacturer receives a lot of 100 parts from a vendor. The lot will be unacceptable if more than five of the parts are defective. The manufacturer is going to select randomly $K$ parts from the lot for inspection and the lot will be accepted if no defective parts are found in the sample.

(a) How large does $K$ have to be to ensure that the probability that the manufacturer accepts an unacceptable lot is less than .10?

(b) Suppose the manufacturer decides to accept the lot if there is at most one defective in the sample. How large does $K$ have to be to ensure that the probability that the manufacturer accepts an unacceptable lot is less than .10?

**3.3** The flow of traffic at certain street corners can sometimes be modeled as a sequence of Bernoulli trials by assuming that the probability of a car passing during any given second is a constant $p$ and that there is no interaction between the passing of cars at different seconds. If we treat seconds as indivisible time units (trials), the Bernoulli model applies. Suppose a pedestrian can cross the street only if no car is to pass during the next 3 seconds. Find the probability that the pedestrian has to wait for exactly 4 seconds before starting to cross.

**3.4** A man with $n$ keys wants to open his door and tries the keys at random. Exactly one key will open the door. Find the mean number of trials if

(a) unsuccessful keys are not eliminated from further selections.

(b) unsuccessful keys are eliminated.

**3.5** A standard drug is known to be effective in 80% of the cases in which it is used. A new drug is tested on 100 patients and found to be effective in 85 cases. Is the new drug superior? (*Hint:* Evaluate the probability of observing 85 or more successes assuming that the new and old drugs are equally effective.)

**3.6** A large number of insects are expected to be attracted to a certain variety of rose plant. A commercial insecticide is advertised as being 99% effective. Suppose 2,000 insects infest a rose garden where the insecticide has been applied, and let $X$ = number of surviving insects.

(a) What probability distribution might provide a reasonable model for this experiment?

(b) Write down, but do not evaluate, an expression for the probability that fewer than 100 insects survive, using the model in part (a).

(c) Evaluate an approximation to the probability in part (b).

**3.7** Let the number of chocolate chips in a certain type of cookie have a Poisson distribution. We want the probability that a randomly chosen cookie has at least two chocolate chips to be greater than .99. Find the smallest value of the mean of the distribution that ensures this probability.

**3.8** Two movie theaters compete for the business of 1,000 customers. Assume that each customer chooses between the movie theaters independently and with "indifference." Let $N$ denote the number of seats in each theater.

(a) Using a binomial model, find an expression for $N$ that will guarantee that the probability of turning away a customer (because of a full house) is less than 1%.

(b) Use the normal approximation to get a numerical value for $N$.

**3.9** Often, news stories that are reported as startling "one-in-a-million" coincidences are actually, upon closer examination, not rare events and can even be expected to occur. A few years ago an elementary school in New York state reported that its incoming kindergarten class contained five sets of twins. This, of course, was reported throughout the state, with a quote from the principal that this was a "statistical impossibility". Was it? Or was it an instance of what Diaconis and Mosteller (1989) call the "law of truly large numbers"? Let's do some calculations.

(a) The probability of a twin birth is approximately 1/90, and we can assume that an elementary school will have approximately 60 children entering kindergarten (three classes of 20 each). Explain how our "statistically impossible" event can be thought of as the probability of 5 or more successes from a binomial(60, 1/90). Is this even rare enough to be newsworthy?

(b) Even if the probability in part (a) is rare enough to be newsworthy, consider that this could have happened in any school in the county, and in any county in the state, and it still would have been reported exactly the same. (The "law of truly large numbers" is starting to come into play.) New York state has 62 counties, and it is reasonable to assume that each county has five elementary schools. Does the event still qualify as a "statistical impossibility", or is it becoming something that could be expected to occur?

(c) If the probability in part (b) still seems small, consider further that this event could have happened in any one of the 50 states, during any of the last 10 years, and still would have received the same news coverage.

In addition to Diaconis and Mosteller (1989), see Hanley (1992) for more on coincidences.

**3.10** Shuster (1991) describes a number of probability calculations that he did for a court case involving the sale of cocaine. A Florida police department seized 496 *suspected* packets of cocaine, of which four were randomly selected and tested and found to actually be cocaine. The police then chose two more packets at random and, posing as drug dealers, sold the packets to the defendant. These last two packets were lost before they could be tested to verify that they were, indeed, cocaine.

(a) If the original 496 packets were composed of $N$ packets of cocaine and $M = 496 - N$ noncocaine, show that the probability of selecting 4 cocaine packets and then 2 noncocaine packets, which is the probability that the defendant is innocent of buying cocaine, is

$$\frac{\binom{N}{4}\binom{M}{2}}{\binom{N+M}{4}\binom{N+M-4}{2}}.$$

(b) Maximizing (in $M$ and $N$) the probability in part (a) maximizes the defendant's "innocence probability". Show that this probability is .022, attained at $M = 165$ and $N = 331$.

**3.11** The hypergeometric distribution can be approximated by either the binomial or the Poisson distribution. (Of course, it can be approximated by other distributions, but in this exercise we will concentrate on only these two.) Let $X$ have the hypergeometric distribution

$$P(X = x|N, M, K) = \frac{\binom{M}{x}\binom{N-M}{K-x}}{\binom{N}{K}}, \qquad x = 0, 1, \ldots, K.$$

(a) Show that as $N \to \infty$, $M \to \infty$, and $M/N \to p$,

$$P(X = x|N, M, K) \to \binom{K}{x}p^x(1-p)^{K-x}, \qquad x = 0, 1, \ldots, K.$$

(Stirling's Formula from Exercise 1.23 may be helpful.)

(b) Use the fact that the binomial can be approximated by the Poisson to show that if $N \to \infty$, $M \to \infty$, $K \to \infty$, $M/N \to 0$, and $KM/N \to \lambda$, then

$$P(X = x|N, M, K) \to \frac{e^{-\lambda}\lambda^x}{x!}, \qquad x = 0, 1, \ldots.$$

(c) Verify the approximation in part (b) directly, without using the Poisson approximation to the binomial. (Lemma 2.3.14 is helpful.)

**3.12** Suppose $X$ has a binomial$(n, p)$ distribution and let $Y$ have a negative binomial$(r, p)$ distribution. Show that $F_X(r - 1) = 1 - F_Y(n - r)$.

**3.13** A *truncated* discrete distribution is one in which a particular class cannot be observed and is eliminated from the sample space. In particular, if $X$ has range $0, 1, 2, \ldots$ and the 0 class cannot be observed (as is usually the case), the 0-*truncated* random variable $X_T$ has pmf

$$P(X_T = x) = \frac{P(X = x)}{P(X > 0)}, \quad x = 1, 2, \ldots.$$

Find the pmf, mean, and variance of the 0-truncated random variable starting from

(a) $X \sim \text{Poisson}(\lambda)$.

(b) $X \sim$ negative binomial$(r, p)$, as in (3.2.10).

**3.14** Starting from the 0-truncated negative binomial (refer to Exercise 3.13), if we let $r \to 0$, we get an interesting distribution, the *logarithmic series distribution*. A random variable $X$ has a logarithmic series distribution with parameter $p$ if

$$P(X = x) = \frac{-(1 - p)^x}{x \log p}, \quad x = 1, 2, \ldots, \quad 0 < p < 1.$$

(a) Verify that this defines a legitimate probability function.

(b) Find the mean and variance of $X$. (The logarithmic series distribution has proved useful in modeling species abundance. See Stuart and Ord 1987 for a more detailed discussion of this distribution.)

**3.15** In Section 3.2 it was claimed that the Poisson$(\lambda)$ distribution is the limit of the negative binomial$(r, p)$ distribution as $r \to \infty$, $p \to 1$, and $r(1 - p) \to \lambda$. Show that under these conditions the mgf of the negative binomial converges to that of the Poisson.

**3.16** Verify these two identities regarding the gamma function that were given in the text:

(a) $\Gamma(\alpha + 1) = \alpha \Gamma(\alpha)$

(b) $\Gamma(\frac{1}{2}) = \sqrt{\pi}$

**3.17** Establish a formula similar to (3.3.18) for the gamma distribution. If $X \sim \text{gamma}(\alpha, \beta)$, then for any positive constant $\nu$,

$$\text{E}X^\nu = \frac{\beta^\nu \Gamma(\nu + \alpha)}{\Gamma(\alpha)}.$$

**3.18** There is an interesting relationship between negative binomial and gamma random variables, which may sometimes provide a useful approximation. Let $Y$ be a negative binomial random variable with parameters $r$ and $p$, where $p$ is the success probability. Show that as $p \to 0$, the mgf of the random variable $pY$ converges to that of a gamma distribution with parameters $r$ and 1.

**3.19** Show that

$$\int_x^\infty \frac{1}{\Gamma(\alpha)} z^{\alpha-1} e^{-z} dz = \sum_{y=0}^{\alpha-1} \frac{x^y e^{-x}}{y!}, \quad \alpha = 1, 2, 3, \ldots.$$

(*Hint:* Use integration by parts.) Express this formula as a probabilistic relationship between Poisson and gamma random variables.

**3.20** Let the random variable $X$ have the pdf

$$f(x) = \frac{2}{\sqrt{2\pi}} e^{-x^2/2}, \quad 0 < x < \infty.$$

(a) Find the mean and variance of $X$. (This distribution is sometimes called a *folded normal*.)

(b) If $X$ has the folded normal distribution, find the transformation $g(X) = Y$ and values of $\alpha$ and $\beta$ so that $Y \sim \text{gamma}(\alpha, \beta)$.

**3.21** Write the integral that would define the mgf of the pdf

$$f(x) = \frac{1}{\pi} \frac{1}{1 + x^2}.$$

Is the integral finite? (Do you expect it to be?)

**3.22** For each of the following distributions, verify the formulas for $EX$ and $\text{Var}\, X$ given in the text.

(a) Verify $\text{Var}\, X$ if $X$ has a Poisson($\lambda$) distribution. (*Hint:* Compute $EX(X - 1) = EX^2 - EX$.)

(b) Verify $\text{Var}\, X$ if $X$ has a negative binomial$(r, p)$ distribution.

(c) Verify $\text{Var}\, X$ if $X$ has a gamma$(\alpha, \beta)$ distribution.

(d) Verify $EX$ and $\text{Var}\, X$ if $X$ has a beta$(\alpha, \beta)$ distribution.

(e) Verify $EX$ and $\text{Var}\, X$ if $X$ has a double exponential$(\mu, \sigma)$ distribution.

**3.23** The *Pareto distribution*, with parameters $\alpha$ and $\beta$, has pdf

$$f(x) = \frac{\beta \alpha^\beta}{x^{\beta+1}}, \quad \alpha < x < \infty, \quad \alpha > 0, \quad \beta > 0.$$

(a) Verify that $f(x)$ is a pdf.

(b) Derive the mean and variance of this distribution.

(c) Prove that the variance does not exist if $\beta \leq 2$.

**3.24** Many "named" distributions are special cases of the more common distributions already discussed. For each of the following named distributions derive the form of the pdf, verify that it is a pdf, and calculate the mean and variance.

(a) If $X \sim \text{exponential}(\beta)$, then $Y = X^{1/\gamma}$ has the *Weibull*$(\gamma, \beta)$ *distribution*, where $\gamma > 0$ is a constant.

(b) If $X \sim \text{exponential}(\beta)$, then $Y = (2X/\beta)^{1/2}$ has the *Rayleigh distribution*.

(c) If $X \sim \text{gamma}(a, b)$, then $Y = 1/X$ has the *inverted gamma* IG$(a, b)$ distribution. (This distribution is useful in Bayesian estimation of variances; see Exercise 7.23.)

(d) If $X \sim \text{gamma}(\frac{3}{2}, \beta)$, then $Y = (X/\beta)^{1/2}$ has the *Maxwell distribution*.

(e) If $X \sim \text{exponential}(1)$, then $Y = \alpha - \gamma \log X$ has the *Gumbel*$(\alpha, \gamma)$ *distribution*, where $-\infty < \alpha < \infty$ and $\gamma > 0$. (The Gumbel distribution is also known as the *extreme value distribution*.)

**3.25** Suppose the random variable $T$ is the length of life of an object (possibly the lifetime of an electrical component or of a subject given a particular treatment). The *hazard function* $h_T(t)$ associated with the random variable $T$ is defined by

$$h_T(t) = \lim_{\delta \to 0} \frac{P\left(t \leq T < t + \delta | T \geq t\right)}{\delta}.$$

Thus, we can interpret $h_T(t)$ as the rate of change of the probability that the object survives a little past time $t$, given that the object survives to time $t$. Show that if $T$ is a continuous random variable, then

$$h_T(t) = \frac{f_T(t)}{1 - F_T(t)} = -\frac{d}{dt} \log\left(1 - F_T(t)\right).$$

**3.26** Verify that the following pdfs have the indicated hazard functions (see Exercise 3.25).

(a) If $T \sim$ exponential($\beta$), then $h_T(t) = 1/\beta$.

(b) If $T \sim$ Weibull($\gamma, \beta$), then $h_T(t) = (\gamma/\beta)t^{\gamma-1}$.

(c) If $T \sim$ logistic($\mu, \beta$), that is,

$$F_T(t) = \frac{1}{1 + e^{-(t-\mu)/\beta}},$$

then $h_T(t) = (1/\beta)F_T(t)$.

**3.27** For each of the following families, show whether all the pdfs in the family are unimodal (see Exercise 2.27).

(a) uniform($a, b$)

(b) gamma($\alpha, \beta$)

(c) n($\mu, \sigma^2$)

(d) beta($\alpha, \beta$)

**3.28** Show that each of the following families is an exponential family.

(a) normal family with either parameter $\mu$ or $\sigma$ known

(b) gamma family with either parameter $\alpha$ or $\beta$ known or both unknown

(c) beta family with either parameter $\alpha$ or $\beta$ known or both unknown

(d) Poisson family

(e) negative binomial family with $r$ known, $0 < p < 1$

**3.29** For each family in Exercise 3.28, describe the natural parameter space.

**3.30** Use the identities of Theorem 3.4.2 to

(a) calculate the variance of a binomial random variable.

(b) calculate the mean and variance of a Poisson($\lambda$) random variable.

**3.31** In this exercise we will prove Theorem 3.4.2.

(a) Start from the equality

$$\int h(x)c(\boldsymbol{\theta}) \exp\left(\sum_{i=1}^{k} w_i(\boldsymbol{\theta})t_i(x)\right) dx = 1,$$

differentiate both sides, and then rearrange terms to establish (3.4.4). (The fact that $\frac{d}{dx} \log g(x) = g'(x)/g(x)$ will be helpful.)

(b) Differentiate the above equality a second time; then rearrange to establish (3.4.5). (The fact that $\frac{d^2}{dx^2} \log g(x) = (g''(x)/g(x)) - (g'(x)/g(x))^2$ will be helpful.)

**3.32** (a) If an exponential family can be written in the form (3.4.7), show that the identities of Theorem 3.4.2 simplify to

$$E(t_j(X)) = -\frac{\partial}{\partial \eta_j} \log c^*(\boldsymbol{\eta}),$$

$$\text{Var}(t_j(X)) = -\frac{\partial^2}{\partial \eta_j^2} \log c^*(\boldsymbol{\eta}).$$

   (b) Use this identity to calculate the mean and variance of a gamma$(a, b)$ random variable.

**3.33** For each of the following families:

   (i) Verify that it is an exponential family.

   (ii) Describe the curve on which the $\boldsymbol{\theta}$ parameter vector lies.

   (iii) Sketch a graph of the curved parameter space.

   (a) $n(\theta, \theta)$
   (b) $n(\theta, a\theta^2)$, $a$ known
   (c) gamma$(\alpha, 1/\alpha)$
   (d) $f(x|\theta) = C \exp\left(-(x - \theta)^4\right)$, $C$ a normalizing constant

**3.34** In Example 3.4.9 we saw that normal approximations can result in curved exponential families. For each of the following normal approximations:

   (i) Describe the curve on which the $\boldsymbol{\theta}$ parameter vector lies.

   (ii) Sketch a graph of the curved parameter space.

   (a) Poisson approximation: $\bar{X} \sim n(\lambda, \lambda/n)$
   (b) binomial approximation: $\bar{X} \sim n(p, p(1 - p)/n)$
   (c) negative binomial approximation: $\bar{X} \sim n(r(1 - p)/p, r(1 - p)/np^2)$

**3.35** (a) The normal family that approximates a Poisson can also be parameterized as $n(e^\theta, e^\theta)$, where $-\infty < \theta < \infty$. Sketch a graph of the parameter space, and compare with the approximation in Exercise 3.34(a).

   (b) Suppose that $X \sim$ gamma$(\alpha, \beta)$ and we assume that $EX = \mu$. Sketch a graph of the parameter space.

   (c) Suppose that $X_i \sim$ gamma$(\alpha_i, \beta_i)$, $i = 1, 2, \ldots, n$, and we assume that $EX_i = \mu$. Describe the parameter space $(\alpha_1, \ldots, \alpha_n, \beta_1, \ldots, \beta_n)$.

**3.36** Consider the pdf $f(x) = \frac{63}{4}(x^6 - x^8)$, $-1 < x < 1$. Graph $(1/\sigma)f((x - \mu)/\sigma)$ for each of the following on the same axes.

   (a) $\mu = 0$, $\sigma = 1$
   (b) $\mu = 3$, $\sigma = 1$
   (c) $\mu = 3$, $\sigma = 2$

**3.37** Show that if $f(x)$ is a pdf, symmetric about 0, then $\mu$ is the median of the location–scale pdf $(1/\sigma)f((x - \mu)/\sigma)$, $-\infty < x < \infty$.

**3.38** Let $Z$ be a random variable with pdf $f(z)$. Define $z_\alpha$ to be a number that satisfies this relationship:

$$\alpha = P(Z > z_\alpha) = \int_{z_\alpha}^{\infty} f(z)dz.$$

Show that if $X$ is a random variable with pdf $(1/\sigma)f((x-\mu)/\sigma)$ and $x_\alpha = \sigma z_\alpha + \mu$, then $P(X > x_\alpha) = \alpha$. (Thus if a table of $z_\alpha$ values were available, then values of $x_\alpha$ could be easily computed for any member of the location–scale family.)

**3.39** Consider the Cauchy family defined in Section 3.3. This family can be extended to a location–scale family yielding pdfs of the form

$$f(x|\mu,\sigma) = \frac{1}{\sigma\pi\left(1+\left(\frac{x-\mu}{\sigma}\right)^2\right)}, \quad -\infty < x < \infty.$$

The mean and variance do not exist for the Cauchy distribution. So the parameters $\mu$ and $\sigma^2$ are not the mean and variance. But they do have important meaning. Show that if $X$ is a random variable with a Cauchy distribution with parameters $\mu$ and $\sigma$, then:

(a) $\mu$ is the median of the distribution of $X$, that is, $P(X \geq \mu) = P(X \leq \mu) = \frac{1}{2}$.

(b) $\mu+\sigma$ and $\mu-\sigma$ are the quartiles of the distribution of $X$, that is, $P(X \geq \mu+\sigma) = P(X \leq \mu - \sigma) = \frac{1}{4}$. (*Hint:* Prove this first for $\mu = 0$ and $\sigma = 1$ and then use Exercise 3.38.)

**3.40** Let $f(x)$ be any pdf with mean $\mu$ and variance $\sigma^2$. Show how to create a location–scale family based on $f(x)$ such that the standard pdf of the family, say $f^*(x)$, has mean 0 and variance 1.

**3.41** A family of cdfs $\{F(x|\theta), \theta \in \Theta\}$ is *stochastically increasing in $\theta$* if $\theta_1 > \theta_2 \Rightarrow F(x|\theta_1)$ is stochastically greater than $F(x|\theta_2)$. (See Exercise 1.49 for the definition of stochastically greater.)

(a) Show that the $n(\mu,\sigma^2)$ family is stochastically increasing in $\mu$ for fixed $\sigma^2$.

(b) Show that the gamma$(\alpha, \beta)$ family of (3.3.6) is stochastically increasing in $\beta$ (scale parameter) for fixed $\alpha$ (shape parameter).

**3.42** Refer to Exercise 3.41 for the definition of a stochastically increasing family.

(a) Show that a location family is stochastically increasing in its location parameter.

(b) Show that a scale family is stochastically increasing in its scale parameter if the sample space is $[0, \infty)$.

**3.43** A family of cdfs $\{F(x|\theta), \theta \in \theta\}$ is *stochastically decreasing in $\theta$* if $\theta_1 > \theta_2 \Rightarrow F(x|\theta_2)$ is stochastically greater than $F(x|\theta_1)$. (See Exercises 3.41 and 3.42.)

(a) Prove that if $X \sim F_X(x|\theta)$, where the sample space of $X$ is $(0,\infty)$ and $F_X(x|\theta)$ is stochastically increasing in $\theta$, then $F_Y(y|\theta)$ is stochastically decreasing in $\theta$, where $Y = 1/X$.

(b) Prove that if $X \sim F_X(x|\theta)$, where $F_X(x|\theta)$ is stochastically increasing in $\theta$ and $\theta > 0$, then $F_X(x|\frac{1}{\theta})$ is stochastically decreasing in $\theta$.

**3.44** For any random variable $X$ for which $EX^2$ and $E|X|$ exist, show that $P(|X| \geq b)$ does not exceed either $EX^2/b^2$ or $E|X|/b$, where $b$ is a positive constant. If $f(x) = e^{-x}$ for $x > 0$, show that one bound is better when $b = 3$ and the other when $b = \sqrt{2}$. (Notice Markov's Inequality in Miscellanea 3.8.2.)

**3.45** Let $X$ be a random variable with moment-generating function $M_X(t)$, $-h < t < h$.

(a) Prove that $P(X \geq a) \leq e^{-at}M_X(t)$, $0 < t < h$. (A proof similar to that used for Chebychev's Inequality will work.)

(b) Similarly, prove that $P(X \le a) \le e^{-at} M_X(t), \quad -h < t < 0$.

(c) A special case of part (a) is that $P(X \ge 0) \le \mathrm{E} e^{tX}$ for all $t \ge 0$ for which the mgf is defined. What are general conditions on a function $h(t, x)$ such that $P(X \ge 0) \le \mathrm{E} h(t, X)$ for all $t \ge 0$ for which $\mathrm{E} h(t, X)$ exists? (In part (a), $h(t, x) = e^{tx}$.)

**3.46** Calculate $P(|X - \mu_X| \ge k\sigma_X)$ for $X \sim \text{uniform}(0, 1)$ and $X \sim \text{exponential}(\lambda)$, and compare your answers to the bound from Chebychev's Inequality.

**3.47** If $Z$ is a standard normal random variable, prove this companion to the inequality in Example 3.6.3:

$$P(|Z| \ge t) \ge \sqrt{\frac{2}{\pi}} \frac{t}{1 + t^2} e^{-t^2/2}.$$

**3.48** Derive recursion relations, similar to the one given in (3.6.2), for the binomial, negative binomial, and hypergeometric distributions.

**3.49** Prove the following analogs to Stein's Lemma, assuming appropriate conditions on the function $g$.

(a) If $X \sim \text{gamma}(\alpha, \beta)$, then

$$\mathrm{E}\left(g(X)(X - \alpha\beta)\right) = \beta \mathrm{E}\left(X g'(X)\right).$$

(b) If $X \sim \text{beta}(\alpha, \beta)$, then

$$\mathrm{E}\left[g(X)\left(\beta - (\alpha - 1)\frac{(1 - X)}{X}\right)\right] = \mathrm{E}\left((1 - X)g'(X)\right).$$

**3.50** Prove the identity for the negative binomial distribution given in Theorem 3.6.8, part (b).

## 3.8 Miscellanea _____

### 3.8.1 The Poisson Postulates

The Poisson distribution can be derived from a set of basic assumptions, sometimes called the Poisson postulates. These assumptions relate to the physical properties of the process under consideration. While, generally speaking, the assumptions are not very easy to verify, they do provide an experimenter with a set of guidelines for considering whether the Poisson will provide a reasonable model. For a more complete treatment of the Poisson postulates, see the classic text by Feller (1968) or Barr and Zehna (1983).

**Theorem 3.8.1** *For each $t \ge 0$, let $N_t$ be an integer-valued random variable with the following properties. (Think of $N_t$ as denoting the number of arrivals in the time period from time 0 to time $t$.)*

*i)* $N_0 = 0$                                    *(start with no arrivals)*

*ii)* $s < t \Rightarrow N_s$ *and* $N_t - N_s$ *are independent.*    $\left(\begin{array}{l}\text{arrivals in disjoint time}\\ \text{periods are independent}\end{array}\right)$

*iii)* $N_s$ and $N_{t+s} - N_t$ *are identically distributed.* $\left(\begin{array}{c}\textit{number of arrivals depends}\\ \textit{only on period length}\end{array}\right)$

*iv)* $\lim_{t\to 0}\dfrac{P(N_t = 1)}{t} = \lambda$ $\left(\begin{array}{c}\textit{arrival probability proportional}\\ \textit{to period length, if length is small}\end{array}\right)$

*v)* $\lim_{t\to 0}\dfrac{P(N_t > 1)}{t} = 0$ $\hspace{2cm}$ (*no simultaneous arrivals*)

*If i–v hold, then for any integer n,*

$$P(N_t = n) = e^{-\lambda t}\frac{(\lambda t)^n}{n!},$$

*that is,* $N_t \sim Poisson(\lambda t)$.

The postulates may also be interpreted as describing the behavior of objects spatially (for example, movement of insects), giving the Poisson application in spatial distributions.

### 3.8.2 Chebychev and Beyond

Ghosh and Meeden (1977) discuss the fact that Chebychev's Inequality is very conservative and is almost never attained. If we write $\bar{X}_n$ for the mean of the random variables $X_1, X_2, \ldots, X_n$, then Chebychev's Inequality states

$$P\left(|\bar{X}_n - \mu| \geq k\sigma\right) \leq \frac{1}{nk^2}.$$

They prove the following theorem.

**Theorem 3.8.2** *If* $0 < \sigma < \infty$, *then*

**a.** *If* $n = 1$, *the inequality is attainable for* $k \geq 1$ *and unattainable for* $0 < k < 1$.
**b.** *If* $n = 2$, *the inequality is attainable if and only if* $k = 1$.
**c.** *If* $n \geq 3$, *the inequality is not attainable.*

Examples are given for the cases when the inequality is attained. Most of their technical arguments are based on the following inequality, known as Markov's Inequality.

**Lemma 3.8.3 (Markov's Inequality)** *If* $P(Y \geq 0) = 1$ *and* $P(Y = 0) < 1$, *then, for any* $r > 0$,

$$P(Y \geq r) \leq \frac{EY}{r}$$

*with equality if and only if* $P(Y = r) = p = 1 - P(Y = 0), 0 < p \leq 1$.

Markov's Inequality can then be applied to the quantity

$$Y = \frac{(\bar{X}_n - \mu)^2}{\sigma^2}$$

to get the above results.

One reason Chebychev's Inequality is so loose is that it puts no restrictions on the underlying distribution. With the additional restriction of *unimodality*, we can get tighter bounds and the inequalities of Gauss and Vysochanskiĭ-Petunin. (See Pukelsheim 1994 for details and elementary calculus-based proofs of these inequalities.)

**Theorem 3.8.4 (Gauss Inequality)** *Let $X \sim f$, where $f$ is unimodal with mode $\nu$, and define $\tau^2 = E(X - \nu)^2$. Then*

$$P(|X - \nu| > \varepsilon) \le \begin{cases} \frac{4\tau^2}{9\varepsilon^2} & \text{for all } \varepsilon \ge \sqrt{4/3}\tau \\ 1 - \frac{\varepsilon}{\sqrt{3}\tau} & \text{for all } \varepsilon \le \sqrt{4/3}\tau. \end{cases}$$

Although this is a tighter bound than Chebychev, the dependence on the mode limits its usefulness. The extension of Vysochanskiĭ-Petunin removes this limitation.

**Theorem 3.8.5 (Vysochanskiĭ-Petunin Inequality)** *Let $X \sim f$, where $f$ is unimodal, and define $\xi^2 = E(X - \alpha)^2$ for an arbitrary point $\alpha$. Then*

$$P(|X - \alpha| > \varepsilon) \le \begin{cases} \frac{4\xi^2}{9\varepsilon^2} & \text{for all } \varepsilon \ge \sqrt{8/3}\xi \\ \frac{4\xi^2}{9\varepsilon^2} - \frac{1}{3} & \text{for all } \varepsilon \le \sqrt{8/3}\xi. \end{cases}$$

Pukelsheim points out that taking $\alpha = \mu = E(X)$ and $\varepsilon = 3\sigma$, where $\sigma^2 = \text{Var}(X)$, yields

$$P(|X - \mu| > 3\sigma) \le \frac{4}{81} < .05,$$

the so-called *three-sigma rule*, that the probability is less than 5% that $X$ is more than three standard deviations from the mean of the population.

### 3.8.3 More on Exponential Families

Is the lognormal distribution in the exponential family? The density given in (3.3.21) can be put into the form specified by (3.4.1). Hence, we have put the lognormal into the exponential family.

According to Brown (1986, Section 1.1), to define an exponential family of distributions we start with a nonnegative function $\nu(x)$ and define the set $\mathcal{N}$ by

$$\mathcal{N} = \left\{ \theta : \int_{\mathcal{X}} e^{\theta x} \nu(x) \, dx < \infty \right\}.$$

If we let $\lambda(\theta) = \int_{\mathcal{X}} e^{\theta x} \nu(x) \, dx$, the set of probability densities defined by

$$f(x|\theta) = \frac{e^{\theta x} \nu(x)}{\lambda(\theta)}, \quad x \in \mathcal{X}, \quad \theta \in \mathcal{N},$$

is an *exponential family*. The moment-generating function of $f(x|\theta)$ is

$$M_X(t) = \int_\mathcal{X} e^{tx} f(x|\theta)\, dx = \frac{\lambda(t+\theta)}{\lambda(\theta)}$$

and hence exists by construction. If the parameter space $\Theta$ is equal to the set $\mathcal{N}$, the exponential family is called *full*. Cases where $\Theta$ is a lower-dimensional subset of $\mathcal{N}$ give rise to curved exponential families.

Returning to the lognormal distribution, we know that it does not have an mgf, so it can't satisfy Brown's definition of an exponential family. However, the lognormal satisfies the expectation identities of Theorem 3.4.2 and enjoys the sufficiency properties detailed in Section 6.2.1 (Theorem 6.2.10). For our purposes, these are the major properties that we need and the main reasons for identifying a member of the exponential family. More advanced properties, which we will not investigate here, may need the existence of the mgf.

Chapter 4

# Multiple Random Variables

*"I confess that I have been blind as a mole, but it is better to learn wisdom late than never to learn it at all."*

**Sherlock Holmes**
*The Man with the Twisted Lip*

## 4.1 Joint and Marginal Distributions

In previous chapters, we have discussed probability models and computation of probability for events involving only one random variable. These are called *univariate models*. In this chapter, we discuss probability models that involve more than one random variable—naturally enough, called *multivariate models*.

In an experimental situation, it would be very unusual to observe only the value of one random variable. That is, it would be an unusual experiment in which the total data collected consisted of just one numeric value. For example, consider an experiment designed to gain information about some health characteristics of a population of people. It would be a modest experiment indeed if the only datum collected was the body weight of one person. Rather, the body weights of several people in the population might be measured. These different weights would be observations on different random variables, one for each person measured. Multiple observations could also arise because several physical characteristics were measured on each person. For example, temperature, height, and blood pressure, in addition to weight, might be measured. These observations on different characteristics could also be modeled as observations on different random variables. Thus, we need to know how to describe and use probability models that deal with more than one random variable at a time. For the first several sections we will discuss mainly *bivariate models*, models involving two random variables.

Recall that, in Definition 1.4.1, a (univariate) random variable was defined to be a function from a sample space $S$ into the real numbers. A random vector, consisting of several random variables, is defined similarly.

**Definition 4.1.1** An *n-dimensional random vector* is a function from a sample space $S$ into $\Re^n$, $n$-dimensional Euclidean space.

Suppose, for example, that with each point in a sample space we associate an ordered pair of numbers, that is, a point $(x, y) \in \Re^2$, where $\Re^2$ denotes the plane. Then we have defined a two-dimensional (or bivariate) random vector $(X, Y)$. Example 4.1.2 illustrates this.

**Example 4.1.2 (Sample space for dice)**    Consider the experiment of tossing two fair dice. The sample space for this experiment has 36 equally likely points and was introduced in Example 1.3.10. For example, the sample point $(3, 3)$ denotes the outcome in which both dice show a 3; the sample point $(4, 1)$ denotes the outcome in which the first die shows a 4 and the second die a 1; etc. Now, with each of these 36 points associate two numbers, $X$ and $Y$. Let

$$X = \text{ sum of the two dice} \quad \text{and} \quad Y = |\text{difference of the two dice}|.$$

For the sample point $(3, 3)$, $X = 3+3 = 6$ and $Y = |3 - 3| = 0$. For $(4, 1)$, $X = 5$ and $Y = 3$. These are also the values of $X$ and $Y$ for the sample point $(1, 4)$. For each of the 36 sample points we could compute the values of $X$ and $Y$. In this way we have defined the bivariate random vector $(X, Y)$.

Having defined a random vector $(X, Y)$, we can now discuss probabilities of events that are defined in terms of $(X, Y)$. The probabilities of events defined in terms of $X$ and $Y$ are just defined in terms of the probabilities of the corresponding events in the sample space $S$. What is $P(X = 5 \text{ and } Y = 3)$? You can verify that the only two sample points that yield $X = 5$ and $Y = 3$ are $(4, 1)$ and $(1, 4)$. Thus the event "$X = 5$ and $Y = 3$" will occur if and only if the event $\{(4, 1), (1, 4)\}$ occurs. Since each of the 36 sample points in $S$ is equally likely,

$$P\left(\{(4, 1), (1, 4)\}\right) = \frac{2}{36} = \frac{1}{18}.$$

Thus,

$$P(X = 5 \text{ and } Y = 3) = \frac{1}{18}.$$

Henceforth, we will write $P(X = 5, Y = 3)$ for $P(X = 5 \text{ and } Y = 3)$. Read the comma as "and." Similarly, $P(X = 6, Y = 0) = \frac{1}{36}$ because the only sample point that yields these values of $X$ and $Y$ is $(3, 3)$. For more complicated events, the technique is the same. For example, $P(X = 7, Y \le 4) = \frac{4}{36} = \frac{1}{9}$ because the only four sample points that yield $X = 7$ and $Y \le 4$ are $(4, 3)$, $(3, 4)$, $(5, 2)$, and $(2, 5)$.    ‖

The random vector $(X, Y)$ defined above is called a *discrete random vector* because it has only a countable (in this case, finite) number of possible values. For a discrete random vector, the function $f(x, y)$ defined by $f(x, y) = P(X = x, Y = y)$ can be used to compute any probabilities of events defined in terms of $(X, Y)$.

**Definition 4.1.3**    Let $(X, Y)$ be a discrete bivariate random vector. Then the function $f(x, y)$ from $\Re^2$ into $\Re$ defined by $f(x, y) = P(X = x, Y = y)$ is called the *joint probability mass function* or *joint pmf* of $(X, Y)$. If it is necessary to stress the fact that $f$ is the joint pmf of the vector $(X, Y)$ rather than some other vector, the notation $f_{X,Y}(x, y)$ will be used.

$x$

| | | 2 | 3 | 4 | 5 | 6 | 7 | 8 | 9 | 10 | 11 | 12 |
|---|---|---|---|---|---|---|---|---|---|---|---|---|
| | 0 | $\frac{1}{36}$ | | $\frac{1}{36}$ | | $\frac{1}{36}$ | | $\frac{1}{36}$ | | $\frac{1}{36}$ | | $\frac{1}{36}$ |
| | 1 | | $\frac{1}{18}$ | | $\frac{1}{18}$ | | $\frac{1}{18}$ | | $\frac{1}{18}$ | | $\frac{1}{18}$ | |
| $y$ | 2 | | | $\frac{1}{18}$ | | $\frac{1}{18}$ | | $\frac{1}{18}$ | | $\frac{1}{18}$ | | |
| | 3 | | | | $\frac{1}{18}$ | | $\frac{1}{18}$ | | $\frac{1}{18}$ | | | |
| | 4 | | | | | $\frac{1}{18}$ | | $\frac{1}{18}$ | | | | |
| | 5 | | | | | | $\frac{1}{18}$ | | | | | |

Table 4.1.1. *Values of the joint pmf* $f(x,y)$

The joint pmf of $(X, Y)$ completely defines the probability distribution of the random vector $(X, Y)$, just as the pmf of a discrete univariate random variable completely defines its distribution. For the $(X, Y)$ defined in Example 4.1.2 in terms of the roll of a pair of dice, there are 21 possible values of $(X, Y)$. The value of $f(x, y)$ for each of these 21 possible values is given in Table 4.1.1. Two of these values, $f(5, 3) = \frac{1}{18}$ and $f(6, 0) = \frac{1}{36}$, were computed above and the rest are obtained by similar reasoning. The joint pmf $f(x, y)$ is defined for all $(x, y) \in \Re^2$, not just the 21 pairs in Table 4.1.1. For any other $(x, y)$, $f(x, y) = P(X = x, Y = y) = 0$.

The joint pmf can be used to compute the probability of any event defined in terms of $(X, Y)$. Let $A$ be any subset of $\Re^2$. Then

$$P((X, Y) \in A) = \sum_{(x,y) \in A} f(x, y).$$

Since $(X, Y)$ is discrete, $f(x, y)$ is nonzero for at most a countable number of points $(x, y)$. Thus, the sum can be interpreted as a countable sum even if $A$ contains an uncountable number of points. For example, let $A = \{(x, y) : x = 7 \text{ and } y \leq 4\}$. This is a half-infinite line in $\Re^2$. But from Table 4.1.1 we see that the only $(x, y) \in A$ for which $f(x, y)$ is nonzero are $(x, y) = (7, 1)$ and $(x, y) = (7, 3)$. Thus,

$$P(X = 7, Y \leq 4) = P((X, Y) \in A) = f(7, 1) + f(7, 3) = \frac{1}{18} + \frac{1}{18} = \frac{1}{9}.$$

This, of course, is the same value computed in Example 4.1.2 by considering the definition of $(X, Y)$ and sample points in $S$. It is usually simpler to work with the joint pmf than it is to work with the fundamental definition.

Expectations of functions of random vectors are computed just as with univariate random variables. Let $g(x, y)$ be a real-valued function defined for all possible values $(x, y)$ of the discrete random vector $(X, Y)$. Then $g(X, Y)$ is itself a random variable and its expected value $Eg(X, Y)$ is given by

$$Eg(X, Y) = \sum_{(x,y) \in \Re^2} g(x, y) f(x, y).$$

**Example 4.1.4 (Continuation of Example 4.1.2)** For the $(X, Y)$ whose joint pmf is given in Table 4.1.1, what is the average value of $XY$? Letting $g(x, y) = xy$, we compute $\mathrm{E}XY = \mathrm{E}g(X, Y)$ by computing $xyf(x, y)$ for each of the 21 $(x, y)$ points in Table 4.1.1 and summing these 21 terms. Thus,

$$\mathrm{E}XY = (2)(0)\frac{1}{36} + (4)(0)\frac{1}{36} + \cdots + (8)(4)\frac{1}{18} + (7)(5)\frac{1}{18} = 13\frac{11}{18}. \qquad \|$$

The expectation operator continues to have the properties listed in Theorem 2.2.5 when the random variable $X$ is replaced by the random vector $(X, Y)$. For example, if $g_1(x, y)$ and $g_2(x, y)$ are two functions and $a$, $b$, and $c$ are constants, then

$$\mathrm{E}(ag_1(X, Y) + bg_2(X, Y) + c) = a\mathrm{E}g_1(X, Y) + b\mathrm{E}g_2(X, Y) + c.$$

These properties follow from the properties of sums exactly as in the univariate case (see Exercise 4.2).

The joint pmf for any discrete bivariate random vector $(X, Y)$ must have certain properties. For any $(x, y)$, $f(x, y) \geq 0$ since $f(x, y)$ is a probability. Also, since $(X, Y)$ is certain to be in $\Re^2$,

$$\sum_{(x,y) \in \Re^2} f(x, y) = P\left((X, Y) \in \Re^2\right) = 1.$$

It turns out that any nonnegative function from $\Re^2$ into $\Re$ that is nonzero for at most a countable number of $(x, y)$ pairs and sums to 1 is the joint pmf for some bivariate discrete random vector $(X, Y)$. Thus, by defining $f(x, y)$, we can define a probability model for $(X, Y)$ without ever working with the fundamental sample space $S$.

**Example 4.1.5 (Joint pmf for dice)**  Define $f(x, y)$ by

$$f(0, 0) = f(0, 1) = \frac{1}{6},$$

$$f(1, 0) = f(1, 1) = \frac{1}{3},$$

$$f(x, y) = 0 \quad \text{for any other } (x, y).$$

Then $f(x, y)$ is nonnegative and sums to 1, so $f(x, y)$ is the joint pmf for some bivariate random vector $(X, Y)$. We can use $f(x, y)$ to compute probabilities such as $P(X = Y) = f(0, 0) + f(1, 1) = \frac{1}{2}$. All this can be done without reference to the sample space $S$. Indeed, there are many sample spaces and functions thereon that lead to this joint pmf for $(X, Y)$. Here is one. Let $S$ be the 36-point sample space for the experiment of tossing two fair dice. Let $X = 0$ if the first die shows at most 2 and $X = 1$ if the first die shows more than 2. Let $Y = 0$ if the second die shows an odd number and $Y = 1$ if the second die shows an even number. It is left as Exercise 4.3 to show that this definition leads to the above probability distribution for $(X, Y)$. $\|$

Even if we are considering a probability model for a random vector $(X, Y)$, there may be probabilities or expectations of interest that involve only one of the random variables in the vector. We may wish to know $P(X = 2)$, for instance. The variable $X$ is itself a random variable, in the sense of Chapter 1, and its probability distribution is described by its pmf, namely, $f_X(x) = P(X = x)$. (As mentioned earlier, we now use the subscript to distinguish $f_X(x)$ from the joint pmf $f_{X,Y}(x,y)$.) We now call $f_X(x)$ the *marginal pmf of X* to emphasize the fact that it is the pmf of $X$ but in the context of the probability model that gives the joint distribution of the vector $(X,Y)$. The marginal pmf of $X$ or $Y$ is easily calculated from the joint pmf of $(X,Y)$ as Theorem 4.1.6 indicates.

**Theorem 4.1.6**   *Let $(X,Y)$ be a discrete bivariate random vector with joint pmf $f_{X,Y}(x,y)$. Then the marginal pmfs of $X$ and $Y$, $f_X(x) = P(X = x)$ and $f_Y(y) = P(Y = y)$, are given by*

$$f_X(x) = \sum_{y \in \Re} f_{X,Y}(x,y) \quad and \quad f_Y(y) = \sum_{x \in \Re} f_{X,Y}(x,y).$$

**Proof:** We will prove the result for $f_X(x)$. The proof for $f_Y(y)$ is similar. For any $x \in \Re$, let $A_x = \{(x,y) : -\infty < y < \infty\}$. That is, $A_x$ is the line in the plane with first coordinate equal to $x$. Then, for any $x \in \Re$,

$$f_X(x) = P(X = x)$$
$$= P(X = x, -\infty < Y < \infty) \qquad (P(-\infty < Y < \infty) = 1)$$
$$= P((X,Y) \in A_x) \qquad \text{(definition of } A_x\text{)}$$
$$= \sum_{(x,y) \in A_x} f_{X,Y}(x,y)$$
$$= \sum_{y \in \Re} f_{X,Y}(x,y). \qquad \qquad \Box$$

**Example 4.1.7 (Marginal pmf for dice)**   Using the result of Theorem 4.1.6, we can compute the marginal distributions for $X$ and $Y$ from the joint distribution given in Table 4.1.1. To compute the marginal pmf of $Y$, for each possible value of $Y$ we sum over the possible values of $X$. In this way we obtain

$$f_Y(0) = f_{X,Y}(2,0) + f_{X,Y}(4,0) + f_{X,Y}(6,0)$$
$$+ f_{X,Y}(8,0) + f_{X,Y}(10,0) + f_{X,Y}(12,0)$$
$$= \frac{1}{6}.$$

Similarly, we obtain

$$f_Y(1) = \tfrac{5}{18}, \quad f_Y(2) = \tfrac{2}{9}, \quad f_Y(3) = \tfrac{1}{6}, \quad f_Y(4) = \tfrac{1}{9}, \quad f_Y(5) = \tfrac{1}{18}.$$

Notice that $f_Y(0) + f_Y(1) + f_Y(2) + f_Y(3) + f_Y(4) + f_Y(5) = 1$, as it must, since these are the only six possible values of $Y$.  ‖

The marginal pmf of $X$ or $Y$ is the same as the pmf of $X$ or $Y$ defined in Chapter 1. The marginal pmf of $X$ or $Y$ can be used to compute probabilities or expectations that involve only $X$ or $Y$. But to compute a probability or expectation that simultaneously involves both $X$ and $Y$, we must use the joint pmf of $X$ and $Y$.

**Example 4.1.8 (Dice probabilities)** Using the marginal pmf of $Y$ computed in Example 4.1.7, we can compute

$$P(Y < 3) = f_Y(0) + f_Y(1) + f_Y(2) = \frac{1}{6} + \frac{5}{18} + \frac{2}{9} = \frac{2}{3}.$$

Also,

$$EY^3 = 0^3 f_Y(0) + \cdots + 5^3 f_Y(5) = 20\frac{11}{18}. \qquad \|$$

The marginal distributions of $X$ and $Y$, described by the marginal pmfs $f_X(x)$ and $f_Y(y)$, do not completely describe the joint distribution of $X$ and $Y$. Indeed, there are many different joint distributions that have the same marginal distributions. Thus, it is hopeless to try to determine the joint pmf, $f_{X,Y}(x,y)$, from knowledge of only the marginal pmfs, $f_X(x)$ and $f_Y(y)$. The next example illustrates the point.

**Example 4.1.9 (Same marginals, different joint pmf)** Define a joint pmf by

$$f(0,0) = \tfrac{1}{12}, \quad f(1,0) = \tfrac{5}{12}, \quad f(0,1) = f(1,1) = \tfrac{3}{12},$$
$$f(x,y) = 0 \quad \text{for all other values.}$$

The marginal pmf of $Y$ is $f_Y(0) = f(0,0) + f(1,0) = \frac{1}{2}$ and $f_Y(1) = f(0,1) + f(1,1) = \frac{1}{2}$. The marginal pmf of $X$ is $f_X(0) = \frac{1}{3}$ and $f_X(1) = \frac{2}{3}$. Now check that for the joint pmf given in Example 4.1.5, which is obviously different from the one given here, the marginal pmfs of both $X$ and $Y$ are exactly the same as the ones just computed. Thus, we cannot determine what the joint pmf is if we know only the marginal pmfs. The joint pmf tells us additional information about the distribution of $(X,Y)$ that is not found in the marginal distributions. $\qquad \|$

To this point we have discussed discrete bivariate random vectors. We can also consider random vectors whose components are continuous random variables. The probability distribution of a continuous random vector is usually described using a density function, as in the univariate case.

**Definition 4.1.10** A function $f(x,y)$ from $\Re^2$ into $\Re$ is called a *joint probability density function* or *joint pdf of the continuous bivariate random vector* $(X,Y)$ if, for every $A \subset \Re^2$,

$$P((X,Y) \in A) = \int_A \int f(x,y)\, dx\, dy.$$

A joint pdf is used just like a univariate pdf except now the integrals are double integrals over sets in the plane. The notation $\int \int_A$ simply means that the limits of integration are set so that the function is integrated over all $(x,y) \in A$. Expectations of functions of continuous random vectors are defined as in the discrete case with integrals replacing sums and the pdf replacing the pmf. That is, if $g(x,y)$ is a real-valued function, then the *expected value of* $g(X,Y)$ is defined to be

(4.1.2)                    $$Eg(X,Y) = \int_{-\infty}^{\infty} \int_{-\infty}^{\infty} g(x,y) f(x,y)\, dx\, dy.$$

It is important to realize that the joint pdf *is* defined for all $(x, y) \in \Re^2$. The pdf may equal 0 on a large set $A$ if $P((X, Y) \in A) = 0$ but the pdf *is* defined for the points in $A$.

The *marginal probability density functions* of $X$ and $Y$ are also defined as in the discrete case with integrals replacing sums. The marginal pdfs may be used to compute probabilities or expectations that involve only $X$ or $Y$. Specifically, the marginal pdfs of $X$ and $Y$ are given by

(4.1.3)
$$f_X(x) = \int_{-\infty}^{\infty} f(x, y)\, dy, \quad -\infty < x < \infty,$$

$$f_Y(y) = \int_{-\infty}^{\infty} f(x, y)\, dx, \quad -\infty < y < \infty.$$

Any function $f(x, y)$ satisfying $f(x, y) \geq 0$ for all $(x, y) \in \Re^2$ and

$$1 = \int_{-\infty}^{\infty} \int_{-\infty}^{\infty} f(x, y)\, dx\, dy$$

is the joint pdf of some continuous bivariate random vector $(X, Y)$. All of these concepts regarding joint pdfs are illustrated in the following two examples.

**Example 4.1.11 (Calculating joint probabilities–I)** Define a joint pdf by

$$f(x, y) = \begin{cases} 6xy^2 & 0 < x < 1 \text{ and } 0 < y < 1 \\ 0 & \text{otherwise.} \end{cases}$$

(Henceforth, it will be understood that $f(x, y) = 0$ for $(x, y)$ values not specifically mentioned in the definition.) First, we might check that $f(x, y)$ is indeed a joint pdf. That $f(x, y) \geq 0$ for all $(x, y)$ in the defined range is fairly obvious. To compute the integral of $f(x, y)$ over the whole plane, note that, since $f(x, y)$ is 0 except on the unit square, the integral over the plane is the same as the integral over the square. Thus we have

$$\int_{-\infty}^{\infty} \int_{-\infty}^{\infty} f(x, y)\, dx\, dy = \int_{0}^{1} \int_{0}^{1} 6xy^2\, dx\, dy = \int_{0}^{1} 3x^2 y^2 \Big|_{0}^{1}\, dy$$

$$= \int_{0}^{1} 3y^2\, dy = y^3 \Big|_{0}^{1} = 1.$$

Now, consider calculating a probability such as $P(X + Y \geq 1)$. Letting $A = \{(x, y) : x + y \geq 1\}$, we can re-express this as $P((X, Y) \in A)$. From Definition 4.1.10, to calculate the probability we integrate the joint pdf over the set $A$. But the joint pdf is 0 except on the unit square. So integrating over $A$ is the same as integrating over only that part of $A$ which is in the unit square. The set $A$ is a half-plane in the northeast part of the plane, and the part of $A$ in the unit square is the triangular region bounded by the lines $x = 1$, $y = 1$, and $x + y = 1$. We can write

$$A = \{(x, y) : x + y \geq 1, 0 < x < 1, 0 < y < 1\}$$
$$= \{(x, y) : x \geq 1 - y, 0 < x < 1, 0 < y < 1\}$$
$$= \{(x, y) : 1 - y \leq x < 1, 0 < y < 1\}.$$

This gives us the limits of integration we need to calculate the probability. We have

$$P(X + Y \geq 1) = \int_A \int f(x, y)\, dx\, dy = \int_0^1 \int_{1-y}^1 6xy^2\, dx\, dy = \frac{9}{10}.$$

Using (4.1.3), we can calculate the marginal pdf of $X$ or $Y$. For example, to calculate $f_X(x)$, we note that for $x \geq 1$ or $x \leq 0$, $f(x, y) = 0$ for all values of $y$. Thus for $x \geq 1$ or $x \leq 0$,

$$f_X(x) = \int_{-\infty}^{\infty} f(x, y)\, dy = 0.$$

For $0 < x < 1$, $f(x, y)$ is nonzero only if $0 < y < 1$. Thus for $0 < x < 1$,

$$f_X(x) = \int_{-\infty}^{\infty} f(x, y)\, dy = \int_0^1 6xy^2\, dy = 2xy^3 \Big|_0^1 = 2x.$$

This marginal pdf of $X$ can now be used to calculate probabilities involving only $X$. For example,

$$P\left(\frac{1}{2} < X < \frac{3}{4}\right) = \int_{\frac{1}{2}}^{\frac{3}{4}} 2x\, dx = \frac{5}{16}. \qquad \|$$

**Example 4.1.12 (Calculating joint probabilities–II)**  As another example of a joint pdf, let $f(x, y) = e^{-y}, 0 < x < y < \infty$. Although $e^{-y}$ does not depend on $x$, $f(x, y)$ certainly is a function of $x$ since the set where $f(x, y)$ is nonzero depends on $x$. This is made more obvious by using an indicator function to write

$$f(x, y) = e^{-y} I_{\{(u,v):0<u<v<\infty\}}(x, y).$$

To calculate $P(X + Y \geq 1)$, we could integrate the joint pdf over the region that is the intersection of the set $A = \{(x, y) : x + y \geq 1\}$ and the set where $f(x, y)$ is nonzero. Graph these sets and notice that this region is an unbounded region (lighter shading in Figure 4.1.1) with three sides given by the lines $x = y, x + y = 1$, and $x = 0$. To integrate over this region we would have to break the region into at least two parts in order to write the appropriate limits of integration.

The integration is easier over the intersection of the set $B = \{(x, y) : x + y < 1\}$ and the set where $f(x, y)$ is nonzero, the triangular region (darker shading in Figure 4.1.1) bounded by the lines $x = y, x + y = 1$, and $x = 0$. Thus

$$P(X + Y \geq 1) = 1 - P(X + Y < 1) = 1 - \int_0^{\frac{1}{2}} \int_x^{1-x} e^{-y}\, dy\, dx$$

$$= 1 - \int_0^{\frac{1}{2}} \left(e^{-x} - e^{-(1-x)}\right) dx = 2e^{-1/2} - e^{-1}.$$

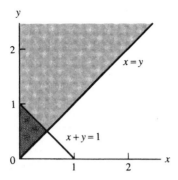

Figure 4.1.1. *Regions for Example 4.1.12*

This illustrates that it is almost always helpful to graph the sets of interest in determining the appropriate limits of integration for problems such as this.          ‖

The joint probability distribution of $(X, Y)$ can be completely described with the *joint cdf* (cumulative distribution function) rather than with the joint pmf or joint pdf. The joint cdf is the function $F(x, y)$ defined by

$$F(x, y) = P(X \le x, Y \le y)$$

for all $(x, y) \in \Re^2$. The joint cdf is usually not very handy to use for a discrete random vector. But for a continuous bivariate random vector we have the important relationship, as in the univariate case,

$$F(x, y) = \int_{-\infty}^{x} \int_{-\infty}^{y} f(s, t) \, dt \, ds.$$

From the bivariate Fundamental Theorem of Calculus, this implies that

(4.1.4) $$\frac{\partial^2 F(x, y)}{\partial x \, \partial y} = f(x, y)$$

at continuity points of $f(x, y)$. This relationship is useful in situations where an expression for $F(x, y)$ can be found. The mixed partial derivative can be computed to find the joint pdf.

## 4.2 Conditional Distributions and Independence

Oftentimes when two random variables, $(X, Y)$, are observed, the values of the two variables are related. For example, suppose that, in sampling from a human population, $X$ denotes a person's height and $Y$ denotes the same person's weight. Surely we would think it more likely that $Y > 200$ pounds if we were told that $X = 73$ inches than if we were told that $X = 41$ inches. Knowledge about the value of $X$ gives us some information about the value of $Y$ even if it does not tell us the value of $Y$ exactly. Conditional probabilities regarding $Y$ given knowledge of the $X$ value can be computed using the joint distribution of $(X, Y)$. Sometimes, however, knowledge about $X$ gives us no information about $Y$. We will discuss these topics concerning conditional probabilities in this section.

If $(X, Y)$ is a discrete random vector, then a conditional probability of the form $P(Y = y | X = x)$ is interpreted exactly as in Definition 1.3.2. For a countable (maybe finite) number of $x$ values, $P(X = x) > 0$. For these values of $x$, $P(Y = y | X = x)$ is simply $P(X = x, Y = y)/P(X = x)$, according to the definition. The event $\{Y = y\}$ is the event $A$ in the formula and the event $\{X = x\}$ is the event $B$. For a fixed value of $x$, $P(Y = y | X = x)$ could be computed for all possible values of $y$. In this way the probability of various values of $y$ could be assessed given the knowledge that $X = x$ was observed. This computation can be simplified by noting that in terms of the joint and marginal pmfs of $X$ and $Y$, the above probabilities are $P(X = x, Y = y) = f(x, y)$ and $P(X = x) = f_X(x)$. This leads to the following definition.

**Definition 4.2.1**  Let $(X, Y)$ be a discrete bivariate random vector with joint pmf $f(x, y)$ and marginal pmfs $f_X(x)$ and $f_Y(y)$. For any $x$ such that $P(X = x) = f_X(x) > 0$, the *conditional pmf of Y given that $X = x$* is the function of $y$ denoted by $f(y|x)$ and defined by

$$f(y|x) = P(Y = y | X = x) = \frac{f(x, y)}{f_X(x)}.$$

For any $y$ such that $P(Y = y) = f_Y(y) > 0$, the *conditional pmf of X given that $Y = y$* is the function of $x$ denoted by $f(x|y)$ and defined by

$$f(x|y) = P(X = x | Y = y) = \frac{f(x, y)}{f_Y(y)}.$$

Since we have called $f(y|x)$ a pmf, we should verify that this function of $y$ does indeed define a pmf for a random variable. First, $f(y|x) \geq 0$ for every $y$ since $f(x, y) \geq 0$ and $f_X(x) > 0$. Second,

$$\sum_y f(y|x) = \frac{\sum_y f(x, y)}{f_X(x)} = \frac{f_X(x)}{f_X(x)} = 1.$$

Thus, $f(y|x)$ is indeed a pmf and can be used in the usual way to compute probabilities involving $Y$ given the knowledge that $X = x$ occurred.

**Example 4.2.2 (Calculating conditional probabilities)**  Define the joint pmf of $(X, Y)$ by

$$f(0, 10) = f(0, 20) = \tfrac{2}{18}, \quad f(1, 10) = f(1, 30) = \tfrac{3}{18},$$
$$f(1, 20) = \tfrac{4}{18}, \quad \text{and} \quad f(2, 30) = \tfrac{4}{18}.$$

We can use Definition 4.2.1 to compute the conditional pmf of $Y$ given $X$ for each of the possible values of $X$, $x = 0, 1, 2$. First, the marginal pmf of $X$ is

$$f_X(0) = f(0, 10) + f(0, 20) = \frac{4}{18},$$

$$f_X(1) = f(1, 10) + f(1, 20) + f(1, 30) = \frac{10}{18},$$

$$f_X(2) = f(2, 30) = \frac{4}{18}.$$

For $x = 0$, $f(0, y)$ is positive only for $y = 10$ and $y = 20$. Thus $f(y|0)$ is positive only for $y = 10$ and $y = 20$, and

$$f(10|0) = \frac{f(0, 10)}{f_X(0)} = \frac{\frac{2}{18}}{\frac{4}{18}} = \frac{1}{2},$$

$$f(20|0) = \frac{f(0, 20)}{f_X(0)} = \frac{1}{2}.$$

That is, given the knowledge that $X = 0$, the conditional probability distribution for $Y$ is the discrete distribution that assigns probability $\frac{1}{2}$ to each of the two points $y = 10$ and $y = 20$.

For $x = 1$, $f(y|1)$ is positive for $y = 10$, 20, and 30, and

$$f(10|1) = f(30|1) = \frac{\frac{3}{18}}{\frac{10}{18}} = \frac{3}{10},$$

$$f(20|1) = \frac{\frac{4}{18}}{\frac{10}{18}} = \frac{4}{10},$$

and for $x = 2$,

$$f(30|2) = \frac{\frac{4}{18}}{\frac{4}{18}} = 1.$$

The latter result reflects a fact that is also apparent from the joint pmf. If we know that $X = 2$, then we know that $Y$ must be 30.

Other conditional probabilities can be computed using these conditional pmfs. For example,

$$P(Y > 10|X = 1) = f(20|1) + f(30|1) = \tfrac{7}{10}$$

or

$$P(Y > 10|X = 0) = f(20|0) = \tfrac{1}{2}. \qquad \|$$

If $X$ and $Y$ are continuous random variables, then $P(X = x) = 0$ for every value of $x$. To compute a conditional probability such as $P(Y > 200|X = 73)$, Definition 1.3.2 cannot be used since the denominator, $P(X = 73)$, is 0. Yet in actuality a value of $X$ is observed. If, to the limit of our measurement, we see $X = 73$, this knowledge might give us information about $Y$ (as the height and weight example at the beginning of this section indicated). It turns out that the appropriate way to define a conditional probability distribution for $Y$ given $X = x$, when $X$ and $Y$ are both continuous, is analogous to the discrete case with pdfs replacing pmfs (see Miscellanea 4.9.3).

**Definition 4.2.3**   Let $(X, Y)$ be a continuous bivariate random vector with joint pdf $f(x, y)$ and marginal pdfs $f_X(x)$ and $f_Y(y)$. For any $x$ such that $f_X(x) > 0$, the *conditional pdf of Y given that X = x* is the function of $y$ denoted by $f(y|x)$ and defined by

$$f(y|x) = \frac{f(x, y)}{f_X(x)}.$$

For any $y$ such that $f_Y(y) > 0$, the *conditional pdf of X given that Y = y* is the function of $x$ denoted by $f(x|y)$ and defined by

$$f(x|y) = \frac{f(x, y)}{f_Y(y)}.$$

To verify that $f(x|y)$ and $f(y|x)$ are indeed pdfs, the same steps can be used as in the earlier verification that Definition 4.2.1 had defined true pmfs with integrals now replacing sums.

In addition to their usefulness for calculating probabilities, the conditional pdfs or pmfs can also be used to calculate expected values. Just remember that $f(y|x)$ as a function of $y$ is a pdf or pmf and use it in the same way that we have previously used unconditional pdfs or pmfs. If $g(Y)$ is a function of $Y$, then the *conditional expected value of g(Y) given that X = x* is denoted by $\mathrm{E}(g(Y)|x)$ and is given by

$$\mathrm{E}(g(Y)|x) = \sum_y g(y) f(y|x) \quad \text{and} \quad \mathrm{E}(g(Y)|x) = \int_{-\infty}^{\infty} g(y) f(y|x) \, dy$$

in the discrete and continuous cases, respectively. The conditional expected value has all of the properties of the usual expected value listed in Theorem 2.2.5. Moreover, $\mathrm{E}(Y|X)$ provides the best guess at $Y$ based on knowledge of $X$, extending the result in Example 2.2.6. (See Exercise 4.13.)

**Example 4.2.4 (Calculating conditional pdfs)**   As in Example 4.1.12, let the continuous random vector $(X, Y)$ have joint pdf $f(x, y) = e^{-y}$, $0 < x < y < \infty$. Suppose we wish to compute the conditional pdf of $Y$ given $X = x$. The marginal pdf of $X$ is computed as follows. If $x \leq 0$, $f(x, y) = 0$ for all values of $y$, so $f_X(x) = 0$. If $x > 0$, $f(x, y) > 0$ only if $y > x$. Thus

$$f_X(x) = \int_{-\infty}^{\infty} f(x, y) \, dy = \int_x^{\infty} e^{-y} \, dy = e^{-x}.$$

Thus, marginally, $X$ has an exponential distribution. From Definition 4.2.3, the conditional distribution of $Y$ given $X = x$ can be computed for any $x > 0$ (since these are the values for which $f_X(x) > 0$). For any such $x$,

$$f(y|x) = \frac{f(x, y)}{f_X(x)} = \frac{e^{-y}}{e^{-x}} = e^{-(y-x)}, \quad \text{if } y > x,$$

and

$$f(y|x) = \frac{f(x, y)}{f_X(x)} = \frac{0}{e^{-x}} = 0, \quad \text{if } y \leq x.$$

Thus, given $X = x$, $Y$ has an exponential distribution, where $x$ is the location parameter in the distribution of $Y$ and $\beta = 1$ is the scale parameter. The conditional distribution of $Y$ is different for every value of $x$. It then follows that

$$E(Y|X = x) = \int_x^\infty y e^{-(y-x)}\, dy = 1 + x.$$

The variance of the probability distribution described by $f(y|x)$ is called the *conditional variance of $Y$ given $X = x$*. Using the notation $\text{Var}(Y|x)$ for this, we have, using the ordinary definition of variance,

$$\text{Var}(Y|x) = E(Y^2|x) - (E(Y|x))^2.$$

Applying this definition to our example, we obtain

$$\text{Var}(Y|x) = \int_x^\infty y^2 e^{-(y-x)}\, dy - \left( \int_x^\infty y e^{-(y-x)}\, dy \right)^2 = 1.$$

In this case the conditional variance of $Y$ given $X = x$ is the same for all values of $x$. In other situations, however, it may be different for different values of $x$. This conditional variance might be compared to the unconditional variance of $Y$. The marginal distribution of $Y$ is gamma$(2, 1)$, which has $\text{Var}\, Y = 2$. Given the knowledge that $X = x$, the variability in $Y$ is considerably reduced.                    ‖

A physical situation for which the model in Example 4.2.4 might be used is this. Suppose we have two light bulbs. The lengths of time each will burn are random variables denoted by $X$ and $Z$. The lifelengths $X$ and $Z$ are independent and both have pdf $e^{-x}, x > 0$. The first bulb will be turned on. As soon as it burns out, the second bulb will be turned on. Now consider observing $X$, the time when the first bulb burns out, and $Y = X + Z$, the time when the second bulb burns out. Given that $X = x$ is when the first burned out and the second is started, $Y = Z + x$. This is like Example 3.5.3. The value $x$ is acting as a location parameter, and the pdf of $Y$, in this case the *conditional* pdf of $Y$ given $X = x$, is $f(y|x) = f_Z(y - x) = e^{-(y-x)}, y > x$.

The conditional distribution of $Y$ given $X = x$ is possibly a different probability distribution for each value of $x$. Thus we really have a family of probability distributions for $Y$, one for each $x$. When we wish to describe this entire family, we will use the phrase "the distribution of $Y|X$." If, for example, $X$ is a positive integer-valued random variable and the conditional distribution of $Y$ given $X = x$ is binomial$(x, p)$, then we might say the distribution of $Y|X$ is binomial$(X, p)$ or write $Y|X \sim$ binomial$(X, p)$. Whenever we use the symbol $Y|X$ or have a random variable as the parameter of a probability distribution, we are describing the family of conditional probability distributions. Joint pdfs or pmfs are sometimes defined by specifying the conditional $f(y|x)$ and the marginal $f_X(x)$. Then the definition yields $f(x, y) = f(y|x)f_X(x)$. These types of models are discussed more in Section 4.4.

Notice also that $E(g(Y)|x)$ is a function of $x$. That is, for each value of $x$, $E(g(Y)|x)$ is a real number obtained by computing the appropriate integral or sum. Thus, $E(g(Y)|X)$ is a random variable whose value depends on the value of $X$. If $X = x$, the value of the random variable $E(g(Y)|X)$ is $E(g(Y)|x)$. Thus, in Example 4.2.4, we can write $E(Y|X) = 1 + X$.

In all the previous examples, the conditional distribution of $Y$ given $X = x$ was different for different values of $x$. In some situations, the knowledge that $X = x$ does not give us any more information about $Y$ than what we already had. This important relationship between $X$ and $Y$ is called *independence*. Just as with independent events in Chapter 1, it is more convenient to define independence in a symmetric fashion and then derive conditional properties like those we just mentioned. This we now do.

**Definition 4.2.5**   Let $(X, Y)$ be a bivariate random vector with joint pdf or pmf $f(x, y)$ and marginal pdfs or pmfs $f_X(x)$ and $f_Y(y)$. Then $X$ and $Y$ are called *independent random variables* if, for every $x \in \Re$ and $y \in \Re$,

$$(4.2.1) \qquad\qquad f(x, y) = f_X(x) f_Y(y).$$

If $X$ and $Y$ are independent, the conditional pdf of $Y$ given $X = x$ is

$$f(y|x) = \frac{f(x, y)}{f_X(x)} \qquad\qquad \text{(definition)}$$

$$= \frac{f_X(x) f_Y(y)}{f_X(x)} \qquad\qquad \text{(from (4.2.1))}$$

$$= f_Y(y),$$

regardless of the value of $x$. Thus, for any $A \subset \Re$ and $x \in \Re, P(Y \in A|x) = \int_A f(y|x)\, dy = \int_A f_Y(y)\, dy = P(Y \in A)$. The knowledge that $X = x$ gives us no additional information about $Y$.

Definition 4.2.5 is used in two different ways. We might start with a joint pdf or pmf and then check whether $X$ and $Y$ are independent. To do this we must verify that (4.2.1) is true for every value of $x$ and $y$. Or we might wish to define a model in which $X$ and $Y$ are independent. Consideration of what $X$ and $Y$ represent might indicate that knowledge that $X = x$ should give us no information about $Y$. In this case we could specify the marginal distributions of $X$ and $Y$ and then define the joint distribution as the product given in (4.2.1).

**Example 4.2.6 (Checking independence–I)**   Consider the discrete bivariate random vector $(X, Y)$, with joint pmf given by

$$f(10, 1) = f(20, 1) = f(20, 2) = \tfrac{1}{10},$$
$$f(10, 2) = f(10, 3) = \tfrac{1}{5}, \quad \text{and} \quad f(20, 3) = \tfrac{3}{10}.$$

The marginal pmfs are easily calculated to be

$$f_X(10) = f_X(20) = \frac{1}{2} \quad \text{and} \quad f_Y(1) = \frac{1}{5},\ f_Y(2) = \frac{3}{10},\ \text{and } f_Y(3) = \frac{1}{2}.$$

The random variables $X$ and $Y$ are not independent because (4.2.1) is not true for every $x$ and $y$. For example,

$$f(10, 3) = \frac{1}{5} \neq \frac{1}{2}\frac{1}{2} = f_X(10) f_Y(3).$$

The relationship (4.2.1) must hold for every choice of $x$ and $y$ if $X$ and $Y$ are to be independent. Note that $f(10,1) = \frac{1}{10} = \frac{1}{2}\frac{1}{5} = f_X(10)f_Y(1)$. That (4.2.1) holds for *some* values of $x$ and $y$ does not ensure that $X$ and $Y$ are independent. All values must be checked.                                                                                 ‖

The verification that $X$ and $Y$ are independent by direct use of (4.2.1) would require the knowledge of $f_X(x)$ and $f_Y(y)$. The following lemma makes the verification somewhat easier.

**Lemma 4.2.7** *Let $(X,Y)$ be a bivariate random vector with joint pdf or pmf $f(x,y)$. Then $X$ and $Y$ are independent random variables if and only if there exist functions $g(x)$ and $h(y)$ such that, for every $x \in \Re$ and $y \in \Re$,*

$$f(x,y) = g(x)h(y).$$

**Proof:** The "only if" part is proved by defining $g(x) = f_X(x)$ and $h(y) = f_Y(y)$ and using (4.2.1). To prove the "if" part for continuous random variables, suppose that $f(x,y) = g(x)h(y)$. Define

$$\int_{-\infty}^{\infty} g(x)\,dx = c \quad \text{and} \quad \int_{-\infty}^{\infty} h(y)\,dy = d,$$

where the constants $c$ and $d$ satisfy

$$
\begin{aligned}
cd &= \left( \int_{-\infty}^{\infty} g(x)\,dx \right)\left( \int_{-\infty}^{\infty} h(y)\,dy \right) \\
(4.2.2) \qquad &= \int_{-\infty}^{\infty}\int_{-\infty}^{\infty} g(x)h(y)\,dx\,dy \\
&= \int_{-\infty}^{\infty}\int_{-\infty}^{\infty} f(x,y)\,dx\,dy \\
&= 1. \qquad\qquad\qquad\qquad\qquad (f(x,y) \text{ is a joint pdf})
\end{aligned}
$$

Furthermore, the marginal pdfs are given by

(4.2.3)

$$f_X(x) = \int_{-\infty}^{\infty} g(x)h(y)\,dy = g(x)d \quad \text{and} \quad f_Y(y) = \int_{-\infty}^{\infty} g(x)h(y)\,dx = h(y)c.$$

Thus, using (4.2.2) and (4.2.3), we have

$$f(x,y) = g(x)h(y) = g(x)h(y)cd = f_X(x)f_Y(y),$$

showing that $X$ and $Y$ are independent. Replacing integrals with sums proves the lemma for discrete random vectors.                                                              □

**Example 4.2.8 (Checking independence–II)** Consider the joint pdf $f(x,y) = \frac{1}{384}x^2y^4e^{-y-(x/2)}$, $x > 0$ and $y > 0$. If we define

$$g(x) = \begin{cases} x^2e^{-x/2} & x > 0 \\ 0 & x \leq 0 \end{cases} \quad \text{and} \quad h(y) = \begin{cases} y^4e^{-y}/384 & y > 0 \\ 0 & y \leq 0, \end{cases}$$

then $f(x,y) = g(x)h(y)$ for all $x \in \Re$ and all $y \in \Re$. By Lemma 4.2.7, we conclude that $X$ and $Y$ are independent random variables. We do not have to compute marginal pdfs.                                                                                                    ‖

If $X$ and $Y$ are independent random variables, then from $(4.2.1)$ it is clear that $f(x,y) > 0$ on the set $\{(x,y) : x \in A \text{ and } y \in B\}$, where $A = \{x : f_X(x) > 0\}$ and $B = \{y : f_Y(y) > 0\}$. A set of this form is called a cross-product and is usually denoted by $A \times B$. Membership in a cross-product can be checked by considering the $x$ and $y$ values separately. If $f(x,y)$ is a joint pdf or pmf and the set where $f(x,y) > 0$ is not a cross-product, then the random variables $X$ and $Y$ with joint pdf or pmf $f(x,y)$ are not independent. In Example 4.2.4, the set $0 < x < y < \infty$ is not a cross-product. To check membership in this set we must check that not only $0 < x < \infty$ and $0 < y < \infty$ but also $x < y$. Thus the random variables in Example 4.2.4 are not independent. Example 4.2.2 gives an example of a joint pmf that is positive on a set that is not a cross-product.

**Example 4.2.9 (Joint probability model)**  As an example of using independence to define a joint probability model, consider this situation. A student from an elementary school in Kansas City is randomly selected and $X = $ the number of living parents of the student is recorded. Suppose the marginal distribution of $X$ is

$$f_X(0) = .01, \quad f_X(1) = .09, \quad \text{and} \quad f_X(2) = .90.$$

A retiree from Sun City is randomly selected and $Y = $ the number of living parents of the retiree is recorded. Suppose the marginal distribution of $Y$ is

$$f_Y(0) = .70, \quad f_Y(1) = .25, \quad \text{and} \quad f_Y(2) = .05.$$

It seems reasonable to assume that these two random variables are independent. Knowledge of the number of parents of the student tells us nothing about the number of parents of the retiree. The only joint distribution of $X$ and $Y$ that reflects this independence is the one defined by $(4.2.1)$. Thus, for example,

$$f(0,0) = f_X(0)f_Y(0) = .0070 \quad \text{and} \quad f(0,1) = f_X(0)f_Y(1) = .0025.$$

This joint distribution can be used to calculate quantities such as

$$P(X = Y) = f(0,0) + f(1,1) + f(2,2)$$
$$= (.01)(.70) + (.09)(.25) + (.90)(.05) = .0745. \qquad ‖$$

Certain probabilities and expectations are easy to calculate if $X$ and $Y$ are independent, as the next theorem indicates.

**Theorem 4.2.10**  *Let $X$ and $Y$ be independent random variables.*

**a.** *For any $A \subset \Re$ and $B \subset \Re$, $P(X \in A, Y \in B) = P(X \in A)P(Y \in B)$; that is, the events $\{X \in A\}$ and $\{Y \in B\}$ are independent events.*

**b.** *Let $g(x)$ be a function only of $x$ and $h(y)$ be a function only of $y$. Then*

$$\mathrm{E}\left(g(X)h(Y)\right) = \left(\mathrm{E}g(X)\right)\left(\mathrm{E}h(Y)\right).$$

**Proof:** For continuous random variables, part (b) is proved by noting that

$$
\begin{aligned}
\mathrm{E}\left(g(X)h(Y)\right) &= \int_{-\infty}^{\infty}\int_{-\infty}^{\infty} g(x)h(y)f(x,y)\,dx\,dy \\
&= \int_{-\infty}^{\infty}\int_{-\infty}^{\infty} g(x)h(y)f_X(x)f_Y(y)\,dx\,dy \qquad \text{(by (4.2.1))} \\
&= \int_{-\infty}^{\infty} h(y)f_Y(y)\int_{-\infty}^{\infty} g(x)f_X(x)\,dx\,dy \\
&= \left(\int_{-\infty}^{\infty} g(x)f_X(x)\,dx\right)\left(\int_{-\infty}^{\infty} h(y)f_Y(y)\,dy\right) \\
&= \left(\mathrm{E}g(X)\right)\left(\mathrm{E}h(Y)\right).
\end{aligned}
$$

The result for discrete random variables is proved by replacing integrals by sums. Part (a) can be proved by a series of steps similar to those above or by the following argument. Let $g(x)$ be the indicator function of the set $A$. Let $h(y)$ be the indicator function of the set $B$. Note that $g(x)h(y)$ is the indicator function of the set $C \subset \Re^2$ defined by $C = \{(x,y) : x \in A, y \in B\}$. Also note that for an indicator function such as $g(x)$, $\mathrm{E}g(X) = P(X \in A)$. Thus using the expectation equality just proved, we have

$$
\begin{aligned}
P(X \in A, Y \in B) &= P\left((X,Y) \in C\right) = \mathrm{E}\left(g(X)h(Y)\right) \\
&= \left(\mathrm{E}g(X)\right)\left(\mathrm{E}h(Y)\right) = P(X \in A)P(Y \in B). \qquad \Box
\end{aligned}
$$

**Example 4.2.11 (Expectations of independent variables)**  Let $X$ and $Y$ be independent exponential(1) random variables. From Theorem 4.2.10 we have

$$P(X \geq 4, Y < 3) = P(X \geq 4)P(Y < 3) = e^{-4}(1 - e^{-3}).$$

Letting $g(x) = x^2$ and $h(y) = y$, we see that

$$\mathrm{E}\left(X^2 Y\right) = \left(\mathrm{E}X^2\right)\left(\mathrm{E}Y\right) = \left(\mathrm{Var}\,X + (\mathrm{E}X)^2\right)\mathrm{E}Y = (1 + 1^2)1 = 2. \qquad \|$$

The following result concerning sums of independent random variables is a simple consequence of Theorem 4.2.10.

**Theorem 4.2.12** *Let $X$ and $Y$ be independent random variables with moment generating functions $M_X(t)$ and $M_Y(t)$. Then the moment generating function of the random variable $Z = X + Y$ is given by*

$$M_Z(t) = M_X(t)M_Y(t).$$

**Proof:** Using the definition of the mgf and Theorem 4.2.10, we have

$$M_Z(t) = \mathrm{E}e^{tZ} = \mathrm{E}e^{t(X+Y)} = \mathrm{E}(e^{tX}e^{tY}) = (\mathrm{E}e^{tX})(\mathrm{E}e^{tY}) = M_X(t)M_Y(t). \qquad \Box$$

**Example 4.2.13 (Mgf of a sum of normal variables)**  Sometimes Theorem 4.2.12 can be used to easily derive the distribution of $Z$ from knowledge of the distribution of $X$ and $Y$. For example, let $X \sim n(\mu, \sigma^2)$ and $Y \sim n(\gamma, \tau^2)$ be independent normal random variables. From Exercise 2.33, the mgfs of $X$ and $Y$ are

$$M_X(t) = \exp(\mu t + \sigma^2 t^2/2) \quad \text{and} \quad M_Y(t) = \exp(\gamma t + \tau^2 t^2/2).$$

Thus, from Theorem 4.2.12, the mgf of $Z = X + Y$ is

$$M_Z(t) = M_X(t)M_Y(t) = \exp\left((\mu + \gamma)t + (\sigma^2 + \tau^2)t^2/2\right).$$

This is the mgf of a normal random variable with mean $\mu + \gamma$ and variance $\sigma^2 + \tau^2$. This result is important enough to be stated as a theorem.                    ‖

**Theorem 4.2.14**  *Let $X \sim n(\mu, \sigma^2)$ and $Y \sim n(\gamma, \tau^2)$ be independent normal random variables. Then the random variable $Z = X + Y$ has a $n(\mu + \gamma, \sigma^2 + \tau^2)$ distribution.*

If $f(x, y)$ is the joint pdf for the continuous random vector $(X, Y)$, (4.2.1) may fail to hold on a set $A$ of $(x, y)$ values for which $\int_A \int dx\, dy = 0$. In such a case $X$ and $Y$ are still called independent random variables. This reflects the fact that two pdfs that differ only on a set such as $A$ define the same probability distribution for $(X, Y)$. To see this, suppose $f(x, y)$ and $f^*(x, y)$ are two pdfs that are equal everywhere except on a set $A$ for which $\int_A \int dx\, dy = 0$. Let $(X, Y)$ have pdf $f(x, y)$, let $(X^*, Y^*)$ have pdf $f^*(x, y)$, and let $B$ be any subset of $\Re^2$. Then

$$P((X, Y) \in B) = \int_B \int f(x, y)\, dx\, dy$$

$$= \int_{B \cap A^c} \int f(x, y)\, dx\, dy$$

$$= \int_{B \cap A^c} \int f^*(x, y)\, dx\, dy$$

$$= \int_B \int f^*(x, y)\, dx\, dy = P((X^*, Y^*) \in B).$$

Thus $(X, Y)$ and $(X^*, Y^*)$ have the same probability distribution. So, for example, $f(x, y) = e^{-x-y}$, $x > 0$ and $y > 0$, is a pdf for two independent exponential random variables and satisfies (4.2.1). But, $f^*(x, y)$, which is equal to $f(x, y)$ except that $f^*(x, y) = 0$ if $x = y$, is also the pdf for two independent exponential random variables even though (4.2.1) is not true on the set $A = \{(x, x) : x > 0\}$.

## 4.3 Bivariate Transformations

In Section 2.1, methods of finding the distribution of a function of a random variable were discussed. In this section we extend these ideas to the case of bivariate random vectors.

Let $(X, Y)$ be a bivariate random vector with a known probability distribution. Now consider a new bivariate random vector $(U, V)$ defined by $U = g_1(X, Y)$ and $V =$

$g_2(X, Y)$, where $g_1(x, y)$ and $g_2(x, y)$ are some specified functions. If $B$ is any subset of $\Re^2$, then $(U, V) \in B$ if and only if $(X, Y) \in A$, where $A = \{(x, y) : (g_1(x, y), g_2(x, y)) \in B\}$. Thus $P((U, V) \in B) = P((X, Y) \in A)$, and the probability distribution of $(U, V)$ is completely determined by the probability distribution of $(X, Y)$.

If $(X, Y)$ is a discrete bivariate random vector, then there is only a countable set of values for which the joint pmf of $(X, Y)$ is positive. Call this set $\mathcal{A}$. Define the set $\mathcal{B} = \{(u, v) : u = g_1(x, y) \text{ and } v = g_2(x, y) \text{ for some } (x, y) \in \mathcal{A}\}$. Then $\mathcal{B}$ is the countable set of possible values for the discrete random vector $(U, V)$. And if, for any $(u, v) \in \mathcal{B}$, $A_{uv}$ is defined to be $\{(x, y) \in \mathcal{A} : g_1(x, y) = u \text{ and } g_2(x, y) = v\}$, then the joint pmf of $(U, V)$, $f_{U,V}(u, v)$, can be computed from the joint pmf of $(X, Y)$ by

$$(4.3.1) \quad f_{U,V}(u, v) = P(U = u, V = v) = P((X, Y) \in A_{uv}) = \sum_{(x,y) \in A_{uv}} f_{X,Y}(x, y).$$

**Example 4.3.1 (Distribution of the sum of Poisson variables)** Let $X$ and $Y$ be independent Poisson random variables with parameters $\theta$ and $\lambda$, respectively. Thus the joint pmf of $(X, Y)$ is

$$f_{X,Y}(x, y) = \frac{\theta^x e^{-\theta}}{x!} \frac{\lambda^y e^{-\lambda}}{y!}, \quad x = 0, 1, 2, \ldots, \; y = 0, 1, 2, \ldots.$$

The set $\mathcal{A}$ is $\{(x, y) : x = 0, 1, 2, \ldots \text{ and } y = 0, 1, 2, \ldots\}$. Now define $U = X + Y$ and $V = Y$. That is, $g_1(x, y) = x + y$ and $g_2(x, y) = y$. We will describe the set $\mathcal{B}$, the set of possible $(u, v)$ values. The possible values for $v$ are the nonnegative integers. The variable $v = y$ and thus has the same set of possible values. For a given value of $v$, $u = x + y = x + v$ must be an integer greater than or equal to $v$ since $x$ is a nonnegative integer. The set of all possible $(u, v)$ values is thus given by $\mathcal{B} = \{(u, v) : v = 0, 1, 2, \ldots \text{ and } u = v, v + 1, v + 2, \ldots\}$. For any $(u, v) \in \mathcal{B}$, the only $(x, y)$ value satisfying $x + y = u$ and $y = v$ is $x = u - v$ and $y = v$. Thus, in this example, $A_{uv}$ always consists of only the single point $(u - v, v)$. From (4.3.1) we thus obtain the joint pmf of $(U, V)$ as

$$f_{U,V}(u, v) = f_{X,Y}(u - v, v) = \frac{\theta^{u-v} e^{-\theta}}{(u - v)!} \frac{\lambda^v e^{-\lambda}}{v!}, \quad \begin{matrix} v = 0, 1, 2, \ldots, \\ u = v, v + 1, v + 2, \ldots. \end{matrix}$$

In this example it is interesting to compute the marginal pmf of $U$. For any fixed nonnegative integer $u$, $f_{U,V}(u, v) > 0$ only for $v = 0, 1, \ldots, u$. This gives the set of $v$ values to sum over to obtain the marginal pmf of $U$. It is

$$f_U(u) = \sum_{v=0}^{u} \frac{\theta^{u-v} e^{-\theta}}{(u - v)!} \frac{\lambda^v e^{-\lambda}}{v!} = e^{-(\theta+\lambda)} \sum_{v=0}^{u} \frac{\theta^{u-v}}{(u - v)!} \frac{\lambda^v}{v!}, \quad u = 0, 1, 2, \ldots.$$

This can be simplified by noting that, if we multiply and divide each term by $u!$, we can use the Binomial Theorem to obtain

$$f_U(u) = \frac{e^{-(\theta+\lambda)}}{u!} \sum_{v=0}^{u} \binom{u}{v} \lambda^v \theta^{u-v} = \frac{e^{-(\theta+\lambda)}}{u!} (\theta + \lambda)^u, \quad u = 0, 1, 2, \ldots.$$

This is the pmf of a Poisson random variable with parameter $\theta + \lambda$. This result is significant enough to be stated as a theorem.    ‖

**Theorem 4.3.2** *If $X \sim$ Poisson$(\theta)$ and $Y \sim$ Poisson$(\lambda)$ and $X$ and $Y$ are independent, then $X + Y \sim$ Poisson$(\theta + \lambda)$.*

If $(X, Y)$ is a continuous random vector with joint pdf $f_{X,Y}(x, y)$, then the joint pdf of $(U, V)$ can be expressed in terms of $f_{X,Y}(x, y)$ in a manner analogous to (2.1.8). As before, $\mathcal{A} = \{(x, y) : f_{X,Y}(x, y) > 0\}$ and $\mathcal{B} = \{(u, v) : u = g_1(x, y)$ and $v = g_2(x, y)$ for some $(x, y) \in \mathcal{A}\}$. The joint pdf $f_{U,V}(u, v)$ will be positive on the set $\mathcal{B}$. For the simplest version of this result we assume that the transformation $u = g_1(x, y)$ and $v = g_2(x, y)$ defines a one-to-one transformation of $\mathcal{A}$ onto $\mathcal{B}$. The transformation is onto because of the definition of $\mathcal{B}$. We are assuming that for each $(u, v) \in \mathcal{B}$ there is only one $(x, y) \in \mathcal{A}$ such that $(u, v) = (g_1(x, y), g_2(x, y))$. For such a one-to-one, onto transformation, we can solve the equations $u = g_1(x, y)$ and $v = g_2(x, y)$ for $x$ and $y$ in terms of $u$ and $v$. We will denote this inverse transformation by $x = h_1(u, v)$ and $y = h_2(u, v)$. The role played by a derivative in the univariate case is now played by a quantity called the *Jacobian of the transformation*. This function of $(u, v)$, denoted by $J$, is the *determinant of a matrix* of partial derivatives. It is defined by

$$J = \begin{vmatrix} \dfrac{\partial x}{\partial u} & \dfrac{\partial x}{\partial v} \\ \dfrac{\partial y}{\partial u} & \dfrac{\partial y}{\partial v} \end{vmatrix} = \frac{\partial x}{\partial u}\frac{\partial y}{\partial v} - \frac{\partial y}{\partial u}\frac{\partial x}{\partial v},$$

where

$$\frac{\partial x}{\partial u} = \frac{\partial h_1(u, v)}{\partial u}, \quad \frac{\partial x}{\partial v} = \frac{\partial h_1(u, v)}{\partial v}, \quad \frac{\partial y}{\partial u} = \frac{\partial h_2(u, v)}{\partial u}, \quad \text{and} \quad \frac{\partial y}{\partial v} = \frac{\partial h_2(u, v)}{\partial v}.$$

We assume that $J$ is not identically 0 on $\mathcal{B}$. Then the joint pdf of $(U, V)$ is 0 outside the set $\mathcal{B}$ and on the set $\mathcal{B}$ is given by

$$(4.3.2) \qquad f_{U,V}(u, v) = f_{X,Y}(h_1(u, v),\ h_2(u, v))|J|,$$

where $|J|$ is the absolute value of $J$. When we use (4.3.2), it is sometimes just as difficult to determine the set $\mathcal{B}$ and verify that the transformation is one-to-one as it is to substitute into formula (4.3.2). Note these parts of the explanations in the following examples.

**Example 4.3.3 (Distribution of the product of beta variables)**   Let $X \sim$ beta$(\alpha, \beta)$ and $Y \sim$ beta$(\alpha + \beta, \gamma)$ be independent random variables. The joint pdf of $(X, Y)$ is

$$f_{X,Y}(x, y) = \frac{\Gamma(\alpha + \beta)}{\Gamma(\alpha)\Gamma(\beta)}x^{\alpha-1}(1 - x)^{\beta-1}\frac{\Gamma(\alpha + \beta + \gamma)}{\Gamma(\alpha + \beta)\Gamma(\gamma)}y^{\alpha+\beta-1}(1 - y)^{\gamma-1},$$
$$0 < x < 1, \quad 0 < y < 1.$$

Consider the transformation $U = XY$ and $V = X$. The set of possible values for $V$ is $0 < v < 1$ since $V = X$. For a fixed value of $V = v$, $U$ must be between 0 and $v$ since $X = V = v$ and $Y$ is between 0 and 1. Thus, this transformation maps the set

$\mathcal{A}$ onto the set $\mathcal{B} = \{(u, v) : 0 < u < v < 1\}$. For any $(u, v) \in \mathcal{B}$, the equations $u = xy$ and $v = x$ can be uniquely solved for $x = h_1(u, v) = v$ and $y = h_2(u, v) = u/v$. Note that if considered as a transformation defined on all of $\Re^2$, this transformation is not one-to-one. Any point $(0, y)$ is mapped into the point $(0, 0)$. But as a function defined only on $\mathcal{A}$, it is a one-to-one transformation onto $\mathcal{B}$. The Jacobian is given by

$$J = \begin{vmatrix} \dfrac{\partial x}{\partial u} & \dfrac{\partial x}{\partial v} \\ \dfrac{\partial y}{\partial u} & \dfrac{\partial y}{\partial v} \end{vmatrix} = \begin{vmatrix} 0 & 1 \\ \dfrac{1}{v} & -\dfrac{u}{v^2} \end{vmatrix} = -\dfrac{1}{v}.$$

Thus, from (4.3.2) we obtain the joint pdf as

$$(4.3.3) \quad f_{U,V}(u, v) = \frac{\Gamma(\alpha + \beta + \gamma)}{\Gamma(\alpha)\Gamma(\beta)\Gamma(\gamma)} v^{\alpha-1}(1 - v)^{\beta-1} \left(\frac{u}{v}\right)^{\alpha+\beta-1} \left(1 - \frac{u}{v}\right)^{\gamma-1} \frac{1}{v},$$

$$0 < u < v < 1.$$

The marginal distribution of $V = X$ is, of course, a beta$(\alpha, \beta)$ distribution. But the distribution of $U$ is also a beta distribution:

$$f_U(u) = \int_u^1 f_{U,V}(u, v) dv$$

$$= \frac{\Gamma(\alpha + \beta + \gamma)}{\Gamma(\alpha)\Gamma(\beta)\Gamma(\gamma)} u^{\alpha-1} \int_u^1 \left(\frac{u}{v} - u\right)^{\beta-1} \left(1 - \frac{u}{v}\right)^{\gamma-1} \left(\frac{u}{v^2}\right) dv.$$

The expression (4.3.3) was used but some terms have been rearranged. Now make the univariate change of variable $y = (u/v - u)/(1 - u)$ so that $dy = -u/[v^2(1 - u)]dv$ to obtain

$$f_U(u) = \frac{\Gamma(\alpha + \beta + \gamma)}{\Gamma(\alpha)\Gamma(\beta)\Gamma(\gamma)} u^{\alpha-1}(1 - u)^{\beta+\gamma-1} \int_0^1 y^{\beta-1}(1 - y)^{\gamma-1} dy$$

$$= \frac{\Gamma(\alpha + \beta + \gamma)}{\Gamma(\alpha)\Gamma(\beta)\Gamma(\gamma)} u^{\alpha-1}(1 - u)^{\beta+\gamma-1} \frac{\Gamma(\beta)\Gamma(\gamma)}{\Gamma(\beta + \gamma)}$$

$$= \frac{\Gamma(\alpha + \beta + \gamma)}{\Gamma(\alpha)\Gamma(\beta + \gamma)} u^{\alpha-1}(1 - u)^{\beta+\gamma-1}, \quad 0 < u < 1.$$

To obtain the second identity we recognized the integrand as the kernel of a beta pdf and used (3.3.17). Thus we see that the marginal distribution of $U$ is beta$(\alpha, \beta+\gamma)$. ‖

## Example 4.3.4 (Sum and difference of normal variables)

Let $X$ and $Y$ be independent, standard normal random variables. Consider the transformation $U = X+Y$ and $V = X-Y$. In the notation used above, $U = g_1(X, Y)$ where $g_1(x, y) = x+y$ and $V = g_2(X, Y)$ where $g_2(x, y) = x - y$. The joint pdf of $X$ and $Y$ is, of course, $f_{X,Y}(x, y) = (2\pi)^{-1} \exp(-x^2/2) \exp(-y^2/2)$, $-\infty < x < \infty, -\infty < y < \infty$. So the set

$\mathcal{A} = \Re^2$. To determine the set $\mathcal{B}$ on which $f_{U,V}(u, v)$ is positive, we must determine all the values that

(4.3.4) $$u = x + y \quad \text{and} \quad v = x - y$$

take on as $(x, y)$ range over the set $\mathcal{A} = \Re^2$. But we can set $u$ to be any number and $v$ to be any number and uniquely solve equations (4.3.4) for $x$ and $y$ to obtain

(4.3.5) $$x = h_1(u, v) = \frac{u + v}{2} \quad \text{and} \quad y = h_2(u, v) = \frac{u - v}{2}.$$

This shows two things. For any $(u, v) \in \Re^2$ there is an $(x, y) \in \mathcal{A}$ (defined by (4.3.5)) such that $u = x + y$ and $v = x - y$. So $\mathcal{B}$, the set of all possible $(u, v)$ values, is $\Re^2$. Since the solution (4.3.5) is unique, this also shows that the transformation we have considered is one-to-one. Only the $(x, y)$ given in (4.3.5) will yield $u = x + y$ and $v = x - y$. From (4.3.5) the partial derivatives of $x$ and $y$ are easy to compute. We obtain

$$J = \begin{vmatrix} \dfrac{\partial x}{\partial u} & \dfrac{\partial x}{\partial v} \\ \dfrac{\partial y}{\partial u} & \dfrac{\partial y}{\partial v} \end{vmatrix} = \begin{vmatrix} \dfrac{1}{2} & \dfrac{1}{2} \\ \dfrac{1}{2} & -\dfrac{1}{2} \end{vmatrix} = -\frac{1}{2}.$$

Substituting the expressions (4.3.5) for $x$ and $y$ into $f_{X,Y}(x, y)$ and using $|J| = \frac{1}{2}$, we obtain the joint pdf of $(U, V)$ from (4.3.2) as

$$f_{U,V}(u, v) = f_{X,Y}(h_1(u, v), h_2(u, v))|J| = \frac{1}{2\pi} e^{-((u+v)/2)^2/2} e^{-((u-v)/2)^2/2} \frac{1}{2}$$

for $-\infty < u < \infty$ and $-\infty < v < \infty$. Multiplying out the squares in the exponentials, we see that the terms involving $uv$ cancel. Thus after some simplification and rearrangement we obtain

$$f_{U,V}(u, v) = \left( \frac{1}{\sqrt{2\pi}\sqrt{2}} e^{-u^2/4} \right) \left( \frac{1}{\sqrt{2\pi}\sqrt{2}} e^{-v^2/4} \right).$$

The joint pdf has factored into a function of $u$ and a function of $v$. By Lemma 4.2.7, $U$ and $V$ are independent. From Theorem 4.2.14, the marginal distribution of $U = X + Y$ is $n(0, 2)$. Similarly, Theorem 4.2.12 could be used to find that the marginal distribution of $V$ is also $n(0, 2)$. This important fact, that sums and differences of independent normal random variables are independent normal random variables, is true regardless of the means of $X$ and $Y$, so long as $\text{Var}\, X = \text{Var}\, Y$. This result is left as Exercise 4.27. Theorems 4.2.12 and 4.2.14 give us the marginal distributions of $U$ and $V$. But the more involved analysis here is required to determine that $U$ and $V$ are independent. ‖

In Example 4.3.4, we found that $U$ and $V$ are independent random variables. There is a much simpler, but very important, situation in which new variables $U$ and $V$, defined in terms of original variables $X$ and $Y$, are independent. Theorem 4.3.5 describes this.

**Theorem 4.3.5** *Let $X$ and $Y$ be independent random variables. Let $g(x)$ be a function only of $x$ and $h(y)$ be a function only of $y$. Then the random variables $U = g(X)$ and $V = h(Y)$ are independent.*

**Proof:** We will prove the theorem assuming $U$ and $V$ are continuous random variables. For any $u \in \Re$ and $v \in \Re$ , define

$$A_u = \{x : g(x) \le u\} \quad \text{and} \quad B_v = \{y : h(y) \le v\}.$$

Then the joint cdf of $(U, V)$ is

$$
\begin{aligned}
F_{U,V}(u, v) &= P(U \le u, V \le v) && \text{(definition of cdf)}\\
&= P(X \in A_u, Y \in B_v) && \text{(definition of } U \text{ and } V)\\
&= P(X \in A_u)P(Y \in B_v). && \text{(Theorem 4.2.10)}
\end{aligned}
$$

The joint pdf of $(U, V)$ is

$$
\begin{aligned}
f_{U,V}(u, v) &= \frac{\partial^2}{\partial u \partial v} F_{U,V}(u, v) && \text{(by (4.1.4))}\\[2mm]
&= \left( \frac{d}{du} P(X \in A_u) \right) \left( \frac{d}{dv} P(Y \in B_v) \right),
\end{aligned}
$$

where, as the notation indicates, the first factor is a function only of $u$ and the second factor is a function only of $v$. Hence, by Lemma 4.2.7, $U$ and $V$ are independent. □

It may be that there is only one function, say $U = g_1(X, Y)$, of interest. In such cases, this method may still be used to find the distribution of $U$. If another convenient function, $V = g_2(X, Y)$, can be chosen so that the resulting transformation from $(X, Y)$ to $(U, V)$ is one-to-one on $\mathcal{A}$, then the joint pdf of $(U, V)$ can be derived using (4.3.2) and the marginal pdf of $U$ can be obtained from the joint pdf. In the previous example, perhaps we were interested only in $U = XY$. We could choose to define $V = X$, recognizing that the resulting transformation is one-to-one on $\mathcal{A}$. Then we would proceed as in the example to obtain the marginal pdf of $U$. But other choices, such as $V = Y$, would work as well (see Exercise 4.23).

Of course, in many situations, the transformation of interest is not one-to-one. Just as Theorem 2.1.8 generalized the univariate method to many-to-one functions, the same can be done here. As before, $\mathcal{A} = \{(x, y) : f_{X,Y}(x, y) > 0\}$. Suppose $A_0, A_1, \ldots, A_k$ form a partition of $\mathcal{A}$ with these properties. The set $A_0$, which may be empty, satisfies $P((X, Y) \in A_0) = 0$. The transformation $U = g_1(X, Y)$ and $V = g_2(X, Y)$ is a one-to-one transformation from $A_i$ onto $\mathcal{B}$ for each $i = 1, 2, \ldots, k$. Then for each $i$, the inverse functions from $\mathcal{B}$ to $A_i$ can be found. Denote the $i$th inverse by $x = h_{1i}(u, v)$ and $y = h_{2i}(u, v)$. This $i$th inverse gives, for $(u, v) \in \mathcal{B}$, the unique $(x, y) \in A_i$ such that $(u, v) = (g_1(x, y), g_2(x, y))$. Let $J_i$ denote the Jacobian computed from the $i$th inverse. Then assuming that these Jacobians do not vanish identically on $\mathcal{B}$, we have the following representation of the joint pdf, $f_{U,V}(u, v)$:

$$
(4.3.6) \qquad f_{U,V}(u, v) = \sum_{i=1}^{k} f_{X,Y}(h_{1i}(u, v), h_{2i}(u, v)) |J_i|.
$$

**Example 4.3.6 (Distribution of the ratio of normal variables)** Let $X$ and $Y$ be independent $n(0,1)$ random variables. Consider the transformation $U = X/Y$ and $V = |Y|$. ($U$ and $V$ can be defined to be any value, say $(1,1)$, if $Y = 0$ since $P(Y = 0) = 0$.) This transformation is not one-to-one since the points $(x,y)$ and $(-x,-y)$ are both mapped into the same $(u,v)$ point. But if we restrict consideration to either positive or negative values of $y$, then the transformation is one-to-one. In the above notation, let

$$A_1 = \{(x,y) : y > 0\}, \quad A_2 = \{(x,y) : y < 0\}, \quad \text{and} \quad A_0 = \{(x,y) : y = 0\}.$$

$A_0, A_1$, and $A_2$ form a partition of $\mathcal{A} = \Re^2$ and $P((X,Y) \in A_0) = P(Y = 0) = 0$. For either $A_1$ or $A_2$, if $(x,y) \in A_i$, $v = |y| > 0$, and for a fixed value of $v = |y|$, $u = x/y$ can be any real number since $x$ can be any real number. Thus, $\mathcal{B} = \{(u,v) : v > 0\}$ is the image of both $A_1$ and $A_2$ under the transformation. Furthermore, the inverse transformations from $\mathcal{B}$ to $A_1$ and $\mathcal{B}$ to $A_2$ are given by $x = h_{11}(u,v) = uv$, $y = h_{21}(u,v) = v$, and $x = h_{12}(u,v) = -uv$, $y = h_{22}(u,v) = -v$. Note that the first inverse gives positive values of $y$ and the second gives negative values of $y$. The Jacobians from the two inverses are $J_1 = J_2 = v$. Using

$$f_{X,Y}(x,y) = \frac{1}{2\pi} e^{-x^2/2} e^{-y^2/2},$$

from (4.3.6) we obtain

$$f_{U,V}(u,v) = \frac{1}{2\pi} e^{-(uv)^2/2} e^{-v^2/2} |v| + \frac{1}{2\pi} e^{-(-uv)^2/2} e^{-(-v)^2/2} |v|$$

$$= \frac{v}{\pi} e^{-(u^2+1)v^2/2}, \quad -\infty < u < \infty, \quad 0 < v < \infty.$$

From this the marginal pdf of $U$ can be computed to be

$$f_U(u) = \int_0^\infty \frac{v}{\pi} e^{-(u^2+1)v^2/2} \, dv$$

$$= \frac{1}{2\pi} \int_0^\infty e^{-(u^2+1)z/2} \, dz \quad z = v^2) \quad \text{(change of variable)}$$

$$= \frac{1}{2\pi} \frac{2}{(u^2+1)} \quad \left( \begin{array}{c} \text{integrand is kernel of} \\ \text{exponential } (\beta = 2/(u^2+1)) \text{ pdf} \end{array} \right)$$

$$= \frac{1}{\pi(u^2+1)}, \quad -\infty < u < \infty.$$

So we see that the ratio of two independent standard normal random variables is a Cauchy random variable. (See Exercise 4.28 for more relationships between normal and Cauchy random variables.) ‖

## 4.4 Hierarchical Models and Mixture Distributions

In the cases we have seen thus far, a random variable has a single distribution, possibly depending on parameters. While, in general, a random variable can have only one distribution, it is often easier to model a situation by thinking of things in a hierarchy.

**Example 4.4.1 (Binomial-Poisson hierarchy)** Perhaps the most classic hierarchical model is the following. An insect lays a large number of eggs, each surviving with probability $p$. On the average, how many eggs will survive?

The "large number" of eggs laid is a random variable, often taken to be Poisson($\lambda$). Furthermore, if we assume that each egg's survival is independent, then we have Bernoulli trials. Therefore, if we let $X$ = number of survivors and $Y$ = number of eggs laid, we have

$$X|Y \sim \text{binomial}(Y, p),$$

$$Y \sim \text{Poisson}(\lambda),$$

a hierarchical model. (Recall that we use notation such as $X|Y \sim \text{binomial}(Y, p)$ to mean that the conditional distribution of $X$ given $Y = y$ is binomial$(y, p)$.)      ‖

The advantage of the hierarchy is that complicated processes may be modeled by a sequence of relatively simple models placed in a hierarchy. Also, dealing with the hierarchy is no more difficult than dealing with conditional and marginal distributions.

**Example 4.4.2 (Continuation of Example 4.4.1)** The random variable of interest, $X$ = number of survivors, has the distribution given by

$$P(X = x) = \sum_{y=0}^{\infty} P(X = x, Y = y)$$

$$= \sum_{y=0}^{\infty} P(X = x|Y = y)P(Y = y) \qquad \begin{pmatrix} \text{definition of} \\ \text{conditional probability} \end{pmatrix}$$

$$= \sum_{y=x}^{\infty} \left[ \binom{y}{x} p^x (1-p)^{y-x} \right] \left[ \frac{e^{-\lambda}\lambda^y}{y!} \right], \qquad \begin{pmatrix} \text{conditional probability} \\ \text{is 0 if } y < x \end{pmatrix}$$

since $X|Y = y$ is binomial$(y, p)$ and $Y$ is Poisson($\lambda$). If we now simplify this last expression, canceling what we can and multiplying by $\lambda^x / \lambda^x$, we get

$$P(X = x) = \frac{(\lambda p)^x e^{-\lambda}}{x!} \sum_{y=x}^{\infty} \frac{((1-p)\lambda)^{y-x}}{(y-x)!}$$

$$= \frac{(\lambda p)^x e^{-\lambda}}{x!} \sum_{t=0}^{\infty} \frac{((1-p)\lambda)^t}{t!} \qquad (t = y - x)$$

$$= \frac{(\lambda p)^x e^{-\lambda}}{x!} e^{(1-p)\lambda} \qquad \begin{pmatrix} \text{sum is a kernel for} \\ \text{a Poisson distribution} \end{pmatrix}$$

$$= \frac{(\lambda p)^x}{x!} e^{-\lambda p},$$

so $X \sim \text{Poisson}(\lambda p)$. Thus, any marginal inference on $X$ is with respect to a Poisson($\lambda p$) distribution, with $Y$ playing no part at all. Introducing $Y$ in the hierarchy was mainly to aid our understanding of the model. There was an added bonus in that the parameter of the distribution of $X$ is the product of two parameters, each relatively simple to understand.

The answer to the original question is now easy to compute:

$$EX = \lambda p,$$

so, on the average, $\lambda p$ eggs will survive. If we were interested only in this mean and did not need the distribution, we could have used properties of conditional expectations.

$\parallel$

Sometimes, calculations can be greatly simplified be using the following theorem. Recall from Section 4.2 that $E(X|y)$ is a function of $y$ and $E(X|Y)$ is a random variable whose value depends on the value of $Y$.

**Theorem 4.4.3**  *If $X$ and $Y$ are any two random variables, then*

(4.4.1)                                     $$EX = E\left(E(X|Y)\right),$$

*provided that the expectations exist.*

**Proof:** Let $f(x, y)$ denote the joint pdf of $X$ and $Y$. By definition, we have

(4.4.2)        $$EX = \int\int x f(x, y)\, dx\, dy = \int \left[\int x f(x|y)\, dx\right] f_Y(y)\, dy,$$

where $f(x|y)$ and $f_Y(y)$ are the conditional pdf of $X$ given $Y = y$ and the marginal pdf of $Y$, respectively. But now notice that the inner integral in (4.4.2) is the conditional expectation $E(X|y)$, and we have

$$EX = \int E(X|y) f_Y(y)\, dy = E\left(E(X|Y)\right),$$

as desired. Replace integrals by sums to prove the discrete case.                    $\square$

Note that equation (4.4.1) contains an abuse of notation, since we have used the "E" to stand for different expectations in the same equation. The "E" in the left-hand side of (4.4.1) is expectation with respect to the marginal distribution of $X$. The first "E" in the right-hand side of (4.4.1) is expectation with respect to the marginal distribution of $Y$, while the second one stands for expectation with respect to the conditional distribution of $X|Y$. However, there is really no cause for confusion because these interpretations are the only ones that the symbol "E" can take!

We can now easily compute the expected number of survivors in Example 4.4.1. From Theorem 4.4.3 we have

$$EX = E\left(E(X|Y)\right)$$
$$= E(pY) \qquad \text{(since } X|Y \sim \text{binomial}(Y, p))$$
$$= p\lambda. \qquad \text{(since } Y \sim \text{Poisson}(\lambda))$$

The term *mixture distribution* in the title of this section refers to a distribution arising from a hierarchical structure. Although there is no standardized definition for this term, we will use the following definition, which seems to be a popular one.

**Definition 4.4.4**  A random variable $X$ is said to have a *mixture distribution* if the distribution of $X$ depends on a quantity that also has a distribution.

Thus, in Example 4.4.1 the Poisson($\lambda p$) distribution is a mixture distribution since it is the result of combining a binomial($Y, p$) with $Y \sim$ Poisson($\lambda$). In general, we can say that hierarchical models lead to mixture distributions.

There is nothing to stop the hierarchy at two stages, but it should be easy to see that any more complicated hierarchy can be treated as a two-stage hierarchy theoretically. There may be advantages, however, in modeling a phenomenon as a multistage hierarchy. It may be easier to understand.

**Example 4.4.5 (Generalization of Example 4.4.1)**  Consider a generalization of Example 4.4.1, where instead of one mother insect there are a large number of mothers and one mother is chosen at random. We are still interested in knowing the average number of survivors, but it is no longer clear that the number of eggs laid follows the same Poisson distribution for each mother. The following three-stage hierarchy may be more appropriate. Let $X =$ number of survivors in a litter; then

$$X|Y \sim \text{binomial}(Y, p),$$

$$Y|\Lambda \sim \text{Poisson}(\Lambda),$$

$$\Lambda \sim \text{exponential}(\beta),$$

where the last stage of the hierarchy accounts for the variability across different mothers.

The mean of $X$ can easily be calculated as

$$
\begin{aligned}
\text{E}X &= \text{E}\left(\text{E}(X|Y)\right) \\
&= \text{E}(pY) && \text{(as before)} \\
&= \text{E}\left(\text{E}(pY|\Lambda)\right) \\
&= \text{E}(p\Lambda) \\
&= p\beta, && \text{(exponential expectation)}
\end{aligned}
$$

completing the calculation.                                                   ‖

In this example we have used a slightly different type of model than before in that two of the random variables are discrete and one is continuous. Using these models should present no problems. We can define a joint density, $f(x, y, \lambda)$; conditional densities, $f(x|y)$, $f(x|y, \lambda)$, etc.; and marginal densities, $f(x)$, $f(x, y)$, etc. as before. Simply understand that, when probabilities or expectations are calculated, discrete variables are summed and continuous variables are integrated.

Note that this three-stage model can also be thought of as a two-stage hierarchy by combining the last two stages. If $Y|\Lambda \sim \text{Poisson}(\Lambda)$ and $\Lambda \sim \text{exponential}(\beta)$, then

$$P(Y = y) = P(Y = y, 0 < \Lambda < \infty)$$

$$= \int_0^\infty f(y, \lambda) \, d\lambda$$

$$= \int_0^\infty f(y|\lambda) f(\lambda) \, d\lambda$$

$$= \int_0^\infty \left[ \frac{e^{-\lambda} \lambda^y}{y!} \right] \frac{1}{\beta} e^{-\lambda/\beta} \, d\lambda$$

$$= \frac{1}{\beta y!} \int_0^\infty \lambda^y e^{-\lambda(1+\beta^{-1})} \, d\lambda \qquad \left( \begin{array}{c} \text{gamma} \\ \text{pdf kernel} \end{array} \right)$$

$$= \frac{1}{\beta y!} \Gamma(y+1) \left( \frac{1}{1 + \beta^{-1}} \right)^{y+1}$$

$$= \frac{1}{(1 + \beta)} \left( \frac{1}{1 + \beta^{-1}} \right)^y.$$

This expression for the pmf of $Y$ is the form (3.2.10) of the negative binomial pmf. Therefore, our three-stage hierarchy in Example 4.4.5 is equivalent to the two-stage hierarchy

$$X|Y \sim \text{binomial}(Y, p),$$

$$Y \sim \text{negative binomial} \left( p = \frac{1}{1 + \beta}, r = 1 \right).$$

However, in terms of understanding the model, the three-stage model is much easier to understand!

A useful generalization is a Poisson–gamma mixture, which is a generalization of a part of the previous model. If we have the hierarchy

$$Y|\Lambda \sim \text{Poisson}(\Lambda),$$

$$\Lambda \sim \text{gamma}(\alpha, \beta),$$

then the marginal distribution of $Y$ is negative binomial (see Exercise 4.32). This model for the negative binomial distribution shows that it can be considered to be a "more variable" Poisson. Solomon (1983) explains these and other biological and mathematical models that lead to the negative binomial distribution. (See Exercise 4.33.)

Aside from the advantage in aiding understanding, hierarchical models can often make calculations easier. For example, a distribution that often occurs in statistics is the *noncentral chi squared distribution*. With $p$ degrees of freedom and noncentrality parameter $\lambda$, the pdf is given by

(4.4.3) $$f(x|\lambda, p) = \sum_{k=0}^\infty \frac{x^{p/2+k-1} e^{-x/2}}{\Gamma(p/2+k) 2^{p/2+k}} \frac{\lambda^k e^{-\lambda}}{k!},$$

an extremely messy expression. Calculating $EX$, for example, looks like quite a chore. However, if we examine the pdf closely, we see that this is a mixture distribution, made up of central chi squared densities (like those given in (3.2.10)) and Poisson distributions. That is, if we set up the hierarchy

$$X|K \sim \chi^2_{p+2K},$$
$$K \sim \text{Poisson}(\lambda),$$

then the marginal distribution of $X$ is given by (4.4.3). Hence

$$EX = E(E(X|K))$$
$$= E(p + 2K)$$
$$= p + 2\lambda,$$

a relatively simple calculation. $\text{Var}\, X$ can also be calculated in this way.

  We close this section with one more hierarchical model and illustrate one more conditional expectation calculation.

**Example 4.4.6 (Beta-binomial hierarchy)**  One generalization of the binomial distribution is to allow the success probability to vary according to a distribution. A standard model for this situation is

$$X|P \sim \text{binomial}(P), \quad i = 1, \ldots, n,$$
$$P \sim \text{beta}(\alpha, \beta).$$

By iterating the expectation, we calculate the mean of $X$ as

$$EX = E[E(X|P)] = E[nP] = n\frac{\alpha}{\alpha + \beta}. \qquad \|$$

  Calculating the variance of $X$ is only slightly more involved. We can make use of a formula for conditional variances, similar in spirit to the expected value identity of Theorem 4.4.3.

**Theorem 4.4.7 (Conditional variance identity)** *For any two random variables $X$ and $Y$,*

(4.4.4) $$\text{Var}\, X = E\left(\text{Var}(X|Y)\right) + \text{Var}\left(E(X|Y)\right),$$

*provided that the expectations exist.*

**Proof:** By definition, we have

$$\text{Var}\, X = E\left([X - EX]^2\right) = E\left([X - E(X|Y) + E(X|Y) - EX]^2\right),$$

where in the last step we have added and subtracted $E(X|Y)$. Expanding the square in this last expectation now gives

$$\text{Var}\, X = E\left([X - E(X|Y)]^2\right) + E\left([E(X|Y) - EX]^2\right)$$
(4.4.5) $$+ 2E\left([X - E(X|Y)][E(X|Y) - EX]\right).$$

The last term in this expression is equal to 0, however, which can easily be seen by iterating the expectation:

(4.4.6)
$$\mathrm{E}\left([X - \mathrm{E}(X|Y)][\mathrm{E}(X|Y) - \mathrm{E}X]\right) = \mathrm{E}\left(\mathrm{E}\left\{[X - \mathrm{E}(X|Y)][\mathrm{E}(X|Y) - \mathrm{E}X]|Y\right\}\right).$$

In the conditional distribution $X|Y$, $X$ is the random variable. So in the expression
$$\mathrm{E}\left\{[X - \mathrm{E}(X|Y)][\mathrm{E}(X|Y) - \mathrm{E}X]|Y\right\},$$
$\mathrm{E}(X|Y)$ and $\mathrm{E}X$ are constants. Thus,
$$\begin{aligned}\mathrm{E}\left\{[X - \mathrm{E}(X|Y)][\mathrm{E}(X|Y) - \mathrm{E}X]|Y\right\} &= (\mathrm{E}(X|Y) - \mathrm{E}X)\left(\mathrm{E}\left\{[X - \mathrm{E}(X|Y)]|Y\right\}\right)\\ &= (\mathrm{E}(X|Y) - \mathrm{E}X)\left(\mathrm{E}(X|Y) - \mathrm{E}(X|Y)\right)\\ &= (\mathrm{E}(X|Y) - \mathrm{E}X)\,(0)\\ &= 0.\end{aligned}$$

Thus, from (4.4.6), we have that $\mathrm{E}((X - \mathrm{E}(X|Y))(\mathrm{E}(X|Y) - \mathrm{E}X)) = \mathrm{E}(0) = 0$. Referring back to equation (4.4.5), we see that
$$\begin{aligned}\mathrm{E}\left([X - \mathrm{E}(X|Y)]^2\right) &= \mathrm{E}\left(\mathrm{E}\left\{[X - \mathrm{E}(X|Y)]^2|Y\right\}\right)\\ &= \mathrm{E}\left(\mathrm{Var}(X|Y)\right)\end{aligned}$$

and
$$\mathrm{E}\left([\mathrm{E}(X|Y) - \mathrm{E}X]^2\right) = \mathrm{Var}\left(\mathrm{E}(X|Y)\right),$$
establishing (4.4.4). □

**Example 4.4.8 (Continuation of Example 4.4.6)** To calculate the variance of $X$, we have from (4.4.4),
$$\mathrm{Var}\,X = \mathrm{Var}\left(\mathrm{E}(X|P)\right) + \mathrm{E}\left(\mathrm{Var}(X|P)\right).$$

Now $\mathrm{E}(X|P) = nP$, and since $P \sim \mathrm{beta}(\alpha, \beta)$,
$$\mathrm{Var}\left(\mathrm{E}(X|P)\right) = \mathrm{Var}(nP) = n^2 \frac{\alpha\beta}{(\alpha + \beta)^2(\alpha + \beta + 1)}.$$

Also, since $X|P$ is binomial$(n, P)$, $\mathrm{Var}(X|P) = nP(1 - P)$. We then have
$$\mathrm{E}\left[\mathrm{Var}(X|P)\right] = n\mathrm{E}\left[P(1 - P)\right] = n\frac{\Gamma(\alpha + \beta)}{\Gamma(\alpha)\Gamma(\beta)}\int_0^1 p(1 - p)p^{\alpha - 1}(1 - p)^{\beta - 1}dp.$$

Notice that the integrand is the kernel of another beta pdf (with parameters $\alpha + 1$ and $\beta + 1$) so
$$\mathrm{E}\left(\mathrm{Var}(X|P)\right) = n\frac{\Gamma(\alpha + \beta)}{\Gamma(\alpha)\Gamma(\beta)}\left[\frac{\Gamma(\alpha + 1)\Gamma(\beta + 1)}{\Gamma(\alpha + \beta + 2)}\right] = n\frac{\alpha\beta}{(\alpha + \beta)(\alpha + \beta + 1)}.$$

Adding together the two pieces and simplifying, we get
$$\mathrm{Var}\,X = n\frac{\alpha\beta(\alpha + \beta + n)}{(\alpha + \beta)^2(\alpha + \beta + 1)}.$$
‖

## 4.5 Covariance and Correlation

In earlier sections, we have discussed the absence or presence of a relationship between two random variables, independence or nonindependence. But if there is a relationship, the relationship may be strong or weak. In this section we discuss two numerical measures of the strength of a relationship between two random variables, the covariance and correlation.

To illustrate what we mean by the strength of a relationship between two random variables, consider two different experiments. In the first, random variables $X$ and $Y$ are measured, where $X$ is the weight of a sample of water and $Y$ is the volume of the same sample of water. Clearly there is a strong relationship between $X$ and $Y$. If $(X, Y)$ pairs are measured on several samples and the observed data pairs are plotted, the data points should fall on a straight line because of the physical relationship between $X$ and $Y$. This will not be exactly the case because of measurement errors, impurities in the water, etc. But with careful laboratory technique, the data points will fall very nearly on a straight line. Now consider another experiment in which $X$ and $Y$ are measured, where $X$ is the body weight of a human and $Y$ is the same human's height. Clearly there is also a relationship between $X$ and $Y$ here but the relationship is not nearly as strong. We would not expect a plot of $(X, Y)$ pairs measured on different people to form a straight line, although we might expect to see an upward trend in the plot. The covariance and correlation are two measures that quantify this difference in the strength of a relationship between two random variables.

Throughout this section we will frequently be referring to the mean and variance of $X$ and the mean and variance of $Y$. For these we will use the notation $EX = \mu_X$, $EY = \mu_Y$, $\operatorname{Var} X = \sigma_X^2$, and $\operatorname{Var} Y = \sigma_Y^2$. We will assume throughout that $0 < \sigma_X^2 < \infty$ and $0 < \sigma_Y^2 < \infty$.

**Definition 4.5.1** The *covariance of $X$ and $Y$* is the number defined by

$$\operatorname{Cov}(X, Y) = E\left((X - \mu_X)(Y - \mu_Y)\right).$$

**Definition 4.5.2** The *correlation of $X$ and $Y$* is the number defined by

$$\rho_{XY} = \frac{\operatorname{Cov}(X, Y)}{\sigma_X \sigma_Y}.$$

The value $\rho_{XY}$ is also called the *correlation coefficient*.

If large values of $X$ tend to be observed with large values of $Y$ and small values of $X$ with small values of $Y$, then $\operatorname{Cov}(X, Y)$ will be positive. If $X > \mu_X$, then $Y > \mu_Y$ is likely to be true and the product $(X - \mu_X)(Y - \mu_Y)$ will be positive. If $X < \mu_X$, then $Y < \mu_Y$ is likely to be true and the product $(X - \mu_X)(Y - \mu_Y)$ will again be positive. Thus $\operatorname{Cov}(X, Y) = E(X - \mu_X)(Y - \mu_Y) > 0$. If large values of $X$ tend to be observed with small values of $Y$ and small values of $X$ with large values of $Y$, then $\operatorname{Cov}(X, Y)$ will be negative because when $X > \mu_X$, $Y$ will tend to be less than $\mu_Y$ and vice versa, and hence $(X - \mu_X)(Y - \mu_Y)$ will tend to be negative. Thus the sign of $\operatorname{Cov}(X, Y)$ gives information regarding the relationship between $X$ and $Y$.

But $\text{Cov}(X, Y)$ can be any number and a given value of $\text{Cov}(X, Y)$, say $\text{Cov}(X, Y) = 3$, does not in itself give information about the strength of the relationship between $X$ and $Y$. On the other hand, the correlation is always between $-1$ and $1$, with the values $-1$ and $1$ indicating a perfect *linear* relationship between $X$ and $Y$. This is proved in Theorem 4.5.7.

Before investigating these properties of covariance and correlation, we will first calculate these measures in a given example. This calculation will be simplified by the following result.

**Theorem 4.5.3** *For any random variables $X$ and $Y$,*

$$\text{Cov}(X, Y) = \text{E}XY - \mu_X\mu_Y.$$

**Proof:** $\text{Cov}(X, Y) = \text{E}\,(X - \mu_X)(Y - \mu_Y)$

$$= \text{E}\,(XY - \mu_X Y - \mu_Y X + \mu_X\mu_Y) \qquad \text{(expanding the product)}$$

$$= \text{E}XY - \mu_X\text{E}Y - \mu_Y\text{E}X + \mu_X\mu_Y \quad (\mu_X \text{ and } \mu_Y \text{ are constants})$$

$$= \text{E}XY - \mu_X\mu_Y - \mu_Y\mu_X + \mu_X\mu_Y$$

$$= \text{E}XY - \mu_X\mu_Y. \qquad\qquad\qquad\qquad\qquad \square$$

**Example 4.5.4 (Correlation–I)** Let the joint pdf of $(X, Y)$ be $f(x, y) = 1$, $0 < x < 1$, $x < y < x+1$. See Figure 4.5.1. The marginal distribution of $X$ is uniform$(0, 1)$ so $\mu_X = \frac{1}{2}$ and $\sigma_X^2 = \frac{1}{12}$. The marginal pdf of $Y$ is $f_Y(y) = y$, $0 < y < 1$, and $f_Y(y) = 2 - y$, $1 \le y < 2$, with $\mu_Y = 1$ and $\sigma_Y^2 = \frac{1}{6}$. We also have

$$\text{E}XY = \int_0^1 \int_x^{x+1} xy \, dy \, dx = \int_0^1 \frac{1}{2}xy^2\Big|_x^{x+1} dx$$

$$= \int_0^1 \left(x^2 + \frac{1}{2}x\right) dx = \frac{7}{12}.$$

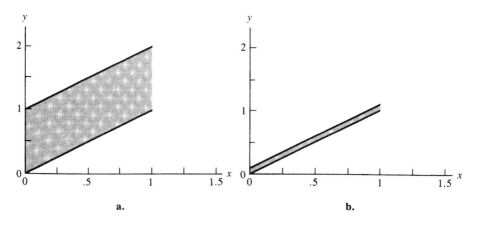

Figure 4.5.1. **(a)** *Region where $f(x, y) > 0$ for Example 4.5.4;* **(b)** *region where $f(x, y) > 0$ for Example 4.5.8*

Using Theorem 4.5.3, we have $\text{Cov}(X,Y) = \frac{7}{12} - \left(\frac{1}{2}\right)(1) = \frac{1}{12}$. The correlation is

$$\rho_{XY} = \frac{\text{Cov}(X,Y)}{\sigma_X \sigma_Y} = \frac{1/12}{\sqrt{1/12}\sqrt{1/6}} = \frac{1}{\sqrt{2}}. \qquad \|$$

In the next three theorems we describe some of the fundamental properties of covariance and correlation.

**Theorem 4.5.5** *If $X$ and $Y$ are independent random variables, then $\text{Cov}(X,Y) = 0$ and $\rho_{XY} = 0$.*

**Proof:** Since $X$ and $Y$ are independent, from Theorem 4.2.10 we have $\text{E}XY = (\text{E}X)(\text{E}Y)$. Thus

$$\text{Cov}(X,Y) = \text{E}XY - (\text{E}X)(\text{E}Y) = (\text{E}X)(\text{E}Y) - (\text{E}X)(\text{E}Y) = 0$$

and

$$\rho_{XY} = \frac{\text{Cov}(X,Y)}{\sigma_X \sigma_Y} = \frac{0}{\sigma_X \sigma_Y} = 0. \qquad \Box$$

Thus, the values $\text{Cov}(X,Y) = \rho_{XY} = 0$ in some sense indicate that there is no relationship between $X$ and $Y$. It is important to note, however, that Theorem 4.5.5 does *not* say that if $\text{Cov}(X,Y) = 0$, then $X$ and $Y$ are independent. For example, if $X \sim f(x - \theta)$, symmetric around 0 with $\text{E}X = \theta$, and $Y$ is the indicator function $Y = I(|X - \theta| < 2)$, then $X$ and $Y$ are obviously not independent. However,

$$\text{E}(XY) = \int_{-\infty}^{\infty} xI(|x-\theta| < 2)f(x-\theta)\,dx = \int_{-2}^{2}(t+\theta)f(t)\,dt = \theta \int_{-2}^{2} f(t)\,dt = \text{E}X\text{E}Y,$$

where we used the fact that, by symmetry, $\int_{-2}^{2} tf(t)\,dt = 0$. So it is easy to find uncorrelated, dependent random variables.

Covariance and correlation measure only a particular kind of *linear* relationship that will be described further in Theorem 4.5.7. Also see Example 4.5.9, which discusses two random variables that have a strong relationship but whose covariance and correlation are 0 because the relationship is not linear.

Covariance also plays an important role in understanding the variation in sums of random variables, as the next theorem, a generalization of Theorem 2.3.4, indicates. (See Exercise 4.44 for a further generalization.)

**Theorem 4.5.6** *If $X$ and $Y$ are any two random variables and $a$ and $b$ are any two constants, then*

$$\text{Var}(aX + bY) = a^2\text{Var}\,X + b^2\text{Var}\,Y + 2ab\,\text{Cov}(X,Y).$$

*If $X$ and $Y$ are independent random variables, then*

$$\text{Var}(aX + bY) = a^2\text{Var}\,X + b^2\text{Var}\,Y.$$

**Proof:** The mean of $aX + bY$ is $E(aX + bY) = aEX + bEY = a\mu_X + b\mu_Y$. Thus

$$
\begin{aligned}
\text{Var}(aX + bY) &= E\left((aX + bY) - (a\mu_X + b\mu_Y)\right)^2 \\
&= E\left(a(X - \mu_X) + b(Y - \mu_Y)\right)^2 \\
&= E\left(a^2(X - \mu_X)^2 + b^2(Y - \mu_Y)^2 + 2ab(X - \mu_X)(Y - \mu_Y)\right) \\
&= a^2 E(X - \mu_X)^2 + b^2 E(Y - \mu_Y)^2 + 2ab E(X - \mu_X)(Y - \mu_Y) \\
&= a^2 \text{Var}\, X + b^2 \text{Var}\, Y + 2ab \text{Cov}(X, Y).
\end{aligned}
$$

If $X$ and $Y$ are independent, then, from Theorem 4.5.5, $\text{Cov}(X, Y) = 0$ and the second equality is immediate from the first. $\qquad\square$

From Theorem 4.5.6 we see that if $X$ and $Y$ are positively correlated $(\text{Cov}(X, Y) > 0)$, then the variation in $X + Y$ is greater than the sum of the variations in $X$ and $Y$. But if they are negatively correlated, then the variation in $X + Y$ is less than the sum. For negatively correlated random variables, large values of one tend to be observed with small values of the other and in the sum these two extremes cancel. The result, $X + Y$, tends not to have as many extreme values and hence has smaller variance. By choosing $a = 1$ and $b = -1$ we get an expression for the variance of the difference of two random variables, and similar arguments apply.

The nature of the linear relationship measured by covariance and correlation is somewhat explained by the following theorem.

**Theorem 4.5.7** *For any random variables $X$ and $Y$,*

**a.** $-1 \leq \rho_{XY} \leq 1$.
**b.** $|\rho_{XY}| = 1$ *if and only if there exist numbers $a \neq 0$ and $b$ such that $P(Y = aX + b) = 1$. If $\rho_{XY} = 1$, then $a > 0$, and if $\rho_{XY} = -1$, then $a < 0$.*

**Proof:** Consider the function $h(t)$ defined by

$$
h(t) = E\left((X - \mu_X)t + (Y - \mu_Y)\right)^2.
$$

Expanding this expression, we obtain

$$
\begin{aligned}
h(t) &= t^2 E(X - \mu_X)^2 + 2t E(X - \mu_X)(Y - \mu_Y) + E(Y - \mu_Y)^2 \\
&= t^2 \sigma_X^2 + 2t \text{Cov}(X, Y) + \sigma_Y^2.
\end{aligned}
$$

This quadratic function of $t$ is greater than or equal to 0 for all values of $t$ since it is the expected value of a nonnegative random variable. Thus, this quadratic function can have at most one real root and thus must have a nonpositive discriminant. That is,

$$
(2\text{Cov}(X, Y))^2 - 4\sigma_X^2\sigma_Y^2 \leq 0.
$$

This is equivalent to

$$
-\sigma_X\sigma_Y \leq \text{Cov}(X, Y) \leq \sigma_X\sigma_Y.
$$

Dividing by $\sigma_X \sigma_Y$ yields

$$-1 \leq \frac{\text{Cov}(X, Y)}{\sigma_X \sigma_Y} = \rho_{XY} \leq 1.$$

Also, $|\rho_{XY}| = 1$ if and only if the discriminant is equal to 0. That is, $|\rho_{XY}| = 1$ if and only if $h(t)$ has a single root. But since $((X - \mu_X)t + (Y - \mu_Y))^2 \geq 0$, the expected value $h(t) = \text{E}((X - \mu_X)t + (Y - \mu_Y))^2 = 0$ if and only if

$$P\big([(X - \mu_X)t + (Y - \mu_Y)]^2 = 0\big) = 1.$$

This is equivalent to

$$P\big((X - \mu_X)t + (Y - \mu_Y) = 0\big) = 1.$$

This is $P(Y = aX + b) = 1$ with $a = -t$ and $b = \mu_X t + \mu_Y$, where $t$ is the root of $h(t)$. Using the quadratic formula, we see that this root is $t = -\text{Cov}(X, Y)/\sigma_X^2$. Thus $a = -t$ has the same sign as $\rho_{XY}$, proving the final assertion.                    □

In Section 4.7 we will prove a theorem called the Cauchy–Schwarz Inequality. This theorem has as a direct consequence that $\rho_{XY}$ is bounded between $-1$ and $1$, and we will see that, with this inequality, the preceding proof can be shortened.

If there is a line $y = ax + b$, with $a \neq 0$, such that the values of $(X, Y)$ have a high probability of being near this line, then the correlation between $X$ and $Y$ will be near $1$ or $-1$. But if no such line exists, the correlation will be near $0$. This is an intuitive notion of the linear relationship that is being measured by correlation. This idea will be illustrated further in the next two examples.

**Example 4.5.8 (Correlation–II)** This example is similar to Example 4.5.4, but we develop it differently to illustrate other model building and computational techniques. Let $X$ have a uniform$(0, 1)$ distribution and $Z$ have a uniform$(0, \frac{1}{10})$ distribution. Suppose $X$ and $Z$ are independent. Let $Y = X + Z$ and consider the random vector $(X, Y)$. The joint distribution of $(X, Y)$ can be derived from the joint distribution of $(X, Z)$ using the techniques of Section 4.3. The joint pdf of $(X, Y)$ is

$$f(x, y) = 10, \quad 0 < x < 1, \quad x < y < x + \frac{1}{10}.$$

Rather than using the formal techniques of Section 4.3, we can justify this as follows. Given $X = x$, $Y = x + Z$. The conditional distribution of $Z$ given $X = x$ is just uniform$(0, \frac{1}{10})$ since $X$ and $Z$ are independent. Thus $x$ serves as a location parameter in the conditional distribution of $Y$ given $X = x$, and this conditional distribution is just uniform$(x, x + \frac{1}{10})$. Multiplying this conditional pdf by the marginal pdf of $X$ (uniform$(0, 1)$) yields the joint pdf above. This representation of $Y = X + Z$ makes the computation of the covariance and correlation easy. The expected values of $X$ and $Y$ are $\text{E}X = \frac{1}{2}$ and $\text{E}Y = \text{E}(X + Z) = \text{E}X + \text{E}Z = \frac{1}{2} + \frac{1}{20} = \frac{11}{20}$, giving

$$\text{Cov}(X, Y) = \text{E}XY - (\text{E}X)(\text{E}Y)$$

$$= \text{E}X(X + Z) - (\text{E}X)(\text{E}(X + Z))$$

$$= EX^2 + EXZ - (EX)^2 - (EX)(EZ)$$

$$= EX^2 - (EX)^2 + (EX)(EZ) - (EX)(EZ) \quad \begin{pmatrix} \text{independence of} \\ X \text{ and } Z \end{pmatrix}$$

$$= \sigma_X^2 = \frac{1}{12}.$$

From Theorem 4.5.6, the variance of $Y$ is $\sigma_Y^2 = \mathrm{Var}(X + Z) = \mathrm{Var}\, X + \mathrm{Var}\, Z = \frac{1}{12} + \frac{1}{1200}$. Thus

$$\rho_{XY} = \frac{\frac{1}{12}}{\sqrt{\frac{1}{12}}\sqrt{\frac{1}{12} + \frac{1}{1200}}} = \sqrt{\frac{100}{101}}.$$

This is much larger than the value of $\rho_{XY} = 1/\sqrt{2}$ obtained in Example 4.5.4. The sets on which $f(x, y)$ is positive for Example 4.5.4 and this example are illustrated in Figure 4.5.1. (Recall that this set is called the support of a distribution.) In each case, $(X, Y)$ is a random point from the set. In both cases there is a linearly increasing relationship between $X$ and $Y$, but the relationship is much stronger in Figure 4.5.1b. Another way to see this is by noting that in this example, the conditional distribution of $Y$ given $X = x$ is uniform$(x, x + \frac{1}{10})$. In Example 4.5.4, the conditional distribution of $Y$ given $X = x$ is uniform$(x, x + 1)$. The knowledge that $X = x$ gives us much more information about the value of $Y$ in this model than in the one in Example 4.5.4. Hence the correlation is nearer to 1 in this example.                                                    ‖

The next example illustrates that there may be a strong relationship between $X$ and $Y$, but if the relationship is not linear, the correlation may be small.

**Example 4.5.9 (Correlation–III)**  In this example, let $X$ have a uniform$(-1, 1)$ distribution and let $Z$ have a uniform$(0, \frac{1}{10})$ distribution. Let $X$ and $Z$ be independent. Let $Y = X^2 + Z$ and consider the random vector $(X, Y)$. As in Example 4.5.8, given $X = x$, $Y = x^2 + Z$ and the conditional distribution of $Y$ given $X = x$ is uniform$(x^2, x^2 + \frac{1}{10})$. The joint pdf of $X$ and $Y$, the product of this conditional pdf and the marginal pdf of $X$, is thus

$$f(x, y) = 5, \quad -1 < x < 1, \quad x^2 < y < x^2 + \frac{1}{10}.$$

The set on which $f(x, y) > 0$ is illustrated in Figure 4.5.2. There is a strong relationship between $X$ and $Y$, as indicated by the conditional distribution of $Y$ given $X = x$. But the relationship is not linear. The possible values of $(X, Y)$ cluster around a parabola rather than a straight line. The correlation does not measure this non-linear relationship. In fact, $\rho_{XY} = 0$. Since $X$ has a uniform$(-1, 1)$ distribution, $EX = EX^3 = 0$, and since $X$ and $Z$ are independent, $EXZ = (EX)(EZ)$. Thus,

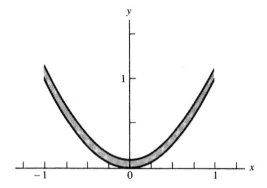

Figure 4.5.2. *Region where $f(x,y) > 0$ for Example 4.5.9*

$$\text{Cov}(X,Y) = E(X(X^2 + Z)) - (EX)(E(X^2 + Z))$$
$$= EX^3 + EXZ - 0E(X^2 + Z)$$
$$= 0 + (EX)(EZ) = 0(EZ) = 0,$$

and $\rho_{XY} = \text{Cov}(X,Y)/(\sigma_X \sigma_Y) = 0.$                    ‖

We close this section by introducing a very important bivariate distribution in which the correlation coefficient arises naturally as a parameter.

**Definition 4.5.10**   Let $-\infty < \mu_X < \infty, -\infty < \mu_Y < \infty, 0 < \sigma_X, 0 < \sigma_Y,$ and $-1 < \rho < 1$ be five real numbers. The *bivariate normal pdf with means $\mu_X$ and $\mu_Y$, variances $\sigma_X^2$ and $\sigma_Y^2$, and correlation $\rho$* is the bivariate pdf given by

$$f(x,y) = \left(2\pi\sigma_X\sigma_Y\sqrt{1-\rho^2}\right)^{-1}$$
$$\times \exp\left(-\frac{1}{2(1-\rho^2)}\left(\left(\frac{x-\mu_X}{\sigma_X}\right)^2\right.\right.$$
$$\left.\left.- 2\rho\left(\frac{x-\mu_X}{\sigma_X}\right)\left(\frac{y-\mu_Y}{\sigma_Y}\right) + \left(\frac{y-\mu_Y}{\sigma_Y}\right)^2\right)\right)$$

for $-\infty < x < \infty$ and $-\infty < y < \infty$.

Although the formula for the bivariate normal pdf looks formidable, this bivariate distribution is one of the most frequently used. (In fact, the derivation of the formula need not be formidable at all. See Exercise 4.46.)

The many nice properties of this distribution include these:

a. The marginal distribution of $X$ is $n(\mu_X, \sigma_X^2)$.

b. The marginal distribution of $Y$ is $n(\mu_Y, \sigma_Y^2)$.

c. The correlation between $X$ and $Y$ is $\rho_{XY} = \rho$.

d. For any constants $a$ and $b$, the distribution of $aX + bY$ is $n(a\mu_X + b\mu_Y, a^2\sigma_X^2 + b^2\sigma_Y^2 + 2ab\rho\sigma_X\sigma_Y)$.

We will leave the verification of properties (a), (b), and (d) as exercises (Exercise 4.45). Assuming (a) and (b) are true, we will prove (c). We have by definition

$$\rho_{XY} = \frac{\text{Cov}(X, Y)}{\sigma_X\sigma_Y}$$

$$= \frac{\text{E}(X - \mu_X)(Y - \mu_Y)}{\sigma_X\sigma_Y}$$

$$= \text{E}\left(\frac{X - \mu_X}{\sigma_X}\right)\left(\frac{Y - \mu_Y}{\sigma_Y}\right)$$

$$= \int_{-\infty}^{\infty}\int_{-\infty}^{\infty}\left(\frac{x - \mu_X}{\sigma_X}\right)\left(\frac{y - \mu_Y}{\sigma_Y}\right)f(x, y)\,dx\,dy.$$

Make the change of variable

$$s = \left(\frac{x - \mu_X}{\sigma_X}\right)\left(\frac{y - \mu_Y}{\sigma_Y}\right) \quad \text{and} \quad t = \left(\frac{x - \mu_X}{\sigma_X}\right).$$

Then $x = \sigma_X t + \mu_X$, $y = (\sigma_Y s/t) + \mu_Y$, and the Jacobian of the transformation is $J = \sigma_X\sigma_Y/t$. With this change of variable, we obtain

$$\rho_{XY} = \int_{-\infty}^{\infty}\int_{-\infty}^{\infty} sf\left(\sigma_X t + \mu_X, \frac{\sigma_Y s}{t} + \mu_Y\right)\left|\frac{\sigma_X\sigma_Y}{t}\right|\,ds\,dt$$

$$= \int_{-\infty}^{\infty}\int_{-\infty}^{\infty} s\left(2\pi\sigma_X\sigma_Y\sqrt{1 - \rho^2}\right)^{-1}$$

$$\times \exp\left(-\frac{1}{2(1 - \rho^2)}\left(t^2 - 2\rho s + \left(\frac{s}{t}\right)^2\right)\right)\frac{\sigma_X\sigma_Y}{|t|}\,ds\,dt.$$

Noting that $|t| = \sqrt{t^2}$ and $t^2 - 2\rho s + \left(\frac{s}{t}\right)^2 = \left(\frac{s - \rho t^2}{t}\right)^2 + (1 - \rho^2)t^2$, we can rewrite this as

$$\rho_{XY} = \int_{-\infty}^{\infty}\frac{1}{\sqrt{2\pi}}\exp\left(-\frac{t^2}{2}\right)\left[\int_{-\infty}^{\infty}\frac{s}{\sqrt{2\pi}\sqrt{(1 - \rho^2)t^2}}\exp\left(-\frac{(s - \rho t^2)^2}{2(1 - \rho^2)t^2}\right)\,ds\right]\,dt.$$

The inner integral is $\text{E}S$, where $S$ is a normal random variable with $\text{E}S = \rho t^2$ and $\text{Var}\,S = (1 - \rho^2)t^2$. Thus the inner integral is $\rho t^2$. Hence we have

$$\rho_{XY} = \int_{-\infty}^{\infty}\frac{\rho t^2}{\sqrt{2\pi}}\exp\left(-\frac{t^2}{2}\right)\,dt.$$

But this integral is $\rho\text{E}T^2$, where $T$ is a $n(0, 1)$ random variable. Hence $\text{E}T^2 = 1$ and $\rho_{XY} = \rho$.

All the conditional distributions of $Y$ given $X = x$ and of $X$ given $Y = y$ are also normal distributions. Using the joint and marginal pdfs given above, it is straightforward to verify that the conditional distribution of $Y$ given $X = x$ is

$$n(\mu_Y + \rho(\sigma_Y/\sigma_X)(x - \mu_X), \sigma_Y^2(1 - \rho^2)).$$

As $\rho$ converges to 1 or $-1$, the conditional variance $\sigma_Y^2(1-\rho^2)$ converges to 0. Thus, the conditional distribution of $Y$ given $X = x$ becomes more concentrated about the point $\mu_Y + \rho(\sigma_Y/\sigma_X)(x - \mu_X)$, and the joint probability distribution of $(X, Y)$ becomes more concentrated about the line $y = \mu_Y + \rho(\sigma_Y/\sigma_X)(x - \mu_X)$. This illustrates again the point made earlier that a correlation near 1 or $-1$ means that there is a line $y = ax + b$ about which the values of $(X, Y)$ cluster with high probability.

Note one important fact: All of the normal marginal and conditional pdfs are derived from the starting point of bivariate normality. The derivation does not go in the opposite direction. That is, marginal normality does not imply joint normality. See Exercise 4.47 for an illustration of this.

## 4.6 Multivariate Distributions

At the beginning of this chapter, we discussed observing more than two random variables in an experiment. In the previous sections our discussions have concentrated on a bivariate random vector $(X, Y)$. In this section we discuss a multivariate random vector $(X_1, \ldots, X_n)$. In the example at the beginning of this chapter, temperature, height, weight, and blood pressure were observed on an individual. In this example, $n = 4$ and the observed random vector is $(X_1, X_2, X_3, X_4)$, where $X_1$ is temperature, $X_2$ is height, etc. The concepts from the earlier sections, including marginal and conditional distributions, generalize from the bivariate to the multivariate setting. We introduce some of these generalizations in this section.

*A note on notation*: We will use boldface letters to denote multiple variates. Thus, we write $\mathbf{X}$ to denote the random variables $X_1, \ldots, X_n$ and $\mathbf{x}$ to denote the sample $x_1, \ldots, x_n$.

The random vector $\mathbf{X} = (X_1, \ldots, X_n)$ has a sample space that is a subset of $\Re^n$. If $(X_1, \ldots, X_n)$ is a discrete random vector (the sample space is countable), then the *joint pmf of* $(X_1, \ldots, X_n)$ is the function defined by $f(\mathbf{x}) = f(x_1, \ldots, x_n) = P(X_1 = x_1, \ldots, X_n = x_n)$ for each $(x_1, \ldots, x_n) \in \Re^n$. Then for any $A \subset \Re^n$,

$$(4.6.1) \qquad\qquad P(\mathbf{X} \in A) = \sum_{\mathbf{x} \in A} f(\mathbf{x}).$$

If $(X_1, \ldots, X_n)$ is a continuous random vector, the *joint pdf of* $(X_1, \ldots, X_n)$ is a function $f(x_1, \ldots, x_n)$ that satisfies

$$(4.6.2) \quad P(\mathbf{X} \in A) = \int \cdots \int_A f(\mathbf{x}) d\mathbf{x} = \int \cdots \int_A f(x_1, \ldots, x_n)\, dx_1 \cdots dx_n.$$

These integrals are $n$-fold integrals with limits of integration set so that the integration is over all points $\mathbf{x} \in A$.

Let $g(\mathbf{x}) = g(x_1, \ldots, x_n)$ be a real-valued function defined on the sample space of $\mathbf{X}$. Then $g(\mathbf{X})$ is a random variable and the *expected value of* $g(\mathbf{X})$ is

$$(4.6.3) \quad \mathrm{E}g(\mathbf{X}) = \int_{-\infty}^{\infty} \cdots \int_{-\infty}^{\infty} g(\mathbf{x}) f(\mathbf{x}) d\mathbf{x} \quad \text{and} \quad \mathrm{E}g(\mathbf{X}) = \sum_{\mathbf{x} \in \Re^n} g(\mathbf{x}) f(\mathbf{x})$$

in the continuous and discrete cases, respectively. These and other definitions are analogous to the bivariate definitions except that now the integrals or sums are over the appropriate subset of $\Re^n$ rather than $\Re^2$.

The *marginal pdf or pmf* of any subset of the coordinates of $(X_1, \ldots, X_n)$ can be computed by integrating or summing the joint pdf or pmf over all possible values of the other coordinates. Thus, for example, the marginal distribution of $(X_1, \ldots, X_k)$, the first $k$ coordinates of $(X_1, \ldots, X_n)$, is given by the pdf or pmf

$$(4.6.4) \qquad f(x_1, \ldots, x_k) = \int_{-\infty}^{\infty} \cdots \int_{-\infty}^{\infty} f(x_1, \ldots, x_n) \, dx_{k+1} \cdots dx_n$$

or

$$(4.6.5) \qquad f(x_1, \ldots, x_k) = \sum_{(x_{k+1}, \ldots, x_n) \in \Re^{n-k}} f(x_1, \ldots, x_n)$$

for every $(x_1, \ldots, x_k) \in \Re^k$. The *conditional pdf or pmf* of a subset of the coordinates of $(X_1, \ldots, X_n)$ given the values of the remaining coordinates is obtained by dividing the joint pdf or pmf by the marginal pdf or pmf of the remaining coordinates. Thus, for example, if $f(x_1, \ldots, x_k) > 0$, the conditional pdf or pmf of $(X_{k+1}, \ldots, X_n)$ given $X_1 = x_1, \ldots, X_k = x_k$ is the function of $(x_{k+1}, \ldots, x_n)$ defined by

$$(4.6.6) \qquad f(x_{k+1}, \ldots, x_n | x_1, \ldots, x_k) = \frac{f(x_1, \ldots, x_n)}{f(x_1, \ldots, x_k)}.$$

These ideas are illustrated in the following example.

**Example 4.6.1 (Multivariate pdfs)** Let $n = 4$ and

$$f(x_1, x_2, x_3, x_4) = \begin{cases} \frac{3}{4}(x_1^2 + x_2^2 + x_3^2 + x_4^2) & 0 < x_i < 1, i = 1, 2, 3, 4 \\ 0 & \text{otherwise.} \end{cases}$$

This nonnegative function is the joint pdf of a random vector $(X_1, X_2, X_3, X_4)$ and it can be verified that

$$\int_{-\infty}^{\infty} \int_{-\infty}^{\infty} \int_{-\infty}^{\infty} \int_{-\infty}^{\infty} f(x_1, x_2, x_3, x_4) \, dx_1 \, dx_2 \, dx_3 \, dx_4$$

$$= \int_0^1 \int_0^1 \int_0^1 \int_0^1 \frac{3}{4}(x_1^2 + x_2^2 + x_3^2 + x_4^2) \, dx_1 \, dx_2 \, dx_3 \, dx_4$$

$$= 1.$$

This joint pdf can be used to compute probabilities such as

$$P\left(X_1 < \frac{1}{2}, X_2 < \frac{3}{4}, X_4 > \frac{1}{2}\right)$$

$$= \int_{\frac{1}{2}}^{1} \int_{0}^{1} \int_{0}^{\frac{3}{4}} \int_{0}^{\frac{1}{2}} \frac{3}{4}\left(x_1^2 + x_2^2 + x_3^2 + x_4^2\right) dx_1\, dx_2\, dx_3\, dx_4.$$

Note how the limits of integration restrict the integration to those values of $(x_1, x_2, x_3, x_4)$ that are in the event in question and for which $f(x_1, x_2, x_3, x_4) > 0$. Each of the four terms, $\frac{3}{4}x_1^2$, $\frac{3}{4}x_2^2$, etc., can be integrated separately and the results summed. For example,

$$\int_{\frac{1}{2}}^{1} \int_{0}^{1} \int_{0}^{\frac{3}{4}} \int_{0}^{\frac{1}{2}} \frac{3}{4}x_1^2\, dx_1\, dx_2\, dx_3\, dx_4 = \frac{3}{256}.$$

The other three integrals are $\frac{7}{1024}$, $\frac{3}{64}$, and $\frac{21}{256}$. Thus

$$P\left(X_1 < \frac{1}{2}, X_2 < \frac{3}{4}, X_4 > \frac{1}{2}\right) = \frac{3}{256} + \frac{7}{1024} + \frac{3}{64} + \frac{21}{256} = \frac{151}{1024}.$$

Using (4.6.4), we can obtain the marginal pdf of $(X_1, X_2)$ by integrating out the variables $x_3$ and $x_4$ to obtain

$$f(x_1, x_2) = \int_{-\infty}^{\infty} \int_{-\infty}^{\infty} f(x_1, x_2, x_3, x_4)\, dx_3\, dx_4$$

$$= \int_{0}^{1} \int_{0}^{1} \frac{3}{4}\left(x_1^2 + x_2^2 + x_3^2 + x_4^2\right) dx_3\, dx_4 = \frac{3}{4}\left(x_1^2 + x_2^2\right) + \frac{1}{2}$$

for $0 < x_1 < 1$ and $0 < x_2 < 1$. Any probability or expected value that involves only $X_1$ and $X_2$ can be computed using this marginal pdf. For example,

$$EX_1 X_2 = \int_{-\infty}^{\infty} \int_{-\infty}^{\infty} x_1 x_2 f(x_1, x_2)\, dx_1\, dx_2$$

$$= \int_{0}^{1} \int_{0}^{1} x_1 x_2 \left(\frac{3}{4}\left(x_1^2 + x_2^2\right) + \frac{1}{2}\right) dx_1\, dx_2$$

$$= \int_{0}^{1} \int_{0}^{1} \left(\frac{3}{4}x_1^3 x_2 + \frac{3}{4}x_1 x_2^3 + \frac{1}{2}x_1 x_2\right) dx_1\, dx_2$$

$$= \int_{0}^{1} \left(\frac{3}{16}x_2 + \frac{3}{8}x_2^3 + \frac{1}{4}x_2\right) dx_2 = \frac{3}{32} + \frac{3}{32} + \frac{1}{8} = \frac{5}{16}.$$

For any $(x_1, x_2)$ with $0 < x_1 < 1$ and $0 < x_2 < 1$, $f(x_1, x_2) > 0$ and the conditional pdf of $(X_3, X_4)$ given $X_1 = x_1$ and $X_2 = x_2$ can be found using (4.6.6). For any such $(x_1, x_2)$, $f(x_1, x_2, x_3, x_4) > 0$ if $0 < x_3 < 1$ and $0 < x_4 < 1$, and for these values of $(x_3, x_4)$, the conditional pdf is

$$f(x_3, x_4 | x_1, x_2) = \frac{f(x_1, x_2, x_3, x_4)}{f(x_1, x_2)}$$

$$= \frac{\frac{3}{4}(x_1^2 + x_2^2 + x_3^2 + x_4^2)}{\frac{3}{4}(x_1^2 + x_2^2) + \frac{1}{2}}$$

$$= \frac{x_1^2 + x_2^2 + x_3^2 + x_4^2}{x_1^2 + x_2^2 + \frac{2}{3}}.$$

For example, the conditional pdf of $(X_3, X_4)$ given $X_1 = \frac{1}{3}$ and $X_2 = \frac{2}{3}$ is

$$f\left(x_3, x_4 \middle| x_1 = \frac{1}{3}, x_2 = \frac{2}{3}\right) = \frac{\left(\frac{1}{3}\right)^2 + \left(\frac{2}{3}\right)^2 + x_3^2 + x_4^2}{\left(\frac{1}{3}\right)^2 + \left(\frac{2}{3}\right)^2 + \frac{2}{3}} = \frac{5}{11} + \frac{9}{11}x_3^2 + \frac{9}{11}x_4^2.$$

This can be used to compute

$$P\left(X_3 > \frac{3}{4}, X_4 < \frac{1}{2} \middle| X_1 = \frac{1}{3}, X_2 = \frac{2}{3}\right) = \int_0^{\frac{1}{2}} \int_{\frac{3}{4}}^1 \left(\frac{5}{11} + \frac{9}{11}x_3^2 + \frac{9}{11}x_4^2\right) dx_3\, dx_4$$

$$= \int_0^{\frac{1}{2}} \left(\frac{5}{44} + \frac{111}{704} + \frac{9}{44}x_4^2\right) dx_4$$

$$= \frac{5}{88} + \frac{111}{1408} + \frac{3}{352} = \frac{203}{1408}. \qquad \|$$

Before giving examples of computations with conditional and marginal distributions for a discrete multivariate random vector, we will introduce an important family of discrete multivariate distributions. This family generalizes the binomial family to the situation in which each trial has $n$ (rather than two) distinct possible outcomes.

**Definition 4.6.2**   Let $n$ and $m$ be positive integers and let $p_1, \ldots, p_n$ be numbers satisfying $0 \le p_i \le 1$, $i = 1, \ldots, n$, and $\sum_{i=1}^n p_i = 1$. Then the random vector $(X_1, \ldots, X_n)$ has a *multinomial distribution with $m$ trials and cell probabilities* $p_1, \ldots, p_n$ if the joint pmf of $(X_1, \ldots, X_n)$ is

$$f(x_1, \ldots, x_n) = \frac{m!}{x_1! \cdot \cdots \cdot x_n!} p_1^{x_1} \cdot \cdots \cdot p_n^{x_n} = m! \prod_{i=1}^n \frac{p_i^{x_i}}{x_i!}$$

on the set of $(x_1, \ldots, x_n)$ such that each $x_i$ is a nonnegative integer and $\sum_{i=1}^n x_i = m$.

The multinomial distribution is a model for the following kind of experiment. The experiment consists of $m$ independent trials. Each trial results in one of $n$ distinct possible outcomes. The probability of the $i$th outcome is $p_i$ on every trial. And $X_i$ is the count of the number of times the $i$th outcome occurred in the $m$ trials. For $n = 2$, this is just a binomial experiment in which each trial has $n = 2$ possible outcomes and $X_1$ counts the number of "successes" and $X_2 = m - X_1$ counts the number of "failures" in $m$ trials. In a general multinomial experiment, there are $n$ different possible outcomes to count.

**Example 4.6.3 (Multivariate pmf)**  Consider tossing a six-sided die ten times. Suppose the die is unbalanced so that the probability of observing a 1 is $\frac{1}{21}$, the probability of observing a 2 is $\frac{2}{21}$, and, in general, the probability of observing an $i$ is $\frac{i}{21}$. Now consider the random vector $(X_1, \ldots, X_6)$, where $X_i$ counts the number of times $i$ comes up in the ten tosses. Then $(X_1, \ldots, X_6)$ has a multinomial distribution with $m = 10$ trials, $n = 6$ possible outcomes, and cell probabilities $p_1 = \frac{1}{21}, p_2 = \frac{2}{21}, \ldots, p_6 = \frac{6}{21}$. The formula in Definition 4.6.2 may be used to calculate the probability of rolling four 6s, three 5s, two 4s, and one 3 to be

$$
f(0,0,1,2,3,4) = \frac{10!}{0!0!1!2!3!4!} \left(\frac{1}{21}\right)^0 \left(\frac{2}{21}\right)^0 \left(\frac{3}{21}\right)^1 \left(\frac{4}{21}\right)^2 \left(\frac{5}{21}\right)^3 \left(\frac{6}{21}\right)^4
$$

$$
= .0059. \qquad\qquad\qquad \|
$$

The factor $m!/(x_1! \cdots \cdots x_n!)$ is called a *multinomial coefficient*. It is the number of ways that $m$ objects can be divided into $n$ groups with $x_1$ in the first group, $x_2$ in the second group, $\ldots$, and $x_n$ in the $n$th group. A generalization of the Binomial Theorem 3.2.2 is the Multinomial Theorem.

**Theorem 4.6.4 (Multinomial Theorem)** *Let $m$ and $n$ be positive integers. Let $\mathcal{A}$ be the set of vectors $\mathbf{x} = (x_1, \ldots, x_n)$ such that each $x_i$ is a nonnegative integer and $\sum_{i=1}^{n} x_i = m$. Then, for any real numbers $p_1, \ldots, p_n$,*

$$
(p_1 + \cdots + p_n)^m = \sum_{\mathbf{x} \in \mathcal{A}} \frac{m!}{x_1! \cdots \cdots x_n!} p_1^{x_1} \cdots \cdots p_n^{x_n}.
$$

Theorem 4.6.4 shows that a multinomial pmf sums to 1. The set $\mathcal{A}$ is the set of points with positive probability in Definition 4.6.2. The sum of the pmf over all those points is, by Theorem 4.6.4, $(p_1 + \cdots + p_n)^m = 1^m = 1$.

Now we consider some marginal and conditional distributions for the multinomial model. Consider a single coordinate $X_i$. If the occurrence of the $i$th outcome is labeled a "success" and anything else is labeled a "failure," then $X_i$ is the count of the number of successes in $m$ independent trials where the probability of a success is $p_i$ on each trial. Thus $X_i$ should have a binomial$(m, p_i)$ distribution. To verify this the marginal distribution of $X_i$ should be computed using (4.6.5). For example, consider the marginal pmf of $X_n$. For a fixed value of $x_n \in \{0, 1, \ldots, n\}$, to compute the marginal pmf $f(x_n)$, we must sum over all possible values of $(x_1, \ldots, x_{n-1})$. That is, we must sum over all $(x_1, \ldots, x_{n-1})$ such that the $x_i$s are all nonnegative integers and $\sum_{i=1}^{n-1} x_i = m - x_n$. Denote this set by $\mathcal{B}$. Then

$$
f(x_n) = \sum_{(x_1, \ldots, x_{n-1}) \in \mathcal{B}} \frac{m!}{x_1! \cdots \cdots x_n!} (p_1)^{x_1} \cdots \cdots (p_n)^{x_n}
$$

$$
= \sum_{(x_1, \ldots, x_{n-1}) \in \mathcal{B}} \frac{m!}{x_1! \cdots \cdots x_n!} p_1^{x_1} \cdots \cdots p_n^{x_n} \frac{(m - x_n)! \, (1 - p_n)^{m - x_n}}{(m - x_n)! \, (1 - p_n)^{m - x_n}}
$$

$$= \frac{m!}{x_n!(m-x_n)!} p_n^{x_n} (1-p_n)^{m-x_n}$$

$$\times \sum_{(x_1,\ldots,x_{n-1})\in\mathcal{B}} \frac{(m-x_n)!}{x_1!\cdot\cdots\cdot x_{n-1}!} \left(\frac{p_1}{1-p_n}\right)^{x_1} \cdots \cdot \left(\frac{p_{n-1}}{1-p_n}\right)^{x_{n-1}}.$$

But using the facts that $x_1 + \cdots + x_{n-1} = m - x_n$ and $p_1 + \cdots + p_{n-1} = 1 - p_n$ and Theorem 4.6.4, we see that the last summation is 1. Hence the marginal distribution of $X_n$ is binomial$(m, p_n)$. Similar arguments show that each of the other coordinates is marginally binomially distributed.

Given that $X_n = x_n$, there must have been $m - x_n$ trials that resulted in one of the first $n - 1$ outcomes. The vector $(X_1, \ldots, X_{n-1})$ counts the number of these $m - x_n$ trials that are of each type. Thus it seems that given $X_n = x_n$, $(X_1, \ldots, X_{n-1})$ might have a multinomial distribution. This is true. From (4.6.6), the conditional pmf of $(X_1, \ldots, X_{n-1})$ given $X_n = x_n$ is

$$f(x_1, \ldots, x_{n-1}|x_n) = \frac{f(x_1, \ldots, x_n)}{f(x_n)}$$

$$= \frac{\frac{m!}{x_1!\cdot\cdots\cdot x_n!}(p_1)^{x_1}\cdot\cdots\cdot(p_n)^{x_n}}{\frac{m!}{x_n!(m-x_n)!}(p_n)^{x_n}(1-p_n)^{m-x_n}}$$

$$= \frac{(m-x_n)!}{x_1!\cdot\cdots\cdot x_{n-1}!} \left(\frac{p_1}{1-p_n}\right)^{x_1} \cdots \cdot \left(\frac{p_{n-1}}{1-p_n}\right)^{x_{n-1}}.$$

This is the pmf of a multinomial distribution with $m - x_n$ trials and cell probabilities $p_1/(1-p_n), \ldots, p_{n-1}/(1-p_n)$. In fact, the conditional distribution of any subset of the coordinates of $(X_1, \ldots, X_n)$ given the values of the rest of the coordinates is a multinomial distribution.

We see from the conditional distributions that the coordinates of the vector $(X_1, \ldots, X_n)$ are related. In particular, there must be some negative correlation. It turns out that all of the pairwise covariances are negative and are given by (Exercise 4.39)

$$\text{Cov}(X_i, X_j) = \text{E}[(X_i - p_i)(X_j - p_j)] = -mp_ip_j.$$

Thus, the negative correlation is greater for variables with higher success probabilities. This makes sense, as the variable total is constrained at $m$, so if one starts to get big, the other tends not to.

**Definition 4.6.5**    Let $\mathbf{X}_1, \ldots, \mathbf{X}_n$ be random vectors with joint pdf or pmf $f(\mathbf{x}_1, \ldots, \mathbf{x}_n)$. Let $f_{\mathbf{X}_i}(\mathbf{x}_i)$ denote the marginal pdf or pmf of $\mathbf{X}_i$. Then $\mathbf{X}_1, \ldots, \mathbf{X}_n$ are called *mutually independent random vectors* if, for every $(\mathbf{x}_1, \ldots, \mathbf{x}_n)$,

$$f(\mathbf{x}_1, \ldots, \mathbf{x}_n) = f_{\mathbf{X}_1}(\mathbf{x}_1)\cdot\cdots\cdot f_{\mathbf{X}_n}(\mathbf{x}_n) = \prod_{i=1}^{n} f_{\mathbf{X}_i}(\mathbf{x}_i).$$

If the $X_i$s are all one-dimensional, then $X_1, \ldots, X_n$ are called *mutually independent random variables*.

If $X_1, \ldots, X_n$ are mutually independent, then knowledge about the values of some coordinates gives us no information about the values of the other coordinates. Using Definition 4.6.5, one can show that the conditional distribution of any subset of the coordinates, given the values of the rest of the coordinates, is the same as the marginal distribution of the subset. Mutual independence implies that any pair, say $X_i$ and $X_j$, are pairwise independent. That is, the bivariate marginal pdf or pmf, $f(x_i, x_j)$, satisfies Definition 4.2.5. But mutual independence implies more than pairwise independence. As in Example 1.3.11, it is possible to specify a probability distribution for $(X_1, \ldots, X_n)$ with the property that each pair, $(X_i, X_j)$, is pairwise independent but $X_1, \ldots, X_n$ are not mutually independent.

Mutually independent random variables have many nice properties. The proofs of the following theorems are analogous to the proofs of their counterparts in Sections 4.2 and 4.3.

**Theorem 4.6.6 (Generalization of Theorem 4.2.10)** *Let $X_1, \ldots, X_n$ be mutually independent random variables. Let $g_1, \ldots, g_n$ be real-valued functions such that $g_i(x_i)$ is a function only of $x_i$, $i = 1, \ldots, n$. Then*

$$E(g_1(X_1) \cdots g_n(X_n)) = (Eg_1(X_1)) \cdots (Eg_n(X_n)).$$

**Theorem 4.6.7 (Generalization of Theorem 4.2.12)** *Let $X_1, \ldots, X_n$ be mutually independent random variables with mgfs $M_{X_1}(t), \ldots, M_{X_n}(t)$. Let $Z = X_1 + \cdots + X_n$. Then the mgf of $Z$ is*

$$M_Z(t) = M_{X_1}(t) \cdots M_{X_n}(t).$$

*In particular, if $X_1, \ldots, X_n$ all have the same distribution with mgf $M_X(t)$, then*

$$M_Z(t) = (M_X(t))^n.$$

**Example 4.6.8 (Mgf of a sum of gamma variables)** Suppose $X_1, \ldots, X_n$ are mutually independent random variables, and the distribution of $X_i$ is gamma$(\alpha_i, \beta)$. From Example 2.3.8, the mgf of a gamma$(\alpha, \beta)$ distribution is $M(t) = (1 - \beta t)^{-\alpha}$. Thus, if $Z = X_1 + \cdots + X_n$, the mgf of $Z$ is

$$M_Z(t) = M_{X_1}(t) \cdots M_{X_n}(t) = (1 - \beta t)^{-\alpha_1} \cdots (1 - \beta t)^{-\alpha_n} = (1 - \beta t)^{-(\alpha_1 + \cdots + \alpha_n)}.$$

This is the mgf of a gamma$(\alpha_1 + \cdots + \alpha_n, \beta)$ distribution. Thus, the sum of independent gamma random variables that have a common scale parameter $\beta$ also has a gamma distribution. ‖

A generalization of Theorem 4.6.7 is obtained if we consider a sum of linear functions of independent random variables.

**Corollary 4.6.9** *Let $X_1, \ldots, X_n$ be mutually independent random variables with mgfs $M_{X_1}(t), \ldots, M_{X_n}(t)$. Let $a_1, \ldots, a_n$ and $b_1, \ldots, b_n$ be fixed constants. Let $Z = (a_1 X_1 + b_1) + \cdots + (a_n X_n + b_n)$. Then the mgf of $Z$ is*

$$M_Z(t) = (e^{t(\Sigma b_i)}) M_{X_1}(a_1 t) \cdots M_{X_n}(a_n t).$$

**Proof:** From the definition, the mgf of $Z$ is

$$M_Z(t) = \mathrm{E}e^{tZ}$$

$$= \mathrm{E}e^{t\Sigma(a_iX_i+b_i)}$$

$$= (e^{t(\Sigma b_i)})\mathrm{E}(e^{ta_1X_1}\cdot\cdots\cdot e^{ta_nX_n}) \quad \binom{\text{properties of exponentials}}{\text{and expectations}}$$

$$= (e^{t(\Sigma b_i)})M_{X_1}(a_1t)\cdot\cdots\cdot M_{X_n}(a_nt), \qquad \text{(Theorem 4.6.6)}$$

as was to be shown. $\qquad\square$

Undoubtedly, the most important application of Corollary 4.6.9 is to the case of normal random variables. *A linear combination of independent normal random variables is normally distributed.*

**Corollary 4.6.10**  *Let $X_1,\ldots,X_n$ be mutually independent random variables with $X_i \sim n(\mu_i, \sigma_i^2)$. Let $a_1,\ldots,a_n$ and $b_1,\ldots,b_n$ be fixed constants. Then*

$$Z = \sum_{i=1}^{n}(a_iX_i + b_i) \sim n\left(\sum_{i=1}^{n}(a_i\mu_i + b_i), \sum_{i=1}^{n}a_i^2\sigma_i^2\right).$$

**Proof:** Recall that the mgf of a $n(\mu, \sigma^2)$ random variable is $M(t) = e^{\mu t + \sigma^2 t^2/2}$. Substituting into the expression in Corollary 4.6.9 yields

$$M_Z(t) = (e^{t(\Sigma b_i)})e^{\mu_1 a_1 t + \sigma_1^2 a_1^2 t^2/2}\cdot\cdots\cdot e^{\mu_n a_n t + \sigma_n^2 a_n^2 t^2/2}$$

$$= e^{\left((\Sigma(a_i\mu_i+b_i))t + (\Sigma a_i^2\sigma_i^2)t^2/2\right)},$$

the mgf of the indicated normal distribution. $\qquad\square$

**Theorem 4.6.11 (Generalization of Lemma 4.2.7)** *Let $\mathbf{X}_1,\ldots,\mathbf{X}_n$ be random vectors. Then $\mathbf{X}_1,\ldots,\mathbf{X}_n$ are mutually independent random vectors if and only if there exist functions $g_i(\mathbf{x}_i), i = 1,\ldots,n$, such that the joint pdf or pmf of $(\mathbf{X}_1,\ldots,\mathbf{X}_n)$ can be written as*

$$f(\mathbf{x}_1,\ldots,\mathbf{x}_n) = g_1(\mathbf{x}_1)\cdot\cdots\cdot g_n(\mathbf{x}_n).$$

**Theorem 4.6.12 (Generalization of Theorem 4.3.5)** *Let $\mathbf{X}_1,\ldots,\mathbf{X}_n$ be independent random vectors. Let $g_i(\mathbf{x}_i)$ be a function only of $\mathbf{x}_i, i = 1, \ldots,n$. Then the random variables $U_i = g_i(\mathbf{X}_i), i = 1,\ldots,n$, are mutually independent.*

We close this section by describing the generalization of a technique for finding the distribution of a transformation of a random vector. We will present the generalization of formula (4.3.6) that gives the pdf of the new random vector in terms of the pdf of the original random vector. Note that to fully understand the remainder of this section, some knowledge of matrix algebra is required. (See, for example, Searle 1982.) In particular, we will need to compute the determinant of a matrix. This is the only place in the book where such knowledge is required.

Let $(X_1, \ldots, X_n)$ be a random vector with pdf $f_{\mathbf{X}}(x_1, \ldots, x_n)$. Let $\mathcal{A} = \{\mathbf{x} : f_{\mathbf{X}}(\mathbf{x}) > 0\}$. Consider a new random vector $(U_1, \ldots, U_n)$, defined by $U_1 = g_1(X_1, \ldots, X_n), U_2 = g_2(X_1, \ldots, X_n), \ldots, U_n = g_n(X_1, \ldots, X_n)$. Suppose that $A_0, A_1, \ldots, A_k$ form a partition of $\mathcal{A}$ with these properties. The set $A_0$, which may be empty, satisfies $P((X_1, \ldots, X_n) \in A_0) = 0$. The transformation $(U_1, \ldots, U_n) = (g_1(\mathbf{X}), \ldots, g_n(\mathbf{X}))$ is a one-to-one transformation from $A_i$ onto $\mathcal{B}$ for each $i = 1, 2, \ldots, k$. Then for each $i$, the inverse functions from $\mathcal{B}$ to $A_i$ can be found. Denote the $i$th inverse by $x_1 = h_{1i}(u_1, \ldots, u_n), x_2 = h_{2i}(u_1, \ldots, u_n), \ldots, x_n = h_{ni}(u_1, \ldots, u_n)$. This $i$th inverse gives, for $(u_1, \ldots, u_n) \in \mathcal{B}$, the unique $(x_1, \ldots, x_n) \in A_i$ such that $(u_1, \ldots, u_n) = (g_1(x_1, \ldots, x_n), \ldots, g_n(x_1, \ldots, x_n))$. Let $J_i$ denote the Jacobian computed from the $i$th inverse. That is,

$$
J_i = \begin{vmatrix} \dfrac{\partial x_1}{\partial u_1} & \dfrac{\partial x_1}{\partial u_2} & \cdots & \dfrac{\partial x_1}{\partial u_n} \\ \dfrac{\partial x_2}{\partial u_1} & \dfrac{\partial x_2}{\partial u_2} & \cdots & \dfrac{\partial x_2}{\partial u_n} \\ \vdots & \vdots & \ddots & \vdots \\ \dfrac{\partial x_n}{\partial u_1} & \dfrac{\partial x_n}{\partial u_2} & \cdots & \dfrac{\partial x_n}{\partial u_n} \end{vmatrix} = \begin{vmatrix} \dfrac{\partial h_{1i}(\mathbf{u})}{\partial u_1} & \dfrac{\partial h_{1i}(\mathbf{u})}{\partial u_2} & \cdots & \dfrac{\partial h_{1i}(\mathbf{u})}{\partial u_n} \\ \dfrac{\partial h_{2i}(\mathbf{u})}{\partial u_1} & \dfrac{\partial h_{2i}(\mathbf{u})}{\partial u_2} & \cdots & \dfrac{\partial h_{2i}(\mathbf{u})}{\partial u_n} \\ \vdots & \vdots & \ddots & \vdots \\ \dfrac{\partial h_{ni}(\mathbf{u})}{\partial u_1} & \dfrac{\partial h_{ni}(\mathbf{u})}{\partial u_2} & \cdots & \dfrac{\partial h_{ni}(\mathbf{u})}{\partial u_n} \end{vmatrix},
$$

the determinant of an $n \times n$ matrix. Assuming that these Jacobians do not vanish identically on $\mathcal{B}$, we have the following representation of the joint pdf, $f_{\mathbf{U}}(u_1, \ldots, u_n)$, for $\mathbf{u} \in \mathcal{B}$:

$$(4.6.7) \qquad f_{\mathbf{U}}(u_1, \ldots, u_n) = \sum_{i=1}^{k} f_{\mathbf{X}}(h_{1i}(u_1, \ldots, u_n), \ldots, h_{ni}(u_1, \ldots, u_n))|J_i|.$$

**Example 4.6.13 (Multivariate change of variables)** Let $(X_1, X_2, X_3, X_4)$ have joint pdf

$$f_{\mathbf{X}}(x_1, x_2, x_3, x_4) = 24e^{-x_1-x_2-x_3-x_4}, \quad 0 < x_1 < x_2 < x_3 < x_4 < \infty.$$

Consider the transformation

$$U_1 = X_1, \quad U_2 = X_2 - X_1, \quad U_3 = X_3 - X_2, \quad U_4 = X_4 - X_3.$$

This transformation maps the set $\mathcal{A}$ onto the set $\mathcal{B} = \{\mathbf{u} : 0 < u_i < \infty, i = 1, 2, 3, 4\}$. The transformation is one-to-one, so $k = 1$, and the inverse is

$$X_1 = U_1, \quad X_2 = U_1 + U_2, \quad X_3 = U_1 + U_2 + U_3, \quad X_4 = U_1 + U_2 + U_3 + U_4.$$

The Jacobian of the inverse is

$$J = \begin{vmatrix} 1 & 0 & 0 & 0 \\ 1 & 1 & 0 & 0 \\ 1 & 1 & 1 & 0 \\ 1 & 1 & 1 & 1 \end{vmatrix} = 1.$$

Since the matrix is triangular, the determinant is equal to the product of the diagonal elements. Thus, from (4.6.7) we obtain

$$f_{\mathbf{U}}(u_1, \ldots, u_4) = 24e^{-u_1-(u_1+u_2)-(u_1+u_2+u_3)-(u_1+u_2+u_3+u_4)}$$
$$= 24e^{-4u_1-3u_2-2\mu_3-u_4}$$

on $\mathcal{B}$. From this the marginal pdfs of $U_1, U_2, U_3$, and $U_4$ can be calculated. It turns out that $f_U(u_i) = (5-i)e^{-(5-i)u_i}, 0 < u_i$; that is, $U_i \sim$ exponential$(1/(5-i))$. From Theorem 4.6.11 we see that $U_1, U_2$, $U_3$, and $U_4$ are mutually independent random variables.                                                                                        ‖

The model in Example 4.6.13 can arise in the following way. Suppose $Y_1, Y_2, Y_3$, and $Y_4$ are mutually independent random variables, each with an exponential(1) distribution. Define $X_1 = \min(Y_1, Y_2, Y_3, Y_4)$, $X_2 =$ second smallest value of $(Y_1, Y_2, Y_3, Y_4)$, $X_3 =$ second largest value of $(Y_1, Y_2, Y_3, Y_4)$, and $X_4 = \max(Y_1, Y_2, Y_3, Y_4)$. These variables will be called *order statistics* in Section 5.5. There we will see that the joint pdf of $(X_1, X_2, X_3, X_4)$ is the pdf given in Example 4.6.13. Now the variables $U_2$, $U_3$, and $U_4$ defined in the example are called the *spacings* between the order statistics. The example showed that, for these exponential random variables $(Y_1, \ldots, Y_n)$, the spacings between the order statistics are mutually independent and also have exponential distributions.

## 4.7 Inequalities

In Section 3.6 we saw inequalities that were derived using probabilistic arguments. In this section we will see inequalities that apply to probabilities and expectations but are based on arguments that use properties of functions and numbers.

### 4.7.1 Numerical Inequalities

The inequalities in this subsection, although often stated in terms of expectations, rely mainly on properties of numbers. In fact, they are all based on the following simple lemma.

**Lemma 4.7.1**  *Let $a$ and $b$ be any positive numbers, and let $p$ and $q$ be any positive numbers (necessarily greater than 1) satisfying*

(4.7.1)
$$\frac{1}{p} + \frac{1}{q} = 1.$$

*Then*

(4.7.2)
$$\frac{1}{p}a^p + \frac{1}{q}b^q \geq ab$$

*with equality if and only if $a^p = b^q$.*

**Proof:** Fix $b$, and consider the function

$$g(a) = \frac{1}{p}a^p + \frac{1}{q}b^q - ab.$$

To minimize $g(a)$, differentiate and set equal to 0:

$$\frac{d}{da}g(a) = 0 \Rightarrow a^{p-1} - b = 0 \Rightarrow b = a^{p-1}.$$

A check of the second derivative will establish that this is indeed a minimum. The value of the function at the minimum is

$$\frac{1}{p}a^p + \frac{1}{q}\left(a^{p-1}\right)^q - aa^{p-1} = \frac{1}{p}a^p + \frac{1}{q}a^p - a^p \quad \left(\begin{array}{c}(p-1)q = p \text{ follows}\\ \text{from (4.7.1)}\end{array}\right)$$

$$= 0. \qquad\qquad \text{(again from (4.7.1))}$$

Hence the minimum is 0 and (4.7.2) is established. Since the minimum is unique (why?), equality holds only if $a^{p-1} = b$, which is equivalent to $a^p = b^q$, again from (4.7.1). $\qquad\qquad\square$

The first of our expectation inequalities, one of the most used and most important, follows easily from the lemma.

**Theorem 4.7.2 (Hölder's Inequality)** *Let $X$ and $Y$ be any two random variables, and let $p$ and $q$ satisfy (4.7.1). Then*

(4.7.3) $\qquad\qquad |EXY| \leq E|XY| \leq (E|X|^p)^{1/p} (E|Y|^q)^{1/q}.$

**Proof:** The first inequality follows from $-|XY| \leq XY \leq |XY|$ and Theorem 2.2.5. To prove the second inequality, define

$$a = \frac{|X|}{(E|X|^p)^{1/p}} \quad \text{and} \quad b = \frac{|Y|}{(E|Y|^q)^{1/q}}.$$

Applying Lemma 4.7.1, we get

$$\frac{1}{p}\frac{|X|^p}{E|X|^p} + \frac{1}{q}\frac{|Y|^q}{E|Y|^q} \geq \frac{|XY|}{(E|X|^p)^{1/p}(E|Y|^q)^{1/q}}.$$

Now take expectations of both sides. The expectation of the left-hand side is 1, and rearrangement gives (4.7.3). $\qquad\qquad\square$

Perhaps the most famous special case of Hölder's Inequality is that for which $p = q = 2$. This is called the Cauchy–Schwarz Inequality.

**Theorem 4.7.3 (Cauchy–Schwarz Inequality)** *For any two random variables $X$ and $Y$,*

(4.7.4) $\qquad\qquad |EXY| \leq E|XY| \leq (E|X|^2)^{1/2}(E|Y|^2)^{1/2}.$

**Example 4.7.4 (Covariance Inequality-I)** If $X$ and $Y$ have means $\mu_X$ and $\mu_Y$ and variances $\sigma_X^2$ and $\sigma_Y^2$, respectively, we can apply the Cauchy–Schwarz Inequality to get

$$|\mathrm{E}(X - \mu_X)(Y - \mu_Y)| \leq \left\{\mathrm{E}(X - \mu_X)^2\right\}^{1/2} \left\{\mathrm{E}(Y - \mu_Y)^2\right\}^{1/2}.$$

Squaring both sides and using statistical notation, we have

$$(\mathrm{Cov}(X, Y))^2 \leq \sigma_X^2 \sigma_Y^2.$$

Recalling the definition of the correlation coefficient, $\rho$, we have proved that $0 \leq \rho^2 \leq 1$. Furthermore, the condition for equality in Lemma 4.7.1 still carries over, and equality is attained here only if $X - \mu_X = c(Y - \mu_Y)$, for some constant $c$. That is, the correlation is $\pm 1$ if and only if $X$ and $Y$ are linearly related. Compare the ease of this proof to the one used in Theorem 4.5.7, before we had the Cauchy–Schwarz Inequality.                                                                                      ‖

Some other special cases of Hölder's Inequality are often useful. If we set $Y \equiv 1$ in (4.7.3), we get

$$(4.7.5) \qquad \mathrm{E}|X| \leq \{\mathrm{E}(|X|^p)\}^{1/p}, \quad 1 < p < \infty.$$

For $1 < r < p$, if we replace $|X|$ by $|X|^r$ in (4.7.5), we obtain

$$\mathrm{E}|X|^r \leq \{\mathrm{E}(|X|^{pr})\}^{1/p}.$$

Now write $s = pr$ (note that $s > r$) and rearrange terms to get

$$(4.7.6) \qquad \{\mathrm{E}|X|^r\}^{1/r} \leq \{\mathrm{E}|X|^s\}^{1/s}, \quad 1 < r < s < \infty,$$

which is known as *Liapounov's Inequality*.

Our next named inequality is similar in spirit to Hölder's Inequality and, in fact, follows from it.

**Theorem 4.7.5 (Minkowski's Inequality)** *Let $X$ and $Y$ be any two random variables. Then for $1 \leq p < \infty$,*

$$(4.7.7) \qquad [\mathrm{E}|X + Y|^p]^{1/p} \leq [\mathrm{E}|X|^p]^{1/p} + [\mathrm{E}|Y|^p]^{1/p}.$$

**Proof:** Write

$$\mathrm{E}|X + Y|^p = \mathrm{E}\left(|X + Y||X + Y|^{p-1}\right)$$

$$(4.7.8) \qquad \leq \mathrm{E}\left(|X||X + Y|^{p-1}\right) + \mathrm{E}\left(|Y||X + Y|^{p-1}\right),$$

where we have used the fact that $|X + Y| \leq |X| + |Y|$ (the *triangle inequality*; see Exercise 4.64). Now apply Hölder's Inequality to each expectation on the right-hand side of (4.7.8) to get

$$\mathrm{E}(|X + Y|^p) \leq [\mathrm{E}(|X|^p)]^{1/p}[\mathrm{E}|X + Y|^{q(p-1)}]^{1/q}$$

$$+ [\mathrm{E}(|Y|^p)]^{1/p}[\mathrm{E}|X + Y|^{q(p-1)}]^{1/q},$$

where $q$ satisfies $1/p + 1/q = 1$. Now divide through by $\left[E(|X+Y|^{q(p-1)})\right]^{1/q}$. Noting that $q(p-1) = p$ and $1 - 1/q = 1/p$, we obtain (4.7.7). $\qquad\square$

The preceding theorems also apply to numerical sums where there is no explicit reference to an expectation. For example, for numbers $a_i, b_i, i = 1, \ldots, n$, the inequality

$$(4.7.9) \qquad \sum_{i=1}^{n} |a_i b_i| \leq \left( \sum_{i=1}^{n} a_i^p \right)^{1/p} \left( \sum_{i=1}^{n} b_i^q \right)^{1/q}, \qquad \frac{1}{p} + \frac{1}{q} = 1,$$

is a version of Hölder's Inequality. To establish (4.7.9) we can formally set up an expectation with respect to random variables taking values $a_1, \ldots, a_n$ and $b_1, \ldots, b_n$. (This is done in Example 4.7.8.)

An important special case of (4.7.9) occurs when $b_i \equiv 1, p = q = 2$. We then have

$$\frac{1}{n} \left( \sum_{i=1}^{n} |a_i| \right)^2 \leq \sum_{i=1}^{n} a_i^2.$$

### 4.7.2 Functional Inequalities

The inequalities in this section rely on properties of real-valued functions rather than on any statistical properties. In many cases, however, they prove to be very useful. One of the most useful is Jensen's Inequality, which applies to convex functions.

**Definition 4.7.6** A function $g(x)$ is *convex* if $g(\lambda x + (1-\lambda)y) \leq \lambda g(x) + (1-\lambda)g(y)$, for all $x$ and $y$, and $0 < \lambda < 1$. The function $g(x)$ is *concave* if $-g(x)$ is convex.

Informally, we can think of convex functions as functions that "hold water"—that is, they are bowl-shaped ($g(x) = x^2$ is convex), while concave functions "spill water" ($g(x) = \log x$ is concave). More formally, convex functions lie below lines connecting any two points (see Figure 4.7.1). As $\lambda$ goes from 0 to 1, $\lambda g(x_1) + (1 - \lambda)g(x_2)$

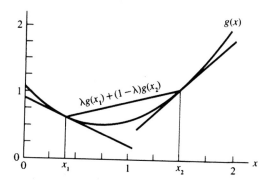

Figure 4.7.1. *Convex function and tangent lines at $x_1$ and $x_2$*

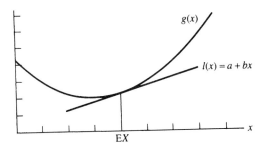

Figure 4.7.2. *Graphical illustration of Jensen's Inequality*

defines a line connecting $g(x_1)$ and $g(x_2)$. This line lies above $g(x)$ if $g(x)$ is convex. Furthermore, a convex function lies above all of its tangent lines (also shown in Figure 4.7.1), and that fact is the basis of Jensen's Inequality.

**Theorem 4.7.7 (Jensen's Inequality)** *For any random variable $X$, if $g(x)$ is a convex function, then*

$$\mathrm{E}g(X) \geq g(\mathrm{E}X).$$

*Equality holds if and only if, for every line $a + bx$ that is tangent to $g(x)$ at $x = EX$,*
$$P(g(X) = a + bX) = 1.$$

**Proof:** To establish the inequality, let $l(x)$ be a tangent line to $g(x)$ at the point $g(\mathrm{E}X)$. (Recall that $\mathrm{E}X$ is a constant.) Write $l(x) = a + bx$ for some $a$ and $b$. The situation is illustrated in Figure 4.7.2.

Now, by the convexity of $g$ we have $g(x) \geq a + bx$. Since expectations preserve inequalities,

$$\mathrm{E}g(X) \geq \mathrm{E}(a + bX)$$

$$= a + b\mathrm{E}X \qquad \left(\begin{array}{c}\text{linearity of expectation,}\\ \text{Theorem 2.2.5}\end{array}\right)$$

$$= l(\mathrm{E}X) \qquad\qquad (\text{definition of } l(x))$$

$$= g(\mathrm{E}X), \qquad\qquad (l \text{ is tangent at } \mathrm{E}X)$$

as was to be shown.

If $g(x)$ is linear, equality follows from properties of expectations (Theorem 2.2.5). For the "only if" part see Exercise 4.62. □

One immediate application of Jensen's Inequality shows that $\mathrm{E}X^2 \geq (\mathrm{E}X)^2$, since $g(x) = x^2$ is convex. Also, if $x$ is positive, then $1/x$ is convex; hence $\mathrm{E}(1/X) \geq 1/\mathrm{E}X$, another useful application.

To check convexity of a twice differentiable function is quite easy. The function $g(x)$ is convex if $g''(x) \geq 0$, for all $x$, and $g(x)$ is concave if $g''(x) \leq 0$, for all $x$. Jensen's Inequality applies to concave functions as well. If $g$ is concave, then $\mathrm{E}g(X) \leq g(\mathrm{E}X)$.

**Example 4.7.8 (An inequality for means)** Jensen's Inequality can be used to prove an inequality between three different kinds of means. If $a_1, \ldots, a_n$ are positive numbers, define

$$a_A = \frac{1}{n}(a_1 + a_2 + \cdots + a_n), \qquad \text{(arithmetic mean)}$$

$$a_G = [a_1 a_2 \cdots \cdot a_n]^{1/n}, \qquad \text{(geometric mean)}$$

$$a_H = \frac{1}{\frac{1}{n}\left(\frac{1}{a_1} + \frac{1}{a_2} + \cdots + \frac{1}{a_n}\right)}. \qquad \text{(harmonic mean)}$$

An inequality relating these means is

$$a_H \leq a_G \leq a_A.$$

To apply Jensen's Inequality, let $X$ be a random variable with range $a_1, \ldots, a_n$ and $P(X = a_i) = 1/n, i = 1, \ldots, n$. Since $\log x$ is a concave function, Jensen's Inequality shows that $E(\log X) \leq \log(EX)$; hence,

$$\log a_G = \frac{1}{n}\sum_{i=1}^{n} \log a_i = E(\log X) \leq \log(EX) = \log\left(\frac{1}{n}\sum_{i=1}^{n} a_i\right) = \log a_A,$$

so $a_G \leq a_A$. Now again use the fact that $\log x$ is concave to get

$$\log \frac{1}{a_H} = \log\left(\frac{1}{n}\sum_{i=1}^{n} \frac{1}{a_i}\right) = \log E\frac{1}{X} \geq E\left(\log \frac{1}{X}\right) = -E(\log X).$$

Since $E(\log X) = \log a_G$, it then follows that $\log(1/a_H) \geq \log(1/a_G)$, or $a_G \geq a_H$.   ‖

The next inequality merely exploits the definition of covariance, but sometimes proves to be useful. If $X$ is a random variable with finite mean $\mu$ and $g(x)$ is a nondecreasing function, then

$$E\left(g(X)(X - \mu)\right) \geq 0,$$

since

$$E(g(X)(X - \mu))$$
$$= E\left(g(X)(X - \mu)I_{(-\infty,0)}(X - \mu)\right) + E\left(g(X)(X - \mu)I_{[0,\infty)}(X - \mu)\right)$$
$$\geq E\left(g(\mu)(X - \mu)I_{(-\infty,0)}(X - \mu)\right)$$
$$\quad + E\left(g(\mu)(X - \mu)I_{[0,\infty)}(X - \mu)\right) \qquad \text{(since $g$ is nondecreasing)}$$
$$= g(\mu)E(X - \mu)$$
$$= 0.$$

A generalization of this argument can be used to establish the following inequality (see Exercise 4.65).

**Theorem 4.7.9 (Covariance Inequality-II)** *Let $X$ be any random variable and $g(x)$ and $h(x)$ any functions such that $\mathrm{E}g(X)$, $\mathrm{E}h(X)$, and $\mathrm{E}(g(X)h(X))$ exist.*

**a.** *If $g(x)$ is a nondecreasing function and $h(x)$ is a nonincreasing function, then*

$$\mathrm{E}\left(g(X)h(X)\right) \leq \left(\mathrm{E}g(X)\right)\left(\mathrm{E}h(X)\right).$$

**b.** *If $g(x)$ and $h(x)$ are either both nondecreasing or both nonincreasing, then*

$$\mathrm{E}\left(g(X)h(X)\right) \geq \left(\mathrm{E}g(X)\right)\left(\mathrm{E}h(X)\right).$$

The intuition behind the inequality is easy. In case (a) there is negative correlation between $g$ and $h$, while in case (b) there is positive correlation. The inequalities merely reflect this fact. The usefulness of the Covariance Inequality is that it allows us to bound an expectation without using higher-order moments.

## 4.8 Exercises ───────────────────────────────────────────

**4.1** A random point $(X,Y)$ is distributed uniformly on the square with vertices $(1,1)$, $(1,-1)$, $(-1,1)$, and $(-1,-1)$. That is, the joint pdf is $f(x,y) = \frac{1}{4}$ on the square. Determine the probabilities of the following events.

(a) $X^2 + Y^2 < 1$
(b) $2X - Y > 0$
(c) $|X + Y| < 2$

**4.2** Prove the following properties of bivariate expectations (the bivariate analog to Theorem 2.2.5). For random variables $X$ and $Y$, functions $g_1(x,y)$ and $g_2(x,y)$, and constants $a$, $b$, and $c$:

(a) $\mathrm{E}(ag_1(X,Y) + bg_2(X,Y) + c) = a\mathrm{E}(g_1(X,Y)) + b\mathrm{E}(g_2(X,Y)) + c$.
(b) If $g_1(x,y) \geq 0$, then $\mathrm{E}(g_1(X,Y)) \geq 0$.
(c) If $g_1(x,y) \geq g_2(x,y)$, then $\mathrm{E}(g_1(X,Y)) \geq \mathrm{E}(g_2(X,Y))$.
(d) If $a \leq g_1(x,y) \leq b$, then $a \leq \mathrm{E}(g_1(X,Y)) \leq b$.

**4.3** Using Definition 4.1.1, show that the random vector $(X,Y)$ defined at the end of Example 4.1.5 has the pmf given in that example.

**4.4** A pdf is defined by

$$f(x,y) = \begin{cases} C(x + 2y) & \text{if } 0 < y < 1 \text{ and } 0 < x < 2 \\ 0 & \text{otherwise.} \end{cases}$$

(a) Find the value of $C$.
(b) Find the marginal distribution of $X$.
(c) Find the joint cdf of $X$ and $Y$.
(d) Find the pdf of the random variable $Z = 9/(X+1)^2$.

**4.5** (a) Find $P(X > \sqrt{Y})$ if $X$ and $Y$ are jointly distributed with pdf

$$f(x,y) = x + y, \quad 0 \leq x \leq 1, \quad 0 \leq y \leq 1.$$

(b) Find $P(X^2 < Y < X)$ if $X$ and $Y$ are jointly distributed with pdf

$$f(x,y) = 2x, \quad 0 \leq x \leq 1, \quad 0 \leq y \leq 1.$$

**4.6** A and B agree to meet at a certain place between 1 PM and 2 PM. Suppose they arrive at the meeting place independently and randomly during the hour. Find the distribution of the length of time that A waits for B. (If B arrives before A, define A's waiting time as 0.)

**4.7** A woman leaves for work between 8 AM and 8:30 AM and takes between 40 and 50 minutes to get there. Let the random variable $X$ denote her time of departure, and the random variable $Y$ the travel time. Assuming that these variables are independent and uniformly distributed, find the probability that the woman arrives at work before 9 AM.

**4.8** Referring to Miscellanea 4.9.1.

   (a) Show that $P(X = m|M = m) = P(X = 2m|M = m) = 1/2$, and verify the expressions for $P(M = x|X = x)$ and $P(M = x/2|X = x)$.

   (b) Verify that one should trade only if $\pi(x/2) < 2\pi(x)$, and if $\pi$ is the exponential($\lambda$) density, show that it is optimal to trade if $x < 2\log 2/\lambda$.

   (c) For the classical approach, show that $P(Y = 2x|X = m) = 1$ and $P(Y = x/2|X = 2m) = 1$ and that your expected winning if you trade or keep your envelope is $E(Y) = 3m/2$.

**4.9** Prove that if the joint cdf of $X$ and $Y$ satisfies

$$F_{X,Y}(x, y) = F_X(x)F_Y(y),$$

then for any pair of intervals $(a, b)$, and $(c, d)$,

$$P(a \leq X \leq b, c \leq Y \leq d) = P(a \leq X \leq b)P(c \leq Y \leq d).$$

**4.10** The random pair $(X, Y)$ has the distribution

|       |   | $X$ |   |   |
|-------|---|-----|---|---|
|       |   | 1   | 2 | 3 |
|       | 2 | $\frac{1}{12}$ | $\frac{1}{6}$ | $\frac{1}{12}$ |
| $Y$   | 3 | $\frac{1}{6}$ | 0 | $\frac{1}{6}$ |
|       | 4 | 0 | $\frac{1}{3}$ | 0 |

   (a) Show that $X$ and $Y$ are dependent.

   (b) Give a probability table for random variables $U$ and $V$ that have the same marginals as $X$ and $Y$ but are independent.

**4.11** Let $U$ = the number of trials needed to get the first head and $V$ = the number of trials needed to get two heads in repeated tosses of a fair coin. Are $U$ and $V$ independent random variables?

**4.12** If a stick is broken at random into three pieces, what is the probability that the pieces can be put together in a triangle? (See Gardner 1961 for a complete discussion of this problem.)

**4.13** Let $X$ and $Y$ be random variables with finite means.

   (a) Show that

$$\min_{g(x)} E\left(Y - g(X)\right)^2 = E\left(Y - E(Y|X)\right)^2,$$

where $g(x)$ ranges over all functions. ($E(Y|X)$ is sometimes called the *regression of $Y$ on $X$*, the "best" predictor of $Y$ conditional on $X$.)

   (b) Show that equation (2.2.3) can be derived as a special case of part (a).

**4.14** Suppose $X$ and $Y$ are independent $n(0,1)$ random variables.

(a) Find $P(X^2 + Y^2 < 1)$.

(b) Find $P(X^2 < 1)$, after verifying that $X^2$ is distributed $\chi_1^2$.

**4.15** Let $X \sim \text{Poisson}(\theta)$, $Y \sim \text{Poisson}(\lambda)$, independent. It was shown in Theorem 4.3.2 that the distribution of $X + Y$ is $\text{Poisson}(\theta + \lambda)$. Show that the distribution of $X|X + Y$ is binomial with success probability $\theta/(\theta + \lambda)$. What is the distribution of $Y|X + Y$?

**4.16** Let $X$ and $Y$ be independent random variables with the same geometric distribution.

(a) Show that $U$ and $V$ are independent, where $U$ and $V$ are defined by

$$U = \min(X, Y) \quad \text{and} \quad V = X - Y.$$

(b) Find the distribution of $Z = X/(X + Y)$, where we define $Z = 0$ if $X + Y = 0$.

(c) Find the joint pmf of $X$ and $X + Y$.

**4.17** Let $X$ be an exponential(1) random variable, and define $Y$ to be the integer part of $X + 1$, that is

$$Y = i + 1 \quad \text{if and only if} \quad i \le X < i + 1, \quad i = 0, 1, 2, \ldots.$$

(a) Find the distribution of $Y$. What well-known distribution does $Y$ have?

(b) Find the conditional distribution of $X - 4$ given $Y \ge 5$.

**4.18** Given that $g(x) \ge 0$ has the property that

$$\int_0^\infty g(x)\,dx = 1,$$

show that

$$f(x, y) = \frac{2g\left(\sqrt{x^2 + y^2}\right)}{\pi\sqrt{x^2 + y^2}}, \quad x, y > 0,$$

is a pdf.

**4.19** (a) Let $X_1$ and $X_2$ be independent $n(0,1)$ random variables. Find the pdf of $(X_1 - X_2)^2/2$.

(b) If $X_i, i = 1, 2$, are independent gamma$(\alpha_i, 1)$ random variables, find the marginal distributions of $X_1/(X_1 + X_2)$ and $X_2/(X_1 + X_2)$.

**4.20** $X_1$ and $X_2$ are independent $n(0, \sigma^2)$ random variables.

(a) Find the joint distribution of $Y_1$ and $Y_2$, where

$$Y_1 = X_1^2 + X_2^2 \quad \text{and} \quad Y_2 = \frac{X_1}{\sqrt{Y_1}}.$$

(b) Show that $Y_1$ and $Y_2$ are independent, and interpret this result geometrically.

**4.21** A point is generated at random in the plane according to the following polar scheme. A radius $R$ is chosen, where the distribution of $R^2$ is $\chi^2$ with 2 degrees of freedom. Independently, an angle $\theta$ is chosen, where $\theta \sim \text{uniform}(0, 2\pi)$. Find the joint distribution of $X = R\cos\theta$ and $Y = R\sin\theta$.

**4.22** Let $(X, Y)$ be a bivariate random vector with joint pdf $f(x, y)$. Let $U = aX + b$ and $V = cY + d$, where $a$, $b$, $c$, and $d$ are fixed constants with $a > 0$ and $c > 0$. Show that the joint pdf of $(U, V)$ is

$$f_{U,V}(u, v) = \frac{1}{ac} f\left(\frac{u - b}{a}, \frac{v - d}{c}\right).$$

**4.23** For $X$ and $Y$ as in Example 4.3.3, find the distribution of $XY$ by making the transformations given in (a) and (b) and integrating out $V$.

(a) $U = XY$, $V = Y$
(b) $U = XY$, $V = X/Y$

**4.24** Let $X$ and $Y$ be independent random variables with $X \sim \text{gamma}(r, 1)$ and $Y \sim \text{gamma}(s, 1)$. Show that $Z_1 = X + Y$ and $Z_2 = X/(X + Y)$ are independent, and find the distribution of each. ($Z_1$ is gamma and $Z_2$ is beta.)

**4.25** Use the techniques of Section 4.3 to derive the joint distribution of $(X, Y)$ from the joint distribution of $(X, Z)$ in Examples 4.5.8 and 4.5.9.

**4.26** $X$ and $Y$ are independent random variables with $X \sim \text{exponential}(\lambda)$ and $Y \sim \text{exponential}(\mu)$. It is impossible to obtain direct observations of $X$ and $Y$. Instead, we observe the random variables $Z$ and $W$, where

$$Z = \min\{X, Y\} \qquad \text{and} \qquad W = \begin{cases} 1 & \text{if } Z = X \\ 0 & \text{if } Z = Y. \end{cases}$$

(This is a situation that arises, in particular, in medical experiments. The $X$ and $Y$ variables are *censored*.)

(a) Find the joint distribution of $Z$ and $W$.
(b) Prove that $Z$ and $W$ are independent. (*Hint:* Show that $P(Z \leq z | W = i) = P(Z \leq z)$ for $i = 0$ or 1.)

**4.27** Let $X \sim \text{n}(\mu, \sigma^2)$ and let $Y \sim \text{n}(\gamma, \sigma^2)$. Suppose $X$ and $Y$ are independent. Define $U = X + Y$ and $V = X - Y$. Show that $U$ and $V$ are independent normal random variables. Find the distribution of each of them.

**4.28** Let $X$ and $Y$ be independent standard normal random variables.

(a) Show that $X/(X + Y)$ has a Cauchy distribution.
(b) Find the distribution of $X/|Y|$.
(c) Is the answer to part (b) surprising? Can you formulate a general theorem?

**4.29** Jones (1999) looked at the distribution of functions of $X$ and $Y$ when $X = R\cos\theta$ and $Y = R\sin\theta$, where $\theta \sim U(0, 2\pi)$ and $R$ is a positive random variable. Here are two of the many situations that he considered.

(a) Show that $X/Y$ has a Cauchy distribution.
(b) Show that the distribution of $(2XY)/\sqrt{X^2 + Y^2}$ is the same as the distribution of $X$. Specialize this result to one about $\text{n}(0, \sigma^2)$ random variables.

**4.30** Suppose the distribution of $Y$, conditional on $X = x$, is $\text{n}(x, x^2)$ and that the marginal distribution of $X$ is uniform$(0, 1)$.

(a) Find $EY$, $\text{Var}\,Y$, and $\text{Cov}(X, Y)$.
(b) Prove that $Y/X$ and $X$ are independent.

**4.31** Suppose that the random variable $Y$ has a binomial distribution with $n$ trials and success probability $X$, where $n$ is a given constant and $X$ is a uniform$(0,1)$ random variable.

(a) Find $EY$ and $\operatorname{Var} Y$.
(b) Find the joint distribution of $X$ and $Y$.
(c) Find the marginal distribution of $Y$.

**4.32** (a) For the hierarchical model

$$Y|\Lambda \sim \text{Poisson}(\Lambda) \quad \text{and} \quad \Lambda \sim \text{gamma}(\alpha, \beta)$$

find the marginal distribution, mean, and variance of $Y$. Show that the marginal distribution of $Y$ is a negative binomial if $\alpha$ is an integer.

(b) Show that the three-stage model

$$Y|N \sim \text{binomial}(N, p), \quad N|\Lambda \sim \text{Poisson}(\Lambda), \quad \text{and} \quad \Lambda \sim \text{gamma}(\alpha, \beta)$$

leads to the same marginal (unconditional) distribution of $Y$.

**4.33** (*Alternative derivation of the negative binomial distribution*)    Solomon (1983) details the following biological model. Suppose that each of a random number, $N$, of insects lays $X_i$ eggs, where the $X_i$s are independent, identically distributed random variables. The total number of eggs laid is $H = X_1 + \cdots + X_N$. What is the distribution of $H$? It is common to assume that $N$ is Poisson$(\lambda)$. Furthermore, if we assume that each $X_i$ has the logarithmic series distribution (see Exercise 3.14) with success probability $p$, we have the hierarchical model

$$H|N = X_1 + \cdots + X_N, \quad P(X_i = t) = \frac{-1}{\log(p)} \frac{(1-p)^t}{t},$$

$$N \sim \text{Poisson}(\lambda).$$

Show that the marginal distribution of $H$ is negative binomial$(r, p)$, where $r = -\lambda/\log(p)$. (It is easiest to calculate and identify the mgf of $H$ using Theorems 4.4.3 and 4.6.7. Stuart and Ord 1987, Section 5.21, also mention this derivation of the logarithmic series distribution. They refer to $H$ as a *randomly stopped sum*.)

**4.34** (a) For the hierarchy in Example 4.4.6, show that the marginal distribution of $X$ is given by the *beta-binomial distribution*,

$$P(X = x) = \binom{n}{x} \frac{\Gamma(\alpha + \beta)}{\Gamma(\alpha)\Gamma(\beta)} \frac{\Gamma(x + \alpha)\Gamma(n - x + \beta)}{\Gamma(\alpha + \beta + n)}.$$

(b) A variation on the hierarchical model in part (a) is

$$X|P \sim \text{negative binomial}(r, P) \quad \text{and} \quad P \sim \text{beta}(\alpha, \beta).$$

Find the marginal pmf of $X$ and its mean and variance. (This distribution is the *beta-Pascal*.)

**4.35** (a) For the hierarchy in Example 4.4.6, show that the variance of $X$ can be written

$$\operatorname{Var} X = nEP(1 - EP) + n(n - 1)\operatorname{Var} P.$$

(The first term reflects binomial variation with success probability $EP$, and the second term is often called "extra-binomial" variation, showing how the hierarchical model has a variance that is larger than the binomial alone.)

(b) For the hierarchy in Exercise 4.32, show that the variance of $Y$ can be written

$$\text{Var}\, Y = \text{E}\Lambda + \text{Var}\,\Lambda = \mu + \frac{1}{\alpha}\mu^2,$$

where $\mu = \text{E}\Lambda$. Identify the "extra-Poisson" variation induced by the hierarchy.

**4.36** One generalization of the Bernoulli trials hierarchy in Example 4.4.6 is to allow the success probability to vary from trial to trial, keeping the trials independent. A standard model for this situation is this. Let $(X_1, P_1), \ldots, (X_n, P_n)$ be independent random vectors with

$$X_i | P_i \sim \text{Bernoulli}(P_i), \quad i = 1, \ldots, n,$$

$$P_i \sim \text{beta}(\alpha, \beta).$$

This model might be appropriate, for example, if we are measuring the success of a drug on $n$ patients and, because the patients are different, we are reluctant to assume that the success probabilities are constant. (This can be thought of as an *empirical Bayes model*; see Miscellanea 7.5.6.)

A random variable of interest is $Y = \sum_{i=1}^{n} X_i$, the total number of successes.

(a) Show that $\text{E}Y = n\alpha/(\alpha + \beta)$.

(b) Show that $\text{Var}\, Y = n\alpha\beta/(\alpha + \beta)^2$, and hence $Y$ has the same mean and variance as a binomial$(n, \frac{\alpha}{\alpha+\beta})$ random variable. What is the distribution of $Y$?

(c) Suppose now that the model is

$$X_i | P_i \sim \text{binomial}(n_i, P_i), \quad i = 1, \ldots, k,$$

$$P_i \sim \text{beta}(\alpha, \beta).$$

Show that for $Y = \sum_{i=1}^{k} X_i$, $\text{E}Y = \frac{\alpha}{\alpha+\beta} \sum_{i=1}^{k} n_i$ and $\text{Var}\, Y = \sum_{i=1}^{k} \text{Var}\, X_i$, where

$$\text{Var}\, X_i = n_i \frac{\alpha\beta(\alpha + \beta + n_i)}{(\alpha + \beta)^2(\alpha + \beta + 1)}.$$

**4.37** A generalization of the hierarchy in Exercise 4.34 is described by D. G. Morrison (1978), who gives a model for *forced binary choices*. A forced binary choice occurs when a person is forced to choose between two alternatives, as in a taste test. It may be that a person cannot actually discriminate between the two choices (can you tell Coke from Pepsi?), but the setup of the experiment is such that a choice must be made. Therefore, there is a confounding between discriminating correctly and guessing correctly. Morrison modeled this by defining the following parameters:

$$p = \text{probability that a person can actually discriminate},$$

$$c = \text{probability that a person discriminates correctly}.$$

Then

$$c = p + \frac{1}{2}(1 - p) = \frac{1}{2}(1 + p), \quad \frac{1}{2} < c < 1,$$

where $\frac{1}{2}(1 - p)$ is the probability that a person guesses correctly. We now run the experiment and observe $X_1, \ldots, X_n \sim \text{Bernoulli}(c)$, so

$$P(\Sigma X_i = k | c) = \binom{n}{k} c^k (1 - c)^{n-k}.$$

However, it is probably the case that $p$ is not constant from person to person, so $p$ is allowed to vary according to a beta distribution,

$$P \sim \text{beta}(a, b).$$

(a) Show that the distribution of $\Sigma X_i$ is a mixture of beta-binomials.
(b) Find the mean and variance of $\Sigma X_i$.

**4.38** (*The gamma as a mixture of exponentials*)   Gleser (1989) shows that, in certain cases, the gamma distribution can be written as a scale mixture of exponentials, an identity suggested by different analyses of the same data. Let $f(x)$ be a gamma$(r, \lambda)$ pdf.

(a) Show that if $r \leq 1$, then $f(x)$ can be written

$$f(x) = \int_0^\lambda \frac{1}{\nu} e^{-x/\nu} p_\lambda(\nu) \, d\nu,$$

where

$$p_\lambda(\nu) = \frac{1}{\Gamma(r)\Gamma(1-r)} \frac{\nu^{r-1}}{(\lambda - \nu)^r}, \quad 0 < \nu < \lambda.$$

(*Hint:* Make a change of variable from $\nu$ to $u$, where $u = x/\nu - x/\lambda$.)
(b) Show that $p_\lambda(\nu)$ is a pdf, for $r \leq 1$, by showing that

$$\int_0^\lambda p_\lambda(\nu) \, d\nu = 1.$$

(c) Show that the restriction $r \leq 1$ is necessary for the representation in part (a) to be valid; that is, there is no such representation if $r > 1$. (*Hint:* Suppose $f(x)$ can be written $f(x) = \int_0^\infty (e^{-x/\nu}/\nu) q_\lambda(\nu) d\nu$ for some pdf $q_\lambda(\nu)$. Show that $\frac{\partial}{\partial x} \log (f(x)) > 0$ but $\frac{\partial}{\partial x} \log \left( \int_0^\infty (e^{-x/\nu}/\nu) q_\lambda(\nu) d\nu \right) < 0$, a contradiction.)

**4.39** Let $(X_1, \ldots, X_n)$ have a multinomial distribution with $m$ trials and cell probabilities $p_1, \ldots, p_n$ (see Definition 4.6.2). Show that, for every $i$ and $j$,

$$X_i | X_j = x_j \sim \text{binomial} \left( m - x_j, \frac{p_i}{1 - p_j} \right)$$

$$X_j \sim \text{binomial} (m, p_j)$$

and that $\text{Cov}(X_i, X_j) = -m p_i p_j$.

**4.40** A generalization of the beta distribution is the *Dirichlet* distribution. In its bivariate version, $(X, Y)$ have pdf

$$f(x, y) = C x^{a-1} y^{b-1} (1 - x - y)^{c-1}, \quad 0 < x < 1, \quad 0 < y < 1, \quad 0 < y < 1 - x < 1,$$

where $a > 0$, $b > 0$, and $c > 0$ are constants.

(a) Show that $C = \frac{\Gamma(a+b+c)}{\Gamma(a)\Gamma(b)\Gamma(c)}$.
(b) Show that, marginally, both $X$ and $Y$ are beta.
(c) Find the conditional distribution of $Y|X = x$, and show that $Y/(1-x)$ is beta$(b, c)$.
(d) Show that $\text{E}(XY) = \frac{ab}{(a+b+c+1)(a+b+c)}$, and find their covariance.

**4.41** Show that $\text{Cov}(X, c) = 0$ for any random variable $X$ with finite mean and any constant $c$.

**4.42** Let $X$ and $Y$ be independent random variables with means $\mu_X, \mu_Y$ and variances $\sigma_X^2$, $\sigma_Y^2$. Find an expression for the correlation of $XY$ and $Y$ in terms of these means and variances.

**4.43** Let $X_1, X_2$, and $X_3$ be uncorrelated random variables, each with mean $\mu$ and variance $\sigma^2$. Find, in terms of $\mu$ and $\sigma^2$, $\text{Cov}(X_1 + X_2, X_2 + X_3)$ and $\text{Cov}(X_1 + X_2, X_1 - X_2)$.

**4.44** Prove the following generalization of Theorem 4.5.6: For any random vector $(X_1, \ldots, X_n)$,

$$\text{Var}\left(\sum_{i=1}^{n} X_i\right) = \sum_{i=1}^{n} \text{Var } X_i + 2 \sum_{1 \le i < j \le n} \text{Cov}(X_i, X_j).$$

**4.45** Show that if $(X, Y) \sim$ bivariate normal$(\mu_X, \mu_Y, \sigma_X^2, \sigma_Y^2, \rho)$, then the following are true.

(a) The marginal distribution of $X$ is $\text{n}(\mu_X, \sigma_X^2)$ and the marginal distribution of $Y$ is $\text{n}(\mu_Y, \sigma_Y^2)$.

(b) The conditional distribution of $Y$ given $X = x$ is

$$\text{n}(\mu_Y + \rho(\sigma_Y/\sigma_X)(x - \mu_X), \sigma_Y^2(1 - \rho^2)).$$

(c) For any constants $a$ and $b$, the distribution of $aX + bY$ is

$$\text{n}(a\mu_X + b\mu_Y, a^2\sigma_X^2 + b^2\sigma_Y^2 + 2ab\rho\sigma_X\sigma_Y).$$

**4.46** (*A derivation of the bivariate normal distribution*) Let $Z_1$ and $Z_2$ be independent $\text{n}(0, 1)$ random variables, and define new random variables $X$ and $Y$ by

$$X = a_X Z_1 + b_X Z_2 + c_X \quad \text{and} \quad Y = a_Y Z_1 + b_Y Z_2 + c_Y,$$

where $a_X, b_X, c_X, a_Y, b_Y$, and $c_Y$ are constants.

(a) Show that

$$\text{E}X = c_X, \quad \text{Var } X = a_X^2 + b_X^2,$$
$$\text{E}Y = c_Y, \quad \text{Var } Y = a_Y^2 + b_Y^2,$$
$$\text{Cov}(X, Y) = a_X a_Y + b_X b_Y.$$

(b) If we define the constants $a_X, b_X, c_X, a_Y, b_Y$, and $c_Y$ by

$$a_X = \sqrt{\frac{1 + \rho}{2}}\sigma_X, \quad b_X = \sqrt{\frac{1 - \rho}{2}}\sigma_X, \quad c_X = \mu_X,$$

$$a_Y = \sqrt{\frac{1 + \rho}{2}}\sigma_Y, \quad b_Y = -\sqrt{\frac{1 - \rho}{2}}\sigma_Y, \quad c_Y = \mu_Y,$$

where $\mu_X, \mu_Y, \sigma_X^2, \sigma_Y^2$, and $\rho$ are constants, $-1 \le \rho \le 1$, then show that

$$\text{E}X = \mu_X, \quad \text{Var } X = \sigma_X^2,$$
$$\text{E}Y = \mu_Y, \quad \text{Var } Y = \sigma_Y^2,$$
$$\rho_{XY} = \rho.$$

(c) Show that $(X, Y)$ has the bivariate normal pdf with parameters $\mu_X, \mu_Y, \sigma_X^2, \sigma_Y^2$, and $\rho$.

(d) If we start with bivariate normal parameters $\mu_X, \mu_Y, \sigma_X^2, \sigma_Y^2$, and $\rho$, we can define constants $a_X, b_X, c_X, a_Y, b_Y$, and $c_Y$ as the solutions to the equations

$$\mu_X = c_X, \quad \sigma_X^2 = a_X^2 + b_X^2,$$
$$\mu_Y = c_Y, \quad \sigma_Y^2 = a_Y^2 + b_Y^2,$$
$$\rho\sigma_X\sigma_Y = a_X a_Y + b_X b_Y.$$

Show that the solution given in part (b) is not unique by exhibiting another solution to these equations. How many solutions are there?

**4.47** (*Marginal normality does not imply bivariate normality.*)  Let $X$ and $Y$ be independent $n(0, 1)$ random variables, and define a new random variable $Z$ by

$$Z = \begin{cases} X & \text{if } XY > 0 \\ -X & \text{if } XY < 0. \end{cases}$$

(a) Show that $Z$ has a normal distribution.
(b) Show that the joint distribution of $Z$ and $Y$ is not bivariate normal. (*Hint*: Show that $Z$ and $Y$ always have the same sign.)

**4.48** Gelman and Meng (1991) give an example of a bivariate family of distributions that are not bivariate normal but have normal conditionals. Define the joint pdf of $(X, Y)$ as

$$f(x, y) \propto \exp\left\{ -\frac{1}{2} \left[ Ax^2y^2 + x^2 + y^2 - 2Bxy - 2Cx - 2Dy \right] \right\},$$

where $A, B, C, D$ are constants.

(a) Show that the distribution of $X|Y = y$ is normal with mean $\frac{By+C}{Ay^2+1}$ and variance $\frac{1}{Ay^2+1}$. Derive a corresponding result for the distribution of $Y|X = x$.
(b) A most interesting configuration is $A = 1, B = 0, C = D = 8$. Show that this joint distribution is bimodal.

**4.49** Behboodian (1990) illustrates how to construct bivariate random variables that are uncorrelated but dependent. Suppose that $f_1, f_2, g_1, g_2$ are univariate densities with means $\mu_1, \mu_2, \xi_1, \xi_2$, respectively, and the bivariate random variable $(X, Y)$ has density

$$(X, Y) \sim a f_1(x)g_1(y) + (1 - a)f_2(x)g_2(y),$$

where $0 < a < 1$ is known.

(a) Show that the marginal distributions are given by $f_X(x) = a f_1(x) + (1 - a)f_2(x)$ and $f_Y(x) = a g_1(y) + (1 - a)g_2(y)$.
(b) Show that $X$ and $Y$ are independent if and only if $[f_1(x) - f_2(x)][g_1(y) - g_2(y)] = 0$.
(c) Show that $\text{Cov}(X, Y) = a(1-a)[\mu_1 - \mu_2][\xi_1 - \xi_2]$, and thus explain how to construct dependent uncorrelated random variables.
(d) Letting $f_1, f_2, g_1, g_2$ be binomial pmfs, give examples of combinations of parameters that lead to independent $(X, Y)$ pairs, correlated $(X, Y)$ pairs, and uncorrelated but dependent $(X, Y)$ pairs.

**4.50** If $(X, Y)$ has the bivariate normal pdf

$$f(x, y) = \frac{1}{2\pi(1 - \rho^2)^{1/2}} \exp\left(\frac{-1}{2(1 - \rho^2)}(x^2 - 2\rho xy + y^2)\right),$$

show that $\text{Corr}(X, Y) = \rho$ and $\text{Corr}(X^2, Y^2) = \rho^2$. (Conditional expectations will simplify calculations.)

**4.51** Let $X$, $Y$, and $Z$ be independent uniform$(0, 1)$ random variables.

(a) Find $P(X/Y \leq t)$ and $P(XY \leq t)$. (Pictures will help.)
(b) Find $P(XY/Z \leq t)$.

**4.52** Bullets are fired at the origin of an $(x, y)$ coordinate system, and the point hit, say $(X, Y)$, is a random variable. The variables $X$ and $Y$ are taken to be independent n$(0, 1)$ random variables. If two bullets are fired independently, what is the distribution of the distance between them?

**4.53** Let $A$, $B$, and $C$ be independent random variables, uniformly distributed on $(0, 1)$. What is the probability that $Ax^2 + Bx + C$ has real roots? (*Hint:* If $X \sim$ uniform$(0, 1)$, then $-\log X \sim$ exponential. The sum of two independent exponentials is gamma.)

**4.54** Find the pdf of $\prod_{i=1}^{n} X_i$, where the $X_i$s are independent uniform$(0, 1)$ random variables. (*Hint:* Try to calculate the cdf, and remember the relationship between uniforms and exponentials.)

**4.55** A *parallel system* is one that functions as long as at least one component of it functions. A particular parallel system is composed of three independent components, each of which has a lifelength with an exponential$(\lambda)$ distribution. The lifetime of the system is the maximum of the individual lifelengths. What is the distribution of the lifetime of the system?

**4.56** A large number, $N = mk$, of people are subject to a blood test. This can be administered in two ways.

(i) Each person can be tested separately. In this case $N$ tests are required.
(ii) The blood samples of $k$ people can be pooled and analyzed together. If the test is negative, this *one* test suffices for $k$ people. If the test is positive, each of the $k$ persons must be tested separately, and, in all, $k + 1$ tests are required for the $k$ people.

Assume that the probability, $p$, that the test is positive is the same for all people and that the test results for different people are statistically independent.

(a) What is the probability that the test for a pooled sample of $k$ people will be positive?
(b) Let $X =$ number of blood tests necessary under plan (ii). Find $\text{E}X$.
(c) In terms of minimizing the expected number of blood tests to be performed on the $N$ people, which plan [(i) or (ii)] would be preferred if it is known that $p$ is close to 0? Justify your answer using the expression derived in part (b).

**4.57** Refer to Miscellanea 4.9.2.

(a) Show that $A_1$ is the arithmetic mean, $A_{-1}$ is the harmonic mean, and $A_0 = \lim_{r \to 0} A_r$ is the geometric mean.
(b) The arithmetic-geometric-harmonic mean inequality will follow if it can be established that $A_r$ is a nondecreasing function of $r$ over the range $-\infty < r < \infty$.

(i) Verify that if $\log A_r$ is nondecreasing in $r$, then it will follow that $A_r$ is non-decreasing in $r$.

(ii) Show that

$$\frac{d}{dr} \log A_r = \frac{1}{r^2} \left\{ \frac{r \sum_i x_i^r \log x_i}{\sum_i x_i^r} - \log \left( \frac{1}{n} \sum_i x_i^r \right) \right\}.$$

(iii) Define $a_i = x_i^r / \sum_i x_i^r$ and write the quantity in braces as

$$\log(n) - \sum a_i \log(1/a_i),$$

where $\sum a_i = 1$. Now prove that this quantity is nonnegative, establishing the monotonicity of $A_r$ and the arithmetic-geometric-harmonic mean inequality as a special case.

The quantity $\sum_i a_i \log(1/a_i)$ is called *entropy*, sometimes considered an absolute measure of uncertainty (see Bernardo and Smith 1994, Section 2.7). The result of part (iii) states that the maximum entropy is attained when all probabilities are the same (randomness).

(*Hint*: To prove the inequality note that the $a_i$ are a probability distribution, and we can write

$$\mathrm{E} \log \left( \frac{1}{a} \right) = \sum_i a_i \log \left( \frac{1}{a_i} \right),$$

and Jensen's Inequality shows that $\mathrm{E} \log \left( \frac{1}{a} \right) \leq \log \left( \mathrm{E} \frac{1}{a} \right)$.)

**4.58** For any two random variables $X$ and $Y$ with finite variances, prove that

(a) $\mathrm{Cov}(X, Y) = \mathrm{Cov}(X, \mathrm{E}(Y|X))$.

(b) $X$ and $Y - \mathrm{E}(Y|X)$ are uncorrelated.

(c) $\mathrm{Var}(Y - \mathrm{E}(Y|X)) = \mathrm{E}(\mathrm{Var}(Y|X))$.

**4.59** For any three random variables $X$, $Y$, and $Z$ with finite variances, prove (in the sprit of Theorem 4.4.7) the covariance identity

$$\mathrm{Cov}(X, Y) = \mathrm{E}(\mathrm{Cov}(X, Y|Z)) + \mathrm{Cov}(\mathrm{E}(X|Z), \mathrm{E}(Y|Z)),$$

where $\mathrm{Cov}(X, Y|Z)$ is the covariance of $X$ and $Y$ under the pdf $f(x, y|z)$.

**4.60** Referring to Miscellanea 4.9.3, find the conditional distribution of $Y$ given that $Y = X$ for each of the three interpretations given for the condition $Y = X$.

**4.61** DeGroot (1986) gives the following example of the Borel Paradox (Miscellanea 4.9.3): Suppose that $X_1$ and $X_2$ are iid exponential(1) random variables, and define $Z = (X_2 - 1)/X_1$. The probability-zero sets $\{Z = 0\}$ and $\{X_2 = 1\}$ seem to be giving us the same information but lead to different conditional distributions.

(a) Find the distribution of $X_1|Z = 0$, and compare it to the distribution of $X_1|X_2 = 1$.

(b) For small $\varepsilon > 0$ and $x_1 > 0, x_2 > 0$, consider the sets

$$B_1 = \{(x_1, x_2) : -\varepsilon < \frac{x_2 - 1}{x_1} < \varepsilon\} \quad \text{and} \quad B_2 = \{(x_1, x_2) : 1 - \varepsilon < x_2 < 1 + \varepsilon\}.$$

Draw these sets and support the argument that $B_1$ is informative about $X_1$ but $B_2$ is not.

(c) Calculate $P(X_1 \leq x | B_1)$ and $P(X_1 \leq x | B_2)$, and show that their limits (as $\varepsilon \to 0$) agree with part (a).

(*Communicated by L. Mark Berliner, Ohio State University.*)

**4.62** Finish the proof of the equality in Jensen's Inequality (Theorem 4.7.7). Let $g(x)$ be a convex function. Suppose $a + bx$ is a line tangent to $g(x)$ at $x = \mathrm{E}\,X$, and $g(x) > a + bx$ except at $x = \mathrm{E}\,X$. Then $\mathrm{E}\,g(X) > g(\mathrm{E}\,X)$ unless $P(X = \mathrm{E}\,X) = 1$.

**4.63** A random variable $X$ is defined by $Z = \log X$, where $\mathrm{E}Z = 0$. Is $\mathrm{E}X$ greater than, less than, or equal to 1?

**4.64** This exercise involves a well-known inequality known as the *triangle inequality* (a special case of Minkowski's Inequality).

(a) Prove (without using Minkowski's Inequality) that for any numbers $a$ and $b$

$$|a + b| \leq |a| + |b|.$$

(b) Use part (a) to establish that for any random variables $X$ and $Y$ with finite expectations,

$$\mathrm{E}|X + Y| \leq \mathrm{E}|X| + \mathrm{E}|Y|.$$

**4.65** Prove the Covariance Inequality-II by generalizing the argument given in the text immediately preceding the inequality.

## 4.9 Miscellanea

### 4.9.1 *The Exchange Paradox*

The *"Exchange Paradox"* (Christensen and Utts 1992) has generated a lengthy dialog among statisticians. The problem (or the paradox) goes as follows:

> A swami puts $m$ dollars in one envelope and $2m$ dollars in another. You and your opponent each get one of the envelopes (at random). You open your envelope and find $x$ dollars, and then the swami asks you if you want to trade envelopes. You reason that if you switch, you will get either $x/2$ or $2x$ dollars, each with probability $1/2$. This makes the expected value of a switch equal to $(1/2)(x/2) + (1/2)(2x) = 5x/4$, which is greater than the $x$ dollars that you hold in your hand. So you offer to trade.
>
> The paradox is that your opponent has done the same calculation. How can the trade be advantageous for both of you?

(i) Christensen and Utts say, "The conclusion that trading envelopes is always optimal is based on the assumption that there is no information obtained by observing the contents of the envelope," and they offer the following resolution. Let $M \sim \pi(m)$ be the pdf for the amount of money placed in the first envelope, and let $X$ be the amount of money in your envelope. Then $P(X = m | M = m) = P(X = 2m | M = m) = 1/2$, and hence

$$P(M = x | X = x) = \frac{\pi(x)}{\pi(x) + \pi(x/2)} \quad \text{and} \quad P(M = x/2 | X = x) = \frac{\pi(x/2)}{\pi(x) + \pi(x/2)}.$$

It then follows that the expected winning from a trade is

$$\frac{\pi(x)}{\pi(x) + \pi(x/2)} 2x + \frac{\pi(x/2)}{\pi(x) + \pi(x/2)} \frac{x}{2},$$

and thus you should trade only if $\pi(x/2) < 2\pi(x)$. If $\pi$ is the exponential($\lambda$) density, it is optimal to trade if $x < 2\log 2/\lambda$.

(ii) A more classical approach does not assume that there is a pdf on the amount of money placed in the first envelope. Christensen and Utts also offer an explanation here, noting that the paradox occurs if one incorrectly assumes that $P(Y = y|X = x) = 1/2$ for all values of $X$ and $Y$, where $X$ is the amount in your envelope and $Y$ is the amount in your opponent's envelope. They argue that the correct conditional distributions are $P(Y = 2x|X = m) = 1$ and $P(Y = x/2|X = 2m) = 1$ and that your expected winning if you trade is $E(Y) = 3m/2$, which is the same as your expected winning if you keep your envelope.

This paradox is often accompanied with arguments for or against the Bayesian methodology of inference (see Chapter 7), but these arguments are somewhat tangential to the underlying probability calculations. For comments, criticisms, and other analyses see the letters to the editor from Binder (1993), Ridgeway (1993) (which contains a solution by Marilyn vos Savant), Ross (1994), and Blachman (1996) and the accompanying responses from Christensen and Utts.

### 4.9.2 More on the Arithmetic-Geometric-Harmonic Mean Inequality

The arithmetic-geometric-harmonic mean inequality is a special case of a general result about *power means*, which are defined by

$$A_r = \left[\frac{1}{n}\sum_{i=1}^{n} x_i^r\right]^{1/r}$$

for $x_i \geq 0$. Shier (1988) shows that $A_r$ is a nondecreasing function of $r$; that is, $A_r \leq A_{r'}$ if $r \leq r'$ or

$$\left[\frac{1}{n}\sum_{i=1}^{n} x_i^r\right]^{1/r} \leq \left[\frac{1}{n}\sum_{i=1}^{n} x_i^{r'}\right]^{1/r'} \qquad \text{for } r \leq r'.$$

It should be clear that $A_1$ is the arithmetic mean and $A_{-1}$ is the harmonic mean. What is less clear, but true, is that $A_0 = \lim_{r\to 0} A_r$ is the geometric mean. Thus, the arithmetic-geometric-harmonic mean inequality follows as a special case of the power mean inequality (see Exercise 4.57).

### 4.9.3 The Borel Paradox

Throughout this chapter, for continuous random variables $X$ and $Y$, we have been writing expressions such as $E(Y|X = x)$ and $P(Y \leq y|X = x)$. Thus far, we have not gotten into trouble. However, we might have.

Formally, the conditioning in a conditional expectation is done with respect to a sub sigma-algebra (Definition 1.2.1), and the conditional expectation $E(Y|\mathcal{G})$ is defined as a random variable whose integral, over any set in the sub sigma-algebra $\mathcal{G}$, agrees with that of $X$. This is quite an advanced concept in probability theory (see Billingsley 1995, Section 34).

Since the conditional expectation is only defined in terms of its integral, it may not be unique even if the conditioning is well-defined. However, when we condition on sets of probability 0 (such as $\{X = x\}$), conditioning may not be well defined, so different conditional expectations are more likely to appear. To see how this could affect us, it is easiest to look at conditional distributions, which amounts to calculating $E[I(Y \leq y)|X = x]$.

Proschan and Presnell (1998) tell the story of a statistics exam that had the question "If $X$ and $Y$ are independent standard normals, what is the conditional distribution of $Y$ given that $Y = X$?" Different students interpreted the condition $Y = X$ in the following ways:

(1) $Z_1 = 0$, where $Z_1 = Y - X$;
(2) $Z_2 = 1$, where $Z_2 = Y/X$;
(3) $Z_3 = 1$, where $Z_3 = I(Y = X)$.

Each condition is a correct interpretation of the condition $Y = X$, and each leads to a different conditional distribution (see Exercise 4.60).

This is the *Borel Paradox* and arises because different (correct) interpretations of the probability 0 conditioning sets result in different conditional expectations. How can we avoid the paradox? One way is to avoid conditioning on sets of probability 0. That is, compute only $E(Y|X \in B)$, where $B$ is a set with $P(X \in B) > 0$. So to compute something like $E(Y|X = x)$, take a sequence $B_n \downarrow x$, and define $E(Y|X = x) = \lim_{n \to \infty} E(Y|X \in B_n)$. We now avoid the paradox, as the different answers for $E(Y|X = x)$ will arise from different sequences, so there should be no surprises (Exercise 4.61).

# Chapter 5

# Properties of a Random Sample

*"I'm afraid that I rather give myself away when I explain," said he. "Results
without causes are much more impressive."*

**Sherlock Holmes**
*The Stock-Broker's Clerk*

## 5.1 Basic Concepts of Random Samples

Often, the data collected in an experiment consist of several observations on a variable
of interest. We discussed examples of this at the beginning of Chapter 4. In this
chapter, we present a model for data collection that is often used to describe this
situation, a model referred to as random sampling. The following definition explains
mathematically what is meant by the random sampling method of data collection.

**Definition 5.1.1** The random variables $X_1, \ldots, X_n$ are called a *random sample of
size n from the population* $f(x)$ *if* $X_1, \ldots, X_n$ are mutually independent random variables and the marginal pdf or pmf of each $X_i$ is the same function $f(x)$. Alternatively,
$X_1, \ldots, X_n$ are called *independent and identically distributed random variables with
pdf or pmf* $f(x)$. This is commonly abbreviated to iid random variables.

The random sampling model describes a type of experimental situation in which
the variable of interest has a probability distribution described by $f(x)$. If only one
observation $X$ is made on this variable, then probabilities regarding $X$ can be calculated using $f(x)$. In most experiments there are $n > 1$ (a fixed, positive integer)
repeated observations made on the variable, the first observation is $X_1$, the second is
$X_2$, and so on. Under the random sampling model each $X_i$ is an observation on the
same variable and each $X_i$ has a marginal distribution given by $f(x)$. Furthermore,
the observations are taken in such a way that the value of one observation has no
effect on or relationship with any of the other observations; that is, $X_1, \ldots, X_n$ are
*mutually independent*. (See Exercise 5.4 for a generalization of independence.)

From Definition 4.6.5, the joint pdf or pmf of $X_1, \ldots, X_n$ is given by

$$(5.1.1) \qquad f(x_1, \ldots, x_n) = f(x_1)f(x_2) \cdots f(x_n) = \prod_{i=1}^{n} f(x_i).$$

This joint pdf or pmf can be used to calculate probabilities involving the sample.
Since $X_1, \ldots, X_n$ are identically distributed, all the marginal densities $f(x)$ are the

same function. In particular, if the population pdf or pmf is a member of a parametric family, say one of those introduced in Chapter 3, with pdf or pmf given by $f(x|\theta)$, then the joint pdf or pmf is

$$(5.1.2) \qquad f(x_1, \ldots, x_n|\theta) = \prod_{i=1}^{n} f(x_i|\theta),$$

where the same parameter value $\theta$ is used in each of the terms in the product. If, in a statistical setting, we assume that the population we are observing is a member of a specified parametric family but the true parameter value is unknown, then a random sample from this population has a joint pdf or pmf of the above form with the value of $\theta$ unknown. By considering different possible values of $\theta$, we can study how a random sample would behave for different populations.

**Example 5.1.2 (Sample pdf–exponential)** Let $X_1, \ldots, X_n$ be a random sample from an exponential($\beta$) population. Specifically, $X_1, \ldots, X_n$ might correspond to the times until failure (measured in years) for $n$ identical circuit boards that are put on test and used until they fail. The joint pdf of the sample is

$$f(x_1, \ldots, x_n|\beta) = \prod_{i=1}^{n} f(x_i|\beta) = \prod_{i=1}^{n} \frac{1}{\beta} e^{-x_i/\beta} = \frac{1}{\beta^n} e^{-(x_1+\cdots+x_n)/\beta}.$$

This pdf can be used to answer questions about the sample. For example, what is the probability that all the boards last more than 2 years? We can compute

$$P(X_1 > 2, \ldots, X_n > 2)$$
$$= \int_{2}^{\infty} \cdots \int_{2}^{\infty} \prod_{i=1}^{n} \frac{1}{\beta} e^{-x_i/\beta} dx_1 \cdots dx_n$$
$$= e^{-2/\beta} \int_{2}^{\infty} \cdots \int_{2}^{\infty} \prod_{i=2}^{n} \frac{1}{\beta} e^{-x_i/\beta} \, dx_2 \cdots dx_n \quad \text{(integrate out } x_1)$$
$$\vdots \qquad\qquad \text{(integrate out the remaining } x_i\text{s successively)}$$
$$= (e^{-2/\beta})^n$$
$$= e^{-2n/\beta}.$$

If $\beta$, the average lifelength of a circuit board, is large relative to $n$, we see that this probability is near 1.

The previous calculation illustrates how the pdf of a random sample defined by (5.1.1) or, more specifically, by (5.1.2) can be used to calculate probabilities about the sample. Realize that the independent and identically distributed property of a random sample can also be used directly in such calculations. For example, the above calculation can be done like this:

$$P(X_1 > 2, \ldots, X_n > 2)$$

$$
\begin{aligned}
&= P(X_1 > 2) \cdots P(X_n > 2) && \text{(independence)}\\
&= [P(X_1 > 2)]^n && \text{(identical distributions)}\\
&= (e^{-2/\beta})^n && \text{(exponential calculation)}\\
&= e^{-2n/\beta}. && \|
\end{aligned}
$$

The random sampling model in Definition 5.1.1 is sometimes called sampling from an *infinite* population. Think of obtaining the values of $X_1, \ldots, X_n$ sequentially. First, the experiment is performed and $X_1 = x_1$ is observed. Then, the experiment is repeated and $X_2 = x_2$ is observed. The assumption of independence in random sampling implies that the probability distribution for $X_2$ is unaffected by the fact that $X_1 = x_1$ was observed first. "Removing" $x_1$ from the infinite population does not change the population, so $X_2 = x_2$ is still a random observation from the same population.

When sampling is from a *finite* population, Definition 5.1.1 may or may not be relevant depending on how the data collection is done. A finite population is a finite set of numbers, $\{x_1, \ldots, x_N\}$. A sample $X_1, \ldots, X_n$ is to be drawn from this population. Four ways of drawing this sample are described in Section 1.2.3. We will discuss the first two.

Suppose a value is chosen from the population in such a way that each of the $N$ values is equally likely (probability $= 1/N$) to be chosen. (Think of drawing numbers from a hat.) This value is recorded as $X_1 = x_1$. Then the process is repeated. Again, each of the $N$ values is equally likely to be chosen. The second value chosen is recorded as $X_2 = x_2$. (If the same number is chosen, then $x_1 = x_2$.) This process of drawing from the $N$ values is repeated $n$ times, yielding the sample $X_1, \ldots, X_n$. This kind of sampling is called *with replacement* because the value chosen at any stage is "replaced" in the population and is available for choice again at the next stage. For this kind of sampling, the conditions of Definition 5.1.1 are met. Each $X_i$ is a discrete random variable that takes on each of the values $x_1, \ldots, x_N$ with equal probability. The random variables $X_1, \ldots, X_n$ are independent because the process of choosing any $X_i$ is the same, regardless of the values that are chosen for any of the other variables. (This type of sampling is used in the *bootstrap*—see Section 10.1.4.)

A second method for drawing a random sample from a finite population is called sampling *without replacement*. Sampling without replacement is done as follows. A value is chosen from $\{x_1, \ldots, x_N\}$ in such a way that each of the $N$ values has probability $1/N$ of being chosen. This value is recorded as $X_1 = x_1$. Now a second value is chosen from the remaining $N - 1$ values. Each of the $N - 1$ values has probability $1/(N-1)$ of being chosen. The second chosen value is recorded as $X_2 = x_2$. Choice of the remaining values continues in this way, yielding the sample $X_1, \ldots, X_n$. But once a value is chosen, it is unavailable for choice at any later stage.

A sample drawn from a finite population without replacement does not satisfy all the conditions of Definition 5.1.1. The random variables $X_1, \ldots, X_n$ are not mutually independent. To see this, let $x$ and $y$ be distinct elements of $\{x_1, \ldots, x_N\}$. Then $P(X_2 = y | X_1 = y) = 0$, since the value $y$ cannot be chosen at the second stage if it was already chosen at the first. However, $P(X_2 = y | X_1 = x) = 1/(N-1)$. The probability distribution for $X_2$ depends on the value of $X_1$ that is observed and, hence, $X_1$ and $X_2$ are not independent. However, it is interesting to note that $X_1, \ldots, X_n$ are identically distributed. That is, the marginal distribution of $X_i$ is the same for each $i = 1, \ldots, n$. For $X_1$ it is clear that the marginal distribution is $P(X_1 = x) = 1/N$ for each $x \in \{x_1, \ldots, x_N\}$. To compute the marginal distribution for $X_2$, use Theorem 1.2.11(a) and the definition of conditional probability to write

$$P(X_2 = x) = \sum_{i=1}^{N} P(X_2 = x | X_1 = x_i) P(X_1 = x_i).$$

For one value of the index, say $k$, $x = x_k$ and $P(X_2 = x | X_1 = x_k) = 0$. For all other $j \neq k$, $P(X_2 = x | X_1 = x_j) = 1/(N-1)$. Thus,

(5.1.3)                    $$P(X_2 = x) = (N-1) \left( \frac{1}{N-1} \frac{1}{N} \right) = \frac{1}{N}.$$

Similar arguments can be used to show that each of the $X_i$s has the same marginal distribution.

Sampling without replacement from a finite population is sometimes called *simple random sampling*. It is important to realize that this is not the same sampling situation as that described in Definition 5.1.1. However, if the population size $N$ is large compared to the sample size $n$, $X_1, \ldots, X_n$ are nearly independent and some approximate probability calculations can be made assuming they are independent. By saying they are "nearly independent" we simply mean that the conditional distribution of $X_i$ given $X_1, \ldots, X_{i-1}$ is not too different from the marginal distribution of $X_i$. For example, the conditional distribution of $X_2$ given $X_1$ is

$$P(X_2 = x_1 | X_1 = x_1) = 0 \quad \text{and} \quad P(X_2 = x | X_1 = x_1) = \frac{1}{N-1} \quad \text{for } x \neq x_1.$$

This is not too different from the marginal distribution of $X_2$ given in (5.1.3) if $N$ is large. The nonzero probabilities in the conditional distribution of $X_i$ given $X_1, \ldots, X_{i-1}$ are $1/(N-i+1)$, which are close to $1/N$ if $i \leq n$ is small compared with $N$.

**Example 5.1.3 (Finite population model)**  As an example of an approximate calculation using independence, suppose $\{1, \ldots, 1000\}$ is the finite population, so $N = 1000$. A sample of size $n = 10$ is drawn without replacement. What is the probability that all ten sample values are greater than 200? If $X_1, \ldots, X_{10}$ were mutually independent we would have

$$P(X_1 > 200, \ldots, X_{10} > 200) = P(X_1 > 200) \cdots P(X_{10} > 200)$$

(5.1.4)                    $$= \left( \frac{800}{1000} \right)^{10} = .107374.$$

To calculate this probability exactly, let $Y$ be a random variable that counts the number of items in the sample that are greater than 200. Then $Y$ has a hypergeometric ($N = 1000$, $M = 800$, $K = 10$) distribution. So

$$P(X_1 > 200, \ldots, X_{10} > 200) = P(Y = 10)$$

$$= \frac{\binom{800}{10}\binom{200}{0}}{\binom{1000}{10}}$$

$$= .106164.$$

Thus, (5.1.4) is a reasonable approximation to the true value.                    ‖

   Throughout the remainder of the book, we will use Definition 5.1.1 as our definition of a random sample from a population.

## 5.2 Sums of Random Variables from a Random Sample

When a sample $X_1, \ldots, X_n$ is drawn, some summary of the values is usually computed. Any well-defined summary may be expressed mathematically as a function $T(x_1, \ldots, x_n)$ whose domain includes the sample space of the random vector $(X_1, \ldots, X_n)$. The function $T$ may be real-valued or vector-valued; thus the summary is a random variable (or vector), $Y = T(X_1, \ldots, X_n)$. This definition of a random variable as a function of others was treated in detail in Chapter 4, and the techniques in Chapter 4 can be used to describe the distribution of $Y$ in terms of the distribution of the population from which the sample was obtained. Since the random sample $X_1, \ldots, X_n$ has a simple probabilistic structure (because the $X_i$s are independent and identically distributed), the distribution of $Y$ is particularly tractable. Because this distribution is usually derived from the distribution of the variables in the random sample, it is called the *sampling distribution* of $Y$. This distinguishes the probability distribution of $Y$ from the distribution of the population, that is, the marginal distribution of each $X_i$. In this section, we will discuss some properties of sampling distributions, especially for functions $T(x_1, \ldots, x_n)$ defined by sums of random variables.

**Definition 5.2.1**   Let $X_1, \ldots, X_n$ be a random sample of size $n$ from a population and let $T(x_1, \ldots, x_n)$ be a real-valued or vector-valued function whose domain includes the sample space of $(X_1, \ldots, X_n)$. Then the random variable or random vector $Y = T(X_1, \ldots, X_n)$ is called a *statistic*. The probability distribution of a statistic $Y$ is called the *sampling distribution of Y*.

   The definition of a statistic is very broad, with the only restriction being that a statistic cannot be a function of a parameter. The sample summary given by a statistic can include many types of information. For example, it may give the smallest or largest value in the sample, the average sample value, or a measure of the variability in the sample observations. Three statistics that are often used and provide good summaries of the sample are now defined.

**Definition 5.2.2**  The *sample mean* is the arithmetic average of the values in a random sample. It is usually denoted by

$$\bar{X} = \frac{X_1 + \cdots + X_n}{n} = \frac{1}{n} \sum_{i=1}^{n} X_i.$$

**Definition 5.2.3**  The *sample variance* is the statistic defined by

$$S^2 = \frac{1}{n-1} \sum_{i=1}^{n} (X_i - \bar{X})^2.$$

The *sample standard deviation* is the statistic defined by $S = \sqrt{S^2}$.

As is commonly done, we have suppressed the functional notation in the above definitions of these statistics. That is, we have written $S$ rather than $S(X_1, \ldots, X_n)$. The dependence of the statistic on the sample is understood. As before, we will denote observed values of statistics with lowercase letters. So $\bar{x}$, $s^2$, and $s$ denote observed values of $\bar{X}$, $S^2$, and $S$.

The sample mean is certainly familiar to all. The sample variance and standard deviation are measures of variability in the sample that are related to the population variance and standard deviation in ways that we shall see below. We begin by deriving some properties of the sample mean and variance. In particular, the relationship for the sample variance given in Theorem 5.2.4 is related to (2.3.1), a similar relationship for the population variance.

**Theorem 5.2.4**  Let $x_1, \ldots, x_n$ be any numbers and $\bar{x} = (x_1 + \cdots + x_n)/n$. Then
**a.** $\min_a \sum_{i=1}^{n} (x_i - a)^2 = \sum_{i=1}^{n} (x_i - \bar{x})^2,$
**b.** $(n-1)s^2 = \sum_{i=1}^{n} (x_i - \bar{x})^2 = \sum_{i=1}^{n} x_i^2 - n\bar{x}^2.$

**Proof:**  To prove part (a), add and subtract $\bar{x}$ to get

$$\sum_{i=1}^{n} (x_i - a)^2 = \sum_{i=1}^{n} (x_i - \bar{x} + \bar{x} - a)^2$$

$$= \sum_{i=1}^{n} (x_i - \bar{x})^2 + 2 \sum_{i=1}^{n} (x_i - \bar{x})(\bar{x} - a) + \sum_{i=1}^{n} (\bar{x} - a)^2$$

$$= \sum_{i=1}^{n} (x_i - \bar{x})^2 + \sum_{i=1}^{n} (\bar{x} - a)^2. \qquad \text{(cross term is 0)}$$

It is now clear that the right-hand side is minimized at $a = \bar{x}$. (Notice the similarity to Example 2.2.6 and Exercise 4.13.)

To prove part (b), take $a = 0$ in the above.  $\square$

The expression in Theorem 5.2.4(b) is useful both computationally and theoretically because it allows us to express $s^2$ in terms of sums that are easy to handle.

We will begin our study of sampling distributions by considering the expected values of some statistics. The following result is quite useful.

**Lemma 5.2.5**  *Let $X_1, \ldots, X_n$ be a random sample from a population and let $g(x)$ be a function such that $Eg(X_1)$ and $\mathrm{Var}\, g(X_1)$ exist. Then*

$$(5.2.1) \qquad\qquad E\left(\sum_{i=1}^{n} g(X_i)\right) = n\left(Eg(X_1)\right)$$

*and*

$$(5.2.2) \qquad\qquad \mathrm{Var}\left(\sum_{i=1}^{n} g(X_i)\right) = n\left(\mathrm{Var}\, g(X_1)\right).$$

**Proof:** To prove (5.2.1), note that

$$E\left(\sum_{i=1}^{n} g(X_i)\right) = \sum_{i=1}^{n} Eg(X_i) = n\left(Eg(X_1)\right).$$

Since the $X_i$s are identically distributed, the second equality is true because $Eg(X_i)$ is the same for all $i$. Note that the independence of $X_1, \ldots, X_n$ is not needed for (5.2.1) to hold. Indeed, (5.2.1) is true for any collection of $n$ identically distributed random variables.

To prove (5.2.2), note that

$$\mathrm{Var}\left(\sum_{i=1}^{n} g(X_i)\right) = E\left[\sum_{i=1}^{n} g(X_i) - E\left(\sum_{i=1}^{n} g(X_i)\right)\right]^2 \qquad \text{(definition of variance)}$$

$$= E\left[\sum_{i=1}^{n} (g(X_i) - Eg(X_i))\right]^2. \qquad \left(\begin{array}{l}\text{expectation property and} \\ \text{rearrangement of terms}\end{array}\right)$$

In this last expression there are $n^2$ terms. First, there are $n$ terms $(g(X_i) - Eg(X_i))^2$, $i = 1, \ldots, n$, and for each, we have

$$E\left(g(X_i) - Eg(X_i)\right)^2 = \mathrm{Var}\, g(X_i) \qquad \text{(definition of variance)}$$

$$= \mathrm{Var}\, g(X_1). \qquad \text{(identically distributed)}$$

The remaining $n(n-1)$ terms are all of the form $\big(g(X_i) - Eg(X_i)\big)\big(g(X_j) - Eg(X_j)\big)$, with $i \neq j$. For each term,

$$E\left[(g(X_i) - Eg(X_i))(g(X_j) - Eg(X_j))\right] = \mathrm{Cov}\left(g(X_i), g(X_j)\right) \qquad \left(\begin{array}{l}\text{definition of} \\ \text{covariance}\end{array}\right)$$

$$= 0. \qquad \left(\begin{array}{l}\text{independence} \\ \text{Theorem 4.5.5}\end{array}\right)$$

Thus, we obtain equation (5.2.2).                                            □

**Theorem 5.2.6**  *Let $X_1, \ldots, X_n$ be a random sample from a population with mean $\mu$ and variance $\sigma^2 < \infty$. Then*

**a.** $E\bar{X} = \mu$,

**b.** $\text{Var }\bar{X} = \dfrac{\sigma^2}{n}$,

**c.** $ES^2 = \sigma^2$.

**Proof:** To prove (a), let $g(X_i) = X_i/n$, so $Eg(X_i) = \mu/n$. Then, by Lemma 5.2.5,

$$E\bar{X} = E\left(\frac{1}{n}\sum_{i=1}^{n}X_i\right) = \frac{1}{n}E\left(\sum_{i=1}^{n}X_i\right) = \frac{1}{n}nEX_1 = \mu.$$

Similarly for (b), we have

$$\text{Var }\bar{X} = \text{Var}\left(\frac{1}{n}\sum_{i=1}^{n}X_i\right) = \frac{1}{n^2}\text{Var}\left(\sum_{i=1}^{n}X_i\right) = \frac{1}{n^2}n\,\text{Var }X_1 = \frac{\sigma^2}{n}.$$

For the sample variance, using Theorem 5.2.4, we have

$$ES^2 = E\left(\frac{1}{n-1}\left[\sum_{i=1}^{n}X_i^2 - n\bar{X}^2\right]\right)$$

$$= \frac{1}{n-1}\left(nEX_1^2 - nE\bar{X}^2\right)$$

$$= \frac{1}{n-1}\left(n(\sigma^2 + \mu^2) - n\left(\frac{\sigma^2}{n} + \mu^2\right)\right) = \sigma^2,$$

establishing part (c) and proving the theorem. $\qquad\square$

The relationships (a) and (c) in Theorem 5.2.6, relationships between a statistic and a population parameter, are examples of *unbiased* statistics. These are discussed in Chapter 7. The statistic $\bar{X}$ is an *unbiased estimator* of $\mu$, and $S^2$ is an *unbiased estimator* of $\sigma^2$. The use of $n-1$ in the definition of $S^2$ may have seemed unintuitive. Now we see that, with this definition, $ES^2 = \sigma^2$. If $S^2$ were defined as the usual average of the squared deviations with $n$ rather than $n-1$ in the denominator, then $ES^2$ would be $\frac{n-1}{n}\sigma^2$ and $S^2$ would not be an unbiased estimator of $\sigma^2$.

We now discuss in more detail the sampling distribution of $\bar{X}$. The methods from Sections 4.3 and 4.6 can be used to derive this sampling distribution from the population distribution. But because of the special probabilistic structure of a random sample (iid random variables), the resulting sampling distribution of $\bar{X}$ is simply expressed.

First we note some simple relationships. Since $\bar{X} = \frac{1}{n}(X_1 + \cdots + X_n)$, if $f(y)$ is the pdf of $Y = (X_1 + \cdots + X_n)$, then $f_{\bar{X}}(x) = nf(nx)$ is the pdf of $\bar{X}$ (see Exercise 5.5). Thus, a result about the pdf of $Y$ is easily transformed into a result about the pdf of $\bar{X}$. A similar relationship holds for mgfs:

$$M_{\bar{X}}(t) = Ee^{t\bar{X}} = Ee^{t(X_1+\cdots+X_n)/n} = Ee^{(t/n)Y} = M_Y(t/n).$$

Since $X_1, \ldots, X_n$ are identically distributed, $M_{X_i}(t)$ is the same function for each $i$. Thus, by Theorem 4.6.7, we have the following.

**Theorem 5.2.7**   *Let $X_1, \ldots, X_n$ be a random sample from a population with mgf $M_X(t)$. Then the mgf of the sample mean is*

$$M_{\bar{X}}(t) = [M_X(t/n)]^n.$$

Of course, Theorem 5.2.7 is useful only if the expression for $M_{\bar{X}}(t)$ is a familiar mgf. Cases when this is true are somewhat limited, but the following example illustrates that, when this method works, it provides a very slick derivation of the sampling distribution of $\bar{X}$.

**Example 5.2.8 (Distribution of the mean)**   Let $X_1, \ldots, X_n$ be a random sample from a n$(\mu, \sigma^2)$ population. Then the mgf of the sample mean is

$$M_{\bar{X}}(t) = \left[ \exp\left( \mu \frac{t}{n} + \frac{\sigma^2(t/n)^2}{2} \right) \right]^n$$

$$= \exp\left( n\left( \mu \frac{t}{n} + \frac{\sigma^2(t/n)^2}{2} \right) \right) = \exp\left( \mu t + \frac{(\sigma^2/n)t^2}{2} \right).$$

Thus, $\bar{X}$ has a n$(\mu, \sigma^2/n)$ distribution.

Another simple example is given by a gamma$(\alpha, \beta)$ random sample (see Example 4.6.8). Here, we can also easily derive the distribution of the sample mean. The mgf of the sample mean is

$$M_{\bar{X}}(t) = \left[ \left( \frac{1}{1 - \beta(t/n)} \right)^\alpha \right]^n = \left( \frac{1}{1 - (\beta/n)t} \right)^{n\alpha},$$

which we recognize as the mgf of a gamma$(n\alpha, \beta/n)$, the distribution of $\bar{X}$.        ‖

If Theorem 5.2.7 is not applicable, because either the resulting mgf of $\bar{X}$ is unrecognizable or the population mgf does not exist, then the transformation method of Sections 4.3 and 4.6 might be used to find the pdf of $Y = (X_1 + \cdots + X_n)$ and $\bar{X}$. In such cases, the following *convolution formula* is useful.

**Theorem 5.2.9**   *If $X$ and $Y$ are independent continuous random variables with pdfs $f_X(x)$ and $f_Y(y)$, then the pdf of $Z = X + Y$ is*

(5.2.3)
$$f_Z(z) = \int_{-\infty}^{\infty} f_X(w) f_Y(z - w)\, dw.$$

**Proof:** Let $W = X$. The Jacobian of the transformation from $(X, Y)$ to $(Z, W)$ is 1. So using (4.3.2), we obtain the joint pdf of $(Z, W)$ as

$$f_{Z,W}(z, w) = f_{X,Y}(w, z - w) = f_X(w) f_Y(z - w).$$

Integrating out $w$, we obtain the marginal pdf of $Z$ as given in (5.2.3).        □

The limits of integration in (5.2.3) might be modified if $f_X$ or $f_Y$ or both are positive for only some values. For example, if $f_X$ and $f_Y$ are positive for only positive values, then the limits of integration are 0 and $z$ because the integrand is 0 for values of $w$ outside this range. Equations similar to the convolution formula of (5.2.3) can be derived for operations other than summing; for example, formulas for differences, products, and quotients are also obtainable (see Exercise 5.6).

**Example 5.2.10 (Sum of Cauchy random variables)** As an example of a situation where the mgf technique fails, consider sampling from a Cauchy distribution. We will eventually derive the distribution of $\bar{Z}$, the mean of $Z_1, \ldots, Z_n$, iid Cauchy$(0, 1)$ observations. We start, however, with the distribution of the sum of two independent Cauchy random variables and apply formula (5.2.3).

Let $U$ and $V$ be independent Cauchy random variables, $U \sim$ Cauchy$(0, \sigma)$ and $V \sim$ Cauchy$(0, \tau)$; that is,

$$f_U(u) = \frac{1}{\pi\sigma}\frac{1}{1 + (u/\sigma)^2}, \qquad f_V(v) = \frac{1}{\pi\tau}\frac{1}{1 + (v/\tau)^2}, \qquad \begin{array}{l} -\infty < u < \infty, \\ -\infty < v < \infty. \end{array}$$

Based on formula (5.2.3), the pdf of $Z = U + V$ is given by

$$(5.2.4)\ f_Z(z) = \int_{-\infty}^{\infty} \frac{1}{\pi\sigma}\frac{1}{1 + (w/\sigma)^2}\frac{1}{\pi\tau}\frac{1}{1 + ((z - w)/\tau)^2}\,dw, \quad -\infty < z < \infty.$$

This integral is somewhat involved but can be solved by a partial fraction decomposition and some careful antidifferentiation (see Exercise 5.7). The result is

$$(5.2.5) \qquad f_Z(z) = \frac{1}{\pi(\sigma + \tau)}\frac{1}{1 + (z/(\sigma + \tau))^2}, \quad -\infty < z < \infty.$$

Thus, the sum of two independent Cauchy random variables is again a Cauchy, with the scale parameters adding. It therefore follows that if $Z_1, \ldots, Z_n$ are iid Cauchy$(0, 1)$ random variables, then $\sum Z_i$ is Cauchy$(0, n)$ and also $\bar{Z}$ is Cauchy$(0, 1)$! The sample mean has the same distribution as the individual observations. (See Example A.0.5 in Appendix A for a computer algebra version of this calculation.) ‖

If we are sampling from a location–scale family or if we are sampling from certain types of exponential families, the sampling distribution of sums of random variables, and in particular of $\bar{X}$, is easy to derive. We will close this section by discussing these two situations.

We first treat the location–scale case discussed in Section 3.5. Suppose $X_1, \ldots, X_n$ is a random sample from $(1/\sigma)f((x-\mu)/\sigma)$, a member of a location–scale family. Then the distribution of $\bar{X}$ has a simple relationship to the distribution of $\bar{Z}$, the sample mean from a random sample from the standard pdf $f(z)$. To see the nature of this relationship, note that from Theorem 3.5.6 there exist random variables $Z_1, \ldots, Z_n$ such that $X_i = \sigma Z_i + \mu$ and the pdf of each $Z_i$ is $f(z)$. Furthermore, we see that $Z_1, \ldots, Z_n$ are mutually independent. Thus $Z_1, \ldots, Z_n$ is a random sample from $f(z)$. The sample means $\bar{X}$ and $\bar{Z}$ are related by

$$\bar{X} = \frac{1}{n}\sum_{i=1}^{n} X_i = \frac{1}{n}\sum_{i=1}^{n}(\sigma Z_i + \mu) = \frac{1}{n}\left(\sigma\sum_{i=1}^{n} Z_i + n\mu\right) = \sigma\bar{Z} + \mu.$$

Thus, again applying Theorem 3.5.6, we find that if $g(z)$ is the pdf of $\bar{Z}$, then $(1/\sigma)g((x-\mu)/\sigma)$ is the pdf of $\bar{X}$. It may be easier to work first with $Z_1, \ldots, Z_n$ and $f(z)$ to find the pdf $g(z)$ of $\bar{Z}$. If this is done, the parameters $\mu$ and $\sigma$ do not have to be dealt with, which may make the computations less messy. Then we immediately know that the pdf of $\bar{X}$ is $(1/\sigma)g((x-\mu)/\sigma)$.

In Example 5.2.10, we found that if $Z_1, \ldots, Z_n$ is a random sample from a Cauchy$(0,1)$ distribution, then $\bar{Z}$ also has a Cauchy$(0,1)$ distribution. Now we can conclude that if $X_1, \ldots, X_n$ is a random sample from a Cauchy$(\mu, \sigma)$ distribution, then $\bar{X}$ also has a Cauchy$(\mu, \sigma)$ distribution. It is important to note that the dispersion in the distribution of $\bar{X}$, as measured by $\sigma$, is the same, regardless of the sample size $n$. This is in sharp contrast to the more common situation in Theorem 5.2.6 (the population has finite variance), where Var $\bar{X} = \sigma^2/n$ decreases as the sample size increases.

When sampling is from an exponential family, some sums from a random sample have sampling distributions that are easy to derive. The statistics $T_1, \ldots, T_k$ in the next theorem are important summary statistics, as will be seen in Section 6.2.

**Theorem 5.2.11** *Suppose $X_1, \ldots, X_n$ is a random sample from a pdf or pmf $f(x|\theta)$, where*

$$f(x|\theta) = h(x)c(\theta)\exp\left(\sum_{i=1}^{k} w_i(\theta)t_i(x)\right)$$

*is a member of an exponential family. Define statistics $T_1, \ldots, T_k$ by*

$$T_i(X_1, \ldots, X_n) = \sum_{j=1}^{n} t_i(X_j), \quad i = 1, \ldots, k.$$

*If the set $\{(w_1(\theta), w_2(\theta), \ldots, w_k(\theta)), \theta \in \Theta\}$ contains an open subset of $\Re^k$, then the distribution of $(T_1, \ldots, T_k)$ is an exponential family of the form*

$$(5.2.6) \qquad f_T(u_1, \ldots, u_k|\theta) = H(u_1, \ldots, u_k)[c(\theta)]^n \exp\left(\sum_{i=1}^{k} w_i(\theta)u_i\right).$$

The open set condition eliminates a density such as the n$(\theta, \theta^2)$ and, in general, eliminates curved exponential families from Theorem 5.2.11. Note that in the pdf or pmf of $(T_1, \ldots, T_k)$, the functions $c(\theta)$ and $w_i(\theta)$ are the same as in the original family although the function $H(u_1, \ldots, u_k)$ is, of course, different from $h(x)$. We will not prove this theorem but will only illustrate the result in a simple case.

**Example 5.2.12 (Sum of Bernoulli random variables)**  Suppose $X_1, \ldots, X_n$ is a random sample from a Bernoulli$(p)$ distribution. From Example 3.4.1 (with $n = 1$) we see that a Bernoulli$(p)$ distribution is an exponential family with $k = 1, c(p) = (1-p), w_1(p) = \log(p/(1-p))$, and $t_1(x) = x$. Thus, in the previous theorem, $T_1 = T_1(X_1, \ldots, X_n) = X_1 + \cdots + X_n$. From the definition of the binomial distribution in Section 3.2, we know that $T_1$ has a binomial$(n, p)$ distribution. From Example 3.4.1 we also see that a binomial$(n, p)$ distribution is an exponential family with the same $w_1(p)$ and $c(p) = (1-p)^n$. Thus expression (5.2.6) is verified for this example.    ‖

## 5.3 Sampling from the Normal Distribution

This section deals with the properties of sample quantities drawn from a normal population—still one of the most widely used statistical models. Sampling from a normal population leads to many useful properties of sample statistics and also to many well-known sampling distributions.

### 5.3.1 Properties of the Sample Mean and Variance

We have already seen how to calculate the means and variances of $\bar{X}$ and $S^2$ in general. Now, under the additional assumption of normality, we can derive their full distributions, and more. The properties of $\bar{X}$ and $S^2$ are summarized in the following theorem.

**Theorem 5.3.1** *Let $X_1, \ldots, X_n$ be a random sample from a $n(\mu, \sigma^2)$ distribution, and let $\bar{X} = (1/n)\sum_{i=1}^{n} X_i$ and $S^2 = [1/(n-1)]\sum_{i=1}^{n}(X_i - \bar{X})^2$. Then*

**a.** $\bar{X}$ *and $S^2$ are independent random variables,*

**b.** $\bar{X}$ *has a $n(\mu, \sigma^2/n)$ distribution,*

**c.** $(n-1)S^2/\sigma^2$ *has a chi squared distribution with $n-1$ degrees of freedom.*

**Proof:** First note that, from Section 3.5 on location–scale families, we can assume, without loss of generality, that $\mu = 0$ and $\sigma = 1$. (Also see the discussion preceding Theorem 5.2.11.) Furthermore, part (b) has already been established in Example 5.2.8, leaving us to prove parts (a) and (c).

To prove part (a) we will apply Theorem 4.6.12, and show that $\bar{X}$ and $S^2$ are functions of independent random vectors. Note that we can write $S^2$ as a function of $n-1$ deviations as follows:

$$S^2 = \frac{1}{n-1} \sum_{i=1}^{n}(X_i - \bar{X})^2$$

$$= \frac{1}{n-1}\left( (X_1 - \bar{X})^2 + \sum_{i=2}^{n}(X_i - \bar{X})^2 \right)$$

$$= \frac{1}{n-1}\left( \left[\sum_{i=2}^{n}(X_i - \bar{X})\right]^2 + \sum_{i=2}^{n}(X_i - \bar{X})^2 \right) \cdot \left( \begin{matrix} \text{since} \\ \sum_{i=1}^{n}(X_i - \bar{X}) = 0 \end{matrix} \right)$$

Thus, $S^2$ can be written as a function only of $(X_2 - \bar{X}, \ldots, X_n - \bar{X})$. We will now show that these random variables are independent of $\bar{X}$. The joint pdf of the sample $X_1, \ldots, X_n$ is given by

$$f(x_1, \ldots, x_n) = \frac{1}{(2\pi)^{n/2}} e^{-(1/2)\sum_{i=1}^{n} x_i^2}, \quad -\infty < x_i < \infty.$$

Make the transformation

$$y_1 = \bar{x},$$

$$y_2 = x_2 - \bar{x},$$

$$\vdots$$

$$y_n = x_n - \bar{x}.$$

This is a linear transformation with a Jacobian equal to $1/n$. We have

$$f(y_1, \ldots, y_n)$$

$$= \frac{n}{(2\pi)^{n/2}} e^{-(1/2)(y_1 - \Sigma_{i=2}^n y_i)^2} e^{-(1/2)\Sigma_{i=2}^n (y_i + y_1)^2}, \quad -\infty < y_i < \infty$$

$$= \left[ \left( \frac{n}{2\pi} \right)^{1/2} e^{(-ny_1^2)/2} \right] \left[ \frac{n^{1/2}}{(2\pi)^{(n-1)/2}} e^{-(1/2)\left[ \Sigma_{i=2}^n y_i^2 + (\Sigma_{i=2}^n y_i)^2 \right]} \right], \quad -\infty < y_i < \infty.$$

Since the joint pdf of $Y_1, \ldots, Y_n$ factors, it follows from Theorem 4.6.11 that $Y_1$ is independent of $Y_2, \ldots, Y_n$ and, hence, from Theorem 4.6.12 that $\bar{X}$ is independent of $S^2$. □

To finish the proof of the theorem we must now derive the distribution of $S^2$. Before doing so, however, we digress a little and discuss the chi squared distribution, whose properties play an important part in the derivation of the pdf of $S^2$. Recall from Section 3.3 that the chi squared pdf is a special case of the gamma pdf and is given by

$$f(x) = \frac{1}{\Gamma(p/2) \, 2^{p/2}} x^{(p/2)-1} e^{-x/2}, \quad 0 < x < \infty,$$

where $p$ is called the *degrees of freedom*. We now summarize some pertinent facts about the chi squared distribution.

**Lemma 5.3.2 (Facts about chi squared random variables)** *We use the notation $\chi_p^2$ to denote a chi squared random variable with $p$ degrees of freedom.*

**a.** *If $Z$ is a $n(0,1)$ random variable, then $Z^2 \sim \chi_1^2$; that is, the square of a standard normal random variable is a chi squared random variable.*

**b.** *If $X_1, \ldots, X_n$ are independent and $X_i \sim \chi_{p_i}^2$, then $X_1 + \cdots + X_n \sim \chi_{p_1 + \cdots + p_n}^2$; that is, independent chi squared variables add to a chi squared variable, and the degrees of freedom also add.*

**Proof:** We have encountered these facts already. Part (a) was established in Example 2.1.7. Part (b) is a special case of Example 4.6.8, which has to do with sums of independent gamma random variables. Since a $\chi_p^2$ random variable is a gamma$(p/2, 2)$, application of the example gives part (b). □

**Proof of Theorem 5.3.1(c):** We will employ an induction argument to establish the distribution of $S^2$, using the notation $\bar{X}_k$ and $S_k^2$ to denote the sample mean and variance based on the first $k$ observations. (Note that the actual ordering of the observations is immaterial—we are just considering them to be ordered to facilitate the proof.) It is straightforward to establish (see Exercise 5.15) that

$$(5.3.1) \qquad (n-1)S_n^2 = (n-2)S_{n-1}^2 + \left(\frac{n-1}{n}\right)(X_n - \bar{X}_{n-1})^2.$$

Now consider $n = 2$. Defining $0 \times S_1^2 = 0$, we have from (5.3.1) that

$$S_2^2 = \frac{1}{2}(X_2 - X_1)^2.$$

Since the distribution of $(X_2 - X_1)/\sqrt{2}$ is $n(0,1)$, part (a) of Lemma 5.3.2 shows that $S_2^2 \sim \chi_1^2$. Proceeding with the induction, we assume that for $n = k$, $(k-1)S_k^2 \sim \chi_{k-1}^2$. For $n = k+1$ we have from (5.3.1)

$$(5.3.2) \qquad kS_{k+1}^2 = (k-1)S_k^2 + \left(\frac{k}{k+1}\right)(X_{k+1} - \bar{X}_k)^2.$$

According to the induction hypothesis, $(k-1)S_k^2 \sim \chi_{k-1}^2$. If we can establish that $(k/(k+1))(X_{k+1} - \bar{X}_k)^2 \sim \chi_1^2$, independent of $S_k^2$, it will follow from part (b) of Lemma 5.3.2 that $kS_{k+1}^2 \sim \chi_k^2$, and the theorem will be proved.

The independence of $(X_{k+1} - \bar{X}_k)^2$ and $S_k^2$ again follows from Theorem 4.6.12. The vector $(X_{k+1}, \bar{X}_k)$ is independent of $S_k^2$ and so is any function of the vector. Furthermore, $X_{k+1} - \bar{X}_k$ is a normal random variable with mean 0 and variance

$$\text{Var}\,(X_{k+1} - \bar{X}_k) = \frac{k+1}{k},$$

and therefore $(k/(k+1))(X_{k+1} - \bar{X}_k)^2 \sim \chi_1^2$, and the theorem is established.    □

The independence of $\bar{X}$ and $S^2$ can be established in a manner different from that used in the proof of Theorem 5.3.1. Rather than show that the joint pdf factors, we can use the following lemma, which ties together independence and correlation for normal samples.

**Lemma 5.3.3**  Let $X_j \sim n(\mu_j, \sigma_j^2), j = 1, \ldots, n$, independent. For constants $a_{ij}$ and $b_{rj}$ $(j = 1, \ldots, n; i = 1, \ldots, k; r = 1, \ldots, m)$, where $k + m \le n$, define

$$U_i = \sum_{j=1}^{n} a_{ij}X_j, \quad i = 1, \ldots, k,$$

$$V_r = \sum_{j=1}^{n} b_{rj}X_j, \quad r = 1, \ldots, m.$$

**a.** The random variables $U_i$ and $V_r$ are independent if and only if $\text{Cov}(U_i, V_r) = 0$. Furthermore, $\text{Cov}(U_i, V_r) = \sum_{j=1}^{n} a_{ij}b_{rj}\sigma_j^2$.

**b.** The random vectors $(U_1, \ldots, U_k)$ and $(V_1, \ldots, V_m)$ are independent if and only if $U_i$ is independent of $V_r$ for all pairs $i, r$ $(i = 1, \ldots, k; r = 1, \ldots, m)$.

**Proof:** It is sufficient to prove the lemma for $\mu_i = 0$ and $\sigma_i^2 = 1$, since the general statement of the lemma then follows quickly. Furthermore, the implication from independence to 0 covariance is immediate (Theorem 4.5.5) and the expression for the covariance is easily verified (Exercise 5.14). Note also that Corollary 4.6.10 shows that $U_i$ and $V_r$ are normally distributed.

Thus, we are left with proving that if the constants satisfy the above restriction (equivalently, the covariance is 0), then we have independence under normality. We prove the lemma only for $n = 2$, since the proof for general $n$ is similar but necessitates a detailed $n$-variate transformation.

To prove part (a) start with the joint pdf of $X_1$ and $X_2$,

$$f_{X_1,X_2}(x_1, x_2) = \frac{1}{2\pi} e^{-(1/2)(x_1^2 + x_2^2)}, \quad -\infty < x_1, x_2 < \infty.$$

Make the transformation (we can suppress the double subscript in the $n = 2$ case)

$$u = a_1 x_1 + a_2 x_2, \qquad v = b_1 x_1 + b_2 x_2,$$

so

$$x_1 = \frac{b_2 u - a_2 v}{a_1 b_2 - b_1 a_2}, \qquad x_2 = \frac{a_1 v - b_1 u}{a_1 b_2 - b_1 a_2},$$

with Jacobian

$$J = \begin{vmatrix} \dfrac{\partial x_1}{\partial u} & \dfrac{\partial x_1}{\partial v} \\[2mm] \dfrac{\partial x_2}{\partial u} & \dfrac{\partial x_2}{\partial v} \end{vmatrix} = \frac{1}{a_1 b_2 - b_1 a_2}.$$

Thus, the pdf of $U$ and $V$ is

$$f_{U,V}(u, v) = f_{X_1,X_2}\left( \frac{b_2 u - a_2 v}{a_1 b_2 - b_1 a_2}, \frac{a_1 v - b_1 u}{a_1 b_2 - b_1 a_2} \right) |J|$$

$$= \frac{1}{2\pi} \exp\left\{ \frac{-1}{2(a_1 b_2 - b_1 a_2)^2} \left[ (b_2 u - a_2 v)^2 + (a_1 v - b_1 u)^2 \right] \right\} |J|,$$

$-\infty < u, v < \infty$. Expanding the squares in the exponent, we can write

$$(b_2 u - a_2 v)^2 + (a_1 v - b_1 u)^2 = (b_1^2 + b_2^2)u^2 + (a_1^2 + a_2^2)v^2 - 2(a_1 b_1 + a_2 b_2)uv.$$

The assumption on the constants shows that the cross-term is identically 0. Hence, the pdf factors so, by Lemma 4.2.7, $U$ and $V$ are independent and part (a) is established.

A similar type of argument will work for part (b), the details of which we will not go into. If the appropriate transformation is made, the joint pdf of the vectors $(U_1, \ldots, U_k)$ and $(V_1, \ldots, V_m)$ can be obtained. By an application of Theorem 4.6.11, the vectors are independent if the joint pdf factors. From the form of the normal pdf, this will happen if and only if $U_i$ is independent of $V_r$ for all pairs $i, r$ ($i = 1, \ldots, k; r = 1, \ldots, m$). $\qquad\Box$

This lemma shows that, if we start with independent normal random variables, covariance and independence are equivalent for linear functions of these random variables. Thus, we can check independence for normal variables by merely checking the covariance term, a much simpler calculation. There is nothing magic about this; it just follows from the form of the normal pdf. Furthermore, part (b) allows us to infer overall independence of normal vectors by just checking pairwise independence, a property that does not hold for general random variables.

We can use Lemma 5.3.3 to provide an alternative proof of the independence of $\bar{X}$ and $S^2$ in normal sampling. Since we can write $S^2$ as a function of $n-1$ deviations $(X_2 - \bar{X}, \ldots, X_n - \bar{X})$, we must show that these random variables are uncorrelated with $\bar{X}$. The normality assumption, together with Lemma 5.3.3, will then allow us to conclude independence.

As an illustration of the application of Lemma 5.3.3, write

$$\bar{X} = \sum_{i=1}^{n} \left(\frac{1}{n}\right) X_i,$$

$$X_j - \bar{X} = \sum_{i=1}^{n} \left(\delta_{ij} - \frac{1}{n}\right) X_i,$$

where $\delta_{ij} = 1$ if $i = j$ and $\delta_{ij} = 0$ otherwise. It is then easy to show that

$$\text{Cov}\left(\bar{X}, X_j - \bar{X}\right) = \sum_{i=1}^{n} \left(\frac{1}{n}\right) \left(\delta_{ij} - \frac{1}{n}\right) = 0,$$

showing that $\bar{X}$ and $X_j - \bar{X}$ are independent (as long as the $X_i$s have the same variance).

### 5.3.2 The Derived Distributions: Student's t and Snedecor's F

The distributions derived in Section 5.3.1 are, in a sense, the first step in a statistical analysis that assumes normality. In particular, in most practical cases the variance, $\sigma^2$, is unknown. Thus, to get any idea of the variability of $\bar{X}$ (as an estimate of $\mu$), it is necessary to estimate this variance. This topic was first addressed by W. S. Gosset (who published under the pseudonym of Student) in the early 1900s. The landmark work of Student resulted in Student's $t$ distribution or, more simply, the $t$ distribution.

If $X_1, \ldots, X_n$ are a random sample from a $n(\mu, \sigma^2)$, we know that the quantity

(5.3.3)
$$\frac{\bar{X} - \mu}{\sigma/\sqrt{n}}$$

is distributed as a $n(0,1)$ random variable. If we knew the value of $\sigma$ and we measured $\bar{X}$, then we could use (5.3.3) as a basis for inference about $\mu$, since $\mu$ would then be the only unknown quantity. Most of the time, however, $\sigma$ is unknown. Student did the obvious thing—he looked at the distribution of

(5.3.4)
$$\frac{\bar{X} - \mu}{S/\sqrt{n}},$$

a quantity that could be used as a basis for inference about $\mu$ when $\sigma$ was unknown.

The distribution of (5.3.4) is easy to derive, provided that we first notice a few simplifying maneuvers. Multiply (5.3.4) by $\sigma/\sigma$ and rearrange slightly to obtain

$$(5.3.5) \qquad \frac{\bar{X} - \mu}{S/\sqrt{n}} = \frac{(\bar{X} - \mu)/(\sigma/\sqrt{n})}{\sqrt{S^2/\sigma^2}}.$$

The numerator of (5.3.5) is a $n(0,1)$ random variable, and the denominator is $\sqrt{\chi_{n-1}^2/(n-1)}$, *independent* of the numerator. Thus, the distribution of (5.3.4) can be found by solving the simplified problem of finding the distribution of $U/\sqrt{V/p}$, where $U$ is $n(0,1)$, $V$ is $\chi_p^2$, and $U$ and $V$ are independent. This gives us Student's $t$ distribution.

**Definition 5.3.4**  Let $X_1, \ldots, X_n$ be a random sample from a $n(\mu, \sigma^2)$ distribution. The quantity $(\bar{X} - \mu)/(S/\sqrt{n})$ has *Student's t distribution with $n-1$ degrees of freedom.* Equivalently, a random variable $T$ has Student's $t$ distribution with $p$ degrees of freedom, and we write $T \sim t_p$ if it has pdf

$$(5.3.6) \qquad f_T(t) = \frac{\Gamma\left(\frac{p+1}{2}\right)}{\Gamma\left(\frac{p}{2}\right)} \frac{1}{(p\pi)^{1/2}} \frac{1}{(1 + t^2/p)^{(p+1)/2}}, \qquad -\infty < t < \infty.$$

Notice that if $p = 1$, then (5.3.6) becomes the pdf of the Cauchy distribution, which occurs for samples of size 2. Once again the Cauchy distribution has appeared in an ordinary situation.

The derivation of the $t$ pdf is straightforward. If we start with $U$ and $V$ defined above, it follows from (5.3.5) that the joint pdf of $U$ and $V$ is

$$f_{U,V}(u,v) = \frac{1}{(2\pi)^{1/2}} e^{-u^2/2} \frac{1}{\Gamma\left(\frac{p}{2}\right) 2^{p/2}} v^{(p/2)-1} e^{-v/2}, \qquad -\infty < u < \infty, \quad 0 < v < \infty.$$

(Recall that $U$ and $V$ are independent.) Now make the transformation

$$t = \frac{u}{\sqrt{v/p}}, \qquad w = v.$$

The Jacobian of the transformation is $(w/p)^{1/2}$, and the marginal pdf of $T$ is given by

$$f_T(t) = \int_0^\infty f_{U,V}\left(t\left(\frac{w}{p}\right)^{1/2}, w\right) \left(\frac{w}{p}\right)^{1/2} dw$$

$$= \frac{1}{(2\pi)^{1/2}} \frac{1}{\Gamma\left(\frac{p}{2}\right) 2^{p/2}} \int_0^\infty e^{-(1/2)t^2 w/p} w^{(p/2)-1} e^{-w/2} \left(\frac{w}{p}\right)^{1/2} dw$$

$$= \frac{1}{(2\pi)^{1/2}} \frac{1}{\Gamma\left(\frac{p}{2}\right) 2^{p/2} p^{1/2}} \int_0^\infty e^{-(1/2)(1 + t^2/p)w} w^{((p+1)/2)-1} dw.$$

Recognize the integrand as the kernel of a gamma$((p+1)/2, 2/(1+t^2/p))$ pdf. We therefore have

$$f_T(t) = \frac{1}{(2\pi)^{1/2}} \frac{1}{\Gamma\left(\frac{p}{2}\right) 2^{p/2} p^{1/2}} \Gamma\left(\frac{p+1}{2}\right) \left[\frac{2}{1+t^2/p}\right]^{(p+1)/2},$$

which is equal to (5.3.6).

Student's $t$ has no mgf because it does not have moments of all orders. In fact, if there are $p$ degrees of freedom, then there are only $p-1$ moments. Hence, a $t_1$ has no mean, a $t_2$ has no variance, etc. It is easy to check (see Exercise 5.18) that if $T_p$ is a random variable with a $t_p$ distribution, then

$$\mathrm{E}T_p = 0, \quad \text{if } p > 1,$$

(5.3.7)

$$\mathrm{Var}\, T_p = \frac{p}{p-2}, \quad \text{if } p > 2.$$

Another important derived distribution is Snedecor's $F$, whose derivation is quite similar to that of Student's $t$. Its motivation, however, is somewhat different. The $F$ distribution, named in honor of Sir Ronald Fisher, arises naturally as the distribution of a ratio of variances.

**Example 5.3.5 (Variance ratio distribution)** Let $X_1, \ldots, X_n$ be a random sample from a n$(\mu_X, \sigma_X^2)$ population, and let $Y_1, \ldots, Y_m$ be a random sample from an independent n$(\mu_Y, \sigma_Y^2)$ population. If we were interested in comparing the variability of the populations, one quantity of interest would be the ratio $\sigma_X^2/\sigma_Y^2$. Information about this ratio is contained in $S_X^2/S_Y^2$, the ratio of sample variances. The $F$ distribution allows us to compare these quantities by giving us a distribution of

(5.3.8)
$$\frac{S_X^2/S_Y^2}{\sigma_X^2/\sigma_Y^2} = \frac{S_X^2/\sigma_X^2}{S_Y^2/\sigma_Y^2}.$$

Examination of (5.3.8) shows us how the $F$ distribution is derived. The ratios $S_X^2/\sigma_X^2$ and $S_Y^2/\sigma_Y^2$ are each scaled chi squared variates, and they are independent.  ‖

**Definition 5.3.6** Let $X_1, \ldots, X_n$ be a random sample from a n$(\mu_X, \sigma_X^2)$ population, and let $Y_1, \ldots, Y_m$ be a random sample from an independent n$(\mu_Y, \sigma_Y^2)$ population. The random variable $F = (S_X^2/\sigma_X^2)/(S_Y^2/\sigma_Y^2)$ has *Snedecor's F distribution with* $n-1$ *and* $m-1$ *degrees of freedom.* Equivalently, the random variable $F$ has the $F$ distribution with $p$ and $q$ degrees of freedom if it has pdf

(5.3.9)   $$f_F(x) = \frac{\Gamma\left(\frac{p+q}{2}\right)}{\Gamma\left(\frac{p}{2}\right)\Gamma\left(\frac{q}{2}\right)} \left(\frac{p}{q}\right)^{p/2} \frac{x^{(p/2)-1}}{[1+(p/q)x]^{(p+q)/2}}, \quad 0 < x < \infty.$$

The $F$ distribution can be derived in a more general setting than is done here. A variance ratio may have an $F$ distribution even if the parent populations are not normal. Kelker (1970) has shown that as long as the parent populations have a certain type of symmetry (*spherical symmetry*), then the variance ratio will have an $F$ distribution.

The derivation of the $F$ pdf, starting from normal distributions, is similar to the derivation of Student's $t$. In fact, in one special case the $F$ is a transform of the $t$. (See Theorem 5.3.8.) Similar to what we did for the $t$, we can reduce the task of deriving the $F$ pdf to that of finding the pdf of $(U/p)/(V/q)$, where $U$ and $V$ are independent, $U \sim \chi_p^2$ and $V \sim \chi_q^2$. (See Exercise 5.17.)

**Example 5.3.7 (Continuation of Example 5.3.5)** To see how the $F$ distribution may be used for inference about the true ratio of population variances, consider the following. The quantity $(S_X^2/\sigma_X^2)/(S_Y^2/\sigma_Y^2)$ has an $F_{n-1,m-1}$ distribution. (In general, we use the notation $F_{p,q}$ to denote an $F$ random variable with $p$ and $q$ degrees of freedom.) We can calculate

$$\mathrm{E}F_{n-1,m-1} = \mathrm{E}\left(\frac{\chi_{n-1}^2/(n-1)}{\chi_{m-1}^2/(m-1)}\right) \qquad \text{(by definition)}$$

$$= \mathrm{E}\left(\frac{\chi_{n-1}^2}{n-1}\right)\mathrm{E}\left(\frac{m-1}{\chi_{m-1}^2}\right) \qquad \text{(independence)}$$

$$= \left(\frac{n-1}{n-1}\right)\left(\frac{m-1}{m-3}\right) \quad \text{(chi squared calculations)}$$

$$= \frac{m-1}{m-3}.$$

Note that this last expression is finite and positive only if $m > 3$. We have that

$$\mathrm{E}\left(\frac{S_X^2/\sigma_X^2}{S_Y^2/\sigma_Y^2}\right) = \mathrm{E}F_{n-1,m-1} = \frac{m-1}{m-3},$$

and, removing expectations, we have for reasonably large $m$,

$$\frac{S_X^2/S_Y^2}{\sigma_X^2/\sigma_Y^2} \approx \frac{m-1}{m-3} \approx 1,$$

as we might expect. ‖

The $F$ distribution has many interesting properties and is related to a number of other distributions. We summarize some of these facts in the next theorem, whose proof is left as an exercise. (See Exercises 5.17 and 5.18.)

**Theorem 5.3.8**

**a.** If $X \sim F_{p,q}$, then $1/X \sim F_{q,p}$; that is, the reciprocal of an $F$ random variable is again an $F$ random variable.

**b.** If $X \sim t_q$, then $X^2 \sim F_{1,q}$.

**c.** If $X \sim F_{p,q}$, then $(p/q)X/(1 + (p/q)X) \sim beta(p/2, q/2)$.

## 5.4 Order Statistics

Sample values such as the smallest, largest, or middle observation from a random sample can provide additional summary information. For example, the highest flood waters or the lowest winter temperature recorded during the last 50 years might be useful data when planning for future emergencies. The median price of houses sold during the previous month might be useful for estimating the cost of living. These are all examples of *order statistics*.

**Definition 5.4.1** The *order statistics* of a random sample $X_1, \ldots, X_n$ are the sample values placed in ascending order. They are denoted by $X_{(1)}, \ldots, X_{(n)}$.

The order statistics are random variables that satisfy $X_{(1)} \leq \cdots \leq X_{(n)}$. In particular,

$$X_{(1)} = \min_{1 \leq i \leq n} X_i,$$

$$X_{(2)} = \text{second smallest } X_i,$$

$$\vdots$$

$$X_{(n)} = \max_{1 \leq i \leq n} X_i.$$

Since they are random variables, we can discuss the probabilities that they take on various values. To calculate these probabilities we need the pdfs or pmfs of the order statistics. The formulas for the pdfs of the order statistics of a random sample from a continuous population will be the main topic later in this section, but first, we will mention some statistics that are easily defined in terms of the order statistics.

The *sample range*, $R = X_{(n)} - X_{(1)}$, is the distance between the smallest and largest observations. It is a measure of the dispersion in the sample and should reflect the dispersion in the population.

The *sample median*, which we will denote by $M$, is a number such that approximately one-half of the observations are less than $M$ and one-half are greater. In terms of the order statistics, $M$ is defined by

$$(5.4.1) \qquad M = \begin{cases} X_{((n+1)/2)} & \text{if } n \text{ is odd} \\ \left(X_{(n/2)} + X_{(n/2+1)}\right)/2 & \text{if } n \text{ is even.} \end{cases}$$

The median is a measure of location that might be considered an alternative to the sample mean. One advantage of the sample median over the sample mean is that it is less affected by extreme observations. (See Section 10.2 for details.)

Although related, the mean and median usually measure different things. For example, in recent baseball salary negotiations a major point of contention was the owners' contributions to the players' pension fund. The owners' view could be paraphrased as, "The average baseball player's annual salary is $433,659 so, with that kind of money, the current pension is adequate." But the players' view was, "Over half of the players make less than $250,000 annually and, because of the short professional life of most players, need the security of a larger pension." (These figures are for the 1988 season, not the year of the dispute.) Both figures were correct, but the owners were discussing the mean while the players were discussing the median. About a dozen players with salaries over $2 million can raise the average salary to $433,659 while the majority of the players make less than $250,000, including all rookies who make $62,500. When discussing salaries, prices, or any variable with a few extreme values, the median gives a better indication of "typical" values than the mean. Other statistics that can be defined in terms of order statistics and are less sensitive to extreme values (such as the $\alpha$-*trimmed* mean discussed in Exercise 10.20) are discussed in texts such as Tukey (1977).

For any number $p$ between 0 and 1, the $(100p)$th sample percentile is the observation such that approximately $np$ of the observations are less than this observation and $n(1 - p)$ of the observations are greater. The 50th sample percentile $(p = .5)$ is the sample median. For other values of $p$, we can more precisely define the sample percentiles in terms of the order statistics in the following way.

**Definition 5.4.2** The notation $\{b\}$, when appearing in a subscript, is defined to be *the number b rounded to the nearest integer in the usual way.* More precisely, if $i$ is an integer and $i - .5 \leq b < i + .5$, then $\{b\} = i$.

The $(100p)$th sample percentile is $X_{(\{np\})}$ if $\frac{1}{2n} < p < .5$ and $X_{(n+1-\{n(1-p)\})}$ if $.5 < p < 1 - \frac{1}{2n}$. For example, if $n = 12$ and the 65th percentile is wanted, we note that $12 \times (1 - .65) = 4.2$ and $12 + 1 - 4 = 9$. Thus the 65th percentile is $X_{(9)}$. There is a restriction on the range of $p$ because the size of the sample limits the range of sample percentiles.

The cases $p < .5$ and $p > .5$ are defined separately so that the sample percentiles exhibit the following symmetry. If the $(100p)$th sample percentile is the $i$th smallest observation, then the $(100(1-p))$th sample percentile should be the $i$th largest observation and the above definition achieves this. For example, if $n = 11$, the 30th sample percentile is $X_{(3)}$ and the 70th sample percentile is $X_{(9)}$.

In addition to the median, two other sample percentiles are commonly identified. These are the *lower quartile* (25th percentile) and *upper quartile* (75th percentile). A measure of dispersion that is sometimes used is the *interquartile range*, the distance between the lower and upper quartiles.

Since the order statistics are functions of the sample, probabilities concerning order statistics can be computed in terms of probabilities for the sample. If $X_1, \ldots, X_n$ are iid discrete random variables, then the calculation of probabilities for the order statistics is mainly a counting task. These formulas are derived in Theorem 5.4.3. If $X_1, \ldots, X_n$ are a random sample from a continuous population, then convenient expressions for the pdf of one or more order statistics are derived in Theorems 5.4.4 and 5.4.6. These can then be used to derive the distribution of functions of the order statistics.

**Theorem 5.4.3** *Let $X_1, \ldots, X_n$ be a random sample from a discrete distribution with pmf $f_X(x_i) = p_i$, where $x_1 < x_2 < \cdots$ are the possible values of $X$ in ascending* *order. Define*

$$P_0 = 0$$

$$P_1 = p_1$$

$$P_2 = p_1 + p_2$$

$$\vdots$$

$$P_i = p_1 + p_2 + \cdots + p_i$$

$$\vdots$$

*Let $X_{(1)}, \ldots, X_{(n)}$ denote the order statistics from the sample. Then*

$$(5.4.2) \qquad P\big(X_{(j)} \le x_i\big) = \sum_{k=j}^{n} \binom{n}{k} P_i^k (1 - P_i)^{n-k}$$

*and*

$$(5.4.3) \qquad P\big(X_{(j)} = x_i\big) = \sum_{k=j}^{n} \binom{n}{k} \left[ P_i^k (1 - P_i)^{n-k} - P_{i-1}^k (1 - P_{i-1})^{n-k} \right].$$

**Proof:** Fix $i$, and let $Y$ be a random variable that counts the number of $X_1, \ldots, X_n$ that are less than or equal to $x_i$. For each of $X_1, \ldots, X_n$, call the event $\{X_j \le x_i\}$ a "success" and $\{X_j > x_i\}$ a "failure." Then $Y$ is the number of successes in $n$ trials. The probability of a success is the same value, namely $P_i = P(X_j \le x_i)$, for each trial, since $X_1, \ldots, X_n$ are identically distributed. The success or failure of the $j$th trial is independent of the outcome of any other trial, since $X_j$ is independent of the other $X_i$s. Thus, $Y \sim \text{binomial}(n, P_i)$.

The event $\{X_{(j)} \le x_i\}$ is equivalent to the event $\{Y \ge j\}$; that is, at least $j$ of the sample values are less than or equal to $x_i$. Equation (5.4.2) expresses this binomial probability,

$$P\big(X_{(j)} \le x_i\big) = P(Y \ge j).$$

Equation (5.4.3) simply expresses the difference,

$$P\big(X_{(j)} = x_i\big) = P\big(X_{(j)} \le x_i\big) - P\big(X_{(j)} \le x_{i-1}\big).$$

The case $i = 1$ is exceptional in that $P\big(X_{(j)} = x_1\big) = P\big(X_{(j)} \le x_1\big)$. The definition of $P_0 = 0$ takes care of this exception in (5.4.3). $\qquad\square$

If $X_1, \ldots, X_n$ are a random sample from a continuous population, then the situation is simplified slightly by the fact that the probability is 0 that any two $X_j$s are equal, freeing us from worrying about ties. Thus $P\big(X_{(1)} < X_{(2)} < \cdots < X_{(n)}\big) = 1$ and the sample space for $\big(X_{(1)}, \ldots, X_{(n)}\big)$ is $\{(x_1, \ldots, x_n) : x_1 < x_2 < \cdots < x_n\}$. In Theorems 5.4.4 and 5.4.6 we derive the pdf for one and the joint pdf for two order statistics, again using binomial arguments.

**Theorem 5.4.4** *Let* $X_{(1)}, \ldots, X_{(n)}$ *denote the order statistics of a random sample,* $X_1, \ldots, X_n$, *from a continuous population with cdf* $F_X(x)$ *and pdf* $f_X(x)$. *Then the pdf of* $X_{(j)}$ *is*

$$(5.4.4) \qquad f_{X_{(j)}}(x) = \frac{n!}{(j-1)!(n-j)!} f_X(x) [F_X(x)]^{j-1} [1 - F_X(x)]^{n-j}.$$

**Proof:** We first find the cdf of $X_{(j)}$ and then differentiate it to obtain the pdf. As in Theorem 5.4.3, let $Y$ be a random variable that counts the number of $X_1, \ldots, X_n$ less than or equal to $x$. Then, defining a "success" as the event $\{X_j \leq x\}$, we see that $Y \sim \text{binomial}(n, F_X(x))$. (Note that we can write $P_i = F_X(x_i)$ in Theorem 5.4.3. Also, although $X_1, \ldots, X_n$ are continuous random variables, the counting variable $Y$ is discrete.) Thus,

$$F_{X_{(j)}}(x) = P(Y \geq j) = \sum_{k=j}^{n} \binom{n}{k} [F_X(x)]^k [1 - F_X(x)]^{n-k},$$

and the pdf of $X_{(j)}$ is

$$f_{X_{(j)}}(x) = \frac{d}{dx} F_{X_{(j)}}(x)$$

$$= \sum_{k=j}^{n} \binom{n}{k} \left( k [F_X(x)]^{k-1} [1 - F_X(x)]^{n-k} f_X(x) \right.$$

$$\left. - (n-k)[F_X(x)]^k [1 - F_X(x)]^{n-k-1} f_X(x) \right) \qquad \text{(chain rule)}$$

$$= \binom{n}{j} j f_X(x) [F_X(x)]^{j-1} [1 - F_X(x)]^{n-j}$$

$$+ \sum_{k=j+1}^{n} \binom{n}{k} k [F_X(x)]^{k-1} [1 - F_X(x)]^{n-k} f_X(x)$$

$$- \sum_{k=j}^{n-1} \binom{n}{k} (n-k)[F_X(x)]^k [1 - F_X(x)]^{n-k-1} f_X(x) \qquad \begin{pmatrix} k = n \text{ term} \\ \text{is } 0 \end{pmatrix}$$

$$= \frac{n!}{(j-1)!(n-j)!} f_X(x) [F_X(x)]^{j-1} [1 - F_X(x)]^{n-j}$$

$$(5.4.5) \qquad + \sum_{k=j}^{n-1} \binom{n}{k+1} (k+1) [F_X(x)]^k [1 - F_X(x)]^{n-k-1} f_X(x) \qquad \begin{pmatrix} \text{change} \\ \text{dummy} \\ \text{variable} \end{pmatrix}$$

$$- \sum_{k=j}^{n-1} \binom{n}{k} (n-k) [F_X(x)]^k [1 - F_X(x)]^{n-k-1} f_X(x).$$

Noting that

$$(5.4.6) \qquad \binom{n}{k+1} (k+1) = \frac{n!}{k!(n-k-1)!} = \binom{n}{k} (n-k),$$

we see that the last two sums in (5.4.5) cancel. Thus, the pdf $f_{X_{(j)}}(x)$ is given by the expression in (5.4.4).                                                                                          $\square$

**Example 5.4.5 (Uniform order statistic pdf)**     Let $X_1, \ldots, X_n$ be iid uniform$(0, 1)$, so $f_X(x) = 1$ for $x \in (0, 1)$ and $F_X(x) = x$ for $x \in (0, 1)$. Using (5.4.4), we see that the pdf of the $j$th order statistic is

$$f_{X_{(j)}}(x) = \frac{n!}{(j-1)!(n-j)!} x^{j-1}(1-x)^{n-j} \quad \text{for } x \in (0, 1)$$

$$= \frac{\Gamma(n+1)}{\Gamma(j)\Gamma(n-j+1)} x^{j-1}(1-x)^{(n-j+1)-1}.$$

Thus, the $j$th order statistic from a uniform$(0, 1)$ sample has a beta$(j, n-j+1)$ distribution. From this we can deduce that

$$\mathrm{E}X_{(j)} = \frac{j}{n+1} \quad \text{and} \quad \mathrm{Var}\, X_{(j)} = \frac{j(n-j+1)}{(n+1)^2(n+2)}. \qquad \|$$

The joint distribution of two or more order statistics can be used to derive the distribution of some of the statistics mentioned at the beginning of this section. The joint pdf of any two order statistics is given in the following theorem, whose proof is left to Exercise 5.26.

**Theorem 5.4.6**  *Let* $X_{(1)}, \ldots, X_{(n)}$ *denote the order statistics of a random sample,* $X_1, \ldots, X_n$, *from a continuous population with cdf* $F_X(x)$ *and pdf* $f_X(x)$. *Then the joint pdf of* $X_{(i)}$ *and* $X_{(j)}, 1 \le i < j \le n$, *is*

$$(5.4.7) \quad f_{X_{(i)}, X_{(j)}}(u, v) = \frac{n!}{(i-1)!(j-1-i)!(n-j)!} f_X(u) f_X(v) [F_X(u)]^{i-1}$$

$$\times [F_X(v) - F_X(u)]^{j-1-i} [1 - F_X(v)]^{n-j}$$

*for* $-\infty < u < v < \infty$.

The joint pdf of three or more order statistics could be derived using similar but even more involved arguments. Perhaps the other most useful pdf is $f_{X_{(1)}, \ldots, X_{(n)}}$ $(x_1, \ldots, x_n)$, the joint pdf of all the order statistics, which is given by

$$f_{X_{(1)}, \ldots, X_{(n)}}(x_1, \ldots, x_n) = \begin{cases} n! f_X(x_1) \cdot \cdots \cdot f_X(x_n) & -\infty < x_1 < \cdots < x_n < \infty \\ 0 & \text{otherwise.} \end{cases}$$

The $n!$ naturally comes into this formula because, for any set of values $x_1, \ldots, x_n$, there are $n!$ equally likely assignments for these values to $X_1, \ldots, X_n$ that all yield the same values for the order statistics. This joint pdf and the techniques from Chapter 4 can be used to derive marginal and conditional distributions and distributions of other functions of the order statistics. (See Exercises 5.27 and 5.28.)

We now use the joint pdf (5.4.7) to derive the distribution of some of the functions mentioned at the beginning of this section.

**Example 5.4.7 (Distribution of the midrange and range)** Let $X_1, \ldots, X_n$ be iid uniform$(0, a)$ and let $X_{(1)}, \ldots, X_{(n)}$ denote the order statistics. The range was earlier defined as $R = X_{(n)} - X_{(1)}$. The *midrange*, a measure of location like the sample median or the sample mean, is defined by $V = (X_{(1)} + X_{(n)})/2$. We will derive the joint pdf of $R$ and $V$ from the joint pdf of $X_{(1)}$ and $X_{(n)}$. From (5.4.7) we have that

$$f_{X_{(1)}, X_{(n)}}(x_1, x_n) = \frac{n(n-1)}{a^2}\left(\frac{x_n}{a} - \frac{x_1}{a}\right)^{n-2}$$

$$= \frac{n(n-1)(x_n - x_1)^{n-2}}{a^n}, \qquad 0 < x_1 < x_n < a.$$

Solving for $X_{(1)}$ and $X_{(n)}$, we obtain $X_{(1)} = V - R/2$ and $X_{(n)} = V + R/2$. The Jacobian for this transformation is $-1$. The transformation from $(X_{(1)}, X_{(n)})$ to $(R, V)$ maps $\{(x_1, x_n): 0 < x_1 < x_n < a\}$ onto the set $\{(r, v): 0 < r < a, r/2 < v < a - r/2\}$. To see this, note that obviously $0 < r < a$ and for a fixed value of $r$, $v$ ranges from $r/2$ (corresponding to $x_1 = 0, x_n = r$) to $a - r/2$ (corresponding to $x_1 = a - r, x_n = a$). Thus, the joint pdf of $(R, V)$ is

$$f_{R,V}(r, v) = \frac{n(n-1)r^{n-2}}{a^n}, \qquad 0 < r < a, \quad r/2 < v < a - r/2.$$

The marginal pdf of $R$ is thus

$$f_R(r) = \int_{r/2}^{a-r/2} \frac{n(n-1)r^{n-2}}{a^n}\, dv$$

(5.4.8)

$$= \frac{n(n-1)r^{n-2}(a - r)}{a^n}, \qquad 0 < r < a.$$

If $a = 1$, we see that $R$ has a beta$(n-1, 2)$ distribution. Or, for arbitrary $a$, it is easy to deduce from (5.4.8) that $R/a$ has a beta distribution. Note that the constant $a$ is a scale parameter.

The set where $f_{R,V}(r, v) > 0$ is shown in Figure 5.4.1, where we see that the range of integration of $r$ depends on whether $v > a/2$ or $v \leq a/2$. Thus, the marginal pdf of $V$ is given by

$$f_V(v) = \int_0^{2v} \frac{n(n-1)r^{n-2}}{a^n}\, dr = \frac{n(2v)^{n-1}}{a^n}, \qquad 0 < v \leq a/2,$$

and

$$f_V(v) = \int_0^{2(a-v)} \frac{n(n-1)r^{n-2}}{a^n}\, dr = \frac{n[2(a - v)]^{n-1}}{a^n}, \qquad a/2 < v \leq a.$$

This pdf is symmetric about $a/2$ and has a peak at $a/2$.                    ‖

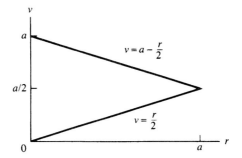

Figure 5.4.1. *Region on which $f_{R,V}(r,v) > 0$ for Example 5.4.7*

## 5.5 Convergence Concepts

This section treats the somewhat fanciful idea of allowing the sample size to approach infinity and investigates the behavior of certain sample quantities as this happens. Although the notion of an infinite sample size is a theoretical artifact, it can often provide us with some useful approximations for the finite-sample case, since it usually happens that expressions become simplified in the limit.

We are mainly concerned with three types of convergence, and we treat them in varying amounts of detail. (A full treatment of convergence is given in Billingsley 1995 or Resnick 1999, for example.) In particular, we want to look at the behavior of $\bar{X}_n$, the mean of $n$ observations, as $n \to \infty$.

### 5.5.1 Convergence in Probability

This type of convergence is one of the weaker types and, hence, is usually quite easy to verify.

**Definition 5.5.1**   A sequence of random variables, $X_1, X_2, \ldots$, *converges in probability* to a random variable $X$ if, for every $\epsilon > 0$,

$$\lim_{n \to \infty} P(|X_n - X| \geq \epsilon) = 0 \quad \text{or, equivalently,} \quad \lim_{n \to \infty} P(|X_n - X| < \epsilon) = 1.$$

The $X_1, X_2, \ldots$ in Definition 5.5.1 (and the other definitions in this section) are typically not independent and identically distributed random variables, as in a random sample. The distribution of $X_n$ changes as the subscript changes, and the convergence concepts discussed in this section describe different ways in which the distribution of $X_n$ converges to some limiting distribution as the subscript becomes large.

Frequently, statisticians are concerned with situations in which the limiting random variable is a constant and the random variables in the sequence are sample means (of some sort). The most famous result of this type is the following.

**Theorem 5.5.2 (Weak Law of Large Numbers)** *Let $X_1, X_2, \ldots$ be iid random variables with $\mathrm{E}X_i = \mu$ and $\mathrm{Var}\, X_i = \sigma^2 < \infty$. Define $\bar{X}_n = (1/n)\sum_{i=1}^n X_i$. Then,* *for every $\epsilon > 0$,*

$$\lim_{n \to \infty} P(|\bar{X}_n - \mu| < \epsilon) = 1;$$

*that is, $\bar{X}_n$ converges in probability to $\mu$.*

**Proof:** The proof is quite simple, being a straightforward application of Chebychev's Inequality. We have, for every $\epsilon > 0$,

$$P(|\bar{X}_n - \mu| \geq \epsilon) = P((\bar{X}_n - \mu)^2 \geq \epsilon^2) \leq \frac{\mathrm{E}(\bar{X}_n - \mu)^2}{\epsilon^2} = \frac{\mathrm{Var}\ \bar{X}}{\epsilon^2} = \frac{\sigma^2}{n\epsilon^2}.$$

Hence, $P(|\bar{X}_n - \mu| < \epsilon) = 1 - P(|\bar{X}_n - \mu| \geq \epsilon) \geq 1 - \sigma^2/(n\epsilon^2) \to 1$, as $n \to \infty$. $\qquad\square$

The Weak Law of Large Numbers (WLLN) quite elegantly states that, under general conditions, the sample mean approaches the population mean as $n \to \infty$. In fact, there are more general versions of the WLLN, where we need assume only that the mean is finite. However, the version stated in Theorem 5.5.2 is applicable in most practical situations.

The property summarized by the WLLN, that a sequence of the "same" sample quantity approaches a constant as $n \to \infty$, is known as *consistency*. We will examine this property more closely in Chapter 7.

**Example 5.5.3 (Consistency of $S^2$)** Suppose we have a sequence $X_1, X_2, \ldots$ of iid random variables with $\mathrm{E}X_i = \mu$ and $\mathrm{Var}\ X_i = \sigma^2 < \infty$. If we define

$$S_n^2 = \frac{1}{n-1} \sum_{i=1}^{n} (X_i - \bar{X}_n)^2,$$

can we prove a WLLN for $S_n^2$? Using Chebychev's Inequality, we have

$$P(|S_n^2 - \sigma^2| \geq \epsilon) \leq \frac{\mathrm{E}(S_n^2 - \sigma^2)^2}{\epsilon^2} = \frac{\mathrm{Var}\ S_n^2}{\epsilon^2}$$

and thus, a sufficient condition that $S_n^2$ converges in probability to $\sigma^2$ is that $\mathrm{Var}\ S_n^2 \to 0$ as $n \to \infty$. $\qquad\|$

A natural extension of Definition 5.5.1 relates to functions of random variables. That is, if the sequence $X_1, X_2, \ldots$ converges in probability to a random variable $X$ or to a constant $a$, can we make any conclusions about the sequence of random variables $h(X_1), h(X_2), \ldots$ for some reasonably behaved function $h$? This next theorem shows that we can. (See Exercise 5.39 for a proof.)

**Theorem 5.5.4** *Suppose that $X_1, X_2, \ldots$ converges in probability to a random variable $X$ and that $h$ is a continuous function. Then $h(X_1), h(X_2), \ldots$ converges in probability to $h(X)$.*

**Example 5.5.5 (Consistency of $S$)** If $S_n^2$ is a consistent estimator of $\sigma^2$, then by Theorem 5.5.4, the sample standard deviation $S_n = \sqrt{S_n^2} = h(S_n^2)$ is a consistent estimator of $\sigma$. Note that $S_n$ is, in fact, a biased estimator of $\sigma$ (see Exercise 5.11), but the bias disappears asymptotically. $\qquad\|$

### 5.5.2 Almost Sure Convergence

A type of convergence that is stronger than convergence in probability is almost sure convergence (sometimes confusingly known as *convergence with probability 1*). This type of convergence is similar to pointwise convergence of a sequence of functions, except that the convergence need not occur on a set with probability 0 (hence the "almost" sure).

**Definition 5.5.6**    A sequence of random variables, $X_1, X_2, \ldots$, *converges almost surely* to a random variable $X$ if, for every $\epsilon > 0$,

$$P(\lim_{n \to \infty} |X_n - X| < \epsilon) = 1.$$

Notice the similarity in the statements of Definitions 5.5.1 and 5.5.6. Although they look similar, they are very different statements, with Definition 5.5.6 much stronger. To understand almost sure convergence, we must recall the basic definition of a random variable as given in Definition 1.4.1. A random variable is a real-valued function defined on a sample space $S$. If a sample space $S$ has elements denoted by $s$, then $X_n(s)$ and $X(s)$ are all functions defined on $S$. Definition 5.5.6 states that $X_n$ converges to $X$ almost surely if the functions $X_n(s)$ converge to $X(s)$ for all $s \in S$ except perhaps for $s \in N$, where $N \subset S$ and $P(N) = 0$. Example 5.5.7 illustrates almost sure convergence. Example 5.5.8 illustrates the difference between convergence in probability and almost sure convergence.

**Example 5.5.7 (Almost sure convergence)**    Let the sample space $S$ be the closed interval $[0, 1]$ with the uniform probability distribution. Define random variables $X_n(s) = s + s^n$ and $X(s) = s$. For every $s \in [0, 1)$, $s^n \to 0$ as $n \to \infty$ and $X_n(s) \to s = X(s)$. However, $X_n(1) = 2$ for every $n$ so $X_n(1)$ does not converge to $1 = X(1)$. But since the convergence occurs on the set $[0, 1)$ and $P([0, 1)) = 1, X_n$ converges to $X$ almost surely.                                                                           ‖

**Example 5.5.8 (Convergence in probability, not almost surely)**    In this example we describe a sequence that converges in probability, but not almost surely. Again, let the sample space $S$ be the closed interval $[0,1]$ with the uniform probability distribution. Define the sequence $X_1, X_2, \ldots$ as follows:

$$X_1(s) = s + I_{[0,1]}(s), \qquad X_2(s) = s + I_{[0,\frac{1}{2}]}(s), \qquad X_3(s) = s + I_{[\frac{1}{2},1]}(s),$$

$$X_4(s) = s + I_{[0,\frac{1}{3}]}(s), \qquad X_5(s) = s + I_{[\frac{1}{3},\frac{2}{3}]}(s), \qquad X_6(s) = s + I_{[\frac{2}{3},1]}(s),$$

etc. Let $X(s) = s$. It is straightforward to see that $X_n$ converges to $X$ in probability. As $n \to \infty, P(|X_n - X| \geq \epsilon)$ is equal to the probability of an interval of $s$ values whose length is going to 0. However, $X_n$ does not converge to $X$ almost surely. Indeed, there is no value of $s \in S$ for which $X_n(s) \to s = X(s)$. For every $s$, the value $X_n(s)$ alternates between the values $s$ and $s + 1$ infinitely often. For example, if $s = \frac{3}{8}, X_1(s) = 1\frac{3}{8}, X_2(s) = 1\frac{3}{8}, X_3(s) = \frac{3}{8}, X_4(s) = \frac{3}{8}, X_5(s) = 1\frac{3}{8}, X_6(s) = \frac{3}{8}$, etc. No pointwise convergence occurs for this sequence.                                                                           ‖

As might be guessed, convergence almost surely, being the stronger criterion, implies convergence in probability. The converse is, of course, false, as Example 5.5.8 shows. However, if a sequence converges in probability, it is possible to find a *subsequence* that converges almost surely. (Resnick 1999, Section 6.3, has a thorough treatment of the connections between the two types of convergence.)

Again, statisticians are often concerned with convergence to a constant. We now state, without proof, the stronger analog of the WLLN, the Strong Law of Large Numbers (SLLN). See Miscellanea 5.8.4 for an outline of a proof.

**Theorem 5.5.9 (Strong Law of Large Numbers)** *Let $X_1, X_2, \ldots$ be iid random variables with $\mathrm{E}X_i = \mu$ and $\mathrm{Var}\, X_i = \sigma^2 < \infty$, and define $\bar{X}_n = (1/n) \sum_{i=1}^n X_i$. Then, for every $\epsilon > 0$,*

$$P(\lim_{n \to \infty} |\bar{X}_n - \mu| < \epsilon) = 1;$$

*that is, $\bar{X}_n$ converges almost surely to $\mu$.*

For both the Weak and Strong Laws of Large Numbers we had the assumption of a finite variance. Although such an assumption is true (and desirable) in most applications, it is, in fact, a stronger assumption than is needed. Both the weak and strong laws hold without this assumption. The only moment condition needed is that $\mathrm{E}|X_i| < \infty$ (see Resnick 1999, Chapter 7, or Billingsley 1995, Section 22).

### 5.5.3 Convergence in Distribution

We have already encountered the idea of convergence in distribution in Chapter 2. Remember the properties of moment generating functions (mgfs) and how their convergence implies convergence in distribution (Theorem 2.3.12).

**Definition 5.5.10** A sequence of random variables, $X_1, X_2, \ldots$, *converges in distribution* to a random variable $X$ if

$$\lim_{n \to \infty} F_{X_n}(x) = F_X(x)$$

at all points $x$ where $F_X(x)$ is continuous.

**Example 5.5.11 (Maximum of uniforms)** If $X_1, X_2, \ldots$ are iid uniform$(0, 1)$ and $X_{(n)} = \max_{1 \le i \le n} X_i$, let us examine if (and to where) $X_{(n)}$ converges in distribution.

As $n \to \infty$, we expect $X_{(n)}$ to get close to 1 and, as $X_{(n)}$ must necessarily be less than 1, we have for any $\varepsilon > 0$,

$$P(|X_{(n)} - 1| \ge \varepsilon) = P(X_{(n)} \ge 1 + \varepsilon) + P(X_{(n)} \le 1 - \varepsilon)$$
$$= 0 + P(X_{(n)} \le 1 - \varepsilon).$$

Next using the fact that we have an *iid* sample, we can write

$$P(X_{(n)} \le 1 - \varepsilon) = P(X_i \le 1 - \varepsilon, i = 1, \ldots n) = (1 - \varepsilon)^n,$$

which goes to 0. So we have proved that $X_{(n)}$ converges to 1 in probability. However, if we take $\varepsilon = t/n$, we then have

$$P(X_{(n)} \leq 1 - t/n) = (1 - t/n)^n \to e^{-t},$$

which, upon rearranging, yields

$$P(n(1 - X_{(n)}) \leq t) \to 1 - e^{-t};$$

that is, the random variable $n(1 - X_{(n)})$ converges in distribution to an exponential(1) random variable.                                                                           ‖

Note that although we talk of a sequence of random variables converging in distribution, it is really the cdfs that converge, not the random variables. In this very fundamental way convergence in distribution is quite different from convergence in probability or convergence almost surely. However, it is implied by the other types of convergence.

**Theorem 5.5.12** *If the sequence of random variables, $X_1, X_2, \ldots$, converges in probability to a random variable $X$, the sequence also converges in distribution to $X$.*

See Exercise 5.40 for a proof. Note also that, from Section 5.5.2, convergence in distribution is also implied by almost sure convergence.

In a special case, Theorem 5.5.12 has a converse that turns out to be useful. See Example 10.1.13 for an illustration and Exercise 5.41 for a proof.

**Theorem 5.5.13** *The sequence of random variables, $X_1, X_2, \ldots$, converges in probability to a constant $\mu$ if and only if the sequence also converges in distribution to $\mu$. That is, the statement*

$$P\left(|X_n - \mu| > \varepsilon\right) \to 0 \text{ for every } \varepsilon > 0$$

*is equivalent to*

$$P\left(X_n \leq x\right) \to \begin{cases} 0 & \text{if } x < \mu \\ 1 & \text{if } x > \mu. \end{cases}$$

The sample mean is one statistic whose large-sample behavior is quite important. In particular, we want to investigate its limiting distribution. This is summarized in one of the most startling theorems in statistics, the Central Limit Theorem (CLT).

**Theorem 5.5.14 (Central Limit Theorem)** *Let $X_1, X_2, \ldots$ be a sequence of iid random variables whose mgfs exist in a neighborhood of 0 (that is, $M_{X_i}(t)$ exists for $|t| < h$, for some positive $h$). Let $\mathrm{E}X_i = \mu$ and $\mathrm{Var}\, X_i = \sigma^2 > 0$. (Both $\mu$ and $\sigma^2$ are finite since the mgf exists.) Define $\bar{X}_n = (1/n) \sum_{i=1}^n X_i$. Let $G_n(x)$ denote the cdf of $\sqrt{n}(\bar{X}_n - \mu)/\sigma$. Then, for any $x$, $-\infty < x < \infty$,*

$$\lim_{n \to \infty} G_n(x) = \int_{-\infty}^x \frac{1}{\sqrt{2\pi}} e^{-y^2/2} \, dy;$$

*that is, $\sqrt{n}(\bar{X}_n - \mu)/\sigma$ has a limiting standard normal distribution.*

Before we prove this theorem (the proof is somewhat anticlimactic) we first look at its implications. Starting from virtually no assumptions (other than independence and finite variances), we end up with normality! The point here is that normality comes from sums of "small" (finite variance), independent disturbances. The assumption of finite variances is essentially necessary for convergence to normality. Although it can be relaxed somewhat, it cannot be eliminated. (Recall Example 5.2.10, dealing with the Cauchy distribution, where there is no convergence to normality.)

While we revel in the wonder of the CLT, it is also useful to reflect on its limitations. Although it gives us a useful general approximation, we have no automatic way of knowing how good the approximation is in general. In fact, the goodness of the approximation is a function of the original distribution, and so must be checked case by case. Furthermore, with the current availability of cheap, plentiful computing power, the importance of approximations like the Central Limit Theorem is somewhat lessened. However, despite its limitations, it is still a marvelous result.

**Proof of Theorem 5.5.14:** We will show that, for $|t| < h$, the mgf of $\sqrt{n}(\bar{X}_n - \mu)/\sigma$ converges to $e^{t^2/2}$, the mgf of a $n(0,1)$ random variable.

Define $Y_i = (X_i - \mu)/\sigma$, and let $M_Y(t)$ denote the common mgf of the $Y_i$s, which exists for $|t| < \sigma h$ and is given by Theorem 2.3.15. Since

$$(5.5.1) \qquad \frac{\sqrt{n}(\bar{X}_n - \mu)}{\sigma} = \frac{1}{\sqrt{n}} \sum_{i=1}^{n} Y_i,$$

we have, from the properties of mgfs (see Theorems 2.3.15 and 4.6.7),

$$(5.5.2) \qquad M_{\sqrt{n}(\bar{X}_n - \mu)/\sigma}(t) = M_{\sum_{i=1}^{n} Y_i/\sqrt{n}}(t)$$

$$= M_{\sum_{i=1}^{n} Y_i}\left(\frac{t}{\sqrt{n}}\right) \qquad \text{(Theorem 2.3.15)}$$

$$= \left(M_Y\left(\frac{t}{\sqrt{n}}\right)\right)^n. \qquad \text{(Theorem 4.6.7)}$$

We now expand $M_Y(t/\sqrt{n})$ in a Taylor series (power series) around 0. (See Definition 5.5.20.) We have

$$(5.5.3) \qquad M_Y\left(\frac{t}{\sqrt{n}}\right) = \sum_{k=0}^{\infty} M_Y^{(k)}(0)\frac{(t/\sqrt{n})^k}{k!},$$

where $M_Y^{(k)}(0) = (d^k/dt^k) M_Y(t)|_{t=0}$. Since the mgfs exist for $|t| < h$, the power series expansion is valid if $t < \sqrt{n}\sigma h$.

Using the facts that $M_Y^{(0)} = 1$, $M_Y^{(1)} = 0$, and $M_Y^{(2)} = 1$ (by construction, the mean and variance of $Y$ are 0 and 1), we have

$$(5.5.4) \qquad M_Y\left(\frac{t}{\sqrt{n}}\right) = 1 + \frac{(t/\sqrt{n})^2}{2!} + R_Y\left(\frac{t}{\sqrt{n}}\right),$$

where $R_Y$ is the remainder term in the Taylor expansion,

$$R_Y\left(\frac{t}{\sqrt{n}}\right) = \sum_{k=3}^{\infty} M_Y^{(k)}(0)\frac{(t/\sqrt{n})^k}{k!}.$$

An application of Taylor's Theorem (Theorem 5.5.21) shows that, for fixed $t \neq 0$, we have

$$\lim_{n\to\infty} \frac{R_Y(t/\sqrt{n})}{(t/\sqrt{n})^2} = 0.$$

Since $t$ is fixed, we also have

(5.5.5) $$\lim_{n\to\infty} \frac{R_Y(t/\sqrt{n})}{(1/\sqrt{n})^2} = \lim_{n\to\infty} nR_Y\left(\frac{t}{\sqrt{n}}\right) = 0,$$

and (5.5.5) is also true at $t = 0$ since $R_Y(0/\sqrt{n}) = 0$. Thus, for any fixed $t$, we can write

(5.5.6) $$\lim_{n\to\infty}\left(M_Y\left(\frac{t}{\sqrt{n}}\right)\right)^n = \lim_{n\to\infty}\left[1 + \frac{(t/\sqrt{n})^2}{2!} + R_Y\left(\frac{t}{\sqrt{n}}\right)\right]^n$$

$$= \lim_{n\to\infty}\left[1 + \frac{1}{n}\left(\frac{t^2}{2} + nR_Y\left(\frac{t}{\sqrt{n}}\right)\right)\right]^n$$

$$= e^{t^2/2}$$

by an application of Lemma 2.3.14, where we set $a_n = (t^2/2) + nR_Y(t/\sqrt{n})$. (Note that (5.5.5) implies that $a_n \to t^2/2$ as $n \to \infty$.) Since $e^{t^2/2}$ is the mgf of the n(0, 1) distribution, the theorem is proved.            □

The Central Limit Theorem is valid in much more generality than is stated in Theorem 5.5.14 (see Miscellanea 5.8.1). In particular, all of the assumptions about mgfs are not needed—the use of characteristic functions (see Miscellanea 2.6.2) can replace them. We state the next theorem without proof. It is a version of the Central Limit Theorem that is general enough for almost all statistical purposes. Notice that the only assumption on the parent distribution is that it has finite variance.

**Theorem 5.5.15 (Stronger form of the Central Limit Theorem)** *Let $X_1, X_2, \ldots$ be a sequence of iid random variables with $\mathrm{E}X_i = \mu$ and $0 < \mathrm{Var}\, X_i = \sigma^2 < \infty$. Define $\bar{X}_n = (1/n)\sum_{i=1}^{n} X_i$. Let $G_n(x)$ denote the cdf of $\sqrt{n}(\bar{X}_n - \mu)/\sigma$. Then, for any $x$, $-\infty < x < \infty$,*

$$\lim_{n\to\infty} G_n(x) = \int_{-\infty}^{x} \frac{1}{\sqrt{2\pi}}e^{-y^2/2}\, dy;$$

*that is, $\sqrt{n}(\bar{X}_n - \mu)/\sigma$ has a limiting standard normal distribution.*

The proof is almost identical to that of Theorem 5.5.14, except that characteristic functions are used instead of mgfs. Since the characteristic function of a distribution always exists, it is not necessary to mention them in the assumptions of the theorem. The proof is more delicate, however, since functions of *complex variables* must be dealt with. Details can be found in Billingsley (1995, Section 27).

The Central Limit Theorem provides us with an all-purpose approximation (but remember the warning about the goodness of the approximation). In practice, it can always be used for a first, rough calculation.

**Example 5.5.16 (Normal approximation to the negative binomial)** Suppose $X_1, \ldots, X_n$ are a random sample from a negative binomial$(r, p)$ distribution. Recall that

$$EX = \frac{r(1-p)}{p}, \quad \text{Var } X = \frac{r(1-p)}{p^2},$$

and the Central Limit Theorem tells us that

$$\frac{\sqrt{n}(\bar{X} - r(1-p)/p)}{\sqrt{r(1-p)/p^2}}$$

is approximately n$(0, 1)$. The approximate probability calculations are much easier than the exact calculations. For example, if $r = 10$, $p = \frac{1}{2}$, and $n = 30$, an exact calculation would be

$$P(\bar{X} \leq 11) = P\left(\sum_{i=1}^{30} X_i \leq 330\right)$$

$$= \sum_{x=0}^{330} \binom{300 + x - 1}{x} \left(\frac{1}{2}\right)^{300} \left(\frac{1}{2}\right)^x \quad \left(\begin{array}{c}\sum X \text{ is negative} \\ \text{binomial}(nr, p)\end{array}\right)$$

$$= .8916,$$

which is a very difficult calculation. (Note that this calculation is difficult even with the aid of a computer—the magnitudes of the factorials cause great difficulty. Try it if you don't believe it!) The CLT gives us the approximation

$$P(\bar{X} \leq 11) = P\left(\frac{\sqrt{30}(\bar{X} - 10)}{\sqrt{20}} \leq \frac{\sqrt{30}(11 - 10)}{\sqrt{20}}\right)$$

$$\approx P(Z \leq 1.2247) = .8888.$$

See Exercise 5.37 for some further refinement.                                    ‖

An approximation tool that can be used in conjunction with the Central Limit Theorem is known as Slutsky's Theorem.

**Theorem 5.5.17 (Slutsky's Theorem)** *If $X_n \to X$ in distribution and $Y_n \to a$, a constant, in probability, then*

**a.** $Y_n X_n \to aX$ *in distribution.*

**b.** $X_n + Y_n \to X + a$ *in distribution.*

The proof of Slutsky's Theorem is omitted, since it relies on a characterization of convergence in distribution that we have not discussed. A typical application is illustrated by the following example.

**Example 5.5.18 (Normal approximation with estimated variance)** Suppose that

$$\frac{\sqrt{n}(\bar{X}_n - \mu)}{\sigma} \to n(0, 1),$$

but the value of $\sigma$ is unknown. We have seen in Example 5.5.3 that, if $\lim_{n \to \infty} \mathrm{Var}\, S_n^2 = 0$, then $S_n^2 \to \sigma^2$ in probability. By Exercise 5.32, $\sigma/S_n \to 1$ in probability. Hence, Slutsky's Theorem tells us

$$\frac{\sqrt{n}(\bar{X}_n - \mu)}{S_n} = \frac{\sigma}{S_n} \frac{\sqrt{n}(\bar{X}_n - \mu)}{\sigma} \to n(0, 1). \qquad \|$$

*5.5.4 The Delta Method*

The previous section gives conditions under which a standardized random variable has a limit normal distribution. There are many times, however, when we are not specifically interested in the distribution of the random variable itself, but rather some function of the random variable.

**Example 5.5.19 (Estimating the odds)** Suppose we observe $X_1, X_2, \ldots, X_n$ independent Bernoulli($p$) random variables. The typical parameter of interest is $p$, the success probability, but another popular parameter is $\frac{p}{1-p}$, the *odds*. For example, if the data represent the outcomes of a medical treatment with $p = 2/3$, then a person has odds $2:1$ of getting better. Moreover, if there were another treatment with success probability $r$, biostatisticians often estimate the *odds ratio* $\frac{p}{1-p}/\frac{r}{1-r}$, giving the relative odds of one treatment over another.

As we would typically estimate the success probability $p$ with the observed success probability $\hat{p} = \sum_i X_i/n$, we might consider using $\frac{\hat{p}}{1-\hat{p}}$ as an estimate of $\frac{p}{1-p}$. But what are the properties of this estimator? How might we estimate the variance of $\frac{\hat{p}}{1-\hat{p}}$? Moreover, how can we approximate its sampling distribution?

Intuition abandons us, and exact calculation is relatively hopeless, so we have to rely on an approximation. The Delta Method will allow us to obtain reasonable, approximate answers to our questions. $\qquad \|$

One method of proceeding is based on using a Taylor series approximation, which allows us to approximate the mean and variance of a function of a random variable. We will also see that these rather straightforward approximations are good enough to obtain a CLT. We begin with a short review of Taylor series.

**Definition 5.5.20**  If a function $g(x)$ has derivatives of order $r$, that is, $g^{(r)}(x) = \frac{d^r}{dx^r}g(x)$ exists, then for any constant $a$, the *Taylor polynomial of order $r$ about $a$* is

$$T_r(x) = \sum_{i=0}^{r} \frac{g^{(i)}(a)}{i!}(x-a)^i.$$

Taylor's major theorem, which we will not prove here, is that the *remainder* from the approximation, $g(x) - T_r(x)$, always tends to 0 faster than the highest-order explicit term.

**Theorem 5.5.21 (Taylor)**  *If* $g^{(r)}(a) = \frac{d^r}{dx^r}g(x)\big|_{x=a}$ *exists, then*

$$\lim_{x \to a} \frac{g(x) - T_r(x)}{(x-a)^r} = 0.$$

In general, we will not be concerned with the explicit form of the remainder. Since we are interested in approximations, we are just going to ignore the remainder. There are, however, many explicit forms, one useful one being

$$g(x) - T_r(x) = \int_a^x \frac{g^{(r+1)}(t)}{r!}(x-t)^r dt.$$

For the statistical application of Taylor's Theorem, we are most concerned with the *first-order* Taylor series, that is, an approximation using just the first derivative (taking $r = 1$ in the above formulas). Furthermore, we will also find use for a multivariate Taylor series. Since the above detail is univariate, some of the following will have to be accepted on faith.

Let $T_1, \ldots, T_k$ be random variables with means $\theta_1, \ldots, \theta_k$, and define $\mathbf{T} = (T_1, \ldots, T_k)$ and $\boldsymbol{\theta} = (\theta_1, \ldots, \theta_k)$. Suppose there is a differentiable function $g(\mathbf{T})$ (an estimator of some parameter) for which we want an approximate estimate of variance. Define

$$g_i'(\boldsymbol{\theta}) = \frac{\partial}{\partial t_i}g(\mathbf{t})\big|_{t_1 = \theta_1, \ldots, t_k = \theta_k}.$$

The first-order Taylor series expansion of $g$ about $\boldsymbol{\theta}$ is

$$g(\mathbf{t}) = g(\boldsymbol{\theta}) + \sum_{i=1}^{k} g_i'(\boldsymbol{\theta})(t_i - \theta_i) + \text{Remainder}.$$

For our statistical approximation we forget about the remainder and write

(5.5.7)                    $$g(\mathbf{t}) \approx g(\boldsymbol{\theta}) + \sum_{i=1}^{k} g_i'(\boldsymbol{\theta})(t_i - \theta_i).$$

Now, take expectations on both sides of (5.5.7) to get

(5.5.8)                    $$E_{\boldsymbol{\theta}} g(\mathbf{T}) \approx g(\boldsymbol{\theta}) + \sum_{i=1}^{k} g_i'(\boldsymbol{\theta}) E_{\boldsymbol{\theta}}(T_i - \theta_i)$$

$$= g(\boldsymbol{\theta}). \qquad\qquad (T_i \text{ has mean } \theta_i)$$

We can now approximate the variance of $g(\mathbf{T})$ by

$$\text{Var}_{\boldsymbol{\theta}}\, g(\mathbf{T}) \approx \text{E}_{\boldsymbol{\theta}}\left([g(\mathbf{T}) - g(\boldsymbol{\theta})]^2\right) \qquad \text{(using (5.5.8))}$$

$$\approx \text{E}_{\boldsymbol{\theta}}\left(\left(\sum_{i=1}^{k} g_i'(\boldsymbol{\theta})(T_i - \theta_i)\right)^2\right) \qquad \text{(using (5.5.7))}$$

$$(5.5.9) \qquad = \sum_{i=1}^{k}[g_i'(\boldsymbol{\theta})]^2 \text{Var}_{\boldsymbol{\theta}}\, T_i + 2\sum_{i>j} g_i'(\boldsymbol{\theta})g_j'(\boldsymbol{\theta})\text{Cov}_{\boldsymbol{\theta}}(T_i, T_j),$$

where the last equality comes from expanding the square and using the definition of variance and covariance (similar to Exercise 4.44). Approximation (5.5.9) is very useful because it gives us a variance formula for a general function, using only simple variances and covariances. Here are two examples.

**Example 5.5.22 (Continuation of Example 5.5.19)** Recall that we are interested in the properties of $\frac{\hat{p}}{1-\hat{p}}$ as an estimate of $\frac{p}{1-p}$, where $p$ is a binomial success probability. In our above notation, take $g(p) = \frac{p}{1-p}$ so $g'(p) = \frac{1}{(1-p)^2}$ and

$$\text{Var}\left(\frac{\hat{p}}{1 - \hat{p}}\right) \approx [g'(p)]^2\,\text{Var}(\hat{p})$$

$$= \left[\frac{1}{(1-p)^2}\right]^2 \frac{p(1-p)}{n} = \frac{p}{n(1-p)^3}, \qquad \|$$

giving us an approximation for the variance of our estimator.

**Example 5.5.23 (Approximate mean and variance)** Suppose $X$ is a random variable with $\text{E}_{\mu}X = \mu \neq 0$. If we want to estimate a function $g(\mu)$, a first-order approximation would give us

$$g(X) = g(\mu) + g'(\mu)(X - \mu).$$

If we use $g(X)$ as an estimator of $g(\mu)$, we can say that approximately

$$\text{E}_{\mu}g(X) \approx g(\mu),$$

$$\text{Var}_{\mu}\, g(X) \approx [g'(\mu)]^2 \text{Var}_{\mu}\, X.$$

For a specific example, take $g(\mu) = 1/\mu$. We estimate $1/\mu$ with $1/X$, and we can say

$$\text{E}_{\mu}\left(\frac{1}{X}\right) \approx \frac{1}{\mu},$$

$$\text{Var}_{\mu}\left(\frac{1}{X}\right) \approx \left(\frac{1}{\mu}\right)^4 \text{Var}_{\mu}X. \qquad \|$$

Using these Taylor series approximations for the mean and variance, we get the following useful generalization of the Central Limit Theorem, known as the *Delta Method*.

**Theorem 5.5.24 (Delta Method)** *Let $Y_n$ be a sequence of random variables that satisfies $\sqrt{n}(Y_n - \theta) \to n(0, \sigma^2)$ in distribution. For a given function $g$ and a specific value of $\theta$, suppose that $g'(\theta)$ exists and is not 0. Then*

$$(5.5.10) \qquad \sqrt{n}[g(Y_n) - g(\theta)] \to n(0, \sigma^2[g'(\theta)]^2) \text{ in distribution.}$$

**Proof:** The Taylor expansion of $g(Y_n)$ around $Y_n = \theta$ is

$$(5.5.11) \qquad g(Y_n) = g(\theta) + g'(\theta)(Y_n - \theta) + \text{ Remainder,}$$

where the remainder $\to 0$ as $Y_n \to \theta$. Since $Y_n \to \theta$ in probability it follows that the remainder $\to 0$ in probability. By applying Slutsky's Theorem (Theorem 5.5.17) to

$$\sqrt{n}[g(Y_n) - g(\theta)] = g'(\theta)\sqrt{n}(Y_n - \theta),$$

the result now follows. See Exercise 5.43 for details.  $\square$

**Example 5.5.25 (Continuation of Example 5.5.23)** Suppose now that we have the mean of a random sample $\bar{X}$. For $\mu \neq 0$, we have

$$\sqrt{n}\left(\frac{1}{\bar{X}} - \frac{1}{\mu}\right) \to n\left(0, \left(\frac{1}{\mu}\right)^4 \text{Var}_\mu X_1\right)$$

in distribution.

If we do not know the variance of $X_1$, to use the above approximation requires an estimate, say $S^2$. Moreover, there is the question of what to do with the $1/\mu$ term, as we also do not know $\mu$. We can estimate everything, which gives us the approximate variance

$$\widehat{\text{Var}}\left(\frac{1}{\bar{X}}\right) \approx \left(\frac{1}{\bar{X}}\right)^4 S^2.$$

Furthermore, as both $\bar{X}$ and $S^2$ are consistent estimators, we can again apply Slutsky's Theorem to conclude that for $\mu \neq 0$,

$$\frac{\sqrt{n}\left(\frac{1}{\bar{X}} - \frac{1}{\mu}\right)}{\left(\frac{1}{\bar{X}}\right)^2 S} \to n(0, 1)$$

in distribution.

Note how we wrote this latter quantity, dividing through by the estimated standard deviation and making the limiting distribution a standard normal. This is the only way that makes sense if we need to estimate any parameters in the limiting distribution. We also note that there is an alternative approach when there are parameters to estimate, and here we can actually avoid using an estimate for $\mu$ in the variance (see the score test in Section 10.3.2).  $\|$

There are two extensions of the basic Delta Method that we need to deal with to complete our treatment. The first concerns the possibility that $g'(\mu) = 0$. This could happen, for example, if we were interested in estimating the variance of a binomial variance (see Exercise 5.44).

If $g'(\theta) = 0$, we take one more term in the Taylor expansion to get

$$g(Y_n) = g(\theta) + g'(\theta)(Y_n - \theta) + \frac{g''(\theta)}{2}(Y_n - \theta)^2 + \text{ Remainder.}$$

If we do some rearranging (setting $g' = 0$), we have

(5.5.12) $$g(Y_n) - g(\theta) = \frac{g''(\theta)}{2}(Y_n - \theta)^2 + \text{ Remainder.}$$

Now recall that the square of a $n(0, 1)$ is a $\chi_1^2$ (Example 2.1.9), which implies that

$$\frac{n(Y_n - \theta)^2}{\sigma^2} \to \chi_1^2$$

in distribution. Therefore, an argument similar to that used in Theorem 5.5.24 will establish the following theorem.

**Theorem 5.5.26 (Second-order Delta Method)** *Let $Y_n$ be a sequence of random variables that satisfies $\sqrt{n}(Y_n - \theta) \to n(0, \sigma^2)$ in distribution. For a given function $g$ and a specific value of $\theta$, suppose that $g'(\theta) = 0$ and $g''(\theta)$ exists and is not 0. Then*

(5.5.13) $$n[g(Y_n) - g(\theta)] \to \sigma^2 \frac{g''(\theta)}{2}\chi_1^2 \text{ in distribution.}$$

Approximation techniques are very useful when more than one parameter makes up the function to be estimated and more than one random variable is used in the estimator. One common example is in growth studies, where a ratio of weight/height is a variable of interest. (Recall that in Chapter 3 we saw that a ratio of two *normal* random variables has a Cauchy distribution. The ratio problem, while being important to experimenters, is nasty in theory.)

This brings us to the second extension of the Delta Method, to the multivariate case. As we already have Taylor's Theorem for the multivariate case, this extension contains no surprises.

**Example 5.5.27 (Moments of a ratio estimator)** Suppose $X$ and $Y$ are random variables with nonzero means $\mu_X$ and $\mu_Y$, respectively. The parametric function to be estimated is $g(\mu_X, \mu_Y) = \mu_X/\mu_Y$. It is straightforward to calculate

$$\frac{\partial}{\partial \mu_X} g(\mu_X, \mu_Y) = \frac{1}{\mu_Y}$$

and

$$\frac{\partial}{\partial \mu_Y} g(\mu_X, \mu_Y) = \frac{-\mu_X}{\mu_Y^2}.$$

The first-order Taylor approximations (5.5.8) and (5.5.9) give

$$\mathrm{E}\left(\frac{X}{Y}\right) \approx \frac{\mu_X}{\mu_Y}$$

and

$$\mathrm{Var}\left(\frac{X}{Y}\right) \approx \frac{1}{\mu_Y^2}\mathrm{Var}\,X + \frac{\mu_X^2}{\mu_Y^4}\mathrm{Var}\,Y - 2\frac{\mu_X}{\mu_Y^3}\mathrm{Cov}(X,Y)$$

$$= \left(\frac{\mu_X}{\mu_Y}\right)^2\left(\frac{\mathrm{Var}\,X}{\mu_X^2} + \frac{\mathrm{Var}\,Y}{\mu_Y^2} - 2\frac{\mathrm{Cov}(X,Y)}{\mu_X\mu_Y}\right).$$

Thus, we have an approximation for the mean and variance of the ratio estimator, and the approximations use only the means, variances, and covariance of $\bar{X}$ and $Y$. Exact calculations would be quite hopeless, with closed-form expressions being unattainable.

$\|$

We next present a CLT to cover an estimator such as the ratio estimator. Note that we must deal with multiple random variables although the ultimate CLT is a univariate one. Suppose the vector-valued random variable $\mathbf{X} = (X_1, \ldots, X_p)$ has mean $\boldsymbol{\mu} = (\mu_1, \ldots, \mu_p)$ and covariances $\mathrm{Cov}(X_i, X_j) = \sigma_{ij}$, and we observe an independent random sample $\mathbf{X}_1, \ldots, \mathbf{X}_n$ and calculate the means $\bar{X}_i = \sum_{k=1}^{n} X_{ik}$, $i = 1, \ldots, p$. For a function $g(\mathbf{x}) = g(x_1, \ldots, x_p)$ we can use the development after (5.5.7) to write

$$g(\bar{x}_1, \ldots, \bar{x}_p) \approx g(\mu_1, \ldots, \mu_p) + \sum_{k=1}^{p} g_k'(\boldsymbol{\mu})(\bar{x}_k - \mu_k),$$

and we then have the following theorem.

**Theorem 5.5.28 (Multivariate Delta Method)** *Let $\mathbf{X}_1, \ldots, \mathbf{X}_n$ be a random sample with $\mathrm{E}(X_{ij}) = \mu_i$ and $\mathrm{Cov}(X_{ik}, X_{jk}) = \sigma_{ij}$. For a given function $g$ with continuous first partial derivatives and a specific value of $\boldsymbol{\mu} = (\mu_1, \ldots, \mu_p)$ for which $\tau^2 = \Sigma\Sigma\sigma_{ij}\frac{\partial g(\boldsymbol{\mu})}{\partial \mu_i} \cdot \frac{\partial g(\boldsymbol{\mu})}{\partial \mu_j} > 0,$*

$$\sqrt{n}[g(\bar{X}_1, \ldots, \bar{X}_s) - g(\mu_1, \ldots, \mu_p)] \to \mathrm{n}(0, \tau^2) \text{ in distribution .}$$

The proof necessitates dealing with the convergence of multivariate random variables, and we will not deal with such multivariate intricacies here, but will take Theorem 5.5.28 on faith. The interested reader can find more details in Lehmann and Casella (1998, Section 1.8).

## 5.6 Generating a Random Sample

Thus far we have been concerned with the many methods of describing the behavior of random variables—transformations, distributions, moment calculations, limit theorems. In practice, these random variables are used to describe and model real phenomena, and observations on these random variables are the data that we collect.

Thus, typically, we observe random variables $X_1, \ldots, X_n$ from a distribution $f(x|\theta)$ and are most concerned with using properties of $f(x|\theta)$ to describe the behavior of the random variables. In this section we are, in effect, going to turn that strategy around. Here we are concerned with *generating* a random sample $X_1, \ldots, X_n$ from a given distribution $f(x|\theta)$.

**Example 5.6.1 (Exponential lifetime)** Suppose that a particular electrical component is to be modeled with an exponential($\lambda$) lifetime. The manufacturer is interested in determining the probability that, out of $c$ components, at least $t$ of them will last $h$ hours. Taking this one step at a time, we have

$$p_1 = P(\text{component lasts at least } h \text{ hours})$$

(5.6.1)
$$= P(X \geq h|\lambda),$$

and assuming that the components are independent, we can model the outcomes of the $c$ components as Bernoulli trials, so

$$p_2 = P(\text{at least } t \text{ components last } h \text{ hours})$$

(5.6.2)
$$= \sum_{k=t}^{c} \binom{c}{k} p_1^k (1 - p_1)^{c-k}.$$

Although calculation of (5.6.2) is straightforward, it can be computationally burdensome, especially if both $t$ and $c$ are large numbers. Moreover, the exponential model has the advantage that $p_1$ can be expressed in closed form, that is,

(5.6.3)
$$p_1 = \int_h^\infty \frac{1}{\lambda} e^{-x/\lambda} \, dx = e^{-h/\lambda}.$$

However, if each component were modeled with, say, a gamma distribution, then $p_1$ may not be expressible in closed form. This would make calculation of $p_2$ even more involved. ‖

A simulation approach to the calculation of expressions such as (5.6.2) is to generate random variables with the desired distribution and then use the Weak Law of Large Numbers (Theorem 5.5.2) to validate the simulation. That is, if $Y_i$, $i = 1, \ldots, n$, are iid, then a consequence of that theorem (provided the assumptions hold) is

(5.6.4)
$$\frac{1}{n} \sum_{i=1}^{n} h(Y_i) \longrightarrow Eh(Y)$$

in probability, as $n \to \infty$. (Expression (5.6.4) also holds almost everywhere, a consequence of Theorem 5.5.9, the Strong Law of Large Numbers.)

**Example 5.6.2 (Continuation of Example 5.6.1)** The probability $p_2$ can be calculated using the following steps. For $j = 1, \ldots, n$:

**a.** Generate $X_1, \ldots, X_c$ iid $\sim$ exponential($\lambda$).

**b.** Set $Y_j = 1$ if at least $t$ $X_i$s are $\geq h$; otherwise, set $Y_j = 0$.

Then, because $Y_j \sim$ Bernoulli($p_2$) and $EY_j = p_2$,

$$\frac{1}{n} \sum_{j=1}^{n} Y_j \to p_2 \text{ as } n \to \infty. \qquad \|$$

Examples 5.6.1 and 5.6.2 highlight the major concerns of this section. First, we must examine how to generate the random variables that we need, and second, we then use a version of the Law of Large Numbers to validate the simulation approximation.

Since we have to start somewhere, we start with the assumption that we can generate iid uniform random variables $U_1, \ldots, U_m$. (This problem of generating uniform random numbers has been worked on, with great success, by computer scientists.) There exist many algorithms for generating *pseudo-random* numbers that will pass almost all tests for uniformity. Moreover, most good statistical packages have a reasonable uniform random number generator. (See Devroye 1985 or Ripley 1987 for more on generating pseudo-random numbers.)

Since we are starting from the uniform random variables, our problem here is really not the problem of generating the desired random variables, but rather of *transforming* the uniform random variables to the desired distribution. In essence, there are two general methodologies for doing this, which we shall (noninformatively) call *direct* and *indirect* methods.

### 5.6.1 Direct Methods

A *direct method* of generating a random variable is one for which there exists a closed-form function $g(u)$ such that the transformed variable $Y = g(U)$ has the desired distribution when $U \sim$ uniform($0, 1$). As might be recalled, this was already accomplished for continuous random variables in Theorem 2.1.10, the Probability Integral Transform, where any distribution was transformed to the uniform. Hence the inverse transformation solves our problem.

**Example 5.6.3 (Probability Integral Transform)** If $Y$ is a continuous random variable with cdf $F_Y$, then Theorem 2.1.10 implies that the random variable $F_Y^{-1}(U)$, where $U \sim$ uniform($0, 1$), has distribution $F_Y$. If $Y \sim$ exponential($\lambda$), then

$$F_Y^{-1}(U) = -\lambda \log(1 - U)$$

is an exponential($\lambda$) random variable (see Exercise 5.49).

Thus, if we generate $U_1, \ldots, U_n$ as iid uniform random variables, $Y_i = -\lambda \log(1 - U_i)$, $i = 1, \ldots, n$, are iid exponential($\lambda$) random variables. As an example, for $n = 10,000$, we generate $u_1, u_2, \ldots, u_{10,000}$ and calculate

$$\frac{1}{n} \sum u_i = .5019 \quad \text{and} \quad \frac{1}{n-1} \sum (u_i - \bar{u})^2 = .0842.$$

From (5.6.4), which follows from the WLLN (Theorem 5.5.2), we know that $\bar{U} \to EU = 1/2$ and, from Example 5.5.3, $S^2 \to \text{Var}\, U = 1/12 = .0833$, so our estimates

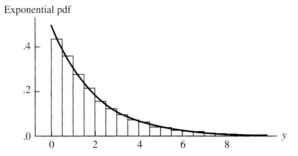

Figure 5.6.1. *Histogram of 10,000 observations from an exponential pdf with $\lambda = 2$, together with the pdf*

are quite close to the true parameters. The transformed variables $Y_i = -2 \log(1 - u_i)$ have an exponential(2) distribution, and we find that

$$\frac{1}{n} \sum y_i = 2.0004 \quad \text{and} \quad \frac{1}{n-1} \sum (y_i - \bar{y})^2 = 4.0908,$$

in close agreement with $EY = 2$ and $\text{Var}\, Y = 4$. Figure 5.6.1 illustrates the agreement between the sample histogram and the population pdf.                    ‖

The relationship between the exponential and other distributions allows the quick generation of many random variables. For example, if $U_j$ are iid uniform$(0, 1)$ random variables, then $Y_j = -\lambda \log(u_j)$ are iid exponential $(\lambda)$ random variables and

$$Y = -2 \sum_{j=1}^{\nu} \log(U_j) \sim \chi^2_{2\nu},$$

(5.6.5)
$$Y = -\beta \sum_{j=1}^{a} \log(U_j) \sim \text{gamma}(a, \beta),$$

$$Y = \frac{\sum_{j=1}^{a} \log(U_j)}{\sum_{j=1}^{a+b} \log(U_j)} \sim \text{beta}(a, b).$$

Many other variations are possible (see Exercises 5.47–5.49), but all are being driven by the exponential-uniform transformation.

Unfortunately, there are limits to this transformation. For example, we cannot use it to generate $\chi^2$ random variables with odd degrees of freedom. Hence, we cannot get a $\chi^2_1$, which would in turn get us a normal$(0, 1)$ — an extremely useful variable to be able to generate. We will return to this problem in the next subsection.

Recall that the basis of Example 5.6.3 and hence the transformations in (5.6.5) was the Probability Integral Transform, which, in general, can be written as

(5.6.6)
$$F_Y^{-1}(u) = y \leftrightarrow u = \int_{-\infty}^{y} f_y(t) dt.$$

Application of this formula to the exponential distribution was particularly handy, as the integral equation had a simple solution (see also Exercise 5.56). However, in many cases no closed-form solution for (5.6.6) will exist. Thus, each random variable generation will necessitate a solution of an integral equation, which, in practice, could be prohibitively long and complicated. This would be the case, for example, if (5.6.6) were used to generate a $\chi_1^2$.

When no closed-form solution for (5.6.6) exists, other options should be explored. These include other types of generation methods and indirect methods. As an example of the former, consider the following.

**Example 5.6.4 (Box-Muller algorithm)** Generate $U_1$ and $U_2$, two independent uniform$(0, 1)$ random variables, and set

$$R = \sqrt{-2\log U_2} \quad \text{and} \quad \theta = 2\pi U_1.$$

Then

$$X = R\cos\theta \quad \text{and} \quad Y = R\sin\theta$$

are independent normal$(0, 1)$ random variables. Thus, although we had no quick transformation for generating a single $n(0, 1)$ random variable, there is such a method for generating two variables. (See Exercise 5.50.) ‖

Unfortunately, solutions such as those in Example 5.6.4 are not plentiful. Moreover, they take advantage of the specific structure of certain distributions and are, thus, less applicable as general strategies. It turns out that, for the most part, generation of other continuous distributions (than those already considered) is best accomplished through indirect methods. Before exploring these, we end this subsection with a look at where (5.6.6) is quite useful: the case of discrete random variables.

If $Y$ is a discrete random variable taking on values $y_1 < y_2 < \cdots < y_k$, then analogous to (5.6.6) we can write

$$(5.6.7) \qquad P[F_Y(y_i) < U \le F_Y(y_{i+1})] = F_Y(y_{i+1}) - F_Y(y_i)$$

$$= P(Y = y_{i+1}).$$

Implementation of (5.6.7) to generate discrete random variables is actually quite straightforward and can be summarized as follows. To generate $Y_i \sim F_Y(y)$,

**a.** Generate $U \sim$ uniform$(0, 1)$.

**b.** If $F_y(y_i) < U \le F_y(y_{i+1})$, set $Y = y_{i+1}$.

We define $y_0 = -\infty$ and $F_Y(y_0) = 0$.

**Example 5.6.5 (Binomial random variable generation)** To generate a $Y \sim$ binomial$(4, \frac{5}{8})$, for example, generate $U \sim$ uniform$(0, 1)$ and set

$$(5.6.8) \qquad Y = \begin{cases} 0 & \text{if } 0 \quad < U \le .020, \\ 1 & \text{if } .020 < U \le .152 \\ 2 & \text{if } .152 < U \le .481 \\ 3 & \text{if } .481 < U \le .847 \\ 4 & \text{if } .847 < U \le 1. \end{cases}$$

‖

The algorithm (5.6.8) also works if the range of the discrete random variable is infinite, say Poisson or negative binomial. Although, theoretically, this could require a large number of evaluations, in practice this does not happen because there are simple and clever ways of speeding up the algorithm. For example, instead of checking each $y_i$ in the order $1, 2, \ldots$, it can be much faster to start checking $y_i$s near the mean. (See Ripley 1987, Section 3.3, and Exercise 5.55.)

We will see many uses of simulation methodology. To start off, consider the following exploration of the Poisson distribution, which is a version of the *parametric bootstrap* that we will see in Section 10.1.4.

**Example 5.6.6 (Distribution of the Poisson variance)**  If $X_1, \ldots, X_n$ are iid Poisson($\lambda$), then by either Theorem 5.2.7 or 5.2.11 the distribution of $\sum X_i$ is Poisson($n\lambda$). Thus, it is quite easy to describe the distribution of the sample mean $\bar{X}$. However, describing the distribution of the sample variance, $S^2 = \frac{1}{n-1}\sum(X_i - \bar{X})^2$, is not a simple task.

The distribution of $S^2$ is quite simple to simulate, however. Figure 5.6.2 shows such a histogram. Moreover, the simulated samples can also be used to calculate probabilities about $S^2$. If $S_i^2$ is the value calculated from the $i$th simulated sample, then

$$\frac{1}{M}\sum_{i=1}^{M} I(S_i^2 \geq a) \to P_\lambda(S^2 \geq a)$$

as $M \to \infty$.

To illustrate the use of such methodology consider the following sample of bay anchovy larvae counts taken from the Hudson River in late August 1984:

(5.6.9)          $19, 32, 29, 13, 8, 12, 16, 20, 14, 17, 22, 18, 23.$

If it is assumed that the larvae are distributed randomly and uniformly in the river, then the number that are collected in a fixed size net should follow a Poisson distribution. Such an argument follows from a spatial version of the Poisson postulates (see the Miscellanea of Chapter 2). To see if such an assumption is tenable, we can check whether the mean and variance of the observed data are consistent with the Poisson assumptions.

For the data in (5.6.9) we calculate $\bar{x} = 18.69$ and $s^2 = 44.90$. Under the Poisson assumptions we expect these values to be the same. Of course, due to sampling variability, they will not be exactly the same, and we can use a simulation to get some idea of what to expect. In Figure 5.6.2 we simulated 5,000 samples of size $n = 13$ from a Poisson distribution with $\lambda = 18.69$, and constructed the relative frequency histogram of $S^2$. Note that the observed value of $S^2 = 44.90$ falls into the tail of the distribution. In fact, since 27 of the values of $S^2$ were greater than 44.90, we can estimate

$$P(S^2 > 44.90 | \lambda = 18.69) \approx \frac{1}{5000}\sum_{i=1}^{5000} I(S_i^2 > 44.90) = \frac{27}{5000} = .0054,$$

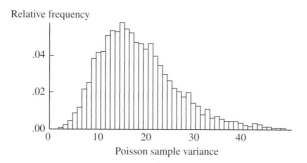

Figure 5.6.2. *Histogram of the sample variances, $S^2$, of 5,000 samples of size 13 from a Poisson distribution with $\lambda = 18.69$. The mean and standard deviation of the 5,000 values are 18.86 and 7.68.*

which leads us to question the Poisson assumption; see Exercise 5.54. (Such findings spawned the extremely bad bilingual pun "Something is fishy in the Hudson—the Poisson has failed.") ‖

## 5.6.2 Indirect Methods

When no easily found direct transformation is available to generate the desired random variable, an extremely powerful indirect method, the Accept/Reject Algorithm, can often provide a solution. The idea behind the Accept/Reject Algorithm is, perhaps, best explained through a simple example.

**Example 5.6.7 (Beta random variable generation—I)** Suppose the goal is to generate $Y \sim \text{beta}(a, b)$. If both $a$ and $b$ are integers, then the direct transformation method (5.6.5) can be used. However, if $a$ and $b$ are not integers, then that method will not work. For definiteness, set $a = 2.7$ and $b = 6.3$. In Figure 5.6.3 we have put the beta density $f_Y(y)$ inside a box with sides 1 and $c \geq \max_y f_Y(y)$. Now consider the following method of calculating $P(Y \leq y)$. If $(U, V)$ are independent uniform$(0, 1)$ random variables, then the probability of the shaded area is

$$
P\left(V \leq y, U \leq \frac{1}{c} f_Y(V)\right) = \int_0^y \int_0^{f_Y(v)/c} du\, dv
$$

(5.6.10)
$$
= \frac{1}{c} \int_0^y f_Y(v)\, dv
$$

$$
= \frac{1}{c} P(Y \leq y).
$$

So we can calculate the beta probabilities from the uniform probabilities, which suggests that we can generate the beta random variable from the uniform random variables.

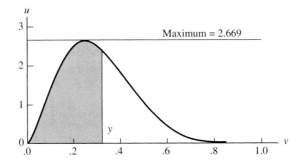

Figure 5.6.3. *The beta distribution with $a = 2.7$ and $b = 6.3$ with $c = \max_y f_Y(y) = 2.669$. The uniform random variable $V$ gives the $x$-coordinate, and we use $U$ to test if we are under the density.*

From (5.6.10), if we set $y = 1$, then we have $\frac{1}{c} = P(U < \frac{1}{c}f_Y(V))$, so

$$P(Y \le y) = \frac{P(V \le y, U \le \frac{1}{c}f_Y(V))}{P(U \le \frac{1}{c}f_Y(V))}$$

(5.6.11)
$$= P\left(V \le y | U \le \frac{1}{c}f_Y(V)\right),$$

which suggests the following algorithm.

To generate $Y \sim \text{beta}(a, b)$:

**a.** Generate $(U, V)$ independent uniform$(0, 1)$.

**b.** If $U < \frac{1}{c}f_Y(V)$, set $Y = V$; otherwise, return to step (a).

This algorithm generates a beta$(a, b)$ random variable as long as $c \ge \max_y f_Y(y)$ and, in fact, can be generalized to any bounded density with bounded support (Exercises 5.59 and 5.60).                                                                                    ‖

It should be clear that the optimal choice of $c$ is $c = \max_y f_Y(y)$. To see why this is so, note that the algorithm is open-ended in the sense that we do not know how many $(U, V)$ pairs will be needed in order to get one $Y$ variable. However, if we define the random variable

(5.6.12)              $N = $ number of $(U, V)$ pairs required for one $Y$,

then, recalling that $\frac{1}{c} = P(U \le \frac{1}{c}f_Y(V))$, we see that $N$ is a geometric$(1/c)$ random variable. Thus to generate one $Y$ we expect to need $E(N) = c$ pairs $(U, V)$, and in this sense minimizing $c$ will optimize the algorithm.

Examining Figure 5.6.3 we see that the algorithm is wasteful in the area where $U > \frac{1}{c}f_Y(V)$. This is because we are using a uniform random variable $V$ to get a beta random variable $Y$. To improve, we might start with something closer to the beta.

The testing step, step (b) of the algorithm, can be thought of as testing whether the random variable $V$ "looks like" it could come from the density $f_Y$. Suppose that $V \sim f_V$, and we compute

$$M = \sup_y \frac{f_Y(y)}{f_V(y)} < \infty.$$

A generalization of step (b) is to compare $U \sim \text{uniform}(0,1)$ to $\frac{1}{M} f_Y(V)/f_V(V)$. The larger this ratio is, the more $V$ "looks like" a random variable from the density $f_Y$, and the more likely it is that $U < \frac{1}{M} f_Y(V)/f_V(V)$. This is the basis of the general *Accept/Reject Algorithm*.

### 5.6.3 The Accept/Reject Algorithm

**Theorem 5.6.8**  *Let $Y \sim f_Y(y)$ and $V \sim f_V(V)$, where $f_Y$ and $f_V$ have common support with*

$$M = \sup_y f_Y(y)/f_V(y) < \infty.$$

*To generate a random variable $Y \sim f_Y$:*
**a.** *Generate $U \sim \text{uniform}(0,1)$, $V \sim f_V$, independent.*
**b.** *If $U < \frac{1}{M} f_Y(V)/f_V(V)$, set $Y = V$; otherwise, return to step (a).*

**Proof:** The generated random variable $Y$ has cdf

$$P(Y \leq y) = P(V \leq y | \text{stop})$$

$$= P\left(V \leq y \Big| U < \frac{1}{M} f_Y(V)/f_V(V)\right)$$

$$= \frac{P(V \leq y, U < \frac{1}{M} f_Y(V)/f_V(V))}{P(U < \frac{1}{M} f_Y(V)/f_V(V))}$$

$$= \frac{\int_{-\infty}^{y} \int_{0}^{\frac{1}{M} f_Y(v)/f_V(v)} du f_V(v) dv}{\int_{-\infty}^{\infty} \int_{0}^{\frac{1}{M} f_Y(v)/f_V(v)} du f_V(v) dv}$$

$$= \int_{-\infty}^{y} f_Y(v) dv,$$

which is the desired cdf.                                                  □

Note also that

$$M = \sup_y f_Y(y)/f_V(y)$$

$$= \left[P\left(U < \frac{1}{M} f_Y(V)/f_V(V)\right)\right]^{-1}$$

$$= \frac{1}{P(\text{Stop})},$$

so the number of trials needed to generate one $Y$ is a geometric$(1/M)$ random variable, and $M$ is the expected number of trials.

**Example 5.6.9 (Beta random variable generation–II)**    To generate $Y \sim$ beta$(2.7, 6.3)$ consider the algorithm:

**a.** Generate $U \sim$ uniform$(0, 1)$, $V \sim$ beta$(2, 6)$.

**b.** If $U < \frac{1}{M} \frac{f_Y(V)}{f_V(V)}$, set $Y = V$; otherwise, return to step (a).

This Accept/Reject Algorithm will generate the required $Y$ as long as $\sup_y f_Y(y)/f_V(y) \leq M < \infty$. For the given densities we have $M = 1.67$, so the requirement is satisfied. (See Exercise 5.63.)

For this algorithm $EN = 1.67$, while for the algorithm of Example 5.6.7, which uses the uniform $V$, we have $EN = 2.67$. Although this seems to indicate that the latter algorithm is faster, remember that generating a beta$(2, 6)$ random variable will need eight uniform random variables. Thus, comparison of algorithms is not always straightforward and will include consideration of both computer speed and programming ease.                                                                    ‖

The importance of the requirement that $M < \infty$ should be stressed. This can be interpreted as requiring the density of $V$ (often called the *candidate density*) to have heavier tails than the density of $Y$ (often called the *target density*). This requirement tends to ensure that we will obtain a good representation of the values of $Y$, even those values that are in the tails. For example, if $V \sim$ Cauchy and $Y \sim$ n$(0, 1)$, then we expect the range of $V$ samples to be wider than that of $Y$ samples, and we should get good performance from an Accept/Reject Algorithm based on these densities. However, it is much more difficult to change n$(0, 1)$ random variables into Cauchy random variables because the extremes will be underrepresented.

There are cases, however, where the target density has heavy tails, and it is difficult to get candidate densities that will result in finite values of $M$. In such cases the Accept/Reject Algorithm will no longer apply, and one is led to another class of methods known as Markov Chain Monte Carlo (MCMC) methods. Special cases of such methods are known as the *Gibbs Sampler* and the *Metropolis Algorithm*. We state the latter.

**Metropolis Algorithm** Let $Y \sim f_Y(y)$ and $V \sim f_V(v)$, where $f_Y$ and $f_V$ have common support. To generate $Y \sim f_Y$:

0. Generate $V \sim f_V$. Set $Z_0 = V$.

For $i = 1, 2, \ldots$:

1. Generate $U_i \sim$ uniform$(0, 1)$ , $V_i \sim f_V$, and calculate

$$\rho_i = \min\left\{ \frac{f_Y(V_i)}{f_V(V_i)} \cdot \frac{f_V(Z_{i-1})}{f_Y(Z_{i-1})}, 1 \right\}.$$

2. Set

$$Z_i = \begin{cases} V_i & \text{if } U_i \leq \rho_i \\ Z_{i-1} & \text{if } U_i > \rho_i. \end{cases}$$

Then, as $i \to \infty$, $Z_i$ converges to $Y$ in distribution.

Although the algorithm does not require a finite $M$, it does not produce a random variable with exactly the density $f_Y$, but rather a convergent sequence. In practice, after the algorithm is run for a while ($i$ gets big), the $Z$s that are produced behave very much like variables from $f_Y$. (See Chib and Greenberg 1995 for an elementary introduction to the Metropolis Algorithm.)

Although MCMC methods can be traced back to at least Metropolis *et al.* (1953), they entered real prominence with the work of Gelfand and Smith (1990), building on that of Geman and Geman (1984). See Miscellanea 5.8.5 for more details.

## 5.7 Exercises

**5.1** Color blindness appears in 1% of the people in a certain population. How large must a sample be if the probability of its containing a color-blind person is to be .95 or more? (Assume that the population is large enough to be considered infinite, so that sampling can be considered to be with replacement.)

**5.2** Suppose $X_1, X_2, \ldots$ are jointly continuous and independent, each distributed with marginal pdf $f(x)$, where each $X_i$ represents annual rainfall at a given location.

    (a) Find the distribution of the number of years until the first year's rainfall, $X_1$, is exceeded for the first time.

    (b) Show that the mean number of years until $X_1$ is exceeded for the first time is infinite.

**5.3** Let $X_1, \ldots, X_n$ be iid random variables with continuous cdf $F_X$, and suppose $EX_i = \mu$. Define the random variables $Y_1, \ldots, Y_n$ by

$$Y_i = \begin{cases} 1 & \text{if } X_i > \mu \\ 0 & \text{if } X_i \leq \mu. \end{cases}$$

Find the distribution of $\sum_{i=1}^{n} Y_i$.

**5.4** A generalization of iid random variables is *exchangeable* random variables, an idea due to deFinetti (1972). A discussion of exchangeability can also be found in Feller (1971). The random variables $X_1, \ldots, X_n$ are *exchangeable* if any permutation of any subset of them of size $k$ ($k \leq n$) has the same distribution. In this exercise we will see an example of random variables that are exchangeable but not iid. Let $X_i | P \sim$ iid Bernoulli($P$), $i = 1, \ldots, n$, and let $P \sim$ uniform$(0, 1)$.

    (a) Show that the marginal distribution of any $k$ of the $X$s is the same as

$$P(X_1 = x_1, \ldots, X_k = x_k) = \int_0^1 p^t(1-p)^{k-t} dp = \frac{t!(k-t)!}{(k+1)!},$$

    where $t = \sum_{i=1}^{k} x_i$. Hence, the $X$s are exchangeable.

    (b) Show that, marginally,

$$P(X_1 = x_1, \ldots, X_n = x_n) \neq \Pi_{i=1}^{n} P(X_i = x_i),$$

    so the distribution of the $X$s is exchangeable but not iid.

(deFinetti proved an elegant characterization theorem for an infinite sequence of exchangeable random variables. He proved that any such sequence of exchangeable random variables is a mixture of iid random variables.)

**5.5** Let $X_1, \ldots, X_n$ be iid with pdf $f_X(x)$, and let $\bar{X}$ denote the sample mean. Show that

$$f_{\bar{X}}(x) = n f_{X_1 + \cdots + X_n}(nx),$$

even if the mgf of $X$ does not exist.

**5.6** If $X$ has pdf $f_X(x)$ and $Y$, independent of $X$, has pdf $f_Y(y)$, establish formulas, similar to (5.2.3), for the random variable $Z$ in each of the following situations.

(a) $Z = X - Y$

(b) $Z = XY$

(c) $Z = X/Y$

**5.7** In Example 5.2.10, a partial fraction decomposition is needed to derive the distribution of the sum of two independent Cauchy random variables. This exercise provides the details that are skipped in that example.

(a) Find the constants $A$, $B$, $C$, and $D$ that satisfy

$$\frac{1}{1 + (w/\sigma)^2} \frac{1}{1 + ((z-w)/\tau)^2} =$$

$$\frac{Aw}{1 + (w/\sigma)^2} + \frac{B}{1 + (w/\sigma)^2} - \frac{Cw}{1 + ((z-w)/\tau)^2} - \frac{D}{1 + ((z-w)/\tau)^2},$$

where $A$, $B$, $C$, and $D$ may depend on $z$ but not on $w$.

(b) Using the facts that

$$\int \frac{t}{1+t^2}\, dt = \frac{1}{2} \log(1+t^2) + constant \quad \text{and} \quad \int \frac{1}{1+t^2}\, dt = \arctan(t) + constant,$$

evaluate (5.2.4) and hence verify (5.2.5).

(Note that the integration in part (b) is quite delicate. Since the mean of a Cauchy does not exist, the integrals $\int_{-\infty}^{\infty} \frac{Aw}{1+(w/\sigma)^2}\, dw$ and $\int_{-\infty}^{\infty} \frac{Cw}{1+((z-w)/\tau)^2}\, dw$ do not exist. However, the integral of the difference *does exist*, which is all that is needed.)

**5.8** Let $X_1, \ldots, X_n$ be a random sample, where $\bar{X}$ and $S^2$ are calculated in the usual way.

(a) Show that

$$S^2 = \frac{1}{2n(n-1)} \sum_{i=1}^{n} \sum_{j=1}^{n} (X_i - X_j)^2.$$

Assume now that the $X_i$s have a finite fourth moment, and denote $\theta_1 = EX_i$, $\theta_j = E(X_i - \theta_1)^j$, $j = 2, 3, 4$.

(b) Show that Var $S^2 = \frac{1}{n}(\theta_4 - \frac{n-3}{n-1}\theta_2^2)$.

(c) Find $\text{Cov}(\bar{X}, S^2)$ in terms of $\theta_1, \ldots, \theta_4$. Under what conditions is $\text{Cov}(\bar{X}, S^2) = 0$?

**5.9** Establish the *Lagrange Identity*, that for any numbers $a_1, a_2, \ldots, a_n$ and $b_1, b_2, \ldots, b_n$,

$$\left( \sum_{i=1}^{n} a_i^2 \right) \left( \sum_{i=1}^{n} b_i^2 \right) - \left( \sum_{i=1}^{n} a_i b_i \right)^2 = \sum_{i=1}^{n-1} \sum_{j=i+1}^{n} (a_i b_j - a_j b_i)^2.$$

Use the identity to show that the squared correlation coefficient is equal to 1 if and only if all of the sample points lie on a straight line (Wright 1992). (*Hint*: Establish the identity for $n = 2$; then induct.)

**5.10** Let $X_1, \ldots, X_n$ be a random sample from a $n(\mu, \sigma^2)$ population.

(a) Find expressions for $\theta_1, \ldots, \theta_4$, as defined in Exercise 5.8, in terms of $\mu$ and $\sigma^2$.

(b) Use the results of Exercise 5.8, together with the results of part (a), to calculate Var $S^2$.

(c) Calculate Var $S^2$ a completely different (and easier) way: Use the fact that $(n-1)S^2/\sigma^2 \sim \chi_{n-1}^2$.

**5.11** Suppose $\bar{X}$ and $S^2$ are calculated from a random sample $X_1, \ldots, X_n$ drawn from a population with finite variance $\sigma^2$. We know that $ES^2 = \sigma^2$. Prove that $ES \leq \sigma$, and if $\sigma^2 > 0$, then $ES < \sigma$.

**5.12** Let $X_1, \ldots, X_n$ be a random sample from a $n(0, 1)$ population. Define

$$Y_1 = \left| \frac{1}{n} \sum_{i=1}^{n} X_i \right|, \qquad Y_2 = \frac{1}{n} \sum_{i=1}^{n} |X_i|.$$

Calculate $EY_1$ and $EY_2$, and establish an inequality between them.

**5.13** Let $X_1, \ldots, X_n$ be iid $n(\mu, \sigma^2)$. Find a function of $S^2$, the sample variance, say $g(S^2)$, that satisfies $Eg(S^2) = \sigma$. (*Hint*: Try $g(S^2) = c\sqrt{S^2}$, where $c$ is a constant.)

**5.14** (a) Prove that the statement of Lemma 5.3.3 follows from the special case of $\mu_i = 0$ and $\sigma_i^2 = 1$. That is, show that if $X_j = \sigma_j Z_j + \mu_j$ and $Z_j \sim n(0, 1)$, $j = 1, \ldots, n$, all independent, $a_{ij}, b_{rj}$ are constants, and

$$\text{Cov}\left( \sum_{j=1}^{n} a_{ij} Z_j, \sum_{j=1}^{n} b_{rj} Z_j \right) = 0 \Rightarrow \sum_{j=1}^{n} a_{ij} Z_j \text{ and } \sum_{j=1}^{n} b_{rj} Z_j \text{ are independent,}$$

then

$$\text{Cov}\left( \sum_{j=1}^{n} a_{ij} X_j, \sum_{j=1}^{n} b_{rj} X_j \right) = 0 \Rightarrow \sum_{j=1}^{n} a_{ij} X_j \text{ and } \sum_{j=1}^{n} b_{rj} X_j \text{ are independent.}$$

(b) Verify the expression for $\text{Cov}\left( \sum_{j=1}^{n} a_{ij} X_j, \sum_{j=1}^{n} b_{rj} X_j \right)$ in Lemma 5.3.3.

**5.15** Establish the following recursion relations for means and variances. Let $\bar{X}_n$ and $S_n^2$ be the mean and variance, respectively, of $X_1, \ldots, X_n$. Then suppose another observation, $X_{n+1}$, becomes available. Show that

(a) $\bar{X}_{n+1} = \frac{X_{n+1} + n\bar{X}_n}{n+1}$.

(b) $nS_{n+1}^2 = (n-1)S_n^2 + \left(\frac{n}{n+1}\right)(X_{n+1} - \bar{X}_n)^2$.

**5.16** Let $X_i, i = 1, 2, 3$, be independent with $n(i, i^2)$ distributions. For each of the following situations, use the $X_i$s to construct a statistic with the indicated distribution.

(a) chi squared with 3 degrees of freedom

(b) $t$ distribution with 2 degrees of freedom

(c) $F$ distribution with 1 and 2 degrees of freedom

**5.17** Let $X$ be a random variable with an $F_{p,q}$ distribution.

(a) Derive the pdf of $X$.

(b) Derive the mean and variance of $X$.

(c) Show that $1/X$ has an $F_{q,p}$ distribution.

(d) Show that $(p/q)X/[1 + (p/q)X]$ has a beta distribution with parameters $p/2$ and $q/2$.

**5.18** Let $X$ be a random variable with a Student's $t$ distribution with $p$ degrees of freedom.

(a) Derive the mean and variance of $X$.

(b) Show that $X^2$ has an $F$ distribution with 1 and $p$ degrees of freedom.

(c) Let $f(x|p)$ denote the pdf of $X$. Show that

$$\lim_{p \to \infty} f(x|p) \to \frac{1}{\sqrt{2\pi}} e^{-x^2/2}$$

at each value of $x$, $-\infty < x < \infty$. This correctly suggests that as $p \to \infty$, $X$ converges in distribution to a $n(0, 1)$ random variable. (*Hint*: Use Stirling's Formula.)

(d) Use the results of parts (a) and (b) to argue that, as $p \to \infty$, $X^2$ converges in distribution to a $\chi_1^2$ random variable.

(e) What might you conjecture about the distributional limit, as $p \to \infty$, of $qF_{q,p}$?

**5.19** (a) Prove that the $\chi^2$ distribution is *stochastically increasing* in its degrees of freedom; that is, if $p > q$, then for any $a$, $P(\chi_p^2 > a) \geq P(\chi_q^2 > a)$, with strict inequality for some $a$.

(b) Use the results of part (a) to prove that for any $\nu$, $kF_{k,\nu}$ is stochastically increasing in $k$.

(c) Show that for any $k$, $\nu$, and $\alpha$, $kF_{\alpha,k,\nu} > (k-1)F_{\alpha,k-1,\nu}$. (The notation $F_{\alpha,k-1,\nu}$ denotes a level-$\alpha$ *cutoff point*; see Section 8.3.1. Also see Miscellanea 8.5.1 and Exercise 11.15.)

**5.20** (a) We can see that the $t$ distribution is a mixture of normals using the following argument:

$$P(T_\nu \leq t) = P\left(\frac{Z}{\sqrt{\chi_\nu^2/\nu}} \leq t\right) = \int_0^\infty P\left(Z \leq t\sqrt{x}/\sqrt{\nu}\right) P\left(\chi_\nu^2 = x\right) dx,$$

where $T_\nu$ is a $t$ random variable with $\nu$ degrees of freedom. Using the Fundamental Theorem of Calculus and interpreting $P(\chi_\nu^2 = \nu x)$ as a pdf, we obtain

$$f_{T_\nu}(t) = \int_0^\infty \frac{1}{\sqrt{2\pi}} e^{-t^2 x/2\nu} \sqrt{x} \, \frac{1}{\sqrt{\nu}} \frac{1}{\Gamma(\nu/2)2^{\nu/2}} (x)^{(\nu/2)-1} e^{-x/2} \, dx,$$

a scale mixture of normals. Verify this formula by direct integration.

(b) A similar formula holds for the $F$ distribution; that is, it can be written as a mixture of chi squareds. If $F_{1,\nu}$ is an $F$ random variable with 1 and $\nu$ degrees of freedom, then we can write

$$P(F_{1,\nu} \le \nu t) = \int_0^\infty P(\chi_1^2 \le ty) \, f_\nu(y) \, dy,$$

where $f_\nu(y)$ is a $\chi_\nu^2$ pdf. Use the Fundamental Theorem of Calculus to obtain an integral expression for the pdf of $F_{1,\nu}$, and show that the integral equals the pdf.

(c) Verify that the generalization of part (b),

$$P\left(F_{m,\nu} \le \frac{\nu}{m} t\right) = \int_0^\infty P(\chi_m^2 \le ty) \, f_\nu(y) \, dy,$$

is valid for all integers $m > 1$.

**5.21** What is the probability that the larger of two continuous iid random variables will exceed the population median? Generalize this result to samples of size $n$.

**5.22** Let $X$ and $Y$ be iid n$(0, 1)$ random variables, and define $Z = \min(X, Y)$. Prove that $Z^2 \sim \chi_1^2$.

**5.23** Let $U_i, i = 1, 2, \ldots,$ be independent uniform$(0, 1)$ random variables, and let $X$ have distribution

$$P(X = x) = \frac{c}{x!}, \quad x = 1, 2, 3, \ldots,$$

where $c = 1/(e - 1)$. Find the distribution of

$$Z = \min\{U_1, \ldots, U_X\}.$$

(*Hint:* Note that the distribution of $Z|X = x$ is that of the first-order statistic from a sample of size $x$.)

**5.24** Let $X_1, \ldots, X_n$ be a random sample from a population with pdf

$$f_X(x) = \begin{cases} 1/\theta & \text{if } 0 < x < \theta \\ 0 & \text{otherwise.} \end{cases}$$

Let $X_{(1)} < \cdots < X_{(n)}$ be the order statistics. Show that $X_{(1)}/X_{(n)}$ and $X_{(n)}$ are independent random variables.

**5.25** As a generalization of the previous exercise, let $X_1, \ldots, X_n$ be iid with pdf

$$f_X(x) = \begin{cases} \frac{a}{\theta^a} x^{a-1} & \text{if } 0 < x < \theta \\ 0 & \text{otherwise.} \end{cases}$$

Let $X_{(1)} < \cdots < X_{(n)}$ be the order statistics. Show that $X_{(1)}/X_{(2)}, X_{(2)}/X_{(3)}, \ldots,$ $X_{(n-1)}/X_{(n)}$, and $X_{(n)}$ are mutually independent random variables. Find the distribution of each of them.

**5.26** Complete the proof of Theorem 5.4.6.

(a) Let $U$ be a random variable that counts the number of $X_1, \ldots, X_n$ less than or equal to $u$, and let $V$ be a random variable that counts the number of $X_1, \ldots, X_n$ greater than $u$ and less than or equal to $v$. Show that $(U, V, n - U - V)$ is a multinomial random vector with $n$ trials and cell probabilities $(F_X(u), F_X(v) - F_X(u), 1 - F_X(v))$.

(b) Show that the joint cdf of $X_{(i)}$ and $X_{(j)}$ can be expressed as

$$F_{X_{(i)}, X_{(j)}}(u, v) = P(U \geq i, U + V \geq j)$$

$$= \sum_{k=i}^{j-1} \sum_{m=j-k}^{n-k} P(U = k, V = m) + P(U \geq j)$$

$$= \sum_{k=i}^{j-1} \sum_{m=j-k}^{n-k} \frac{n!}{k! m! (n-k-m)!} [F_X(u)]^k [F_X(v) - F_X(u)]^m$$

$$\times [1 - F_X(v)]^{n-k-m} + P(U \geq j).$$

(c) Find the joint pdf by computing the mixed partial as indicated in (4.1.4). (The mixed partial of $P(U \geq j)$ is 0 since this term depends only on $u$, not $v$. For the other terms, there is much cancellation using relationships like (5.4.6).)

**5.27** Let $X_1, \ldots, X_n$ be iid with pdf $f_X(x)$ and cdf $F_X(x)$, and let $X_{(1)} < \cdots < X_{(n)}$ be the order statistics.

(a) Find an expression for the conditional pdf of $X_{(i)}$ given $X_{(j)}$ in terms of $f_X$ and $F_X$.

(b) Find the pdf of $V | R = r$, where $V$ and $R$ are defined in Example 5.4.7.

**5.28** Let $X_1, \ldots, X_n$ be iid with pdf $f_X(x)$ and cdf $F_X(x)$, and let $X_{(i_1)} < \cdots < X_{(i_l)}$ and $X_{(j_1)} < \cdots < X_{(j_m)}$ be any two disjoint groups of order statistics. In terms of the pdf $f_X(x)$ and the cdf $F_X(x)$, find expressions for

(a) The marginal cdf and pdf of $X_{(i_1)}, \ldots, X_{(i_l)}$.

(b) The conditional cdf and pdf of $X_{(i_1)}, \ldots, X_{(i_l)}$ given $X_{(j_1)}, \ldots, X_{(j_m)}$.

**5.29** A manufacturer of booklets packages them in boxes of 100. It is known that, on the average, the booklets weigh 1 ounce, with a standard deviation of .05 ounce. The manufacturer is interested in calculating

$$P(100 \text{ booklets weigh more than } 100.4 \text{ ounces}),$$

a number that would help detect whether too many booklets are being put in a box. Explain how you would calculate the (approximate?) value of this probability. Mention any relevant theorems or assumptions needed.

**5.30** If $\bar{X}_1$ and $\bar{X}_2$ are the means of two independent samples of size $n$ from a population with variance $\sigma^2$, find a value for $n$ so that $P(|\bar{X}_1 - \bar{X}_2| < \sigma/5) \approx .99$. Justify your calculations.

**5.31** Suppose $\bar{X}$ is the mean of 100 observations from a population with mean $\mu$ and variance $\sigma^2 = 9$. Find limits between which $\bar{X} - \mu$ will lie with probability at least .90. Use both Chebychev's Inequality and the Central Limit Theorem, and comment on each.

**5.32** Let $X_1, X_2, \ldots$ be a sequence of random variables that converges in probability to a constant $a$. Assume that $P(X_i > 0) = 1$ for all $i$.

(a) Verify that the sequences defined by $Y_i = \sqrt{X_i}$ and $Y_i' = a/X_i$ converge in probability.

(b) Use the results in part (a) to prove the fact used in Example 5.5.18, that $\sigma/S_n$ converges in probability to 1.

**5.33** Let $X_n$ be a sequence of random variables that converges in distribution to a random variable $X$. Let $Y_n$ be a sequence of random variables with the property that for any finite number $c$,

$$\lim_{n \to \infty} P(Y_n > c) = 1.$$

Show that for any finite number $c$,

$$\lim_{n \to \infty} P(X_n + Y_n > c) = 1.$$

(This is the type of result used in the discussion of the power properties of the tests described in Section 10.3.2.)

**5.34** Let $X_1, \ldots, X_n$ be a random sample from a population with mean $\mu$ and variance $\sigma^2$. Show that

$$\mathrm{E}\frac{\sqrt{n}(\bar{X}_n - \mu)}{\sigma} = 0 \quad \text{and} \quad \mathrm{Var}\,\frac{\sqrt{n}(\bar{X}_n - \mu)}{\sigma} = 1.$$

Thus, the normalization of $\bar{X}_n$ in the Central Limit Theorem gives random variables that have the same mean and variance as the limiting $n(0, 1)$ distribution.

**5.35** Stirling's Formula (derived in Exercise 1.28), which gives an approximation for factorials, can be easily derived using the CLT.

(a) Argue that, if $X_i \sim$ exponential$(1), i = 1, 2, \ldots$, all independent, then for every $x$,

$$P\left(\frac{\bar{X}_n - 1}{1/\sqrt{n}} \le x\right) \to P(Z \le x),$$

where $Z$ is a standard normal random variable.

(b) Show that differentiating both sides of the approximation in part (a) suggests

$$\frac{\sqrt{n}}{\Gamma(n)}(x\sqrt{n} + n)^{n-1}e^{-(x\sqrt{n}+n)} \approx \frac{1}{\sqrt{2\pi}}e^{-x^2/2}$$

and that $x = 0$ gives Stirling's Formula.

**5.36** Given that $N = n$, the conditional distribution of $Y$ is $\chi^2_{2n}$. The unconditional distribution of $N$ is Poisson$(\theta)$.

(a) Calculate $\mathrm{E}Y$ and $\mathrm{Var}\,Y$ (unconditional moments).

(b) Show that, as $\theta \to \infty$, $(Y - \mathrm{E}Y)/\sqrt{\mathrm{Var}\,Y} \to n(0, 1)$ in distribution.

**5.37** In Example 5.5.16, a normal approximation to the negative binomial distribution was given. Just as with the normal approximation to the binomial distribution given in Example 3.3.2, the approximation might be improved with a "continuity correction." For $X_i$s defined as in Example 5.5.16, let $V_n = \sum_{i=1}^{n} X_i$. For $n = 10, p = .7$, and $r = 2$, calculate $P(V_n = v)$ for $v = 0, 1, \ldots, 10$ using each of the following three methods.

(a) exact calculations

(b) normal approximation as given in Example 5.5.16

(c) normal approximation with continuity correction

**5.38** The following extensions of the inequalities established in Exercise 3.45 are useful in establishing a SLLN (see Miscellanea 5.8.4). Let $X_1, X_2, \ldots, X_n$ be iid with mgf $M_X(t)$, $-h < t < h$, and let $S_n = \sum_{i=1}^{n} X_i$ and $\bar{X}_n = S_n/n$.

(a) Show that $P(S_n > a) \le e^{-at}[M_X(t)]^n$, for $0 < t < h$, and $P(S_n \le a) \le e^{-at}[M_X(t)]^n$, for $-h < t < 0$.

(b) Use the facts that $M_X(0) = 1$ and $M_X'(0) = \mathrm{E}(X)$ to show that, if $\mathrm{E}(X) < 0$, then there is a $0 < c < 1$ with $P(S_n > a) \le c^n$. Establish a similar bound for $P(S_n \le a)$.

(c) Define $Y_i = X_i - \mu - \varepsilon$ and use the above argument, with $a = 0$, to establish that $P(\bar{X}_n - \mu > \varepsilon) \le c^n$.

(d) Now define $Y_i = -X_i + \mu - \varepsilon$, establish an equality similar to part (c), and combine the two to get

$$P(|\bar{X}_n - \mu| > \varepsilon) \le 2c^n \text{ for some } 0 < c < 1.$$

**5.39** This exercise, and the two following, will look at some of the mathematical details of convergence.

(a) Prove Theorem 5.5.4. (*Hint:* Since $h$ is continuous, given $\varepsilon > 0$ we can find a $\delta$ such that $|h(x_n) - h(x)| < \varepsilon$ whenever $|x_n - x| < \delta$. Translate this into probability statements.)

(b) In Example 5.5.8, find a subsequence of the $X_i$s that converges almost surely, that is, that converges pointwise.

**5.40** Prove Theorem 5.5.12 for the case where $X_n$ and $X$ are continuous random variables.

(a) Given $t$ and $\varepsilon$, show that $P(X \le t - \varepsilon) \le P(X_n \le t) + P(|X_n - X| \ge \varepsilon)$. This gives a lower bound on $P(X_n \le t)$.

(b) Use a similar strategy to get an upper bound on $P(X_n \le t)$.

(c) By pinching, deduce that $P(X_n \le t) \to P(X \le t)$.

**5.41** Prove Theorem 5.5.13; that is, show that

$$P(|X_n - \mu| > \varepsilon) \to 0 \text{ for every } \varepsilon \quad \Leftrightarrow \quad P(X_n \le x) \to \begin{cases} 0 & \text{if } x < \mu \\ 1 & \text{if } x > \mu. \end{cases}$$

(a) Set $\varepsilon = |x - \mu|$ and show that if $x > \mu$, then $P(X_n \le x) \ge P(|X_n - \mu| \le \varepsilon)$, while if $x < \mu$, then $P(X_n \le x) \le P(|X_n - \mu| \ge \varepsilon)$. Deduce the $\Rightarrow$ implication.

(b) Use the fact that $\{x : |x - \mu| > \varepsilon\} = \{x : x - \mu < -\varepsilon\} \cup \{x : x - \mu > \varepsilon\}$ to deduce the $\Leftarrow$ implication.

(See Billingsley 1995, Section 25, for a detailed treatment of the above results.)

**5.42** Similar to Example 5.5.11, let $X_1, X_2, \ldots$ be iid and $X_{(n)} = \max_{1 \le i \le n} X_i$.

(a) If $X_i$ beta$(1, \beta)$, find a value of $\nu$ so that $n^\nu(1 - X_{(n)})$ converges in distribution.

(b) If $X_i$ exponential(1), find a sequence $a_n$ so that $X_{(n)} - a_n$ converges in distribution.

**5.43** Fill in the details in the proof of Theorem 5.5.24.

(a) Show that if $\sqrt{n}(Y_n - \mu) \to n(0, \sigma^2)$ in distribution, then $Y_n \to \mu$ in probability.

(b) Give the details for the application of Slutsky's Theorem (Theorem 5.5.17).

**5.44** Let $X_i, i = 1, 2, \ldots,$ be independent Bernoulli($p$) random variables and let $Y_n = \frac{1}{n} \sum_{i=1}^n X_i$.

(a) Show that $\sqrt{n}\,(Y_n - p) \to n\,[0, p(1 - p)]$ in distribution.

(b) Show that for $p \neq 1/2$, the estimate of variance $Y_n(1 - Y_n)$ satisfies $\sqrt{n}[Y_n(1 - Y_n) - p(1 - p)] \to n\,[0, (1 - 2p)^2 p(1 - p)]$ in distribution.

(c) Show that for $p = 1/2$, $n\left[Y_n(1 - Y_n) - \frac{1}{4}\right] \to -\frac{1}{4}\chi_1^2$ in distribution. (If this appears strange, note that $Y_n(1 - Y_n) \leq 1/4$, so the left-hand side is always negative. An equivalent form is $4n\left[\frac{1}{4} - Y_n(1 - Y_n)\right] \to \chi_1^2$.)

**5.45** For the situation of Example 5.6.1, calculate the probability that at least 50% of the components last 100 hours when

(a) $c = 300, X \sim \text{gamma}(a, b), a = 4, b = 25.$

(b) $c = 100, X \sim \text{gamma}(a, b), a = 20, b = 5.$

(c) $c = 100, X \sim \text{gamma}(a, b), a = 20.7, b = 5.$

(Hint: In parts (a) and (b) it is possible to evaluate the gamma integral in closed form, although it probably isn't worth the effort in (b). There is no closed-form expression for the integral in part (c), which has to be evaluated through either numerical integration or simulation.)

**5.46** Referring to Exercise 5.45, compare your answers to what is obtained from a normal approximation to the binomial (see Example 3.3.2).

**5.47** Verify the distributions of the random variables in (5.6.5).

**5.48** Using strategies similar to (5.6.5), show how to generate an $F_{m,n}$ random variable, where both $m$ and $n$ are even integers.

**5.49** Let $U \sim \text{uniform}(0, 1)$.

(a) Show that both $-\log U$ and $-\log(1 - U)$ are exponential random variables.

(b) Show that $X = \log \frac{u}{1-u}$ is a logistic(0, 1) random variable.

(c) Show how to generate a logistic($\mu, \beta$) random variable.

**5.50** The Box-Muller method for generating normal *pseudo-random variables* (Example 5.6.4) is based on the transformation

$$X_1 = \cos(2\pi U_1)\sqrt{-2\log U_2}, \qquad X_2 = \sin(2\pi U_1)\sqrt{-2\log U_2},$$

where $U_1$ and $U_2$ are iid uniform(0,1). Prove that $X_1$ and $X_2$ are independent n(0, 1) random variables.

**5.51** One of the earlier methods (not one of the better ones) of generating pseudo-random standard normal random variables from uniform random variables is to take $X = \sum_{i=1}^{12} U_i - 6$, where the $U_i$s are iid uniform(0, 1).

(a) Justify the fact that $X$ is approximately n(0, 1).

(b) Can you think of any obvious way in which the approximation fails?

(c) Show how good (or bad) the approximation is by comparing the first four moments. (The fourth moment is 29/10 and is a lengthy calculation—mgfs and computer algebra would help; see Example A.0.6 in Appendix A.)

**5.52** For each of the following distributions write down an algorithm for generating the indicated random variables.

(a) $Y \sim$ binomial$(8, \frac{2}{3})$
(b) $Y \sim$ hypergeometric $N = 10, M = 8, K = 4$
(c) $Y \sim$ negative binomial$(5, \frac{1}{3})$

**5.53** For each of the distributions in the previous exercise:

(a) Generate 1,000 variables from the indicated distribution.
(b) Compare the mean, variance, and histogram of the generated random variables with the theoretical values.

**5.54** Refer to Example 5.6.6. Another sample of bay anchovy larvae counts yielded the data

$$158, 143, 106, 57, 97, 80, 109, 109, 350, 224, 109, 214, 84.$$

(a) Use the technique of Example 5.6.6 to construct a simulated distribution of $S^2$ to see if the assumption of Poisson counts is tenable.
(b) A possible explanation of the failure of the Poisson assumptions (and increased variance) is the failure of the assumption that the larvae are uniformly distributed in the river. If the larvae tend to clump, the negative binomial$(r, p)$ distribution (with mean $\mu = r\frac{1-p}{p}$ and variance $\mu + \frac{\mu^2}{r}$) is a reasonable alternative model. For $\mu = \bar{x}$, what values of $r$ lead to simulated distributions that are consistent with the data?

**5.55** Suppose the method of (5.6.7) is used to generate the random variable $Y$, where $y_i = i$, $i = 0, 1, 2, \ldots$. Show that the expected number of comparisons is $E(Y + 1)$. (*Hint:* See Exercise 2.14.)

**5.56** Let $Y$ have the Cauchy distribution, $f_Y(y) = \frac{1}{1+y^2}, -\infty < y < \infty$.

(a) Show that $F_Y(y) = \tan^{-1}(y)$.
(b) Show how to simulate a Cauchy$(a, b)$ random variable starting from a uniform$(0, 1)$ random variable.

(See Exercise 2.12 for a related result.)

**5.57** Park *et al.* (1996) describe a method for generating correlated binary variables based on the follow scheme. Let $X_1, X_2, X_3$ be independent Poisson random variables with mean $\lambda_1, \lambda_2, \lambda_3$, respectively, and create the random variables

$$Y_1 = X_1 + X_3 \quad \text{and} \quad Y_2 = X_2 + X_3.$$

(a) Show that $\text{Cov}(Y_1, Y_2) = \lambda_3$.
(b) Define $Z_i = I(Y_i = 0)$ and $p_i = e^{-(\lambda_i + \lambda_3)}$. Show that $Z_i$ are Bernoulli$(p_i)$ with

$$\text{Corr}(Z_1, Z_2) = \frac{p_1 p_2 (e^{\lambda_3} - 1)}{\sqrt{p_1(1 - p_1)}\sqrt{p_2(1 - p_2)}}.$$

(c) Show that the correlation of $Z_1$ and $Z_2$ is not unrestricted in the range $[-1, 1]$, but

$$\text{Corr}(Z_1, Z_2) \leq \min\left\{\sqrt{\frac{p_2(1 - p_1)}{p_1(1 - p_2)}}, \sqrt{\frac{p_1(1 - p_2)}{p_2(1 - p_1)}}\right\}.$$

**5.58** Suppose that $U_1, U_2, \ldots, U_n$ are iid uniform$(0,1)$ random variables, and let $S_n = \sum_{i=1}^{n} U_i$. Define the random variable $N$ by

$$N = \min\{k : S_k > 1\}.$$

(a) Show that $P(S_k \leq t) = t^k/k!$ if $0 \leq t \leq 1$.

(b) Show that $P(N = n) = P(S_{n-1} < 1) - P(S_n < 1)$ and, surprisingly, that $E(N) = e$, the base of the natural logarithms.

(c) Use the result of part (b) to calculate the value of $e$ by simulation.

(d) How large should $n$ be so that you are 95% confident that you have the first four digits of $e$ correct?

(Russell 1991, who attributes the problem to Gnedenko 1978, describes such a simulation experiment.)

**5.59** Prove that the algorithm of Example 5.6.7 generates a beta$(a, b)$ random variable.

**5.60** Generalize the algorithm of Example 5.6.7 to apply to any bounded pdf; that is, for an arbitrary bounded pdf $f(x)$ on $[a, b]$, define $c = \max_{a \leq x \leq b} f(x)$. Let $X$ and $Y$ be independent, with $X \sim$ uniform$(a, b)$ and $Y \sim$ uniform$(0, c)$. Let $d$ be a number greater than $b$, and define a new random variable

$$W = \begin{cases} X & \text{if } Y < f(X) \\ d & \text{if } Y \geq f(X). \end{cases}$$

(a) Show that $P(W \leq w) = \int_a^w f(t)dt/[c(b-a)]$ for $a \leq w \leq b$.

(b) Using part (a), explain how a random variable with pdf $f(x)$ can be generated. (*Hint:* Use a geometric argument; a picture will help.)

**5.61** (a) Suppose it is desired to generate $Y \sim$ beta$(a, b)$, where $a$ and $b$ are not integers. Show that using $V \sim$ beta$([a], [b])$ will result in a finite value of $M = \sup_y f_Y(y)/f_V(y)$ .

(b) Suppose it is desired to generate $Y \sim$ gamma$(a, b)$, where $a$ and $b$ are not integers. Show that using $V \sim$ gamma$([a], b)$ will result in a finite value of $M = \sup_y f_Y(y)/f_V(y)$ .

(c) Show that, in each of parts (a) and (b), if $V$ had parameter $[a] + 1$, then $M$ would be infinite.

(d) In each of parts (a) and (b) find optimal values for the parameters of $V$ in the sense of minimizing $E(N)$ (see (5.6.12)).

(Recall that $[a] = $ greatest integer $\leq a$ .)

**5.62** Find the values of $M$ so that an Accept/Reject Algorithm can generate $Y \sim$ n$(0,1)$ using $U \sim$ uniform$(0,1)$ and

(a) $V \sim$ Cauchy.

(b) $V \sim$ double exponential.

(c) Compare the algorithms. Which one do you recommend?

**5.63** For generating $Y \sim$ n$(0,1)$ using an Accept/Reject Algorithm, we could generate $U \sim$ uniform, $V \sim$ exponential$(\lambda)$ , and attach a random sign to $V$ ($\pm$ each with equal probability). What value of $\lambda$ will optimize this algorithm?

**5.64** A technique similar to Accept/Reject is *importance sampling*, which is quite useful for calculating features of a distribution. Suppose that $X \sim f$, but the pdf $f$ is difficult to simulate from. Generate $Y_1, Y_2, \ldots, Y_m$, iid from $g$, and, for any function $h$, calculate $\frac{1}{m} \sum_{i=1}^{m} \frac{f(Y_i)}{g(Y_i)} h(Y_i)$. We assume that the supports of $f$ and $g$ are the same and $\text{Var} \, h(X) < \infty$.

(a) Show that $\text{E} \left( \frac{1}{m} \sum_{i=1}^{m} \frac{f(Y_i)}{g(Y_i)} h(Y_i) \right) = \text{E}h(X)$.

(b) Show that $\frac{1}{m} \sum_{i=1}^{m} \frac{f(Y_i)}{g(Y_i)} h(Y_i) \to \text{E}h(X)$ in probability.

(c) Although the estimator of part (a) has the correct expectation, in practice the estimator

$$\sum_{i=1}^{m} \left( \frac{f(Y_i)/g(Y_i)}{\sum_{j=1}^{m} f(Y_j)/g(Y_j)} \right) h(Y_i)$$

is preferred. Show that this estimator converges in probability to $\text{E}h(X)$. Moreover, show that if $h$ is constant, this estimator is superior to the one in part (a). (Casella and Robert 1996 further explore the properties of this estimator.)

**5.65** A variation of the importance sampling algorithm of Exercise 5.64 can actually produce an approximate sample from $f$. Again let $X \sim f$ and generate $Y_1, Y_2, \ldots, Y_m$, iid from $g$. Calculate $q_i = [f(Y_i)/g(Y_i)]/[\sum_{j=1}^{m} f(Y_j)/g(Y_j)]$. Then generate random variables $X^*$ from the discrete distribution on $Y_1, Y_2, \ldots, Y_m$, where $P(X^* = Y_k) = q_k$. Show that $X_1^*, X_2^*, \ldots, X_r^*$ is approximately a random sample from $f$.

(*Hint:* Show that $P(X^* \leq x) = \sum_{i=1}^{m} q_i I(Y_i \leq x)$, let $m \to \infty$, and use the WLLN in the numerator and denominator.)

This algorithm is called the *Sampling/Importance Resampling (SIR)* algorithm by Rubin (1988) and is referred to as the *weighted bootstrap* by Smith and Gelfand (1992).

**5.66** If $X_1, \ldots, X_n$ are iid $\text{n}(\mu, \sigma^2)$, we have seen that the distribution of the sample mean $\bar{X}$ is $\text{n}(\mu, \sigma^2/n)$. If we are interested in using a more robust estimator of location, such as the median (5.4.1), it becomes a more difficult task to derive its distribution.

(a) Show that $M$ is the median of the $X_i$s if and only if $(M - \mu)/\sigma$ is the median of $(X_i - \mu)/\sigma$. Thus, we only need consider the distribution of the median from a $\text{n}(0, 1)$ sample.

(b) For a sample of size $n = 15$ from a $\text{n}(0, 1)$, simulate the distribution of the median $M$.

(c) Compare the distribution in part (b) to the asymptotic distribution of the median $\sqrt{n}(M - \mu) \sim \text{n}[0, 1/4f^2(0)]$, where $f$ is the pdf. Is $n = 15$ large enough for the asymptotics to be valid?

**5.67** In many instances the Metropolis Algorithm is the algorithm of choice because either (i) there are no obvious candidate densities that satisfy the Accept/Reject supremum condition, or (ii) the supremum condition is difficult to verify, or (iii) laziness leads us to substitute computing power for brain power.

For each of the following situations show how to implement the Metropolis Algorithm to generate a sample of size 100 from the specified distribution.

(a) $X \sim \frac{1}{\sigma} f[(x - \mu)/\sigma]$, $f = $ Student's $t$ with $\nu$ degrees of freedom, $\nu$, $\mu$, and $\sigma$ known

(b) $X \sim \text{lognormal}(\mu, \sigma^2)$, $\mu$, $\sigma^2$ known

(c) $X \sim \text{Weibull}(\alpha, \beta)$, $\alpha$, $\beta$ known

**5.68** If we use the Metropolis Algorithm rather than the Accept/Reject Algorithm we are freed from verifying the supremum condition of the Accept/Reject Algorithm. Of course, we give up the property of getting random variables with exactly the distribution we want and must settle for an approximation.

(a) Show how to use the Metropolis Algorithm to generate a random variable with an approximate Student's $t$ distribution with $\nu$ degrees of freedom, starting from $n(0, 1)$ random variables.

(b) Show how to use the Accept/Reject Algorithm to generate a random variable with a Student's $t$ distribution with $\nu$ degrees of freedom, starting from Cauchy random variables.

(c) Show how to use transformations to generate directly a random variable with a Student's $t$ distribution with $\nu$ degrees of freedom.

(d) For $\nu = 2, 10, 25$ compare the methods by generating a sample of size 100. Which method do you prefer? Why?

(Mengersen and Tweedie 1996 show that the convergence of the Metropolis Algorithm is much faster if the supremum condition is satisfied, that is, if $\sup f/g \leq M < \infty$, where $f$ is the target density and $g$ is the candidate density.)

**5.69** Show that the pdf $f_Y(y)$ is a *stable point* of the Metropolis Algorithm. That is, if $Z_i \sim f_Y(y)$, then $Z_{i+1} \sim f_Y(y)$.

## 5.8 Miscellanea

### 5.8.1 More on the Central Limit Theorem

For the case of a sequence of iid random variables, necessary and sufficient conditions for convergence to normality are known, with probably the most famous result due to Lindeberg and Feller. The following special case is due to Lévy. Let $X_1, X_2, \ldots$ be an iid sequence with $EX_i = \mu < \infty$, and let $V_n = \sum_{i=1}^{n} X_i$. The sequence $V_n$ will converge to a $n(0, 1)$ random variable (when suitably normalized) if and only if

$$\lim_{t \to \infty} \frac{t^2 P(|X_1 - \mu| > t)}{E\left((X_1 - \mu)^2 I_{[-t,t]}(X_1 - \mu)\right)} = 0.$$

Note that the condition is a variance condition. While it does not quite require that the variances be finite, it does require that they be "almost" finite. This is an important point in the convergence to normality—normality comes from summing up small disturbances.

Other types of central limit theorems abound—in particular, ones aimed at relaxing the independence assumption. While this assumption cannot be done away with, it can be made less restrictive (see Billingsley 1995, Section 27, or Resnick 1999, Chapter 8).

### 5.8.2 The Bias of $S^2$

Most of the calculations that we have done in this chapter have assumed that the observations are independent, and calculations of some expectations have relied on this fact. David (1985) pointed out that, if the observations are dependent, then $S^2$ may be a biased estimate of $\sigma^2$. That is, it may not happen that $ES^2 = \sigma^2$. However, the range of the possible bias is easily calculated. If $X_1, \ldots, X_n$ are random variables (not necessarily independent) with mean $\mu$ and variance $\sigma^2$, then

$$(n-1)\mathrm{E}S^2 = \mathrm{E}\left(\sum_{i=1}^{n}(X_i - \mu)^2 - n(\bar{X} - \mu)^2\right) = n\sigma^2 - n\operatorname{Var}\bar{X}.$$

Var $\bar{X}$ can vary, according to the amount and type of dependence, from 0 (if all of the variables are constant) to $\sigma^2$ (if all of the variables are copies of $X_1$). Substituting these values in the above equation, we get the range of $\mathrm{E}S^2$ under dependence as

$$0 \le \mathrm{E}S^2 \le \frac{n}{n-1}\sigma^2.$$

### 5.8.3 Chebychev's Inequality Revisited

In Section 3.6 we looked at Chebychev's Inequality (see also Miscellanea 3.8.2), and in Example 3.6.2 we saw a particularly useful form. That form still requires knowledge of the mean and variance of a random variable, and in some cases we might be interested in bounds using estimated values for the mean and variance.

If $X_1, \ldots, X_n$ is a random sample from a population with mean $\mu$ and variance $\sigma^2$, Chebychev's Inequality says

$$P(|X - \mu| \ge k\sigma) \le \frac{1}{k^2}.$$

Saw *et al.* (1984) showed that if we substitute $\bar{X}$ for $\mu$ and $S^2$ for $\sigma^2$, we obtain

$$P\big(|X - \bar{X}| \ge kS\big) \le \frac{1}{n+1}g\left(\frac{n(n+1)k^2}{n-1+(n+1)k^2}\right),$$

where

$$g(t) = \begin{cases} \nu & \text{if } \nu \text{ is even} \\ \nu & \text{if } \nu \text{ is odd and } t < a \\ \nu - 1 & \text{if } \nu \text{ is odd and } t > a \end{cases}$$

and

$$\nu = \text{largest integer} < \frac{n+1}{t}, \quad a = \frac{(n+1)(n+1-\nu)}{1+\nu(n+1-\nu)}.$$

### 5.8.4 More on the Strong Law

As mentioned, the Strong Law of Large Numbers, Theorem 5.5.9, can be proved under the less restrictive condition that the random variables have only a finite mean (see, for example, Resnick 1999, Chapter 7, or Billingsley 1995, Section 22). However, under the assumptions of the existence of an mgf, Koopmans (1993) has presented a proof that uses only calculus.

The type of convergence asserted in the SLLN is the convergence that we are most familiar with; it is *pointwise* convergence of the sequence of random variables $\bar{X}_n$

to their common mean $\mu$. As we saw in Example 5.5.8, this is stronger form of convergence than convergence in probability, the convergence of the weak law. The conclusion of the SLLN is that

$$P(\lim_{n \to \infty} |\bar{X}_n - \mu| < \epsilon) = 1;$$

that is, with probability 1, the limit of the sequence of $\{\bar{X}_n\}$ is $\mu$. Alternatively, the set where the sequence diverges has probability 0. For the sequence to diverge, there must exist $\delta > 0$ such that for every $n$ there is a $k > n$ with $|\bar{X}_k - \mu| > \delta$. The set of all $\bar{X}_k$ that satisfy this is a divergent sequence and is represented by the set

$$A_\delta = \cap_{n=1}^{\infty} \cup_{k=n}^{\infty} \{|\bar{X}_k - \mu| > \delta\}.$$

We can get an upper bound on $P(A_\delta)$ by dropping the intersection term, and then the probability of the set where the sequence $\{\bar{X}_n\}$ diverges is bounded above by

$$P(A_\delta) \le P(\cup_{k=n}^{\infty}\{|\bar{X}_k - \mu| > \delta\})$$

$$\le \sum_{k=n}^{\infty} P(\{|\bar{X}_k - \mu| > \delta\}) \quad \text{(Boole's Inequality, Theorem 1.2.11)}$$

$$\le 2 \sum_{k=n}^{\infty} c^k, \quad 0 < c < 1,$$

where Exercise 5.38(d) can be used to establish the last inequality. We then note that we are summing the geometric series, and it follows from (1.5.4) that

$$P(A_\delta) \le 2 \sum_{k=n}^{\infty} c^k = 2\frac{c^n}{1-c} \to 0 \text{ as } n \to \infty,$$

and, hence, the set where the sequence $\{\bar{X}_n\}$ diverges has probability 0 and the SLLN is established.

### 5.8.5 Markov Chain Monte Carlo

Methods that are collectively known as Markov Chain Monte Carlo (MCMC) methods are used in the generation of random variables and have proved extremely useful for doing complicated calculations, most notably, calculations involving integrations and maximizations. The Metropolis Algorithm (see Section 5.6) is an example of an MCMC method.

As the name suggests, these methods are based on Markov chains, a probabilistic structure that we haven't explored (see Chung 1974 or Ross 1988 for an introduction). The sequence of random variables $X_1, X_2, \ldots$ is a *Markov chain* if

$$P(X_{k+1} \in A | X_1, \ldots, X_k) = P(X_{k+1} \in A | X_k);$$

that is, the distribution of the present random variable depends, at most, on the immediate past random variable. Note that this is a generalization of independence.

The *Ergodic Theorem*, which is a generalization of the Law of Large Numbers, says that if the Markov chain $X_1, X_2, \ldots$ satisfies some regularity conditions (which are often satisfied in statistical problems), then

$$\frac{1}{n} \sum_{i=1}^{n} h(X_i) \to \mathrm{E}h(X) \text{ as } n \to \infty,$$

provided the expectation exists. Thus, the calculations of Section 5.6 can be extended to Markov chains and MCMC methods.

To fully understand MCMC methods it is really necessary to understand more about Markov chains, which we will not do here. There is already a vast literature on MCMC methods, encompassing both theory and applications. Tanner (1996) provides a good introduction to computational methods in statistics, as does Robert (1994, Chapter 9), who provides a more theoretical treatment with a Bayesian flavor. An easier introduction to this topic via the Gibbs sampler (a particular MCMC method) is given by Casella and George (1992). The Gibbs sampler is, perhaps, the MCMC method that is still the most widely used and is responsible for the popularity of this method (due to the seminal work of Gelfand and Smith 1990 expanding on Geman and Geman 1984). The list of references involving MCMC methods is prohibitively long. Some other introductions to this literature are through the papers of Gelman and Rubin (1992), Geyer and Thompson (1992), and Smith and Roberts (1993), with a particularly elegant theoretical introduction given by Tierney (1994). Robert and Casella (1999) is a textbook-length treatment of this field.

Chapter 6

# Principles of Data Reduction

"...we are suffering from a plethora of surmise, conjecture and hypothesis. The difficulty is to detach the framework of fact – of absolute undeniable fact – from the embellishments of theorists and reporters."

**Sherlock Holmes**
*Silver Blaze*

## 6.1 Introduction

An experimenter uses the information in a sample $X_1, \ldots, X_n$ to make inferences about an unknown parameter $\theta$. If the sample size $n$ is large, then the observed sample $x_1, \ldots, x_n$ is a long list of numbers that may be hard to interpret. An experimenter might wish to summarize the information in a sample by determining a few key features of the sample values. This is usually done by computing statistics, functions of the sample. For example, the sample mean, the sample variance, the largest observation, and the smallest observation are four statistics that might be used to summarize some key features of the sample. Recall that we use boldface letters to denote multiple variates, so $\mathbf{X}$ denotes the random variables $X_1, \ldots, X_n$ and $\mathbf{x}$ denotes the sample $x_1, \ldots, x_n$.

Any statistic, $T(\mathbf{X})$, defines a form of data reduction or data summary. An experimenter who uses only the observed value of the statistic, $T(\mathbf{x})$, rather than the entire observed sample, $\mathbf{x}$, will treat as equal two samples, $\mathbf{x}$ and $\mathbf{y}$, that satisfy $T(\mathbf{x}) = T(\mathbf{y})$ even though the actual sample values may be different in some ways.

Data reduction in terms of a particular statistic can be thought of as a partition of the sample space $\mathcal{X}$. Let $\mathcal{T} = \{t : t = T(\mathbf{x}) \text{ for some } \mathbf{x} \in \mathcal{X}\}$ be the image of $\mathcal{X}$ under $T(\mathbf{x})$. Then $T(\mathbf{x})$ partitions the sample space into sets $A_t, t \in \mathcal{T}$, defined by $A_t = \{\mathbf{x} : T(\mathbf{x}) = t\}$. The statistic summarizes the data in that, rather than reporting the entire sample $\mathbf{x}$, it reports only that $T(\mathbf{x}) = t$ or, equivalently, $\mathbf{x} \in A_t$. For example, if $T(\mathbf{x}) = x_1 + \cdots + x_n$, then $T(\mathbf{x})$ does not report the actual sample values but only the sum. There may be many different sample points that have the same sum. The advantages and consequences of this type of data reduction are the topics of this chapter.

We study three principles of data reduction. We are interested in methods of data reduction that do not discard important information about the unknown parameter $\theta$ and methods that successfully discard information that is irrelevant as far as gaining knowledge about $\theta$ is concerned. The Sufficiency Principle promotes a method of data reduction that does not discard information about $\theta$ while achieving some summarization of the data. The Likelihood Principle describes a function of the parameter, determined by the observed sample, that contains all the information about $\theta$ that is available from the sample. The Equivariance Principle prescribes yet another method of data reduction that still preserves some important features of the model.

## 6.2 The Sufficiency Principle

A *sufficient statistic* for a parameter $\theta$ is a statistic that, in a certain sense, captures all the information about $\theta$ contained in the sample. Any additional information in the sample, besides the value of the sufficient statistic, does not contain any more information about $\theta$. These considerations lead to the data reduction technique known as the Sufficiency Principle.

*SUFFICIENCY PRINCIPLE*: If $T(\mathbf{X})$ is a sufficient statistic for $\theta$, then any inference about $\theta$ should depend on the sample $\mathbf{X}$ only through the value $T(\mathbf{X})$. That is, if $\mathbf{x}$ and $\mathbf{y}$ are two sample points such that $T(\mathbf{x}) = T(\mathbf{y})$, then the inference about $\theta$ should be the same whether $\mathbf{X} = \mathbf{x}$ or $\mathbf{X} = \mathbf{y}$ is observed.

In this section we investigate some aspects of sufficient statistics and the Sufficiency Principle.

### 6.2.1 Sufficient Statistics

A sufficient statistic is formally defined in the following way.

**Definition 6.2.1**    A statistic $T(\mathbf{X})$ is a *sufficient statistic for $\theta$* if the conditional distribution of the sample $\mathbf{X}$ given the value of $T(\mathbf{X})$ does not depend on $\theta$.

If $T(\mathbf{X})$ has a continuous distribution, then $P_\theta(T(\mathbf{X}) = t) = 0$ for all values of $t$. A more sophisticated notion of conditional probability than that introduced in Chapter 1 is needed to fully understand Definition 6.2.1 in this case. A discussion of this can be found in more advanced texts such as Lehmann (1986). We will do our calculations in the discrete case and will point out analogous results that are true in the continuous case.

To understand Definition 6.2.1, let $t$ be a possible value of $T(\mathbf{X})$, that is, a value such that $P_\theta(T(\mathbf{X}) = t) > 0$. We wish to consider the conditional probability $P_\theta(\mathbf{X} = \mathbf{x}|T(\mathbf{X}) = t)$. If $\mathbf{x}$ is a sample point such that $T(\mathbf{x}) \neq t$, then clearly $P_\theta(\mathbf{X} = \mathbf{x}|T(\mathbf{X}) = t) = 0$. Thus, we are interested in $P(\mathbf{X} = \mathbf{x}|T(\mathbf{X}) = T(\mathbf{x}))$. By the definition, if $T(\mathbf{X})$ is a sufficient statistic, this conditional probability is the same for all values of $\theta$ so we have omitted the subscript.

A sufficient statistic captures all the information about $\theta$ in this sense. Consider Experimenter 1, who observes $\mathbf{X} = \mathbf{x}$ and, of course, can compute $T(\mathbf{X}) = T(\mathbf{x})$. To make an inference about $\theta$ he can use the information that $\mathbf{X} = \mathbf{x}$ and $T(\mathbf{X}) = T(\mathbf{x})$. Now consider Experimenter 2, who is not told the value of $\mathbf{X}$ but only that $T(\mathbf{X}) = T(\mathbf{x})$. Experimenter 2 knows $P(\mathbf{X} = \mathbf{y}|T(\mathbf{X}) = T(\mathbf{x}))$, a probability distribution on

$A_{T(\mathbf{x})} = \{\mathbf{y}: T(\mathbf{y}) = T(\mathbf{x})\}$, because this can be computed from the model without knowledge of the true value of $\theta$. Thus, Experimenter 2 can use this distribution and a randomization device, such as a random number table, to generate an observation $\mathbf{Y}$ satisfying $P(\mathbf{Y} = \mathbf{y}|T(\mathbf{X}) = T(\mathbf{x})) = P(\mathbf{X} = \mathbf{y}|T(\mathbf{X}) = T(\mathbf{x}))$. It turns out that, for each value of $\theta$, $\mathbf{X}$ and $\mathbf{Y}$ have the same unconditional probability distribution, as we shall see below. So Experimenter 1, who knows $\mathbf{X}$, and Experimenter 2, who knows $\mathbf{Y}$, have equivalent information about $\theta$. But surely the use of the random number table to generate $\mathbf{Y}$ has not added to Experimenter 2's knowledge of $\theta$. All his knowledge about $\theta$ is contained in the knowledge that $T(\mathbf{X}) = T(\mathbf{x})$. So Experimenter 2, who knows only $T(\mathbf{X}) = T(\mathbf{x})$, has just as much information about $\theta$ as does Experimenter 1, who knows the entire sample $\mathbf{X} = \mathbf{x}$.

To complete the above argument, we need to show that $\mathbf{X}$ and $\mathbf{Y}$ have the same unconditional distribution, that is, $P_\theta(\mathbf{X} = \mathbf{x}) = P_\theta(\mathbf{Y} = \mathbf{x})$ for all $\mathbf{x}$ and $\theta$. Note that the events $\{\mathbf{X} = \mathbf{x}\}$ and $\{\mathbf{Y} = \mathbf{x}\}$ are both subsets of the event $\{T(\mathbf{X}) = T(\mathbf{x})\}$. Also recall that

$$P(\mathbf{X} = \mathbf{x}|T(\mathbf{X}) = T(\mathbf{x})) = P(\mathbf{Y} = \mathbf{x}|T(\mathbf{X}) = T(\mathbf{x}))$$

and these conditional probabilities do not depend on $\theta$. Thus we have

$$
\begin{aligned}
P_\theta(\mathbf{X} = \mathbf{x}) \\
&= P_\theta(\mathbf{X} = \mathbf{x} \text{ and } T(\mathbf{X}) = T(\mathbf{x})) \\
&= P(\mathbf{X} = \mathbf{x}|T(\mathbf{X}) = T(\mathbf{x}))P_\theta(T(\mathbf{X}) = T(\mathbf{x})) \quad \left(\begin{array}{c}\text{definition of} \\ \text{conditional probability}\end{array}\right) \\
&= P(\mathbf{Y} = \mathbf{x}|T(\mathbf{X}) = T(\mathbf{x}))P_\theta(T(\mathbf{X}) = T(\mathbf{x})) \\
&= P_\theta(\mathbf{Y} = \mathbf{x} \text{ and } T(\mathbf{X}) = T(\mathbf{x})) \\
&= P_\theta(\mathbf{Y} = \mathbf{x}).
\end{aligned}
$$

To use Definition 6.2.1 to verify that a statistic $T(\mathbf{X})$ is a sufficient statistic for $\theta$, we must verify that for any fixed values of $\mathbf{x}$ and $t$, the conditional probability $P_\theta(\mathbf{X} = \mathbf{x}|T(\mathbf{X}) = t)$ is the same for all values of $\theta$. Now, this probability is 0 for all values of $\theta$ if $T(\mathbf{x}) \neq t$. So, we must verify only that $P_\theta(\mathbf{X} = \mathbf{x}|T(\mathbf{X}) = T(\mathbf{x}))$ does not depend on $\theta$. But since $\{\mathbf{X} = \mathbf{x}\}$ is a subset of $\{T(\mathbf{X}) = T(\mathbf{x})\}$,

$$
\begin{aligned}
P_\theta(\mathbf{X} = \mathbf{x}|T(\mathbf{X}) = T(\mathbf{x})) &= \frac{P_\theta(\mathbf{X} = \mathbf{x} \text{ and } T(\mathbf{X}) = T(\mathbf{x}))}{P_\theta(T(\mathbf{X}) = T(\mathbf{x}))} \\
&= \frac{P_\theta(\mathbf{X} = \mathbf{x})}{P_\theta(T(\mathbf{X}) = T(\mathbf{x}))} \\
&= \frac{p(\mathbf{x}|\theta)}{q(T(\mathbf{x})|\theta)},
\end{aligned}
$$

where $p(\mathbf{x}|\theta)$ is the joint pmf of the sample $\mathbf{X}$ and $q(t|\theta)$ is the pmf of $T(\mathbf{X})$. Thus, $T(\mathbf{X})$ is a sufficient statistic for $\theta$ if and only if, for every $\mathbf{x}$, the above ratio of pmfs is constant as a function of $\theta$. If $\mathbf{X}$ and $T(\mathbf{X})$ have continuous distributions, then the above conditional probabilities cannot be interpreted in the sense of Chapter 1. But it is still appropriate to use the above criterion to determine if $T(\mathbf{X})$ is a sufficient statistic for $\theta$.

**Theorem 6.2.2**    *If $p(\mathbf{x}|\theta)$ is the joint pdf or pmf of $\mathbf{X}$ and $q(t|\theta)$ is the pdf or pmf of $T(\mathbf{X})$, then $T(\mathbf{X})$ is a sufficient statistic for $\theta$ if, for every $\mathbf{x}$ in the sample space, the ratio $p(\mathbf{x}|\theta)/q(T(\mathbf{x})|\theta)$ is constant as a function of $\theta$.*

We now use Theorem 6.2.2 to verify that certain common statistics are sufficient statistics.

**Example 6.2.3 (Binomial sufficient statistic)**    Let $X_1, \ldots, X_n$ be iid Bernoulli random variables with parameter $\theta, 0 < \theta < 1$. We will show that $T(\mathbf{X}) = X_1 + \cdots + X_n$ is a sufficient statistic for $\theta$. Note that $T(\mathbf{X})$ counts the number of $X_i$s that equal 1, so $T(\mathbf{X})$ has a binomial$(n, \theta)$ distribution. The ratio of pmfs is thus

$$
\begin{aligned}
\frac{p(\mathbf{x}|\theta)}{q(T(\mathbf{x})|\theta)} &= \frac{\Pi \theta^{x_i}(1-\theta)^{1-x_i}}{\binom{n}{t}\theta^t(1-\theta)^{n-t}} && \text{(define } t = \Sigma x_i) \\
&= \frac{\theta^{\Sigma x_i}(1-\theta)^{\Sigma(1-x_i)}}{\binom{n}{t}\theta^t(1-\theta)^{n-t}} && (\Pi \theta^{x_i} = \theta^{\Sigma x_i}) \\
&= \frac{\theta^t(1-\theta)^{n-t}}{\binom{n}{t}\theta^t(1-\theta)^{n-t}} \\
&= \frac{1}{\binom{n}{t}} \\
&= \frac{1}{\binom{n}{\Sigma x_i}}.
\end{aligned}
$$

Since this ratio does not depend on $\theta$, by Theorem 6.2.2, $T(\mathbf{X})$ is a sufficient statistic for $\theta$. The interpretation is this: The total number of 1s in this Bernoulli sample contains all the information about $\theta$ that is in the data. Other features of the data, such as the exact value of $X_3$, contain no additional information.    ‖

**Example 6.2.4 (Normal sufficient statistic)**    Let $X_1, \ldots, X_n$ be iid n$(\mu, \sigma^2)$, where $\sigma^2$ is known. We wish to show that the sample mean, $T(\mathbf{X}) = \bar{X} = (X_1 + \cdots + X_n)/n$, is a sufficient statistic for $\mu$. The joint pdf of the sample $\mathbf{X}$ is

$$
\begin{aligned}
f(\mathbf{x}|\mu) &= \prod_{i=1}^{n}(2\pi\sigma^2)^{-1/2}\exp\left(-(x_i-\mu)^2/(2\sigma^2)\right) \\
&= (2\pi\sigma^2)^{-n/2}\exp\left(-\sum_{i=1}^{n}(x_i-\mu)^2/(2\sigma^2)\right)
\end{aligned}
$$

$$= (2\pi\sigma^2)^{-n/2} \exp\left(-\sum_{i=1}^{n}(x_i - \bar{x} + \bar{x} - \mu)^2/(2\sigma^2)\right) \quad \text{(add and subtract } \bar{x})$$

$$(6.2.1) \quad = (2\pi\sigma^2)^{-n/2} \exp\left(-\left(\sum_{i=1}^{n}(x_i - \bar{x})^2 + n(\bar{x} - \mu)^2\right)/(2\sigma^2)\right).$$

The last equality is true because the cross-product term $\sum_{i=1}^{n}(x_i - \bar{x})(\bar{x} - \mu)$ may be rewritten as $(\bar{x} - \mu)\sum_{i=1}^{n}(x_i - \bar{x})$, and $\sum_{i=1}^{n}(x_i - \bar{x}) = 0$. Recall that the sample mean $\bar{X}$ has a $n(\mu, \sigma^2/n)$ distribution. Thus, the ratio of pdfs is

$$\frac{f(\mathbf{x}|\theta)}{q(T(\mathbf{x})|\theta)} = \frac{(2\pi\sigma^2)^{-n/2} \exp\left(-\left(\sum_{i=1}^{n}(x_i - \bar{x})^2 + n(\bar{x} - \mu)^2\right)/(2\sigma^2)\right)}{(2\pi\sigma^2/n)^{-1/2} \exp(-n(\bar{x} - \mu)^2/(2\sigma^2))}$$

$$= n^{-1/2}(2\pi\sigma^2)^{-(n-1)/2} \exp\left(-\sum_{i=1}^{n}(x_i - \bar{x})^2/(2\sigma^2)\right),$$

which does not depend on $\mu$. By Theorem 6.2.2, the sample mean is a sufficient statistic for $\mu$. ‖

In the next example we look at situations in which a substantial reduction of the sample is not possible.

**Example 6.2.5 (Sufficient order statistics)**   Let $X_1, \ldots, X_n$ be iid from a pdf $f$, where we are unable to specify any more information about the pdf (as is the case in *nonparametric* estimation). It then follows that the sample density is given by

$$(6.2.2) \qquad\qquad f(\mathbf{x}) = \prod_{i=1}^{n} f(x_i) = \prod_{i=1}^{n} f(x_{(i)}),$$

where $x_{(1)} \leq x_{(2)} \leq \cdots \leq x_{(n)}$ are the order statistics. By Theorem 6.2.2, we can show that the order statistics are a sufficient statistic. Of course, this is not much of a reduction, but we shouldn't expect more with so little information about the density $f$.

However, even if we do specify more about the density, we still may not be able to get much of a sufficiency reduction. For example, suppose that $f$ is the Cauchy pdf $f(x|\theta) = \frac{1}{\pi(x-\theta)^2}$ or the logistic pdf $f(x|\theta) = \frac{e^{-(x-\theta)}}{\left(1+e^{-(x-\theta)}\right)^2}$. We then have the same reduction as in (6.2.2), and no more. So reduction to the order statistics is the most we can get in these families (see Exercises 6.8 and 6.9 for more examples).

It turns out that outside of the exponential family of distributions, it is rare to have a sufficient statistic of smaller dimension than the size of the sample, so in many cases it will turn out that the order statistics are the best that we can do. (See Lehmann and Casella 1998, Section 1.6, for further details.) ‖

It may be unwieldy to use the definition of a sufficient statistic to find a sufficient statistic for a particular model. To use the definition, we must guess a statistic $T(\mathbf{X})$ to be sufficient, find the pmf or pdf of $T(\mathbf{X})$, and check that the ratio of pdfs or pmfs does not depend on $\theta$. The first step requires a good deal of intuition and the second sometimes requires some tedious analysis. Fortunately, the next theorem, due to Halmos and Savage (1949), allows us to find a sufficient statistic by simple inspection of the pdf or pmf of the sample.[1]

**Theorem 6.2.6 (Factorization Theorem)**  *Let $f(\mathbf{x}|\theta)$ denote the joint pdf or pmf of a sample $\mathbf{X}$. A statistic $T(\mathbf{X})$ is a sufficient statistic for $\theta$ if and only if there exist functions $g(t|\theta)$ and $h(\mathbf{x})$ such that, for all sample points $\mathbf{x}$ and all parameter points $\theta$,*

(6.2.3)                          $$f(\mathbf{x}|\theta) = g(T(\mathbf{x})|\theta)h(\mathbf{x}).$$

**Proof:** We give the proof only for discrete distributions.

Suppose $T(\mathbf{X})$ is a sufficient statistic. Choose $g(t|\theta) = P_\theta(T(\mathbf{X}) = t)$ and $h(\mathbf{x}) = P(\mathbf{X} = \mathbf{x}|T(\mathbf{X}) = T(\mathbf{x}))$. Because $T(\mathbf{X})$ is sufficient, the conditional probability defining $h(\mathbf{x})$ does not depend on $\theta$. Thus this choice of $h(\mathbf{x})$ and $g(t|\theta)$ is legitimate, and for this choice we have

$$f(\mathbf{x}|\theta) = P_\theta(\mathbf{X} = \mathbf{x})$$
$$= P_\theta(\mathbf{X} = \mathbf{x} \text{ and } T(\mathbf{X}) = T(\mathbf{x}))$$
$$= P_\theta(T(\mathbf{X}) = T(\mathbf{x}))P(\mathbf{X} = \mathbf{x}|T(\mathbf{X}) = T(\mathbf{x}))  \qquad \text{(sufficiency)}$$
$$= g(T(\mathbf{x})|\theta)h(\mathbf{x}).$$

So factorization (6.2.3) has been exhibited. We also see from the last two lines above that

$$P_\theta(T(\mathbf{X}) = T(\mathbf{x})) = g(T(\mathbf{x})|\theta),$$

so $g(T(\mathbf{x})|\theta)$ is the pmf of $T(\mathbf{X})$.

Now assume the factorization (6.2.3) exists. Let $q(t|\theta)$ be the pmf of $T(\mathbf{X})$. To show that $T(\mathbf{X})$ is sufficient we examine the ratio $f(\mathbf{x}|\theta)/q(T(\mathbf{x})|\theta)$. Define $A_{T(\mathbf{x})} = \{\mathbf{y}: T(\mathbf{y}) = T(\mathbf{x})\}$. Then

$$\frac{f(\mathbf{x}|\theta)}{q(T(\mathbf{x})|\theta)} = \frac{g(T(\mathbf{x})|\theta)h(\mathbf{x})}{q(T(\mathbf{x})|\theta)} \qquad \text{(since (6.2.3) is satisfied)}$$

$$= \frac{g(T(\mathbf{x})|\theta)h(\mathbf{x})}{\Sigma_{A_{T(\mathbf{x})}} g(T(\mathbf{y})|\theta)h(\mathbf{y})} \qquad \text{(definition of the pmf of } T)$$

$$= \frac{g(T(\mathbf{x})|\theta)h(\mathbf{x})}{g(T(\mathbf{x})|\theta)\Sigma_{A_{T(\mathbf{x})}} h(\mathbf{y})} \qquad \text{(since } T \text{ is constant on } A_{T(\mathbf{x})})$$

$$= \frac{h(\mathbf{x})}{\Sigma_{A_{T(\mathbf{x})}} h(\mathbf{y})}.$$

---

[1] Although, according to Halmos and Savage, their theorem "may be recast in a form more akin in spirit to previous investigations of the concept of sufficiency." The investigations are those of Neyman (1935). (This was pointed out by Prof. J. Beder, University of Wisconsin, Milwaukee.)

Since the ratio does not depend on $\theta$, by Theorem 6.2.2, $T(\mathbf{X})$ is a sufficient statistic for $\theta$. $\qquad\square$

To use the Factorization Theorem to find a sufficient statistic, we factor the joint pdf of the sample into two parts, with one part not depending on $\theta$. The part that does not depend on $\theta$ constitutes the $h(\mathbf{x})$ function. The other part, the one that depends on $\theta$, usually depends on the sample $\mathbf{x}$ only through some function $T(\mathbf{x})$ and this function is a sufficient statistic for $\theta$. This is illustrated in the following example.

**Example 6.2.7 (Continuation of Example 6.2.4)**    For the normal model described earlier, we saw that the pdf could be factored as

$$(6.2.4) \qquad f(\mathbf{x}|\mu) = (2\pi\sigma^2)^{-n/2} \exp\left(-\sum_{i=1}^{n}(x_i - \bar{x})^2/(2\sigma^2)\right) \exp(-n(\bar{x}-\mu)^2/(2\sigma^2)).$$

We can define

$$h(\mathbf{x}) = (2\pi\sigma^2)^{-n/2} \exp\left(-\sum_{i=1}^{n}(x_i-\bar{x})^2/(2\sigma^2)\right),$$

which does not depend on the unknown parameter $\mu$. The factor in (6.2.4) that contains $\mu$ depends on the sample $\mathbf{x}$ only through the function $T(\mathbf{x}) = \bar{x}$, the sample mean. So we have

$$g(t|\mu) = \exp\left(-n(t-\mu)^2/(2\sigma^2)\right)$$

and note that

$$f(\mathbf{x}|\mu) = h(\mathbf{x})g(T(\mathbf{x})|\mu).$$

Thus, by the Factorization Theorem, $T(\mathbf{X}) = \bar{X}$ is a sufficient statistic for $\mu$. $\qquad\|$

The Factorization Theorem requires that the equality $f(\mathbf{x}|\theta) = g(T(\mathbf{x})|\theta)h(\mathbf{x})$ hold for all $\mathbf{x}$ and $\theta$. If the set of $\mathbf{x}$ on which $f(\mathbf{x}|\theta)$ is positive depends on $\theta$, care must be taken in the definition of $h$ and $g$ to ensure that the product is 0 where $f$ is 0. Of course, correct definition of $h$ and $g$ makes the sufficient statistic evident, as the next example illustrates.

**Example 6.2.8 (Uniform sufficient statistic)**    Let $X_1, \ldots, X_n$ be iid observations from the discrete uniform distribution on $1, \ldots, \theta$. That is, the unknown parameter, $\theta$, is a positive integer and the pmf of $X_i$ is

$$f(x|\theta) = \begin{cases} \frac{1}{\theta} & x = 1, 2, \ldots, \theta \\ 0 & \text{otherwise.} \end{cases}$$

Thus the joint pmf of $X_1, \ldots, X_n$ is

$$f(\mathbf{x}|\theta) = \begin{cases} \theta^{-n} & x_i \in \{1, \ldots, \theta\} \text{ for } i = 1, \ldots, n \\ 0 & \text{otherwise.} \end{cases}$$

The restriction "$x_i \in \{1, \ldots, \theta\}$ for $i = 1, \ldots, n$" can be re-expressed as "$x_i \in \{1, 2, \ldots\}$ for $i = 1, \ldots, n$ (note that there is no $\theta$ in this restriction) and $\max_i x_i \leq \theta$." If we define $T(\mathbf{x}) = \max_i x_i$,

$$h(x) = \begin{cases} 1 & x_i \in \{1, 2, \ldots\} \text{ for } i = 1, \ldots, n \\ 0 & \text{otherwise,} \end{cases}$$

and

$$g(t|\theta) = \begin{cases} \theta^{-n} & t \leq \theta \\ 0 & \text{otherwise,} \end{cases}$$

it is easily verified that $f(\mathbf{x}|\theta) = g(T(\mathbf{x})|\theta)h(\mathbf{x})$ for all $\mathbf{x}$ and $\theta$. Thus, the largest order statistic, $T(\mathbf{X}) = \max_i X_i$, is a sufficient statistic in this problem.

This type of analysis can sometimes be carried out more clearly and concisely using indicator functions. Recall that $I_A(x)$ is the indicator function of the set $A$; that is, it is equal to 1 if $x \in A$ and equal to 0 otherwise. Let $\mathcal{N} = \{1, 2, \ldots\}$ be the set of positive integers and let $\mathcal{N}_\theta = \{1, 2, \ldots, \theta\}$. Then the joint pmf of $X_1, \ldots, X_n$ is

$$f(\mathbf{x}|\theta) = \prod_{i=1}^n \theta^{-1} I_{\mathcal{N}_\theta}(x_i) = \theta^{-n} \prod_{i=1}^n I_{\mathcal{N}_\theta}(x_i).$$

Defining $T(\mathbf{x}) = \max_i x_i$, we see that

$$\prod_{i=1}^n I_{\mathcal{N}_\theta}(x_i) = \left(\prod_{i=1}^n I_{\mathcal{N}}(x_i)\right) I_{\mathcal{N}_\theta}(T(\mathbf{x})).$$

Thus we have the factorization

$$f(\mathbf{x}|\theta) = \theta^{-n} I_{\mathcal{N}_\theta}(T(\mathbf{x})) \left(\prod_{i=1}^n I_{\mathcal{N}}(x_i)\right).$$

The first factor depends on $x_1, \ldots, x_n$ only through the value of $T(\mathbf{x}) = \max_i x_i$, and the second factor does not depend on $\theta$. By the Factorization Theorem, $T(\mathbf{X}) = \max_i X_i$ is a sufficient statistic for $\theta$. ‖

In all the previous examples, the sufficient statistic is a real-valued function of the sample. All the information about $\theta$ in the sample $\mathbf{x}$ is summarized in the single number $T(\mathbf{x})$. Sometimes, the information cannot be summarized in one number and several numbers are required instead. In such cases, a sufficient statistic is a vector, say $T(\mathbf{X}) = (T_1(\mathbf{X}), \ldots, T_r(\mathbf{X}))$. This situation often occurs when the parameter is also a vector, say $\boldsymbol{\theta} = (\theta_1, \ldots, \theta_s)$, and it is usually the case that the sufficient statistic and the parameter vectors are of equal length, that is, $r = s$. Different combinations of lengths are possible, however, as the exercises and Examples 6.2.15, 6.2.18, and 6.2.20 illustrate. The Factorization Theorem may be used to find a vector-valued sufficient statistic, as in Example 6.2.9.

**Example 6.2.9 (Normal sufficient statistic, both parameters unknown)**
Again assume that $X_1, \ldots, X_n$ are iid $n(\mu, \sigma^2)$ but, unlike Example 6.2.4, assume that both $\mu$ and $\sigma^2$ are unknown so the parameter vector is $\boldsymbol{\theta} = (\mu, \sigma^2)$. Now when we use the Factorization Theorem, any part of the joint pdf that depends on either $\mu$ or $\sigma^2$ must be included in the $g$ function. From (6.2.1) it is clear that the pdf depends on the sample $\mathbf{x}$ only through the two values $T_1(\mathbf{x}) = \bar{x}$ and $T_2(\mathbf{x}) = s^2 = \sum_{i=1}^{n}(x_i - \bar{x})^2/(n-1)$. Thus we can define $h(\mathbf{x}) = 1$ and

$$g(\mathbf{t}|\theta) = g(t_1, t_2|\mu, \sigma^2)$$
$$= (2\pi\sigma^2)^{-n/2} \exp\left(-\left(n(t_1 - \mu)^2 + (n-1)t_2\right)/(2\sigma^2)\right).$$

Then it can be seen that

(6.2.5) $$f(\mathbf{x}|\mu, \sigma^2) = g(T_1(\mathbf{x}), T_2(\mathbf{x})|\mu, \sigma^2)h(\mathbf{x}).$$

Thus, by the Factorization Theorem, $T(\mathbf{X}) = (T_1(\mathbf{X}), T_2(\mathbf{X})) = (\bar{X}, S^2)$ is a sufficient statistic for $(\mu, \sigma^2)$ in this normal model. ‖

Example 6.2.9 demonstrates that, for the normal model, the common practice of summarizing a data set by reporting only the sample mean and variance is justified. The sufficient statistic $(\bar{X}, S^2)$ contains all the information about $(\mu, \sigma^2)$ that is available in the sample. The experimenter should remember, however, that the definition of a sufficient statistic is model-dependent. For another model, that is, another family of densities, the sample mean and variance may not be a sufficient statistic for the population mean and variance. The experimenter who calculates only $\bar{X}$ and $S^2$ and totally ignores the rest of the data would be placing strong faith in the normal model assumption.

It is easy to find a sufficient statistic for an exponential family of distributions using the Factorization Theorem. The proof of the following important result is left as Exercise 6.4.

**Theorem 6.2.10** *Let $X_1, \ldots, X_n$ be iid observations from a pdf or pmf $f(x|\boldsymbol{\theta})$ that belongs to an exponential family given by*

$$f(x|\boldsymbol{\theta}) = h(x)c(\boldsymbol{\theta}) \exp\left(\sum_{i=1}^{k} w_i(\boldsymbol{\theta})t_i(x)\right),$$

*where $\boldsymbol{\theta} = (\theta_1, \theta_2, \ldots, \theta_d)$, $d \le k$. Then*

$$T(\mathbf{X}) = \left(\sum_{j=1}^{n} t_1(X_j), \ldots, \sum_{j=1}^{n} t_k(X_j)\right)$$

*is a sufficient statistic for $\boldsymbol{\theta}$.*

### 6.2.2 Minimal Sufficient Statistics

In the preceding section we found one sufficient statistic for each model considered. In any problem there are, in fact, many sufficient statistics.

It is always true that the complete sample, $\mathbf{X}$, is a sufficient statistic. We can factor the pdf or pmf of $\mathbf{X}$ as $f(\mathbf{x}|\theta) = f(T(\mathbf{x})|\theta)h(\mathbf{x})$, where $T(\mathbf{x}) = \mathbf{x}$ and $h(\mathbf{x}) = 1$ for all $\mathbf{x}$. By the Factorization Theorem, $T(\mathbf{X}) = \mathbf{X}$ is a sufficient statistic.

Also, it follows that any one-to-one function of a sufficient statistic is a sufficient statistic. Suppose $T(\mathbf{X})$ is a sufficient statistic and define $T^*(\mathbf{x}) = r(T(\mathbf{x}))$ for all $\mathbf{x}$, where $r$ is a one-to-one function with inverse $r^{-1}$. Then by the Factorization Theorem there exist $g$ and $h$ such that

$$f(\mathbf{x}|\theta) = g(T(\mathbf{x})|\theta)h(\mathbf{x}) = g(r^{-1}(T^*(\mathbf{x}))|\theta)h(\mathbf{x}).$$

Defining $g^*(t|\theta) = g(r^{-1}(t)|\theta)$, we see that

$$f(\mathbf{x}|\theta) = g^*(T^*(\mathbf{x})|\theta)h(\mathbf{x}).$$

So, by the Factorization Theorem, $T^*(\mathbf{X})$ is a sufficient statistic.

Because of the numerous sufficient statistics in a problem, we might ask whether one sufficient statistic is any better than another. Recall that the purpose of a sufficient statistic is to achieve data reduction without loss of information about the parameter $\theta$; thus, a statistic that achieves the most data reduction while still retaining all the information about $\theta$ might be considered preferable. The definition of such a statistic is formalized now.

**Definition 6.2.11**   A sufficient statistic $T(\mathbf{X})$ is called a *minimal sufficient statistic* if, for any other sufficient statistic $T'(\mathbf{X})$, $T(\mathbf{x})$ is a function of $T'(\mathbf{x})$.

To say that $T(\mathbf{x})$ is a function of $T'(\mathbf{x})$ simply means that if $T'(\mathbf{x}) = T'(\mathbf{y})$, then $T(\mathbf{x}) = T(\mathbf{y})$. In terms of the partition sets described at the beginning of the chapter, if $\{B_{t'} : t' \in \mathcal{T}'\}$ are the partition sets for $T'(\mathbf{x})$ and $\{A_t : t \in \mathcal{T}\}$ are the partition sets for $T(\mathbf{x})$, then Definition 6.2.11 states that every $B_{t'}$ is a subset of some $A_t$. Thus, the partition associated with a minimal sufficient statistic, is the *coarsest* possible partition for a sufficient statistic, and a minimal sufficient statistic achieves the greatest possible data reduction for a sufficient statistic.

**Example 6.2.12 (Two normal sufficient statistics)**   The model considered in Example 6.2.4 has $X_1, \ldots, X_n$ iid $n(\mu, \sigma^2)$ with $\sigma^2$ known. Using factorization (6.2.4), we concluded that $T(\mathbf{X}) = \bar{X}$ is a sufficient statistic for $\mu$. Instead, we could write down factorization (6.2.5) for this problem ($\sigma^2$ is a known value now) and correctly conclude that $T'(\mathbf{X}) = (\bar{X}, S^2)$ is a sufficient statistic for $\mu$ in this problem. Clearly $T(\mathbf{X})$ achieves a greater data reduction than $T'(\mathbf{X})$ since we do not know the sample variance if we know only $T(\mathbf{X})$. We can write $T(\mathbf{x})$ as a function of $T'(\mathbf{x})$ by defining the function $r(a, b) = a$. Then $T(\mathbf{x}) = \bar{x} = r(\bar{x}, s^2) = r(T'(\mathbf{x}))$. Since $T(\mathbf{X})$ and $T'(\mathbf{X})$ are both sufficient statistics, they both contain the same information about $\mu$. Thus, the additional information about the value of $S^2$, the sample variance, does not add to our knowledge of $\mu$ since the population variance $\sigma^2$ is known. Of course, if $\sigma^2$ is unknown, as in Example 6.2.9, $T(\mathbf{X}) = \bar{X}$ is not a sufficient statistic and $T'(\mathbf{X})$ contains more information about the parameter $(\mu, \sigma^2)$ than does $T(\mathbf{X})$.                    $\|$

Using Definition 6.2.11 to find a minimal sufficient statistic is impractical, as was using Definition 6.2.1 to find sufficient statistics. We would need to guess that $T(\mathbf{X})$

was a minimal sufficient statistic and then verify the condition in the definition. (Note that we did not show that $\overline{X}$ is a minimal sufficient statistic in Example 6.2.12.) Fortunately, the following result of Lehmann and Scheffé (1950, Theorem 6.3) gives an easier way to find a minimal sufficient statistic.

**Theorem 6.2.13**    *Let $f(\mathbf{x}|\theta)$ be the pmf or pdf of a sample $\mathbf{X}$. Suppose there exists a function $T(\mathbf{x})$ such that, for every two sample points $\mathbf{x}$ and $\mathbf{y}$, the ratio $f(\mathbf{x}|\theta)/f(\mathbf{y}|\theta)$ is constant as a function of $\theta$ if and only if $T(\mathbf{x}) = T(\mathbf{y})$. Then $T(\mathbf{X})$ is a minimal sufficient statistic for $\theta$.*

**Proof:** To simplify the proof, we assume $f(\mathbf{x}|\theta) > 0$ for all $\mathbf{x} \in \mathcal{X}$ and $\theta$.

First we show that $T(\mathbf{X})$ is a sufficient statistic. Let $\mathcal{T} = \{t : t = T(\mathbf{x}) \text{ for some } \mathbf{x} \in \mathcal{X}\}$ be the image of $\mathcal{X}$ under $T(\mathbf{x})$. Define the partition sets induced by $T(\mathbf{x})$ as $A_t = \{\mathbf{x} : T(\mathbf{x}) = t\}$. For each $A_t$, choose and fix one element $\mathbf{x}_t \in A_t$. For any $\mathbf{x} \in \mathcal{X}$, $\mathbf{x}_{T(\mathbf{x})}$ is the fixed element that is in the same set, $A_t$, as $\mathbf{x}$. Since $\mathbf{x}$ and $\mathbf{x}_{T(\mathbf{x})}$ are in the same set $A_t$, $T(\mathbf{x}) = T(\mathbf{x}_{T(\mathbf{x})})$ and, hence, $f(\mathbf{x}|\theta)/f(\mathbf{x}_{T(\mathbf{x})}|\theta)$ is constant as a function of $\theta$. Thus, we can define a function on $\mathcal{X}$ by $h(\mathbf{x}) = f(\mathbf{x}|\theta)/f(\mathbf{x}_{T(\mathbf{x})}|\theta)$ and $h$ does not depend on $\theta$. Define a function on $\mathcal{T}$ by $g(t|\theta) = f(\mathbf{x}_t|\theta)$. Then it can be seen that

$$f(\mathbf{x}|\theta) = \frac{f(\mathbf{x}_{T(\mathbf{x})}|\theta)f(\mathbf{x}|\theta)}{f(\mathbf{x}_{T(\mathbf{x})}|\theta)} = g(T(\mathbf{x})|\theta)h(\mathbf{x})$$

and, by the Factorization Theorem, $T(\mathbf{X})$ is a sufficient statistic for $\theta$.

Now to show that $T(\mathbf{X})$ is minimal, let $T'(\mathbf{X})$ be any other sufficient statistic. By the Factorization Theorem, there exist functions $g'$ and $h'$ such that $f(\mathbf{x}|\theta) = g'(T'(\mathbf{x})|\theta)h'(\mathbf{x})$. Let $\mathbf{x}$ and $\mathbf{y}$ be any two sample points with $T'(\mathbf{x}) = T'(\mathbf{y})$. Then

$$\frac{f(\mathbf{x}|\theta)}{f(\mathbf{y}|\theta)} = \frac{g'(T'(\mathbf{x})|\theta)h'(\mathbf{x})}{g'(T'(\mathbf{y})|\theta)h'(\mathbf{y})} = \frac{h'(\mathbf{x})}{h'(\mathbf{y})}.$$

Since this ratio does not depend on $\theta$, the assumptions of the theorem imply that $T(\mathbf{x}) = T(\mathbf{y})$. Thus, $T(\mathbf{x})$ is a function of $T'(\mathbf{x})$ and $T(\mathbf{x})$ is minimal.   □

**Example 6.2.14 (Normal minimal sufficient statistic)**    Let $X_1, \ldots, X_n$ be iid $n(\mu, \sigma^2)$, both $\mu$ and $\sigma^2$ unknown. Let $\mathbf{x}$ and $\mathbf{y}$ denote two sample points, and let $(\bar{x}, s_{\mathbf{x}}^2)$ and $(\bar{y}, s_{\mathbf{y}}^2)$ be the sample means and variances corresponding to the $\mathbf{x}$ and $\mathbf{y}$ samples, respectively. Then, using (6.2.5), we see that the ratio of densities is

$$\frac{f(\mathbf{x}|\mu, \sigma^2)}{f(\mathbf{y}|\mu, \sigma^2)} = \frac{(2\pi\sigma^2)^{-n/2} \exp\left(-\left[n(\bar{x} - \mu)^2 + (n-1)s_{\mathbf{x}}^2\right]/(2\sigma^2)\right)}{(2\pi\sigma^2)^{-n/2} \exp\left(-\left[n(\bar{y} - \mu)^2 + (n-1)s_{\mathbf{y}}^2\right]/(2\sigma^2)\right)}$$

$$= \exp\left(\left[-n(\bar{x}^2 - \bar{y}^2) + 2n\mu(\bar{x} - \bar{y}) - (n-1)(s_{\mathbf{x}}^2 - s_{\mathbf{y}}^2)\right]/(2\sigma^2)\right).$$

This ratio will be constant as a function of $\mu$ and $\sigma^2$ if and only if $\bar{x} = \bar{y}$ and $s_{\mathbf{x}}^2 = s_{\mathbf{y}}^2$. Thus, by Theorem 6.2.13, $(\bar{X}, S^2)$ is a minimal sufficient statistic for $(\mu, \sigma^2)$.   ‖

If the set of $\mathbf{x}$s on which the pdf or pmf is positive depends on the parameter $\theta$, then, for the ratio in Theorem 6.2.13 to be constant as a function of $\theta$, the numerator and denominator must be positive for exactly the same values of $\theta$. This restriction is usually reflected in a minimal sufficient statistic, as the next example illustrates.

**Example 6.2.15 (Uniform minimal sufficient statistic)**    Suppose $X_1, \ldots, X_n$ are iid uniform observations on the interval $(\theta, \theta + 1), -\infty < \theta < \infty$. Then the joint pdf of $\mathbf{X}$ is

$$f(\mathbf{x}|\theta) = \begin{cases} 1 & \theta < x_i < \theta + 1, \ i = 1, \ldots, n, \\ 0 & \text{otherwise}, \end{cases}$$

which can be written as

$$f(\mathbf{x}|\theta) = \begin{cases} 1 & \max_i x_i - 1 < \theta < \min_i x_i \\ 0 & \text{otherwise}. \end{cases}$$

Thus, for two sample points $\mathbf{x}$ and $\mathbf{y}$, the numerator and denominator of the ratio $f(\mathbf{x}|\theta)/f(\mathbf{y}|\theta)$ will be positive for the same values of $\theta$ if and only if $\min_i x_i = \min_i y_i$ and $\max_i x_i = \max_i y_i$. And, if the minima and maxima are equal, then the ratio is constant and, in fact, equals 1. Thus, letting $X_{(1)} = \min_i X_i$ and $X_{(n)} = \max_i X_i$, we have that $T(\mathbf{X}) = (X_{(1)}, X_{(n)})$ is a minimal sufficient statistic. This is a case in which the dimension of a minimal sufficient statistic does not match the dimension of the parameter.                                                                            ‖

A minimal sufficient statistic is not unique. Any one-to-one function of a minimal sufficient statistic is also a minimal sufficient statistic. So, for example, $T'(\mathbf{X}) = (X_{(n)} - X_{(1)}, (X_{(n)} + X_{(1)})/2)$ is also a minimal sufficient statistic in Example 6.2.15 and $T'(\mathbf{X}) = (\sum_{i=1}^n X_i, \sum_{i=1}^n X_i^2)$ is also a minimal sufficient statistic in Example 6.2.14.

### 6.2.3 Ancillary Statistics

In the preceding sections, we considered sufficient statistics. Such statistics, in a sense, contain all the information about $\theta$ that is available in the sample. In this section we introduce a different sort of statistic, one that has a complementary purpose.

**Definition 6.2.16**    A statistic $S(\mathbf{X})$ whose distribution does not depend on the parameter $\theta$ is called an *ancillary statistic*.

Alone, an ancillary statistic contains no information about $\theta$. An ancillary statistic is an observation on a random variable whose distribution is fixed and known, unrelated to $\theta$. Paradoxically, an ancillary statistic, when used in conjunction with other statistics, sometimes does contain valuable information for inferences about $\theta$. We will investigate this behavior in the next section. For now, we just give some examples of ancillary statistics.

**Example 6.2.17 (Uniform ancillary statistic)**    As in Example 6.2.15, let $X_1, \ldots, X_n$ be iid uniform observations on the interval $(\theta, \theta + 1), -\infty < \theta < \infty$. Let $X_{(1)} < \cdots < X_{(n)}$ be the order statistics from the sample. We show below that the range statistic, $R = X_{(n)} - X_{(1)}$, is an ancillary statistic by showing that the pdf of $R$ does not depend on $\theta$. Recall that the cdf of each $X_i$ is

$$F(x|\theta) = \begin{cases} 0 & x \leq \theta \\ x - \theta & \theta < x < \theta + 1 \\ 1 & \theta + 1 \leq x. \end{cases}$$

Thus, the joint pdf of $X_{(1)}$ and $X_{(n)}$, as given by (5.4.7), is

$$g(x_{(1)}, x_{(n)}|\theta) = \begin{cases} n(n-1)(x_{(n)} - x_{(1)})^{n-2} & \theta < x_{(1)} < x_{(n)} < \theta + 1 \\ 0 & \text{otherwise.} \end{cases}$$

Making the transformation $R = X_{(n)} - X_{(1)}$ and $M = (X_{(1)} + X_{(n)})/2$, which has the inverse transformation $X_{(1)} = (2M - R)/2$ and $X_{(n)} = (2M + R)/2$ with Jacobian 1, we see that the joint pdf of $R$ and $M$ is

$$h(r, m|\theta) = \begin{cases} n(n-1)r^{n-2} & 0 < r < 1, \theta + (r/2) < m < \theta + 1 - (r/2) \\ 0 & \text{otherwise.} \end{cases}$$

(Notice the rather involved region of positivity for $h(r, m|\theta)$.) Thus, the pdf for $R$ is

$$h(r|\theta) = \int_{\theta+(r/2)}^{\theta+1-(r/2)} n(n-1)r^{n-2} dm$$

$$= n(n-1)r^{n-2}(1-r), \quad 0 < r < 1.$$

This is a beta pdf with $\alpha = n - 1$ and $\beta = 2$. More important, the pdf is the same for all $\theta$. Thus, the distribution of $R$ does not depend on $\theta$, and $R$ is ancillary.    ‖

In Example 6.2.17 the range statistic is ancillary because the model considered there is a location parameter model. The ancillarity of $R$ does not depend on the uniformity of the $X_i$s, but rather on the parameter of the distribution being a location parameter. We now consider the general location parameter model.

**Example 6.2.18 (Location family ancillary statistic)**    Let $X_1, \ldots, X_n$ be iid observations from a location parameter family with cdf $F(x - \theta)$, $-\infty < \theta < \infty$. We will show that the range, $R = X_{(n)} - X_{(1)}$, is an ancillary statistic. We use Theorem 3.5.6 and work with $Z_1, \ldots, Z_n$ iid observations from $F(x)$ (corresponding to $\theta = 0$) with $X_1 = Z_1 + \theta, \ldots, X_n = Z_n + \theta$. Thus the cdf of the range statistic, $R$, is

$$F_R(r|\theta) = P_\theta(R \leq r)$$

$$= P_\theta(\max_i X_i - \min_i X_i \leq r)$$

$$= P_\theta(\max_i(Z_i + \theta) - \min_i(Z_i + \theta) \leq r)$$

$$= P_\theta(\max_i Z_i - \min_i Z_i + \theta - \theta \leq r)$$

$$= P_\theta(\max_i Z_i - \min_i Z_i \leq r).$$

The last probability does not depend on $\theta$ because the distribution of $Z_1, \ldots, Z_n$ does not depend on $\theta$. Thus, the cdf of $R$ does not depend on $\theta$ and, hence, $R$ is an ancillary statistic.    ‖

**Example 6.2.19 (Scale family ancillary statistic)** Scale parameter families also have certain kinds of ancillary statistics. Let $X_1, \ldots, X_n$ be iid observations from a scale parameter family with cdf $F(x/\sigma), \sigma > 0$. Then any statistic that depends on the sample only through the $n-1$ values $X_1/X_n, \ldots, X_{n-1}/X_n$ is an ancillary statistic. For example,

$$\frac{X_1 + \cdots + X_n}{X_n} = \frac{X_1}{X_n} + \cdots + \frac{X_{n-1}}{X_n} + 1$$

is an ancillary statistic. To see this fact, let $Z_1, \ldots, Z_n$ be iid observations from $F(x)$ (corresponding to $\sigma = 1$) with $X_i = \sigma Z_i$. The joint cdf of $X_1/X_n, \ldots, X_{n-1}/X_n$ is

$$
\begin{aligned}
F(y_1, \ldots, y_{n-1}|\sigma) &= P_\sigma(X_1/X_n \leq y_1, \ldots, X_{n-1}/X_n \leq y_{n-1}) \\
&= P_\sigma(\sigma Z_1/(\sigma Z_n) \leq y_1, \ldots, \sigma Z_{n-1}/(\sigma Z_n) \leq y_{n-1}) \\
&= P_\sigma(Z_1/Z_n \leq y_1, \ldots, Z_{n-1}/Z_n \leq y_{n-1}).
\end{aligned}
$$

The last probability does not depend on $\sigma$ because the distribution of $Z_1, \ldots, Z_n$ does not depend on $\sigma$. So the distribution of $X_1/X_n, \ldots, X_{n-1}/X_n$ is independent of $\sigma$, as is the distribution of any function of these quantities.

In particular, let $X_1$ and $X_2$ be iid $n(0, \sigma^2)$ observations. From the above result, we see that $X_1/X_2$ has a distribution that is the same for every value of $\sigma$. But, in Example 4.3.6, we saw that, if $\sigma = 1$, $X_1/X_2$ has a Cauchy$(0, 1)$ distribution. Thus, for any $\sigma > 0$, the distribution of $X_1/X_2$ is this same Cauchy distribution. ∥

In this section, we have given examples, some rather general, of statistics that are ancillary for various models. In the next section we will consider the relationship between sufficient statistics and ancillary statistics.

### 6.2.4 Sufficient, Ancillary, and Complete Statistics

A minimal sufficient statistic is a statistic that has achieved the maximal amount of data reduction possible while still retaining all the information about the parameter $\theta$. Intuitively, a minimal sufficient statistic eliminates all the extraneous information in the sample, retaining only that piece with information about $\theta$. Since the distribution of an ancillary statistic does not depend on $\theta$, it might be suspected that a minimal sufficient statistic is unrelated to (or mathematically speaking, functionally independent of) an ancillary statistic. However, this is not necessarily the case. In this section, we investigate this relationship in some detail.

We have already discussed a situation in which an ancillary statistic is not independent of a minimal sufficient statistic. Recall Example 6.2.15 in which $X_1, \ldots, X_n$ were iid observations from a uniform$(\theta, \theta + 1)$ distribution. At the end of Section 6.2.2, we noted that the statistic $(X_{(n)} - X_{(1)}, (X_{(n)} + X_{(1)})/2)$ is a minimal sufficient statistic, and in Example 6.2.17, we showed that $X_{(n)} - X_{(1)}$ is an ancillary statistic. Thus, in this case, the ancillary statistic is an important component of the minimal sufficient statistic. Certainly, the ancillary statistic and the minimal sufficient statistic are not independent.

To emphasize the point that an ancillary statistic can sometimes give important information for inferences about $\theta$, we give another example.

**Example 6.2.20 (Ancillary precision)**   Let $X_1$ and $X_2$ be iid observations from the discrete distribution that satisfies

$$P_\theta(X = \theta) = P_\theta(X = \theta + 1) = P_\theta(X = \theta + 2) = \frac{1}{3},$$

where $\theta$, the unknown parameter, is any integer. Let $X_{(1)} \leq X_{(2)}$ be the order statistics for the sample. It can be shown with an argument similar to that in Example 6.2.15 that $(R, M)$, where $R = X_{(2)} - X_{(1)}$ and $M = (X_{(1)} + X_{(2)})/2$, is a minimal sufficient statistic. Since this is a location parameter family, by Example 6.2.17, $R$ is an ancillary statistic. To see how $R$ might give information about $\theta$, even though it is ancillary, consider a sample point $(r, m)$, where $m$ is an integer. First we consider only $m$; for this sample point to have positive probability, $\theta$ must be one of three values. Either $\theta = m$ or $\theta = m - 1$ or $\theta = m - 2$. With only the information that $M = m$, all three $\theta$ values are possible values. But now suppose we get the additional information that $R = 2$. Then it must be the case that $X_{(1)} = m - 1$ and $X_{(2)} = m + 1$. With this additional information, the only possible value for $\theta$ is $\theta = m - 1$. Thus, the knowledge of the value of the ancillary statistic $R$ has increased our knowledge about $\theta$. Of course, the knowledge of $R$ alone would give us no information about $\theta$. (The idea that an ancillary statistic gives information about the *precision* of an estimate of $\theta$ is not new. See Cox 1971 or Efron and Hinkley 1978 for more ideas.)     ‖

For many important situations, however, our intuition that a minimal sufficient statistic is independent of any ancillary statistic is correct. A description of situations in which this occurs relies on the next definition.

**Definition 6.2.21**   Let $f(t|\theta)$ be a family of pdfs or pmfs for a statistic $T(\mathbf{X})$. The family of probability distributions is called *complete* if $E_\theta g(T) = 0$ for all $\theta$ implies $P_\theta(g(T) = 0) = 1$ for all $\theta$. Equivalently, $T(\mathbf{X})$ is called a *complete statistic*.

Notice that completeness is a property of a family of probability distributions, not of a particular distribution. For example, if $X$ has a $n(0, 1)$ distribution, then defining $g(x) = x$, we have that $Eg(X) = EX = 0$. But the function $g(x) = x$ satisfies $P(g(X) = 0) = P(X = 0) = 0$, not 1. However, this is a particular distribution, not a family of distributions. If $X$ has a $n(\theta, 1)$ distribution, $-\infty < \theta < \infty$, we shall see that no function of $X$, except one that is 0 with probability 1 for all $\theta$, satisfies $E_\theta g(X) = 0$ for all $\theta$. Thus, the family of $n(\theta, 1)$ distributions, $-\infty < \theta < \infty$, is complete.

**Example 6.2.22 (Binomial complete sufficient statistic)**   Suppose that $T$ has a binomial$(n, p)$ distribution, $0 < p < 1$. Let $g$ be a function such that $E_p g(T) = 0$.

Then

$$0 = \mathrm{E}_p g(T) = \sum_{t=0}^{n} g(t) \binom{n}{t} p^t (1-p)^{n-t}$$

$$= (1-p)^n \sum_{t=0}^{n} g(t) \binom{n}{t} \left( \frac{p}{1-p} \right)^t$$

for all $p$, $0 < p < 1$. The factor $(1-p)^n$ is not 0 for any $p$ in this range. Thus it must be that

$$0 = \sum_{t=0}^{n} g(t) \binom{n}{t} \left( \frac{p}{1-p} \right)^t = \sum_{t=0}^{n} g(t) \binom{n}{t} r^t$$

for all $r$, $0 < r < \infty$. But the last expression is a polynomial of degree $n$ in $r$, where the coefficient of $r^t$ is $g(t) \binom{n}{t}$. For the polynomial to be 0 for all $r$, each coefficient must be 0. Since none of the $\binom{n}{t}$ terms is 0, this implies that $g(t) = 0$ for $t = 0, 1, \dots, n$. Since $T$ takes on the values $0, 1, \dots, n$ with probability 1, this yields that $P_p(g(T) = 0) = 1$ for all $p$, the desired conclusion. Hence, $T$ is a complete statistic.                    ‖

**Example 6.2.23 (Uniform complete sufficient statistic)**     Let $X_1, \dots, X_n$ be iid uniform$(0, \theta)$ observations, $0 < \theta < \infty$. Using an argument similar to that in Example 6.2.8, we can see that $T(\mathbf{X}) = \max_i X_i$ is a sufficient statistic and, by Theorem 5.4.4, the pdf of $T(\mathbf{X})$ is

$$f(t|\theta) = \begin{cases} nt^{n-1}\theta^{-n} & 0 < t < \theta \\ 0 & \text{otherwise.} \end{cases}$$

Suppose $g(t)$ is a function satisfying $\mathrm{E}_\theta g(T) = 0$ for all $\theta$. Since $\mathrm{E}_\theta g(T)$ is constant as a function of $\theta$, its derivative with respect to $\theta$ is 0. Thus we have that

$$0 = \frac{d}{d\theta} \mathrm{E}_\theta g(T) = \frac{d}{d\theta} \int_0^\theta g(t) n t^{n-1} \theta^{-n} dt$$

$$= (\theta^{-n}) \frac{d}{d\theta} \int_0^\theta ng(t) t^{n-1} dt + \left( \frac{d}{d\theta} \theta^{-n} \right) \int_0^\theta ng(t) t^{n-1} dt$$

$$= \theta^{-n} ng(\theta) \theta^{n-1} + 0 \qquad \begin{pmatrix} \text{applying the product} \\ \text{rule for differentiation} \end{pmatrix}$$

$$= \theta^{-1} ng(\theta).$$

The first term in the next to last line is the result of an application of the Fundamental Theorem of Calculus. The second term is 0 because the integral is, except for a constant, equal to $\mathrm{E}_\theta g(T)$, which is 0. Since $\theta^{-1} ng(\theta) = 0$ and $\theta^{-1} n \neq 0$, it must be that $g(\theta) = 0$. This is true for every $\theta > 0$; hence, $T$ is a complete statistic. (On a somewhat pedantic note, realize that the Fundamental Theorem of Calculus does not apply to all functions, but only to functions that are *Riemann-integrable*. The equation

$$\frac{d}{d\theta} \int_0^\theta g(t)dt = g(\theta)$$

is valid only at points of continuity of Riemann-integrable $g$. Thus, strictly speaking, the above argument does not show that $T$ is a complete statistic, since the condition of completeness applies to all functions, not just Riemann-integrable ones. From a more practical view, however, this distinction is not of concern since the condition of Riemann-integrability is so general that it includes virtually any function we could think of.)                                                                                                ‖

We now use completeness to state a condition under which a minimal sufficient statistic is independent of every ancillary statistic.

**Theorem 6.2.24 (Basu's Theorem)**    *If $T(\mathbf{X})$ is a complete and minimal sufficient statistic, then $T(\mathbf{X})$ is independent of every ancillary statistic.*

**Proof:** We give the proof only for discrete distributions.

Let $S(\mathbf{X})$ be any ancillary statistic. Then $P(S(\mathbf{X}) = s)$ does not depend on $\theta$ since $S(\mathbf{X})$ is ancillary. Also the conditional probability,

$$P(S(\mathbf{X}) = s | T(\mathbf{X}) = t) = P(\mathbf{X} \in \{\mathbf{x} \colon S(\mathbf{x}) = s\} | T(\mathbf{X}) = t),$$

does not depend on $\theta$ because $T(\mathbf{X})$ is a sufficient statistic (recall the definition!). Thus, to show that $S(\mathbf{X})$ and $T(\mathbf{X})$ are independent, it suffices to show that

(6.2.6)                     $$P(S(\mathbf{X}) = s | T(\mathbf{X}) = t) = P(S(\mathbf{X}) = s)$$

for all possible values $t \in \mathcal{T}$. Now,

$$P(S(\mathbf{X}) = s) = \sum_{t \in \mathcal{T}} P(S(\mathbf{X}) = s | T(\mathbf{X}) = t) P_\theta(T(\mathbf{X}) = t).$$

Furthermore, since $\sum_{t \in \mathcal{T}} P_\theta(T(\mathbf{X}) = t) = 1$, we can write

$$P(S(\mathbf{X}) = s) = \sum_{t \in \mathcal{T}} P(S(\mathbf{X}) = s) P_\theta(T(\mathbf{X}) = t).$$

Therefore, if we define the statistic

$$g(t) = P(S(\mathbf{X}) = s | T(\mathbf{X}) = t) - P(S(\mathbf{X}) = s),$$

the above two equations show that

$$\mathrm{E}_\theta g(T) = \sum_{t \in \mathcal{T}} g(t) P_\theta(T(\mathbf{X}) = t) = 0 \quad \text{for all } \theta.$$

Since $T(\mathbf{X})$ is a complete statistic, this implies that $g(t) = 0$ for all possible values $t \in \mathcal{T}$. Hence (6.2.6) is verified.                                                                □

Basu's Theorem is useful in that it allows us to deduce the independence of two statistics without ever finding the joint distribution of the two statistics. To use Basu's Theorem, we need to show that a statistic is complete, which is sometimes a rather difficult analysis problem. Fortunately, most problems we are concerned with are covered by the following theorem. We will not prove this theorem but note that its proof depends on the uniqueness of a Laplace transform, a property that was mentioned in Section 2.3.

**Theorem 6.2.25 (Complete statistics in the exponential family)**     *Let* $X_1, \ldots, X_n$ *be iid observations from an exponential family with pdf or pmf of the form*

$$(6.2.7) \qquad f(x|\boldsymbol{\theta}) = h(x)c(\boldsymbol{\theta}) \exp\left(\sum_{j=1}^{k} w_j(\boldsymbol{\theta})t_j(x)\right),$$

*where* $\boldsymbol{\theta} = (\theta_1, \theta_2, \ldots, \theta_k)$. *Then the statistic*

$$T(\mathbf{X}) = \left(\sum_{i=1}^{n} t_1(X_i), \sum_{i=1}^{n} t_2(X_i), \ldots, \sum_{i=1}^{n} t_k(X_i)\right)$$

*is complete if* $\{(w_1(\boldsymbol{\theta}), \ldots, w_k(\boldsymbol{\theta})): \boldsymbol{\theta} \in \Theta\}$ *contains an open set in* $\Re^k$.

The condition that the parameter space contain an open set is needed to avoid a situation like the following. The $n(\theta, \theta^2)$ distribution can be written in the form (6.2.7); however, the parameter space $(\theta, \theta^2)$ does not contain a two-dimensional open set, as it consists of only the points on a parabola. As a result, we can find a transformation of the statistic $T(\mathbf{X})$ that is an unbiased estimator of 0 (see Exercise 6.15). (Recall that exponential families such as the $n(\theta, \theta^2)$, where the parameter space is a lower-dimensional curve, are called *curved exponential families*; see Section 3.4.) The relationship between sufficiency, completeness, and minimality in exponential families is an interesting one. For a brief introduction, see Miscellanea 6.6.3.

We now give some examples of the use of Basu's Theorem, Theorem 6.2.25, and many of the earlier results in this chapter.

**Example 6.2.26 (Using Basu's Theorem–I)**   Let $X_1, \ldots, X_n$ be iid exponential observations with parameter $\theta$. Consider computing the expected value of

$$g(\mathbf{X}) = \frac{X_n}{X_1 + \cdots + X_n}.$$

We first note that the exponential distributions form a scale parameter family and thus, by Example 6.2.19, $g(\mathbf{X})$ is an ancillary statistic. The exponential distributions also form an exponential family with $t(x) = x$ and so, by Theorem 6.2.25,

$$T(\mathbf{X}) = \sum_{i=1}^{n} X_i$$

is a complete statistic and, by Theorem 6.2.10, $T(\mathbf{X})$ is a sufficient statistic. (As noted below, we need not verify that $T(\mathbf{X})$ is minimal, although it could easily be verified using Theorem 6.2.13.) Hence, by Basu's Theorem, $T(\mathbf{X})$ and $g(\mathbf{X})$ are independent. Thus we have

$$\theta = \mathrm{E}_\theta X_n = \mathrm{E}_\theta T(\mathbf{X}) g(\mathbf{X}) = (\mathrm{E}_\theta T(\mathbf{X}))(\mathrm{E}_\theta g(\mathbf{X})) = n\theta \mathrm{E}_\theta g(\mathbf{X}).$$

Hence, for any $\theta$, $\mathrm{E}_\theta g(\mathbf{X}) = n^{-1}$.          ||

**Example 6.2.27 (Using Basu's Theorem–II)**     As another example of the use of Basu's Theorem, we consider the independence of $\bar{X}$ and $S^2$, the sample mean and variance, when sampling from a $n(\mu, \sigma^2)$ population. We have, of course, already shown that these statistics are independent in Theorem 5.3.1, but we will illustrate the use of Basu's Theorem in this important context. First consider $\sigma^2$ fixed and let $\mu$ vary, $-\infty < \mu < \infty$. By Example 6.2.4, $\bar{X}$ is a sufficient statistic for $\mu$. Theorem 6.2.25 may be used to deduce that the family of $n(\mu, \sigma^2/n)$ distributions, $-\infty < \mu < \infty$, $\sigma^2/n$ known, is a complete family. Since this is the distribution of $\bar{X}$, $\bar{X}$ is a complete statistic. Now consider $S^2$. An argument similar to those used in Examples 6.2.18 and 6.2.19 could be used to show that in any location parameter family (remember $\sigma^2$ is fixed, $\mu$ is the location parameter), $S^2$ is an ancillary statistic. Or, for this normal model, we can use Theorem 5.3.1 to see that the distribution of $S^2$ depends on the fixed quantity $\sigma^2$ but not on the parameter $\mu$. Either way, $S^2$ is ancillary and so, by Basu's Theorem, $S^2$ is independent of the complete sufficient statistic $\bar{X}$. For any $\mu$ and the fixed $\sigma^2$, $\bar{X}$ and $S^2$ are independent. But since $\sigma^2$ was arbitrary, we have that the sample mean and variance are independent for any choice of $\mu$ and $\sigma^2$. Note that neither $\bar{X}$ nor $S^2$ is ancillary in this model when both $\mu$ and $\sigma^2$ are unknown. Yet, by this argument, we are still able to use Basu's Theorem to deduce independence. This kind of argument is sometimes useful, but the fact remains that it is often harder to show that a statistic is complete than it is to show that two statistics are independent.

          ||

It should be noted that the "minimality" of the sufficient statistic was not used in the proof of Basu's Theorem. Indeed, the theorem is true with this word omitted, because a fundamental property of a complete statistic is that it is minimal.

**Theorem 6.2.28**     *If a minimal sufficient statistic exists, then any complete statistic is also a minimal sufficient statistic.*

So even though the word "minimal" is redundant in the statement of Basu's Theorem, it was stated in this way as a reminder that the statistic $T(\mathbf{X})$ in the theorem is a minimal sufficient statistic. (More about the relationship between complete statistics and minimal sufficient statistics can be found in Lehmann and Scheffé 1950 and Schervish 1995, Section 2.1.)

Basu's Theorem gives one relationship between sufficient statistics and ancillary statistics using the concept of complete statistics. There are other possible definitions of ancillarity and completeness. Some relationships between sufficiency and ancillarity for these definitions are discussed by Lehmann (1981).

## 6.3 The Likelihood Principle

In this section we study a specific, important statistic called the likelihood function that also can be used to summarize data. There are many ways to use the likelihood function some of which are mentioned in this section and some in later chapters. But the main consideration in this section is an argument which indicates that, if certain other principles are accepted, the likelihood function *must* be used as a data reduction device.

### 6.3.1 The Likelihood Function

**Definition 6.3.1**    Let $f(\mathbf{x}|\theta)$ denote the joint pdf or pmf of the sample $\mathbf{X} = (X_1, \ldots, X_n)$. Then, given that $\mathbf{X} = \mathbf{x}$ is observed, the function of $\theta$ defined by

$$L(\theta|\mathbf{x}) = f(\mathbf{x}|\theta)$$

is called the *likelihood function*.

If $\mathbf{X}$ is a discrete random vector, then $L(\theta|\mathbf{x}) = P_\theta(\mathbf{X} = \mathbf{x})$. If we compare the likelihood function at two parameter points and find that

$$P_{\theta_1}(\mathbf{X} = \mathbf{x}) = L(\theta_1|\mathbf{x}) > L(\theta_2|\mathbf{x}) = P_{\theta_2}(\mathbf{X} = \mathbf{x}),$$

then the sample we actually observed is more likely to have occurred if $\theta = \theta_1$ than if $\theta = \theta_2$, which can be interpreted as saying that $\theta_1$ is a more plausible value for the true value of $\theta$ than is $\theta_2$. Many different ways have been proposed to use this information, but certainly it seems reasonable to examine the probability of the sample we actually observed under various possible values of $\theta$. This is the information provided by the likelihood function.

If $X$ is a continuous, real-valued random variable and if the pdf of $X$ is continuous in $x$, then, for small $\epsilon$, $P_\theta(x - \epsilon < X < x + \epsilon)$ is approximately $2\epsilon f(x|\theta) = 2\epsilon L(\theta|x)$ (this follows from the definition of a derivative). Thus,

$$\frac{P_{\theta_1}(x - \epsilon < X < x + \epsilon)}{P_{\theta_2}(x - \epsilon < X < x + \epsilon)} \approx \frac{L(\theta_1|x)}{L(\theta_2|x)},$$

and comparison of the likelihood function at two parameter values again gives an approximate comparison of the probability of the observed sample value, $\mathbf{x}$.

Definition 6.3.1 almost seems to be defining the likelihood function to be the same as the pdf or pmf. The only distinction between these two functions is which variable is considered fixed and which is varying. When we consider the pdf or pmf $f(\mathbf{x}|\theta)$, we are considering $\theta$ as fixed and $\mathbf{x}$ as the variable; when we consider the likelihood function $L(\theta|\mathbf{x})$, we are considering $\mathbf{x}$ to be the observed sample point and $\theta$ to be varying over all possible parameter values.

**Example 6.3.2 (Negative binomial likelihood)**    Let $X$ have a negative binomial distribution with $r = 3$ and success probability $p$. If $x = 2$ is observed, then the likelihood function is the fifth-degree polynomial on $0 \le p \le 1$ defined by

$$L(p|2) = P_p(X = 2) = \binom{4}{2} p^3 (1-p)^2.$$

In general, if $X = x$ is observed, then the likelihood function is the polynomial of degree $3 + x$,

$$L(p|x) = \binom{3 + x - 1}{x} p^3 (1 - p)^x.$$                    ||

The Likelihood Principle specifies how the likelihood function should be used as a data reduction device.

*LIKELIHOOD PRINCIPLE:*  If $\mathbf{x}$ and $\mathbf{y}$ are two sample points such that $L(\theta|\mathbf{x})$ is proportional to $L(\theta|\mathbf{y})$, that is, there exists a constant $C(\mathbf{x}, \mathbf{y})$ such that

(6.3.1)                    $L(\theta|\mathbf{x}) = C(\mathbf{x}, \mathbf{y})L(\theta|\mathbf{y})$   for all $\theta$,

then the conclusions drawn from $\mathbf{x}$ and $\mathbf{y}$ should be identical.

Note that the constant $C(\mathbf{x}, \mathbf{y})$ in (6.3.1) may be different for different $(\mathbf{x}, \mathbf{y})$ pairs but $C(\mathbf{x}, \mathbf{y})$ does not depend on $\theta$.

In the special case of $C(\mathbf{x}, \mathbf{y}) = 1$, the Likelihood Principle states that if two sample points result in the same likelihood function, then they contain the same information about $\theta$. But the Likelihood Principle goes further. It states that even if two sample points have only proportional likelihoods, then they contain equivalent information about $\theta$. The rationale is this: The likelihood function is used to compare the plausibility of various parameter values, and if $L(\theta_2|\mathbf{x}) = 2L(\theta_1|\mathbf{x})$, then, in some sense, $\theta_2$ is twice as plausible as $\theta_1$. If (6.3.1) is also true, then $L(\theta_2|\mathbf{y}) = 2L(\theta_1|\mathbf{y})$. Thus, whether we observe $\mathbf{x}$ or $\mathbf{y}$ we conclude that $\theta_2$ is twice as plausible as $\theta_1$.

We carefully used the word "plausible" rather than "probable" in the preceding paragraph because we often think of $\theta$ as a fixed (albeit unknown) value. Furthermore, although $f(\mathbf{x}|\theta)$, as a function of $\mathbf{x}$, is a pdf, there is no guarantee that $L(\theta|\mathbf{x})$, as a function of $\theta$, is a pdf.

One form of inference, called *fiducial inference*, sometimes interprets likelihoods as probabilities for $\theta$. That is, $L(\theta|\mathbf{x})$ is multiplied by $M(\mathbf{x}) = (\int_{-\infty}^{\infty} L(\theta|\mathbf{x})d\theta)^{-1}$ (the integral is replaced by a sum if the parameter space is countable) and then $M(\mathbf{x})L(\theta|\mathbf{x})$ is interpreted as a pdf for $\theta$ (provided, of course, that $M(\mathbf{x})$ is finite!). Clearly, $L(\theta|\mathbf{x})$ and $L(\theta|\mathbf{y})$ satisfying (6.3.1) will yield the same pdf since the constant $C(\mathbf{x}, \mathbf{y})$ will simply be absorbed into the normalizing constant. Most statisticians do not subscribe to the fiducial theory of inference but it has a long history, dating back to the work of Fisher (1930) on what was called *inverse probability* (an application of the probability integral transform). For now, we will for history's sake compute one fiducial distribution.

**Example 6.3.3 (Normal fiducial distribution)**   Let $X_1, \ldots, X_n$ be iid $n(\mu, \sigma^2)$, $\sigma^2$ known. Using expression (6.2.4) for $L(\mu|\mathbf{x})$, we note first that (6.3.1) is satisfied if and only if $\bar{x} = \bar{y}$, in which case

$$C(\mathbf{x}, \mathbf{y}) = \exp\left(-\sum_{i=1}^n (x_i - \bar{x})^2/(2\sigma^2) + \sum_{i=1}^n (y_i - \bar{y})^2/(2\sigma^2)\right).$$

Thus, the Likelihood Principle states that the same conclusion about $\mu$ should be drawn for any two sample points satisfying $\bar{x} = \bar{y}$. To compute the fiducial pdf for $\mu$, we see that if we define $M(\mathbf{x}) = n^{n/2} \exp\left(\sum_{i=1}^{n}(x_i - \bar{x})^2/(2\sigma^2)\right)$, then $M(\mathbf{x})L(\mu|\mathbf{x})$ (as a function of $\mu$) is a $n(\bar{x}, \sigma^2/n)$ pdf. This is the *fiducial distribution* of $\mu$, and a fiducialist can make the following probability calculation regarding $\mu$.

The parameter $\mu$ has a $n(\bar{x}, \sigma^2/n)$ distribution. Hence, $(\mu - \bar{x})/(\sigma/\sqrt{n})$ has a $n(0, 1)$ distribution. Thus we have

$$
\begin{aligned}
.95 &= P\left(-1.96 < \frac{\mu - \bar{x}}{\sigma/\sqrt{n}} < 1.96\right) \\
&= P(-1.96\sigma/\sqrt{n} < \mu - \bar{x} < 1.96\sigma/\sqrt{n}) \\
&= P(\bar{x} - 1.96\sigma/\sqrt{n} < \mu < \bar{x} + 1.96\sigma/\sqrt{n}).
\end{aligned}
$$

This algebra is similar to earlier calculations but the interpretation is quite different. Here $\bar{x}$ is a fixed, known number, the observed data value, and $\mu$ is the variable with the normal probability distribution.                                          ‖

We will discuss other more common uses of the likelihood function in later chapters when we discuss specific methods of inference. But now we consider an argument that shows that the Likelihood Principle is a necessary consequence of two other fundamental principles.

### 6.3.2 The Formal Likelihood Principle

For discrete distributions, the Likelihood Principle can be derived from two intuitively simpler ideas. This is also true, with some qualifications, for continuous distributions. In this subsection we will deal only with discrete distributions. Berger and Wolpert (1984) provide a thorough discussion of the Likelihood Principle in both the discrete and continuous cases. These results were first proved by Birnbaum (1962) in a landmark paper, but our presentation more closely follows that of Berger and Wolpert.

Formally, we define an experiment $E$ to be a triple $(\mathbf{X}, \theta, \{f(\mathbf{x}|\theta)\})$, where $\mathbf{X}$ is a random vector with pmf $f(\mathbf{x}|\theta)$ for some $\theta$ in the parameter space $\Theta$. An experimenter, knowing what experiment $E$ was performed and having observed a particular sample $\mathbf{X} = \mathbf{x}$, will make some inference or draw some conclusion about $\theta$. This conclusion we denote by $\mathrm{Ev}(E, \mathbf{x})$, which stands for the *evidence about $\theta$ arising from $E$ and $\mathbf{x}$*.

**Example 6.3.4 (Evidence function)**     Let $E$ be the experiment consisting of observing $X_1, \ldots, X_n$ iid $n(\mu, \sigma^2)$, $\sigma^2$ known. Since the sample mean, $\bar{X}$, is a sufficient statistic for $\mu$ and $\mathrm{E}\bar{X} = \mu$, we might use the observed value $\bar{X} = \bar{x}$ as an estimate of $\mu$. To give a measure of the accuracy of this estimate, it is common to report the standard deviation of $\bar{X}, \sigma/\sqrt{n}$. Thus we could define $\mathrm{Ev}(E, \mathbf{x}) = (\bar{x}, \sigma/\sqrt{n})$. Here we see that the $\bar{x}$ coordinate depends on the observed sample $\mathbf{x}$, while the $\sigma/\sqrt{n}$ coordinate depends on the knowledge of $E$.                                          ‖

To relate the concept of an evidence function to something familiar we now restate the Sufficiency Principle of Section 6.2 in terms of these concepts.

*FORMAL SUFFICIENCY PRINCIPLE*: Consider experiment $E = (\mathbf{X}, \theta, \{f(\mathbf{x}|\theta)\})$ and suppose $T(\mathbf{X})$ is a sufficient statistic for $\theta$. If $\mathbf{x}$ and $\mathbf{y}$ are sample points satisfying $T(\mathbf{x}) = T(\mathbf{y})$, then $\mathrm{Ev}(E, \mathbf{x}) = \mathrm{Ev}(E, \mathbf{y})$.

Thus, the *Formal* Sufficiency Principle goes slightly further than the Sufficiency Principle of Section 6.2. There no mention was made of the experiment. Here, we are agreeing to equate evidence if the sufficient statistics match. The Likelihood Principle can be derived from the Formal Sufficiency Principle and the following principle, an eminently reasonable one.

*CONDITIONALITY PRINCIPLE*: Suppose that $E_1 = (\mathbf{X}_1, \theta, \{f_1(\mathbf{x}_1|\theta)\})$ and $E_2 = (\mathbf{X}_2, \theta, \{f_2(\mathbf{x}_2|\theta)\})$ are two experiments, where only the unknown parameter $\theta$ need be common between the two experiments. Consider the mixed experiment in which the random variable $J$ is observed, where $P(J = 1) = P(J = 2) = \frac{1}{2}$ (independent of $\theta$, $\mathbf{X}_1$, or $\mathbf{X}_2$), and then experiment $E_J$ is performed. Formally, the experiment performed is $E^* = (\mathbf{X}^*, \theta, \{f^*(\mathbf{x}^*|\theta)\})$, where $\mathbf{X}^* = (j, \mathbf{X}_j)$ and $f^*(\mathbf{x}^*|\theta) = f^*((j, \mathbf{x}_j)|\theta) = \frac{1}{2}f_j(\mathbf{x}_j|\theta)$. Then

$$(6.3.2) \qquad\qquad \mathrm{Ev}(E^*, (j, \mathbf{x}_j)) = \mathrm{Ev}(E_j, \mathbf{x}_j).$$

The Conditionality Principle simply says that if one of two experiments is randomly chosen and the chosen experiment is done, yielding data $\mathbf{x}$, the information about $\theta$ *depends only on the experiment performed*. That is, it is the same information as would have been obtained if it were decided (nonrandomly) to do that experiment from the beginning, and data $\mathbf{x}$ had been observed. The fact that this experiment was performed, rather than some other, has not increased, decreased, or changed knowledge of $\theta$.

**Example 6.3.5 (Binomial/negative binomial experiment)**    Suppose the parameter of interest is the probability $p$, $0 < p < 1$, where $p$ denotes the probability that a particular coin will land "heads" when it is flipped. Let $E_1$ be the experiment consisting of tossing the coin 20 times and recording the number of heads in those 20 tosses. $E_1$ is a binomial experiment and $\{f_1(x_1|p)\}$ is the family of binomial$(20, p)$ pmfs. Let $E_2$ be the experiment consisting of tossing the coin until the seventh head occurs and recording the number of tails before the seventh head. $E_2$ is a negative binomial experiment. Now suppose the experimenter uses a random number table to choose between these two experiments, happens to choose $E_2$, and collects data consisting of the seventh head occurring on trial 20. The Conditionality Principle says that the information about $\theta$ that the experimenter now has, $\mathrm{Ev}(E^*, (2, 13))$, is the same as that which he would have, $\mathrm{Ev}(E_2, 13)$, if he had just chosen to do the negative binomial experiment and had never contemplated the binomial experiment.    ‖

The following Formal Likelihood Principle can now be derived from the Formal Sufficiency Principle and the Conditionality Principle.

*FORMAL LIKELIHOOD PRINCIPLE*: Suppose that we have two experiments, $E_1 = (\mathbf{X}_1, \theta, \{f_1(\mathbf{x}_1|\theta)\})$ and $E_2 = (\mathbf{X}_2, \theta, \{f_2(\mathbf{x}_2|\theta)\})$, where the unknown parameter $\theta$ is the same in both experiments. Suppose $\mathbf{x}_1^*$ and $\mathbf{x}_2^*$ are sample points from $E_1$ and

$E_2$, respectively, such that

(6.3.3)                    $$L(\theta|\mathbf{x}_2^*) = CL(\theta|\mathbf{x}_1^*)$$

for all $\theta$ and for some constant $C$ that may depend on $\mathbf{x}_1^*$ and $\mathbf{x}_2^*$ but not $\theta$. Then

$$\text{Ev}(E_1, \mathbf{x}_1^*) = \text{Ev}(E_2, \mathbf{x}_2^*).$$

The Formal Likelihood Principle is different from the Likelihood Principle in Section 6.3.1 because the Formal Likelihood Principle concerns two experiments, whereas the Likelihood Principle concerns only one. The Likelihood Principle, however, can be derived from the Formal Likelihood Principle by letting $E_2$ be an exact replicate of $E_1$. Thus, the two-experiment setting in the Formal Likelihood Principle is something of an artifact and the important consequence is the following corollary, whose proof is left as an exercise. (See Exercise 6.32.)

*LIKELIHOOD PRINCIPLE COROLLARY*: If $E = (\mathbf{X}, \theta, \{f(\mathbf{x}|\theta)\})$ is an experiment, then $\text{Ev}(E, \mathbf{x})$ should depend on $E$ and $\mathbf{x}$ only through $L(\theta|\mathbf{x})$.

Now we state Birnbaum's Theorem and then investigate its somewhat surprising consequences.

**Theorem 6.3.6 (Birnbaum's Theorem)**    *The Formal Likelihood Principle follows from the Formal Sufficiency Principle and the Conditionality Principle. The converse is also true.*

**Proof:** We only outline the proof, leaving details to Exercise 6.33. Let $E_1$, $E_2$, $\mathbf{x}_1^*$, and $\mathbf{x}_2^*$ be as defined in the Formal Likelihood Principle, and let $E^*$ be the mixed experiment from the Conditionality Principle. On the sample space of $E^*$ define the statistic

$$T(j, \mathbf{x}_j) = \begin{cases} (1, \mathbf{x}_1^*) & \text{if } j = 1 \text{ and } \mathbf{x}_1 = \mathbf{x}_1^* \text{ or if } j = 2 \text{ and } \mathbf{x}_2 = \mathbf{x}_2^* \\ (j, \mathbf{x}_j) & \text{otherwise.} \end{cases}$$

The Factorization Theorem can be used to prove that $T(J, \mathbf{X}_J)$ is a sufficient statistic in the $E^*$ experiment. Then the Formal Sufficiency Principle implies

(6.3.4)                    $$\text{Ev}(E^*, (1, \mathbf{x}_1^*)) = \text{Ev}(E^*, (2, \mathbf{x}_2^*)),$$

the Conditionality Principle implies

(6.3.5)                    $$\text{Ev}(E^*, (1, \mathbf{x}_1^*)) = \text{Ev}(E_1, \mathbf{x}_1^*)$$

$$\text{Ev}(E^*, (2, \mathbf{x}_2^*)) = \text{Ev}(E_2, \mathbf{x}_2^*),$$

and we can deduce that $\text{Ev}(E_1, \mathbf{x}_1^*) = \text{Ev}(E_2, \mathbf{x}_2^*)$, the Formal Likelihood Principle.

To prove the converse, first let one experiment be the $E^*$ experiment and the other $E_j$. It can be shown that $\text{Ev}(E^*, (j, \mathbf{x}_j)) = \text{Ev}(E_j, \mathbf{x}_j)$, the Conditionality Principle. Then, if $T(\mathbf{X})$ is sufficient and $T(\mathbf{x}) = T(\mathbf{y})$, the likelihoods are proportional and the Formal Likelihood Principle implies that $\text{Ev}(E, \mathbf{x}) = \text{Ev}(E, \mathbf{y})$, the Formal Sufficiency Principle.                    □

**Example 6.3.7 (Continuation of Example 6.3.5)**    Consider again the binomial and negative binomial experiments with the two sample points $x_1 = 7$ (7 out of 20 heads in the binomial experiment) and $x_2 = 13$ (the 7th head occurs on the 20th flip of the coin). The likelihood functions are

$$L(p|x_1 = 7) = \binom{20}{7} p^7 (1 - p)^{13} \quad \text{for the binomial experiment}$$

and

$$L(p|x_2 = 13) = \binom{19}{6} p^7 (1 - p)^{13} \quad \text{for the negative binomial experiment.}$$

These are proportional likelihood functions, so the Formal Likelihood Principle states that the same conclusion regarding $p$ should be made in both cases. In particular, the Formal Likelihood Principle asserts that the fact that in the first case sampling ended because 20 trials were completed and in the second case sampling stopped because the 7th head was observed is immaterial as far as our conclusions about $p$ are concerned. Lindley and Phillips (1976) give a thorough discussion of the binomial–negative binomial inference problem.                                                    ‖

   This point, of equivalent inferences from different experiments, may be amplified by considering the sufficient statistic, $T$, defined in the proof of Birnbaum's Theorem and the sample points $\mathbf{x}_1^* = 7$ and $\mathbf{x}_2^* = 13$. For any sample points in the mixed experiment, other than $(1, 7)$ or $(2, 13)$, $T$ tells which experiment, binomial or negative binomial, was performed and the result of the experiment. But for $(1, 7)$ and $(2, 13)$ we have $T(1, 7) = T(2, 13) = (1, 7)$. If we use only the sufficient statistic to make an inference and if $T = (1, 7)$, then all we know is that 7 out of 20 heads were observed. We do not know whether the 7 or the 20 was the fixed quantity.

   Many common statistical procedures violate the Formal Likelihood Principle. With these procedures, different conclusions would be reached for the two experiments discussed in Example 6.3.5. This violation of the Formal Likelihood Principle may seem strange because, by Birnbaum's Theorem, we are then violating either the Sufficiency Principle or the Conditionality Principle. Let us examine these two principles more closely.

   The Formal Sufficiency Principle is, in essence, the same as that discussed in Section 6.1. There, we saw that all the information about $\theta$ is contained in the sufficient statistic, and knowledge of the entire sample cannot add any information. Thus, basing evidence on the sufficient statistic is an eminently plausible principle. One shortcoming of this principle, one that invites violation, is that it is very model-dependent. As mentioned in the discussion after Example 6.2.9, belief in this principle necessitates belief in the model, something that may not be easy to do.

   Most data analysts perform some sort of "model checking" when analyzing a set of data. Most model checking is, necessarily, based on statistics other than a sufficient statistic. For example, it is common practice to examine *residuals* from a model, statistics that measure variation in the data not accounted for by the model. (We will see residuals in more detail in Chapters 11 and 12.) Such a practice immediately violates the Sufficiency Principle, since the residuals are not based on sufficient statistics.

(Of course, such a practice directly violates the Likelihood Principle also.) Thus, it must be realized that *before* considering the Sufficiency Principle (or the Likelihood Principle), we must be comfortable with the model.

The Conditionality Principle, stated informally, says that "only the experiment actually performed matters." That is, in Example 6.3.5, if we did the binomial experiment, and not the negative binomial experiment, then the (not done) negative binomial experiment should in no way influence our conclusion about $\theta$. This principle, also, seems to be eminently plausible.

How, then, can statistical practice violate the Formal Likelihood Principle, when it would mean violating either the Principle of Sufficiency or Conditionality? Several authors have addressed this question, among them Durbin (1970) and Kalbfleisch (1975). One argument, put forth by Kalbfleisch, is that the proof of the Formal Likelihood Principle is not compelling. This is because the Sufficiency Principle is applied in ignorance of the Conditionality Principle. The sufficient statistic, $T(J, \mathbf{X}_J)$, used in the proof of Theorem 6.3.6 is defined on the mixture experiment. If the Conditionality Principle were invoked first, then separate sufficient statistics would have to be defined for each experiment. In this case, the Formal Likelihood Principle would no longer follow. (A key argument in the proof of Birnbaum's Theorem is that $T(J, \mathbf{X}_J)$ *can take on the same value for sample points from each experiment.* This cannot happen with separate sufficient statistics.)

At any rate, since many intuitively appealing inference procedures do violate the Likelihood Principle, it is not universally accepted by all statisticians. Yet it is mathematically appealing and does suggest a useful data reduction technique.

## 6.4  The Equivariance Principle

The previous two sections both describe data reduction principles in the following way. A function $T(\mathbf{x})$ of the sample is specified, and the principle states that if $\mathbf{x}$ and $\mathbf{y}$ are two sample points with $T(\mathbf{x}) = T(\mathbf{y})$, then the same inference about $\theta$ should be made whether $\mathbf{x}$ or $\mathbf{y}$ is observed. The function $T(\mathbf{x})$ is a sufficient statistic when the Sufficiency Principle is used. The "value" of $T(\mathbf{x})$ is the set of all likelihood functions proportional to $L(\theta|\mathbf{x})$ if the Likelihood Principle is used. The Equivariance Principle describes a data reduction technique in a slightly different way. In any application of the Equivariance Principle, a function $T(\mathbf{x})$ is specified, but if $T(\mathbf{x}) = T(\mathbf{y})$, then the Equivariance Principle states that the inference made if $\mathbf{x}$ is observed should have a *certain relationship* to the inference made if $\mathbf{y}$ is observed, although the two inferences may not be the same. This restriction on the inference procedure sometimes leads to a simpler analysis, just as do the data reduction principles discussed in earlier sections.[2]

Although commonly combined into what is called the Equivariance Principle, the data reduction technique we will now describe actually combines two different equivariance considerations.

---

[2] As in many other texts (Schervish 1995; Lehmann and Casella 1998; Stuart, Ord, and Arnold 1999) we distinguish between *equivariance*, in which the estimate changes in a prescribed way as the data are transformed, and *invariance*, in which the estimate remains unchanged as the data are transformed.

The first type of equivariance might be called *measurement equivariance*. It prescribes that the inference made should not depend on the measurement scale that is used. For example, suppose two foresters are going to estimate the average diameter of trees in a forest. The first uses data on tree diameters expressed in inches, and the second uses the same data expressed in meters. Now both are asked to produce an estimate in inches. (The second might conveniently estimate the average diameter in meters and then transform the estimate to inches.) Measurement equivariance requires that both foresters produce the same estimates. No doubt, almost all would agree that this type of equivariance is reasonable.

The second type of equivariance, actually an invariance, might be called *formal invariance*. It states that if two inference problems have the same formal structure in terms of the mathematical model used, then the same inference procedure should be used in both problems. The elements of the model that must be the same are: $\Theta$, the parameter space; $\{f(\mathbf{x}|\theta): \theta \in \Theta\}$, the set of pdfs or pmfs for the sample; and the set of *allowable inferences and consequences of wrong inferences*. This last element has not been discussed much prior to this; for this section we will assume that the set of possible inferences is the same as $\Theta$; that is, an inference is simply a choice of an element of $\Theta$ as an estimate or guess at the true value of $\theta$. Formal invariance is concerned only with the mathematical entities involved, not the physical description of the experiment. For example, $\Theta$ may be $\Theta = \{\theta: \theta > 0\}$ in two problems. But in one problem $\theta$ may be the average price of a dozen eggs in the United States (measured in cents) and in another problem $\theta$ may refer to the average height of giraffes in Kenya (measured in meters). Yet, formal invariance equates these two parameter spaces since they both refer to the same set of real numbers.

*EQUIVARIANCE PRINCIPLE*: If $\mathbf{Y} = g(\mathbf{X})$ is a change of measurement scale such that the model for $\mathbf{Y}$ has the same formal structure as the model for $\mathbf{X}$, then an inference procedure should be both measurement equivariant and formally equivariant.

We will now illustrate how these two concepts of equivariance can work together to provide useful data reduction.

**Example 6.4.1 (Binomial equivariance)**    Let $X$ have a binomial distribution with sample size $n$ known and success probability $p$ unknown. Let $T(x)$ be the estimate of $p$ that is used when $X = x$ is observed. Rather than using the number of successes, $X$, to make an inference about $p$, we could use the number of failures, $Y = n - X$. $Y$ also has a binomial distribution with parameters $(n, q = 1 - p)$. Let $T^*(y)$ be the estimate of $q$ that is used when $Y = y$ is observed, so that $1 - T^*(y)$ is the estimate of $p$ when $Y = y$ is observed. If $x$ successes are observed, then the estimate of $p$ is $T(x)$. But if there are $x$ successes, then there are $n - x$ failures and $1 - T^*(n - x)$ is also an estimate of $p$. Measurement equivariance requires that these two estimates be equal, that is, $T(x) = 1 - T^*(n - x)$, since the change from $X$ to $Y$ is just a change in measurement scale. Furthermore, the formal structures of the inference problems based on $X$ and $Y$ are the same. $X$ and $Y$ both have binomial$(n, \theta)$ distributions, $0 \leq \theta \leq 1$. So formal invariance requires that $T(z) = T^*(z)$ for all $z = 0, \ldots, n$. Thus, measurement and formal invariance together require that

$$(6.4.1) \qquad\qquad T(x) = 1 - T^*(n - x) = 1 - T(n - x).$$

If we consider only estimators satisfying (6.4.1), then we have greatly reduced and simplified the set of estimators we are willing to consider. Whereas the specification of an arbitrary estimator requires the specification of $T(0), T(1), \ldots, T(n)$, the specification of an estimator satisfying (6.4.1) requires the specification only of $T(0), T(1), \ldots, T([n/2])$, where $[n/2]$ is the greatest integer not larger than $n/2$. The remaining values of $T(x)$ are determined by those already specified and (6.4.1). For example, $T(n) = 1 - T(0)$ and $T(n-1) = 1 - T(1)$. This is the type of data reduction that is always achieved by the Equivariance Principle. The inference to be made for some sample points determines the inference to be made for other sample points.

Two estimators that are equivariant for this problem are $T_1(x) = x/n$ and $T_2(x) = .9(x/n) + .1(.5)$. The estimator $T_1(x)$ uses the sample proportion of successes to estimate $p$. $T_2(x)$ "shrinks" the sample proportion toward .5, an estimator that might be sensible if there is reason to think that $p$ is near .5. Condition (6.4.1) is easily verified for both of these estimators and so they are both equivariant. An estimator that is not equivariant is $T_3(x) = .8(x/n) + .2(1)$. Condition (6.4.1) is not satisfied since $T_3(0) = .2 \neq 0 = 1 - T_3(n - 0)$. See Exercise 6.39 for more on measurement vs. formal invariance.                                                                                          ∥

A key to the equivariance argument in Example 6.4.1 and to any equivariance argument is the choice of the transformations. The data transformation used in Example 6.4.1 is $Y = n - X$. The transformations (changes of measurement scale) used in any application of the Equivariance Principle are described by a set of functions on the sample space called a *group of transformations*.

**Definition 6.4.2**    A set of functions $\{g(\mathbf{x}) : g \in \mathcal{G}\}$ from the sample space $\mathcal{X}$ onto $\mathcal{X}$ is called a *group of transformations of $\mathcal{X}$* if

(i) (*Inverse*)   For every $g \in \mathcal{G}$ there is a $g' \in \mathcal{G}$ such that $g'(g(\mathbf{x})) = \mathbf{x}$ for all $\mathbf{x} \in \mathcal{X}$.

(ii) (*Composition*)   For every $g \in \mathcal{G}$ and $g' \in \mathcal{G}$ there exists $g'' \in \mathcal{G}$ such that $g'(g(\mathbf{x})) = g''(\mathbf{x})$ for all $\mathbf{x} \in \mathcal{X}$.

Sometimes the third requirement,

(iii) (*Identity*)   The identity, $e(\mathbf{x})$, defined by $e(\mathbf{x}) = \mathbf{x}$ is an element of $\mathcal{G}$,

is stated as part of the definition of a group. But (iii) is a consequence of (i) and (ii) and need not be verified separately. (See Exercise 6.38.)

**Example 6.4.3 (Continuation of Example 6.4.1)**    For this problem, only two transformations are involved so we may set $\mathcal{G} = \{g_1, g_2\}$, with $g_1(x) = n - x$ and $g_2(x) = x$. Conditions (i) and (ii) are easily verified. The choice of $g' = g$ verifies (i), that is, each element is its own inverse. For example,

$$g_1(g_1(x)) = g_1(n - x) = n - (n - x) = x.$$

In (ii), if $g' = g$, then $g'' = g_2$, while if $g' \neq g$, then $g'' = g_1$ satisfies the equality. For example, take $g' \neq g = g_1$. Then

$$g_2(g_1(x)) = g_2(n - x) = n - x = g_1(x). \qquad \|$$

To use the Equivariance Principle, we must be able to apply formal invariance to the transformed problem. That is, after changing the measurement scale we must still have the same formal structure. As the structure does not change, we want the underlying model, or family of distributions, to be invariant. This requirement is summarized in the next definition.

**Definition 6.4.4**    Let $\mathcal{F} = \{f(\mathbf{x}|\theta) : \theta \in \Theta\}$ be a set of pdfs or pmfs for $\mathbf{X}$, and let $\mathcal{G}$ be a group of transformations of the sample space $\mathcal{X}$. Then $\mathcal{F}$ is *invariant under the group* $\mathcal{G}$ if for every $\theta \in \Theta$ and $g \in \mathcal{G}$ there exists a unique $\theta' \in \Theta$ such that $\mathbf{Y} = g(\mathbf{X})$ has the distribution $f(\mathbf{y}|\theta')$ if $\mathbf{X}$ has the distribution $f(\mathbf{x}|\theta)$.

**Example 6.4.5 (Conclusion of Example 6.4.1)**    In the binomial problem, we must check both $g_1$ and $g_2$. If $\mathbf{X} \sim$ binomial$(n, p)$, then $g_1(X) = n - X \sim$ binomial$(n, 1 - p)$ so $p' = 1 - p$, where $p$ plays the role of $\theta$ in Definition 6.4.4. Also $g_2(X) = X \sim$ binomial$(n, p)$ so $p' = p$ in this case. Thus the set of binomial pmfs is invariant under the group $\mathcal{G} = \{g_1, g_2\}$. $\qquad \|$

In Example 6.4.1, the group of transformations had only two elements. In many cases, the group of transformations is infinite, as the next example illustrates (see also Exercises 6.41 and 6.42).

**Example 6.4.6 (Normal location invariance)**    Let $X_1, \ldots, X_n$ be iid n$(\mu, \sigma^2)$, both $\mu$ and $\sigma^2$ unknown. Consider the group of transformations defined by $\mathcal{G} = \{g_a(\mathbf{x}), -\infty < a < \infty\}$, where $g_a(x_1, \ldots, x_n) = (x_1 + a, \ldots, x_n + a)$. To verify that this set of transformations is a group, conditions (i) and (ii) from Definition 6.4.2 must be verified. For (i) note that

$$g_{-a}(g_a(x_1, \ldots, x_n)) = g_{-a}(x_1 + a, \ldots, x_n + a)$$
$$= (x_1 + a - a, \ldots, x_n + a - a)$$
$$= (x_1, \ldots, x_n).$$

So if $g = g_a$, then $g' = g_{-a}$ satisfies (i). For (ii) note that

$$g_{a_2}(g_{a_1}(x_1, \ldots, x_n)) = g_{a_2}(x_1 + a_1, \ldots, x_n + a_1)$$
$$= (x_1 + a_1 + a_2, \ldots, x_n + a_1 + a_2)$$
$$= g_{a_1+a_2}(x_1, \ldots, x_n).$$

So if $g = g_{a_1}$ and $g' = g_{a_2}$, then $g'' = g_{a_1+a_2}$ satisfies (ii), and Definition 6.4.2 is verified. $\mathcal{G}$ is a group of transformations.

The set $\mathcal{F}$ in this problem is the set of all joint densities $f(x_1, \ldots, x_n|\mu, \sigma^2)$ for $X_1, \ldots, X_n$ defined by "$X_1, \ldots, X_n$ are iid n$(\mu, \sigma^2)$ for some $-\infty < \mu < \infty$ and

$\sigma^2 > 0$." For any $a, -\infty < a < \infty$, the random variables $Y_1, \ldots, Y_n$ defined by

$$(Y_1, \ldots, Y_n) = g_a(X_1, \ldots, X_n) = (X_1 + a, \ldots, X_n + a)$$

are iid $n(\mu + a, \sigma^2)$ random variables. Thus, the joint distribution of $\mathbf{Y} = g_a(\mathbf{X})$ is in $\mathcal{F}$ and hence $\mathcal{F}$ is invariant under $\mathcal{G}$. In terms of the notation in Definition 6.4.4, if $\theta = (\mu, \sigma^2)$, then $\theta' = (\mu + a, \sigma^2)$.                          ‖

Remember, once again, that the Equivariance Principle is composed of two distinct types of equivariance. One type, measurement equivariance, is intuitively reasonable. When many people think of the Equivariance Principle, they think that it refers only to measurement equivariance. If this were the case, the Equivariance Principle would probably be universally accepted. But the other principle, formal invariance, is quite different. It equates any two problems with the same mathematical structure, regardless of the physical reality they are trying to explain. It says that one inference procedure is appropriate *even if the physical realities are quite different*, an assumption that is sometimes difficult to justify.

But like the Sufficiency Principle and the Likelihood Principle, the Equivariance Principle is a data reduction technique that restricts inference by prescribing what other inferences must be made at related sample points. All three principles prescribe relationships between inferences at different sample points, restricting the set of allowable inferences and, in this way, simplifying the analysis of the problem.

## 6.5 Exercises

**6.1** Let $X$ be one observation from a $n(0, \sigma^2)$ population. Is $|X|$ a sufficient statistic?

**6.2** Let $X_1, \ldots, X_n$ be independent random variables with densities

$$f_{X_i}(x|\theta) = \begin{cases} e^{i\theta - x} & x \geq i\theta \\ 0 & x < i\theta. \end{cases}$$

Prove that $T = \min_i(X_i/i)$ is a sufficient statistic for $\theta$.

**6.3** Let $X_1, \ldots, X_n$ be a random sample from the pdf

$$f(x|\mu, \sigma) = \frac{1}{\sigma} e^{-(x-\mu)/\sigma}, \quad \mu < x < \infty, \ 0 < \sigma < \infty.$$

Find a two-dimensional sufficient statistic for $(\mu, \sigma)$.

**6.4** Prove Theorem 6.2.10.

**6.5** Let $X_1, \ldots, X_n$ be independent random variables with pdfs

$$f(x_i|\theta) = \begin{cases} \frac{1}{2i\theta} & -i(\theta - 1) < x_i < i(\theta + 1) \\ 0 & \text{otherwise,} \end{cases}$$

where $\theta > 0$. Find a two-dimensional sufficient statistic for $\theta$.

**6.6** Let $X_1, \ldots, X_n$ be a random sample from a gamma$(\alpha, \beta)$ population. Find a two-dimensional sufficient statistic for $(\alpha, \beta)$.

**6.7** Let $f(x, y|\theta_1, \theta_2, \theta_3, \theta_4)$ be the bivariate pdf for the uniform distribution on the rectangle with lower left corner $(\theta_1, \theta_2)$ and upper right corner $(\theta_3, \theta_4)$ in $\Re^2$. The parameters satisfy $\theta_1 < \theta_3$ and $\theta_2 < \theta_4$. Let $(X_1, Y_1), \ldots, (X_n, Y_n)$ be a random sample from this pdf. Find a four-dimensional sufficient statistic for $(\theta_1, \theta_2, \theta_3, \theta_4)$.

**6.8** Let $X_1, \ldots, X_n$ be a random sample from a population with location pdf $f(x-\theta)$. Show that the order statistics, $T(X_1, \ldots, X_n) = (X_{(1)}, \ldots, X_{(n)})$, are a sufficient statistic for $\theta$ and no further reduction is possible.

**6.9** For each of the following distributions let $X_1, \ldots, X_n$ be a random sample. Find a minimal sufficient statistic for $\theta$.

(a) $f(x|\theta) = \frac{1}{\sqrt{2\pi}} e^{-(x-\theta)^2/2}$,   $-\infty < x < \infty$,   $-\infty < \theta < \infty$   (normal)

(b) $f(x|\theta) = e^{-(x-\theta)}$,   $\theta < x < \infty$,   $-\infty < \theta < \infty$   (location exponential)

(c) $f(x|\theta) = \frac{e^{-(x-\theta)}}{\left(1+e^{-(x-\theta)}\right)^2}$,   $-\infty < x < \infty$,   $-\infty < \theta < \infty$   (logistic)

(d) $f(x|\theta) = \frac{1}{\pi[1+(x-\theta)^2]}$,   $-\infty < x < \infty$,   $-\infty < \theta < \infty$   (Cauchy)

(e) $f(x|\theta) = \frac{1}{2} e^{-|x-\theta|}$,   $-\infty < x < \infty$,   $-\infty < \theta < \infty$   (double exponential)

**6.10** Show that the minimal sufficient statistic for the uniform$(\theta, \theta+1)$, found in Example 6.2.15, is not complete.

**6.11** Refer to the pdfs given in Exercise 6.9. For each, let $X_{(1)} < \cdots < X_{(n)}$ be the ordered sample, and define $Y_i = X_{(n)} - X_{(i)}, i = 1, \ldots, n-1$.

(a) For each of the pdfs in Exercise 6.9, verify that the set $(Y_1, \ldots, Y_{n-1})$ is ancillary for $\theta$. Try to prove a general theorem, like Example 6.2.18, that handles all these families at once.

(b) In each case determine whether the set $(Y_1, \ldots, Y_{n-1})$ is independent of the minimal sufficient statistic.

**6.12** A natural ancillary statistic in most problems is the *sample size*. For example, let $N$ be a random variable taking values $1, 2, \ldots$ with known probabilities $p_1, p_2, \ldots$, where $\Sigma p_i = 1$. Having observed $N = n$, perform $n$ Bernoulli trials with success probability $\theta$, getting $X$ successes.

(a) Prove that the pair $(X, N)$ is minimal sufficient and $N$ is ancillary for $\theta$. (Note the similarity to some of the hierarchical models discussed in Section 4.4.)

(b) Prove that the estimator $X/N$ is unbiased for $\theta$ and has variance $\theta(1-\theta)\mathrm{E}(1/N)$.

**6.13** Suppose $X_1$ and $X_2$ are iid observations from the pdf $f(x|\alpha) = \alpha x^{\alpha-1} e^{-x^\alpha}, x > 0, \alpha > 0$. Show that $(\log X_1)/(\log X_2)$ is an ancillary statistic.

**6.14** Let $X_1, \ldots, X_n$ be a random sample from a location family. Show that $M - \bar{X}$ is an ancillary statistic, where $M$ is the sample median.

**6.15** Let $X_1, \ldots, X_n$ be iid n$(\theta, a\theta^2)$, where $a$ is a known constant and $\theta > 0$.

(a) Show that the parameter space does not contain a two-dimensional open set.

(b) Show that the statistic $T = (\bar{X}, S^2)$ is a sufficient statistic for $\theta$, but the family of distributions is not complete.

**6.16** A famous example in genetic modeling (Tanner, 1996 or Dempster, Laird, and Rubin 1977) is a genetic linkage multinomial model, where we observe the multinomial vector $(x_1, x_2, x_3, x_4)$ with cell probabilities given by $(\frac{1}{2} + \frac{\theta}{4}, \frac{1}{4}(1-\theta), \frac{1}{4}(1-\theta), \frac{\theta}{4})$.

(a) Show that this is a curved exponential family.

(b) Find a sufficient statistic for $\theta$.

(c) Find a minimal sufficient statistic for $\theta$.

**6.17** Let $X_1, \ldots, X_n$ be iid with geometric distribution

$$P_\theta(X = x) = \theta(1 - \theta)^{x-1}, \quad x = 1, 2, \ldots, \quad 0 < \theta < 1.$$

Show that $\Sigma X_i$ is sufficient for $\theta$, and find the family of distributions of $\Sigma X_i$. Is the family complete?

**6.18** Let $X_1, \ldots, X_n$ be iid Poisson($\lambda$). Show that the family of distributions of $\Sigma X_i$ is complete. Prove completeness without using Theorem 6.2.25.

**6.19** The random variable $X$ takes the values 0, 1, 2 according to one of the following distributions:

|  | $P(X = 0)$ | $P(X = 1)$ | $P(X = 2)$ |  |
| --- | --- | --- | --- | --- |
| Distribution 1 | $p$ | $3p$ | $1 - 4p$ | $0 < p < \frac{1}{4}$ |
| Distribution 2 | $p$ | $p^2$ | $1 - p - p^2$ | $0 < p < \frac{1}{2}$ |

In each case determine whether the family of distributions of $X$ is complete.

**6.20** For each of the following pdfs let $X_1, \ldots, X_n$ be iid observations. Find a complete sufficient statistic, or show that one does not exist.

(a) $f(x|\theta) = \frac{2x}{\theta^2}, \quad 0 < x < \theta, \quad \theta > 0$

(b) $f(x|\theta) = \frac{\theta}{(1+x)^{1+\theta}}, \quad 0 < x < \infty, \quad \theta > 0$

(c) $f(x|\theta) = \frac{(\log \theta)\theta^x}{\theta - 1}, \quad 0 < x < 1, \quad \theta > 1$

(d) $f(x|\theta) = e^{-(x-\theta)} \exp(-e^{-(x-\theta)}), \quad -\infty < x < \infty, \quad -\infty < \theta < \infty$

(e) $f(x|\theta) = \binom{2}{x} \theta^x (1 - \theta)^{2-x}, \quad x = 0, 1, 2, \quad 0 \le \theta \le 1$

**6.21** Let $X$ be one observation from the pdf

$$f(x|\theta) = \left(\frac{\theta}{2}\right)^{|x|} (1 - \theta)^{1-|x|}, \quad x = -1, 0, 1, \quad 0 \le \theta \le 1.$$

(a) Is $X$ a complete sufficient statistic?

(b) Is $|X|$ a complete sufficient statistic?

(c) Does $f(x|\theta)$ belong to the exponential class?

**6.22** Let $X_1, \ldots, X_n$ be a random sample from a population with pdf

$$f(x|\theta) = \theta x^{\theta-1}, \quad 0 < x < 1, \quad \theta > 0.$$

(a) Is $\Sigma X_i$ sufficient for $\theta$?

(b) Find a complete sufficient statistic for $\theta$.

**6.23** Let $X_1, \ldots, X_n$ be a random sample from a uniform distribution on the interval $(\theta, 2\theta)$, $\theta > 0$. Find a minimal sufficient statistic for $\theta$. Is the statistic complete?

**6.24** Consider the following family of distributions:

$$\mathcal{P} = \{P_\lambda(X = x) : P_\lambda(X = x) = \lambda^x e^{-\lambda}/x!; x = 0, 1, 2, \ldots; \lambda = 0 \text{ or } 1\}.$$

This is a Poisson family with $\lambda$ restricted to be 0 or 1. Show that the family $\mathcal{P}$ is *not complete*, demonstrating that completeness can be dependent on the range of the parameter. (See Exercises 6.15 and 6.18.)

**6.25** We have seen a number of theorems concerning sufficiency and related concepts for exponential families. Theorem 5.2.11 gave the distribution of a statistic whose sufficiency is characterized in Theorem 6.2.10 and completeness in Theorem 6.2.25. But if the family is curved, the open set condition of Theorem 6.2.25 is not satisfied. In such cases, is the sufficient statistic of Theorem 6.2.10 also minimal? By applying Theorem 6.2.13 to $T(\mathbf{x})$ of Theorem 6.2.10, establish the following:

(a) The statistic $(\sum X_i, \sum X_i^2)$ is sufficient, but not minimal sufficient, in the $n(\mu, \mu)$ family.
(b) The statistic $\sum X_i^2$ is minimal sufficient in the $n(\mu, \mu)$ family.
(c) The statistic $(\sum X_i, \sum X_i^2)$ is minimal sufficient in the $n(\mu, \mu^2)$ family.
(d) The statistic $(\sum X_i, \sum X_i^2)$ is minimal sufficient in the $n(\mu, \sigma^2)$ family.

**6.26** Use Theorem 6.6.5 to establish that, given a sample $X_1, \ldots, X_n$, the following statistics are minimal sufficient.

|       | Statistic                   | Distribution                        |
|-------|-----------------------------|-------------------------------------|
| (a)   | $\bar{X}$                   | $n(\theta, 1)$                      |
| (b)   | $\sum X_i$                  | gamma$(\alpha, \beta)$, $\alpha$ known |
| (c)   | $\max X_i$                  | uniform$(0, \theta)$                |
| (d)   | $X_{(1)}, \ldots, X_{(n)}$  | Cauchy$(\theta, 1)$                 |
| (e)   | $X_{(1)}, \ldots, X_{(n)}$  | logistic$(\mu, \beta)$              |

**6.27** Let $X_1, \ldots, X_n$ be a random sample from the *inverse Gaussian distribution* with pdf

$$f(x|\mu, \lambda) = \left(\frac{\lambda}{2\pi x^3}\right)^{1/2} e^{\frac{-\lambda(x-\mu)^2}{2\mu^2 x}}, \quad 0 < x < \infty.$$

(a) Show that the statistics

$$\bar{X} = \frac{1}{n}\sum_{i=1}^{n} X_i \quad \text{and} \quad T = \frac{n}{\sum_{i=1}^{n}\frac{1}{X_i} - \frac{1}{\bar{X}}}$$

are sufficient and complete.
(b) For $n = 2$, show that $\bar{X}$ has an inverse Gaussian distribution, $n\lambda/T$ has a $\chi^2_{n-1}$ distribution, and they are independent. (Schwarz and Samanta 1991 do the general case.)

The inverse Gaussian distribution has many applications, particularly in modeling of lifetimes. See the books by Chikkara and Folks (1989) and Seshadri (1993).

**6.28** Prove Theorem 6.6.5. (*Hint:* First establish that the minimal sufficiency of $T(\mathbf{X})$ in the family $\{f_0(\mathbf{x}), \ldots, f_k(\mathbf{x})\}$ follows from Theorem 6.2.13. Then argue that any statistic that is sufficient in $\mathcal{F}$ must be a function of $T(\mathbf{x})$.)

**6.29** The concept of minimal sufficiency can be extended beyond parametric families of distributions. Show that if $X_1, \ldots, X_n$ are a random sample from a density $f$ that is unknown, then the order statistics are minimal sufficient.
(*Hint:* Use Theorem 6.6.5, taking the family $\{f_0(\mathbf{x}), \ldots, f_k(\mathbf{x})\}$ to be logistic densities.)

**6.30** Let $X_1, \ldots, X_n$ be a random sample from the pdf $f(x|\mu) = e^{-(x-\mu)}$, where $-\infty < \mu < x < \infty$.

(a) Show that $X_{(1)} = \min_i X_i$ is a complete sufficient statistic.
(b) Use Basu's Theorem to show that $X_{(1)}$ and $S^2$ are independent.

**6.31** Boos and Hughes-Oliver (1998) detail a number of instances where application of Basu's Theorem can simplify calculations. Here are a few.

(a) Let $X_1, \ldots, X_n$ be iid $n(\mu, \sigma^2)$, where $\sigma^2$ is known.

    (i) Show that $\bar{X}$ is complete sufficient for $\mu$, and $S^2$ is ancillary. Hence by Basu's Theorem, $\bar{X}$ and $S^2$ are independent.

    (ii) Show that this independence carries over even if $\sigma^2$ is unknown, as knowledge of $\sigma^2$ has no bearing on the distributions. (Compare this proof to the more involved Theorem 5.3.1(a).)

(b) A *Monte Carlo swindle* is a technique for improving variance estimates. Suppose that $X_1, \ldots, X_n$ are iid $n(\mu, \sigma^2)$ and that we want to compute the variance of the median, $M$.

    (i) Apply Basu's Theorem to show that $\text{Var}(M) = \text{Var}(M - \bar{X}) + \text{Var}(\bar{X})$; thus we only have to simulate the $\text{Var}(M - \bar{X})$ piece of $\text{Var}(M)$ (since $\text{Var}(\bar{X}) = \sigma^2/n$).

    (ii) Show that the swindle estimate is more precise by showing that the variance of $M$ is approximately $2[\text{Var}(M)]^2/(N-1)$ and that of $M - \bar{X}$ is approximately $2[\text{Var}(M - \bar{X})]^2/(N-1)$, where $N$ is the number of Monte Carlo samples.

(c) (i) If $X/Y$ and $Y$ are independent random variables, show that

$$\text{E}\left(\frac{X}{Y}\right)^k = \frac{\text{E}(X^k)}{\text{E}(Y^k)}.$$

    (ii) Use this result and Basu's Theorem to show that if $X_1, \ldots, X_n$ are iid gamma$(\alpha, \beta)$, where $\alpha$ is known, then for $T = \sum_i X_i$,

$$\text{E}\left(X_{(i)} \mid T\right) = \text{E}\left(\frac{X_{(i)}}{T} T \mid T\right) = T\frac{\text{E}(X_{(i)})}{\text{E}T}.$$

**6.32** Prove the Likelihood Principle Corollary. That is, assuming both the Formal Sufficiency Principle and the Conditionality Principle, prove that if $E = (\mathbf{X}, \theta, \{f(\mathbf{x}|\theta)\})$ is an experiment, then $\text{Ev}(E, \mathbf{x})$ should depend on $E$ and $\mathbf{x}$ only through $L(\theta|\mathbf{x})$.

**6.33** Fill in the gaps in the proof of Theorem 6.3.6, Birnbaum's Theorem.

(a) Define $g(\mathbf{t}|\theta) = g((j, \mathbf{x}_j)|\theta) = f^*((j, \mathbf{x}_j)|\theta)$ and

$$h(j, \mathbf{x}_j) = \begin{cases} C & \text{if } (j, \mathbf{x}_j) = (2, \mathbf{x}_2^*) \\ 1 & \text{otherwise.} \end{cases}$$

Show that $T(j, \mathbf{x}_j)$ is a sufficient statistic in the $E^*$ experiment by verifying that

$$g(T(j, \mathbf{x}_j)|\theta)h(j, \mathbf{x}_j) = f^*((j, \mathbf{x}_j)|\theta)$$

for all $(j, \mathbf{x}_j)$.

(b) As $T$ is sufficient, show that the Formal Sufficiency Principle implies (6.3.4). Also the Conditionality Principle implies (6.3.5), and hence deduce the Formal Likelihood Principle.

(c) To prove the converse, first let one experiment be the $E^*$ experiment and the other $E_j$ and deduce that $\text{Ev}(E^*, (j, \mathbf{x}_j)) = \text{Ev}(E_j, \mathbf{x}_j)$, the Conditionality Principle. Then, if $T(\mathbf{X})$ is sufficient and $T(\mathbf{x}) = T(\mathbf{y})$, show that the likelihoods are proportional and then use the Formal Likelihood Principle to deduce $\text{Ev}(E, \mathbf{x}) = \text{Ev}(E, \mathbf{y})$, the Formal Sufficiency Principle.

**6.34** Consider the model in Exercise 6.12. Show that the Formal Likelihood Principle implies that any conclusions about $\theta$ should not depend on the fact that the sample size $n$ was chosen randomly. That is, the likelihood for $(n, x)$, a sample point from Exercise 6.12, is proportional to the likelihood for the sample point $x$, a sample point from a fixed-sample-size binomial$(n, \theta)$ experiment.

**6.35** A risky experimental treatment is to be given to at most three patients. The treatment will be given to one patient. If it is a success, then it will be given to a second. If it is a success, it will be given to a third patient. Model the outcomes for the patients as independent Bernoulli$(p)$ random variables. Identify the four sample points in this model and show that, according to the Formal Likelihood Principle, the inference about $p$ should not depend on the fact that the sample size was determined by the data.

**6.36** One advantage of using a minimal sufficient statistic is that unbiased estimators will have smaller variance, as the following exercise will show. Suppose that $T_1$ is sufficient and $T_2$ is minimal sufficient, $U$ is an unbiased estimator of $\theta$, and define $U_1 = E(U|T_1)$ and $U_2 = E(U|T_2)$.

(a) Show that $U_2 = E(U_1|T_2)$.
(b) Now use the conditional variance formula (Theorem 4.4.7) to show that $\operatorname{Var} U_2 \leq \operatorname{Var} U_1$.

(See Pena and Rohatgi 1994 for more on the relationship between sufficiency and unbiasedness.)

**6.37** Joshi and Nabar (1989) examine properties of linear estimators for the parameter in the so-called "Problem of the Nile," where $(X, Y)$ has the joint density

$$f(x, y|\theta) = \exp\{-(\theta x + y/\theta)\}, \qquad x > 0, \quad y > 0.$$

(a) For an iid sample of size $n$, show that the Fisher information is $I(\theta) = 2n/\theta^2$.
(b) For the estimators

$$T = \sqrt{\sum Y_i / \sum X_i} \quad \text{and} \quad U = \sqrt{\sum X_i \sum Y_i},$$

show that
   (i) the information in $T$ alone is $[2n/(2n+1)]I(\theta)$;
   (ii) the information in $(T, U)$ is $I(\theta)$;
   (iii) $(T, U)$ is jointly sufficient but not complete.

**6.38** In Definition 6.4.2, show that (iii) is implied by (i) and (ii).

**6.39** Measurement equivariance requires the same inference for two equivalent data points: **x**, measurements expressed in one scale, and **y**, *exactly the same measurements* expressed in a different scale. Formal invariance, in the end, leads to a relationship between the inferences at two *different* data points in the same measurement scale. Suppose an experimenter wishes to estimate $\theta$, the mean boiling point of water, based on a single observation $X$, the boiling point measured in degrees Celsius. Because of the altitude and impurities in the water he decides to use the estimate $T(x) = .5x + .5(100)$. If the measurement scale is changed to degrees Fahrenheit, the experimenter would use $T^*(y) = .5y + .5(212)$ to estimate the mean boiling point expressed in degrees Fahrenheit.

(a) The familiar relation between degrees Celsius and degrees Fahrenheit would lead us to convert Fahrenheit to Celsius using the transformation $\frac{5}{9}(T^*(y) - 32)$. Show that this procedure is measurement equivariant in that the same answer will be obtained for the same data; that is, $\frac{5}{9}(T^*(y) - 32) = T(x)$.

(b) Formal invariance would require that $T(x) = T^*(x)$ for all $x$. Show that the estimators we have defined above do not satisfy this. So they are not equivariant in the sense of the Equivariance Principle.

**6.40** Let $X_1, \ldots, X_n$ be iid observations from a location–scale family. Let $T_1(X_1, \ldots, X_n)$ and $T_2(X_1, \ldots, X_n)$ be two statistics that both satisfy

$$T_i(ax_1 + b, \ldots, ax_n + b) = aT_i(x_1, \ldots, x_n)$$

for all values of $x_1, \ldots, x_n$ and $b$ and for any $a > 0$.

(a) Show that $T_1/T_2$ is an ancillary statistic.

(b) Let $R$ be the sample range and $S$ be the sample standard deviation. Verify that $R$ and $S$ satisfy the above condition so that $R/S$ is an ancillary statistic.

**6.41** Suppose that for the model in Example 6.4.6, the inference to be made is an estimate of the mean $\mu$. Let $T(\mathbf{x})$ be the estimate used if $\mathbf{X} = \mathbf{x}$ is observed. If $g_a(\mathbf{X}) = \mathbf{Y} = \mathbf{y}$ is observed, then let $T^*(\mathbf{y})$ be the estimate of $\mu + a$, the mean of each $Y_i$. If $\mu + a$ is estimated by $T^*(\mathbf{y})$, then $\mu$ would be estimated by $T^*(\mathbf{y}) - a$.

(a) Show that measurement equivariance requires that $T(\mathbf{x}) = T^*(\mathbf{y}) - a$ for all $\mathbf{x} = (x_1, \ldots, x_n)$ and all $a$.

(b) Show that formal invariance requires that $T(\mathbf{x}) = T^*(\mathbf{x})$ and hence the Equivariance Principle requires that $T(x_1, \ldots, x_n) + a = T(x_1 + a, \ldots, x_n + a)$ for all $(x_1, \ldots, x_n)$ and all $a$.

(c) If $X_1, \ldots, X_n$ are iid $f(x - \theta)$, show that, as long as $E_0 X_1 = 0$, the estimator $W(X_1, \ldots, X_n) = \bar{X}$ is equivariant for estimating $\theta$ and satisfies $E_\theta W = \theta$.

**6.42** Suppose we have a random sample $X_1, \ldots, X_n$ from $\frac{1}{\sigma} f((x - \theta)/\sigma)$, a location–scale pdf. We want to estimate $\theta$, and we have two groups of transformations under consideration:

$$\mathcal{G}_1 = \{g_{a,c}(\mathbf{x}) : -\infty < a < \infty, \ c > 0\},$$

where $g_{a,c}(x_1, \ldots, x_n) = (cx_1 + a, \ldots, cx_n + a)$, and

$$\mathcal{G}_2 = \{g_a(\mathbf{x}) : -\infty < a < \infty\},$$

where $g_a(x_1, \ldots, x_n) = (x_1 + a, \ldots, x_n + a)$.

(a) Show that estimators of the form

$$W(x_1, \ldots, x_n) = \bar{x} + k,$$

where $k$ is a nonzero constant, are equivariant with respect to the group $\mathcal{G}_2$ but are not equivariant with respect to the group $\mathcal{G}_1$.

(b) For each group, under what conditions does an equivariant estimator $W$ satisfy $E_\theta W = \theta$, that is, it is unbiased for estimating $\theta$?

**6.43** Again, suppose we have a random sample $X_1, \ldots, X_n$ from $\frac{1}{\sigma} f((x - \theta)/\sigma)$, a location–scale pdf, but we are now interested in estimating $\sigma^2$. We can consider three groups of transformations:

$$\mathcal{G}_1 = \{g_{a,c}(\mathbf{x}) : -\infty < a < \infty, \ c > 0\},$$

where $g_{a,c}(x_1, \ldots, x_n) = (cx_1 + a, \ldots, cx_n + a)$;

$$\mathcal{G}_2 = \{g_a(\mathbf{x}): -\infty < a < \infty\},$$

where $g_a(x_1, \ldots, x_n) = (x_1 + a, \ldots, x_n + a)$; and

$$\mathcal{G}_3 = \{g_c(\mathbf{x}): c > 0\},$$

where $g_c(x_1, \ldots, x_n) = (cx_1, \ldots, cx_n)$.

(a) Show that estimators of $\sigma^2$ of the form $kS^2$, where $k$ is a positive constant and $S^2$ is the sample variance, are invariant with respect to $\mathcal{G}_2$ and equivariant with respect to the other two groups.

(b) Show that the larger class of estimators of $\sigma^2$ of the form

$$W(X_1, \ldots, X_n) = \phi\left(\frac{\bar{X}}{S}\right) S^2,$$

where $\phi(x)$ is a function, are equivariant with respect to $\mathcal{G}_3$ but not with respect to either $\mathcal{G}_1$ or $\mathcal{G}_2$, unless $\phi(x)$ is a constant (Brewster and Zidek 1974). Consideration of estimators of this form led Stein (1964) and Brewster and Zidek (1974) to find improved estimators of variance (see Lehmann and Casella 1998, Section 3.3).

## 6.6 Miscellanea

### 6.6.1 The Converse of Basu's Theorem

An interesting statistical fact is that the converse of Basu's Theorem is false. That is, if $T(\mathbf{X})$ is independent of every ancillary statistic, it does not necessarily follow that $T(\mathbf{X})$ is a complete, minimal sufficient statistic. A particularly nice treatment of the topic is given by Lehmann (1981). He makes the point that one reason the converse fails is that ancillarity is a property of the *entire distribution* of a statistic, whereas completeness is a property dealing only with *expectations*. Consider the following modification of the definition of ancillarity.

**Definition 6.6.1**    A statistic $V(\mathbf{X})$ is called *first-order ancillary* if $E_\theta V(\mathbf{X})$ is independent of $\theta$.

Lehmann then proves the following theorem, which is somewhat of a converse to Basu's Theorem.

**Theorem 6.6.2**    *Let $T$ be a statistic with $\mathrm{Var}\, T < \infty$. A necessary and sufficient condition for $T$ to be complete is that every bounded first-order ancillary $V$ is uncorrelated (for all $\theta$) with every bounded real-valued function of $T$.*

Lehmann also notes that a type of converse is also obtainable if, instead of modifying the definition of ancillarity, the definition of completeness is modified.

### 6.6.2 Confusion About Ancillarity

One of the problems with the concept of *ancillarity* is that there are many different definitions of ancillarity, and different properties are given in these definitions. As was seen in this chapter, ancillarity is confusing enough with one definition—with five or six the situation becomes hopeless.

As told by Buehler (1982), the concept of ancillarity goes back to Sir Ronald Fisher (1925), "who left a characteristic trail of intriguing concepts but no definition." Buehler goes on to tell of at least *three* definitions of ancillarity, crediting, among others, Basu (1959) and Cox and Hinkley (1974). Buehler gives eight properties of ancillary statistics and lists 25 examples.

However, it is worth the effort to understand the difficult topic of ancillarity, as it can play an important role in inference. Brown (1996) shows how ancillarity affects inference in regression, and Reid (1995) reviews the role of ancillarity (and other conditioning) in inference. The review article of Lehmann and Scholz (1992) provides a good entry to the topic.

### 6.6.3 More on Sufficiency

1. *Sufficiency and Likelihood*

   There is a striking similarity between the statement of Theorem 6.2.13 and the Likelihood Principle. Both relate to the ratio $L(\theta|\mathbf{x})/L(\theta|\mathbf{y})$, one to describe a minimal sufficient statistic and the other to describe the Likelihood Principle. In fact, these theorems can be combined, with a bit of care, into the fact that a statistic $T(\mathbf{x})$ is a minimal sufficient statistic if and only if it is a one-to-one function of $L(\theta|\mathbf{x})$ (where two sample points that satisfy (6.3.1) are said to have the same likelihood function). Example 6.3.3 and Exercise 6.9 illustrate this point.

2. *Sufficiency and Necessity*

   We may ask, "If there are *sufficient* statistics, why aren't there *necessary* statistics?" In fact, there are. According to Dynkin (1951), we have the following definition.

   **Definition 6.6.3**     A statistic is said to be *necessary* if it can be written as a function of every sufficient statistic.

   If we compare the definition of a necessary statistic and the definition of a minimal sufficient statistic, it should come as no surprise that we have the following theorem.

   **Theorem 6.6.4**     *A statistic is a minimal sufficient statistic if and only if it is a necessary and sufficient statistic.*

3. *Minimal Sufficiency*

   There is an interesting development of minimal sufficiency that actually follows from Theorem 6.2.13 (see Exercise 6.28) and is extremely useful in establishing minimal sufficiency outside of the exponential family.

**Theorem 6.6.5 (Minimal sufficient statistics)** *Suppose that the family of densities* $\{f_0(\mathbf{x}), \ldots, f_k(\mathbf{x})\}$ *all have common support. Then*

**a.** *The statistic*

$$T(\mathbf{X}) = \left( \frac{f_1(\mathbf{X})}{f_0(\mathbf{X})}, \frac{f_2(\mathbf{X})}{f_0(\mathbf{X})}, \ldots, \frac{f_k(\mathbf{X})}{f_0(\mathbf{X})} \right)$$

*is minimal sufficient for the family* $\{f_0(\mathbf{x}), \ldots, f_k(\mathbf{x})\}$.

**b.** *If* $\mathcal{F}$ *is a family of densities with common support, and*

   (i) $f_i(\mathbf{x}) \in \mathcal{F}$, $i = 0, 1, \ldots, k$,
   (ii) $T(\mathbf{x})$ *is sufficient for* $\mathcal{F}$,

   *then* $T(\mathbf{x})$ *is minimal sufficient for* $\mathcal{F}$.

Although Theorem 6.6.5 can be used to establish the minimal sufficiency of $\bar{X}$ in a $n(\theta, 1)$ family, its real usefulness comes when we venture outside of simple situations. For example, Theorem 6.6.5 can be used to show that for samples from distributions like the logistic or double exponential, the order statistics are minimal sufficient (Exercise 6.26). Even further, it can extend to nonparametric families of distributions (Exercise 6.26).

For more on minimal sufficiency and completeness, see Lehmann and Casella (1998, Section 1.6).

# Chapter 7

# Point Estimation

*"What! you have solved it already?"*
*"Well, that would be too much to say. I have discovered a suggestive fact, that is all."*

**Dr. Watson and Sherlock Holmes**
*The Sign of Four*

## 7.1 Introduction

This chapter is divided into two parts. The first part deals with methods for finding estimators, and the second part deals with evaluating these (and other) estimators. In general these two activities are intertwined. Often the methods of evaluating estimators will suggest new ones. However, for the time being, we will make the distinction between finding estimators and evaluating them.

The rationale behind point estimation is quite simple. When sampling is from a population described by a pdf or pmf $f(x|\theta)$, knowledge of $\theta$ yields knowledge of the entire population. Hence, it is natural to seek a method of finding a good estimator of the point $\theta$, that is, a good point estimator. It is also the case that the parameter $\theta$ has a meaningful physical interpretation (as in the case of a population mean) so there is direct interest in obtaining a good point estimate of $\theta$. It may also be the case that some function of $\theta$, say $\tau(\theta)$, is of interest. The methods described in this chapter can also be used to obtain estimators of $\tau(\theta)$.

The following definition of a point estimator may seem unnecessarily vague. However, at this point, we want to be careful not to eliminate any candidates from consideration.

**Definition 7.1.1**   A *point estimator* is any function $W(X_1, \ldots, X_n)$ of a sample; that is, any statistic is a point estimator.

Notice that the definition makes no mention of any correspondence between the estimator and the parameter it is to estimate. While it might be argued that such a statement should be included in the definition, such a statement would restrict the available set of estimators. Also, there is no mention in the definition of the range of the statistic $W(X_1, \ldots, X_n)$. While, in principle, the range of the statistic should coincide with that of the parameter, we will see that this is not always the case.

There is one distinction that must be made clear, the difference between an estimate and an estimator. An *estimator* is a function of the sample, while an *estimate* is the realized value of an estimator (that is, a number) that is obtained when a sample is actually taken. Notationally, when a sample is taken, an estimator is a function of the random variables $X_1, \ldots, X_n$, while an estimate is a function of the realized values $x_1, \ldots, x_n$.

In many cases, there will be an obvious or natural candidate for a point estimator of a particular parameter. For example, the sample mean is a natural candidate for a point estimator of the population mean. However, when we leave a simple case like this, intuition may not only desert us, it may also lead us astray. Therefore, it is useful to have some techniques that will at least give us some reasonable candidates for consideration. Be advised that these techniques do not carry any guarantees with them. The point estimators that they yield still must be evaluated before their worth is established.

## 7.2 Methods of Finding Estimators

In some cases it is an easy task to decide how to estimate a parameter, and often intuition alone can lead us to very good estimators. For example, estimating a parameter with its sample analogue is usually reasonable. In particular, the sample mean is a good estimate for the population mean. In more complicated models, ones that often arise in practice, we need a more methodical way of estimating parameters. In this section we detail four methods of finding estimators.

### 7.2.1 Method of Moments

The method of moments is, perhaps, the oldest method of finding point estimators, dating back at least to Karl Pearson in the late 1800s. It has the virtue of being quite simple to use and almost always yields some sort of estimate. In many cases, unfortunately, this method yields estimators that may be improved upon. However, it is a good place to start when other methods prove intractable.

Let $X_1, \ldots, X_n$ be a sample from a population with pdf or pmf $f(x|\theta_1, \ldots, \theta_k)$. Method of moments estimators are found by equating the first $k$ sample moments to the corresponding $k$ population moments, and solving the resulting system of simultaneous equations. More precisely, define

$$m_1 = \frac{1}{n}\sum_{i=1}^{n} X_i^1, \quad \mu'_1 = \mathrm{E}X^1,$$

$$m_2 = \frac{1}{n}\sum_{i=1}^{n} X_i^2, \quad \mu'_2 = \mathrm{E}X^2,$$

(7.2.1)
$$\vdots$$

$$m_k = \frac{1}{n}\sum_{i=1}^{n} X_i^k, \quad \mu'_k = \mathrm{E}X^k.$$

The population moment $\mu'_j$ will typically be a function of $\theta_1, \ldots, \theta_k$, say $\mu'_j(\theta_1, \ldots, \theta_k)$. The method of moments estimator $(\tilde{\theta}_1, \ldots, \tilde{\theta}_k)$ of $(\theta_1, \ldots, \theta_k)$ is obtained by solving the following system of equations for $(\theta_1, \ldots, \theta_k)$ in terms of $(m_1, \ldots, m_k)$:

$$m_1 = \mu'_1(\theta_1, \ldots, \theta_k),$$
$$m_2 = \mu'_2(\theta_1, \ldots, \theta_k),$$

(7.2.2)
$$\vdots$$

$$m_k = \mu'_k(\theta_1, \ldots, \theta_k).$$

**Example 7.2.1 (Normal method of moments)**     Suppose $X_1, \ldots, X_n$ are iid $n(\theta, \sigma^2)$. In the preceding notation, $\theta_1 = \theta$ and $\theta_2 = \sigma^2$. We have $m_1 = \bar{X}$, $m_2 = (1/n) \sum X_i^2$, $\mu'_1 = \theta$, $\mu'_2 = \theta^2 + \sigma^2$, and hence we must solve

$$\bar{X} = \theta, \quad \frac{1}{n} \sum X_i^2 = \theta^2 + \sigma^2.$$

Solving for $\theta$ and $\sigma^2$ yields the method of moments estimators

$$\tilde{\theta} = \bar{X} \quad \text{and} \quad \tilde{\sigma}^2 = \frac{1}{n} \sum X_i^2 - \bar{X}^2 = \frac{1}{n} \sum (X_i - \bar{X})^2. \qquad \|$$

In this simple example, the method of moments solution coincides with our intuition and perhaps gives some credence to both. The method is somewhat more helpful, however, when no obvious estimator suggests itself.

**Example 7.2.2 (Binomial method of moments)**     Let $X_1, \ldots, X_n$ be iid binomial$(k, p)$, that is,

$$P(X_i = x | k, p) = \binom{k}{x} p^x (1 - p)^{k-x}, \quad x = 0, 1, \ldots, k.$$

Here we assume that both $k$ and $p$ are unknown and we desire point estimators for both parameters. (This somewhat unusual application of the binomial model has been used to estimate crime rates for crimes that are known to have many unreported occurrences. For such a crime, both the true reporting rate, $p$, and the total number of occurrences, $k$, are unknown.)

Equating the first two sample moments to those of the population yields the system of equations

$$\bar{X} = kp,$$

$$\frac{1}{n} \sum X_i^2 = kp(1 - p) + k^2 p^2,$$

which now must be solved for $k$ and $p$. After a little algebra, we obtain the method of moments estimators

$$\tilde{k} = \frac{\bar{X}^2}{\bar{X} - (1/n) \sum (X_i - \bar{X})^2}$$

and

$$\tilde{p} = \frac{\bar{X}}{\tilde{k}}.$$

Admittedly, these are not the best estimators for the population parameters. In particular, it is possible to get negative estimates of $k$ and $p$ which, of course, must be positive numbers. (This is a case where the range of the estimator does not coincide with the range of the parameter it is estimating.) However, in fairness to the method of moments, note that negative estimates will occur only when the sample mean is smaller than the sample variance, indicating a large degree of variability in the data. The method of moments has, in this case, at least given us a set of candidates for point estimators of $k$ and $p$. Although our intuition may have given us a candidate for an estimator of $p$, coming up with an estimator of $k$ is much more difficult.      ‖

The method of moments can be very useful in obtaining approximations to the distributions of statistics. This technique, sometimes called "moment matching," gives us an approximation that is based on matching moments of distributions. In theory, the moments of the distribution of any statistic can be matched to those of any distribution but, in practice, it is best to use distributions that are similar. The following example illustrates one of the most famous uses of this technique, the approximation of Satterthwaite (1946). It is still used today (see Exercise 8.42).

**Example 7.2.3 (Satterthwaite approximation)**      If $Y_i, i = 1, \ldots, k$, are independent $\chi^2_{r_i}$ random variables, we have already seen (Lemma 5.3.2) that the distribution of $\sum Y_i$ is also chi squared, with degrees of freedom equal to $\sum r_i$. Unfortunately, the distribution of $\sum a_i Y_i$, where the $a_i$s are known constants, is, in general, quite difficult to obtain. It does seem reasonable, however, to assume that a $\chi^2_\nu$, for some value of $\nu$, will provide a good approximation.

This is almost Satterthwaite's problem. He was interested in approximating the denominator of a $t$ statistic, and $\sum a_i Y_i$ represented the square of the denominator of his statistic. Hence, for given $a_1, \ldots, a_k$, he wanted to find a value of $\nu$ so that

$$\sum_{i=1}^{k} a_i Y_i \sim \frac{\chi^2_\nu}{\nu} \qquad \text{(approximately)}.$$

Since $\mathrm{E}(\chi^2_\nu / \nu) = 1$, to match first moments we need

$$\mathrm{E}\left( \sum_{i=1}^{k} a_i Y_i \right) = \sum_{i=1}^{k} a_i \mathrm{E} Y_i = \sum_{i=1}^{k} a_i r_i = 1,$$

which gives us a constraint on the $a_i$s but gives us no information on how to estimate $\nu$. To do this we must match second moments, and we need

$$\mathrm{E}\left( \sum_{i=1}^{k} a_i Y_i \right)^2 = \mathrm{E}\left( \frac{\chi^2_\nu}{\nu} \right)^2 = \frac{2}{\nu} + 1.$$

Applying the method of moments, we drop the first expectation and solve for $\nu$, yielding

$$\hat{\nu} = \frac{2}{(\sum_{i=1}^{k} a_i Y_i)^2 - 1}.$$

Thus, straightforward application of the method of moments yields an estimator of $\nu$, but one that can be negative. We might suppose that Satterthwaite was aghast at this possibility, for this is not the estimator he proposed. Working much harder, he customized the method of moments in the following way. Write

$$\text{E}\left(\sum a_i Y_i\right)^2 = \text{Var}\left(\sum a_i Y_i\right) + \left(\text{E}\sum a_i Y_i\right)^2$$

$$= \left(\text{E}\sum a_i Y_i\right)^2 \left[\frac{\text{Var}(\sum a_i Y_i)}{(\text{E}\sum a_i Y_i)^2} + 1\right]$$

$$= \left[\frac{\text{Var}(\sum a_i Y_i)}{(\text{E}\sum a_i Y_i)^2} + 1\right]. \qquad (\text{E }\Sigma a_i Y_i = 1)$$

Now equate second moments to obtain

$$\nu = \frac{2(\text{E}\sum a_i Y_i)^2}{\text{Var}(\sum a_i Y_i)}.$$

Finally, use the fact that $Y_1, \ldots, Y_k$ are independent chi squared random variables to write

$$\text{Var}\left(\sum a_i Y_i\right) = \sum a_i^2 \text{Var } Y_i$$

$$= 2\sum \frac{a_i^2 (\text{E}Y_i)^2}{r_i}. \qquad (\text{Var } Y_i = 2(\text{E}Y_i)^2/r_i)$$

Substituting this expression for the variance and removing the expectations, we obtain Satterthwaite's estimator

$$\hat{\nu} = \frac{(\sum a_i Y_i)^2}{\sum \frac{a_i^2}{r_i} Y_i^2}.$$

This approximation is quite good and is still widely used today. Notice that Satterthwaite succeeded in obtaining an estimator that is always positive, thus alleviating the obvious problems with the straightforward method of moments estimator.    ‖

### 7.2.2 Maximum Likelihood Estimators

The method of maximum likelihood is, by far, the most popular technique for deriving estimators. Recall that if $X_1, \ldots, X_n$ are an iid sample from a population with pdf or pmf $f(x|\theta_1, \ldots, \theta_k)$, the likelihood function is defined by

$$(7.2.3) \qquad L(\theta|\mathbf{x}) = L(\theta_1, \ldots, \theta_k|x_1, \ldots, x_n) = \prod_{i=1}^{n} f(x_i|\theta_1, \ldots, \theta_k).$$

**Definition 7.2.4**   For each sample point $\mathbf{x}$, let $\hat{\theta}(\mathbf{x})$ be a parameter value at which $L(\theta|\mathbf{x})$ attains its maximum as a function of $\theta$, with $\mathbf{x}$ held fixed. A *maximum likelihood estimator* (MLE) of the parameter $\theta$ based on a sample $\mathbf{X}$ is $\hat{\theta}(\mathbf{X})$.

Notice that, by its construction, the range of the MLE coincides with the range of the parameter. We also use the abbreviation MLE to stand for maximum likelihood *estimate* when we are talking of the realized value of the estimator.

Intuitively, the MLE is a reasonable choice for an estimator. The MLE is the parameter point for which the observed sample is most likely. In general, the MLE is a good point estimator, possessing some of the optimality properties discussed later.

There are two inherent drawbacks associated with the general problem of finding the maximum of a function, and hence of maximum likelihood estimation. The first problem is that of actually finding the global maximum and verifying that, indeed, a global maximum has been found. In many cases this problem reduces to a simple differential calculus exercise but, sometimes even for common densities, difficulties do arise. The second problem is that of numerical sensitivity. That is, how sensitive is the estimate to small changes in the data? (Strictly speaking, this is a mathematical rather than statistical problem associated with any maximization procedure. Since an MLE is found through a maximization procedure, however, it is a problem that we must deal with.) Unfortunately, it is sometimes the case that a slightly different sample will produce a vastly different MLE, making its use suspect. We consider first the problem of finding MLEs.

If the likelihood function is differentiable (in $\theta_i$), possible candidates for the MLE are the values of $(\theta_1, \dots, \theta_k)$ that solve

$$(7.2.4) \qquad\qquad \frac{\partial}{\partial \theta_i} L(\theta|\mathbf{x}) = 0, \qquad i = 1, \dots, k.$$

Note that the solutions to (7.2.4) are only *possible candidates* for the MLE since the first derivative being 0 is only a necessary condition for a maximum, not a sufficient condition. Furthermore, the zeros of the first derivative locate only extreme points in the interior of the domain of a function. If the extrema occur on the boundary the first derivative may not be 0. Thus, the boundary must be checked separately for extrema.

Points at which the first derivatives are 0 may be local or global minima, local or global maxima, or inflection points. Our job is to find a global maximum.

**Example 7.2.5 (Normal likelihood)**      Let $X_1, \dots, X_n$ be iid $n(\theta, 1)$, and let $L(\theta|\mathbf{x})$ denote the likelihood function. Then

$$L(\theta|\mathbf{x}) = \prod_{i=1}^{n} \frac{1}{(2\pi)^{1/2}} e^{-(1/2)(x_i - \theta)^2} = \frac{1}{(2\pi)^{n/2}} e^{(-1/2)\Sigma_{i=1}^{n}(x_i - \theta)^2}.$$

The equation $(d/d\theta)L(\theta|\mathbf{x}) = 0$ reduces to

$$\sum_{i=1}^{n} (x_i - \theta) = 0,$$

which has the solution $\hat{\theta} = \bar{x}$. Hence, $\bar{x}$ is a candidate for the MLE. To verify that $\bar{x}$ is, in fact, a global maximum of the likelihood function, we can use the following argument. First, note that $\hat{\theta} = \bar{x}$ is the only solution to $\sum(x_i - \theta) = 0$; hence $\bar{x}$ is the only zero of the first derivative. Second, verify that

$$\frac{d^2}{d\theta^2} L(\theta|\mathbf{x})|_{\theta=\bar{x}} < 0.$$

Thus, $\bar{x}$ is the only extreme point in the interior and it is a maximum. To finally verify that $\bar{x}$ is a global maximum, we must check the boundaries, $\pm\infty$. By taking limits it is easy to establish that the likelihood is 0 at $\pm\infty$. So $\hat{\theta} = \bar{x}$ is a global maximum and hence $\bar{X}$ is the MLE. (Actually, we can be a bit more clever and avoid checking $\pm\infty$. Since we established that $\bar{x}$ is a *unique* interior extremum and is a maximum, there can be no maximum at $\pm\infty$. If there were, then there would have to be an interior minimum, which contradicts uniqueness.) ‖

Another way to find an MLE is to abandon differentiation and proceed with a direct maximization. This method is usually simpler algebraically, especially if the derivatives tend to get messy, but is sometimes harder to implement because there are no set rules to follow. One general technique is to find a global upper bound on the likelihood function and then establish that there is a unique point for which the upper bound is attained.

**Example 7.2.6　(Continuation of Example 7.2.5)** Recall (Theorem 5.2.4) that for any number $a$,

$$\sum_{i=1}^{n}(x_i - a)^2 \geq \sum_{i=1}^{n}(x_i - \bar{x})^2$$

with equality if and only if $a = \bar{x}$. This implies that for any $\theta$,

$$e^{-(1/2)\Sigma(x_i-\theta)^2} \leq e^{-(1/2)\Sigma(x_i-\bar{x})^2}$$

with equality if and only if $\theta = \bar{x}$. Hence $\bar{X}$ is the MLE. ‖

In most cases, especially when differentiation is to be used, it is easier to work with the natural logarithm of $L(\theta|\mathbf{x})$, $\log L(\theta|\mathbf{x})$ (known as the *log likelihood*), than it is to work with $L(\theta|\mathbf{x})$ directly. This is possible because the log function is strictly increasing on $(0, \infty)$, which implies that the extrema of $L(\theta|\mathbf{x})$ and $\log L(\theta|\mathbf{x})$ coincide (see Exercise 7.3).

**Example 7.2.7 (Bernoulli MLE)**　Let $X_1, \ldots, X_n$ be iid Bernoulli($p$). Then the likelihood function is

$$L(p|\mathbf{x}) = \prod_{i=1}^{n} p^{x_i}(1-p)^{1-x_i} = p^y(1-p)^{n-y},$$

where $y = \sum x_i$. While this function is not all that hard to differentiate, it is much easier to differentiate the log likelihood

$$\log L(p|\mathbf{x}) = y \log p + (n - y) \log(1 - p).$$

If $0 < y < n$, differentiating $\log L(p|\mathbf{x})$ and setting the result equal to 0 give the solution, $\hat{p} = y/n$. It is also straightforward to verify that $y/n$ is the global maximum in this case. If $y = 0$ or $y = n$, then

$$\log L(p|\mathbf{x}) = \begin{cases} n \log(1 - p) & \text{if } y = 0 \\ n \log p & \text{if } y = n. \end{cases}$$

In either case $\log L(p|\mathbf{x})$ is a monotone function of $p$, and it is again straightforward to verify that $\hat{p} = y/n$ in each case. Thus, we have shown that $\sum X_i/n$ is the MLE of $p$.                                                                                                      ‖

In this derivation we have assumed that the parameter space is $0 \le p \le 1$. The values $p = 0$ and 1 must be in the parameter space in order for $\hat{p} = y/n$ to be the MLE for $y = 0$ and $n$. Contrast this with Example 3.4.1, where we took $0 < p < 1$ to satisfy the requirements of an exponential family.

One other point to be aware of when finding a maximum likelihood estimator is that the maximization takes place only over the range of parameter values. In some cases this point plays an important part.

**Example 7.2.8 (Restricted range MLE)** Let $X_1, \ldots, X_n$ be iid $n(\theta, 1)$, where it is known that $\theta$ must be nonnegative. With no restrictions on $\theta$, we saw that the MLE of $\theta$ is $\bar{X}$; however, if $\bar{X}$ is negative, it will be outside the range of the parameter.

If $\bar{x}$ is negative, it is easy to check (see Exercise 7.4) that the likelihood function $L(\theta|\mathbf{x})$ is decreasing in $\theta$ for $\theta \ge 0$ and is maximized at $\hat{\theta} = 0$. Hence, in this case, the MLE of $\theta$ is

$$\hat{\theta} = \bar{X} \text{ if } \bar{X} \ge 0 \quad \text{and} \quad \hat{\theta} = 0 \text{ if } \bar{X} < 0. \qquad ‖$$

If $L(\theta|\mathbf{x})$ cannot be maximized analytically, it may be possible to use a computer and maximize $L(\theta|\mathbf{x})$ numerically. In fact, this is one of the most important features of MLEs. If a model (likelihood) can be written down, then there is some hope of maximizing it numerically and, hence, finding MLEs of the parameters. When this is done, there is still always the question of whether a local or global maximum has been found. Thus, it is always important to analyze the likelihood function as much as possible, to find the number and nature of its local maxima, before using numeric maximization.

**Example 7.2.9 (Binomial MLE, unknown number of trials)**   Let $X_1, \ldots, X_n$ be a random sample from a binomial$(k, p)$ population, where $p$ is known and $k$ is unknown. For example, we flip a coin we know to be fair and observe $x_i$ heads but we do not know how many times the coin was flipped. The likelihood function is

$$L(k|\mathbf{x}, p) = \prod_{i=1}^{n} \binom{k}{x_i} p^{x_i} (1 - p)^{k - x_i}.$$

Maximizing $L(k|\mathbf{x}, p)$ by differentiation is difficult because of the factorials and because $k$ must be an integer. Thus we try a different approach.

Of course, $L(k|\mathbf{x}, p) = 0$ if $k < \max_i x_i$. Thus the MLE is an integer $k \geq \max_i x_i$ that satisfies $L(k|\mathbf{x}, p)/L(k - 1|\mathbf{x}, p) \geq 1$ and $L(k + 1|\mathbf{x}, p)/L(k|\mathbf{x}, p) < 1$. We will show that there is only one such $k$. The ratio of likelihoods is

$$\frac{L(k|\mathbf{x}, p)}{L(k - 1|\mathbf{x}, p)} = \frac{(k(1 - p))^n}{\prod_{i=1}^{n}(k - x_i)}.$$

Thus the condition for a maximum is

$$(k(1 - p))^n \geq \prod_{i=1}^{n}(k - x_i) \quad \text{and} \quad ((k + 1)(1 - p))^n < \prod_{i=1}^{n}(k + 1 - x_i).$$

Dividing by $k^n$ and letting $z = 1/k$, we want to solve

$$(1 - p)^n = \prod_{i=1}^{n}(1 - x_i z)$$

for $0 \leq z \leq 1/\max_i x_i$. The right-hand side is clearly a strictly decreasing function of $z$ for $z$ in this range with a value of 1 at $z = 0$ and a value of 0 at $z = 1/\max_i x_i$. Thus there is a unique $z$ (call it $\hat{z}$) that solves the equation. The quantity $1/\hat{z}$ may not be an integer. But the integer $\hat{k}$ that satisfies the inequalities, and is the MLE, is the largest integer less than or equal to $1/\hat{z}$ (see Exercise 7.5). Thus, this analysis shows that there is a unique maximum for the likelihood function and it can be found by numerically solving an $n$th-degree polynomial equality. This description of the MLE for $k$ was found by Feldman and Fox (1968). See Example 7.2.13 for more about estimating $k$.                                                                    ‖

A useful property of maximum likelihood estimators is what has come to be known as the *invariance property of maximum likelihood estimators* (not to be confused with the type of invariance discussed in Chapter 6). Suppose that a distribution is indexed by a parameter $\theta$, but the interest is in finding an estimator for some function of $\theta$, say $\tau(\theta)$. Informally speaking, the invariance property of MLEs says that if $\hat{\theta}$ is the MLE of $\theta$, then $\tau(\hat{\theta})$ is the MLE of $\tau(\theta)$. For example, if $\theta$ is the mean of a normal distribution, the MLE of $\sin(\theta)$ is $\sin(\bar{X})$. We present the approach of Zehna (1966), but see Pal and Berry (1992) for alternative approaches to MLE invariance.

There are, of course, some technical problems to be overcome before we can formalize this notion of invariance of MLEs, and they mostly focus on the function $\tau(\theta)$ that we are trying to estimate. If the mapping $\theta \to \tau(\theta)$ is one-to-one (that is, for each value of $\theta$ there is a unique value of $\tau(\theta)$, and vice versa), then there is no problem. In this case, it is easy to see that it makes no difference whether we maximize the likelihood as a function of $\theta$ or as a function of $\tau(\theta)$ — in each case we get the same answer. If we let $\eta = \tau(\theta)$, then the inverse function $\tau^{-1}(\eta) = \theta$ is well defined and the likelihood function of $\tau(\theta)$, written as a function of $\eta$, is given by

$$L^*(\eta|\mathbf{x}) = \prod_{i=1}^{n} f(x_i|\tau^{-1}(\eta)) = L(\tau^{-1}(\eta)|\mathbf{x})$$

and

$$\sup_{\eta} L^*(\eta|\mathbf{x}) = \sup_{\eta} L(\tau^{-1}(\eta)|\mathbf{x}) = \sup_{\theta} L(\theta|\mathbf{x}).$$

Thus, the maximum of $L^*(\eta|\mathbf{x})$ is attained at $\eta = \tau(\theta) = \tau(\hat{\theta})$, showing that the MLE of $\tau(\theta)$ is $\tau(\hat{\theta})$.

In many cases, this simple version of the invariance of MLEs is not useful because many of the functions we are interested in are not one-to-one. For example, to estimate $\theta^2$, the square of a normal mean, the mapping $\theta \to \theta^2$ is not one-to-one. Thus, we need a more general theorem and, in fact, a more general definition of the likelihood function of $\tau(\theta)$.

If $\tau(\theta)$ is not one-to-one, then for a given value $\eta$ there may be more than one value of $\theta$ that satisfies $\tau(\theta) = \eta$. In such cases, the correspondence between the maximization over $\eta$ and that over $\theta$ can break down. For example, if $\hat{\theta}$ is the MLE of $\theta$, there may be another value of $\theta$, say $\theta_0$, for which $\tau(\hat{\theta}) = \tau(\theta_0)$. We need to avoid such difficulties.

We proceed by defining for $\tau(\theta)$ the *induced likelihood function* $L^*$, given by

$$(7.2.5) \qquad L^*(\eta|\mathbf{x}) = \sup_{\{\theta:\tau(\theta)=\eta\}} L(\theta|\mathbf{x}).$$

The value $\hat{\eta}$ that maximizes $L^*(\eta|\mathbf{x})$ will be called the MLE of $\eta = \tau(\theta)$, and it can be seen from (7.2.5) that the maxima of $L^*$ and $L$ coincide.

**Theorem 7.2.10 (Invariance property of MLEs)**   *If $\hat{\theta}$ is the MLE of $\theta$, then for any function $\tau(\theta)$, the MLE of $\tau(\theta)$ is $\tau(\hat{\theta})$.*

**Proof:** Let $\hat{\eta}$ denote the value that maximizes $L^*(\eta|\mathbf{x})$. We must show that $L^*(\hat{\eta}|\mathbf{x}) = L^*[\tau(\hat{\theta})|\mathbf{x}]$. Now, as stated above, the maxima of $L$ and $L^*$ coincide, so we have

$$
\begin{aligned}
L^*(\hat{\eta}|\mathbf{x}) &= \sup_{\eta}\ \sup_{\{\theta:\tau(\theta)=\eta\}} L(\theta|\mathbf{x}) && (\text{definition of } L^*) \\
&= \sup_{\theta} L(\theta|\mathbf{x}) \\
&= L(\hat{\theta}|\mathbf{x}), && (\text{definition of } \hat{\theta})
\end{aligned}
$$

where the second equality follows because the iterated maximization is equal to the unconditional maximization over $\theta$, which is attained at $\hat{\theta}$. Furthermore

$$
\begin{aligned}
L(\hat{\theta}|\mathbf{x}) &= \sup_{\{\theta:\tau(\theta)=\tau(\hat{\theta})\}} L(\theta|\mathbf{x}) && (\hat{\theta} \text{ is the MLE}) \\
&= L^*[\tau(\hat{\theta})|\mathbf{x}]. && (\text{definition of } L^*)
\end{aligned}
$$

Hence, the string of equalities shows that $L^*(\hat{\eta}|\mathbf{x}) = L^*(\tau(\hat{\theta})|\mathbf{x})$ and that $\tau(\hat{\theta})$ is the MLE of $\tau(\theta)$.                                                           $\square$

Using this theorem, we now see that the MLE of $\theta^2$, the square of a normal mean, is $\bar{X}^2$. We can also apply Theorem 7.2.10 to more complicated functions to see that, for example, the MLE of $\sqrt{p(1-p)}$, where $p$ is a binomial probability, is given by $\sqrt{\hat{p}(1-\hat{p})}$.

Before we leave the subject of finding maximum likelihood estimators, there are a few more points to be mentioned.

The invariance property of MLEs also holds in the multivariate case. There is nothing in the proof of Theorem 7.2.10 that precludes $\theta$ from being a vector. If the MLE of $(\theta_1, \ldots, \theta_k)$ is $(\hat{\theta}_1, \ldots, \hat{\theta}_k)$, and if $\tau(\theta_1, \ldots, \theta_k)$ is any function of the parameters, the MLE of $\tau(\theta_1, \ldots, \theta_k)$ is $\tau(\hat{\theta}_1, \ldots, \hat{\theta}_k)$.

If $\boldsymbol{\theta} = (\theta_1, \ldots, \theta_k)$ is multidimensional, then the problem of finding an MLE is that of maximizing a function of several variables. If the likelihood function is differentiable, setting the first partial derivatives equal to 0 provides a necessary condition for an extremum in the interior. However, in the multidimensional case, using a second derivative condition to check for a maximum is a tedious task, and other methods might be tried first. We first illustrate a technique that usually proves simpler, that of successive maximizations.

**Example 7.2.11 (Normal MLEs, $\mu$ and $\sigma$ unknown)**    Let $X_1, \ldots, X_n$ be iid $n(\theta, \sigma^2)$, with both $\theta$ and $\sigma^2$ unknown. Then

$$L(\theta, \sigma^2|\mathbf{x}) = \frac{1}{(2\pi\sigma^2)^{n/2}} e^{-(1/2)\Sigma_{i=1}^n (x_i-\theta)^2/\sigma^2}$$

and

$$\log L(\theta, \sigma^2|\mathbf{x}) = -\frac{n}{2}\log 2\pi - \frac{n}{2}\log \sigma^2 - \frac{1}{2}\sum_{i=1}^n (x_i - \theta)^2/\sigma^2.$$

The partial derivatives, with respect to $\theta$ and $\sigma^2$, are

$$\frac{\partial}{\partial\theta}\log L(\theta, \sigma^2|\mathbf{x}) = \frac{1}{\sigma^2}\sum_{i=1}^n (x_i - \theta)$$

and

$$\frac{\partial}{\partial\sigma^2}\log L(\theta, \sigma^2|\mathbf{x}) = -\frac{n}{2\sigma^2} + \frac{1}{2\sigma^4}\sum_{i=1}^n (x_i - \theta)^2.$$

Setting these partial derivatives equal to 0 and solving yields the solution $\hat{\theta} = \bar{x}, \hat{\sigma}^2 = n^{-1}\sum_{i=1}^n (x_i - \bar{x})^2$. To verify that this solution is, in fact, a global maximum, recall first that if $\theta \neq \bar{x}$, then $\sum(x_i - \theta)^2 > \sum(x_i - \bar{x})^2$. Hence, for any value of $\sigma^2$,

$$(7.2.6)\qquad \frac{1}{(2\pi\sigma^2)^{n/2}} e^{-(1/2)\Sigma_{i=1}^n (x_i-\bar{x})^2/\sigma^2} \geq \frac{1}{(2\pi\sigma^2)^{n/2}} e^{-(1/2)\Sigma_{i=1}^n (x_i-\theta)^2/\sigma^2}.$$

Therefore, verifying that we have found the maximum likelihood estimators is reduced to a one-dimensional problem, verifying that $(\sigma^2)^{-n/2}\exp(-\frac{1}{2}\sum(x_i-\bar{x})^2/\sigma^2)$ achieves its global maximum at $\sigma^2 = n^{-1}\sum(x_i - \bar{x})^2$. This is straightforward to do using univariate calculus and, in fact, the estimators $(\bar{X}, n^{-1}\sum(X_i - \bar{X})^2)$ are the MLEs.

We note that the left side of the inequality in (7.2.6) is known as the *profile likelihood* for $\sigma^2$. See Miscellanea 7.5.5.                                                              ‖

Now consider the solution to the same problem using two-variate calculus.

**Example 7.2.12 (Continuation of Example 7.2.11)**      To use two-variate calculus to verify that a function $H(\theta_1, \theta_2)$ has a local maximum at $(\hat{\theta}_1, \hat{\theta}_2)$, it must be shown that the following three conditions hold.

a. The first-order partial derivatives are 0,

$$\frac{\partial}{\partial\theta_1}H(\theta_1, \theta_2)|_{\theta_1=\hat{\theta}_1, \theta_2=\hat{\theta}_2} = 0 \quad \text{and} \quad \frac{\partial}{\partial\theta_2}H(\theta_1, \theta_2)|_{\theta_1=\hat{\theta}_1, \theta_2=\hat{\theta}_2} = 0.$$

b. At least one second-order partial derivative is negative,

$$\frac{\partial^2}{\partial\theta_1^2}H(\theta_1, \theta_2)|_{\theta_1=\hat{\theta}_1, \theta_2=\hat{\theta}_2} < 0 \quad \text{or} \quad \frac{\partial^2}{\partial\theta_2^2}H(\theta_1, \theta_2)|_{\theta_1=\hat{\theta}_1, \theta_2=\hat{\theta}_2} < 0.$$

c. The Jacobian of the second-order partial derivatives is positive,

$$\begin{vmatrix} \frac{\partial^2}{\partial\theta_1^2}H(\theta_1, \theta_2) & \frac{\partial^2}{\partial\theta_1\partial\theta_2}H(\theta_1, \theta_2) \\ \frac{\partial^2}{\partial\theta_1\partial\theta_2}H(\theta_1, \theta_2) & \frac{\partial^2}{\partial\theta_2^2}H(\theta_1, \theta_2) \end{vmatrix}_{\theta_1=\hat{\theta}_1, \theta_2=\hat{\theta}_2}$$

$$= \frac{\partial^2}{\partial\theta_1^2}H(\theta_1, \theta_2)\frac{\partial^2}{\partial\theta_2^2}H(\theta_1, \theta_2) - \left(\frac{\partial^2}{\partial\theta_1\partial\theta_2}H(\theta_1, \theta_2)\right)^2\bigg|_{\theta_1=\hat{\theta}_1, \theta_2=\hat{\theta}_2} > 0.$$

For the normal log likelihood, the second-order partial derivatives are

$$\frac{\partial^2}{\partial\theta^2}\log L(\theta, \sigma^2|\mathbf{x}) = \frac{-n}{\sigma^2},$$

$$\frac{\partial^2}{\partial(\sigma^2)^2}\log L(\theta, \sigma^2|\mathbf{x}) = \frac{n}{2\sigma^4} - \frac{1}{\sigma^6}\sum_{i=1}^{n}(x_i - \theta)^2,$$

$$\frac{\partial^2}{\partial\theta\,\partial\sigma^2}\log L(\theta, \sigma^2|\mathbf{x}) = -\frac{1}{\sigma^4}\sum_{i=1}^{n}(x_i - \theta).$$

Properties (a) and (b) are easily seen to hold, and the Jacobian is

$$\begin{vmatrix} \frac{-n}{\sigma^2} & -\frac{1}{\sigma^4}\sum_{i=1}^{n}(x_i - \theta) \\ -\frac{1}{\sigma^4}\sum_{i=1}^{n}(x_i - \theta) & \frac{n}{2\sigma^4} - \frac{1}{\sigma^6}\sum_{i=1}^{n}(x_i - \theta)^2 \end{vmatrix}_{\theta=\bar{x}, \sigma^2=\hat{\sigma}^2}$$

$$= \frac{1}{\sigma^6} \left[ \frac{-n^2}{2} + \frac{n}{\sigma^2} \sum_{i=1}^{n} (x_i - \theta)^2 - \frac{1}{\sigma^2} \left( \sum_{i=1}^{n} (x_i - \theta) \right)^2 \right] \Bigg|_{\theta = \bar{x}, \sigma^2 = \hat{\sigma}^2}$$

$$= \frac{1}{\hat{\sigma}^6} \left[ \frac{-n^2}{2} + \frac{n^2}{\hat{\sigma}^2} \hat{\sigma}^2 - \frac{1}{\hat{\sigma}^2} \left( \sum_{i=1}^{n} (x_i - \bar{x}) \right)^2 \right]$$

$$= \frac{1}{\hat{\sigma}^6} \frac{n^2}{2} > 0.$$

Thus, the calculus conditions are satisfied and we have indeed found a maximum. (Of course, to be really formal, we have verified that $(\bar{x}, \hat{\sigma}^2)$ is *an* interior maximum. We still have to check that it is unique and that there is no maximum at infinity.) The amount of calculation, even in this simple problem, is formidable, and things will only get worse. (Think of what we would have to do for three parameters.) Thus, the moral is that, while we always have to verify that we have, indeed, found a maximum, we should look for ways to do it other than using second derivative conditions.    ‖

Finally, it was mentioned earlier that, since MLEs are found by a maximization process, they are susceptible to the problems associated with that process, among them that of numerical instability. We now look at this problem in more detail.

Recall that the likelihood function is a function of the parameter, $\theta$, with the data, **x**, held constant. However, since the data are measured with error, we might ask how small changes in the data might affect the MLE. That is, we calculate $\hat{\theta}$ based on $L(\theta|\mathbf{x})$, but we might inquire what value we would get for the MLE if we based our calculations on $L(\theta|\mathbf{x} + \epsilon)$, for small $\epsilon$. Intuitively, this new MLE, say $\hat{\theta}_1$, should be close to $\hat{\theta}$ if $\epsilon$ is small. But this is not always the case.

**Example 7.2.13 (Continuation of Example 7.2.2)**  Olkin, Petkau, and Zidek (1981) demonstrate that the MLEs of $k$ and $p$ in binomial sampling can be highly unstable. They illustrate their case with the following example. Five realizations of a binomial$(k, p)$ experiment are observed, where both $k$ and $p$ are unknown. The first data set is (16, 18, 22, 25, 27). (These are the observed numbers of successes from an unknown number of binomial trials.) For this data set, the MLE of $k$ is $\hat{k} = 99$. If a second data set is (16, 18, 22, 25, 28), where the only difference is that the 27 is replaced with 28, then the MLE of $k$ is $\hat{k} = 190$, demonstrating a large amount of variability.    ‖

Such occurrences happen when the likelihood function is very flat in the neighborhood of its maximum or when there is no finite maximum. When the MLEs can be found explicitly, as will often be the case in our examples, this is usually not a problem. However, in many instances, such as in the above example, the MLE cannot be solved for explicitly and must be found by numeric methods. When faced with such a problem, it is often wise to spend a little extra time investigating the stability of the solution.

### 7.2.3 Bayes Estimators

The Bayesian approach to statistics is fundamentally different from the classical approach that we have been taking. Nevertheless, some aspects of the Bayesian approach can be quite helpful to other statistical approaches. Before going into the methods for finding Bayes estimators, we first discuss the Bayesian approach to statistics.

In the classical approach the parameter, $\theta$, is thought to be an unknown, but fixed, quantity. A random sample $X_1, \ldots, X_n$ is drawn from a population indexed by $\theta$ and, based on the observed values in the sample, knowledge about the value of $\theta$ is obtained. In the Bayesian approach $\theta$ is considered to be a quantity whose variation can be described by a probability distribution (called the *prior distribution*). This is a subjective distribution, based on the experimenter's belief, and is formulated before the data are seen (hence the name prior distribution). A sample is then taken from a population indexed by $\theta$ and the prior distribution is updated with this sample information. The updated prior is called the *posterior distribution*. This updating is done with the use of Bayes' Rule (seen in Chapter 1), hence the name Bayesian statistics.

If we denote the prior distribution by $\pi(\theta)$ and the sampling distribution by $f(\mathbf{x}|\theta)$, then the posterior distribution, the conditional distribution of $\theta$ given the sample, $\mathbf{x}$, is

$$(7.2.7) \qquad \pi(\theta|\mathbf{x}) = f(\mathbf{x}|\theta)\pi(\theta)/m(\mathbf{x}), \qquad (f(\mathbf{x}|\theta)\pi(\theta) = f(\mathbf{x}, \theta))$$

where $m(\mathbf{x})$ is the marginal distribution of $\mathbf{X}$, that is,

$$(7.2.8) \qquad m(\mathbf{x}) = \int f(\mathbf{x}|\theta)\pi(\theta)d\theta.$$

Notice that the posterior distribution is a conditional distribution, conditional upon observing the sample. The posterior distribution is now used to make statements about $\theta$, which is still considered a random quantity. For instance, the mean of the posterior distribution can be used as a point estimate of $\theta$.

*A note on notation:* When dealing with distributions on a parameter, $\theta$, we will break our notation convention of using uppercase letters for random variables and lowercase letters for arguments. Thus, we may speak of the random quantity $\theta$ with distribution $\pi(\theta)$. This is more in line with common usage and should not cause confusion.

**Example 7.2.14 (Binomial Bayes estimation)**     Let $X_1, \ldots, X_n$ be iid Bernoulli($p$). Then $Y = \sum X_i$ is binomial($n, p$). We assume the prior distribution on $p$ is beta($\alpha, \beta$). The joint distribution of $Y$ and $p$ is

$$f(y, p) = \left[\binom{n}{y} p^y (1-p)^{n-y}\right]\left[\frac{\Gamma(\alpha+\beta)}{\Gamma(\alpha)\Gamma(\beta)} p^{\alpha-1}(1-p)^{\beta-1}\right] \quad \left(\begin{array}{c}\text{conditional} \times \text{marginal}\\ f(y|p) \times \pi(p)\end{array}\right)$$

$$= \binom{n}{y} \frac{\Gamma(\alpha+\beta)}{\Gamma(\alpha)\Gamma(\beta)} p^{y+\alpha-1}(1-p)^{n-y+\beta-1}.$$

The marginal pdf of $Y$ is

$$(7.2.9) \qquad f(y) = \int_0^1 f(y,p)dp = \binom{n}{y} \frac{\Gamma(\alpha+\beta)}{\Gamma(\alpha)\Gamma(\beta)} \frac{\Gamma(y+\alpha)\Gamma(n-y+\beta)}{\Gamma(n+\alpha+\beta)},$$

a distribution known as the beta-binomial (see Exercise 4.34 and Example 4.4.6). The posterior distribution, the distribution of $p$ given $y$, is

$$f(p|y) = \frac{f(y,p)}{f(y)} = \frac{\Gamma(n+\alpha+\beta)}{\Gamma(y+\alpha)\Gamma(n-y+\beta)} p^{y+\alpha-1}(1-p)^{n-y+\beta-1},$$

which is beta$(y+\alpha, n-y+\beta)$. (Remember that $p$ is the variable and $y$ is treated as fixed.) A natural estimate for $p$ is the mean of the posterior distribution, which would give us as the Bayes estimator of $p$,

$$\hat{p}_{\mathrm{B}} = \frac{y+\alpha}{\alpha+\beta+n}. \qquad\qquad \|$$

Consider how the Bayes estimate of $p$ is formed. The prior distribution has mean $\alpha/(\alpha+\beta)$, which would be our best estimate of $p$ without having seen the data. Ignoring the prior information, we would probably use $p = y/n$ as our estimate of $p$. The Bayes estimate of $p$ combines all of this information. The manner in which this information is combined is made clear if we write $\hat{p}_B$ as

$$\hat{p}_{\mathrm{B}} = \left(\frac{n}{\alpha+\beta+n}\right)\left(\frac{y}{n}\right) + \left(\frac{\alpha+\beta}{\alpha+\beta+n}\right)\left(\frac{\alpha}{\alpha+\beta}\right).$$

Thus $\hat{p}_{\mathrm{B}}$ is a linear combination of the prior mean and the sample mean, with the weights being determined by $\alpha$, $\beta$, and $n$.

When estimating a binomial parameter, it is not necessary to choose a prior distribution from the beta family. However, there was a certain advantage to choosing the beta family, not the least of which being that we obtained a closed-form expression for the estimator. In general, for any sampling distribution, there is a natural family of prior distributions, called the conjugate family.

**Definition 7.2.15** Let $\mathcal{F}$ denote the class of pdfs or pmfs $f(x|\theta)$ (indexed by $\theta$). A class $\prod$ of prior distributions is a *conjugate family* for $\mathcal{F}$ if the posterior distribution is in the class $\prod$ for all $f \in \mathcal{F}$, all priors in $\prod$, and all $x \in \mathcal{X}$.

The beta family is conjugate for the binomial family. Thus, if we start with a beta prior, we will end up with a beta posterior. The updating of the prior takes the form of updating its parameters. Mathematically, this is very convenient, for it usually makes calculation quite easy. Whether or not a conjugate family is a reasonable choice for a particular problem, however, is a question to be left to the experimenter.

We end this section with one more example.

**Example 7.2.16 (Normal Bayes estimators)**    Let $X \sim n(\theta, \sigma^2)$, and suppose that the prior distribution on $\theta$ is $n(\mu, \tau^2)$. (Here we assume that $\sigma^2$, $\mu$, and $\tau^2$ are all known.) The posterior distribution of $\theta$ is also normal, with mean and variance given by

$$E(\theta|x) = \frac{\tau^2}{\tau^2 + \sigma^2} x + \frac{\sigma^2}{\sigma^2 + \tau^2} \mu,$$

(7.2.10)

$$\text{Var}\,(\theta|x) = \frac{\sigma^2 \tau^2}{\sigma^2 + \tau^2}.$$

(See Exercise 7.22 for details.) Notice that the normal family is its own conjugate family. Again using the posterior mean, we have the Bayes estimator of $\theta$ is $E(\theta|X)$.

The Bayes estimator is, again, a linear combination of the prior and sample means. Notice also that as $\tau^2$, the prior variance, is allowed to tend to infinity, the Bayes estimator tends toward the sample mean. We can interpret this as saying that, as the prior information becomes more vague, the Bayes estimator tends to give more weight to the sample information. On the other hand, if the prior information is good, so that $\sigma^2 > \tau^2$, then more weight is given to the prior mean.      ‖

### 7.2.4 The EM Algorithm[1]

A last method that we will look at for finding estimators is inherently different in its approach and specifically designed to find MLEs. Rather than detailing a procedure for solving for the MLE, we specify an algorithm that is guaranteed to converge to the MLE. This algorithm is called the EM (**E**xpectation-**M**aximization) algorithm. It is based on the idea of replacing one difficult likelihood maximization with a sequence of easier maximizations whose limit is the answer to the original problem. It is particularly suited to "missing data" problems, as the very fact that there are missing data can sometimes make calculations cumbersome. However, we will see that filling in the "missing data" will often make the calculation go more smoothly. (We will also see that "missing data" have different interpretations–see, for example, Exercise 7.30.)

In using the EM algorithm we consider two different likelihood problems. The problem that we are interested in solving is the "incomplete-data" problem, and the problem that we actually solve is the "complete-data problem." Depending on the situation, we can start with either problem.

**Example 7.2.17 (Multiple Poisson rates)**   We observe $X_1, \ldots, X_n$ and $Y_1, \ldots, Y_n$, all mutually independent, where $Y_i \sim \text{Poisson}(\beta\tau_i)$ and $X_i \sim \text{Poisson}(\tau_i)$. This would model, for instance, the incidence of a disease, $Y_i$, where the underlying rate is a function of an overall effect $\beta$ and an additional factor $\tau_i$. For example, $\tau_i$ could be a measure of population density in area $i$, or perhaps health status of the population in area $i$. We do not see $\tau_i$ but get information on it through $X_i$.

---

[1] This section contains material that is somewhat specialized and more advanced. It may be skipped without interrupting the flow of the text.

The joint pmf is therefore

$$f((x_1, y_1), (x_2, y_2), \ldots, (x_n, y_n)|\beta, \tau_1, \tau_2, \ldots, \tau_n)$$

(7.2.11)
$$= \prod_{i=1}^{n} \frac{e^{-\beta\tau_i}(\beta\tau_i)^{y_i}}{y_i!} \frac{e^{-\tau_i}(\tau_i)^{x_i}}{x_i!}.$$

The likelihood estimators, which can be found by straightforward differentiation (see Exercise 7.27) are

(7.2.12)
$$\hat{\beta} = \frac{\sum_{i=1}^{n} y_i}{\sum_{i=1}^{n} x_i} \quad \text{and} \quad \hat{\tau}_j = \frac{x_j + y_j}{\hat{\beta} + 1}, \quad j = 1, 2, \ldots, n.$$

The likelihood based on the pmf (7.2.11) is the complete-data likelihood, and $((x_1, y_1), (x_2, y_2), \ldots, (x_n, y_n))$ is called the complete data. Missing data, which is a common occurrence, would make estimation more difficult. Suppose, for example, that the value of $x_1$ was missing. We could also discard $y_1$ and proceed with a sample of size $n - 1$, but this is ignoring the information in $y_1$. Using this information would improve our estimates.

Starting from the pmf (7.2.11), the pmf of the sample with $x_1$ missing is

(7.2.13)
$$\sum_{x_1=0}^{\infty} f((x_1, y_1), (x_2, y_2), \ldots, (x_n, y_n)|\beta, \tau_1, \tau_2, \ldots, \tau_n).$$

The likelihood based on (7.2.13) is the incomplete-data likelihood. This is the likelihood that we need to maximize.                                                              ‖

In general, we can move in either direction, from the complete-data problem to the incomplete-data problem or the reverse. If $\mathbf{Y} = (Y_1, \ldots, Y_n)$ are the *incomplete data*, and $\mathbf{X} = (X_1, \ldots, X_m)$ are the *augmented* data, making $(\mathbf{Y}, \mathbf{X})$ the *complete data*, the densities $g(\cdot|\theta)$ of $\mathbf{Y}$ and $f(\cdot|\theta)$ of $(\mathbf{Y}, \mathbf{X})$ have the relationship

(7.2.14)
$$g(\mathbf{y}|\theta) = \int f(\mathbf{y}, \mathbf{x}|\theta) \, d\mathbf{x}$$

with sums replacing integrals in the discrete case.

If we turn these into likelihoods, $L(\theta|\mathbf{y}) = g(\mathbf{y}|\theta)$ is the *incomplete-data likelihood* and $L(\theta|\mathbf{y}, \mathbf{x}) = f(\mathbf{y}, \mathbf{x}|\theta)$ is the *complete-data likelihood*. If $L(\theta|\mathbf{y})$ is difficult to work with, it will sometimes be the case that the complete-data likelihood will be easier to work with.

**Example 7.2.18 (Continuation of Example 7.2.17)**  The incomplete-data likelihood is obtained from (7.2.11) by summing over $x_1$. This gives

$$L(\beta, \tau_1, \tau_2, \ldots, \tau_n|y_1, (x_2, y_2), \ldots, (x_n, y_n))$$

(7.2.15)
$$= \left[ \prod_{i=1}^{n} \frac{e^{-\beta\tau_i}(\beta\tau_i)^{y_i}}{y_i!} \right] \left[ \prod_{i=2}^{n} \frac{e^{-\tau_i}(\tau_i)^{x_i}}{x_i!} \right],$$

and $(y_1, (x_2, y_2), \ldots, (x_n, y_n))$ is the incomplete data. This is the likelihood that we need to maximize. Differentiation leads to the MLE equations

$$\hat{\beta} = \frac{\sum_{i=1}^{n} y_i}{\sum_{i=1}^{n} \hat{\tau}_i},$$

(7.2.16)
$$y_1 = \hat{\tau}_1 \hat{\beta},$$

$$x_j + y_j = \hat{\tau}_j(\hat{\beta} + 1), \quad j = 2, 3, \ldots, n,$$

which we now solve with the EM algorithm. ‖

The EM algorithm allows us to maximize $L(\theta|\mathbf{y})$ by working with only $L(\theta|\mathbf{y}, \mathbf{x})$ and the conditional pdf or pmf of $\mathbf{X}$ given $\mathbf{y}$ and $\theta$, defined by

(7.2.17) $\quad L(\theta|\mathbf{y}, \mathbf{x}) = f(\mathbf{y}, \mathbf{x}|\theta), \quad L(\theta|\mathbf{y}) = g(\mathbf{y}|\theta), \quad \text{and} \quad k(\mathbf{x}|\theta, \mathbf{y}) = \dfrac{f(\mathbf{y}, \mathbf{x}|\theta)}{g(\mathbf{y}|\theta)}.$

Rearrangement of the last equation in (7.2.17) gives the identity

(7.2.18) $\qquad \log L(\theta|\mathbf{y}) = \log L(\theta|\mathbf{y}, \mathbf{x}) - \log k(\mathbf{x}|\theta, \mathbf{y}).$

As $\mathbf{x}$ is missing data and hence not observed, we replace the right side of (7.2.18) with its expectation under $k(\mathbf{x}|\theta', \mathbf{y})$, creating the new identity

(7.2.19) $\qquad \log L(\theta|\mathbf{y}) = \mathrm{E}\left[\log L(\theta|\mathbf{y}, \mathbf{X})|\theta', \mathbf{y}\right] - \mathrm{E}\left[\log k(\mathbf{X}|\theta, \mathbf{y})|\theta', \mathbf{y}\right].$

Now we start the algorithm: From an initial value $\theta^{(0)}$ we create a sequence $\theta^{(r)}$ according to

(7.2.20) $\qquad \theta^{(r+1)} = $ the value that maximizes $\mathrm{E}\left[\log L(\theta|\mathbf{y}, \mathbf{X})|\theta^{(r)}, \mathbf{y}\right].$

The "E-step" of the algorithm calculates the expected log likelihood, and the "M-step" finds its maximum. Before we look into why this algorithm actually converges to the MLE, let us return to our example.

**Example 7.2.19 (Conclusion of Example 7.2.17)** Let $(\mathbf{x}, \mathbf{y}) = ((x_1, y_1), (x_2, y_2), \ldots, (x_n, y_n))$ denote the complete data and $(\mathbf{x}_{(-1)}, \mathbf{y}) = (y_1, (x_2, y_2), \ldots, (x_n, y_n))$ denote the incomplete data. The expected complete-data log likelihood is

$$\mathrm{E}[\log L(\beta, \tau_1, \tau_2, \ldots, \tau_n|(\mathbf{x}, \mathbf{y}))|\tau^{(r)}, (\mathbf{x}_{(-1)}, \mathbf{y})]$$

$$= \sum_{x_1=0}^{\infty} \log \left( \prod_{i=1}^{n} \frac{e^{-\beta \tau_i}(\beta \tau_i)^{y_i}}{y_i!} \frac{e^{-\tau_i}(\tau_i)^{x_i}}{x_i!} \right) \frac{e^{-\tau_1^{(r)}}(\tau_1^{(r)})^{x_1}}{x_1!}$$

$$= \sum_{i=1}^{n} [-\beta \tau_i + y_i(\log \beta + \log \tau_i) - \log y_i!] + \sum_{i=2}^{n} [-\tau_i + x_i \log \tau_i - \log x_i!]$$

(7.2.21)
$$+ \sum_{x_1=0}^{\infty} [-\tau_1 + x_1 \log \tau_1 - \log x_1!] \frac{e^{-\tau_1^{(r)}}(\tau_1^{(r)})^{x_1}}{x_1!}$$

$$= \left( \sum_{i=1}^{n} [-\beta\tau_i + y_i(\log\beta + \log\tau_i)] + \sum_{i=2}^{n} [-\tau_i + x_i\log\tau_i] + \sum_{x_1=0}^{\infty} [-\tau_1 + x_1\log\tau_1] \frac{e^{-\tau_1^{(r)}}(\tau_1^{(r)})^{x_1}}{x_1!} \right)$$

$$- \left( \sum_{i=1}^{n} \log y_i! + \sum_{i=2}^{n} \log x_i! + \sum_{x_1=0}^{\infty} \log x_1! \frac{e^{-\tau_1^{(r)}}(\tau_1^{(r)})^{x_1}}{x_1!} \right),$$

where in the last equality we have grouped together terms involving $\beta$ and $\tau_i$ and terms that do not involve these parameters. Since we are calculating this expected log likelihood for the purpose of maximizing it in $\beta$ and $\tau_i$, we can ignore the terms in the second set of parentheses. We thus have to maximize only the terms in the first set of parentheses, where we can write the last sum as

$$(7.2.22) \qquad -\tau_1 + \log\tau_1 \sum_{x_1=0}^{\infty} x_1 \frac{e^{-\tau_1^{(r)}}(\tau_1^{(r)})^{x_1}}{x_1!} = -\tau_1 + \tau_1^{(r)}\log\tau_1.$$

When substituting this back into (7.2.21), we see that the expected complete-data likelihood is the same as the original complete-data likelihood, with the exception that $x_1$ is replaced by $\tau_1^{(r)}$. Thus, in the $r$th step the MLEs are only a minor variation of (7.2.12) and are given by

$$(7.2.23) \qquad \hat{\beta}^{(r+1)} = \frac{\sum_{i=1}^{n} y_i}{\tau_1^{(r)} + \sum_{i=2}^{n} x_i}, \qquad \hat{\tau}_1^{(r+1)} = \frac{\hat{\tau}_1^{(r)} + y_1}{\hat{\beta}^{(r+1)} + 1},$$

$$\hat{\tau}_j^{(r+1)} = \frac{x_j + y_j}{\hat{\beta}^{(r+1)} + 1}, \qquad j = 2, 3, \ldots, n.$$

This defines both the E-step (which results in the substitution of $\hat{\tau}_1^{(r)}$ for $x_1$) and the M-step (which results in the calculation in (7.2.23) for the MLEs at the $r$th iteration. The properties of the EM algorithm give us assurance that the sequence $(\hat{\beta}^{(r)}, \hat{\tau}_1^{(r)}, \hat{\tau}_2^{(r)}, \ldots, \hat{\tau}_n^{(r)})$ converges to the incomplete-data MLE as $r \to \infty$. See Exercise 7.27 for more. $\|$

We will not give a complete proof that the EM sequence $\{\hat{\theta}^{(r)}\}$ converges to the incomplete-data MLE, but the following key property suggests that this is true. The proof is left to Exercise 7.31.

**Theorem 7.2.20 (Monotonic EM sequence)** *The sequence $\{\hat{\theta}^{(r)}\}$ defined by (7.2.20) satisfies*

$$(7.2.24) \qquad L\left(\hat{\theta}^{(r+1)}|\mathbf{y}\right) \geq L\left(\hat{\theta}^{(r)}|\mathbf{y}\right),$$

*with equality holding if and only if successive iterations yield the same value of the maximized expected complete-data log likelihood, that is,*

$$E\left[\log L(\hat{\theta}^{(r+1)}|\mathbf{y},\mathbf{X})|\hat{\theta}^{(r)},\mathbf{y}\right] = E\left[\log L\left(\hat{\theta}^{(r)}|\mathbf{y},\mathbf{X}\right)|\hat{\theta}^{(r)},\mathbf{y}\right].$$

## 7.3 Methods of Evaluating Estimators

The methods discussed in the previous section have outlined reasonable techniques for finding point estimators of parameters. A difficulty that arises, however, is that since we can usually apply more than one of these methods in a particular situation, we are often faced with the task of choosing between estimators. Of course, it is possible that different methods of finding estimators will yield the same answer, which makes evaluation a bit easier, but, in many cases, different methods will lead to different estimators.

The general topic of evaluating statistical procedures is part of the branch of statistics known as decision theory, which will be treated in some detail in Section 7.3.4. However, no procedure should be considered until some clues about its performance have been gathered. In this section we will introduce some basic criteria for evaluating estimators, and examine several estimators against these criteria.

### 7.3.1 Mean Squared Error

We first investigate finite-sample measures of the quality of an estimator, beginning with its mean squared error.

**Definition 7.3.1**   The *mean squared error* (MSE) of an estimator $W$ of a parameter $\theta$ is the function of $\theta$ defined by $\mathrm{E}_\theta(W - \theta)^2$.

Notice that the MSE measures the average squared difference between the estimator $W$ and the parameter $\theta$, a somewhat reasonable measure of performance for a point estimator. In general, any increasing function of the absolute distance $|W - \theta|$ would serve to measure the goodness of an estimator (mean absolute error, $\mathrm{E}_\theta(|W - \theta|)$, is a reasonable alternative), but MSE has at least two advantages over other distance measures: First, it is quite tractable analytically and, second, it has the interpretation

$$(7.3.1) \qquad \mathrm{E}_\theta(W - \theta)^2 = \mathrm{Var}_\theta\, W + (\mathrm{E}_\theta W - \theta)^2 = \mathrm{Var}_\theta\, W + (\mathrm{Bias}_\theta\, W)^2,$$

where we define the bias of an estimator as follows.

**Definition 7.3.2**   The *bias* of a point estimator $W$ of a parameter $\theta$ is the difference between the expected value of $W$ and $\theta$; that is, $\mathrm{Bias}_\theta\, W = \mathrm{E}_\theta W - \theta$. An estimator whose bias is identically (in $\theta$) equal to 0 is called *unbiased* and satisfies $\mathrm{E}_\theta W = \theta$ for all $\theta$.

Thus, MSE incorporates two components, one measuring the variability of the estimator (precision) and the other measuring its bias (accuracy). An estimator that has good MSE properties has small combined variance and bias. To find an estimator with good MSE properties, we need to find estimators that control both variance and bias. Clearly, unbiased estimators do a good job of controlling bias.

For an unbiased estimator we have

$$\mathrm{E}_\theta(W - \theta)^2 = \mathrm{Var}_\theta\, W,$$

and so, if an estimator is unbiased, its MSE is equal to its variance.

**Example 7.3.3 (Normal MSE)**   Let $X_1, \ldots, X_n$ be iid $n(\mu, \sigma^2)$. The statistics $\bar{X}$ and $S^2$ are both unbiased estimators since

$$E\bar{X} = \mu, \quad ES^2 = \sigma^2, \quad \text{for all } \mu \text{ and } \sigma^2.$$

(This is true without the normality assumption; see Theorem 5.2.6.) The MSEs of these estimators are given by

$$E(\bar{X} - \mu)^2 = \operatorname{Var}\bar{X} = \frac{\sigma^2}{n},$$

$$E(S^2 - \sigma^2)^2 = \operatorname{Var} S^2 = \frac{2\sigma^4}{n-1}.$$

The MSE of $\bar{X}$ remains $\sigma^2/n$ even if the normality assumption is dropped. However, the above expression for the MSE of $S^2$ does not remain the same if the normality assumption is relaxed (see Exercise 5.8).                                    ∥

Although many unbiased estimators are also reasonable from the standpoint of MSE, be aware that controlling bias does not guarantee that MSE is controlled. In particular, it is sometimes the case that a trade-off occurs between variance and bias in such a way that a small increase in bias can be traded for a larger decrease in variance, resulting in an improvement in MSE.

**Example 7.3.4 (Continuation of Example 7.3.3)**   An alternative estimator for $\sigma^2$ is the maximum likelihood estimator $\hat{\sigma}^2 = \frac{1}{n}\sum_{i=1}^{n}(X_i - \bar{X})^2 = \frac{n-1}{n}S^2$. It is straightforward to calculate

$$E\hat{\sigma}^2 = E\left(\frac{n-1}{n}S^2\right) = \frac{n-1}{n}\sigma^2,$$

so $\hat{\sigma}^2$ is a biased estimator of $\sigma^2$. The variance of $\hat{\sigma}^2$ can also be calculated as

$$\operatorname{Var}\hat{\sigma}^2 = \operatorname{Var}\left(\frac{n-1}{n}S^2\right) = \left(\frac{n-1}{n}\right)^2 \operatorname{Var} S^2 = \frac{2(n-1)\sigma^4}{n^2},$$

and, hence, its MSE is given by

$$E(\hat{\sigma}^2 - \sigma^2)^2 = \frac{2(n-1)\sigma^4}{n^2} + \left(\frac{n-1}{n}\sigma^2 - \sigma^2\right)^2 = \left(\frac{2n-1}{n^2}\right)\sigma^4.$$

We thus have

$$E(\hat{\sigma}^2 - \sigma^2)^2 = \left(\frac{2n-1}{n^2}\right)\sigma^4 < \left(\frac{2}{n-1}\right)\sigma^4 = E(S^2 - \sigma^2)^2,$$

showing that $\hat{\sigma}^2$ has smaller MSE than $S^2$. Thus, by trading off variance for bias, the MSE is improved.                                    ∥

We hasten to point out that the above example does not imply that $S^2$ should be abandoned as an estimator of $\sigma^2$. The above argument shows that, on the average, $\hat{\sigma}^2$ will be closer to $\sigma^2$ than $S^2$ if MSE is used as a measure. However, $\hat{\sigma}^2$ is biased and will, on the average, underestimate $\sigma^2$. This fact alone may make us uncomfortable about using $\hat{\sigma}^2$ as an estimator of $\sigma^2$. Furthermore, it can be argued that MSE, while a reasonable criterion for location parameters, is not reasonable for scale parameters, so the above comparison should not even be made. (One problem is that MSE penalizes equally for overestimation and underestimation, which is fine in the location case. In the scale case, however, 0 is a natural lower bound, so the estimation problem is not symmetric. Use of MSE in this case tends to be forgiving of underestimation.) The end result of this is that no absolute answer is obtained but rather more information is gathered about the estimators in the hope that, for a particular situation, a good estimator is chosen.

In general, since MSE is a function of the parameter, there will not be one "best" estimator. Often, the MSEs of two estimators will cross each other, showing that each estimator is better (with respect to the other) in only a portion of the parameter space. However, even this partial information can sometimes provide guidelines for choosing between estimators.

**Example 7.3.5 (MSE of binomial Bayes estimator)**     Let $X_1, \ldots, X_n$ be iid Bernoulli$(p)$. The MSE of $\hat{p}$, the MLE, as an estimator of $p$, is

$$\mathrm{E}_p(\hat{p} - p)^2 = \mathrm{Var}_p \, \bar{X} = \frac{p(1-p)}{n}.$$

Let $Y = \sum X_i$ and recall the Bayes estimator derived in Example 7.2.14, $\hat{p}_\mathrm{B} = \frac{Y+\alpha}{\alpha+\beta+n}$. The MSE of this Bayes estimator of $p$ is

$$\mathrm{E}_p(\hat{p}_B - p)^2 = \mathrm{Var}_p \, \hat{p}_B + (\mathrm{Bias}_p \, \hat{p}_B)^2$$

$$= \mathrm{Var}_p \left( \frac{Y+\alpha}{\alpha+\beta+n} \right) + \left( \mathrm{E}_p \left( \frac{Y+\alpha}{\alpha+\beta+n} \right) - p \right)^2$$

$$= \frac{np(1-p)}{(\alpha+\beta+n)^2} + \left( \frac{np+\alpha}{\alpha+\beta+n} - p \right)^2.$$

In the absence of good prior information about $p$, we might try to choose $\alpha$ and $\beta$ to make the MSE of $\hat{p}_B$ constant. The details are not too difficult to work out (see Exercise 7.33), and the choice $\alpha = \beta = \sqrt{n}/4$ yields

$$\hat{p}_B = \frac{Y + \sqrt{n}/4}{n + \sqrt{n}} \quad \text{and} \quad \mathrm{E}(\hat{p}_B - p)^2 = \frac{n}{4(n + \sqrt{n})^2}.$$

If we want to choose between $\hat{p}_B$ and $\hat{p}$ on the basis of MSE, Figure 7.3.1 is helpful. For small $n$, $\hat{p}_B$ is the better choice (unless there is a strong belief that $p$ is near 0 or 1). For large $n$, $\hat{p}$ is the better choice (unless there is a strong belief that $p$ is close to $\frac{1}{2}$). Even though the MSE criterion does not show one estimator to be uniformly better than the other, useful information is provided. This information, combined

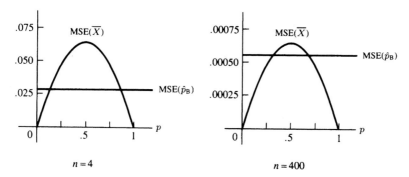

Figure 7.3.1. *Comparison of MSE of $\hat{p}$ and $\hat{p}_B$ for sample sizes $n = 4$ and $n = 400$ in Example 7.3.5*

with the knowledge of the problem at hand, can lead to choosing the better estimator for the situation.                                                                                    ‖

In certain situations, particularly in location parameter estimation, MSE can be a helpful criterion for finding the best estimator in a class of equivariant estimators (see Section 6.4). For a fixed $g$ in the group $\mathcal{G}$, denote the function that takes $\theta \to \theta'$ by $\bar{g}(\theta) = \theta'$. Then if $W(\mathbf{X})$ estimates $\theta$ we have

*Measurement Equivariance:* $W(\mathbf{x})$ estimates $\theta \Rightarrow \bar{g}(W(\mathbf{x}))$ estimates $\bar{g}(\theta) = \theta'$.

*Formal Invariance:* $W(\mathbf{x})$ estimates $\theta \Rightarrow W(g(\mathbf{x}))$ estimates $\bar{g}(\theta) = \theta'$.

Putting these two requirements together gives $W(g(\mathbf{x})) = \bar{g}(W(\mathbf{x}))$.

**Example 7.3.6 (MSE of equivariant estimators)**   Let $X_1, \ldots, X_n$ be iid $f(x - \theta)$. For an estimator $W(X_1, \ldots, X_n)$ to satisfy $W(g_a(\mathbf{x})) = \bar{g}_a(W(\mathbf{x}))$, we must have

$$(7.3.2) \qquad W(x_1, \ldots, x_n) + a = W(x_1 + a, \ldots, x_n + a),$$

which specifies the equivariant estimators with respect to the group of transformations defined by $\mathcal{G} = \{g_a(\mathbf{x}): -\infty < a < \infty\}$, where $g_a(x_1, \ldots, x_n) = (x_1 + a, \ldots, x_n + a)$. For these estimators we have

$$E_\theta(W(X_1, \ldots, X_n) - \theta)^2$$

$$= E_\theta(W(X_1 + a, \ldots, X_n + a) - a - \theta)^2$$

$$= E_\theta(W(X_1 - \theta, \ldots, X_n - \theta))^2 \qquad (a = -\theta)$$

$$= \int_{-\infty}^{\infty} \cdots \int_{-\infty}^{\infty} (W(x_1 - \theta, \ldots, x_n - \theta))^2 \prod_{i=1}^{n} f(x_i - \theta)\, dx_i$$

$$(7.3.3) \qquad = \int_{-\infty}^{\infty} \cdots \int_{-\infty}^{\infty} (W(u_1, \ldots, u_n))^2 \prod_{i=1}^{n} f(u_i)\, du_i. \qquad (u_i = x_i - \theta)$$

This last expression does not depend on $\theta$; hence, the MSEs of these equivariant estimators are not functions of $\theta$. The MSE can therefore be used to order the equivariant estimators, and an equivariant estimator with smallest MSE can be found. In fact, this estimator is the solution to the mathematical problem of finding the function $W$ that minimizes (7.3.3) subject to (7.3.2). (See Exercises 7.35 and 7.36.)                    ‖

### 7.3.2 Best Unbiased Estimators

As noted in the previous section, a comparison of estimators based on MSE considerations may not yield a clear favorite. Indeed, there is no one "best MSE" estimator. Many find this troublesome or annoying, and rather than doing MSE comparisons of candidate estimators, they would rather have a "recommended" one.

The reason that there is no one "best MSE" estimator is that the class of all estimators is too large a class. (For example, the estimator $\hat{\theta} = 17$ cannot be beaten in MSE at $\theta = 17$ but is a terrible estimator otherwise.) One way to make the problem of finding a "best" estimator tractable is to limit the class of estimators. A popular way of restricting the class of estimators, the one we consider in this section, is to consider only unbiased estimators.

If $W_1$ and $W_2$ are both unbiased estimators of a parameter $\theta$, that is, $E_\theta W_1 = E_\theta W_2 = \theta$, then their mean squared errors are equal to their variances, so we should choose the estimator with the smaller variance. If we can find an unbiased estimator with uniformly smallest variance—a best unbiased estimator—then our task is done.

Before proceeding we note that, although we will be dealing with unbiased estimators, the results here and in the next section are actually more general. Suppose that there is an estimator $W^*$ of $\theta$ with $E_\theta W^* = \tau(\theta) \neq \theta$, and we are interested in investigating the worth of $W^*$. Consider the class of estimators

$$\mathcal{C}_\tau = \{W \colon E_\theta W = \tau(\theta)\}.$$

For any $W_1, W_2 \in \mathcal{C}_\tau$, $\text{Bias}_\theta W_1 = \text{Bias}_\theta W_2$, so

$$E_\theta(W_1 - \theta)^2 - E_\theta(W_2 - \theta)^2 = \text{Var}_\theta W_1 - \text{Var}_\theta W_2,$$

and MSE comparisons, within the class $\mathcal{C}_\tau$, can be based on variance alone. Thus, although we speak in terms of unbiased estimators, we really are comparing estimators that have the same expected value, $\tau(\theta)$.

The goal of this section is to investigate a method for finding a "best" unbiased estimator, which we define in the following way.

**Definition 7.3.7**    An estimator $W^*$ is a *best unbiased estimator* of $\tau(\theta)$ if it satisfies $E_\theta W^* = \tau(\theta)$ for all $\theta$ and, for any other estimator $W$ with $E_\theta W = \tau(\theta)$, we have $\text{Var}_\theta W^* \leq \text{Var}_\theta W$ for all $\theta$. $W^*$ is also called a *uniform minimum variance unbiased estimator* (UMVUE) of $\tau(\theta)$.

Finding a best unbiased estimator (if one exists!) is not an easy task for a variety of reasons, two of which are illustrated in the following example.

**Example 7.3.8 (Poisson unbiased estimation)**        Let $X_1, \ldots, X_n$ be iid Poisson($\lambda$), and let $\bar{X}$ and $S^2$ be the sample mean and variance, respectively. Recall that for the Poisson pmf both the mean and variance are equal to $\lambda$. Therefore, applying Theorem 5.2.6, we have

$$\mathrm{E}_\lambda \bar{X} = \lambda, \quad \text{for all } \lambda,$$

and

$$\mathrm{E}_\lambda S^2 = \lambda, \quad \text{for all } \lambda,$$

so both $\bar{X}$ and $S^2$ are unbiased estimators of $\lambda$.

To determine the better estimator, $\bar{X}$ or $S^2$, we should now compare variances. Again from Theorem 5.2.6, we have $\mathrm{Var}_\lambda \bar{X} = \lambda/n$, but $\mathrm{Var}_\lambda S^2$ is quite a lengthy calculation (resembling that in Exercise 5.10(b)). This is one of the first problems in finding a best unbiased estimator. Not only may the calculations be long and involved, but they may be for naught (as in this case), for we will see that $\mathrm{Var}_\lambda \bar{X} \leq \mathrm{Var}_\lambda S^2$ for all $\lambda$.

Even if we can establish that $\bar{X}$ is better than $S^2$, consider the class of estimators

$$W_a(\bar{X}, S^2) = a\bar{X} + (1-a)S^2.$$

For every constant $a$, $\mathrm{E}_\lambda W_a(\bar{X}, S^2) = \lambda$, so we now have infinitely many unbiased estimators of $\lambda$. Even if $\bar{X}$ is better than $S^2$, is it better than every $W_a(\bar{X}, S^2)$? Furthermore, how can we be sure that there are not other, better, unbiased estimators lurking about?        ‖

This example shows some of the problems that might be encountered in trying to find a best unbiased estimator, and perhaps that a more comprehensive approach is desirable. Suppose that, for estimating a parameter $\tau(\theta)$ of a distribution $f(x|\theta)$, we can specify a lower bound, say $B(\theta)$, on the variance of *any* unbiased estimator of $\tau(\theta)$. If we can then find an unbiased estimator $W^*$ satisfying $\mathrm{Var}_\theta W^* = B(\theta)$, we have found a best unbiased estimator. This is the approach taken with the use of the Cramér–Rao Lower Bound.

**Theorem 7.3.9 (Cramér–Rao Inequality)**   *Let $X_1, \ldots, X_n$ be a sample with pdf $f(\mathbf{x}|\theta)$, and let $W(\mathbf{X}) = W(X_1, \ldots, X_n)$ be any estimator satisfying*

$$\frac{d}{d\theta} \mathrm{E}_\theta W(\mathbf{X}) = \int_{\mathcal{X}} \frac{\partial}{\partial\theta} [W(\mathbf{x})f(\mathbf{x}|\theta)] \, d\mathbf{x}$$

(7.3.4)        *and*

$$\mathrm{Var}_\theta W(\mathbf{X}) < \infty.$$

*Then*

(7.3.5)        $$\mathrm{Var}_\theta \left( W(\mathbf{X}) \right) \geq \frac{\left( \frac{d}{d\theta} \mathrm{E}_\theta W(\mathbf{X}) \right)^2}{\mathrm{E}_\theta \left( \left( \frac{\partial}{\partial\theta} \log f(\mathbf{X}|\theta) \right)^2 \right)}.$$

**Proof:** The proof of this theorem is elegantly simple and is a clever application of the Cauchy–Schwarz Inequality or, stated statistically, the fact that for any two random variables $X$ and $Y$,

$$(7.3.6) \qquad [\mathrm{Cov}(X, Y)]^2 \leq (\mathrm{Var}\, X)(\mathrm{Var}\, Y).$$

If we rearrange (7.3.6) we can get a lower bound on the variance of $X$,

$$\mathrm{Var}\, X \geq \frac{[\mathrm{Cov}(X, Y)]^2}{\mathrm{Var}\, Y}.$$

The cleverness in this theorem follows from choosing $X$ to be the estimator $W(\mathbf{X})$ and $Y$ to be the quantity $\frac{\partial}{\partial \theta} \log f(\mathbf{X}|\theta)$ and applying the Cauchy–Schwarz Inequality. First note that

$$
\begin{aligned}
\frac{d}{d\theta} \mathrm{E}_\theta W(\mathbf{X}) &= \int_{\mathcal{X}} W(\mathbf{x}) \left[ \frac{\partial}{\partial \theta} f(\mathbf{x}|\theta) \right] \, d\mathbf{x} \\
(7.3.7) \qquad\qquad &= \mathrm{E}_\theta \left[ W(\mathbf{X}) \frac{\frac{\partial}{\partial \theta} f(\mathbf{X}|\theta)}{f(\mathbf{X}|\theta)} \right] \qquad \text{(multiply by } f(\mathbf{X}|\theta)/f(\mathbf{X}|\theta)) \\
&= \mathrm{E}_\theta \left[ W(\mathbf{X}) \frac{\partial}{\partial \theta} \log f(\mathbf{X}|\theta) \right], \qquad\qquad \text{(property of logs)}
\end{aligned}
$$

which suggests a covariance between $W(\mathbf{X})$ and $\frac{\partial}{\partial \theta} \log f(\mathbf{X}|\theta)$. For it to be a covariance, we need to subtract the product of the expected values, so we calculate $\mathrm{E}_\theta \left( \frac{\partial}{\partial \theta} \log f(\mathbf{X}|\theta) \right)$. But if we apply (7.3.7) with $W(\mathbf{x}) = 1$, we have

$$(7.3.8) \qquad \mathrm{E}_\theta \left( \frac{\partial}{\partial \theta} \log f(\mathbf{X}|\theta) \right) = \frac{d}{d\theta} \mathrm{E}_\theta[1] = 0.$$

Therefore $\mathrm{Cov}_\theta(W(\mathbf{X}), \frac{\partial}{\partial \theta} \log f(\mathbf{X}|\theta))$ is equal to the expectation of the product, and it follows from (7.3.7) and (7.3.8) that

$$(7.3.9) \; \mathrm{Cov}_\theta \left( W(\mathbf{X}), \frac{\partial}{\partial \theta} \log f(\mathbf{X}|\theta) \right) = \mathrm{E}_\theta \left( W(\mathbf{X}) \frac{\partial}{\partial \theta} \log f(\mathbf{X}|\theta) \right) = \frac{d}{d\theta} \mathrm{E}_\theta W(\mathbf{X}).$$

Also, since $\mathrm{E}_\theta(\frac{\partial}{\partial \theta} \log f(\mathbf{X}|\theta)) = 0$ we have

$$(7.3.10) \qquad \mathrm{Var}_\theta \left( \frac{\partial}{\partial \theta} \log f(\mathbf{X}|\theta) \right) = \mathrm{E}_\theta \left( \left( \frac{\partial}{\partial \theta} \log f(\mathbf{X}|\theta) \right)^2 \right).$$

Using the Cauchy–Schwarz Inequality together with (7.3.9) and (7.3.10), we obtain

$$\mathrm{Var}_\theta (W(\mathbf{X})) \geq \frac{\left( \frac{d}{d\theta} \mathrm{E}_\theta W(\mathbf{X}) \right)^2}{\mathrm{E}_\theta \left( \left( \frac{\partial}{\partial \theta} \log f(\mathbf{X}|\theta) \right)^2 \right)},$$

proving the theorem.                                                        □

If we add the assumption of independent samples, then the calculation of the lower bound is simplified. The expectation in the denominator becomes a univariate calculation, as the following corollary shows.

**Corollary 7.3.10 (Cramér–Rao Inequality, iid case)** *If the assumptions of Theorem 7.3.9 are satisfied and, additionally, if $X_1, \ldots, X_n$ are iid with pdf $f(x|\theta)$, then*

$$\operatorname{Var}_\theta W(\mathbf{X}) \geq \frac{\left(\frac{d}{d\theta} \operatorname{E}_\theta W(\mathbf{X})\right)^2}{n \operatorname{E}_\theta \left(\left(\frac{\partial}{\partial\theta} \log f(X|\theta)\right)^2\right)}.$$

**Proof:** We only need to show that

$$\operatorname{E}_\theta \left(\left(\frac{\partial}{\partial\theta} \log f(\mathbf{X}|\theta)\right)^2\right) = n \operatorname{E}_\theta \left(\left(\frac{\partial}{\partial\theta} \log f(X|\theta)\right)^2\right).$$

Since $X_1, \ldots, X_n$ are independent,

$$\operatorname{E}_\theta \left(\frac{\partial}{\partial\theta} \log f(\mathbf{X}|\theta)\right)^2 = \operatorname{E}_\theta \left(\left(\frac{\partial}{\partial\theta} \log \prod_{i=1}^n f(X_i|\theta)\right)^2\right)$$

$$= \operatorname{E}_\theta \left(\left(\sum_{i=1}^n \frac{\partial}{\partial\theta} \log f(X_i|\theta)\right)^2\right) \qquad \text{(property of logs)}$$

$$= \sum_{i=1}^n \operatorname{E}_\theta \left(\left(\frac{\partial}{\partial\theta} \log f(X_i|\theta)\right)^2\right) \qquad \text{(expand the square)}$$

$$(7.3.11) \qquad\qquad + \sum_{i \neq j} \operatorname{E}_\theta \left(\frac{\partial}{\partial\theta} \log f(X_i|\theta) \frac{\partial}{\partial\theta} \log f(X_j|\theta)\right).$$

For $i \neq j$ we have

$$\operatorname{E}_\theta \left(\frac{\partial}{\partial\theta} \log f(X_i|\theta) \frac{\partial}{\partial\theta} \log f(X_j|\theta)\right)$$

$$= \operatorname{E}_\theta \left(\frac{\partial}{\partial\theta} \log f(X_i|\theta)\right) \operatorname{E}_\theta \left(\frac{\partial}{\partial\theta} \log f(X_j|\theta)\right) \qquad \text{(independence)}$$

$$= 0. \qquad\qquad\qquad\qquad\qquad\qquad\qquad\qquad\qquad\qquad\qquad \text{(from (7.3.8))}$$

Therefore the second sum in (7.3.11) is 0, and the first term is

$$\sum_{i=1}^n \operatorname{E}_\theta \left(\left(\frac{\partial}{\partial\theta} \log f(X_i|\theta)\right)^2\right) = n \operatorname{E}_\theta \left(\left(\frac{\partial}{\partial\theta} \log f(X|\theta)\right)^2\right), \quad \text{(identical distributions)}$$

which establishes the corollary. $\qquad\qquad\qquad\qquad\qquad\qquad\qquad\qquad\qquad\qquad\square$

Before going on we note that although the Cramér–Rao Lower Bound is stated for continuous random variables, it also applies to discrete random variables. The key condition, (7.3.4), which allows interchange of integration and differentiation, undergoes the obvious modification. If $f(x|\theta)$ is a pmf, then we must be able to interchange differentiation and summation. (Of course, this assumes that even though $f(x|\theta)$ is a pmf and *not* differentiable in $x$, it *is* differentiable in $\theta$. This is the case for most common pmfs.)

The quantity $\mathrm{E}_\theta\left(\left(\frac{\partial}{\partial\theta}\log f(\mathbf{X}|\theta)\right)^2\right)$ is called the *information number*, or *Fisher information* of the sample. This terminology reflects the fact that the information number gives a bound on the variance of the best unbiased estimator of $\theta$. As the information number gets bigger and we have more information about $\theta$, we have a smaller bound on the variance of the best unbiased estimator.

In fact, the term *Information Inequality* is an alternative to *Cramér–Rao Inequality*, and the Information Inequality exists in much more general forms than is presented here. A key difference of the more general form is that all assumptions about the candidate estimators are removed and are replaced with assumptions on the underlying density. In this form, the Information Inequality becomes very useful in comparing the performance of estimators. See Lehmann and Casella (1998, Section 2.6) for details.

For any differentiable function $\tau(\theta)$ we now have a lower bound on the variance of any estimator $W$ satisfying (7.3.4) and $\mathrm{E}_\theta W = \tau(\theta)$. The bound depends only on $\tau(\theta)$ and $f(x|\theta)$ and is a uniform lower bound on the variance. Any candidate estimator satisfying $\mathrm{E}_\theta W = \tau(\theta)$ and attaining this lower bound is a best unbiased estimator of $\tau(\theta)$.

Before looking at some examples, we present a computational result that aids in the application of this theorem. Its proof is left to Exercise 7.39.

**Lemma 7.3.11** *If $f(x|\theta)$ satisfies*

$$\frac{d}{d\theta}\mathrm{E}_\theta\left(\frac{\partial}{\partial\theta}\log f(X|\theta)\right) = \int \frac{\partial}{\partial\theta}\left[\left(\frac{\partial}{\partial\theta}\log f(x|\theta)\right)f(x|\theta)\right]dx$$

*(true for an exponential family), then*

$$\mathrm{E}_\theta\left(\left(\frac{\partial}{\partial\theta}\log f(X|\theta)\right)^2\right) = -\mathrm{E}_\theta\left(\frac{\partial^2}{\partial\theta^2}\log f(X|\theta)\right).$$

Using the tools just developed, we return to, and settle, the Poisson example.

**Example 7.3.12 (Conclusion of Example 7.3.8)**   Here $\tau(\lambda) = \lambda$, so $\tau'(\lambda) = 1$. Also, since we have an exponential family, using Lemma 7.3.11 gives us

$$\mathrm{E}_\lambda\left(\left(\frac{\partial}{\partial\lambda}\log\prod_{i=1}^n f(X_i|\lambda)\right)^2\right) = -n\mathrm{E}_\lambda\left(\frac{\partial^2}{\partial\lambda^2}\log f(X|\lambda)\right)$$

$$= -n\mathrm{E}_\lambda\left(\frac{\partial^2}{\partial\lambda^2}\log\left(\frac{e^{-\lambda}\lambda^X}{X!}\right)\right)$$

$$= -n\mathrm{E}_\lambda \left( \frac{\partial^2}{\partial \lambda^2} (-\lambda + X \log \lambda - \log X!) \right)$$

$$= -n\mathrm{E}_\lambda \left( -\frac{X}{\lambda^2} \right)$$

$$= \frac{n}{\lambda}.$$

Hence for any unbiased estimator, $W$, of $\lambda$, we must have

$$\mathrm{Var}_\lambda\, W \geq \frac{\lambda}{n}.$$

Since $\mathrm{Var}_\lambda\, \bar{X} = \lambda/n$, $\bar{X}$ is a best unbiased estimator of $\lambda$.　　　$\parallel$

It is important to remember that a key assumption in the Cramér–Rao Theorem is the ability to differentiate under the integral sign, which, of course, is somewhat restrictive. As we have seen, densities in the exponential class will satisfy the assumptions but, in general, such assumptions need to be checked, or contradictions such as the following will arise.

**Example 7.3.13 (Unbiased estimator for the scale uniform)**　　Let $X_1, \ldots,$ $X_n$ be iid with pdf $f(x|\theta) = 1/\theta, 0 < x < \theta$. Since $\frac{\partial}{\partial \theta} \log f(x|\theta) = -1/\theta$, we have

$$\mathrm{E}_\theta \left( \left( \frac{\partial}{\partial \theta} \log f(X|\theta) \right)^2 \right) = \frac{1}{\theta^2}.$$

The Cramér–Rao Theorem would seem to indicate that if $W$ is any unbiased estimator of $\theta$,

$$\mathrm{Var}_\theta\, W \geq \frac{\theta^2}{n}.$$

We would now like to find an unbiased estimator with small variance. As a first guess, consider the sufficient statistic $Y = \max(X_1, \ldots, X_n)$, the largest order statistic. The pdf of $Y$ is $f_Y(y|\theta) = ny^{n-1}/\theta^n$, $0 < y < \theta$, so

$$\mathrm{E}_\theta Y = \int_0^\theta \frac{ny^n}{\theta^n}\, dy = \frac{n}{n+1}\theta,$$

showing that $\frac{n+1}{n}Y$ is an unbiased estimator of $\theta$. We next calculate

$$\mathrm{Var}_\theta \left( \frac{n+1}{n}Y \right) = \left( \frac{n+1}{n} \right)^2 \mathrm{Var}_\theta\, Y$$

$$= \left( \frac{n+1}{n} \right)^2 \left[ \mathrm{E}_\theta Y^2 - \left( \frac{n}{n+1}\theta \right)^2 \right]$$

$$= \left( \frac{n+1}{n} \right)^2 \left[ \frac{n}{n+2}\theta^2 - \left( \frac{n}{n+1}\theta \right)^2 \right]$$

$$= \frac{1}{n(n+2)}\theta^2,$$

which is uniformly smaller than $\theta^2/n$. This indicates that the Cramér–Rao Theorem is not applicable to this pdf. To see that this is so, we can use Leibnitz's Rule (Section 2.4) to calculate

$$\frac{d}{d\theta} \int_0^\theta h(x) f(x|\theta) \, dx = \frac{d}{d\theta} \int_0^\theta h(x) \frac{1}{\theta} \, dx$$

$$= \frac{h(\theta)}{\theta} + \int_0^\theta h(x) \frac{\partial}{\partial\theta} \left( \frac{1}{\theta} \right) \, dx$$

$$\neq \int_0^\theta h(x) \frac{\partial}{\partial\theta} f(x|\theta) \, dx,$$

unless $h(\theta)/\theta = 0$ for all $\theta$. Hence, the Cramér–Rao Theorem does not apply. In general, if the range of the pdf depends on the parameter, the theorem will not be applicable.                                                                                                   ‖

A shortcoming of this approach to finding best unbiased estimators is that, even if the Cramér–Rao Theorem is applicable, there is no guarantee that the bound is sharp. That is to say, the value of the Cramér–Rao Lower Bound may be *strictly smaller* than the variance of *any* unbiased estimator. In fact, in the usually favorable case of $f(x|\theta)$ being a one-parameter exponential family, the most that we can say is that there exists a parameter $\tau(\theta)$ with an unbiased estimator that achieves the Cramér–Rao Lower Bound. However, in other typical situations, for other parameters, the bound may not be attainable. These situations cause concern because, if we cannot find an estimator that attains the lower bound, we have to decide whether no estimator can attain it or whether we must look at more estimators.

**Example 7.3.14 (Normal variance bound)**      Let $X_1, \ldots, X_n$ be iid $n(\mu, \sigma^2)$, and consider estimation of $\sigma^2$, where $\mu$ is unknown. The normal pdf satisfies the assumptions of the Cramér–Rao Theorem and Lemma 7.3.11, so we have

$$\frac{\partial^2}{\partial(\sigma^2)^2} \log \left( \frac{1}{(2\pi\sigma^2)^{1/2}} e^{-(1/2)(x-\mu)^2/\sigma^2} \right) = \frac{1}{2\sigma^4} - \frac{(x-\mu)^2}{\sigma^6}$$

and

$$-\mathrm{E}\left( \frac{\partial^2}{\partial(\sigma^2)^2} \log f(X|\mu,\sigma^2) \,\middle|\, \mu, \sigma^2 \right) = -\mathrm{E}\left( \frac{1}{2\sigma^4} - \frac{(X-\mu)^2}{\sigma^6} \,\middle|\, \mu, \sigma^2 \right)$$

$$= \frac{1}{2\sigma^4}.$$

Thus, any unbiased estimator, $W$, of $\sigma^2$ must satisfy

$$\mathrm{Var}(W|\mu,\sigma^2) \geq \frac{2\sigma^4}{n}.$$

In Example 7.3.3 we saw

$$\text{Var}(S^2|\mu, \sigma^2) = \frac{2\sigma^4}{n-1},$$

so $S^2$ does not attain the Cramér–Rao Lower Bound.                    ‖

In the above example we are left with an incomplete answer; that is, is there a better unbiased estimator of $\sigma^2$ than $S^2$, or is the Cramér–Rao Lower Bound unattainable?

The conditions for attainment of the Cramér–Rao Lower Bound are actually quite simple. Recall that the bound follows from an application of the Cauchy–Schwarz Inequality, so conditions for attainment of the bound are the conditions for equality in the Cauchy–Schwarz Inequality (see Section 4.7). Note also that Corollary 7.3.15 is a useful tool because it implicitly gives us a way of finding a best unbiased estimator.

**Corollary 7.3.15 (Attainment)**   Let $X_1, \ldots, X_n$ be iid $f(x|\theta)$, where $f(x|\theta)$ sat-isfies the conditions of the Cramér–Rao Theorem. Let $L(\theta|\mathbf{x}) = \prod_{i=1}^{n} f(x_i|\theta)$ denote the likelihood function. If $W(\mathbf{X}) = W(X_1, \ldots, X_n)$ is any unbiased estimator of $\tau(\theta)$, then $W(\mathbf{X})$ attains the Cramér–Rao Lower Bound if and only if

$$(7.3.12) \qquad a(\theta)[W(\mathbf{x}) - \tau(\theta)] = \frac{\partial}{\partial\theta} \log L(\theta|\mathbf{x})$$

for some function $a(\theta)$.

**Proof:**  The Cramér–Rao Inequality, as given in (7.3.6), can be written as

$$\left[\text{Cov}_\theta\left(W(\mathbf{X}), \frac{\partial}{\partial\theta} \log \prod_{i=1}^{n} f(X_i|\theta)\right)\right]^2 \le \text{Var}_\theta W(\mathbf{X})\text{Var}_\theta\left(\frac{\partial}{\partial\theta} \log \prod_{i=1}^{n} f(X_i|\theta)\right),$$

and, recalling that $E_\theta W = \tau(\theta)$, $E_\theta(\frac{\partial}{\partial\theta} \log \prod_{i=1}^{n} f(X_i|\theta)) = 0$, and using the results of Theorem 4.5.7, we can have equality if and only if $W(\mathbf{x}) - \tau(\theta)$ is proportional to $\frac{\partial}{\partial\theta} \log \prod_{i=1}^{n} f(x_i|\theta)$. That is exactly what is expressed in (7.3.12).      □

**Example 7.3.16 (Continuation of Example 7.3.14)**   Here we have

$$L(\mu, \sigma^2|\mathbf{x}) = \frac{1}{(2\pi\sigma^2)^{n/2}} e^{-(1/2)\Sigma_{i=1}^{n}(x_i-\mu)^2/\sigma^2},$$

and hence

$$\frac{\partial}{\partial\sigma^2} \log L(\mu, \sigma^2|\mathbf{x}) = \frac{n}{2\sigma^4}\left(\sum_{i=1}^{n} \frac{(x_i-\mu)^2}{n} - \sigma^2\right).$$

Thus, taking $a(\sigma^2) = n/(2\sigma^4)$ shows that the best unbiased estimator of $\sigma^2$ is $\sum_{i=1}^{n}(x_i - \mu)^2/n$, which is calculable only if $\mu$ is known. If $\mu$ is unknown, the bound *cannot be attained.*                    ‖

The theory developed in this section still leaves some questions unanswered. First, what can we do if $f(x|\theta)$ does not satisfy the assumptions of the Cramér–Rao Theorem? (In Example 7.3.13, we still do not know if $\frac{n+1}{n}Y$ is a best unbiased estimator.) Second, what if the bound is unattainable by allowable estimators, as in Example 7.3.14? There, we still do not know if $S^2$ is a best unbiased estimator.

One way of answering these questions is to search for methods that are more widely applicable and yield sharper (that is, greater) lower bounds. Much research has been done on this topic, with perhaps the most well-known bound being that of Chapman and Robbins (1951). Stuart, Ord, and Arnold (1999, Chapter 17) have a good treatment of this subject. Rather than take this approach, however, we will continue the study of best unbiased estimators from another view, using the concept of sufficiency.

### 7.3.3 Sufficiency and Unbiasedness

In the previous section, the concept of sufficiency was not used in our search for unbiased estimates. We will now see that consideration of sufficiency is a powerful tool, indeed.

The main theorem of this section, which relates sufficient statistics to unbiased estimates, is, as in the case of the Cramér–Rao Theorem, another clever application of some well-known theorems. Recall from Chapter 4 that if $X$ and $Y$ are any two random variables, then, provided the expectations exist, we have

(7.3.13)
$$EX = E[E(X|Y)],$$
$$\operatorname{Var} X = \operatorname{Var}[E(X|Y)] + E[\operatorname{Var}(X|Y)].$$

Using these tools we can prove the following theorem.

**Theorem 7.3.17 (Rao–Blackwell)** *Let $W$ be any unbiased estimator of $\tau(\theta)$, and let $T$ be a sufficient statistic for $\theta$. Define $\phi(T) = E(W|T)$. Then $E_\theta\phi(T) = \tau(\theta)$ and $\operatorname{Var}_\theta \phi(T) \leq \operatorname{Var}_\theta W$ for all $\theta$; that is, $\phi(T)$ is a uniformly better unbiased estimator of $\tau(\theta)$.*

**Proof:** From (7.3.13) we have

$$\tau(\theta) = E_\theta W = E_\theta[E(W|T)] = E_\theta\phi(T),$$

so $\phi(T)$ is unbiased for $\tau(\theta)$. Also,

$$\operatorname{Var}_\theta W = \operatorname{Var}_\theta [E(W|T)] + E_\theta [\operatorname{Var}(W|T)]$$
$$= \operatorname{Var}_\theta \phi(T) + E_\theta[\operatorname{Var}(W|T)]$$
$$\geq \operatorname{Var}_\theta \phi(T). \qquad\qquad (\operatorname{Var}(W|T) \geq 0)$$

Hence $\phi(T)$ is uniformly better than $W$, and it only remains to show that $\phi(T)$ is indeed an estimator. That is, we must show that $\phi(T) = E(W|T)$ is a function of only the sample and, in particular, is independent of $\theta$. But it follows from the definition of sufficiency, and the fact that $W$ is a function only of the sample, that the distribution of $W|T$ is independent of $\theta$. Hence $\phi(T)$ is a uniformly better unbiased estimator of $\tau(\theta)$.                                                                                    □

Therefore, conditioning any unbiased estimator on a sufficient statistic will result in a uniform improvement, so we need consider only statistics that are functions of a sufficient statistic in our search for best unbiased estimators.

The identities in (7.3.13) make no mention of sufficiency, so it might at first seem that conditioning on anything will result in an improvement. This is, in effect, true, but the problem is that the resulting quantity will probably depend on $\theta$ and not be an estimator.

**Example 7.3.18 (Conditioning on an insufficient statistic)**    Let $X_1, X_2$ be iid $n(\theta, 1)$. The statistic $\bar{X} = \frac{1}{2}(X_1 + X_2)$ has

$$\mathrm{E}_\theta \bar{X} = \theta \quad \text{and} \quad \mathrm{Var}_\theta \bar{X} = \frac{1}{2}.$$

Consider conditioning on $X_1$, which is not sufficient. Let $\phi(X_1) = \mathrm{E}_\theta(\bar{X}|X_1)$. It follows from (7.3.13) that $\mathrm{E}_\theta \phi(X_1) = \theta$ and $\mathrm{Var}_\theta \phi(X_1) \leq \mathrm{Var}_\theta \bar{X}$, so $\phi(X_1)$ is better than $\bar{X}$. However,

$$\phi(X_1) = \mathrm{E}_\theta(\bar{X}|X_1)$$
$$= \frac{1}{2}\mathrm{E}_\theta(X_1|X_1) + \frac{1}{2}\mathrm{E}_\theta(X_2|X_1)$$
$$= \frac{1}{2}X_1 + \frac{1}{2}\theta,$$

since $\mathrm{E}_\theta(X_2|X_1) = \mathrm{E}_\theta X_2$ by independence. Hence, $\phi(X_1)$ is not an estimator.    ‖

We now know that, in looking for a best unbiased estimator of $\tau(\theta)$, we need consider only estimators based on a sufficient statistic. The question now arises that if we have $\mathrm{E}_\theta \phi = \tau(\theta)$ and $\phi$ is based on a sufficient statistic, that is, $\mathrm{E}(\phi|T) = \phi$, how do we know that $\phi$ is best unbiased? Of course, if $\phi$ attains the Cramér–Rao Lower Bound, then it is best unbiased, but if it does not, have we gained anything? For example, if $\phi^*$ is another unbiased estimator of $\tau(\theta)$, how does $\mathrm{E}(\phi^*|T)$ compare to $\phi$? The next theorem answers this question in part by showing that a best unbiased estimator is unique.

**Theorem 7.3.19**    *If $W$ is a best unbiased estimator of $\tau(\theta)$, then $W$ is unique.*

**Proof:** Suppose $W'$ is another best unbiased estimator, and consider the estimator $W^* = \frac{1}{2}(W + W')$. Note that $\mathrm{E}_\theta W^* = \tau(\theta)$ and

$$\text{Var}_\theta W^* = \text{Var}_\theta \left( \frac{1}{2} W + \frac{1}{2} W' \right)$$

$$= \frac{1}{4} \text{Var}_\theta W + \frac{1}{4} \text{Var}_\theta W' + \frac{1}{2} \text{Cov}_\theta(W, W') \qquad \text{(Exercise 4.44)}$$

(7.3.14)

$$\leq \frac{1}{4} \text{Var}_\theta W + \frac{1}{4} \text{Var}_\theta W' + \frac{1}{2} [(\text{Var}_\theta W)(\text{Var}_\theta W')]^{1/2} \quad \text{(Cauchy–Schwarz)}$$

$$= \text{Var}_\theta W. \qquad\qquad (\text{Var}_\theta W = \text{Var}_\theta W')$$

But if the above inequality is strict, then the best unbiasedness of $W$ is contradicted, so we must have equality for all $\theta$. Since the inequality is an application of Cauchy–Schwarz, we can have equality only if $W' = a(\theta)W + b(\theta)$. Now using properties of covariance, we have

$$\text{Cov}_\theta(W, W') = \text{Cov}_\theta[W, a(\theta)W + b(\theta)]$$

$$= \text{Cov}_\theta[W, a(\theta)W]$$

$$= a(\theta)\text{Var}_\theta W,$$

but $\text{Cov}_\theta(W, W') = \text{Var}_\theta W$ since we had equality in (7.3.14). Hence $a(\theta) = 1$ and, since $\text{E}_\theta W' = \tau(\theta)$, we must have $b(\theta) = 0$ and $W = W'$, showing that $W$ is unique. $\qquad\square$

To see when an unbiased estimator is best unbiased, we might ask how could we improve upon a given unbiased estimator? Suppose that $W$ satisfies $\text{E}_\theta W = \tau(\theta)$, and we have another estimator, $U$, that satisfies $\text{E}_\theta U = 0$ for all $\theta$, that is, $U$ is an *unbiased estimator of 0*. The estimator

$$\phi_a = W + aU,$$

where $a$ is a constant, satisfies $\text{E}_\theta \phi_a = \tau(\theta)$ and hence is also an unbiased estimator of $\tau(\theta)$. Can $\phi_a$ be better than $W$? The variance of $\phi_a$ is

$$\text{Var}_\theta \phi_a = \text{Var}_\theta (W + aU) = \text{Var}_\theta W + 2a\text{Cov}_\theta(W, U) + a^2 \text{Var}_\theta U.$$

Now, if for some $\theta = \theta_0, \text{Cov}_{\theta_0}(W, U) < 0$, then we can make $2a\text{Cov}_{\theta_0}(W, U) + a^2\text{Var}_{\theta_0} U < 0$ by choosing $a \in (0, -2\text{Cov}_{\theta_0}(W, U)/\text{Var}_{\theta_0} U)$. Hence, $\phi_a$ will be better than $W$ at $\theta = \theta_0$ and $W$ cannot be best unbiased. A similar argument will show that if $\text{Cov}_{\theta_0}(W, U) > 0$ for any $\theta_0, W$ also cannot be best unbiased. (See Exercise 7.53.) Thus, the relationship of $W$ with unbiased estimators of 0 is crucial in evaluating whether $W$ is best unbiased. This relationship, in fact, characterizes best unbiasedness.

**Theorem 7.3.20**   *If* $\text{E}_\theta W = \tau(\theta), W$ *is the best unbiased estimator of* $\tau(\theta)$ *if and only if* $W$ *is uncorrelated with all unbiased estimators of 0.*

**Proof:** If $W$ is best unbiased, the above argument shows that $W$ must satisfy $\text{Cov}_\theta(W, U) = 0$ for all $\theta$, for any $U$ satisfying $\text{E}_\theta U = 0$. Hence the necessity is established.

Suppose now that we have an unbiased estimator $W$ that is uncorrelated with all unbiased estimators of 0. Let $W'$ be any other estimator satisfying $E_\theta W' = E_\theta W = \tau(\theta)$. We will show that $W$ is better than $W'$. Write

$$W' = W + (W' - W),$$

and calculate

$$(7.3.15) \qquad \text{Var}_\theta\, W' = \text{Var}_\theta\, W + \text{Var}_\theta\, (W' - W) + 2\text{Cov}_\theta(W, W' - W)$$
$$= \text{Var}_\theta\, W + \text{Var}_\theta\, (W' - W),$$

where the last equality is true because $W' - W$ is an unbiased estimator of 0 and is uncorrelated with $W$ by assumption. Since $\text{Var}_\theta\, (W' - W) \geq 0$, (7.3.15) implies that $\text{Var}_\theta\, W' \geq \text{Var}_\theta\, W$. Since $W'$ is arbitrary, it follows that $W$ is the best unbiased estimator of $\tau(\theta)$.                                                                   □

Note that an unbiased estimator of 0 is nothing more than *random noise*; that is, there is no information in an estimator of 0. (It makes sense that the most sensible way to estimate 0 is with 0, not with random noise.) Therefore, if an estimator could be improved by adding random noise to it, the estimator probably is defective. (Alternatively, we could question the criterion used to evaluate the estimator, but in this case the criterion seems above suspicion.) This intuition is what is formalized in Theorem 7.3.20.

Although we now have an interesting characterization of best unbiased estimators, its usefulness is limited in application. It is often a difficult task to verify that an estimator is uncorrelated with *all* unbiased estimators of 0 because it is usually difficult to describe all unbiased estimators of 0. However, it is sometimes useful in determining that an estimator is not best unbiased.

**Example 7.3.21 (Unbiased estimators of zero)**   Let $X$ be an observation from a uniform$(\theta, \theta + 1)$ distribution. Then

$$E_\theta X = \int_\theta^{\theta+1} x\, dx = \theta + \frac{1}{2},$$

and so $X - \frac{1}{2}$ is an unbiased estimator of $\theta$, and it is easy to check that $\text{Var}_\theta\, X = \frac{1}{12}$.

For this pdf, unbiased estimators of zero are periodic functions with period 1. This follows from the fact that if $h(x)$ satisfies

$$\int_\theta^{\theta+1} h(x)\, dx = 0, \quad \text{for all } \theta,$$

then

$$0 = \frac{d}{d\theta} \int_\theta^{\theta+1} h(x)\, dx = h(\theta + 1) - h(\theta), \quad \text{for all } \theta.$$

Such a function is $h(x) = \sin(2\pi x)$. Now

$$\mathrm{Cov}_\theta(X - \tfrac{1}{2}, \sin(2\pi X)) = \mathrm{Cov}_\theta(X, \sin(2\pi X))$$

$$= \int_\theta^{\theta+1} x \sin(2\pi x)\, dx$$

$$= -\frac{x\cos(2\pi x)}{2\pi}\bigg|_\theta^{\theta+1} + \int_\theta^{\theta+1} \frac{\cos(2\pi x)}{2\pi}\, dx$$

(integration by parts)

$$= -\frac{\cos(2\pi\theta)}{2\pi},$$

where we used $\cos(2\pi(\theta+1)) = \cos(2\pi\theta)$ and $\sin(2\pi(\theta+1)) = \sin(2\pi\theta)$.

Hence $X - \tfrac{1}{2}$ is correlated with an unbiased estimator of zero, and cannot be a best unbiased estimator of $\theta$. In fact, it is straightforward to check that the estimator $X - \tfrac{1}{2} + \sin(2\pi X)/(2\pi)$ is unbiased for $\theta$ and has variance less than $\frac{1}{12}$ for some $\theta$ values.                                                                                                    ‖

To answer the question about existence of a best unbiased estimator, what is needed is some characterization of all unbiased estimators of zero. Given such a characterization, we could then see if our candidate for best unbiased estimator is, in fact, optimal.

Characterizing the unbiased estimators of zero is not an easy task and requires conditions on the pdf (or pmf) with which we are working. Note that, thus far in this section, we have not specified conditions on pdfs (as were needed, for example, in the Cramér–Rao Lower Bound). The price we have paid for this generality is the difficulty in verifying the existence of the best unbiased estimator.

If a family of pdfs or pmfs $f(x|\theta)$ has the property that there are *no* unbiased estimators of zero (other than zero itself), then our search would be ended, since any statistic $W$ satisfies $\mathrm{Cov}_\theta(W, 0) = 0$. Recall that the property of *completeness*, defined in Definition 6.1.4, guarantees such a situation.

**Example 7.3.22 (Continuation of Example 7.3.13)**     For $X_1, \ldots, X_n$ iid uniform$(0, \theta)$, we saw that $\frac{n+1}{n}Y$ is an unbiased estimator of $\theta$, where $Y = \max\{X_1, \ldots, X_n\}$. The conditions of the Cramér–Rao Theorem are not satisfied, and we have not yet established whether this estimator is best unbiased. In Example 6.2.23, however, it was shown that $Y$ is a *complete* sufficient statistic. This means that the family of pdfs of $Y$ is complete, and there are no unbiased estimators of zero that are based on $Y$. (By sufficiency, in the form of the Rao–Blackwell Theorem, we need consider only unbiased estimators of zero based on $Y$.) Therefore, $\frac{n+1}{n}Y$ is uncorrelated with all unbiased estimators of zero (since the only one is zero itself) and thus $\frac{n+1}{n}Y$ is the best unbiased estimator of $\theta$.                                                                 ‖

It is worthwhile to note once again that what is important is the completeness of the family of distributions of the sufficient statistic. Completeness of the original family is of no consequence. This follows from the Rao–Blackwell Theorem, which says that we can restrict attention to functions of a sufficient statistic, so all expectations will be taken with respect to its distribution.

We sum up the relationship between completeness and best unbiasedness in the following theorem.

**Theorem 7.3.23**    *Let $T$ be a complete sufficient statistic for a parameter $\theta$, and let $\phi(T)$ be any estimator based only on $T$. Then $\phi(T)$ is the unique best unbiased estimator of its expected value.*

We close this section with an interesting and useful application of the theory developed here. In many situations, there will be no obvious candidate for an unbiased estimator of a function $\tau(\theta)$, much less a candidate for best unbiased estimator. However, in the presence of completeness, the theory of this section tells us that if we can find any unbiased estimator, we can find the best unbiased estimator. If $T$ is a complete sufficient statistic for a parameter $\theta$ and $h(X_1, \ldots, X_n)$ is *any* unbiased estimator of $\tau(\theta)$, then $\phi(T) = \mathrm{E}(h(X_1, \ldots, X_n)|T)$ is the best unbiased estimator of $\tau(\theta)$ (see Exercise 7.56).

**Example 7.3.24 (Binomial best unbiased estimation)**    Let $X_1, \ldots, X_n$ be iid binomial$(k, \theta)$. The problem is to estimate the probability of exactly one success from a binomial$(k, \theta)$, that is, estimate

$$\tau(\theta) = P_\theta(X = 1) = k\theta(1 - \theta)^{k-1}.$$

Now $\sum_{i=1}^n X_i \sim$ binomial$(kn, \theta)$ is a complete sufficient statistic, but no unbiased estimator based on it is immediately evident. When in this situation, try for the simplest solution. The simple-minded estimator

$$h(X_1) = \begin{cases} 1 & \text{if } X_1 = 1 \\ 0 & \text{otherwise} \end{cases}$$

satisfies

$$\mathrm{E}_\theta h(X_1) = \sum_{x_1=0}^k h(x_1) \binom{k}{x_1} \theta^{x_1} (1 - \theta)^{k-x_1}$$

$$= k\theta(1 - \theta)^{k-1}$$

and hence is an unbiased estimator of $k\theta(1 - \theta)^{k-1}$. Our theory now tells us that the estimator

$$\phi\left(\sum_{i=1}^n X_i\right) = \mathrm{E}\left(h(X_1) \Big| \sum_{i=1}^n X_i\right)$$

is the best unbiased estimator of $k\theta(1-\theta)^{k-1}$. (Notice that we do not need to actually calculate the expectation of $\phi(\sum_{i=1}^n X_i)$; we *know* that it has the correct expected value by properties of iterated expectations.) We must, however, be able to evaluate $\phi$. Suppose that we observe $\sum_{i=1}^n X_i = t$. Then

$$\phi(t) = E\left( h(X_1) \middle| \sum_{i=1}^{n} X_i = t \right) \qquad \left( \begin{array}{c} \text{the expectation does} \\ \text{not depend on } \theta \end{array} \right)$$

$$= P\left( X_1 = 1 \middle| \sum_{i=1}^{n} X_i = t \right) \qquad (h \text{ is 0 or 1})$$

$$= \frac{P_\theta(X_1 = 1, \sum_{i=1}^{n} X_i = t)}{P_\theta(\sum_{i=1}^{n} X_i = t)} \qquad \left( \begin{array}{c} \text{definition of} \\ \text{conditional probability} \end{array} \right)$$

$$= \frac{P_\theta(X_1 = 1, \sum_{i=2}^{n} X_i = t - 1)}{P_\theta(\sum_{i=1}^{n} X_i = t)} \qquad \left( \begin{array}{c} X_1 = 1 \text{ is} \\ \text{redundant} \end{array} \right)$$

$$= \frac{P_\theta(X_1 = 1)P_\theta(\sum_{i=2}^{n} X_i = t - 1)}{P_\theta(\sum_{i=1}^{n} X_i = t)}. \qquad \left( \begin{array}{c} X_1 \text{ is independent} \\ \text{of } X_2, \dots, X_n \end{array} \right)$$

Now $X_1 \sim \text{binomial}(k, \theta)$, $\sum_{i=2}^{n} X_i \sim \text{binomial}(k(n-1), \theta)$, and $\sum_{i=1}^{n} X_i \sim \text{binomial}(kn, \theta)$. Using these facts we have

$$\phi(t) = \frac{\left[ k\theta(1-\theta)^{k-1} \right] \left[ \binom{k(n-1)}{t-1} \theta^{t-1}(1-\theta)^{k(n-1)-(t-1)} \right]}{\binom{kn}{t} \theta^t (1-\theta)^{kn-t}}$$

$$= k\frac{\binom{k(n-1)}{t-1}}{\binom{kn}{t}}.$$

Note that all of the $\theta$s cancel, as they must since $\sum_{i=1}^{n} X_i$ is sufficient. Hence, the best unbiased estimator of $k\theta(1-\theta)^{k-1}$ is

$$\phi\left( \sum_{i=1}^{n} X_i \right) = k\frac{\binom{k(n-1)}{\Sigma X_i - 1}}{\binom{kn}{\Sigma X_i}}.$$

We can assert unbiasedness without performing the difficult evaluation of $E_\theta[\phi\left( \sum_{i=1}^{n} X_i \right)]$.        ‖

### 7.3.4 Loss Function Optimality

Our evaluations of point estimators have been based on their mean squared error performance. Mean squared error is a special case of a function called a *loss function*. The study of the performance, and the optimality, of estimators evaluated through loss functions is a branch of *decision theory*.

After the data $\mathbf{X} = \mathbf{x}$ are observed, where $X \sim f(\mathbf{x}|\theta)$, $\theta \in \Theta$, a decision regarding $\theta$ is made. The set of allowable decisions is the *action space*, denoted by $\mathcal{A}$. Often in point estimation problems $\mathcal{A}$ is equal to $\Theta$, the parameter space, but this will change in other problems (such as hypothesis testing—see Section 8.3.5).

The loss function in a point estimation problem reflects the fact that if an action $a$ is close to $\theta$, then the decision $a$ is reasonable and little loss is incurred. If $a$ is far from $\theta$, then a large loss is incurred. The loss function is a nonnegative function that generally increases as the distance between $a$ and $\theta$ increases. If $\theta$ is real-valued, two commonly used loss functions are

$$\text{absolute error loss,} \quad L(\theta, a) = |a - \theta|,$$

and

$$\text{squared error loss,} \quad L(\theta, a) = (a - \theta)^2.$$

Both of these loss functions increase as the distance between $\theta$ and $a$ increases, with minimum value $L(\theta, \theta) = 0$. That is, the loss is minimum if the action is correct. Squared error loss gives relatively more penalty for large discrepancies, and absolute error loss gives relatively more penalty for small discrepancies. A variation of squared error loss, one that penalizes overestimation more than underestimation, is

$$L(\theta, a) = \begin{cases} (a - \theta)^2 & \text{if } a < \theta \\ 10(a - \theta)^2 & \text{if } a \geq \theta. \end{cases}$$

A loss that penalizes errors in estimation more if $\theta$ is near 0 than if $|\theta|$ is large, a relative squared error loss, is

$$L(\theta, a) = \frac{(a - \theta)^2}{|\theta| + 1}.$$

Notice that both of these last variations of squared error loss could have been based instead on absolute error loss. In general, the experimenter must consider the consequences of various errors in estimation for different values of $\theta$ and specify a loss function that reflects these consequences.

In a loss function or *decision theoretic* analysis, the quality of an estimator is quantified in its *risk function*; that is, for an estimator $\delta(\mathbf{x})$ of $\theta$, the risk function, a function of $\theta$, is

$$(7.3.16) \qquad\qquad R(\theta, \delta) = \mathrm{E}_\theta L(\theta, \delta(\mathbf{X})).$$

At a given $\theta$, the risk function is the average loss that will be incurred if the estimator $\delta(\mathbf{x})$ is used.

Since the true value of $\theta$ is unknown, we would like to use an estimator that has a small value of $R(\theta, \delta)$ for all values of $\theta$. This would mean that, regardless of the true value of $\theta$, the estimator will have a small expected loss. If the qualities of two different estimators, $\delta_1$ and $\delta_2$, are to be compared, then they will be compared by comparing their risk functions, $R(\theta, \delta_1)$ and $R(\theta, \delta_2)$. If $R(\theta, \delta_1) < R(\theta, \delta_2)$ for all $\theta \in \Theta$, then $\delta_1$ is the preferred estimator because $\delta_1$ performs better for all $\theta$. More typically, the two risk functions will cross. Then the judgment as to which estimator is better may not be so clear-cut.

The risk function for an estimator $\delta$ is the expected loss, as defined in (7.3.16). For squared error loss, the risk function is a familiar quantity, the mean squared error (MSE) that was used in Section 7.3.1. There the MSE of an estimator was defined as

$\mathrm{MSE}(\theta) = \mathrm{E}_\theta(\delta(\mathbf{X}) - \theta)^2$, which is just $\mathrm{E}_\theta L(\theta, \delta(\mathbf{X})) = R(\theta, \delta)$ if $L(\theta, a) = (a - \theta)^2$. As in Chapter 7 we have that, for squared error loss,

$$(7.3.17) \qquad R(\theta, \delta) = \mathrm{Var}_\theta \, \delta(\mathbf{X}) + (\mathrm{E}_\theta \delta(\mathbf{X}) - \theta)^2 = \mathrm{Var}_\theta \, \delta(\mathbf{X}) + (\mathrm{Bias}_\theta \, \delta(\mathbf{X}))^2 \, .$$

This risk function for squared error loss clearly indicates that a good estimator should have both a small variance and a small bias. A decision theoretic analysis would judge how well an estimator succeeded in simultaneously minimizing these two quantities.

It would be an atypical decision theoretic analysis in which the set $\mathcal{D}$ of allowable estimators was restricted to the set of unbiased estimators, as was done in Section 7.3.2. Then, minimizing the risk would just be minimizing the variance. A decision theoretic analysis would be more comprehensive in that both the variance and bias are in the risk and will be considered simultaneously. An estimator would be judged good if it had a small, but probably nonzero, bias combined with a small variance.

**Example 7.3.25 (Binomial risk functions)**    In Example 7.3.5 we considered $X_1, \ldots, X_n$, a random sample from a Bernoulli($p$) population. We considered two estimators,

$$\hat{p}_{\mathrm{B}} = \frac{\sum_{i=1}^n X_i + \sqrt{n/4}}{n + \sqrt{n}} \qquad \text{and} \qquad \bar{X} = \frac{1}{n} \sum_{i=1}^n X_i.$$

The risk functions for these two estimators, for $n = 4$ and $n = 400$, were graphed in Figure 7.3.1, and the comparisons of these risk functions are as stated in Example 7.3.5. On the basis of risk comparison, the estimator $\hat{p}_{\mathrm{B}}$ would be preferred for small $n$ and the estimator $\bar{X}$ would be preferred for large $n$.                                    ‖

**Example 7.3.26 (Risk of normal variance)**    Let $X_1, \ldots, X_n$ be a random sample from a n($\mu, \sigma^2$) population. Consider estimating $\sigma^2$ using squared error loss. We will consider estimators of the form $\delta_b(\mathbf{X}) = bS^2$, where $S^2$ is the sample variance and $b$ can be any nonnegative constant. Recall that $\mathrm{E}S^2 = \sigma^2$ and, for a normal sample, $\mathrm{Var}\, S^2 = 2\sigma^4/(n-1)$. Using (7.3.17), we can compute the risk function for $\delta_b$ as

$$\begin{aligned} R((\mu, \sigma^2), \delta_b) &= \mathrm{Var}\, bS^2 + \left(\mathrm{E}bS^2 - \sigma^2\right)^2 \\ &= b^2 \mathrm{Var}\, S^2 + \left(b\mathrm{E}S^2 - \sigma^2\right)^2 \\ &= \frac{b^2 2\sigma^4}{n-1} + (b-1)^2 \sigma^4 \qquad \text{(using Var } S^2) \\ &= \left[\frac{2b^2}{n-1} + (b-1)^2\right] \sigma^4. \end{aligned}$$

The risk function for $\delta_b$ does not depend on $\mu$ and is a quadratic function of $\sigma^2$. This quadratic function is of the form $c_b(\sigma^2)^2$, where $c_b$ is a positive constant. To compare two risk functions, and hence the worth of two estimators, note that if $c_b < c_{b'}$, then

$$R((\mu, \sigma^2), \delta_b) = c_b(\sigma^2)^2 < c_{b'}(\sigma^2)^2 = R((\mu, \sigma^2), \delta_{b'})$$

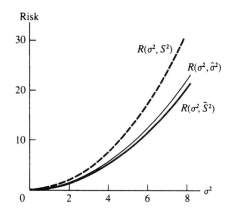

Figure 7.3.2. *Risk functions for three variance estimators in Example 7.3.26*

for all values of $(\mu, \sigma^2)$. Thus $\delta_b$ would be a better estimator than $\delta_{b'}$. The value of $b$ that gives the overall minimum value of

$$(7.3.18) \qquad\qquad c_b = \frac{2b^2}{n-1} + (b-1)^2$$

yields the best estimator $\delta_b$ in this class. Standard calculus methods show that $b = (n-1)/(n+1)$ is the minimizing value. Thus, at every value of $(\mu, \sigma^2)$, the estimator

$$\tilde{S}^2 = \frac{n-1}{n+1}S^2 = \frac{1}{n+1}\sum(X_i - \bar{X})^2$$

has the smallest risk among all estimators of the form $bS^2$. For $n = 5$, the risk functions for this estimator and two other estimators in this class are shown in Figure 7.3.2. The other estimators are $S^2$, the unbiased estimator, and $\hat{\sigma}^2 = \frac{n-1}{n}S^2$, the MLE of $\sigma^2$. It is clear that the risk function for $\tilde{S}^2$ is smallest everywhere.   ‖

**Example 7.3.27 (Variance estimation using Stein's loss)**  Again we consider estimating a population variance $\sigma^2$ with an estimator of the form $bS^2$. In this analysis we can be quite general and assume only that $X_1, \ldots, X_n$ is a random sample from some population with positive, finite variance $\sigma^2$. Now we will use the loss function

$$L(\sigma^2, a) = \frac{a}{\sigma^2} - 1 - \log\frac{a}{\sigma^2},$$

attributed to Stein (James and Stein 1961; see also Brown 1990a). This loss is more complicated than squared error loss but it has some reasonable properties. Note that if $a = \sigma^2$, the loss is 0. Also, for any fixed value of $\sigma^2$, $L(\sigma^2, a) \to \infty$ as $a \to 0$ or $a \to \infty$. That is, gross underestimation is penalized just as heavily as gross overestimation. (A criticism of squared error loss in a variance estimation problem is that underestimation has only a finite penalty, while overestimation has an infinite penalty.) The loss function also arises out of the likelihood function for $\sigma^2$, if this is a sample from a normal population, and thus ties together good decision theoretic properties with good likelihood properties (see Exercise 7.61).

For the estimator $\delta_b = bS^2$, the risk function is

$$R(\sigma^2, \delta_b) = \mathrm{E}\left(\frac{bS^2}{\sigma^2} - 1 - \log \frac{bS^2}{\sigma^2}\right)$$

$$= b\mathrm{E}\frac{S^2}{\sigma^2} - 1 - \mathrm{E}\log \frac{bS^2}{\sigma^2}$$

$$= b - \log b - 1 - \mathrm{E}\log \frac{S^2}{\sigma^2}. \qquad \left(\mathrm{E}\frac{S^2}{\sigma^2} = 1\right)$$

The quantity $\mathrm{E}\log(S^2/\sigma^2)$ may be a function of $\sigma^2$ and other population parameters but it is not a function of $b$. Thus $R(\sigma^2, \delta_b)$ is minimized in $b$, for all $\sigma^2$, by the value of $b$ that minimizes $b - \log b$, that is, $b = 1$. Therefore the estimator of the form $bS^2$ that has the smallest risk for all values of $\sigma^2$ is $\delta_1 = S^2$. ‖

We can also use a Bayesian approach to the problem of loss function optimality, where we would have a prior distribution, $\pi(\theta)$. In a Bayesian analysis we would use this prior distribution to compute an average risk

$$\int_\Theta R(\theta, \delta)\pi(\theta)\, d\theta,$$

known as the *Bayes risk*. Averaging the risk function gives us one number for assessing the performance of an estimator with respect to a given loss function. Moreover, we can attempt to find the estimator that yields the smallest value of the Bayes risk. Such an estimator is called the *Bayes rule with respect to a prior* $\pi$ and is often denoted $\delta^\pi$.

Finding the Bayes decision rule for a given prior $\pi$ may look like a daunting task, but it turns out to be rather mechanical, as the following indicates. (The technique of finding Bayes rules by the method given below works in greater generality than presented here; see Brown and Purves 1973.)

For $\mathbf{X} \sim f(\mathbf{x}|\theta)$ and $\theta \sim \pi$, the Bayes risk of a decision rule $\delta$ can be written as

$$\int_\Theta R(\theta, \delta)\pi(\theta)\, d\theta = \int_\Theta \left(\int_\mathcal{X} L(\theta, \delta(\mathbf{x}))f(\mathbf{x}|\theta)\, d\mathbf{x}\right)\pi(\theta)\, d\theta.$$

Now if we write $f(\mathbf{x}|\theta)\pi(\theta) = \pi(\theta|\mathbf{x})m(\mathbf{x})$, where $\pi(\theta|\mathbf{x})$ is the posterior distribution of $\theta$ and $m(\mathbf{x})$ is the marginal distribution of $\mathbf{X}$, we can write the Bayes risk as

$$(7.3.19) \qquad \int_\Theta R(\theta, \delta)\pi(\theta)\, d\theta = \int_\mathcal{X}\left[\int_\Theta L(\theta, \delta(\mathbf{x}))\pi(\theta|\mathbf{x})\, d\theta\right]m(\mathbf{x})\, d\mathbf{x}.$$

The quantity in square brackets is the expected value of the loss function with respect to the posterior distribution, called the *posterior expected loss*. It is a function only of $\mathbf{x}$, and not a function of $\theta$. Thus, for each $\mathbf{x}$, if we choose the action $\delta(\mathbf{x})$ to minimize the posterior expected loss, we will minimize the Bayes risk.

Notice that we now have a recipe for constructing a Bayes rule. For a given observation $\mathbf{x}$, the Bayes rule should minimize the posterior expected loss. This is quite unlike any prescription we have had in previous sections. For example, consider the methods of finding best unbiased estimators discussed previously. To use Theorem 7.3.23, first we need to find a complete sufficient statistic $T$. Then we need to find a function $\phi(T)$ that is an unbiased estimator of the parameter. The Rao–Blackwell Theorem, Theorem 7.3.17, may be helpful if we know of some unbiased estimator of the parameter. But if we cannot dream up some unbiased estimator, then the method does not tell us how to construct one.

Even if the minimization of the posterior expected loss cannot be done analytically, the integral can be evaluated and the minimization carried out numerically. In fact, having observed $\mathbf{X} = \mathbf{x}$, we need to do the minimization only for this particular $\mathbf{x}$. However, in some problems we can explicitly describe the Bayes rule.

**Example 7.3.28 (Two Bayes rules)**   Consider a point estimation problem for a real-valued parameter $\theta$.

a. For squared error loss, the posterior expected loss is

$$\int_{\Theta} (\theta - a)^2 \pi(\theta|\mathbf{x}) \, d\theta = \mathrm{E}\left((\theta - a)^2 | \mathbf{X} = \mathbf{x}\right).$$

Here $\theta$ is the random variable with distribution $\pi(\theta|\mathbf{x})$. By Example 2.2.6, this expected value is minimized by $\delta^{\pi}(\mathbf{x}) = \mathrm{E}(\theta|\mathbf{x})$. So the Bayes rule is the mean of the posterior distribution.

b. For absolute error loss, the posterior expected loss is $\mathrm{E}\left(|\theta - a||\mathbf{X} = \mathbf{x}\right)$. By applying Exercise 2.18, we see that this is minimized by choosing $\delta^{\pi}(\mathbf{x}) = $ median of $\pi(\theta|\mathbf{x})$.

‖

In Section 7.2.3, the Bayes estimator we discussed was $\delta^{\pi}(\mathbf{x}) = \mathrm{E}(\theta|\mathbf{x})$, the posterior mean. We now see that this is the Bayes estimator with respect to squared error loss. If some other loss function is deemed more appropriate than squared error loss, the Bayes estimator might be a different statistic.

**Example 7.3.29 (Normal Bayes estimates)**   Let $X_1, \ldots, X_n$ be a random sample from a $n(\theta, \sigma^2)$ population and let $\pi(\theta)$ be $n(\mu, \tau^2)$. The values $\sigma^2$, $\mu$, and $\tau^2$ are known. In Example 7.2.16, as extended in Exercise 7.22, we found that the posterior distribution of $\theta$ given $\bar{X} = \bar{x}$ is normal with

$$\mathrm{E}(\theta|\bar{x}) = \frac{\tau^2}{\tau^2 + (\sigma^2/n)} \bar{x} + \frac{\sigma^2/n}{\tau^2 + (\sigma^2/n)} \mu$$

and

$$\mathrm{Var}(\theta|\bar{x}) = \frac{\tau^2 \sigma^2/n}{\tau^2 + (\sigma^2/n)}.$$

For squared error loss, the Bayes estimator is $\delta^{\pi}(\mathbf{x}) = \mathrm{E}(\theta|\bar{x})$. Since the posterior distribution is normal, it is symmetric about its mean and the median of $\pi(\theta|\mathbf{x})$ is equal to $\mathrm{E}(\theta|\bar{x})$. Thus, for absolute error loss, the Bayes estimator is also $\delta^{\pi}(\mathbf{x}) = \mathrm{E}(\theta|\bar{x})$.

‖

Table 7.3.1. *Three estimators for a binomial p*

| $n = 10$ | | prior $\pi(p) \sim$ uniform$(0,1)$ | |
|:---:|:---:|:---:|:---:|
| $y$ | MLE | Bayes absolute error | Bayes squared error |
| 0 | .0000 | .0611 | .0833 |
| 1 | .1000 | .1480 | .1667 |
| 2 | .2000 | .2358 | .2500 |
| 3 | .3000 | .3238 | .3333 |
| 4 | .4000 | .4119 | .4167 |
| 5 | .5000 | .5000 | .5000 |
| 6 | .6000 | .5881 | .5833 |
| 7 | .7000 | .6762 | .6667 |
| 8 | .8000 | .7642 | .7500 |
| 9 | .9000 | .8520 | .8333 |
| 10 | 1.0000 | .9389 | .9137 |

**Example 7.3.30 (Binomial Bayes estimates)**      Let $X_1, \ldots, X_n$ be iid Bernoulli$(p)$ and let $Y = \sum X_i$. Suppose the prior on $p$ is beta$(\alpha, \beta)$. In Example 7.2.14 we found that the posterior distribution depends on the sample only through the observed value of $Y = y$ and is beta$(y + \alpha, n - y + \beta)$. Hence, $\delta^\pi(y) = \mathrm{E}(p|y) = (y + \alpha)/(\alpha + \beta + n)$ is the Bayes estimator of $p$ for squared error loss.

For absolute error loss, we need to find the median of $\pi(p|y) = $ beta$(y+\alpha, n-y+\beta)$. In general, there is no simple expression for this median. The median is implicitly defined to be the number, $m$, that satisfies

$$\int_0^m \frac{\Gamma(\alpha + \beta + n)}{\Gamma(y + \alpha)\Gamma(n - y + \beta)} p^{y+\alpha-1}(1 - p)^{n-y+\beta-1} dp = \frac{1}{2}.$$

This integral can be evaluated numerically to find (approximately) the value $m$ that satisfies the equality. We have done this for $n = 10$ and $\alpha = \beta = 1$, the uniform$(0, 1)$ prior. The Bayes estimator for absolute error loss is given in Table 7.3.1. In the table we have also listed the Bayes estimator for squared error loss, derived above, and the MLE, $\hat{p} = y/n$.

Notice in Table 7.3.1 that, unlike the MLE, neither Bayes estimator estimates $p$ to be 0 or 1, even if $y$ is 0 or $n$. It is typical of Bayes estimators that they would not take on extreme values in the parameter space. No matter how large the sample size, the prior always has some influence on the estimator and tends to draw it away from the extreme values. In the above expression for $\mathrm{E}(p|y)$, you can see that even if $y = 0$ and $n$ is large, the Bayes estimator is a positive number.                    ‖

## 7.4 Exercises

**7.1** One observation is taken on a discrete random variable $X$ with pmf $f(x|\theta)$, where $\theta \in \{1, 2, 3\}$. Find the MLE of $\theta$.

| $x$ | $f(x|1)$ | $f(x|2)$ | $f(x|3)$ |
|-----|----------|----------|----------|
| 0 | $\frac{1}{3}$ | $\frac{1}{4}$ | 0 |
| 1 | $\frac{1}{3}$ | $\frac{1}{4}$ | 0 |
| 2 | 0 | $\frac{1}{4}$ | $\frac{1}{4}$ |
| 3 | $\frac{1}{6}$ | $\frac{1}{4}$ | $\frac{1}{2}$ |
| 4 | $\frac{1}{6}$ | 0 | $\frac{1}{4}$ |

**7.2** Let $X_1, \ldots, X_n$ be a random sample from a gamma$(\alpha, \beta)$ population.

(a) Find the MLE of $\beta$, assuming $\alpha$ is known.

(b) If $\alpha$ and $\beta$ are both unknown, there is no explicit formula for the MLEs of $\alpha$ and $\beta$, but the maximum can be found numerically. The result in part (a) can be used to reduce the problem to the maximization of a univariate function. Find the MLEs for $\alpha$ and $\beta$ for the data in Exercise 7.10(c).

**7.3** Given a random sample $X_1, \ldots, X_n$ from a population with pdf $f(x|\theta)$, show that maximizing the likelihood function, $L(\theta|\mathbf{x})$, as a function of $\theta$ is equivalent to maximizing $\log L(\theta|\mathbf{x})$.

**7.4** Prove the assertion in Example 7.2.8. That is, prove that $\hat{\theta}$ given there is the MLE when the range of $\theta$ is restricted to the positive axis.

**7.5** Consider estimating the binomial parameter $k$ as in Example 7.2.9.

(a) Prove the assertion that the integer $\hat{k}$ that satisfies the inequalities and is the MLE is the largest integer less than or equal to $1/\hat{z}$.

(b) Let $p = \frac{1}{2}$, $n = 4$, and $X_1 = 0$, $X_2 = 20$, $X_3 = 1$, and $X_4 = 19$. What is $\hat{k}$?

**7.6** Let $X_1, \ldots, X_n$ be a random sample from the pdf

$$f(x|\theta) = \theta x^{-2}, \quad 0 < \theta \le x < \infty.$$

(a) What is a sufficient statistic for $\theta$?

(b) Find the MLE of $\theta$.

(c) Find the method of moments estimator of $\theta$.

**7.7** Let $X_1, \ldots, X_n$ be iid with one of two pdfs. If $\theta = 0$, then

$$f(x|\theta) = \begin{cases} 1 & \text{if } 0 < x < 1 \\ 0 & \text{otherwise,} \end{cases}$$

while if $\theta = 1$, then

$$f(x|\theta) = \begin{cases} 1/(2\sqrt{x}) & \text{if } 0 < x < 1 \\ 0 & \text{otherwise.} \end{cases}$$

Find the MLE of $\theta$.

**7.8** One observation, $X$, is taken from a $n(0, \sigma^2)$ population.

    (a) Find an unbiased estimator of $\sigma^2$.

    (b) Find the MLE of $\sigma$.

    (c) Discuss how the method of moments estimator of $\sigma$ might be found.

**7.9** Let $X_1, \ldots, X_n$ be iid with pdf

$$f(x|\theta) = \frac{1}{\theta}, \quad 0 \le x \le \theta, \quad \theta > 0.$$

Estimate $\theta$ using both the method of moments and maximum likelihood. Calculate the means and variances of the two estimators. Which one should be preferred and why?

**7.10** The independent random variables $X_1, \ldots, X_n$ have the common distribution

$$P(X_i \le x | \alpha, \beta) = \begin{cases} 0 & \text{if } x < 0 \\ (x/\beta)^\alpha & \text{if } 0 \le x \le \beta \\ 1 & \text{if } x > \beta, \end{cases}$$

where the parameters $\alpha$ and $\beta$ are positive.

    (a) Find a two-dimensional sufficient statistic for $(\alpha, \beta)$.

    (b) Find the MLEs of $\alpha$ and $\beta$.

    (c) The length (in millimeters) of cuckoos' eggs found in hedge sparrow nests can be modeled with this distribution. For the data

       22.0, 23.9, 20.9, 23.8, 25.0, 24.0, 21.7, 23.8, 22.8, 23.1, 23.1, 23.5, 23.0, 23.0,

    find the MLEs of $\alpha$ and $\beta$.

**7.11** Let $X_1, \ldots, X_n$ be iid with pdf

$$f(x|\theta) = \theta x^{\theta-1}, \quad 0 \le x \le 1, \quad 0 < \theta < \infty.$$

    (a) Find the MLE of $\theta$, and show that its variance $\to 0$ as $n \to \infty$.

    (b) Find the method of moments estimator of $\theta$.

**7.12** Let $X_1, \ldots, X_n$ be a random sample from a population with pmf

$$P_\theta(X = x) = \theta^x (1-\theta)^{1-x}, \quad x = 0 \text{ or } 1, \quad 0 \le \theta \le \frac{1}{2}.$$

    (a) Find the method of moments estimator and MLE of $\theta$.

    (b) Find the mean squared errors of each of the estimators.

    (c) Which estimator is preferred? Justify your choice.

**7.13** Let $X_1, \ldots, X_n$ be a sample from a population with double exponential pdf

$$f(x|\theta) = \frac{1}{2} e^{-|x-\theta|}, \quad -\infty < x < \infty, \quad -\infty < \theta < \infty.$$

Find the MLE of $\theta$. (*Hint:* Consider the case of even $n$ separate from that of odd $n$, and find the MLE in terms of the order statistics. A complete treatment of this problem is given in Norton 1984.)

**7.14** Let $X$ and $Y$ be independent exponential random variables, with

$$f(x|\lambda) = \frac{1}{\lambda} e^{-x/\lambda}, \, x > 0, \quad f(y|\mu) = \frac{1}{\mu} e^{-y/\mu}, \, y > 0.$$

We observe Z and $W$ with

$$Z = \min(X, Y) \quad \text{and} \quad W = \begin{cases} 1 & \text{if } Z = X \\ 0 & \text{if } Z = Y. \end{cases}$$

In Exercise 4.26 the joint distribution of $Z$ and $W$ was obtained. Now assume that $(Z_i, W_i), i = 1, \ldots, n$, are $n$ iid observations. Find the MLEs of $\lambda$ and $\mu$.

**7.15** Let $X_1, X_2, \ldots, X_n$ be a sample from the *inverse Gaussian* pdf,

$$f(x|\mu, \lambda) = \left(\frac{\lambda}{2\pi x^3}\right)^{1/2} \exp\left\{-\lambda(x - \mu)^2/(2\mu^2 x)\right\}, \quad x > 0.$$

(a) Show that the MLEs of $\mu$ and $\lambda$ are

$$\hat{\mu}_n = \bar{X} \quad \text{and} \quad \hat{\lambda}_n = \frac{n}{\sum_i \left(\frac{1}{X_i} - \frac{1}{\bar{X}}\right)}.$$

(b) Tweedie (1957) showed that $\hat{\mu}_n$ and $\hat{\lambda}_n$ are independent, $\hat{\mu}_n$ having an inverse Gaussian distribution with parameters $\mu$ and $n\lambda$, and $n\lambda/\hat{\lambda}_n$ having a $\chi^2_{n-1}$ distribution. Schwarz and Samanta (1991) give a proof of these facts using an induction argument.

   (i) Show that $\hat{\mu}_2$ has an inverse Gaussian distribution with parameters $\mu$ and $2\lambda$, $2\lambda/\hat{\lambda}_2$ has a $\chi^2_1$ distribution, and they are independent.

   (ii) Assume the result is true for $n = k$ and that we get a new, independent observation $x$. Establish the induction step used by Schwarz and Samanta (1991), and transform the pdf $f(x, \hat{\mu}_k, \hat{\lambda}_k)$ to $f(x, \hat{\mu}_{k+1}, \hat{\lambda}_{k+1})$. Show that this density factors in the appropriate way and that the result of Tweedie follows.

**7.16** Berger and Casella (1992) also investigate *power means*, which we have seen in Exercise 4.57. Recall that a power mean is defined as $\left[\frac{1}{n}\sum_{i=1}^n x_i^r\right]^{1/r}$. This definition can be further generalized by noting that the power function $x^r$ can be replaced by any continuous, monotone function $h$, yielding the *generalized mean* $h^{-1}\left(\frac{1}{n}\sum_{i=1}^n h(x_i)\right)$.

(a) The least squares problem $\min_a \sum_i (x_i - a)^2$ is sometimes solved using transformed variables, that is, solving $\min_a \sum_i [h(x_i) - h(a)]^2$. Show that the solution to this latter problem is $a = h^{-1}((1/n)\sum_i h(x_i))$.

(b) Show that the arithmetic mean is the solution to the untransformed least squares problem, the geometric mean is the solution to the problem transformed by $h(x) = \log x$, and the harmonic mean is the solution to the problem transformed by $h(x) = 1/x$.

(c) Show that if the least squares problem is transformed with the *Box-Cox Transformation* (see Exercise 11.3), then the solution is a generalized mean with $h(x) = x^\lambda$.

(d) Let $X_1, \ldots, X_n$ be a sample from a lognormal$(\mu, \sigma^2)$ population. Show that the MLE of $\mu$ is the geometric mean.

(e) Suppose that $X_1, \ldots, X_n$ are a sample from a one-parameter exponential family $f(x|\theta) = \exp\{\theta h(x) - H(\theta)\}g(x)$, where $h = H'$ and $h$ is an increasing function.

   (i) Show that the MLE of $\theta$ is $\hat{\theta} = h^{-1}((1/n)\sum_i h(x_i))$.

   (ii) Show that two densities that satisfy $h = H'$ are the normal and the inverted gamma with pdf $f(x|\theta) = \theta x^{-2} \exp\{-\theta/x\}$ for $x > 0$, and for the normal the MLE is the arithmetic mean and for the inverted gamma it is the harmonic mean.

**7.17** The Borel Paradox (Miscellanea 4.9.3) can also arise in inference problems. Suppose that $X_1$ and $X_2$ are iid exponential($\theta$) random variables.

(a) If we observe only $X_2$, show that the MLE of $\theta$ is $\hat{\theta} = X_2$.

(b) Suppose that we instead observe only $Z = (X_2 - 1)/X_1$. Find the joint distribution of $(X_1, Z)$, and integrate out $X_1$ to get the likelihood function.

(c) Suppose that $X_2 = 1$. Compare the MLEs for $\theta$ from parts (a) and (b).

(d) Bayesian analysis is not immune to the Borel Paradox. If $\pi(\theta)$ is a prior density for $\theta$, show that the posterior distributions, at $X_2 = 1$, are different in parts (a) and (b).

(*Communicated by L. Mark Berliner, Ohio State University.*)

**7.18** Let $(X_1, Y_1), \ldots, (X_n, Y_n)$ be iid bivariate normal random variables (pairs) where all five parameters are unknown.

(a) Show that the method of moments estimators for $\mu_X, \mu_Y, \sigma_X^2, \sigma_Y^2, \rho$ are $\tilde{\mu}_X = \bar{x}, \tilde{\mu}_Y = \bar{y}, \tilde{\sigma}_X^2 = \frac{1}{n}\sum(x_i - \bar{x})^2, \tilde{\sigma}_Y^2 = \frac{1}{n}\sum(y_i - \bar{y})^2, \tilde{\rho} = \frac{1}{n}\sum(x_i - \bar{x})(y_i - \bar{y})/(\tilde{\sigma}_X \tilde{\sigma}_Y)$.

(b) Derive the MLEs of the unknown parameters and show that they are the same as the method of moments estimators. (One attack is to write the joint pdf as the product of a conditional and a marginal, that is, write

$$f(x, y | \mu_X, \mu_Y, \sigma_X^2, \sigma_Y^2, \rho) = f(y | x, \mu_X, \mu_Y, \sigma_X^2, \sigma_Y^2, \rho) f(x | \mu_X, \sigma_X^2),$$

and argue that the MLEs for $\mu_X$ and $\sigma_X^2$ are given by $\bar{x}$ and $\frac{1}{n}\sum(x_i - \bar{x})^2$. Then, turn things around to get the MLEs for $\mu_Y$ and $\sigma_Y^2$. Finally, work with the "partially maximized" likelihood function $L(\bar{x}, \bar{y}, \hat{\sigma}_X^2, \hat{\sigma}_Y^2, \rho | \mathbf{x}, \mathbf{y})$ to get the MLE for $\rho$. As might be guessed, this is a difficult problem.)

**7.19** Suppose that the random variables $Y_1, \ldots, Y_n$ satisfy

$$Y_i = \beta x_i + \epsilon_i, \quad i = 1, \ldots, n,$$

where $x_1, \ldots, x_n$ are fixed constants, and $\epsilon_1, \ldots, \epsilon_n$ are iid n$(0, \sigma^2)$, $\sigma^2$ unknown.

(a) Find a two-dimensional sufficient statistic for $(\beta, \sigma^2)$.

(b) Find the MLE of $\beta$, and show that it is an unbiased estimator of $\beta$.

(c) Find the distribution of the MLE of $\beta$.

**7.20** Consider $Y_1, \ldots, Y_n$ as defined in Exercise 7.19.

(a) Show that $\sum Y_i / \sum x_i$ is an unbiased estimator of $\beta$.

(b) Calculate the exact variance of $\sum Y_i / \sum x_i$ and compare it to the variance of the MLE.

**7.21** Again, let $Y_1, \ldots, Y_n$ be as defined in Exercise 7.19.

(a) Show that $\left[\sum(Y_i/x_i)\right]/n$ is also an unbiased estimator of $\beta$.

(b) Calculate the exact variance of $\left[\sum(Y_i/x_i)\right]/n$ and compare it to the variances of the estimators in the previous two exercises.

**7.22** This exercise will prove the assertions in Example 7.2.16, and more. Let $X_1, \ldots, X_n$ be a random sample from a n$(\theta, \sigma^2)$ population, and suppose that the prior distribution on $\theta$ is n$(\mu, \tau^2)$. Here we assume that $\sigma^2$, $\mu$, and $\tau^2$ are all known.

(a) Find the joint pdf of $\bar{X}$ and $\theta$.

(b) Show that $m(\bar{x}|\sigma^2, \mu, \tau^2)$, the marginal distribution of $\bar{X}$, is n$(\mu, (\sigma^2/n) + \tau^2)$.

(c) Show that $\pi(\theta|\bar{x}, \sigma^2, \mu, \tau^2)$, the posterior distribution of $\theta$, is normal with mean and variance given by (7.2.10).

**7.23** If $S^2$ is the sample variance based on a sample of size $n$ from a normal population, we know that $(n-1)S^2/\sigma^2$ has a $\chi^2_{n-1}$ distribution. The conjugate prior for $\sigma^2$ is the *inverted gamma* pdf, $IG(\alpha, \beta)$, given by

$$\pi(\sigma^2) = \frac{1}{\Gamma(\alpha)\beta^\alpha} \frac{1}{(\sigma^2)^{\alpha+1}} e^{-1/(\beta\sigma^2)}, \quad 0 < \sigma^2 < \infty,$$

where $\alpha$ and $\beta$ are positive constants. Show that the posterior distribution of $\sigma^2$ is $IG(\alpha + \frac{n-1}{2}, [\frac{(n-1)S^2}{2} + \frac{1}{\beta}]^{-1})$. Find the mean of this distribution, the Bayes estimator of $\sigma^2$.

**7.24** Let $X_1, \ldots, X_n$ be iid Poisson($\lambda$), and let $\lambda$ have a gamma($\alpha, \beta$) distribution, the conjugate family for the Poisson.

(a) Find the posterior distribution of $\lambda$.

(b) Calculate the posterior mean and variance.

**7.25** We examine a generalization of the hierarchical (Bayes) model considered in Example 7.2.16 and Exercise 7.22. Suppose that we observe $X_1, \ldots, X_n$, where

$$
\begin{aligned}
X_i|\theta_i &\sim n(\theta_i, \sigma^2), & i &= 1, \ldots, n, & \text{independent},\\
\theta_i &\sim n(\mu, \tau^2), & i &= 1, \ldots, n, & \text{independent}.
\end{aligned}
$$

(a) Show that the marginal distribution of $X_i$ is $n(\mu, \sigma^2 + \tau^2)$ and that, marginally, $X_1, \ldots, X_n$ are iid. (*Empirical Bayes analysis* would use the marginal distribution of the $X_i$s to estimate the prior parameters $\mu$ and $\tau^2$. See Miscellanea 7.5.6.)

(b) Show, in general, that if

$$
\begin{aligned}
X_i|\theta_i &\sim f(x|\theta_i), & i &= 1, \ldots, n, & \text{independent},\\
\theta_i &\sim \pi(\theta|\tau), & i &= 1, \ldots, n, & \text{independent},
\end{aligned}
$$

then marginally, $X_1, \ldots, X_n$ are iid.

**7.26** In Example 7.2.16 we saw that the normal distribution is its own conjugate family. It is sometimes the case, however, that a conjugate prior does not accurately reflect prior knowledge, and a different prior is sought. Let $X_1, \ldots, X_n$ be iid $n(\theta, \sigma^2)$, and let $\theta$ have a double exponential distribution, that is, $\pi(\theta) = e^{-|\theta|/a}/(2a), a$ known. Find the mean of the posterior distribution of $\theta$.

**7.27** Refer to Example 7.2.17.

(a) Show that the likelihood estimators from the complete-data likelihood (7.2.11) are given by (7.2.12).

(b) Show that the limit of the EM sequence in (7.2.23) satisfies (7.2.16)

(c) A direct solution of the original (incomplete-data) likelihood equations is possible. Show that the solution to (7.2.16) is given by

$$\hat{\beta} = \frac{\sum_{i=2}^n y_i}{\sum_{i=2}^n x_i}, \quad \hat{\tau}_1 = \frac{y_1}{\hat{\beta}}, \quad \hat{\tau}_j = \frac{x_j + y_j}{\hat{\beta} + 1}, \quad j = 2, 3, \ldots, n,$$

and that this is the limit of the EM sequence in (7.2.23).

**7.28** Use the model of Example 7.2.17 on the data in the following table adapted from Lange *et al.* (1994). These are leukemia counts and the associated populations for a number of areas in New York State.

*Counts of leukemia cases*

| Population | 3540 | 3560 | 3739 | 2784 | 2571 | 2729 | 3952 | 993 | 1908 |
|---|---|---|---|---|---|---|---|---|---|
| Number of cases | 3 | 4 | 1 | 1 | 3 | 1 | 2 | 0 | 2 |
| Population | 948 | 1172 | 1047 | 3138 | 5485 | 5554 | 2943 | 4969 | 4828 |
| Number of cases | 0 | 1 | 3 | 5 | 4 | 6 | 2 | 5 | 4 |

(a) Fit the Poisson model to these data both to the full data set and to an "incomplete" data set where we suppose that the first population count ($x_1 = 3540$) is missing.

(b) Suppose that instead of having an $x$ value missing, we actually have lost a leukemia count (assume that $y_1 = 3$ is missing). Use the EM algorithm to find the MLEs in this case, and compare your answers to those of part (a).

**7.29** An alternative to the model of Example 7.2.17 is the following, where we observe $(Y_i, X_i)$, $i = 1, 2, \ldots, n$, where $Y_i \sim$ Poisson$(m\beta\tau_i)$ and $(X_1, \ldots, X_n) \sim$ multinomial$(m; \boldsymbol{\tau})$, where $\boldsymbol{\tau} = (\tau_1, \tau_2, \ldots, \tau_n)$ with $\sum_{i=1}^n \tau_i = 1$. So here, for example, we assume that the population counts are multinomial allocations rather than Poisson counts. (Treat $m = \sum x_i$ as known.)

(a) Show that the joint density of $\mathbf{Y} = (Y_1, \ldots, Y_n)$ and $\mathbf{X} = (X_1, \ldots, X_n)$ is

$$f(\mathbf{y}, \mathbf{x}|\beta, \boldsymbol{\tau}) = \prod_{i=1}^n \frac{e^{-m\beta\tau_i}(m\beta\tau_i)^{y_i}}{y_i!} m! \frac{\tau_i^{x_i}}{x_i!}.$$

(b) If the complete data are observed, show that the MLEs are given by

$$\hat{\beta} = \frac{\sum_{i=1}^n y_i}{\sum_{i=1}^n x_i} \quad \text{and} \quad \hat{\tau}_j = \frac{x_j + y_j}{\sum_{i=1}^n x_i + y_i}, \quad j = 1, 2, \ldots, n.$$

(c) Suppose that $x_1$ is missing. Use the fact that $X_1 \sim$ binomial$(m, t_1)$ to calculate the expected complete-data log likelihood. Show that the EM sequence is given by

$$\hat{\beta}^{(r+1)} = \frac{\sum_{i=1}^n y_i}{m\hat{\tau}_1^{(r)} + \sum_{i=2}^n x_i} \quad \text{and} \quad \hat{\tau}_j^{(r+1)} = \frac{x_j + y_j}{m\hat{\tau}_1^{(r)} + \sum_{i=2}^n x_i + \sum_{i=1}^n y_i},$$

$$j = 1, 2, \ldots, n.$$

(d) Use this model to find the MLEs for the data in Exercise 7.28, first assuming that you have all the data and then assuming that $x_1 = 3540$ is missing.

**7.30** The EM algorithm is useful in a variety of situation, and the definition of "missing data" can be stretched to accommodate many different models. Suppose that we have a mixture density $pf(x) + (1-p)g(x)$, where $p$ is unknown. If we observe $\mathbf{X} = (X_1, \ldots, X_n)$, the sample density is

$$\prod_{i=1}^n [pf(x_i) + (1-p)g(x_i)],$$

which could be difficult to deal with. (Actually, a mixture of two is not terrible, but consider what the likelihood would look like with a mixture $\sum_{i=1}^k p_i f_i(x)$ for large $k$.) The EM solution is to augment the observed (or incomplete) data with $\mathbf{Z} = (Z_1, \ldots, Z_n)$, where $Z_i$ tells which component of the mixture $X_i$ came from; that is,

$$X_i|z_i = 1 \sim f(x_i) \quad \text{and} \quad X_i|z_i = 0 \sim g(x_i),$$

and $P(Z_i = 1) = p$.

(a) Show that the joint density of $(\mathbf{X}, \mathbf{Z})$ is given by $\prod_{i=1}^{n}[pf(x_i)^{z_i}][(1-p)g(x_i)^{1-z_i}]$.

(b) Show that the missing data distribution, the distribution of $Z_i|x_i, p$ is Bernoulli with success probability $pf(x_i)/(pf(x_i) + (1-p)g(x_i))$.

(c) Calculate the expected complete-data log likelihood, and show that the EM sequence is given by

$$\hat{p}^{(r+1)} = \frac{1}{n} \sum_{i=1}^{n} \frac{\hat{p}^{(r)} f(x_i)}{\hat{p}^{(r)} f(x_i) + (1 - \hat{p}^{(r)})g(x_i)}.$$

**7.31** Prove Theorem 7.2.20.

(a) Show that, using (7.2.19), we can write

$$\log L(\hat{\theta}^{(r)}|\mathbf{y}) = \mathrm{E}\left[\log L(\hat{\theta}^{(r)}|\mathbf{y}, \mathbf{X})|\hat{\theta}^{(r)}, \mathbf{y}\right] - \mathrm{E}\left[\log k(\mathbf{X}|\hat{\theta}^{(r)}, \mathbf{y})|\hat{\theta}^{(r)}, \mathbf{y}\right],$$

and, since $\hat{\theta}^{(r+1)}$ is a maximum, $\log L(\hat{\theta}^{(r+1)}|\mathbf{y}, \mathbf{X}) \geq \mathrm{E}\left[\log L(\hat{\theta}^{(r)}|\mathbf{y}, \mathbf{X})|\hat{\theta}^{(r)}, \mathbf{y}\right]$. When is the inequality an equality?

(b) Now use Jensen's inequality to show that

$$\mathrm{E}\left[\log k(\mathbf{X}|\hat{\theta}^{(r+1)}, \mathbf{y})|\hat{\theta}^{(r)}, \mathbf{y}\right] \leq \mathrm{E}\left[\log k(\mathbf{X}|\hat{\theta}^{(r)}, \mathbf{y})|\hat{\theta}^{(r)}, \mathbf{y}\right],$$

which together with part (a) proves the theorem.

(*Hint:* If $f$ and $g$ are densities, since log is a concave function, Jensen's inequality (4.7.7) implies

$$\int \log\left(\frac{f(x)}{g(x)}\right) g(x)\, dx \leq \log\left(\int \frac{f(x)}{g(x)} g(x)\, dx\right) = \log\left(\int f(x)\, dx\right) = 0.$$

By the property of logs, this in turn implies that

$$\int \log[f(x)]g(x)\, dx \leq \int \log[g(x)]g(x)\, dx.)$$

**7.32** The algorithm of Exercise 5.65 can be adapted to simulate (approximately) a sample from the posterior distribution using only a sample from the prior distribution. Let $X_1, \ldots, X_n \sim f(x|\theta)$, where $\theta$ has prior distribution $\pi$. Generate $\theta_1, \ldots, \theta_m$ from $\pi$, and calculate $q_i = L(\theta_i|\mathbf{x})/\sum_j L(\theta_j/|\mathbf{x})$, where $L(\theta|\mathbf{x}) = \prod_i f(x_i|\theta)$ is the likelihood function.

(a) Generate $\theta_1^*, \ldots, \theta_r^*$, where $P(\theta^* = \theta_i) = q_i$. Show that this is a (approximate) sample from the posterior in the sense that $P(\theta^* \leq t)$ converges to $\int_{-\infty}^{t} \pi(\theta|\mathbf{x})\, d\theta$.

(b) Show that the estimator $\sum_{j=1}^{r} h(\theta_j^*)/r$ converges to $\mathrm{E}[h(\theta)|\mathbf{x}]$, where the expectation is with respect to the posterior distribution.

(c) Ross (1996) suggests that Rao-Blackwellization can improve the estimate in part (b). Show that for any $j$,

$$\mathrm{E}[h(\theta_j^*)|\theta_1, \ldots, \theta_m] = \frac{1}{\sum_{i=1}^{m} L(\theta_i|\mathbf{x})} \sum_{i=1}^{m} h(\theta_i)L(\theta_i|\mathbf{x})$$

has the same mean and smaller variance than the estimator in part (b).

**7.33** In Example 7.3.5 the MSE of the Bayes estimator, $\hat{p}_B$, of a success probability was calculated (the estimator was derived in Example 7.2.14). Show that the choice $\alpha = \beta = \sqrt{n}/4$ yields a constant MSE for $\hat{p}_B$.

**7.34** Let $X_1, \ldots, X_n$ be a random sample from a binomial$(n, p)$. We want to find equivariant point estimators of $p$ using the group described in Example 6.4.1.

   (a) Find the class of estimators that are equivariant with respect to this group.
   (b) Within the class of Bayes estimators of Example 7.2.14, find the estimators that are equivariant with respect to this group.
   (c) From the equivariant Bayes estimators of part (b), find the one with the smallest MSE.

**7.35** The *Pitman Estimator of Location* (see Lehmann and Casella 1998 Section 3.1, or the original paper by Pitman 1939) is given by

$$d_{\mathrm{P}}(\mathbf{X}) = \frac{\int_{-\infty}^{\infty} t \prod_{i=1}^{n} f(x_i - t)\, dt}{\int_{-\infty}^{\infty} \prod_{i=1}^{n} f(x_i - t)\, dt},$$

where we observe a random sample $X_1, \ldots, X_n$ from $f(x - \theta)$. Pitman showed that this estimator is the location-equivariant estimator with smallest mean squared error (that is, it minimizes (7.3.3)). The goals of this exercise are more modest.

   (a) Show that $d_{\mathrm{P}}(\mathbf{X})$ is invariant with respect to the location group of Example 7.3.6.
   (b) Show that if $f(x - \theta)$ is n$(\theta, 1)$, then $d_{\mathrm{P}}(\mathbf{X}) = \bar{X}$.
   (c) Show that if $f(x - \theta)$ is uniform$(\theta - \frac{1}{2}, \theta + \frac{1}{2})$, then $d_{\mathrm{P}}(\mathbf{X}) = \frac{1}{2}(X_{(1)} + X_{(n)})$.

**7.36** The *Pitman Estimator of Scale* is given by

$$d_{\mathrm{P}}^{r}(\mathbf{X}) = \frac{\int_{0}^{\infty} t^{n+r-1} \prod_{i=1}^{n} f(tx_i)\, dt}{\int_{0}^{\infty} t^{n+2r-1} \prod_{i=1}^{n} f(tx_i)\, dt},$$

where we observe a random sample $X_1, \ldots, X_n$ from $\frac{1}{\sigma} f(x/\sigma)$. Pitman showed that this estimator is the scale-equivariant estimator of $\sigma^r$ with smallest scaled mean squared error (that is, it minimizes $\mathrm{E}(d - \sigma^r)^2 / \sigma^{2r}$).

   (a) Show that $d_{\mathrm{P}}^{r}(\mathbf{X})$ is equivariant with respect to the scale group, that is, it satisfies

$$d_{\mathrm{P}}^{r}(cx_1, \ldots, cx_n) = c^r d_{\mathrm{P}}^{r}(x_1, \ldots, x_n),$$

   for any constant $c > 0$.
   (b) Find the Pitman scale-equivariant estimator for $\sigma^2$ if $X_1, \ldots, X_n$ are iid n$(0, \sigma^2)$.
   (c) Find the Pitman scale-equivariant estimator for $\beta$ if $X_1, \ldots, X_n$ are iid exponential$(\beta)$.
   (d) Find the Pitman scale-equivariant estimator for $\theta$ if $X_1, \ldots, X_n$ are iid uniform$(0, \theta)$.

**7.37** Let $X_1, \ldots, X_n$ be a random sample from a population with pdf

$$f(x|\theta) = \frac{1}{2\theta}, \quad -\theta < x < \theta, \quad \theta > 0.$$

Find, if one exists, a best unbiased estimator of $\theta$.

**7.38** For each of the following distributions, let $X_1, \ldots, X_n$ be a random sample. Is there a function of $\theta$, say $g(\theta)$, for which there exists an unbiased estimator whose variance attains the Cramér–Rao Lower Bound? If so, find it. If not, show why not.

   (a) $f(x|\theta) = \theta x^{\theta-1}, \quad 0 < x < 1, \quad \theta > 0$
   (b) $f(x|\theta) = \frac{\log(\theta)}{\theta-1} \theta^x, \quad 0 < x < 1, \quad \theta > 1$

**7.39** Prove Lemma 7.3.11.

**7.40** Let $X_1, \ldots, X_n$ be iid Bernoulli$(p)$. Show that the variance of $\bar{X}$ attains the Cramér–Rao Lower Bound, and hence $\bar{X}$ is the best unbiased estimator of $p$.

**7.41** Let $X_1, \ldots, X_n$ be a random sample from a population with mean $\mu$ and variance $\sigma^2$.

(a) Show that the estimator $\sum_{i=1}^{n} a_i X_i$ is an unbiased estimator of $\mu$ if $\sum_{i=1}^{n} a_i = 1$.

(b) Among all unbiased estimators of this form (called *linear unbiased estimators*) find the one with minimum variance, and calculate the variance.

**7.42** Let $W_1, \ldots, W_k$ be unbiased estimators of a parameter $\theta$ with $\operatorname{Var} W_i = \sigma_i^2$ and $\operatorname{Cov}(W_i, W_j) = 0$ if $i \neq j$.

(a) Show that, of all estimators of the form $\sum a_i W_i$, where the $a_i$s are constant and $E_\theta(\sum a_i W_i) = \theta$, the estimator $W^* = \dfrac{\sum W_i/\sigma_i^2}{\sum (1/\sigma_i^2)}$ has minimum variance.

(b) Show that $\operatorname{Var} W^* = \dfrac{1}{\sum (1/\sigma_i^2)}$.

**7.43** Exercise 7.42 established that the optimal weights are $q_i^* = (1/\sigma_i^2)/(\sum_j 1/\sigma_j^2)$. A result due to Tukey (see Bloch and Moses 1988) states that if $W = \sum_i q_i W_i$ is an estimator based on another sets of weights $q_i \geq 0$, $\sum_i q_i = 1$, then

$$\frac{\operatorname{Var} W}{\operatorname{Var} W^*} \leq \frac{1}{1 - \lambda^2},$$

where $\lambda$ satisfies $(1 + \lambda)/(1 - \lambda) = b_{\max}/b_{\min}$, and $b_{\max}$ and $b_{\min}$ are the largest and smallest of $b_i = q_i/q_i^*$.

(a) Prove Tukey's inequality.

(b) Use the inequality to assess the performance of the usual mean $\sum_i W_i/k$ as a function of $\sigma_{\max}^2/\sigma_{\min}^2$

**7.44** Let $X_1, \ldots, X_n$ be iid $n(\theta, 1)$. Show that the best unbiased estimator of $\theta^2$ is $\bar{X}^2 - (1/n)$. Calculate its variance (use Stein's Identity from Section 3.6), and show that it is greater than the Cramér–Rao Lower Bound.

**7.45** Let $X_1, X_2, \ldots, X_n$ be iid from a distribution with mean $\mu$ and variance $\sigma^2$, and let $S^2$ be the usual unbiased estimator of $\sigma^2$. In Example 7.3.4 we saw that, under normality, the MLE has smaller MSE than $S^2$. In this exercise will explore variance estimates some more.

(a) Show that, for any estimator of the form $aS^2$, where $a$ is a constant,

$$\operatorname{MSE}(aS^2) = E[aS^2 - \sigma^2]^2 = a^2 \operatorname{Var}(S^2) + (a - 1)^2 \sigma^4.$$

(b) Show that

$$\operatorname{Var}(S^2) = \frac{1}{n}\left(\kappa - \frac{n-3}{n-1}\right)\sigma^4,$$

where $\kappa = E[X - \mu]^4/\sigma^4$ is the *kurtosis*. (You may have already done this in Exercise 5.8(b).)

(c) Show that, under normality, the kurtosis is 3 and establish that, in this case, the estimator of the form $aS^2$ with the minimum MSE is $\frac{n-1}{n+1}S^2$. (Lemma 3.6.5 may be helpful.)

(d) If normality is not assumed, show that $\operatorname{MSE}(aS^2)$ is minimized at

$$a = \frac{n-1}{(n+1) + \frac{(\kappa-3)(n-1)}{n}},$$

which is useless as it depends on a parameter.

(e) Show that

  (i) for distributions with $\kappa > 3$, the optimal $a$ will satisfy $a < \frac{n-1}{n+1}$;

  (ii) for distributions with $\kappa < 3$, the optimal $a$ will satisfy $\frac{n-1}{n+1} < a < 1$.

See Searls and Intarapanich (1990) for more details.

**7.46** Let $X_1, X_2$, and $X_3$ be a random sample of size three from a uniform$(\theta, 2\theta)$ distribution, where $\theta > 0$.

(a) Find the method of moments estimator of $\theta$.

(b) Find the MLE, $\hat{\theta}$, and find a constant $k$ such that $E_\theta(k\hat{\theta}) = \theta$.

(c) Which of the two estimators can be improved by using sufficiency? How?

(d) Find the method of moments estimate and the MLE of $\theta$ based on the data

$$1.29, \ .86, \ 1.33,$$

three observations of average berry sizes (in centimeters) of wine grapes.

**7.47** Suppose that when the radius of a circle is measured, an error is made that has a n$(0, \sigma^2)$ distribution. If $n$ independent measurements are made, find an unbiased estimator of the area of the circle. Is it best unbiased?

**7.48** Suppose that $X_i, i = 1, \ldots, n$, are iid Bernoulli$(p)$.

(a) Show that the variance of the MLE of $p$ attains the Cramér–Rao Lower Bound.

(b) For $n \geq 4$, show that the product $X_1 X_2 X_3 X_4$ is an unbiased estimator of $p^4$, and use this fact to find the best unbiased estimator of $p^4$.

**7.49** Let $X_1, \ldots, X_n$ be iid exponential$(\lambda)$.

(a) Find an unbiased estimator of $\lambda$ based only on $Y = \min\{X_1, \ldots, X_n\}$.

(b) Find a better estimator than the one in part (a). Prove that it is better.

(c) The following data are high-stress failure times (in hours) of Kevlar/epoxy spherical vessels used in a sustained pressure environment on the space shuttle:

$$50.1, \ 70.1, \ 137.0, \ 166.9, \ 170.5, \ 152.8, \ 80.5, \ 123.5, \ 112.6, \ 148.5, \ 160.0, \ 125.4.$$

Failure times are often modeled with the exponential distribution. Estimate the mean failure time using the estimators from parts (a) and (b).

**7.50** Let $X_1, \ldots, X_n$ be iid n$(\theta, \theta^2), \theta > 0$. For this model both $\bar{X}$ and $cS$ are unbiased estimators of $\theta$, where $c = \frac{\sqrt{n-1}\Gamma((n-1)/2)}{\sqrt{2}\Gamma(n/2)}$.

(a) Prove that for any number $a$ the estimator $a\bar{X} + (1-a)(cS)$ is an unbiased estimator of $\theta$.

(b) Find the value of $a$ that produces the estimator with minimum variance.

(c) Show that $(\bar{X}, S^2)$ is a sufficient statistic for $\theta$ but it is not a complete sufficient statistic.

**7.51** Gleser and Healy (1976) give a detailed treatment of the estimation problem in the n$(\theta, a\theta^2)$ family, where $a$ is a known constant (of which Exercise 7.50 is a special case). We explore a small part of their results here. Again let $X_1, \ldots, X_n$ be iid n$(\theta, \theta^2)$, $\theta > 0$, and let $\bar{X}$ and $cS$ be as in Exercise 7.50. Define the class of estimators

$$\mathcal{T} = \left\{ T: T = a_1\bar{X} + a_2(cS) \right\},$$

where we do not assume that $a_1 + a_2 = 1$.

(a) Find the estimator $T \in \mathcal{T}$ that minimizes $E_\theta(\theta - T)^2$; call it $T^*$.

(b) Show that the MSE of $T^*$ is smaller than the MSE of the estimator derived in Exercise 7.50(b).

(c) Show that the MSE of $T^{*^+} = \max\{0, T^*\}$ is smaller than the MSE of $T^*$.

(d) Would $\theta$ be classified as a location parameter or a scale parameter? Explain.

**7.52** Let $X_1, \ldots, X_n$ be iid Poisson($\lambda$), and let $\bar{X}$ and $S^2$ denote the sample mean and variance, respectively. We now complete Example 7.3.8 in a different way. There we used the Cramér–Rao Bound; now we use completeness.

(a) Prove that $\bar{X}$ is the best unbiased estimator of $\lambda$ without using the Cramér–Rao Theorem.

(b) Prove the rather remarkable identity $E(S^2|\bar{X}) = \bar{X}$, and use it to explicitly demonstrate that $\mathrm{Var}\, S^2 > \mathrm{Var}\, \bar{X}$.

(c) Using completeness, can a general theorem be formulated for which the identity in part (b) is a special case?

**7.53** Finish some of the details left out of the proof of Theorem 7.3.20. Suppose $W$ is an unbiased estimator of $\tau(\theta)$, and $U$ is an unbiased estimator of 0. Show that if, for some $\theta = \theta_0$, $\mathrm{Cov}_{\theta_0}(W, U) \neq 0$, then $W$ cannot be the best unbiased estimator of $\tau(\theta)$.

**7.54** Consider the "Problem of the Nile" (see Exercise 6.37).

(a) Show that $T$ is the MLE of $\theta$ and $U$ is ancillary, and

$$ET = \frac{\Gamma(n + 1/2)\Gamma(n - 1/2)}{[\Gamma(n)]^2}\theta \quad \text{and} \quad ET^2 = \frac{\Gamma(n + 1)\Gamma(n - 1)}{[\Gamma(n)]^2}\theta^2.$$

(b) Let $Z_1 = (n-1)/\sum X_i$ and $Z_2 = \sum Y_i/n$. Show that both are unbiased with variances $\theta^2/(n-2)$ and $\theta^2/n$, respectively.

(c) Find the best unbiased estimator of the form $aZ_1 + (1-a)Z_2$, calculate its variance, and compare it to the bias-corrected MLE.

**7.55** For each of the following pdfs, let $X_1, \ldots, X_n$ be a sample from that distribution. In each case, find the best unbiased estimator of $\theta^r$. (See Guenther 1978 for a complete discussion of this problem.)

(a) $f(x|\theta) = \frac{1}{\theta}, \quad 0 < x < \theta,\ r < n$

(b) $f(x|\theta) = e^{-(x-\theta)}, \quad x > \theta$

(c) $f(x|\theta) = \frac{e^{-x}}{e^{-\theta} - e^{-b}}, \quad \theta < x < b, \quad b$ known

**7.56** Prove the assertion made in the text preceding Example 7.3.24: If $T$ is a complete sufficient statistic for a parameter $\theta$, and $h(X_1, \ldots, X_n)$ is *any* unbiased estimator of $\tau(\theta)$, then $\phi(T) = E(h(X_1, \ldots, X_n)|T)$ is *the* best unbiased estimator of $\tau(\theta)$.

**7.57** Let $X_1, \ldots, X_{n+1}$ be iid Bernoulli($p$), and define the function $h(p)$ by

$$h(p) = P\left(\sum_{i=1}^{n} X_i > X_{n+1} \middle| p\right),$$

the probability that the first $n$ observations exceed the $(n+1)$st.

(a) Show that

$$T(X_1, \ldots, X_{n+1}) = \begin{cases} 1 & \text{if } \sum_{i=1}^{n} X_i > X_{n+1} \\ 0 & \text{otherwise} \end{cases}$$

is an unbiased estimator of $h(p)$.

(b) Find the best unbiased estimator of $h(p)$.

**7.58** Let $X$ be an observation from the pdf

$$f(x|\theta) = \left(\frac{\theta}{2}\right)^{|x|} (1-\theta)^{1-|x|}, \quad x = -1, 0, 1; \quad 0 \leq \theta \leq 1.$$

(a) Find the MLE of $\theta$.

(b) Define the estimator $T(X)$ by

$$T(X) = \begin{cases} 2 & \text{if } x = 1 \\ 0 & \text{otherwise.} \end{cases}$$

Show that $T(X)$ is an unbiased estimator of $\theta$.

(c) Find a better estimator than $T(X)$ and prove that it is better.

**7.59** Let $X_1, \ldots, X_n$ be iid $n(\mu, \sigma^2)$. Find the best unbiased estimator of $\sigma^p$, where $p$ is a known positive constant, not necessarily an integer.

**7.60** Let $X_1, \ldots, X_n$ be iid gamma$(\alpha, \beta)$ with $\alpha$ known. Find the best unbiased estimator of $1/\beta$.

**7.61** Show that the log of the likelihood function for estimating $\sigma^2$, based on observing $S^2 \sim \sigma^2 \chi_\nu^2 / \nu$, can be written in the form

$$\log L(\sigma^2 | s^2) = K_1 \frac{s^2}{\sigma^2} - K_2 \log \frac{s^2}{\sigma^2} + K_3,$$

where $K_1, K_2$, and $K_3$ are constants, not dependent on $\sigma^2$. Relate the above log likelihood to the loss function discussed in Example 7.3.27. See Anderson (1984a) for a discussion of this relationship.

**7.62** Let $X_1, \ldots, X_n$ be a random sample from a $n(\theta, \sigma^2)$ population, $\sigma^2$ known. Consider estimating $\theta$ using squared error loss. Let $\pi(\theta)$ be a $n(\mu, \tau^2)$ prior distribution on $\theta$ and let $\delta^\pi$ be the Bayes estimator of $\theta$. Verify the following formulas for the risk function and Bayes risk.

(a) For any constants $a$ and $b$, the estimator $\delta(\mathbf{x}) = a\bar{X} + b$ has risk function

$$R(\theta, \delta) = a^2 \frac{\sigma^2}{n} + (b - (1-a)\theta)^2.$$

(b) Let $\eta = \sigma^2 / (n\tau^2 + \sigma^2)$. The risk function for the Bayes estimator is

$$R(\theta, \delta^\pi) = (1-\eta)^2 \frac{\sigma^2}{n} + \eta^2 (\theta - \mu)^2.$$

(c) The Bayes risk for the Bayes estimator is

$$B(\pi, \delta^\pi) = \tau^2 \eta.$$

**7.63** Let $X \sim n(\mu, 1)$. Let $\delta^\pi$ be the Bayes estimator of $\mu$ for squared error loss. Compute and graph the risk functions, $R(\mu, \delta^\pi)$, for $\pi(\mu) \sim n(0, 1)$ and $\pi(\mu) \sim n(0, 10)$. Comment on how the prior affects the risk function of the Bayes estimator.

**7.64** Let $X_1, \ldots, X_n$ be independent random variables, where $X_i$ has cdf $F(x|\theta_i)$. Show that, for $i = 1, \ldots, n$, if $\delta_i^{\pi_i}(X_i)$ is a Bayes rule for estimating $\theta_i$ using loss $L(\theta_i, a_i)$ and prior $\pi_i(\theta_i)$, then $\delta^\pi(\mathbf{X}) = (\delta^{\pi_1}(X_1), \ldots, \delta^{\pi_n}(X_n))$ is a Bayes rule for estimating $\theta = (\theta_1, \ldots, \theta_n)$ using the loss $\sum_{i=1}^{n} L(\theta_i, a_i)$ and prior $\pi(\theta) = \prod_{i=1}^{n} \pi_i(\theta_i)$.

**7.65** A loss function investigated by Zellner (1986) is the LINEX (LINear–EXponential) loss, a loss function that can handle asymmetries in a smooth way. The LINEX loss is given by

$$L(\theta, a) = e^{c(a-\theta)} - c(a - \theta) - 1,$$

where $c$ is a positive constant. As the constant $c$ varies, the loss function varies from very asymmetric to almost symmetric.

(a) For $c = .2, .5, 1$, plot $L(\theta, a)$ as a function of $a - \theta$.

(b) If $X \sim F(x|\theta)$, show that the Bayes estimator of $\theta$, using a prior $\pi$, is given by $\delta^\pi(X) = \frac{-1}{c} \log \mathrm{E}(e^{-c\theta}|X)$.

(c) Let $X_1, \ldots, X_n$ be iid $\mathrm{n}(\theta, \sigma^2)$, where $\sigma^2$ is known, and suppose that $\theta$ has the noninformative prior $\pi(\theta) = 1$. Show that the Bayes estimator versus LINEX loss is given by $\delta^\mathrm{B}(\bar{X}) = \bar{X} - (c\sigma^2/(2n))$.

(d) Calculate the posterior expected loss for $\delta^\mathrm{B}(\bar{X})$ and $\bar{X}$ using LINEX loss.

(e) Calculate the posterior expected loss for $\delta^\mathrm{B}(\bar{X})$ and $\bar{X}$ using squared error loss.

**7.66** The *jackknife* is a general technique for reducing bias in an estimator (Quenouille, 1956). A one-step jackknife estimator is defined as follows. Let $X_1, \ldots, X_n$ be a random sample, and let $T_n = T_n(X_1, \ldots, X_n)$ be some estimator of a parameter $\theta$. In order to "jackknife" $T_n$ we calculate the $n$ statistics $T_n{}^{(i)}$, $i = 1, \ldots, n$, where $T_n{}^{(i)}$ is calculated just as $T_n$ but using the $n - 1$ observations with $X_i$ removed from the sample. The jackknife estimator of $\theta$, denoted by $\mathrm{JK}(T_n)$, is given by

$$\mathrm{JK}(T_n) = nT_n - \frac{n-1}{n} \sum_{i=1}^{n} T_n{}^{(i)}.$$

(In general, $\mathrm{JK}(T_n)$ will have a smaller bias than $T_n$. See Miller 1974 for a good review of the properties of the jackknife.)

Now, to be specific, let $X_1, \ldots, X_n$ be iid Bernoulli($\theta$). The object is to estimate $\theta^2$.

(a) Show that the MLE of $\theta^2$, $(\sum_{i=1}^{n} X_i/n)^2$, is a biased estimator of $\theta^2$.

(b) Derive the one-step jackknife estimator based on the MLE.

(c) Show that the one-step jackknife estimator is an unbiased estimator of $\theta^2$. (In general, jackknifing only reduces bias. In this special case, however, it removes it entirely.)

(d) Is this jackknife estimator the best unbiased estimator of $\theta^2$? If so, prove it. If not, find the best unbiased estimator.

# 7.5 Miscellanea

## 7.5.1 Moment Estimators and MLEs

In general, method of moments estimators are not functions of sufficient statistics; hence, they can always be improved upon by conditioning on a sufficient statistic. In the case of exponential families, however, there can be a correspondence between a modified method of moments strategy and maximum likelihood estimation. This correspondence is discussed in detail by Davidson and Solomon (1974), who also relate some interesting history.

Suppose that we have a random sample $\mathbf{X} = (X_1, \ldots, X_n)$ from a pdf in the exponential family (see Theorem 5.2.11)

$$f(x|\theta) = h(x)c(\theta) \exp\left(\sum_{i=1}^{k} w_i(\theta)t_i(x)\right),$$

where the range of $f(x|\theta)$ is independent of $\theta$. (Note that $\theta$ may be a vector.) The likelihood function is of the form

$$L(\theta|\mathbf{x}) = H(\mathbf{x})[c(\theta)]^n \exp\left(\sum_{i=1}^{k} w_i(\theta) \sum_{j=1}^{n} t_i(x_j)\right),$$

and a modified method of moments would estimate $w_i(\theta), i = 1, \ldots, k$, by $\hat{w}_i(\theta)$, the solutions to the $k$ equations

$$\sum_{j=1}^{n} t_i(x_j) = \mathrm{E}_\theta\left(\sum_{j=1}^{n} t_i(X_j)\right), \quad i = 1, \ldots, k.$$

Davidson and Solomon, extending work of Huzurbazar (1949), show that the estimators $\hat{w}_i(\theta)$ are, in fact, the MLEs of $w_i(\theta)$. If we define $\eta_i = w_i(\theta), i = 1, \ldots, k$, then the MLE of $g(\eta_i)$ is equal to $g(\hat{\eta}_i) = g(\hat{w}_i(\theta))$ for any one-to-one function $g$. Calculation of the above expectations may be simplified by using the facts (Lehmann 1986, Section 2.7) that

$$\mathrm{E}_\theta(t_i(X_j)) = \frac{\partial}{\partial w_i(\theta)} \log(c(\theta)), \quad i = 1, \ldots, k, \quad j = 1, \ldots, n;$$

$$\mathrm{Cov}_\theta(t_i(X_j), t_{i'}(X_j)) = \frac{\partial^2}{\partial w_i(\theta)\partial w_{i'}(\theta)} \log(c(\theta)), \quad i, i' = 1, \ldots, k, \quad j = 1, \ldots, n.$$

### 7.5.2 Unbiased Bayes Estimates

As was seen in Section 7.2.3, if a Bayesian calculation is done, the mean of the posterior distribution usually is taken as a point estimator. To be specific, if $X$ has pdf $f(x|\theta)$ with $\mathrm{E}_\theta(X) = \theta$ and there is a prior distribution $\pi(\theta)$, then the posterior mean, a Bayesian point estimator of $\theta$, is given by

$$\mathrm{E}(\theta|x) = \int \theta\pi(\theta|x)d\theta.$$

A question that could be asked is whether $\mathrm{E}(\theta|X)$ can be an unbiased estimator of $\theta$ and thus satisfy the equation

$$\mathrm{E}_\theta[\mathrm{E}(\theta|X)] = \int\left[\int \theta\pi(\theta|x)\,d\theta\right]f(x|\theta)\,dx = \theta.$$

The answer is no. That is, posterior means are *never* unbiased estimators. If they were, then taking the expectation over the joint distribution of $X$ and $\theta$, we could write

$$E[(X - \theta)^2] = E[X^2 - 2X\theta + \theta^2] \qquad \text{(expand the square)}$$
$$= E\left(E(X^2 - 2X\theta + \theta^2|\theta)\right) \qquad \text{(iterate the expectation)}$$
$$= E\left(E(X^2|\theta) - 2\theta^2 + \theta^2\right) \qquad (E(X|\theta) = E_\theta X = \theta)$$
$$= E\left(E(X^2|\theta) - \theta^2\right)$$
$$= E(X^2) - E(\theta^2) \qquad \text{(properties of expectations)}$$

doing the conditioning one way, and conditioning on $X$, we could similarly calculate

$$E[(X - \theta)^2] = E\left(E[X^2 - 2X\theta + \theta^2|X]\right)$$
$$= E\left(X^2 - 2X^2 + E(\theta^2|X)\right) \qquad \begin{pmatrix} E(\theta|X) = X \\ \text{by assumption} \end{pmatrix}$$
$$= E(\theta^2) - E(X^2).$$

Comparing the two calculations, we see that the only way that there is no contradiction is if $E(X^2) = E(\theta^2)$, which then implies that $E(X - \theta)^2 = 0$, so $X = \theta$. This occurs only if $P(X = \theta) = 1$, an uninteresting situation, so we have argued to a contradiction. Thus, either $E(X|\theta) \neq \theta$ or $E(\theta|X) \neq X$, showing that posterior means cannot be unbiased estimators. Notice that we have implicitly made the assumption that $E(X^2) < \infty$, but, in fact, this result holds under more general conditions. Bickel and Mallows (1988) have a more thorough development of this topic. At a more advanced level, this connection is characterized by Noorbaloochi and Meeden (1983).

### 7.5.3 The Lehmann–Scheffé Theorem

The Lehmann–Scheffé Theorem represents a major achievement in mathematical statistics, tying together sufficiency, completeness, and uniqueness. The development in the text is somewhat complementary to the Lehmann–Scheffé Theorem, and thus we never stated it in its classical form (which is similar to Theorem 7.3.23). In fact, the Lehmann–Scheffé Theorem is contained in Theorems 7.3.19 and 7.3.23.

**Theorem 7.5.1 (Lehmann–Scheffé)**  *Unbiased estimators based on complete sufficient statistics are unique.*

**Proof:** Suppose $T$ is a complete sufficient statistic, and $\phi(T)$ is an estimator with $E_\theta \phi(T) = \tau(\theta)$. From Theorem 7.3.23 we know that $\phi(T)$ is the best unbiased estimator of $\tau(\theta)$, and from Theorem 7.3.19, best unbiased estimators are unique. $\qquad \square$

This theorem can also be proved without Theorem 7.3.19, using just the consequences of completeness, and provides a slightly different route to Theorem 7.3.23.

### 7.5.4 More on the EM Algorithm

The EM algorithm has its roots in work done in the 1950s (Hartley 1958) but really came into statistical prominence after the seminal work of Dempster, Laird, and Rubin (1977), which detailed the underlying structure of the algorithm and illustrated its use in a wide variety of applications.

One of the strengths of the EM algorithm is that conditions for convergence to the incomplete-data MLEs are known, although this topic has obtained an additional bit of folklore. Dempster, Laird, and Rubin's (1977) original proof of convergence had a flaw, but valid convergence proofs were later given by Boyles (1983) and Wu (1983); see also Finch, Mendell, and Thode (1989).

In our development we stopped with Theorem 7.2.20, which guarantees that the likelihood will increase at each iteration. However, this may not be enough to conclude that the sequence $\{\hat{\theta}^{(r)}\}$ converges to a maximum likelihood estimator. Such a guarantee requires further conditions. The following theorem, due to Wu (1983), guarantees convergence to a *stationary point*, which may be a local maximum or saddlepoint.

**Theorem 7.5.2** *If the expected complete-data log likelihood* $\mathrm{E}\left[\log L(\theta|\mathbf{y}, \mathbf{x})|\theta', \mathbf{y}\right]$ *is continuous in both* $\theta$ *and* $\theta'$, *then all limit points of an EM sequence* $\{\hat{\theta}^{(r)}\}$ *are stationary points of* $L(\theta|\mathbf{y})$, *and* $L(\hat{\theta}^{(r)}|\mathbf{y})$ *converges monotonically to* $L(\hat{\theta}|\mathbf{y})$ *for some stationary point* $\hat{\theta}$.

In an exponential family computations become simplified because the log likelihood will be linear in the missing data. We can write

$$\mathrm{E}\left[\log L(\theta|\mathbf{y}, \mathbf{x})|\theta', \mathbf{y}\right] = \mathrm{E}_{\theta'}\left[\log\left(h(\mathbf{y}, \mathbf{X}) \; e^{\sum \eta_i(\theta)T_i - B(\theta)}\right)|\mathbf{y}\right]$$

$$= \mathrm{E}_{\theta'}\left[\log h(\mathbf{y}, \mathbf{X})\right] + \sum \eta_i(\theta)\mathrm{E}_{\theta'}\left[T_i|\mathbf{y}\right] - B(\theta).$$

Thus, calculating the complete-data MLE involves only the simpler expectation $\mathrm{E}_{\theta'}\left[T_i|\mathbf{y}\right]$.

Good overviews of the EM algorithm are provided by Little and Rubin (1987), Tanner (1996), and Shafer (1997); see also Lehmann and Casella (1998, Section 6.4). McLachlan and Krishnan (1997) provide a book-length treatment of EM.

### 7.5.5 Other Likelihoods

In this chapter we have used the method of maximum likelihood and seen that it not only provides us with a method for finding estimators, but also brings along a large-sample theory that is quite useful for inference.

Likelihood has many modifications. Some are used to deal with nuisance parameters (such as *profile* likelihood); others are used when a more robust specification is desired (such as *quasi* likelihood); and others are useful when the data are censored (such as *partial* likelihood).

There are many other variations, and they all can provide some improvement over the plain likelihood that we have described here. Entries to this wealth of likelihoods can be found in the review article of Hinkley (1980) or the volume of review articles edited by Hinkley, Reid, and Snell (1991).

### 7.5.6 Other Bayes Analyses

1. *Robust Bayes Analysis*    The fact that Bayes rules may be quite sensitive to the (subjective) choice of a prior distribution is a cause of concern for many Bayesian statisticians. The paper of Berger (1984) introduced the idea of a *robust Bayes analysis*. This is a Bayes analysis in which estimators are sought that have good properties for a range of prior distributions. That is, we look for an estimator $\delta^*$ whose performance is robust in that it is not sensitive to which prior $\pi$, in a class of priors, is the correct prior. Robust Bayes estimators can also have good frequentist performance, making then rather attractive procedures. The review papers by Berger (1990, 1994) and Wasserman (1992) provide an entry to this topic.

2. *Empirical Bayes Analysis*    In a standard Bayesian analysis, there are usually parameters in the prior distribution that are to be specified by the experimenter. For example, consider the specification

$$X|\theta \sim \text{n}(\theta, 1),$$

$$\theta|\tau^2 \sim \text{n}(0, \tau^2).$$

The Bayesian experimenter would specify a prior value for $\tau^2$ and a Bayesian analysis can be done. However, as the marginal distribution of $X$ is $\text{n}(0, \tau^2 + 1)$, it contains information about $\tau$ and can be used to estimate $\tau$. This idea of *estimation of prior parameters from the marginal distribution* is what distinguishes empirical Bayes analysis. Empirical Bayes methods are useful in constructing improved procedures, as illustrated in Morris (1983) and Casella and Hwang (1987). Gianola and Fernando (1986) have successfully applied these types of methods to solve practical problems. A comprehensive treatment of empirical Bayes is Carlin and Louis (1996), and less technical introductions are found in Casella (1985, 1992).

3. *Hierarchical Bayes Analysis*    Another way of dealing with the specification above, without giving a prior value to $\tau^2$, is with a hierarchical specification, that is, a specification of a second-stage prior on $\tau^2$. For example, we could use

$$X|\theta \sim \text{n}(\theta, 1),$$

$$\theta|\tau^2 \sim \text{n}(0, \tau^2),$$

$$\tau^2 \sim \text{uniform}(0, \infty) \text{ (improper prior)}.$$

Hierarchical modeling, both Bayes and non-Bayes, is a very effective tool and usually gives answers that are reasonably robust to the underlying model. Their usefulness was demonstrated by Lindley and Smith (1972) and, since then, their use and development have been quite widespread. The seminal paper of Gelfand

and Smith (1990) tied hierarchical models to computing algorithms, and the applicability of Bayesian methods exploded. Lehmann and Casella (1998, Section 4.5) give an introduction to the theory of hierarchical Bayes, and Robert and Casella (1999) cover applications and connections to computational algorithms.

# Chapter 8

# Hypothesis Testing

*"It is a mistake to confound strangeness with mystery."*

**Sherlock Holmes**
*A Study in Scarlet*

## 8.1 Introduction

In Chapter 7 we studied a method of inference called point estimation. Now we move to another inference method, hypothesis testing. Reflecting the need both to find and to evaluate hypothesis tests, this chapter is divided into two parts, as was Chapter 7. We begin with the definition of a statistical hypothesis.

**Definition 8.1.1**   A *hypothesis* is a statement about a population parameter.

The definition of a hypothesis is rather general, but the important point is that a hypothesis makes a statement about the population. The goal of a hypothesis test is to decide, based on a sample from the population, which of two complementary hypotheses is true.

**Definition 8.1.2**   The two complementary hypotheses in a hypothesis testing problem are called the *null hypothesis* and the *alternative hypothesis*. They are denoted by $H_0$ and $H_1$, respectively.

If $\theta$ denotes a population parameter, the general format of the null and alternative hypotheses is $H_0: \theta \in \Theta_0$ and $H_1: \theta \in \Theta_0^c$, where $\Theta_0$ is some subset of the parameter space and $\Theta_0^c$ is its complement. For example, if $\theta$ denotes the average change in a patient's blood pressure after taking a drug, an experimenter might be interested in testing $H_0: \theta = 0$ versus $H_1: \theta \neq 0$. The null hypothesis states that, on the average, the drug has no effect on blood pressure, and the alternative hypothesis states that there is some effect. This common situation, in which $H_0$ states that a treatment has no effect, has led to the term "null" hypothesis. As another example, a consumer might be interested in the proportion of defective items produced by a supplier. If $\theta$ denotes the proportion of defective items, the consumer might wish to test $H_0: \theta \geq \theta_0$ versus $H_1: \theta < \theta_0$. The value $\theta_0$ is the maximum acceptable proportion of defective items, and $H_0$ states that the proportion of defectives is unacceptably high. Problems in which the hypotheses concern the quality of a product are called acceptance sampling problems.

In a hypothesis testing problem, after observing the sample the experimenter must decide either to accept $H_0$ as true or to reject $H_0$ as false and decide $H_1$ is true.

**Definition 8.1.3**    A *hypothesis testing procedure* or *hypothesis test* is a rule that specifies:

i. For which sample values the decision is made to accept $H_0$ as true.

ii. For which sample values $H_0$ is rejected and $H_1$ is accepted as true.

The subset of the sample space for which $H_0$ will be rejected is called the *rejection region* or *critical region*. The complement of the rejection region is called the *acceptance region*.

On a philosophical level, some people worry about the distinction between "rejecting $H_0$" and "accepting $H_1$." In the first case, there is nothing implied about what state the experimenter *is* accepting, only that the state defined by $H_0$ is being rejected. Similarly, a distinction can be made between "accepting $H_0$" and "not rejecting $H_0$." The first phrase implies that the experimenter is willing to assert the state of nature specified by $H_0$, while the second phrase implies that the experimenter really does not believe $H_0$ but does not have the evidence to reject it. For the most part, we will not be concerned with these issues. We view a hypothesis testing problem as a problem in which one of two actions is going to be taken—the actions being the assertion of $H_0$ or $H_1$.

Typically, a hypothesis test is specified in terms of a *test statistic* $W(X_1, \ldots, X_n)$ $= W(\mathbf{X})$, a function of the sample. For example, a test might specify that $H_0$ is to be rejected if $\bar{X}$, the sample mean, is greater than 3. In this case $W(\mathbf{X}) = \bar{X}$ is the test statistic and the rejection region is $\{(x_1, \ldots, x_n) : \bar{x} > 3\}$. In Section 8.2, methods of choosing test statistics and rejection regions are discussed. Criteria for evaluating tests are introduced in Section 8.3. As with point estimators, the methods of finding tests carry no guarantees; the tests they yield must be evaluated before their worth is established.

## 8.2 Methods of Finding Tests

We will detail four methods of finding test procedures, procedures that are useful in different situations and take advantage of different aspects of a problem. We start with a very general method, one that is almost always applicable and is also optimal in some cases.

### 8.2.1 Likelihood Ratio Tests

The likelihood ratio method of hypothesis testing is related to the maximum likelihood estimators discussed in Section 7.2.2, and likelihood ratio tests are as widely applicable as maximum likelihood estimation. Recall that if $X_1, \ldots, X_n$ is a random sample from a population with pdf or pmf $f(x|\theta)$ ($\theta$ may be a vector), the likelihood function is defined as

$$L(\theta|x_1,\ldots,x_n) = L(\theta|\mathbf{x}) = f(\mathbf{x}|\theta) = \prod_{i=1}^{n} f(x_i|\theta).$$

Let $\Theta$ denote the entire parameter space. Likelihood ratio tests are defined as follows.

**Definition 8.2.1**    The *likelihood ratio test statistic* for testing $H_0 : \theta \in \Theta_0$ versus $H_1 : \theta \in \Theta_0^c$ is

$$\lambda(\mathbf{x}) = \frac{\sup_{\Theta_0} L(\theta|\mathbf{x})}{\sup_{\Theta} L(\theta|\mathbf{x})}.$$

A *likelihood ratio test* (LRT) is any test that has a rejection region of the form $\{\mathbf{x}: \lambda(\mathbf{x}) \le c\}$, where $c$ is any number satisfying $0 \le c \le 1$.

   The rationale behind LRTs may best be understood in the situation in which $f(x|\theta)$ is the pmf of a discrete random variable. In this case, the numerator of $\lambda(\mathbf{x})$ is the maximum probability of the observed sample, the maximum being computed over parameters in the null hypothesis. (See Exercise 8.4.) The denominator of $\lambda(\mathbf{x})$ is the maximum probability of the observed sample over all possible parameters. The ratio of these two maxima is small if there are parameter points in the alternative hypothesis for which the observed sample is much more likely than for any parameter point in the null hypothesis. In this situation, the LRT criterion says $H_0$ should be rejected and $H_1$ accepted as true. Methods for selecting the number $c$ are discussed in Section 8.3.

   If we think of doing the maximization over both the entire parameter space (unrestricted maximization) and a subset of the parameter space (restricted maximization), then the correspondence between LRTs and MLEs becomes more clear. Suppose $\hat{\theta}$, an MLE of $\theta$, exists; $\hat{\theta}$ is obtained by doing an unrestricted maximization of $L(\theta|\mathbf{x})$. We can also consider the MLE of $\theta$, call it $\hat{\theta}_0$, obtained by doing a restricted maximization, assuming $\Theta_0$ is the parameter space. That is, $\hat{\theta}_0 = \hat{\theta}_0(\mathbf{x})$ is the value of $\theta \in \Theta_0$ that maximizes $L(\theta|\mathbf{x})$. Then, the LRT statistic is

$$\lambda(\mathbf{x}) = \frac{L(\hat{\theta}_0|\mathbf{x})}{L(\hat{\theta}|\mathbf{x})}.$$

**Example 8.2.2 (Normal LRT)**    Let $X_1,\ldots,X_n$ be a random sample from a $n(\theta, 1)$ population. Consider testing $H_0 : \theta = \theta_0$ versus $H_1 : \theta \ne \theta_0$. Here $\theta_0$ is a number fixed by the experimenter prior to the experiment. Since there is only one value of $\theta$ specified by $H_0$, the numerator of $\lambda(\mathbf{x})$ is $L(\theta_0|\mathbf{x})$. In Example 7.2.5 the (unrestricted) MLE of $\theta$ was found to be $\bar{X}$, the sample mean. Thus the denominator of $\lambda(\mathbf{x})$ is $L(\bar{x}|\mathbf{x})$. So the LRT statistic is

$$(8.2.1) \qquad \lambda(\mathbf{x}) = \frac{(2\pi)^{-n/2} \exp\left[-\sum_{i=1}^{n}(x_i - \theta_0)^2/2\right]}{(2\pi)^{-n/2} \exp\left[-\sum_{i=1}^{n}(x_i - \bar{x})^2/2\right]}$$

$$= \exp\left[\left(-\sum_{i=1}^{n}(x_i - \theta_0)^2 + \sum_{i=1}^{n}(x_i - \bar{x})^2\right)/2\right].$$

The expression for $\lambda(\mathbf{x})$ can be simplified by noting that

$$\sum_{i=1}^{n}(x_i - \theta_0)^2 = \sum_{i=1}^{n}(x_i - \bar{x})^2 + n(\bar{x} - \theta_0)^2.$$

Thus the LRT statistic is

$$(8.2.2) \qquad\qquad \lambda(\mathbf{x}) = \exp\left[-n(\bar{x} - \theta_0)^2/2\right].$$

An LRT is a test that rejects $H_0$ for small values of $\lambda(\mathbf{x})$. From (8.2.2), the rejection region, $\{\mathbf{x} \colon \lambda(\mathbf{x}) \le c\}$, can be written as

$$\{\mathbf{x} \colon |\bar{x} - \theta_0| \ge \sqrt{-2(\log c)/n}\}.$$

As $c$ ranges between 0 and 1, $\sqrt{-2(\log c)/n}$ ranges between 0 and $\infty$. Thus the LRTs are just those tests that reject $H_0 \colon \theta = \theta_0$ if the sample mean differs from the hypothesized value $\theta_0$ by more than a specified amount.  $\|$

The analysis in Example 8.2.2 is typical in that first the expression for $\lambda(\mathbf{X})$ from Definition 8.2.1 is found, as we did in (8.2.1). Then the description of the rejection region is simplified, if possible, to an expression involving a simpler statistic, $|\bar{X} - \theta_0|$ in the example.

**Example 8.2.3 (Exponential LRT)**   Let $X_1, \ldots, X_n$ be a random sample from an exponential population with pdf

$$f(x|\theta) = \begin{cases} e^{-(x-\theta)} & x \ge \theta \\ 0 & x < \theta, \end{cases}$$

where $-\infty < \theta < \infty$. The likelihood function is

$$L(\theta|\mathbf{x}) = \begin{cases} e^{-\Sigma x_i + n\theta} & \theta \le x_{(1)} \\ 0 & \theta > x_{(1)}. \end{cases} \qquad (x_{(1)} = \min x_i)$$

Consider testing $H_0 \colon \theta \le \theta_0$ versus $H_1 \colon \theta > \theta_0$, where $\theta_0$ is a value specified by the experimenter. Clearly $L(\theta|\mathbf{x})$ is an increasing function of $\theta$ on $-\infty < \theta \le x_{(1)}$. Thus, the denominator of $\lambda(\mathbf{x})$, the unrestricted maximum of $L(\theta|\mathbf{x})$, is

$$L(x_{(1)}|\mathbf{x}) = e^{-\Sigma x_i + n x_{(1)}}.$$

If $x_{(1)} \le \theta_0$, the numerator of $\lambda(\mathbf{x})$ is also $L(x_{(1)}|\mathbf{x})$. But since we are maximizing $L(\theta|\mathbf{x})$ over $\theta \le \theta_0$, the numerator of $\lambda(\mathbf{x})$ is $L(\theta_0|\mathbf{x})$ if $x_{(1)} > \theta_0$. Therefore, the likelihood ratio test statistic is

$$\lambda(\mathbf{x}) = \begin{cases} 1 & x_{(1)} \le \theta_0 \\ e^{-n(x_{(1)} - \theta_0)} & x_{(1)} > \theta_0. \end{cases}$$

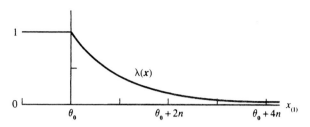

Figure 8.2.1. $\lambda(\mathbf{x})$, a function only of $x_{(1)}$.

A graph of $\lambda(\mathbf{x})$ is shown in Figure 8.2.1. An LRT, a test that rejects $H_0$ if $\lambda(\mathbf{X}) \le c$, is a test with rejection region $\{\mathbf{x} : x_{(1)} \ge \theta_0 - \frac{\log c}{n}\}$. Note that the rejection region depends on the sample only through the sufficient statistic $X_{(1)}$. That this is generally the case will be seen in Theorem 8.2.4.                                                    ‖

Example 8.2.3 again illustrates the point, expressed in Section 7.2.2, that differentiation of the likelihood function is not the only method of finding an MLE. In Example 8.2.3, $L(\theta|\mathbf{x})$ is not differentiable at $\theta = x_{(1)}$.

If $T(\mathbf{X})$ is a sufficient statistic for $\theta$ with pdf or pmf $g(t|\theta)$, then we might consider constructing an LRT based on $T$ and its likelihood function $L^*(\theta|t) = g(t|\theta)$, rather than on the sample $\mathbf{X}$ and its likelihood function $L(\theta|\mathbf{x})$. Let $\lambda^*(t)$ denote the likelihood ratio test statistic based on $T$. Given the intuitive notion that all the information about $\theta$ in $\mathbf{x}$ is contained in $T(\mathbf{x})$, the test based on $T$ should be as good as the test based on the complete sample $\mathbf{X}$. In fact the tests are equivalent.

**Theorem 8.2.4**    *If $T(\mathbf{X})$ is a sufficient statistic for $\theta$ and $\lambda^*(t)$ and $\lambda(\mathbf{x})$ are the LRT statistics based on $T$ and $\mathbf{X}$, respectively, then $\lambda^*(T(\mathbf{x})) = \lambda(\mathbf{x})$ for every $\mathbf{x}$ in the sample space.*

**Proof:** From the Factorization Theorem (Theorem 6.2.6), the pdf or pmf of $\mathbf{X}$ can be written as $f(\mathbf{x}|\theta) = g(T(\mathbf{x})|\theta)h(\mathbf{x})$, where $g(t|\theta)$ is the pdf or pmf of $T$ and $h(\mathbf{x})$ does not depend on $\theta$. Thus

$$\lambda(\mathbf{x}) = \frac{\sup_{\Theta_0} L(\theta|\mathbf{x})}{\sup_{\Theta} L(\theta|\mathbf{x})}$$

$$= \frac{\sup_{\Theta_0} f(\mathbf{x}|\theta)}{\sup_{\Theta} f(\mathbf{x}|\theta)}$$

$$= \frac{\sup_{\Theta_0} g(T(\mathbf{x})|\theta)h(\mathbf{x})}{\sup_{\Theta} g(T(\mathbf{x})|\theta)h(\mathbf{x})} \qquad (T \text{ is sufficient})$$

$$= \frac{\sup\limits_{\Theta_0} g(T(\mathbf{x})|\theta)}{\sup\limits_{\Theta} g(T(\mathbf{x})|\theta)} \qquad (h \text{ does not depend on } \theta)$$

$$= \frac{\sup\limits_{\Theta_0} L^*(\theta|T(\mathbf{x}))}{\sup\limits_{\Theta} L^*(\theta|T(\mathbf{x}))} \qquad (g \text{ is the pdf or pmf of } T)$$

$$= \lambda^*(T(\mathbf{x})). \qquad \qquad \square$$

The comment after Example 8.2.2 was that, after finding an expression for $\lambda(\mathbf{x})$, we try to simplify that expression. In light of Theorem 8.2.4, one interpretation of this comment is that the simplified expression for $\lambda(\mathbf{x})$ should depend on $\mathbf{x}$ only through $T(\mathbf{x})$ if $T(\mathbf{X})$ is a sufficient statistic for $\theta$.

**Example 8.2.5 (LRT and sufficiency)**  In Example 8.2.2, we can recognize that $\bar{X}$ is a sufficient statistic for $\theta$. We could use the likelihood function associated with $\bar{X}$ ($\bar{X} \sim \text{n}(\theta, \frac{1}{n})$) to more easily reach the conclusion that a likelihood ratio test of $H_0: \theta = \theta_0$ versus $H_1: \theta \neq \theta_0$ rejects $H_0$ for large values of $|\bar{X} - \theta_0|$.

Similarly in Example 8.2.3, $X_{(1)} = \min X_i$ is a sufficient statistic for $\theta$. The likelihood function of $X_{(1)}$ (the pdf of $X_{(1)}$) is

$$L^*(\theta|x_{(1)}) = \begin{cases} ne^{-n(x_{(1)}-\theta)} & \theta \leq x_{(1)} \\ 0 & \theta > x_{(1)}. \end{cases}$$

This likelihood could also be used to derive the fact that a likelihood ratio test of $H_0: \theta \leq \theta_0$ versus $H_1: \theta > \theta_0$ rejects $H_0$ for large values of $X_{(1)}$.  ‖

Likelihood ratio tests are also useful in situations where there are *nuisance* parameters, that is, parameters that are present in a model but are not of direct inferential interest. The presence of such nuisance parameters does not affect the LRT construction method but, as might be expected, the presence of nuisance parameters might lead to a different test.

**Example 8.2.6 (Normal LRT with unknown variance)**  Suppose $X_1, \ldots, X_n$ are a random sample from a $\text{n}(\mu, \sigma^2)$, and an experimenter is interested only in inferences about $\mu$, such as testing $H_0: \mu \leq \mu_0$ versus $H_1: \mu > \mu_0$. Then the parameter $\sigma^2$ is a nuisance parameter. The LRT statistic is

$$\lambda(\mathbf{x}) = \frac{\max\limits_{\{\mu,\sigma^2: \mu \leq \mu_0, \sigma^2 \geq 0\}} L(\mu, \sigma^2|\mathbf{x})}{\max\limits_{\{\mu,\sigma^2: -\infty < \mu < \infty, \sigma^2 \geq 0\}} L(\mu, \sigma^2|\mathbf{x})}$$

$$= \frac{\max\limits_{\{\mu,\sigma^2: \mu \leq \mu_0, \sigma^2 \geq 0\}} L(\mu, \sigma^2|\mathbf{x})}{L(\hat{\mu}, \hat{\sigma}^2|\mathbf{x})}.$$

where $\hat{\mu}$ and $\hat{\sigma}^2$ are the MLEs of $\mu$ and $\sigma^2$ (see Example 7.2.11). Furthermore, if $\hat{\mu} \leq \mu_0$, then the restricted maximum is the same as the unrestricted maximum, while if $\hat{\mu} > \mu_0$, the restricted maximum is $L(\mu_0, \hat{\sigma}_0^2|\mathbf{x})$, where $\hat{\sigma}_0^2 = \Sigma (x_i - \mu_0)^2 / n$. Thus

$$\lambda(\mathbf{x}) = \begin{cases} 1 & \text{if } \hat{\mu} \leq \mu_0 \\ \frac{L(\mu_0, \hat{\sigma}_0^2|\mathbf{x})}{L(\hat{\mu}, \hat{\sigma}^2|\mathbf{x})} & \text{if } \hat{\mu} > \mu_0. \end{cases}$$

With some algebra, it can be shown that the test based on $\lambda(\mathbf{x})$ is equivalent to a test based on Student's $t$ statistic. Details are left to Exercise 8.37. (Exercises 8.38–8.42 also deal with nuisance parameter problems.)  ∥

### 8.2.2 Bayesian Tests

Hypothesis testing problems may also be formulated in a Bayesian model. Recall from Section 7.2.3 that a Bayesian model includes not only the sampling distribution $f(\mathbf{x}|\theta)$ but also the prior distribution $\pi(\theta)$, with the prior distribution reflecting the experimenter's opinion about the parameter $\theta$ prior to sampling.

The Bayesian paradigm prescribes that the sample information be combined with the prior information using Bayes' Theorem to obtain the posterior distribution $\pi(\theta|\mathbf{x})$. All inferences about $\theta$ are now based on the posterior distribution.

In a hypothesis testing problem, the posterior distribution may be used to calculate the probabilities that $H_0$ and $H_1$ are true. Remember, $\pi(\theta|\mathbf{x})$ is a probability distribution for a random variable. Hence, the posterior probabilities $P(\theta \in \Theta_0|\mathbf{x}) = P(H_0$ is true$|\mathbf{x})$ and $P(\theta \in \Theta_0^c|\mathbf{x}) = P(H_1$ is true$|\mathbf{x})$ may be computed.

The probabilities $P(H_0$ is true$|\mathbf{x})$ and $P(H_1$ is true$|\mathbf{x})$ are not meaningful to the classical statistician. The classical statistician considers $\theta$ to be a fixed number. Consequently, a hypothesis is *either true or false*. If $\theta \in \Theta_0$, $P(H_0$ is true$|\mathbf{x}) = 1$ and $P(H_1$ is true$|\mathbf{x}) = 0$ for all values of $\mathbf{x}$. If $\theta \in \Theta_0^c$, these values are reversed. Since these probabilities are unknown (since $\theta$ is unknown) and do not depend on the sample $\mathbf{x}$, they are not used by the classical statistician. In a Bayesian formulation of a hypothesis testing problem, these probabilities depend on the sample $\mathbf{x}$ and can give useful information about the veracity of $H_0$ and $H_1$.

One way a Bayesian hypothesis tester may choose to use the posterior distribution is to decide to accept $H_0$ as true if $P(\theta \in \Theta_0|\mathbf{X}) \geq P(\theta \in \Theta_0^c|\mathbf{X})$ and to reject $H_0$ otherwise. In the terminology of the previous sections, the test statistic, a function of the sample, is $P(\theta \in \Theta_0^c|\mathbf{X})$ and the rejection region is $\{\mathbf{x}: P(\theta \in \Theta_0^c|\mathbf{x}) > \frac{1}{2}\}$. Alternatively, if the Bayesian hypothesis tester wishes to guard against falsely rejecting $H_0$, he may decide to reject $H_0$ only if $P(\theta \in \Theta_0^c|\mathbf{X})$ is greater than some large number, .99 for example.

**Example 8.2.7 (Normal Bayesian test)** Let $X_1, \ldots, X_n$ be iid $n(\theta, \sigma^2)$ and let the prior distribution on $\theta$ be $n(\mu, \tau^2)$, where $\sigma^2$, $\mu$, and $\tau^2$ are known. Consider testing $H_0: \theta \leq \theta_0$ versus $H_1: \theta > \theta_0$. From Example 7.2.16, the posterior $\pi(\theta|\bar{x})$ is normal with mean $(n\tau^2\bar{x} + \sigma^2\mu)/(n\tau^2 + \sigma^2)$ and variance $\sigma^2\tau^2/(n\tau^2 + \sigma^2)$.

If we decide to accept $H_0$ if and only if $P(\theta \in \Theta_0|\mathbf{X}) \geq P(\theta \in \Theta_0^c|\mathbf{X})$, then we will accept $H_0$ if and only if

$$\frac{1}{2} \leq P(\theta \in \Theta_0|\mathbf{X}) = P(\theta \leq \theta_0|\mathbf{X}).$$

Since $\pi(\theta|\mathbf{x})$ is symmetric, this is true if and only if the mean of $\pi(\theta|\mathbf{x})$ is less than or equal to $\theta_0$. Therefore $H_0$ will be accepted as true if

$$\bar{X} \leq \theta_0 + \frac{\sigma^2(\theta_0 - \mu)}{n\tau^2}$$

and $H_1$ will be accepted as true otherwise. In particular, if $\mu = \theta_0$ so that prior to experimentation probability $\frac{1}{2}$ is assigned to both $H_0$ and $H_1$, then $H_0$ will be accepted as true if $\bar{x} \leq \theta_0$ and $H_1$ accepted otherwise. ‖

Other methods that use the posterior distribution to make inferences in hypothesis testing problems are discussed in Section 8.3.5.

### 8.2.3 Union–Intersection and Intersection–Union Tests

In some situations, tests for complicated null hypotheses can be developed from tests for simpler null hypotheses. We discuss two related methods.

The *union–intersection method* of test construction might be useful when the null hypothesis is conveniently expressed as an intersection, say

(8.2.3)
$$H_0\colon \theta \in \bigcap_{\gamma \in \Gamma} \Theta_\gamma.$$

Here $\Gamma$ is an arbitrary index set that may be finite or infinite, depending on the problem. Suppose that tests are available for each of the problems of testing $H_{0\gamma}\colon \theta \in \Theta_\gamma$ versus $H_{1\gamma}\colon \theta \in \Theta_\gamma^c$. Say the rejection region for the test of $H_{0\gamma}$ is $\{\mathbf{x}\colon T_\gamma(\mathbf{x}) \in R_\gamma\}$. Then the rejection region for the union–intersection test is

(8.2.4)
$$\bigcup_{\gamma \in \Gamma} \{\mathbf{x}\colon\ T_\gamma(\mathbf{x}) \in R_\gamma\}.$$

The rationale is simple. If any one of the hypotheses $H_{0\gamma}$ is rejected, then $H_0$, which, by (8.2.3), is true only if $H_{0\gamma}$ is true for every $\gamma$, must also be rejected. Only if each of the hypotheses $H_{0\gamma}$ is accepted as true will the intersection $H_0$ be accepted as true.

In some situations a simple expression for the rejection region of a union–intersection test can be found. In particular, suppose that each of the individual tests has a rejection region of the form $\{\mathbf{x}\colon T_\gamma(\mathbf{x}) > c\}$, where $c$ does not depend on $\gamma$. The rejection region for the union–intersection test, given in (8.2.4), can be expressed as

$$\bigcup_{\gamma \in \Gamma} \{\mathbf{x}\colon T_\gamma(\mathbf{x}) > c\} = \{\mathbf{x}\colon \sup_{\gamma \in \Gamma} T_\gamma(\mathbf{x}) > c\}.$$

Thus the test statistic for testing $H_0$ is $T(\mathbf{x}) = \sup_{\gamma \in \Gamma} T_\gamma(\mathbf{x})$. Some examples in which $T(\mathbf{x})$ has a simple formula may be found in Chapter 11.

**Example 8.2.8 (Normal union–intersection test)**    Let $X_1, \ldots, X_n$ be a random sample from a $n(\mu, \sigma^2)$ population. Consider testing $H_0\colon \mu = \mu_0$ versus $H_1\colon \mu \neq \mu_0$, where $\mu_0$ is a specified number. We can write $H_0$ as the intersection of two sets,

$$H_0\colon \{\mu\colon \mu \leq \mu_0\} \cap \{\mu\colon \mu \geq \mu_0\}.$$

The LRT of $H_{0L} : \mu \leq \mu_0$ versus $H_{1L} : \mu > \mu_0$ is

$$\text{reject } H_{0L} : \mu \leq \mu_0 \text{ in favor of } H_{1L} : \mu > \mu_0 \text{ if } \frac{\bar{X} - \mu_0}{S/\sqrt{n}} \geq t_L$$

(see Exercise 8.37). Similarly, the LRT of $H_{0U} : \mu \geq \mu_0$ versus $H_{1U} : \mu < \mu_0$ is

$$\text{reject } H_{0U} : \mu \geq \mu_0 \text{ in favor of } H_{1U} : \mu < \mu_0 \text{ if } \frac{\bar{X} - \mu_0}{S/\sqrt{n}} \leq t_U.$$

Thus the union–intersection test of $H_0 : \mu = \mu_0$ versus $H_1 : \mu \neq \mu_0$ formed from these two LRTs is

$$\text{reject } H_0 \text{ if } \frac{\bar{X} - \mu_0}{S/\sqrt{n}} \geq t_L \text{ or } \frac{\bar{X} - \mu_0}{S/\sqrt{n}} \leq t_U.$$

If $t_L = -t_U \geq 0$, the union–intersection test can be more simply expressed as

$$\text{reject } H_0 \text{ if } \frac{|\bar{X} - \mu_0|}{S/\sqrt{n}} \geq t_L.$$

It turns out that this union–intersection test is also the LRT for this problem (see Exercise 8.38) and is called the two-sided $t$ test.                    ‖

The union–intersection method of test construction is useful if the null hypothesis is conveniently expressed as an intersection. Another method, the *intersection–union method*, may be useful if the null hypothesis is conveniently expressed as a union. Suppose we wish to test the null hypothesis

$$(8.2.5) \qquad\qquad H_0 : \theta \in \bigcup_{\gamma \in \Gamma} \Theta_\gamma.$$

Suppose that for each $\gamma \in \Gamma, \{\mathbf{x} : T_\gamma(\mathbf{x}) \in R_\gamma\}$ is the rejection region for a test of $H_{0\gamma} : \theta \in \Theta_\gamma$ versus $H_{1\gamma} : \theta \in \Theta_\gamma^c$. Then the rejection region for the intersection–union test of $H_0$ versus $H_1$ is

$$(8.2.6) \qquad\qquad \bigcap_{\gamma \in \Gamma} \{\mathbf{x} : T_\gamma(\mathbf{x}) \in R_\gamma\}.$$

From (8.2.5), $H_0$ is false if and only if *all* of the $H_{0\gamma}$ are false, so $H_0$ can be rejected if and only if each of the individual hypotheses $H_{0\gamma}$ can be rejected. Again, the test can be greatly simplified if the rejection regions for the individual hypotheses are all of the form $\{\mathbf{x} : T_\gamma(\mathbf{x}) \geq c\}$ ($c$ independent of $\gamma$). In such cases, the rejection region for $H_0$ is

$$\bigcap_{\gamma \in \Gamma} \{\mathbf{x} : T_\gamma(\mathbf{x}) \geq c\} = \{\mathbf{x} : \inf_{\gamma \in \Gamma} T_\gamma(\mathbf{x}) \geq c\}.$$

Here, the intersection-union test statistic is $\inf_{\gamma \in \Gamma} T_\gamma(\mathbf{X})$, and the test rejects $H_0$ for large values of this statistic.

**Example 8.2.9 (Acceptance sampling)**   The topic of acceptance sampling provides an extremely useful application of an intersection–union test, as this example will illustrate. (See Berger 1982 for a more detailed treatment of this problem.)

Two parameters that are important in assessing the quality of upholstery fabric are $\theta_1$, the mean breaking strength, and $\theta_2$, the probability of passing a flammability test. Standards may dictate that $\theta_1$ should be over 50 pounds and $\theta_2$ should be over .95, and the fabric is acceptable only if it meets both of these standards. This can be modeled with the hypothesis test

$$H_0 \colon \{\theta_1 \leq 50 \text{ or } \theta_2 \leq .95\} \qquad \text{versus} \qquad H_1 \colon \{\theta_1 > 50 \text{ and } \theta_2 > .95\},$$

where a batch of material is acceptable only if $H_1$ is accepted.

Suppose $X_1, \ldots, X_n$ are measurements of breaking strength for $n$ samples and are assumed to be iid $\text{n}(\theta_1, \sigma^2)$. The LRT of $H_{01} \colon \theta_1 \leq 50$ will reject $H_{01}$ if $(\bar{X} - 50)/(S/\sqrt{n}) > t$. Suppose that we also have the results of $m$ flammability tests, denoted by $Y_1, \ldots, Y_m$, where $Y_i = 1$ if the $i$th sample passes the test and $Y_i = 0$ otherwise. If $Y_1, \ldots, Y_m$ are modeled as iid Bernoulli($\theta_2$) random variables, the LRT will reject $H_{02} \colon \theta_2 \leq .95$ if $\sum_{i=1}^{m} Y_i > b$ (see Exercise 8.3). Putting all of this together, the rejection region for the intersection–union test is given by

$$\left\{ (\mathbf{x}, \mathbf{y}) \colon \frac{\bar{x} - 50}{s/\sqrt{n}} > t \text{ and } \sum_{i=1}^{m} y_i > b \right\}.$$

Thus the intersection–union test decides the product is acceptable, that is, $H_1$ is true, if and only if it decides that each of the individual parameters meets its standard, that is, $H_{1i}$ is true. If more than two parameters define a product's quality, individual tests for each parameter can be combined, by means of the intersection–union method, to yield an overall test of the product's quality.                                               ‖

## 8.3 Methods of Evaluating Tests

In deciding to accept or reject the null hypothesis $H_0$, an experimenter might be making a mistake. Usually, hypothesis tests are evaluated and compared through their probabilities of making mistakes. In this section we discuss how these error probabilities can be controlled. In some cases, it can even be determined which tests have the smallest possible error probabilities.

### 8.3.1 Error Probabilities and the Power Function

A hypothesis test of $H_0 : \theta \in \Theta_0$ versus $H_1 : \theta \in \Theta_0^c$ might make one of two types of errors. These two types of errors traditionally have been given the non-mnemonic names, Type I Error and Type II Error. If $\theta \in \Theta_0$ but the hypothesis test incorrectly decides to reject $H_0$, then the test has made a *Type I Error*. If, on the other hand, $\theta \in \Theta_0^c$ but the test decides to accept $H_0$, a *Type II Error* has been made. These two different situations are depicted in Table 8.3.1.

Suppose $R$ denotes the rejection region for a test. Then for $\theta \in \Theta_0$, the test will make a mistake if $\mathbf{x} \in R$, so the probability of a Type I Error is $P_\theta(\mathbf{X} \in R)$. For

Table 8.3.1. *Two types of errors in hypothesis testing*

|  |  | Decision | |
| --- | --- | --- | --- |
|  |  | Accept $H_0$ | Reject $H_0$ |
|  | $H_0$ | Correct decision | Type I Error |
| Truth | | | |
|  | $H_1$ | Type II Error | Correct decision |

$\theta \in \Theta_0^c$, the probability of a Type II Error is $P_\theta(\mathbf{X} \in R^c)$. This switching from $R$ to $R^c$ is a bit confusing, but, if we realize that $P_\theta(\mathbf{X} \in R^c) = 1 - P_\theta(\mathbf{X} \in R)$, then the function of $\theta$, $P_\theta(\mathbf{X} \in R)$, contains all the information about the test with rejection region $R$. We have

$$P_\theta(\mathbf{X} \in R) = \begin{cases} \text{probability of a Type I Error} & \text{if } \theta \in \Theta_0 \\ \text{one minus the probability of a Type II Error} & \text{if } \theta \in \Theta_0^c. \end{cases}$$

This consideration leads to the following definition.

**Definition 8.3.1**     The *power function* of a hypothesis test with rejection region $R$ is the function of $\theta$ defined by $\beta(\theta) = P_\theta(\mathbf{X} \in R)$.

The ideal power function is 0 for all $\theta \in \Theta_0$ and 1 for all $\theta \in \Theta_0^c$. Except in trivial situations, this ideal cannot be attained. Qualitatively, a good test has power function near 1 for most $\theta \in \Theta_0^c$ and near 0 for most $\theta \in \Theta_0$.

**Example 8.3.2 (Binomial power function)**     Let $X \sim \text{binomial}(5, \theta)$. Consider testing $H_0: \theta \leq \frac{1}{2}$ versus $H_1: \theta > \frac{1}{2}$. Consider first the test that rejects $H_0$ if and only if all "successes" are observed. The power function for this test is

$$\beta_1(\theta) = P_\theta(X \in R) = P_\theta(X = 5) = \theta^5.$$

The graph of $\beta_1(\theta)$ is in Figure 8.3.1. In examining this power function, we might decide that although the probability of a Type I Error is acceptably low ($\beta_1(\theta) \leq (\frac{1}{2})^5 = .0312$) for all $\theta \leq \frac{1}{2}$, the probability of a Type II Error is too high ($\beta_1(\theta)$ is too small) for most $\theta > \frac{1}{2}$. The probability of a Type II Error is less than $\frac{1}{2}$ only if $\theta > (\frac{1}{2})^{1/5} = .87$. To achieve smaller Type II Error probabilities, we might consider using the test that rejects $H_0$ if $X = 3$, 4, or 5. The power function for this test is

$$\beta_2(\theta) = P_\theta(X = 3, 4, \text{ or } 5) = \binom{5}{3}\theta^3(1-\theta)^2 + \binom{5}{4}\theta^4(1-\theta)^1 + \binom{5}{5}\theta^5(1-\theta)^0.$$

The graph of $\beta_2(\theta)$ is also in Figure 8.3.1. It can be seen in Figure 8.3.1 that the second test has achieved a smaller Type II Error probability in that $\beta_2(\theta)$ is larger for $\theta > \frac{1}{2}$. But the Type I Error probability is larger for the second test; $\beta_2(\theta)$ is larger for $\theta \leq \frac{1}{2}$. If a choice is to be made between these two tests, the researcher must decide which error structure, that described by $\beta_1(\theta)$ or that described by $\beta_2(\theta)$, is more acceptable.     ‖

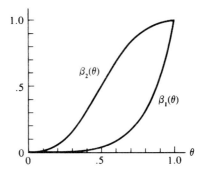

Figure 8.3.1. *Power functions for Example 8.3.2*

**Example 8.3.3 (Normal power function)**  Let $X_1, \ldots, X_n$ be a random sample from a $n(\theta, \sigma^2)$ population, $\sigma^2$ known. An LRT of $H_0 : \theta \leq \theta_0$ versus $H_1 : \theta > \theta_0$ is a test that rejects $H_0$ if $(\bar{X} - \theta_0)/(\sigma/\sqrt{n}) > c$ (see Exercise 8.37). The constant $c$ can be any positive number. The power function of this test is

$$\beta(\theta) = P_\theta\left(\frac{\bar{X} - \theta_0}{\sigma/\sqrt{n}} > c\right)$$

$$= P_\theta\left(\frac{\bar{X} - \theta}{\sigma/\sqrt{n}} > c + \frac{\theta_0 - \theta}{\sigma/\sqrt{n}}\right)$$

$$= P\left(Z > c + \frac{\theta_0 - \theta}{\sigma/\sqrt{n}}\right),$$

where $Z$ is a standard normal random variable, since $(\bar{X} - \theta)/(\sigma/\sqrt{n}) \sim n(0, 1)$. As $\theta$ increases from $-\infty$ to $\infty$, it is easy to see that this normal probability increases from 0 to 1. Therefore, it follows that $\beta(\theta)$ is an increasing function of $\theta$, with

$$\lim_{\theta \to -\infty} \beta(\theta) = 0, \quad \lim_{\theta \to \infty} \beta(\theta) = 1, \quad \text{and} \quad \beta(\theta_0) = \alpha \text{ if } P(Z > c) = \alpha.$$

A graph of $\beta(\theta)$ for $c = 1.28$ is given in Figure 8.3.2.                                    $\|$

Typically, the power function of a test will depend on the sample size $n$. If $n$ can be chosen by the experimenter, consideration of the power function might help determine what sample size is appropriate in an experiment.

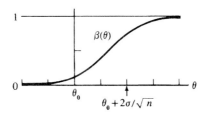

Figure 8.3.2. *Power function for Example 8.3.3*

**Example 8.3.4 (Continuation of Example 8.3.3)**     Suppose the experimenter wishes to have a maximum Type I Error probability of .1. Suppose, in addition, the experimenter wishes to have a maximum Type II Error probability of .2 if $\theta \geq \theta_0 + \sigma$. We now show how to choose $c$ and $n$ to achieve these goals, using a test that rejects $H_0 : \theta \leq \theta_0$ if $(\bar{X} - \theta_0)/(\sigma/\sqrt{n}) > c$. As noted above, the power function of such a test is

$$\beta(\theta) = P\left(Z > c + \frac{\theta_0 - \theta}{\sigma/\sqrt{n}}\right).$$

Because $\beta(\theta)$ is increasing in $\theta$, the requirements will be met if

$$\beta(\theta_0) = .1 \quad \text{and} \quad \beta(\theta_0 + \sigma) = .8.$$

By choosing $c = 1.28$, we achieve $\beta(\theta_0) = P(Z > 1.28) = .1$, regardless of $n$. Now we wish to choose $n$ so that $\beta(\theta_0 + \sigma) = P(Z > 1.28 - \sqrt{n}) = .8$. But, $P(Z > -.84) = .8$. So setting $1.28 - \sqrt{n} = -.84$ and solving for $n$ yield $n = 4.49$. Of course $n$ must be an integer. So choosing $c = 1.28$ and $n = 5$ yield a test with error probabilities controlled as specified by the experimenter.                    ‖

For a fixed sample size, it is usually impossible to make both types of error probabilities arbitrarily small. In searching for a good test, it is common to restrict consideration to tests that control the Type I Error probability at a specified level. Within this class of tests we then search for tests that have Type II Error probability that is as small as possible. The following two terms are useful when discussing tests that control Type I Error probabilities.

**Definition 8.3.5**     For $0 \leq \alpha \leq 1$, a test with power function $\beta(\theta)$ is a *size $\alpha$ test* if $\sup_{\theta \in \Theta_0} \beta(\theta) = \alpha$.

**Definition 8.3.6**     For $0 \leq \alpha \leq 1$, a test with power function $\beta(\theta)$ is a *level $\alpha$ test* if $\sup_{\theta \in \Theta_0} \beta(\theta) \leq \alpha$.

Some authors do not make the distinction between the terms *size* and *level* that we have made, and sometimes these terms are used interchangeably. But according to our definitions, the set of level $\alpha$ tests contains the set of size $\alpha$ tests. Moreover, the distinction becomes important in complicated models and complicated testing situations, where it is often computationally impossible to construct a size $\alpha$ test. In such situations, an experimenter must be satisfied with a level $\alpha$ test, realizing that some compromises may be made. We will see some examples, especially in conjunction with union–intersection and intersection–union tests.

Experimenters commonly specify the level of the test they wish to use, with typical choices being $\alpha = .01, .05$, and $.10$. Be aware that, in fixing the level of the test, the experimenter is controlling only the Type I Error probabilities, not the Type II Error. If this approach is taken, the experimenter should specify the null and alternative hypotheses so that it is most important to control the Type I Error probability. For example, suppose an experimenter expects an experiment to give support to a particular hypothesis, but she does not wish to make the assertion unless the data really do give convincing support. The test can be set up so that the alternative hypothesis is the one that she expects the data to support, and hopes to prove. (The alternative hypothesis is sometimes called the *research hypothesis* in this context.) By using a level $\alpha$ test with small $\alpha$, the experimenter is guarding against saying the data support the research hypothesis when it is false.

The methods of Section 8.2 usually yield test statistics and general forms for rejection regions. However, they do not generally lead to one specific test. For example, an LRT (Definition 8.2.1) is one that rejects $H_0$ if $\lambda(\mathbf{X}) \leq c$, but $c$ was unspecified, so not one but an entire class of LRTs is defined, one for each value of $c$. The restriction to size $\alpha$ tests may now lead to the choice of one out of the class of tests.

**Example 8.3.7 (Size of LRT)**   In general, a size $\alpha$ LRT is constructed by choosing $c$ such that $\sup_{\theta \in \Theta_0} P_\theta(\lambda(\mathbf{X}) \leq c) = \alpha$. How that $c$ is determined depends on the particular problem. For example, in Example 8.2.2, $\Theta_0$ consists of the single point $\theta = \theta_0$ and $\sqrt{n}(\bar{X} - \theta_0) \sim n(0, 1)$ if $\theta = \theta_0$. So the test

$$\text{reject } H_0 \text{ if } |\bar{X} - \theta_0| \geq z_{\alpha/2}/\sqrt{n},$$

where $z_{\alpha/2}$ satisfies $P(Z \geq z_{\alpha/2}) = \alpha/2$ with $Z \sim n(0, 1)$, is the size $\alpha$ LRT. Specifically, this corresponds to choosing $c = \exp(-z_{\alpha/2}^2/2)$, but this is not an important point.

For the problem described in Example 8.2.3, finding a size $\alpha$ LRT is complicated by the fact that the null hypothesis $H_0: \theta \leq \theta_0$ consists of more than one point. The LRT rejects $H_0$ if $X_{(1)} \geq c$, where $c$ is chosen so that this is a size $\alpha$ test. But if $c = (-\log \alpha)/n + \theta_0$, then

$$P_{\theta_0}\left(X_{(1)} \geq c\right) = e^{-n(c-\theta_0)} = \alpha.$$

Since $\theta$ is a location parameter for $X_{(1)}$,

$$P_\theta\left(X_{(1)} \geq c\right) \leq P_{\theta_0}\left(X_{(1)} \geq c\right) \quad \text{for any } \theta \leq \theta_0.$$

Thus

$$\sup_{\theta \in \Theta_0} \beta(\theta) = \sup_{\theta \leq \theta_0} P_\theta(X_{(1)} \geq c) = P_{\theta_0}(X_{(1)} \geq c) = \alpha$$

and this $c$ yields the size $\alpha$ LRT.                                        ∥

*A note on notation:*   In the above example we used the notation $z_{\alpha/2}$ to denote the point having probability $\alpha/2$ to the right of it for a standard normal pdf. We will use this notation in general, not just for the normal but for other distributions as well (defining what we need to for clarity's sake). For example, the point $z_\alpha$ satisfies $P(Z > z_\alpha) = \alpha$, where $Z \sim n(0, 1)$; $t_{n-1,\alpha/2}$ satisfies $P(T_{n-1} > t_{n-1,\alpha/2}) = \alpha/2$, where $T_{n-1} \sim t_{n-1}$; and $\chi_{p,1-\alpha}^2$ satisfies $P(\chi_p^2 > \chi_{p,1-\alpha}^2) = 1 - \alpha$, where $\chi_p^2$ is a chi squared random variable with $p$ degrees of freedom. Points like $z_{\alpha/2}, z_\alpha, t_{n-1,\alpha/2}$, and $\chi_{p,1-\alpha}^2$ are known as *cutoff points*.

**Example 8.3.8 (Size of union–intersection test)**    The problem of finding a size $\alpha$ union–intersection test in Example 8.2.8 involves finding constants $t_L$ and $t_U$ such that

$$\sup_{\theta \in \Theta_0} P_\theta \left( \frac{\bar{X} - \mu_0}{\sqrt{S^2/n}} \geq t_L \quad \text{or} \quad \frac{\bar{X} - \mu_0}{\sqrt{S^2/n}} \leq t_U \right) = \alpha.$$

But for any $(\mu, \sigma^2) = \theta \in \Theta_0$, $\mu = \mu_0$ and thus $(\bar{X} - \mu_0)/\sqrt{S^2/n}$ has a Student's $t$ distribution with $n - 1$ degrees of freedom. So any choice of $t_U = t_{n-1,1-\alpha_1}$ and $t_L = t_{n-1,\alpha_2}$, with $\alpha_1 + \alpha_2 = \alpha$, will yield a test with Type I Error probability of exactly $\alpha$ for all $\theta \in \Theta_0$. The usual choice is $t_L = -t_U = t_{n-1,\alpha/2}$.    ‖

Other than $\alpha$ levels, there are other features of a test that might also be of concern. For example, we would like a test to be more likely to reject $H_0$ if $\theta \in \Theta_0^c$ than if $\theta \in \Theta_0$. All of the power functions in Figures 8.3.1 and 8.3.2 have this property, yielding tests that are called unbiased.

**Definition 8.3.9**    A test with power function $\beta(\theta)$ is *unbiased* if $\beta(\theta') \geq \beta(\theta'')$ for every $\theta' \in \Theta_0^c$ and $\theta'' \in \Theta_0$.

**Example 8.3.10 (Conclusion of Example 8.3.3)**    An LRT of $H_0 : \theta \leq \theta_0$ versus $H_1 : \theta > \theta_0$ has power function

$$\beta(\theta) = P\left( Z > c + \frac{\theta_0 - \theta}{\sigma/\sqrt{n}} \right),$$

where $Z \sim n(0, 1)$. Since $\beta(\theta)$ is an increasing function of $\theta$ (for fixed $\theta_0$), it follows that

$$\beta(\theta) > \beta(\theta_0) = \max_{t \leq \theta_0} \beta(t) \quad \text{for all } \theta > \theta_0$$

and, hence, that the test is unbiased.    ‖

In most problems there are many unbiased tests. (See Exercise 8.45.) Likewise, there are many size $\alpha$ tests, likelihood ratio tests, etc. In some cases we have imposed enough restrictions to narrow consideration to one test. For the two problems in Example 8.3.7, there is only one size $\alpha$ likelihood ratio test. In other cases there remain many tests from which to choose. We discussed only the one that rejects $H_0$ for large values of $T$. In the following sections we will discuss other criteria for selecting one out of a class of tests, criteria that are all related to the power functions of the tests.

### 8.3.2 Most Powerful Tests

In previous sections we have described various classes of hypothesis tests. Some of these classes control the probability of a Type I Error; for example, level $\alpha$ tests have Type I Error probabilities at most $\alpha$ for all $\theta \in \Theta_0$. A good test in such a class would also have a small Type II Error probability, that is, a large power function for $\theta \in \Theta_0^c$. If one test had a smaller Type II Error probability than all other tests in the class, it would certainly be a strong contender for the best test in the class, a notion that is formalized in the next definition.

**Definition 8.3.11**     Let $\mathcal{C}$ be a class of tests for testing $H_0 : \theta \in \Theta_0$ versus $H_1 : \theta \in \Theta_0^c$. A test in class $\mathcal{C}$, with power function $\beta(\theta)$, is a *uniformly most powerful* (UMP) *class $\mathcal{C}$ test* if $\beta(\theta) \geq \beta'(\theta)$ for every $\theta \in \Theta_0^c$ and every $\beta'(\theta)$ that is a power function of a test in class $\mathcal{C}$.

In this section, the class $\mathcal{C}$ will be the class of *all* level $\alpha$ tests. The test described in Definition 8.3.11 is then called a UMP level $\alpha$ test. For this test to be interesting, restriction to the class $\mathcal{C}$ must involve some restriction on the Type I Error probability. A minimization of the Type II Error probability without some control of the Type I Error probability is not very interesting. (For example, a test that rejects $H_0$ with probability 1 will never make a Type II Error. See Exercise 8.16.)

The requirements in Definition 8.3.11 are so strong that UMP tests do not exist in many realistic problems. But in problems that have UMP tests, a UMP test might well be considered the best test in the class. Thus, we would like to be able to identify UMP tests if they exist. The following famous theorem clearly describes which tests are UMP level $\alpha$ tests in the situation where the null and alternative hypotheses both consist of only one probability distribution for the sample (that is, when both $H_0$ and $H_1$ are *simple* hypotheses).

**Theorem 8.3.12 (Neyman–Pearson Lemma)**     *Consider testing $H_0 : \theta = \theta_0$ versus $H_1 : \theta = \theta_1$, where the pdf or pmf corresponding to $\theta_i$ is $f(\mathbf{x}|\theta_i), i = 0, 1$, using a test with rejection region $R$ that satisfies*

$$\mathbf{x} \in R \quad if \quad f(\mathbf{x}|\theta_1) > k f(\mathbf{x}|\theta_0)$$

(8.3.1)                    *and*

$$\mathbf{x} \in R^c \quad if \quad f(\mathbf{x}|\theta_1) < k f(\mathbf{x}|\theta_0),$$

*for some $k \geq 0$, and*

(8.3.2)                              $\alpha = P_{\theta_0}(\mathbf{X} \in R).$

*Then*

**a.** *(Sufficiency)   Any test that satisfies (8.3.1) and (8.3.2) is a UMP level $\alpha$ test.*

**b.** *(Necessity)   If there exists a test satisfying (8.3.1) and (8.3.2) with $k > 0$, then every UMP level $\alpha$ test is a size $\alpha$ test (satisfies (8.3.2)) and every UMP level $\alpha$ test satisfies (8.3.1) except perhaps on a set $A$ satisfying $P_{\theta_0}(\mathbf{X} \in A) = P_{\theta_1}(\mathbf{X} \in A) = 0.$*

**Proof:** We will prove the theorem for the case that $f(\mathbf{x}|\theta_0)$ and $f(\mathbf{x}|\theta_1)$ are pdfs of continuous random variables. The proof for discrete random variables can be accomplished by replacing integrals with sums. (See Exercise 8.21.)

Note first that any test satisfying (8.3.2) is a size $\alpha$ and, hence, a level $\alpha$ test because $\sup_{\theta \in \Theta_0} P_\theta(\mathbf{X} \in R) = P_{\theta_0}(\mathbf{X} \in R) = \alpha$, since $\Theta_0$ has only one point.

To ease notation, we define a *test function*, a function on the sample space that is 1 if $\mathbf{x} \in R$ and 0 if $\mathbf{x} \in R^c$. That is, it is the indicator function of the rejection region. Let $\phi(\mathbf{x})$ be the test function of a test satisfying (8.3.1) and (8.3.2). Let $\phi'(\mathbf{x})$ be the test function of any other level $\alpha$ test, and let $\beta(\theta)$ and $\beta'(\theta)$ be the power functions corresponding to the tests $\phi$ and $\phi'$, respectively. Because $0 \leq \phi'(\mathbf{x}) \leq 1$, (8.3.1) implies that $(\phi(\mathbf{x}) - \phi'(\mathbf{x}))(f(\mathbf{x}|\theta_1) - kf(\mathbf{x}|\theta_0)) \geq 0$ for every $\mathbf{x}$ (since $\phi = 1$ if $f(\mathbf{x}|\theta_1) > kf(\mathbf{x}|\theta_0)$ and $\phi = 0$ if $f(\mathbf{x}|\theta_1) < kf(\mathbf{x}|\theta_0)$). Thus

$$(8.3.3) \qquad 0 \leq \int [\phi(\mathbf{x}) - \phi'(\mathbf{x})][f(\mathbf{x}|\theta_1) - kf(\mathbf{x}|\theta_0)]\, d\mathbf{x}$$

$$= \beta(\theta_1) - \beta'(\theta_1) - k(\beta(\theta_0) - \beta'(\theta_0)).$$

Statement (a) is proved by noting that, since $\phi'$ is a level $\alpha$ test and $\phi$ is a size $\alpha$ test, $\beta(\theta_0) - \beta'(\theta_0) = \alpha - \beta'(\theta_0) \geq 0$. Thus (8.3.3) and $k \geq 0$ imply that

$$0 \leq \beta(\theta_1) - \beta'(\theta_1) - k(\beta(\theta_0) - \beta'(\theta_0)) \leq \beta(\theta_1) - \beta'(\theta_1),$$

showing that $\beta(\theta_1) \geq \beta'(\theta_1)$ and hence $\phi$ has greater power than $\phi'$. Since $\phi'$ was an arbitrary level $\alpha$ test and $\theta_1$ is the only point in $\Theta_0^c$, $\phi$ is a UMP level $\alpha$ test.

To prove statement (b), let $\phi'$ now be the test function for any UMP level $\alpha$ test. By part (a), $\phi$, the test satisfying (8.3.1) and (8.3.2), is also a UMP level $\alpha$ test, thus $\beta(\theta_1) = \beta'(\theta_1)$. This fact, (8.3.3), and $k > 0$ imply

$$\alpha - \beta'(\theta_0) = \beta(\theta_0) - \beta'(\theta_0) \leq 0.$$

Now, since $\phi'$ is a level $\alpha$ test, $\beta'(\theta_0) \leq \alpha$. Thus $\beta'(\theta_0) = \alpha$, that is, $\phi'$ is a size $\alpha$ test, and this also implies that (8.3.3) is an equality in this case. But the nonnegative integrand $(\phi(\mathbf{x}) - \phi'(\mathbf{x}))(f(\mathbf{x}|\theta_1) - kf(\mathbf{x}|\theta_0))$ will have a zero integral only if $\phi'$ satisfies (8.3.1), except perhaps on a set $A$ with $\int_A f(\mathbf{x}|\theta_i)\, d\mathbf{x} = 0$. This implies that the last assertion in statement (b) is true. $\qquad\qquad\square$

The following corollary connects the Neyman–Pearson Lemma to sufficiency.

**Corollary 8.3.13** *Consider the hypothesis problem posed in Theorem 8.3.12. Suppose $T(\mathbf{X})$ is a sufficient statistic for $\theta$ and $g(t|\theta_i)$ is the pdf or pmf of $T$ corresponding to $\theta_i$, $i = 0, 1$. Then any test based on $T$ with rejection region $S$ (a subset of the sample space of $T$) is a UMP level $\alpha$ test if it satisfies*

$$t \in S \quad \text{if} \quad g(t|\theta_1) > kg(t|\theta_0)$$

$$(8.3.4) \qquad\qquad and$$

$$t \in S^c \quad \text{if} \quad g(t|\theta_1) < kg(t|\theta_0),$$

*for some $k \geq 0$, where*

$$(8.3.5) \qquad\qquad\qquad \alpha = P_{\theta_0}(T \in S).$$

**Proof:** In terms of the original sample $\mathbf{X}$, the test based on $T$ has the rejection region $R = \{\mathbf{x} : T(\mathbf{x}) \in S\}$. By the Factorization Theorem, the pdf or pmf of $\mathbf{X}$ can be written as $f(\mathbf{x}|\theta_i) = g(T(\mathbf{x})|\theta_i)h(\mathbf{x}), i = 0, 1$, for some nonnegative function $h(\mathbf{x})$. Multiplying the inequalities in (8.3.4) by this nonnegative function, we see that $R$ satisfies

$$\mathbf{x} \in R \text{ if } f(\mathbf{x}|\theta_1) = g(T(\mathbf{x})|\theta_1)h(\mathbf{x}) > kg(T(\mathbf{x})|\theta_0)h(\mathbf{x}) = kf(\mathbf{x}|\theta_0)$$

and

$$\mathbf{x} \in R^c \text{ if } f(\mathbf{x}|\theta_1) = g(T(\mathbf{x})|\theta_1)h(\mathbf{x}) < kg(T(\mathbf{x})|\theta_0)h(\mathbf{x}) = kf(\mathbf{x}|\theta_0).$$

Also, by (8.3.5),

$$P_{\theta_0}(\mathbf{X} \in R) = P_{\theta_0}(T(\mathbf{X}) \in S) = \alpha.$$

So, by the sufficiency part of the Neyman–Pearson Lemma, the test based on $T$ is a UMP level $\alpha$ test. $\qquad\square$

When we derive a test that satisfies the inequalities (8.3.1) or (8.3.4), and hence is a UMP level $\alpha$ test, it is usually easier to rewrite the inequalities as $f(\mathbf{x}|\theta_1)/f(\mathbf{x}|\theta_0) > k$. (We must be careful about dividing by 0.) This method is used in the following examples.

**Example 8.3.14 (UMP binomial test)**      Let $X \sim$ binomial$(2, \theta)$. We want to test $H_0: \theta = \frac{1}{2}$ versus $H_1: \theta = \frac{3}{4}$. Calculating the ratios of the pmfs gives

$$\frac{f(0|\theta = \frac{3}{4})}{f(0|\theta = \frac{1}{2})} = \frac{1}{4}, \quad \frac{f(1|\theta = \frac{3}{4})}{f(1|\theta = \frac{1}{2})} = \frac{3}{4}, \quad \text{and} \quad \frac{f(2|\theta = \frac{3}{4})}{f(2|\theta = \frac{1}{2})} = \frac{9}{4}.$$

If we choose $\frac{3}{4} < k < \frac{9}{4}$, the Neyman–Pearson Lemma says that the test that rejects $H_0$ if $X = 2$ is the UMP level $\alpha = P(X = 2|\theta = \frac{1}{2}) = \frac{1}{4}$ test. If we choose $\frac{1}{4} < k < \frac{3}{4}$, the Neyman–Pearson Lemma says that the test that rejects $H_0$ if $X = 1$ or 2 is the UMP level $\alpha = P(X = 1 \text{ or } 2|\theta = \frac{1}{2}) = \frac{3}{4}$ test. Choosing $k < \frac{1}{4}$ or $k > \frac{9}{4}$ yields the UMP level $\alpha = 1$ or level $\alpha = 0$ test.

Note that if $k = \frac{3}{4}$, then (8.3.1) says we must reject $H_0$ for the sample point $x = 2$ and accept $H_0$ for $x = 0$ but leaves our action for $x = 1$ undetermined. But if we accept $H_0$ for $x = 1$, we get the UMP level $\alpha = \frac{1}{4}$ test as above. If we reject $H_0$ for $x = 1$, we get the UMP level $\alpha = \frac{3}{4}$ test as above. $\qquad\|$

Example 8.3.14 also shows that for a discrete distribution, the $\alpha$ level at which a test can be done is a function of the particular pmf with which we are dealing. (No such problem arises in the continuous case. Any $\alpha$ level can be attained.)

**Example 8.3.15 (UMP normal test)**      Let $X_1, \ldots, X_n$ be a random sample from a n$(\theta, \sigma^2)$ population, $\sigma^2$ known. The sample mean $\bar{X}$ is a sufficient statistic for $\theta$. Consider testing $H_0: \theta = \theta_0$ versus $H_1: \theta = \theta_1$, where $\theta_0 > \theta_1$. The inequality (8.3.4), $g(\bar{x}|\theta_1) > kg(\bar{x}|\theta_0)$, is equivalent to

$$\bar{x} < \frac{(2\sigma^2 \log k)/n - \theta_0^2 + \theta_1^2}{2(\theta_1 - \theta_0)}.$$

The fact that $\theta_1 - \theta_0 < 0$ was used to obtain this inequality. The right-hand side increases from $-\infty$ to $\infty$ as $k$ increases from 0 to $\infty$. Thus, by Corollary 8.3.13, the test with rejection region $\bar{x} < c$ is the UMP level $\alpha$ test, where $\alpha = P_{\theta_0}(\bar{X} < c)$. If a particular $\alpha$ is specified, then the UMP test rejects $H_0$ if $\bar{X} < c = -\sigma z_\alpha / \sqrt{n} + \theta_0$. This choice of $c$ ensures that (8.3.5) is true. $\qquad\qquad\qquad\qquad\qquad\qquad\qquad\qquad\qquad\qquad\qquad\quad\|$

Hypotheses, such as $H_0$ and $H_1$ in the Neyman–Pearson Lemma, that specify only one possible distribution for the sample $\mathbf{X}$ are called simple hypotheses. In most realistic problems, the hypotheses of interest specify more than one possible distribution for the sample. Such hypotheses are called *composite hypotheses*. Since Definition 8.3.11 requires a UMP test to be most powerful against *each* individual $\theta \in \Theta_0^c$, the Neyman–Pearson Lemma can be used to find UMP tests in problems involving composite hypotheses.

In particular, hypotheses that assert that a univariate parameter is large, for example, $H: \theta \geq \theta_0$, or small, for example, $H: \theta < \theta_0$, are called *one-sided hypotheses*. Hypotheses that assert that a parameter is either large or small, for example, $H: \theta \neq \theta_0$, are called *two-sided hypotheses*. A large class of problems that admit UMP level $\alpha$ tests involve one-sided hypotheses and pdfs or pmfs with the monotone likelihood ratio property.

**Definition 8.3.16**   A family of pdfs or pmfs $\{g(t|\theta): \theta \in \Theta\}$ for a univariate random variable $T$ with real-valued parameter $\theta$ has a *monotone likelihood ratio* (MLR) if, for every $\theta_2 > \theta_1$, $g(t|\theta_2)/g(t|\theta_1)$ is a monotone (nonincreasing or nondecreasing) function of $t$ on $\{t: g(t|\theta_1) > 0 \text{ or } g(t|\theta_2) > 0\}$. Note that $c/0$ is defined as $\infty$ if $0 < c$.

Many common families of distributions have an MLR. For example, the normal (known variance, unknown mean), Poisson, and binomial all have an MLR. Indeed, any regular exponential family with $g(t|\theta) = h(t)c(\theta)e^{w(\theta)t}$ has an MLR if $w(\theta)$ is a nondecreasing function (see Exercise 8.25).

**Theorem 8.3.17 (Karlin–Rubin)**    *Consider testing $H_0 : \theta \leq \theta_0$ versus $H_1 : \theta > \theta_0$. Suppose that $T$ is a sufficient statistic for $\theta$ and the family of pdfs or pmfs $\{g(t|\theta): \theta \in \Theta\}$ of $T$ has an MLR. Then for any $t_0$, the test that rejects $H_0$ if and only if $T > t_0$ is a UMP level $\alpha$ test, where $\alpha = P_{\theta_0}(T > t_0)$.*

**Proof:** Let $\beta(\theta) = P_\theta(T > t_0)$ be the power function of the test. Fix $\theta' > \theta_0$ and consider testing $H_0': \theta = \theta_0$ versus $H_1': \theta = \theta'$. Since the family of pdfs or pmfs of $T$ has an MLR, $\beta(\theta)$ is nondecreasing (see Exercise 8.34), so

i.  $\sup_{\theta \leq \theta_0} \beta(\theta) = \beta(\theta_0) = \alpha$, and this is a level $\alpha$ test.

ii. If we define

$$k' = \inf_{t \in \mathcal{T}} \frac{g(t|\theta')}{g(t|\theta_0)},$$

where $\mathcal{T} = \{t: t > t_0 \text{ and either } g(t|\theta') > 0 \text{ or } g(t|\theta_0) > 0\}$, it follows that

$$T > t_0 \Leftrightarrow \frac{g(t|\theta')}{g(t|\theta_0)} > k'.$$

Together with Corollary 8.3.13, (i) and (ii) imply that $\beta(\theta') \geq \beta^*(\theta')$, where $\beta^*(\theta)$ is the power function for any other level $\alpha$ test of $H_0'$, that is, any test satisfying $\beta(\theta_0) \leq \alpha$. However, any level $\alpha$ test of $H_0$ satisfies $\beta^*(\theta_0) \leq \sup_{\theta \in \Theta_0} \beta^*(\theta) \leq \alpha$. Thus, $\beta(\theta') \geq \beta^*(\theta')$ for any level $\alpha$ test of $H_0$. Since $\theta'$ was arbitrary, the test is a UMP level $\alpha$ test.                                                                                    □

By an analogous argument, it can be shown that under the conditions of Theorem 8.3.17, the test that rejects $H_0: \theta \geq \theta_0$ in favor of $H_1: \theta < \theta_0$ if and only if $T < t_0$ is a UMP level $\alpha = P_{\theta_0}(T < t_0)$ test.

**Example 8.3.18 (Continuation of Example 8.3.15)** Consider testing $H_0': \theta \geq \theta_0$ versus $H_1': \theta < \theta_0$ using the test that rejects $H_0'$ if

$$\bar{X} < -\frac{\sigma z_\alpha}{\sqrt{n}} + \theta_0.$$

As $\bar{X}$ is sufficient and its distribution has an MLR (see Exercise 8.25), it follows from Theorem 8.3.17 that the test is a UMP level $\alpha$ test in this problem.

As the power function of this test,

$$\beta(\theta) = P_\theta\left(\bar{X} < -\frac{\sigma z_\alpha}{\sqrt{n}} + \theta_0\right),$$

is a decreasing function of $\theta$ (since $\theta$ is a location parameter in the distribution of $\bar{X}$), the value of $\alpha$ is given by $\sup_{\theta \geq \theta_0} \beta(\theta) = \beta(\theta_0) = \alpha$.                          ‖

Although most experimenters would choose to use a UMP level $\alpha$ test if they knew of one, unfortunately, for many problems there is no UMP level $\alpha$ test. That is, no UMP test exists because the class of level $\alpha$ tests is so large that no one test dominates all the others in terms of power. In such cases, a common method of continuing the search for a good test is to consider some subset of the class of level $\alpha$ tests and attempt to find a UMP test in this subset. This tactic should be reminiscent of what we did in Chapter 7, when we restricted attention to unbiased point estimators in order to investigate optimality. We illustrate how restricting attention to the subset consisting of unbiased tests can result in finding a best test.

First we consider an example that illustrates a typical situation in which a UMP level $\alpha$ test does not exist.

**Example 8.3.19 (Nonexistence of UMP test)** Let $X_1, \ldots, X_n$ be iid $n(\theta, \sigma^2)$, $\sigma^2$ known. Consider testing $H_0: \theta = \theta_0$ versus $H_1: \theta \neq \theta_0$. For a specified value of $\alpha$, a level $\alpha$ test in this problem is any test that satisfies

(8.3.6)                          $P_{\theta_0}(\text{reject } H_0) \leq \alpha$.

Consider an alternative parameter point $\theta_1 < \theta_0$. The analysis in Example 8.3.18 shows that, among all tests that satisfy (8.3.6), the test that rejects $H_0$ if $\bar{X} < -\sigma z_\alpha/\sqrt{n} + \theta_0$ has the highest possible power at $\theta_1$. Call this Test 1. Furthermore, by part (b) (necessity) of the Neyman–Pearson Lemma, any other level $\alpha$ test that has as high a power as Test 1 at $\theta_1$ must have the same rejection region as Test 1 except possibly for a set $A$ satisfying $\int_A f(\mathbf{x}|\theta_i)\, d\mathbf{x} = 0$. Thus, if a UMP level $\alpha$ test exists for this problem, it must be Test 1 because no other test has as high a power as Test 1 at $\theta_1$.

Now consider Test 2, which rejects $H_0$ if $\bar{X} > \sigma z_\alpha/\sqrt{n} + \theta_0$. Test 2 is also a level $\alpha$ test. Let $\beta_i(\theta)$ denote the power function of Test $i$. For any $\theta_2 > \theta_0$,

$$\beta_2(\theta_2) = P_{\theta_2}\left(\bar{X} > \frac{\sigma z_\alpha}{\sqrt{n}} + \theta_0\right)$$

$$= P_{\theta_2}\left(\frac{\bar{X} - \theta_2}{\sigma/\sqrt{n}} > z_\alpha + \frac{\theta_0 - \theta_2}{\sigma/\sqrt{n}}\right)$$

$$> P(Z > z_\alpha) \qquad \left(\begin{array}{l} Z \sim \mathrm{n}(0,1), \\ > \text{ since } \theta_0 - \theta_2 < 0 \end{array}\right)$$

$$= P(Z < -z_\alpha)$$

$$> P_{\theta_2}\left(\frac{\bar{X} - \theta_2}{\sigma/\sqrt{n}} < -z_\alpha + \frac{\theta_0 - \theta_2}{\sigma/\sqrt{n}}\right) \qquad \left(\begin{array}{l} \text{again, } > \text{ since} \\ \theta_0 - \theta_2 < 0 \end{array}\right)$$

$$= P_{\theta_2}\left(\bar{X} < -\frac{\sigma z_\alpha}{\sqrt{n}} + \theta_0\right)$$

$$= \beta_1(\theta_2).$$

Thus Test 1 is not a UMP level $\alpha$ test because Test 2 has a higher power than Test 1 at $\theta_2$. Earlier we showed that if there were a UMP level $\alpha$ test, it would have to be Test 1. Therefore, no UMP level $\alpha$ test exists in this problem.    ‖

Example 8.3.19 illustrates again the usefulness of the Neyman–Pearson Lemma. Previously, the *sufficiency* part of the lemma was used to construct UMP level $\alpha$ tests, but to show the nonexistence of a UMP level $\alpha$ test, the *necessity* part of the lemma is used.

**Example 8.3.20 (Unbiased test)**   When no UMP level $\alpha$ test exists within the class of all tests, we might try to find a UMP level $\alpha$ test within the class of unbiased tests. The power function, $\beta_3(\theta)$, of Test 3, which rejects $H_0 : \theta = \theta_0$ in favor of $H_1 : \theta \neq \theta_0$ if and only if

$$\bar{X} > \sigma z_{\alpha/2}/\sqrt{n} + \theta_0 \quad \text{or} \quad \bar{X} < -\sigma z_{\alpha/2}/\sqrt{n} + \theta_0,$$

as well as $\beta_1(\theta)$ and $\beta_2(\theta)$ from Example 8.3.19, is shown in Figure 8.3.3. Test 3 is actually a UMP unbiased level $\alpha$ test; that is, it is UMP in the class of unbiased tests.

Note that although Test 1 and Test 2 have slightly higher powers than Test 3 for some parameter points, Test 3 has much higher power than Test 1 and Test 2 at other parameter points. For example, $\beta_3(\theta_2)$ is near 1, whereas $\beta_1(\theta_2)$ is near 0. If the interest is in rejecting $H_0$ for both large and small values of $\theta$, Figure 8.3.3 shows that Test 3 is better overall than either Test 1 or Test 2.    ‖

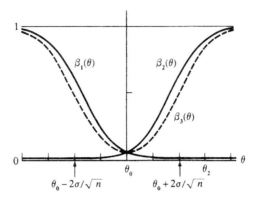

Figure 8.3.3. *Power functions for three tests in Example 8.3.19; $\beta_3(\theta)$ is the power function of an unbiased level $\alpha = .05$ test*

### 8.3.3 Sizes of Union–Intersection and Intersection–Union Tests

Because of the simple way in which they are constructed, the sizes of union–intersection tests (UIT) and intersection–union tests (IUT) can often be bounded above by the sizes of some other tests. Such bounds are useful if a level $\alpha$ test is wanted, but the size of the UIT or IUT is too difficult to evaluate. In this section we discuss these bounds and give examples in which the bounds are sharp, that is, the size of the test is equal to the bound.

First consider UITs. Recall that, in this situation, we are testing a null hypothesis of the form $H_0: \theta \in \Theta_0$, where $\Theta_0 = \bigcap_{\gamma \in \Gamma} \Theta_\gamma$. To be specific, let $\lambda_\gamma(\mathbf{x})$ be the LRT statistic for testing $H_{0\gamma}: \theta \in \Theta_\gamma$ versus $H_{1\gamma}: \theta \in \Theta_\gamma^c$, and let $\lambda(\mathbf{x})$ be the LRT statistic for testing $H_0: \theta \in \Theta_0$ versus $H_1: \theta \in \Theta_0^c$. Then we have the following relationships between the overall LRT and the UIT based on $\lambda_\gamma(\mathbf{x})$.

**Theorem 8.3.21**    *Consider testing $H_0: \theta \in \Theta_0$ versus $H_1: \theta \in \Theta_0^c$, where $\Theta_0 = \bigcap_{\gamma \in \Gamma} \Theta_\gamma$ and $\lambda_\gamma(\mathbf{x})$ is defined in the previous paragraph. Define $T(\mathbf{x}) = \inf_{\gamma \in \Gamma} \lambda_\gamma(\mathbf{x})$, and form the UIT with rejection region*

$$\{\mathbf{x} \colon \lambda_\gamma(\mathbf{x}) < c \text{ for some } \gamma \in \Gamma\} = \{\mathbf{x} \colon T(\mathbf{x}) < c\}.$$

*Also consider the usual LRT with rejection region $\{\mathbf{x} \colon \lambda(\mathbf{x}) < c\}$. Then*

**a.** $T(\mathbf{x}) \geq \lambda(\mathbf{x})$ *for every $\mathbf{x}$;*

**b.** *If $\beta_T(\theta)$ and $\beta_\lambda(\theta)$ are the power functions for the tests based on $T$ and $\lambda$, respectively, then $\beta_T(\theta) \leq \beta_\lambda(\theta)$ for every $\theta \in \Theta$;*

**c.** *If the LRT is a level $\alpha$ test, then the UIT is a level $\alpha$ test.*

**Proof:** Since $\Theta_0 = \bigcap_{\gamma \in \Gamma} \Theta_\gamma \subset \Theta_\gamma$ for any $\gamma$, from Definition 8.2.1 we see that for any $\mathbf{x}$,

$$\lambda_\gamma(\mathbf{x}) \geq \lambda(\mathbf{x}) \quad \text{for each } \gamma \in \Gamma$$

because the region of maximization is bigger for the individual $\lambda_\gamma$. Thus $T(\mathbf{x}) = \inf_{\gamma \in \Gamma} \lambda_\gamma(\mathbf{x}) \geq \lambda(\mathbf{x})$, proving (a). By (a), $\{\mathbf{x} : T(\mathbf{x}) < c\} \subset \{\mathbf{x} : \lambda(\mathbf{x}) < c\}$, so

$$\beta_T(\theta) = P_\theta (T(\mathbf{X}) < c) \leq P_\theta (\lambda(\mathbf{X}) < c) = \beta_\lambda(\theta),$$

proving (b). Since (b) holds for every $\theta$, $\sup_{\theta \in \Theta_0} \beta_T(\theta) \leq \sup_{\theta \in \Theta_0} \beta_\lambda(\theta) \leq \alpha$, proving (c). □

**Example 8.3.22 (An equivalence)**  In some situations, $T(\mathbf{x}) = \lambda(\mathbf{x})$ in Theorem 8.3.21. The UIT built up from individual LRTs is the same as the overall LRT. This was the case in Example 8.2.8. There the UIT formed from two one-sided $t$ tests was equivalent to the two-sided LRT.  ‖

Since the LRT is uniformly more powerful than the UIT in Theorem 8.3.21, we might ask why we should use the UIT. One reason is that the UIT has a smaller Type I Error probability for every $\theta \in \Theta_0$. Furthermore, if $H_0$ is rejected, we may wish to look at the individual tests of $H_{0\gamma}$ to see why. As yet, we have not discussed inferences for the individual $H_{0\gamma}$. The error probabilities for such inferences would have to be examined before such an inference procedure were adopted. But the possibility of gaining additional information by looking at the $H_{0\gamma}$ individually, rather than looking only at the overall LRT, is evident.

Now we investigate the sizes of IUTs. A simple bound for the size of an IUT is related to the sizes of the individual tests that are used to define the IUT. Recall that in this situation the null hypothesis is expressible as a *union*; that is, we are testing

$$H_0 : \theta \in \Theta_0 \qquad \text{versus} \qquad H_1 : \theta \in \Theta_0^c, \quad \text{where } \Theta_0 = \bigcup_{\gamma \in \Gamma} \Theta_\gamma.$$

An IUT has a rejection region of the form $R = \bigcap_{\gamma \in \Gamma} R_\gamma$, where $R_\gamma$ is the rejection region for a test of $H_{0\gamma} : \theta \in \Theta_\gamma$.

**Theorem 8.3.23**  *Let $\alpha_\gamma$ be the size of the test of $H_{0\gamma}$ with rejection region $R_\gamma$. Then the IUT with rejection region $R = \bigcap_{\gamma \in \boldsymbol{\Gamma}} R_\gamma$ is a level $\alpha = \sup_{\gamma \in \Gamma} \alpha_\gamma$ test.*

**Proof:** Let $\theta \in \Theta_0$. Then $\theta \in \Theta_\gamma$ for some $\gamma$ and

$$P_\theta(\mathbf{X} \in R) \leq P_\theta(\mathbf{X} \in R_\gamma) \leq \alpha_\gamma \leq \alpha.$$

Since $\theta \in \Theta_0$ was arbitrary, the IUT is a level $\alpha$ test.  □

Typically, the individual rejection regions $R_\gamma$ are chosen so that $\alpha_\gamma = \alpha$ for all $\gamma$. In such a case, Theorem 8.3.23 states that the resulting IUT is a level $\alpha$ test.

Theorem 8.3.23, which provides an upper bound for the size of an IUT, is somewhat more useful than Theorem 8.3.21, which provides an upper bound for the size of a UIT. Theorem 8.3.21 applies only to UITs constructed from likelihood ratio tests. In contrast, Theorem 8.3.23 applies to any IUT.

The bound in Theorem 8.3.21 is the size of the LRT, which, in a complicated problem, may be difficult to compute. In Theorem 8.3.23, however, the LRT need not be used to obtain the upper bound. Any test of $H_{0\gamma}$ with known size $\alpha_\gamma$ can be used, and then the upper bound on the size of the IUT is given in terms of the known sizes $\alpha_\gamma, \gamma \in \Gamma$.

The IUT in Theorem 8.3.23 is a level $\alpha$ test. But the size of the IUT may be much less than $\alpha$; the IUT may be very conservative. The following theorem gives conditions under which the size of the IUT is exactly $\alpha$ and the IUT is not too conservative.

**Theorem 8.3.24**    *Consider testing $H_0 : \theta \in \bigcup_{j=1}^k \Theta_j$, where $k$ is a finite positive integer. For each $j = 1, \ldots, k$, let $R_j$ be the rejection region of a level $\alpha$ test of $H_{0j}$. Suppose that for some $i = 1, \ldots, k$, there exists a sequence of parameter points, $\theta_l \in \Theta_i$, $l = 1, 2, \ldots$, such that*

i. $\lim_{l \to \infty} P_{\theta_l}(\mathbf{X} \in R_i) = \alpha$,

ii. *for each $j = 1, \ldots, k$, $j \neq i$, $\lim_{l \to \infty} P_{\theta_l}(\mathbf{X} \in R_j) = 1$.*

*Then, the IUT with rejection region $R = \bigcap_{j=1}^k R_j$ is a size $\alpha$ test.*

**Proof:** By Theorem 8.3.23, $R$ is a level $\alpha$ test, that is,

$$(8.3.7) \qquad\qquad\qquad \sup_{\theta \in \Theta_0} P_\theta(\mathbf{X} \in R) \leq \alpha.$$

But, because all the parameter points $\theta_l$ satisfy $\theta_l \in \Theta_i \subset \Theta_0$,

$$\sup_{\theta \in \Theta_0} P_\theta(\mathbf{X} \in R) \geq \lim_{l \to \infty} P_{\theta_l}(\mathbf{X} \in R)$$

$$= \lim_{l \to \infty} P_{\theta_l}\left(\mathbf{X} \in \bigcap_{j=1}^k R_j\right)$$

$$\geq \lim_{l \to \infty} \sum_{j=1}^k P_{\theta_l}(\mathbf{X} \in R_j) - (k-1) \quad \begin{pmatrix} \text{Bonferroni's} \\ \text{Inequality} \end{pmatrix}$$

$$= (k-1) + \alpha - (k-1) \qquad\qquad \text{(by (i) and (ii))}$$

$$= \alpha.$$

This and (8.3.7) imply the test has size exactly equal to $\alpha$.                    □

**Example 8.3.25 (Intersection–union test)**    In Example 8.2.9, let $n = m = 58$, $t = 1.672$, and $b = 57$. Then each of the individual tests has size $\alpha = .05$ (approximately). Therefore, by Theorem 8.3.23, the IUT is a level $\alpha = .05$ test; that is, the probability of deciding the product is good, when in fact it is not, is no more than .05. In fact, this test is a size $\alpha = .05$ test. To see this consider a sequence of parameter points $\theta_l = (\theta_{1l}, \theta_2)$, with $\theta_{1l} \to \infty$ as $l \to \infty$ and $\theta_2 = .95$. All such parameter points are in $\Theta_0$ because $\theta_2 \leq .95$. Also, $P_{\theta_l}(\mathbf{X} \in R_1) \to 1$ as $\theta_{1l} \to \infty$, while $P_{\theta_l}(\mathbf{X} \in R_2) = .05$ for all $l$ because $\theta_2 = .95$. Thus, by Theorem 8.3.24, the IUT is a size $\alpha$ test.                    ‖

Note that, in Example 8.3.25, only the marginal distributions of the $X_1, \ldots, X_n$ and $Y_1, \ldots, Y_m$ were used to find the size of the test. This point is extremely important and directly relates to the usefulness of IUTs, because the joint distribution is often difficult to know and, if known, often difficult to work with. For example, $X_i$ and $Y_i$ may be related if they are measurements on the same piece of fabric, but this relationship would have to be modeled and used to calculate the exact power of the IUT at any particular parameter value.

### 8.3.4 p-Values

After a hypothesis test is done, the conclusions must be reported in some statistically meaningful way. One method of reporting the results of a hypothesis test is to report the size, $\alpha$, of the test used and the decision to reject $H_0$ or accept $H_0$. The size of the test carries important information. If $\alpha$ is small, the decision to reject $H_0$ is fairly convincing, but if $\alpha$ is large, the decision to reject $H_0$ is not very convincing because the test has a large probability of incorrectly making that decision. Another way of reporting the results of a hypothesis test is to report the value of a certain kind of test statistic called a *p-value*.

**Definition 8.3.26** A *p-value* $p(\mathbf{X})$ is a test statistic satisfying $0 \leq p(\mathbf{x}) \leq 1$ for every sample point $\mathbf{x}$. Small values of $p(\mathbf{X})$ give evidence that $H_1$ is true. A p-value is *valid* if, for every $\theta \in \Theta_0$ and every $0 \leq \alpha \leq 1$,

$$(8.3.8) \qquad\qquad P_\theta \left( p(\mathbf{X}) \leq \alpha \right) \leq \alpha.$$

If $p(\mathbf{X})$ is a valid p-value, it is easy to construct a level $\alpha$ test based on $p(\mathbf{X})$. The test that rejects $H_0$ if and only if $p(\mathbf{X}) \leq \alpha$ is a level $\alpha$ test because of (8.3.8). An advantage to reporting a test result via a p-value is that each reader can choose the $\alpha$ he or she considers appropriate and then can compare the reported $p(\mathbf{x})$ to $\alpha$ and know whether these data lead to acceptance or rejection of $H_0$. Furthermore, the smaller the p-value, the stronger the evidence for rejecting $H_0$. Hence, a p-value reports the results of a test on a more continuous scale, rather than just the dichotomous decision "Accept $H_0$" or "Reject $H_0$."

The most common way to define a valid p-value is given in Theorem 8.3.27.

**Theorem 8.3.27** *Let $W(\mathbf{X})$ be a test statistic such that large values of $W$ give evidence that $H_1$ is true. For each sample point $\mathbf{x}$, define*

$$(8.3.9) \qquad\qquad p(\mathbf{x}) = \sup_{\theta \in \Theta_0} P_\theta \left( W(\mathbf{X}) \geq W(\mathbf{x}) \right).$$

*Then, $p(\mathbf{X})$ is a valid p-value.*

**Proof:** Fix $\theta \in \Theta_0$. Let $F_\theta(w)$ denote the cdf of $-W(\mathbf{X})$. Define

$$p_\theta(\mathbf{x}) = P_\theta \left( W(\mathbf{X}) \geq W(\mathbf{x}) \right) = P_\theta \left( -W(\mathbf{X}) \leq -W(\mathbf{x}) \right) = F_\theta \left( -W(\mathbf{x}) \right).$$

Then the random variable $p_\theta(\mathbf{X})$ is equal to $F_\theta(-W(\mathbf{X}))$. Hence, by the Probability Integral Transformation or Exercise 2.10, the distribution of $p_\theta(\mathbf{X})$ is stochastically greater than or equal to a uniform$(0, 1)$ distribution. That is, for every $0 \le \alpha \le 1$, $P_\theta(p_\theta(\mathbf{X}) \le \alpha) \le \alpha$. Because $p(\mathbf{x}) = \sup_{\theta' \in \Theta_0} p_{\theta'}(\mathbf{x}) \ge p_\theta(\mathbf{x})$ for every $\mathbf{x}$,

$$P_\theta\left(p(\mathbf{X}) \le \alpha\right) \le P_\theta\left(p_\theta(\mathbf{X}) \le \alpha\right) \le \alpha.$$

This is true for every $\theta \in \Theta_0$ and for every $0 \le \alpha \le 1$; $p(\mathbf{X})$ is a valid p-value.  □

The calculation of the supremum in (8.3.9) might be difficult. The next two examples illustrate common situations in which it is not too difficult. In the first, no supremum is necessary; in the second, it is easy to determine the $\theta$ value at which the supremum occurs.

**Example 8.3.28 (Two-sided normal p-value)**    Let $X_1, \ldots, X_n$ be a random sample from a n$(\mu, \sigma^2)$ population. Consider testing $H_0 : \mu = \mu_0$ versus $H_1 : \mu \ne \mu_0$. By Exercise 8.38, the LRT rejects $H_0$ for large values of $W(\mathbf{X}) = |\bar{X} - \mu_0|/(S/\sqrt{n})$. If $\mu = \mu_0$, regardless of the value of $\sigma$, $(\bar{X} - \mu_0)/(S/\sqrt{n})$ has a Student's $t$ distribution with $n - 1$ degrees of freedom. Thus, in calculating (8.3.9), the probability is the same for all values of $\theta$, that is, all values of $\sigma$. Thus, the p-value from (8.3.9) for this two-sided $t$ test is $p(\mathbf{x}) = 2P(T_{n-1} \ge |\bar{x} - \mu_0|/(s/\sqrt{n}))$, where $T_{n-1}$ has a Student's $t$ distribution with $n - 1$ degrees of freedom.  ‖

**Example 8.3.29 (One-sided normal p-value)**    Again consider the normal model of Example 8.3.28, but consider testing $H_0 : \mu \le \mu_0$ versus $H_1 : \mu > \mu_0$. By Exercise 8.37, the LRT rejects $H_0$ for large values of $W(\mathbf{X}) = (\bar{X} - \mu_0)/(S/\sqrt{n})$. The following argument shows that, for this statistic, the supremum in (8.3.9) always occurs at a parameter $(\mu_0, \sigma)$, and the value of $\sigma$ used does not matter. Consider any $\mu \le \mu_0$ and any $\sigma$:

$$P_{\mu,\sigma}\left(W(\mathbf{X}) \ge W(\mathbf{x})\right) = P_{\mu,\sigma}\left(\frac{\bar{X} - \mu_0}{S/\sqrt{n}} \ge W(\mathbf{x})\right)$$

$$= P_{\mu,\sigma}\left(\frac{\bar{X} - \mu}{S/\sqrt{n}} \ge W(\mathbf{x}) + \frac{\mu_0 - \mu}{S/\sqrt{n}}\right)$$

$$= P_{\mu,\sigma}\left(T_{n-1} \ge W(\mathbf{x}) + \frac{\mu_0 - \mu}{S/\sqrt{n}}\right)$$

$$\le P\left(T_{n-1} \ge W(\mathbf{x})\right).$$

Here again, $T_{n-1}$ has a Student's $t$ distribution with $n - 1$ degrees of freedom. The inequality in the last line is true because $\mu_0 \ge \mu$ and $(\mu_0 - \mu)/(S/\sqrt{n})$ is a nonnegative random variable. The subscript on $P$ is dropped here, because this probability does not depend on $(\mu, \sigma)$. Furthermore,

$$P\left(T_{n-1} \ge W(\mathbf{x})\right) = P_{\mu_0,\sigma}\left(\frac{\bar{X} - \mu_0}{S/\sqrt{n}} \ge W(\mathbf{x})\right) = P_{\mu_0,\sigma}\left(W(\mathbf{X}) \ge W(\mathbf{x})\right),$$

and this probability is one of those considered in the calculation of the supremum in (8.3.9) because $(\mu_0, \sigma) \in \Theta_0$. Thus, the p-value from (8.3.9) for this one-sided $t$ test is $p(\mathbf{x}) = P(T_{n-1} \ge W(\mathbf{x})) = P(T_{n-1} \ge (\bar{x} - \mu_0)/(s/\sqrt{n}))$.  ‖

Another method for defining a valid p-value, an alternative to using (8.3.9), involves conditioning on a sufficient statistic. Suppose $S(\mathbf{X})$ is a sufficient statistic for the model $\{f(\mathbf{x}|\theta) : \theta \in \Theta_0\}$. (To avoid tests with low power it is important that $S$ is sufficient only for the null model, not the entire model $\{f(\mathbf{x}|\theta) : \theta \in \Theta\}$.) If the null hypothesis is true, the conditional distribution of $\mathbf{X}$ given $S = s$ does not depend on $\theta$. Again, let $W(\mathbf{X})$ denote a test statistic for which large values give evidence that $H_1$ is true. Then, for each sample point $\mathbf{x}$ define

(8.3.10) $$p(\mathbf{x}) = P\left(W(\mathbf{X}) \geq W(\mathbf{x})|S = S(\mathbf{x})\right).$$

Arguing as in Theorem 8.3.27, but considering only the single distribution that is the conditional distribution of $\mathbf{X}$ given $S = s$, we see that, for any $0 \leq \alpha \leq 1$,

$$P\left(p(\mathbf{X}) \leq \alpha|S = s\right) \leq \alpha.$$

Then, for any $\theta \in \Theta_0$, unconditionally we have

$$P_\theta\left(p(\mathbf{X}) \leq \alpha\right) = \sum_s P\left(p(\mathbf{X}) \leq \alpha|S = s\right) P_\theta\left(S = s\right) \leq \sum_s \alpha P_\theta\left(S = s\right) \leq \alpha.$$

Thus, $p(\mathbf{X})$ defined by (8.3.10) is a valid p-value. Sums can be replaced by integrals for continuous $S$, but this method is usually used for discrete $S$, as in the next example.

**Example 8.3.30 (Fisher's Exact Test)**      Let $S_1$ and $S_2$ be independent observations with $S_1 \sim \text{binomial}(n_1, p_1)$ and $S_2 \sim \text{binomial}(n_2, p_2)$. Consider testing $H_0: p_1 = p_2$ versus $H_1: p_1 > p_2$. Under $H_0$, if we let $p$ denote the common value of $p_1 = p_2$, the joint pmf of $(S_1, S_2)$ is

$$f(s_1, s_2|p) = \binom{n_1}{s_1} p^{s_1}(1-p)^{n_1-s_1} \binom{n_2}{s_2} p^{s_2}(1-p)^{n_2-s_2}$$

$$= \binom{n_1}{s_1}\binom{n_2}{s_2} p^{s_1+s_2}(1-p)^{n_1+n_2-(s_1+s_2)}.$$

Thus, $S = S_1 + S_2$ is a sufficient statistic under $H_0$. Given the value of $S = s$, it is reasonable to use $S_1$ as a test statistic and reject $H_0$ in favor of $H_1$ for large values of $S_1$, because large values of $S_1$ correspond to small values of $S_2 = s - S_1$. The conditional distribution $S_1$ given $S = s$ is hypergeometric$(n_1 + n_2, n_1, s)$ (see Exercise 8.48). Thus the conditional p-value in (8.3.10) is

$$p(s_1, s_2) = \sum_{j=s_1}^{\min\{n_1, s\}} f(j|s),$$

the sum of hypergeometric probabilities. The test defined by this p-value is called Fisher's Exact Test.      ‖

### 8.3.5 Loss Function Optimality

A decision theoretic analysis, as in Section 7.3.4, may be used to compare hypothesis tests, rather than just comparing them via their power functions. To carry out this kind of analysis, we must specify the action space and loss function for our hypothesis testing problem.

In a hypothesis testing problem, only two actions are allowable, "accept $H_0$" or "reject $H_0$." These two actions might be denoted $a_0$ and $a_1$, respectively. The action space in hypothesis testing is the two-point set $\mathcal{A} = \{a_0, a_1\}$. A decision rule $\delta(\mathbf{x})$ (a hypothesis test) is a function on $\mathcal{X}$ that takes on only two values, $a_0$ and $a_1$. The set $\{\mathbf{x}\colon \delta(\mathbf{x}) = a_0\}$ is the acceptance region for the test, and the set $\{\mathbf{x}\colon \delta(\mathbf{x}) = a_1\}$ is the rejection region, just as in Definition 8.1.3.

The loss function in a hypothesis testing problem should reflect the fact that, if $\theta \in \Theta_0$ and decision $a_1$ is made, or if $\theta \in \Theta_0^c$ and decision $a_0$ is made, a mistake has been made. But in the other two possible cases, the correct decision has been made. Since there are only two possible actions, the loss function $L(\theta, a)$ in a hypothesis testing problem is composed of only two parts. The function $L(\theta, a_0)$ is the loss incurred for various values of $\theta$ if the decision to accept $H_0$ is made, and $L(\theta, a_1)$ is the loss incurred for various values of $\theta$ if the decision to reject $H_0$ is made.

The simplest kind of loss in a testing problem is called *0–1 loss* and is defined by

$$L(\theta, a_0) = \begin{cases} 0 & \theta \in \Theta_0 \\ 1 & \theta \in \Theta_0^c \end{cases} \quad \text{and} \quad L(\theta, a_1) = \begin{cases} 1 & \theta \in \Theta_0 \\ 0 & \theta \in \Theta_0^c. \end{cases}$$

With 0–1 loss, the value 0 is lost if a correct decision is made and the value 1 is lost if an incorrect decision is made. This is a particularly simple situation in which both types of error have the same consequence. A slightly more realistic loss, one that gives different costs to the two types of error, is *generalized 0–1 loss*,

$$(8.3.11) \qquad L(\theta, a_0) = \begin{cases} 0 & \theta \in \Theta_0 \\ c_{\text{II}} & \theta \in \Theta_0^c \end{cases} \quad \text{and} \quad L(\theta, a_1) = \begin{cases} c_{\text{I}} & \theta \in \Theta_0 \\ 0 & \theta \in \Theta_0^c. \end{cases}$$

In this loss, $c_{\text{I}}$ is the cost of a Type I Error, the error of falsely rejecting $H_0$, and $c_{\text{II}}$ is the cost of a Type II Error, the error of falsely accepting $H_0$. (Actually, when we compare tests, all that really matters is the ratio $c_{\text{II}}/c_{\text{I}}$, not the two individual values. If $c_{\text{I}} = c_{\text{II}}$, we essentially have 0–1 loss.)

In a decision theoretic analysis, the risk function (the expected loss) is used to evaluate a hypothesis testing procedure. The risk function of a test is closely related to its power function, as the following analysis shows.

Let $\beta(\theta)$ be the power function of the test based on the decision rule $\delta$. That is, if $R = \{\mathbf{x}\colon \delta(\mathbf{x}) = a_1\}$ denotes the rejection region of the test, then

$$\beta(\theta) = P_\theta(\mathbf{X} \in R) = P_\theta(\delta(\mathbf{X}) = a_1).$$

The risk function associated with (8.3.11) and, in particular, 0–1 loss is very simple. For any value of $\theta \in \Theta$, $L(\theta, a)$ takes on only two values, 0 and $c_{\text{I}}$ if $\theta \in \Theta_0$ and 0 and $c_{\text{II}}$ if $\theta \in \Theta_0^c$. Thus the risk is

$$R(\theta, \delta) = 0P_\theta(\delta(\mathbf{X}) = a_0) + c_{\mathrm{I}}P_\theta(\delta(\mathbf{X}) = a_1) = c_{\mathrm{I}}\beta(\theta) \quad \text{if } \theta \in \Theta_0,$$

(8.3.12)

$$R(\theta, \delta) = c_{\mathrm{II}}P_\theta(\delta(\mathbf{X}) = a_0) + 0P_\theta(\delta(\mathbf{X}) = a_1) = c_{\mathrm{II}}(1 - \beta(\theta)) \quad \text{if } \theta \in \Theta_0^c.$$

This similarity between a decision theoretic approach and a more traditional power approach is due, in part, to the form of the loss function. But in all hypothesis testing problems, as we shall see below, the power function plays an important role in the risk function.

**Example 8.3.31 (Risk of UMP test)** Let $X_1, \ldots, X_n$ be a random sample from a $n(\mu, \sigma^2)$ population, $\sigma^2$ known. The UMP level $\alpha$ test of $H_0: \theta \geq \theta_0$ versus $H_1: \theta < \theta_0$ is the test that rejects $H_0$ if $(\bar{X} - \theta_0)/(\sigma/\sqrt{n}) < -z_\alpha$ (see Example 8.3.15). The power function for this test is

$$\beta(\theta) = P_\theta\left(Z < -z_\alpha - \frac{\theta - \theta_0}{\sigma/\sqrt{n}}\right),$$

where $Z$ has a $n(0, 1)$ distribution. For $\alpha = .10$, the risk function (8.3.12) for $c_{\mathrm{I}} = 8$ and $c_{\mathrm{II}} = 3$ is shown in Figure 8.3.4. Notice the discontinuity in the risk function at $\theta = \theta_0$. This is due to the fact that at $\theta_0$ the expression in the risk function changes from $\beta(\theta)$ to $1 - \beta(\theta)$ as well as to the difference between $c_{\mathrm{I}}$ and $c_{\mathrm{II}}$. ‖

The 0–1 loss judges only whether a decision is right or wrong. It may be the case that some wrong decisions are more serious than others and the loss function should reflect this. When we test $H_0: \theta \geq \theta_0$ versus $H_1: \theta < \theta_0$, it is a Type I Error to reject $H_0$ if $\theta$ is slightly bigger than $\theta_0$, but it may not be a very serious mistake. The adverse consequences of rejecting $H_0$ may be much worse if $\theta$ is much larger than $\theta_0$. A loss function that reflects this is

(8.3.13) $\quad L(\theta, a_0) = \begin{cases} 0 & \theta \geq \theta_0 \\ b(\theta_0 - \theta) & \theta < \theta_0 \end{cases}$ and $\quad L(\theta, a_1) = \begin{cases} c(\theta - \theta_0)^2 & \theta \geq \theta_0 \\ 0 & \theta < \theta_0, \end{cases}$

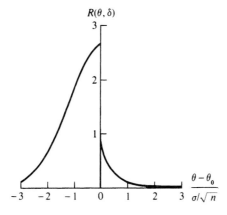

Figure 8.3.4. *Risk function for test in Example 8.3.31*

where $b$ and $c$ are positive constants. For example, if an experimenter is testing whether a drug lowers cholesterol level, $H_0$ and $H_1$ might be set up like this with $\theta_0$ = standard acceptable cholesterol level. Since a high cholesterol level is associated with heart disease, the consequences of rejecting $H_0$ when $\theta$ is large are quite serious. A loss function like (8.3.13) reflects such a consequence. A similar type of loss function is advocated by Vardeman (1987).

Even for a general loss function like (8.3.13), the risk function and the power function are closely related. For any fixed value of $\theta$, the loss is either $L(\theta, a_0)$ or $L(\theta, a_1)$. Thus the expected loss is

$$R(\theta, \delta) = L(\theta, a_0) P_\theta(\delta(\mathbf{X}) = a_0) + L(\theta, a_1) P_\theta(\delta(\mathbf{X}) = a_1)$$

(8.3.14)
$$= L(\theta, a_0)(1 - \beta(\theta)) + L(\theta, a_1)\beta(\theta).$$

The power function of a test is always important when evaluating a hypothesis test. But in a decision theoretic analysis, the weights given by the loss function are also important.

## 8.4 Exercises

**8.1** In 1,000 tosses of a coin, 560 heads and 440 tails appear. Is it reasonable to assume that the coin is fair? Justify your answer.

**8.2** In a given city it is assumed that the number of automobile accidents in a given year follows a Poisson distribution. In past years the average number of accidents per year was 15, and this year it was 10. Is it justified to claim that the accident rate has dropped?

**8.3** Here, the LRT alluded to in Example 8.2.9 will be derived. Suppose that we observe $m$ iid Bernoulli($\theta$) random variables, denoted by $Y_1, \ldots, Y_m$. Show that the LRT of $H_0: \theta \le \theta_0$ versus $H_1: \theta > \theta_0$ will reject $H_0$ if $\sum_{i=1}^m Y_i > b$.

**8.4** Prove the assertion made in the text after Definition 8.2.1. If $f(x|\theta)$ is the pmf of a discrete random variable, then the numerator of $\lambda(\mathbf{x})$, the LRT statistic, is the maximum probability of the observed sample when the maximum is computed over parameters in the null hypothesis. Furthermore, the denominator of $\lambda(\mathbf{x})$ is the maximum probability of the observed sample over all possible parameters.

**8.5** A random sample, $X_1, \ldots, X_n$, is drawn from a Pareto population with pdf

$$f(x|\theta, \nu) = \frac{\theta \nu^\theta}{x^{\theta+1}} I_{[\nu, \infty)}(x), \quad \theta > 0, \quad \nu > 0.$$

(a) Find the MLEs of $\theta$ and $\nu$.

(b) Show that the LRT of

$$H_0: \theta = 1, \ \nu \text{ unknown}, \qquad \text{versus} \qquad H_1: \theta \ne 1, \ \nu \text{ unknown},$$

has critical region of the form $\{\mathbf{x}: T(\mathbf{x}) \le c_1 \text{ or } T(\mathbf{x}) \ge c_2\}$, where $0 < c_1 < c_2$ and

$$T = \log \left[ \frac{\prod_{i=1}^n X_i}{(\min_i X_i)^n} \right].$$

(c) Show that, under $H_0$, $2T$ has a chi squared distribution, and find the number of degrees of freedom. (*Hint*: Obtain the joint distribution of the $n - 1$ nontrivial terms $X_i/(\min_i X_i)$ conditional on $\min_i X_i$. Put these $n - 1$ terms together, and notice that the distribution of $T$ given $\min_i X_i$ does not depend on $\min_i X_i$, so it is the unconditional distribution of $T$.)

**8.6** Suppose that we have two independent random samples: $X_1, \ldots, X_n$ are exponential($\theta$), and $Y_1, \ldots, Y_m$ are exponential($\mu$).

(a) Find the LRT of $H_0: \theta = \mu$ versus $H_1: \theta \neq \mu$.

(b) Show that the test in part (a) can be based on the statistic

$$T = \frac{\Sigma X_i}{\Sigma X_i + \Sigma Y_i}.$$

(c) Find the distribution of $T$ when $H_0$ is true.

**8.7** We have already seen the usefulness of the LRT in dealing with problems with nuisance parameters. We now look at some other nuisance parameter problems.

(a) Find the LRT of

$$H_0: \theta \leq 0 \qquad \text{versus} \qquad H_1: \theta > 0$$

based on a sample $X_1, \ldots, X_n$ from a population with probability density function $f(x|\theta, \lambda) = \frac{1}{\lambda} e^{-(x-\theta)/\lambda} I_{[\theta, \infty)}(x)$, where both $\theta$ and $\lambda$ are unknown.

(b) We have previously seen that the exponential pdf is a special case of a gamma pdf. Generalizing in another way, the exponential pdf can be considered as a special case of the Weibull($\gamma, \beta$). The Weibull pdf, which reduces to the exponential if $\gamma = 1$, is very important in modeling reliability of systems. Suppose that $X_1, \ldots, X_n$ is a random sample from a Weibull population with both $\gamma$ and $\beta$ unknown. Find the LRT of $H_0: \gamma = 1$ versus $H_1: \gamma \neq 1$.

**8.8** A special case of a normal family is one in which the mean and the variance are related, the $n(\theta, a\theta)$ family. If we are interested in testing this relationship, regardless of the value of $\theta$, we are again faced with a nuisance parameter problem.

(a) Find the LRT of $H_0: a = 1$ versus $H_1: a \neq 1$ based on a sample $X_1, \ldots, X_n$ from a $n(\theta, a\theta)$ family, where $\theta$ is unknown.

(b) A similar question can be asked about a related family, the $n(\theta, a\theta^2)$ family. Thus, if $X_1, \ldots, X_n$ are iid $n(\theta, a\theta^2)$, where $\theta$ is unknown, find the LRT of $H_0: a = 1$ versus $H_1: a \neq 1$.

**8.9** Stefanski (1996) establishes the arithmetic-geometric-harmonic mean inequality (see Example 4.7.8 and Miscellanea 4.9.2) using a proof based on likelihood ratio tests. Suppose that $Y_1, \ldots, Y_n$ are independent with pdfs $\lambda_i e^{-\lambda_i y_i}$, and we want to test $H_0: \lambda_1 = \cdots = \lambda_n$ vs. $H_1: \lambda_i$ are not all equal.

(a) Show that the LRT statistic is given by $(\bar{Y})^{-n}/(\prod_i Y_i)^{-1}$ and hence deduce the arithmetic-geometric mean inequality.

(b) Make the transformation $X_i = 1/Y_i$, and show that the LRT statistic based on $X_1, \ldots, X_n$ is given by $[n/\sum_i(1/X_i)]^n/\prod_i X_i$ and hence deduce the geometric-harmonic mean inequality.

**8.10** Let $X_1, \ldots, X_n$ be iid Poisson($\lambda$), and let $\lambda$ have a gamma($\alpha, \beta$) distribution, the conjugate family for the Poisson. In Exercise 7.24 the posterior distribution of $\lambda$ was found, including the posterior mean and variance. Now consider a Bayesian test of $H_0: \lambda \leq \lambda_0$ versus $H_1: \lambda > \lambda_0$.

(a) Calculate expressions for the posterior probabilities of $H_0$ and $H_1$.

(b) If $\alpha = \frac{5}{2}$ and $\beta = 2$, the prior distribution is a chi squared distribution with 5 degrees of freedom. Explain how a chi squared table could be used to perform a Bayesian test.

**8.11** In Exercise 7.23 the posterior distribution of $\sigma^2$, the variance of a normal population, given $S^2$, the sample variance based on a sample of size $n$, was found using a conjugate prior for $\sigma^2$ (the inverted gamma pdf with parameters $\alpha$ and $\beta$). Based on observing $S^2$, a decision about the hypotheses $H_0: \sigma \leq 1$ versus $H_1: \sigma > 1$ is to be made.

(a) Find the region of the sample space for which $P(\sigma \leq 1|s^2) > P(\sigma > 1|s^2)$, the region for which a Bayes test will decide that $\sigma \leq 1$.

(b) Compare the region in part (a) with the acceptance region of an LRT. Is there any choice of prior parameters for which the regions agree?

**8.12** For samples of size $n = 1, 4, 16, 64, 100$ from a normal population with mean $\mu$ and known variance $\sigma^2$, plot the power function of the following LRTs. Take $\alpha = .05$.

(a) $H_0: \mu \leq 0$ versus $H_1: \mu > 0$

(b) $H_0: \mu = 0$ versus $H_1: \mu \neq 0$

**8.13** Let $X_1, X_2$ be iid uniform$(\theta, \theta + 1)$. For testing $H_0: \theta = 0$ versus $H_1: \theta > 0$, we have two competing tests:

$$\phi_1(X_1) : \text{Reject } H_0 \text{ if } X_1 > .95,$$

$$\phi_2(X_1, X_2) : \text{Reject } H_0 \text{ if } X_1 + X_2 > C.$$

(a) Find the value of $C$ so that $\phi_2$ has the same size as $\phi_1$.

(b) Calculate the power function of each test. Draw a well-labeled graph of each power function.

(c) Prove or disprove: $\phi_2$ is a more powerful test than $\phi_1$.

(d) Show how to get a test that has the same size but is more powerful than $\phi_2$.

**8.14** For a random sample $X_1, \ldots, X_n$ of Bernoulli$(p)$ variables, it is desired to test

$$H_0: p = .49 \qquad \text{versus} \qquad H_1: p = .51.$$

Use the Central Limit Theorem to determine, approximately, the sample size needed so that the two probabilities of error are both about .01. Use a test function that rejects $H_0$ if $\sum_{i=1}^{n} X_i$ is large.

**8.15** Show that for a random sample $X_1, \ldots, X_n$ from a n$(0, \sigma^2)$ population, the most powerful test of $H_0: \sigma = \sigma_0$ versus $H_1: \sigma = \sigma_1$, where $\sigma_0 < \sigma_1$, is given by

$$\phi(\Sigma X_i^2) = \begin{cases} 1 & \text{if } \Sigma X_i^2 > c \\ 0 & \text{if } \Sigma X_i^2 \leq c. \end{cases}$$

For a given value of $\alpha$, the size of the Type I Error, show how the value of $c$ is explicitly determined.

**8.16** One very striking abuse of $\alpha$ levels is to choose them *after* seeing the data and to choose them in such a way as to force rejection (or acceptance) of a null hypothesis. To see what the *true* Type I and Type II Error probabilities of such a procedure are, calculate size and power of the following two trivial tests:

(a) Always reject $H_0$, no matter what data are obtained (equivalent to the practice of choosing the $\alpha$ level to force rejection of $H_0$).

(b) Always accept $H_0$, no matter what data are obtained (equivalent to the practice of choosing the $\alpha$ level to force acceptance of $H_0$).

**8.17** Suppose that $X_1, \ldots, X_n$ are iid with a beta$(\mu, 1)$ pdf and $Y_1, \ldots, Y_m$ are iid with a beta$(\theta, 1)$ pdf. Also assume that the $X$s are independent of the $Y$s.

(a) Find an LRT of $H_0: \theta = \mu$ versus $H_1: \theta \neq \mu$.

(b) Show that the test in part (a) can be based on the statistic

$$T = \frac{\Sigma \log X_i}{\Sigma \log X_i + \Sigma \log Y_i}.$$

(c) Find the distribution of $T$ when $H_0$ is true, and then show how to get a test of size $\alpha = .10$.

**8.18** Let $X_1, \ldots, X_n$ be a random sample from a n$(\theta, \sigma^2)$ population, $\sigma^2$ known. An LRT of $H_0: \theta = \theta_0$ versus $H_1: \theta \neq \theta_0$ is a test that rejects $H_0$ if $|\bar{X} - \theta_0|/(\sigma/\sqrt{n}) > c$.

(a) Find an expression, in terms of standard normal probabilities, for the power function of this test.

(b) The experimenter desires a Type I Error probability of .05 and a maximum Type II Error probability of .25 at $\theta = \theta_0 + \sigma$. Find values of $n$ and $c$ that will achieve this.

**8.19** The random variable $X$ has pdf $f(x) = e^{-x}, x > 0$. One observation is obtained on the random variable $Y = X^\theta$, and a test of $H_0: \theta = 1$ versus $H_1: \theta = 2$ needs to be constructed. Find the UMP level $\alpha = .10$ test and compute the Type II Error probability.

**8.20** Let $X$ be a random variable whose pmf under $H_0$ and $H_1$ is given by

| $x$ | 1 | 2 | 3 | 4 | 5 | 6 | 7 |
|-----|-----|-----|-----|-----|-----|-----|-----|
| $f(x\|H_0)$ | .01 | .01 | .01 | .01 | .01 | .01 | .94 |
| $f(x\|H_1)$ | .06 | .05 | .04 | .03 | .02 | .01 | .79 |

Use the Neyman–Pearson Lemma to find the most powerful test for $H_0$ versus $H_1$ with size $\alpha = .04$. Compute the probability of Type II Error for this test.

**8.21** In the proof of Theorem 8.3.12 (Neyman–Pearson Lemma), it was stated that the proof, which was given for continuous random variables, can easily be adapted to cover discrete random variables. Provide the details; that is, prove the Neyman–Pearson Lemma for discrete random variables. Assume that the $\alpha$ level is attainable.

**8.22** Let $X_1, \ldots, X_{10}$ be iid Bernoulli$(p)$.

(a) Find the most powerful test of size $\alpha = .0547$ of the hypotheses $H_0: p = \frac{1}{2}$ versus $H_1: p = \frac{1}{4}$. Find the power of this test.

(b) For testing $H_0: p \leq \frac{1}{2}$ versus $H_1: p > \frac{1}{2}$, find the size and sketch the power function of the test that rejects $H_0$ if $\sum_{i=1}^{10} X_i \geq 6$.

(c) For what $\alpha$ levels does there exist a UMP test of the hypotheses in part (a)?

**8.23** Suppose $X$ is one observation from a population with beta$(\theta, 1)$ pdf.

(a) For testing $H_0: \theta \leq 1$ versus $H_1: \theta > 1$, find the size and sketch the power function of the test that rejects $H_0$ if $X > \frac{1}{2}$.

(b) Find the most powerful level $\alpha$ test of $H_0: \theta = 1$ versus $H_1: \theta = 2$.

(c) Is there a UMP test of $H_0: \theta \leq 1$ versus $H_1: \theta > 1$? If so, find it. If not, prove so.

**8.24** Find the LRT of a *simple* $H_0$ versus a *simple* $H_1$. Is this test equivalent to the one obtained from the Neyman–Pearson Lemma? (This relationship is treated in some detail by Solomon 1975.)

**8.25** Show that each of the following families has an MLR.

(a) $n(\theta, \sigma^2)$ family with $\sigma^2$ known

(b) Poisson($\theta$) family

(c) binomial($n, \theta$) family with $n$ known

**8.26** (a) Show that if a family of pdfs $\{f(x|\theta): \theta \in \Theta\}$ has an MLR, then the corresponding family of cdfs is *stochastically increasing* in $\theta$. (See the Miscellanea section.)

(b) Show that the converse of part (a) is false; that is, give an example of a family of cdfs that is stochastically increasing in $\theta$ for which the corresponding family of pdfs does not have an MLR.

**8.27** Suppose $g(t|\theta) = h(t)c(\theta)e^{w(\theta)t}$ is a one-parameter exponential family for the random variable $T$. Show that this family has an MLR if $w(\theta)$ is an increasing function of $\theta$. Give three examples of such a family.

**8.28** Let $f(x|\theta)$ be the logistic location pdf

$$f(x|\theta) = \frac{e^{(x-\theta)}}{(1 + e^{(x-\theta)})^2}, \quad -\infty < x < \infty, \quad -\infty < \theta < \infty.$$

(a) Show that this family has an MLR.

(b) Based on one observation, $X$, find the most powerful size $\alpha$ test of $H_0: \theta = 0$ versus $H_1: \theta = 1$. For $\alpha = .2$, find the size of the Type II Error.

(c) Show that the test in part (b) is UMP size $\alpha$ for testing $H_0: \theta \leq 0$ versus $H_1: \theta > 0$. What can be said about UMP tests in general for the logistic location family?

**8.29** Let $X$ be one observation from a Cauchy($\theta$) distribution.

(a) Show that this family does not have an MLR.

(b) Show that the test

$$\phi(x) = \begin{cases} 1 & \text{if } 1 < x < 3 \\ 0 & \text{otherwise} \end{cases}$$

is most powerful of its size for testing $H_0: \theta = 0$ versus $H_1: \theta = 1$. Calculate the Type I and Type II Error probabilities.

(c) Prove or disprove: The test in part (b) is UMP for testing $H_0: \theta \leq 0$ versus $H_1: \theta > 0$. What can be said about UMP tests in general for the Cauchy location family?

**8.30** Let $f(x|\theta)$ be the Cauchy *scale* pdf

$$f(x|\theta) = \frac{\theta}{\pi} \frac{1}{\theta^2 + x^2}, \quad -\infty < x < \infty, \quad \theta > 0.$$

(a) Show that this family does not have an MLR.

(b) If $X$ is one observation from $f(x|\theta)$, show that $|X|$ is sufficient for $\theta$ and that the distribution of $|X|$ *does* have an MLR.

**8.31** Let $X_1, \ldots, X_n$ be iid Poisson($\lambda$).

(a) Find a UMP test of $H_0: \lambda \leq \lambda_0$ versus $H_1: \lambda > \lambda_0$.

(b) Consider the specific case $H_0: \lambda \leq 1$ versus $H_1: \lambda > 1$. Use the Central Limit Theorem to determine the sample size $n$ so a UMP test satisfies $P(\text{reject } H_0|\lambda = 1) = .05$ and $P(\text{reject } H_0|\lambda = 2) = .9$.

**8.32** Let $X_1, \ldots, X_n$ be iid $n(\theta, 1)$, and let $\theta_0$ be a specified value of $\theta$.

    (a) Find the UMP, size $\alpha$, test of $H_0: \theta \geq \theta_0$ versus $H_1: \theta < \theta_0$.

    (b) Show that there does not exist a UMP, size $\alpha$, test of $H_0: \theta = \theta_0$ versus $H_1: \theta \neq \theta_0$.

**8.33** Let $X_1, \ldots, X_n$ be a random sample from the uniform$(\theta, \theta+1)$ distribution. To test $H_0: \theta = 0$ versus $H_1: \theta > 0$, use the test

$$\text{reject } H_0 \text{ if } Y_n \geq 1 \text{ or } Y_1 \geq k,$$

where $k$ is a constant, $Y_1 = \min\{X_1, \ldots, X_n\}, Y_n = \max\{X_1, \ldots, X_n\}$.

    (a) Determine $k$ so that the test will have size $\alpha$.

    (b) Find an expression for the power function of the test in part (a).

    (c) Prove that the test is UMP size $\alpha$.

    (d) Find values of $n$ and $k$ so that the UMP .10 level test will have power at least .8 if $\theta > 1$.

**8.34** In each of the following two situations, show that for any number $c$, if $\theta_1 \leq \theta_2$, then

$$P_{\theta_1}(T > c) \leq P_{\theta_2}(T > c).$$

    (a) $\theta$ is a location parameter in the distribution of the random variable $T$.

    (b) The family of pdfs of $T$, $\{g(t|\theta): \theta \in \Theta\}$, has an MLR.

**8.35** The usual $t$ distribution, as derived in Section 5.3.2, is also known as a *central t distribution.* It can be thought of as the pdf of a random variable of the form $T = n(0,1)/\sqrt{\chi_\nu^2/\nu}$, where the normal and the chi squared random variables are independent. A generalization of the $t$ distribution, the *noncentral t*, is of the form $T' = n(\mu,1)/\sqrt{\chi_\nu^2/\nu}$, where the normal and the chi squared random variables are independent and we can have $\mu \neq 0$. (We have already seen a noncentral pdf, the noncentral chi squared, in (4.4.3).) Formally, if $X \sim n(\mu, 1)$ and $Y \sim \chi_\nu^2$, independent of $X$, then $T' = X/\sqrt{Y/\nu}$ has a noncentral $t$ distribution with $\nu$ degrees of freedom and noncentrality parameter $\delta = \sqrt{\mu^2}$.

    (a) Calculate the mean and variance of $T'$.

    (b) The pdf of $T'$ is given by

$$f_{T'}(t|\delta) = \frac{e^{-\delta^2/2}}{\Gamma(\frac{1}{2})\Gamma(\frac{\nu}{2})\sqrt{\nu}} \sum_{k=0}^{\infty} \frac{(2/\nu)^{k/2}(\delta t)^k}{k!} \frac{\Gamma([\nu+k+1]/2)}{(1+(t^2/\nu))^{(\nu+k+1)/2}}.$$

    Show that this pdf reduces to that of a central $t$ if $\delta = 0$.

    (c) Show that the pdf of $T'$ has an MLR in its noncentrality parameter.

**8.36** We have one observation from a beta$(1, \theta)$ population.

    (a) To test $H_0: \theta_1 \leq \theta \leq \theta_2$ versus $H_1: \theta < \theta_1$ or $\theta > \theta_2$, where $\theta_1 = 1$ and $\theta_2 = 2$, a test satisfies $E_{\theta_1}\phi = .5$ and $E_{\theta_2}\phi = .3$. Find a test that is as good, and explain why it is as good.

    (b) For testing $H_0: \theta = \theta_1$ versus $H_1: \theta \neq \theta_1$, with $\theta_1 = 1$, find a two-sided test (other than $\phi \equiv .1$) that satisfies $E_{\theta_1}\phi = .1$ and $\frac{d}{d\theta}E_\theta(\phi)\big|_{\theta=\theta_1} = 0$.

**8.37** Let $X_1, \ldots, X_n$ be a random sample from a $n(\theta, \sigma^2)$ population. Consider testing

$$H_0: \theta \leq \theta_0 \qquad \text{versus} \qquad H_1: \theta > \theta_0.$$

(a) If $\sigma^2$ is known, show that the test that rejects $H_0$ when

$$\bar{X} > \theta_0 + z_\alpha \sqrt{\sigma^2/n}$$

is a test of size $\alpha$. Show that the test can be derived as an LRT.

(b) Show that the test in part (a) is a UMP test.

(c) If $\sigma^2$ is unknown, show that the test that rejects $H_0$ when

$$\bar{X} > \theta_0 + t_{n-1,\alpha} \sqrt{S^2/n}$$

is a test of size $\alpha$. Show that the test can be derived as an LRT.

**8.38** Let $X_1, \ldots, X_n$ be iid $n(\theta, \sigma^2)$, where $\theta_0$ is a specified value of $\theta$ and $\sigma^2$ is unknown. We are interested in testing

$$H_0: \theta = \theta_0 \qquad \text{versus} \qquad H_1: \theta \neq \theta_0.$$

(a) Show that the test that rejects $H_0$ when

$$|\bar{X} - \theta_0| > t_{n-1,\alpha/2} \sqrt{S^2/n}$$

is a test of size $\alpha$.

(b) Show that the test in part (a) can be derived as an LRT.

**8.39** Let $(X_1, Y_1), \ldots, (X_n, Y_n)$ be a random sample from a bivariate normal distribution with parameters $\mu_X, \mu_Y, \sigma_X^2, \sigma_Y^2, \rho$. We are interested in testing

$$H_0: \mu_X = \mu_Y \qquad \text{versus} \qquad H_1: \mu_X \neq \mu_Y.$$

(a) Show that the random variables $W_i = X_i - Y_i$ are iid $n(\mu_W, \sigma_W^2)$.

(b) Show that the above hypothesis can be tested with the statistic

$$T_W = \frac{\bar{W}}{\sqrt{\frac{1}{n} S_W^2}},$$

where $\bar{W} = \frac{1}{n} \sum_{i=1}^n W_i$ and $S_W^2 = \frac{1}{(n-1)} \sum_{i=1}^n (W_i - \bar{W})^2$. Furthermore, show that, under $H_0$, $T_W \sim$ Student's $t$ with $n-1$ degrees of freedom. (This test is known as the *paired-sample t test.*)

**8.40** Let $(X_1, Y_1), \ldots, (X_n, Y_n)$ be a random sample from a bivariate normal distribution with parameters $\mu_X, \mu_Y, \sigma_X^2, \sigma_Y^2, \rho$.

(a) Derive the LRT of

$$H_0: \mu_X = \mu_Y \qquad \text{versus} \qquad H_1: \mu_X \neq \mu_Y,$$

where $\sigma_X^2, \sigma_Y^2$, and $\rho$ are unspecified and unknown.

(b) Show that the test derived in part (a) is equivalent to the paired $t$ test of Exercise 8.39.

(*Hint*: Straightforward maximization of the bivariate likelihood is possible but somewhat nasty. Filling in the gaps of the following argument gives a more elegant proof.)

Make the transformation $u = x - y, v = x + y$. Let $f(x, y)$ denote the bivariate normal pdf, and write

$$f(x, y) = g(v|u)h(u),$$

where $g(v|u)$ is the conditional pdf of $V$ given $U$, and $h(u)$ is the marginal pdf of $U$. Argue that (1) the likelihood can be equivalently factored and (2) the piece involving $g(v|u)$ has the same maximum whether or not the means are restricted. Thus, it can be ignored (since it will cancel) and the LRT is based only on $h(u)$. However, $h(u)$ is a normal pdf with mean $\mu_X - \mu_Y$, and the LRT is the usual one-sample $t$ test, as derived in Exercise 8.38.

**8.41** Let $X_1, \ldots, X_n$ be a random sample from a $n(\mu_X, \sigma_X^2)$, and let $Y_1, \ldots, Y_m$ be an independent random sample from a $n(\mu_Y, \sigma_Y^2)$. We are interested in testing

$$H_0 : \mu_X = \mu_Y \quad \text{versus} \quad H_1 : \mu_X \neq \mu_Y$$

with the assumption that $\sigma_X^2 = \sigma_Y^2 = \sigma^2$.

(a) Derive the LRT for these hypotheses. Show that the LRT can be based on the statistic

$$T = \frac{\bar{X} - \bar{Y}}{\sqrt{S_p^2 \left( \frac{1}{n} + \frac{1}{m} \right)}},$$

where

$$S_p^2 = \frac{1}{(n + m - 2)} \left( \sum_{i=1}^{n} (X_i - \bar{X})^2 + \sum_{i=1}^{m} (Y_i - \bar{Y})^2 \right).$$

(The quantity $S_p^2$ is sometimes referred to as a *pooled variance estimate*. This type of estimate will be used extensively in Section 11.2.)

(b) Show that, under $H_0$, $T \sim t_{n+m-2}$. (This test is known as the *two-sample t test*.)

(c) Samples of wood were obtained from the core and periphery of a certain Byzantine church. The date of the wood was determined, giving the following data.

| Core | | Periphery | |
|------|------|------|------|
| 1294 | 1251 | 1284 | 1274 |
| 1279 | 1248 | 1272 | 1264 |
| 1274 | 1240 | 1256 | 1256 |
| 1264 | 1232 | 1254 | 1250 |
| 1263 | 1220 | 1242 | |
| 1254 | 1218 | | |
| 1251 | 1210 | | |

Use the two-sample $t$ test to determine if the mean age of the core is the same as the mean age of the periphery.

**8.42** The assumption of equal variances, which was made in Exercise 8.41, is not always tenable. In such a case, the distribution of the statistic is no longer a $t$. Indeed, there is doubt as to the wisdom of calculating a pooled variance estimate. (This problem, of making inference on means when variances are unequal, is, in general, quite a difficult one. It is known as the *Behrens–Fisher Problem.*) A natural test to try is the following modification of the two-sample $t$ test: Test

$$H_0 : \mu_X = \mu_Y \quad \text{versus} \quad H_1 : \mu_X \neq \mu_Y,$$

where we do not assume that $\sigma_X^2 = \sigma_Y^2$, using the statistic

$$T' = \frac{\bar{X} - \bar{Y}}{\sqrt{\left(\frac{S_X^2}{n} + \frac{S_Y^2}{m}\right)}},$$

where

$$S_X^2 = \frac{1}{n-1} \sum_{i=1}^{n} (X_i - \bar{X})^2 \quad \text{and} \quad S_Y^2 = \frac{1}{m-1} \sum_{i=1}^{m} (Y_i - \bar{Y})^2.$$

The exact distribution of $T'$ is not pleasant, but we can approximate the distribution using Satterthwaite's approximation (Example 7.2.3).

(a) Show that

$$\frac{\frac{S_X^2}{n} + \frac{S_Y^2}{m}}{\frac{\sigma_X^2}{n} + \frac{\sigma_Y^2}{m}} \sim \frac{\chi_\nu^2}{\nu} \qquad \text{(approximately)},$$

where $\nu$ can be estimated with

$$\hat{\nu} = \frac{\left(\frac{S_X^2}{n} + \frac{S_Y^2}{m}\right)^2}{\frac{S_X^4}{n^2(n-1)} + \frac{S_Y^4}{m^2(m-1)}}.$$

(b) Argue that the distribution of $T'$ can be approximated by a $t$ distribution with $\hat{\nu}$ degrees of freedom.

(c) Re-examine the data from Exercise 8.41 using the approximate $t$ test of this exercise; that is, test if the mean age of the core is the same as the mean age of the periphery using the $T'$ statistic.

(d) Is there any statistical evidence that the variance of the data from the core may be different from the variance of the data from the periphery? (Recall Example 5.4.1.)

**8.43** Sprott and Farewell (1993) note that in the two-sample $t$ test, a valid $t$ statistic can be derived as long as the ratio of variances is known. Let $X_1, \ldots, X_{n_1}$ be a sample from a $n(\mu_1, \sigma^2)$ and $Y_1, \ldots, Y_{n_2}$ a sample from a $n(\mu_2, \rho^2\sigma^2)$, where $\rho^2$ is known. Show that

$$\frac{(\bar{X} - \bar{Y}) - (\mu_1 - \mu_2)}{\sqrt{\frac{1}{n_1} + \frac{\rho^2}{n_2}} \sqrt{\frac{(n_1-1)s_X^2 + (n_2-1)s_Y^2/\rho^2}{n_1+n_2-2}}}$$

has Student's $t$ distribution with $n_1 + n_2 - 2$ degrees of freedom and $\frac{s_Y^2}{\rho^2 s_X^2}$ has an $F$ distribution with $n_2 - 1$ and $n_1 - 1$ degrees of freedom.

Sprott and Farewell also note that the $t$ statistic is maximized at $\rho^2 = \frac{n_1\sqrt{n_1 - 1}s_X^2}{n_2\sqrt{n_2 - 1}s_Y^2}$, and they suggest plotting the statistic for plausible values of $\rho^2$, possibly those in a confidence interval.

**8.44** Verify that Test 3 in Example 8.3.20 is an unbiased level $\alpha$ test.

**8.45** Let $X_1, \ldots, X_n$ be a random sample from a $n(\theta, \sigma^2)$ population. Consider testing

$$H_0: \theta \leq \theta_0 \qquad \text{versus} \qquad H_1: \theta > \theta_0.$$

Let $\bar{X}_m$ denote the sample mean of the first $m$ observations, $X_1, \ldots, X_m$, for $m = 1, \ldots, n$. If $\sigma^2$ is known, show that for each $m = 1, \ldots, n$, the test that rejects $H_0$ when

$$\bar{X}_m > \theta_0 + z_\alpha \sqrt{\sigma^2/m}$$

is an unbiased size $\alpha$ test. Graph the power function for each of these tests if $n = 4$.

**8.46** Let $X_1, \ldots, X_n$ be a random sample from a $n(\theta, \sigma^2)$ population. Consider testing

$$H_0: \theta_1 \leq \theta \leq \theta_2 \qquad \text{versus} \qquad H_1: \theta < \theta_1 \text{ or } \theta > \theta_2.$$

(a) Show that the test

$$\text{reject } H_0 \text{ if } \bar{X} > \theta_2 + t_{n-1,\alpha/2}\sqrt{S^2/n} \quad \text{or} \quad \bar{X} < \theta_1 - t_{n-1,\alpha/2}\sqrt{S^2/n}$$

is not a size $\alpha$ test.

(b) Show that, for an appropriately chosen constant $k$, a size $\alpha$ test is given by

$$\text{reject } H_0 \text{ if } |\bar{X} - \bar{\theta}| > k\sqrt{S^2/n},$$

where $\bar{\theta} = (\theta_1 + \theta_2)/2$.

(c) Show that the tests in parts (a) and (b) are unbiased of their size. (Assume that the noncentral $t$ distribution has an MLR.)

**8.47** Consider two independent normal samples with equal variances, as in Exercise 8.41. Consider testing $H_0: \mu_X - \mu_Y \leq -\delta$ or $\mu_X - \mu_Y \geq \delta$ versus $H_1: -\delta < \mu_X - \mu_Y < \delta$, where $\delta$ is a specified positive constant. (This is called an *equivalence testing* problem.)

(a) Show that the size $\alpha$ LRT of $H_0^-: \mu_X - \mu_Y \leq -\delta$ versus $H_1^-: \mu_X - \mu_Y > -\delta$ rejects $H_0^-$ if

$$T^- = \frac{\bar{X} - \bar{Y} - (-\delta)}{\sqrt{S_p^2\left(\frac{1}{n} + \frac{1}{m}\right)}} \geq t_{n+m-2,\alpha}.$$

(b) Find the size $\alpha$ LRT of $H_0^+: \mu_X - \mu_Y \geq \delta$ versus $H_1^+: \mu_X - \mu_Y < \delta$.

(c) Explain how the tests in (a) and (b) can be combined into a level $\alpha$ test of $H_0$ versus $H_1$.

(d) Show that the test in (c) is a size $\alpha$ test. (*Hint:* Consider $\sigma \to 0$.)

This procedure is sometimes known as the *two one-sided tests procedure* and was derived by Schuirmann (1987) (see also Westlake 1981) for the problem of testing *bioequivalence*. See also the review article by Berger and Hsu (1996) and Exercise 9.33 for a confidence interval counterpart.

**8.48** Prove the assertion in Example 8.3.30 that the conditional distribution of $S_1$ given $S$ is hypergeometric.

**8.49** In each of the following situations, calculate the p-value of the observed data.

(a) For testing $H_0: \theta \leq \frac{1}{2}$ versus $H_1: \theta > \frac{1}{2}$, 7 successes are observed out of 10 Bernoulli trials.

(b) For testing $H_0 : \lambda \leq 1$ versus $H_1 : \lambda > 1, X = 3$ are observed, where $X \sim$ Poisson($\lambda$).

(c) For testing $H_0 : \lambda \leq 1$ versus $H_1 : \lambda > 1, X_1 = 3, X_2 = 5$, and $X_3 = 1$ are observed, where $X_i \sim$ Poisson($\lambda$), independent.

**8.50** Let $X_1, \ldots, X_n$ be iid n($\theta, \sigma^2$), $\sigma^2$ known, and let $\theta$ have a double exponential distribution, that is, $\pi(\theta) = e^{-|\theta|/a}/(2a)$, $a$ known. A Bayesian test of the hypotheses $H_0 : \theta \leq 0$ versus $H_1 : \theta > 0$ will decide in favor of $H_1$ if its posterior probability is large.

(a) For a given constant $K$, calculate the posterior probability that $\theta > K$, that is, $P(\theta > K | x_1, \ldots, x_n, a)$.

(b) Find an expression for $\lim_{a \to \infty} P(\theta > K | x_1, \ldots, x_n, a)$.

(c) Compare your answer in part (b) to the p-value associated with the classical hypothesis test.

**8.51** Here is another common interpretation of p-values. Consider a problem of testing $H_0$ versus $H_1$. Let $W(\mathbf{X})$ be a test statistic. Suppose that for each $\alpha$, $0 \leq \alpha \leq 1$, a critical value $c_\alpha$ can be chosen so that $\{\mathbf{x} : W(\mathbf{x}) \geq c_\alpha\}$ is the rejection region of a size $\alpha$ test of $H_0$. Using this family of tests, show that the usual p-value $p(\mathbf{x})$, defined by (8.3.9), is the smallest $\alpha$ level at which we could reject $H_0$, having observed the data $\mathbf{x}$.

**8.52** Consider testing $H_0 : \theta \in \bigcup_{j=1}^{k} \Theta_j$. For each $j = 1, \ldots, k$, let $p_j(\mathbf{x})$ denote a valid p-value for testing $H_{0j} : \theta \in \Theta_j$. Let $p(\mathbf{x}) = \max_{1 \leq j \leq k} p_j(\mathbf{x})$.

(a) Show that $p(\mathbf{X})$ is a valid p-value for testing $H_0$.

(b) Show that the $\alpha$ level test defined by $p(\mathbf{X})$ is the same as an $\alpha$ level IUT defined in terms of individual tests based on the $p_j(\mathbf{x})$s.

**8.53** In Example 8.2.7 we saw an example of a one-sided Bayesian hypothesis test. Now we will consider a similar situation, but with a two-sided test. We want to test

$$H_0 : \theta = 0 \qquad \text{versus} \qquad H_1 : \theta \neq 0,$$

and we observe $X_1, \ldots, X_n$, a random sample from a n($\theta, \sigma^2$) population, $\sigma^2$ known. A type of prior distribution that is often used in this situation is a mixture of a point mass on $\theta = 0$ and a pdf spread out over $H_1$. A typical choice is to take $P(\theta = 0) = \frac{1}{2}$, and if $\theta \neq 0$, take the prior distribution to be $\frac{1}{2}$n($0, \tau^2$), where $\tau^2$ is known.

(a) Show that the prior defined above is proper, that is, $P(-\infty < \theta < \infty) = 1$.

(b) Calculate the posterior probability that $H_0$ is true, $P(\theta = 0 | x_1, \ldots, x_n)$.

(c) Find an expression for the p-value corresponding to a value of $\bar{x}$.

(d) For the special case $\sigma^2 = \tau^2 = 1$, compare $P(\theta = 0 | x_1, \ldots, x_n)$ and the p-value for a range of values of $\bar{x}$. In particular,

(i) For $n = 9$, plot the p-value and posterior probability as a function of $\bar{x}$, and show that the Bayes probability is greater than the p-value for moderately large values of $\bar{x}$.

(ii) Now, for $\alpha = .05$, set $\bar{x} = Z_{\alpha/2}/\sqrt{n}$, fixing the p-value at $\alpha$ for all $n$. Show that the posterior probability at $\bar{x} = Z_{\alpha/2}/\sqrt{n}$ goes to 1 as $n \to \infty$. This is *Lindley's Paradox*.

Note that small values of $P(\theta = 0 | x_1, \ldots, x_n)$ are evidence *against* $H_0$, and thus this quantity is similar in spirit to a p-value. The fact that these two quantities can have very different values was noted by Lindley (1957) and is also examined by Berger and Sellke (1987). (See the Miscellanea section.)

**8.54** The discrepancies between p-values and Bayes posterior probabilities are not as dramatic in the one-sided problem, as is discussed by Casella and Berger (1987) and also mentioned in the Miscellanea section. Let $X_1, \ldots, X_n$ be a random sample from a $n(\theta, \sigma^2)$ population, and suppose that the hypotheses to be tested are

$$H_0: \theta \leq 0 \qquad \text{versus} \qquad H_1: \theta > 0.$$

The prior distribution on $\theta$ is $n(0, \tau^2)$, $\tau^2$ known, which is symmetric about the hypotheses in the sense that $P(\theta \leq 0) = P(\theta > 0) = \frac{1}{2}$.

(a) Calculate the posterior probability that $H_0$ is true, $P(\theta \leq 0 | x_1, \ldots, x_n)$.

(b) Find an expression for the p-value corresponding to a value of $\bar{x}$, using tests that reject for large values of $\bar{X}$.

(c) For the special case $\sigma^2 = \tau^2 = 1$, compare $P(\theta \leq 0 | x_1, \ldots, x_n)$ and the p-value for values of $\bar{x} > 0$. Show that the Bayes probability is always greater than the p-value.

(d) Using the expression derived in parts (a) and (b), show that

$$\lim_{\tau^2 \to \infty} P(\theta \leq 0 | x_1, \ldots, x_n) = \text{p-value},$$

an equality that does not occur in the two-sided problem.

**8.55** Let $X$ have a $n(\theta, 1)$ distribution, and consider testing $H_0: \theta \geq \theta_0$ versus $H_1: \theta < \theta_0$. Use the loss function (8.3.13) and investigate the three tests that reject $H_0$ if $X < -z_\alpha + \theta_0$ for $\alpha = .1, .3$, and .5.

(a) For $b = c = 1$, graph and compare their risk functions.

(b) For $b = 3, c = 1$, graph and compare their risk functions.

(c) Graph and compare the power functions of the three tests to the risk functions in parts (a) and (b).

**8.56** Consider testing $H_0: p \leq \frac{1}{3}$ versus $H_1: p > \frac{1}{3}$, where $X \sim \text{binomial}(5, p)$, using 0–1 loss. Graph and compare the risk functions for the following two tests. Test I rejects $H_0$ if $X = 0$ or 1. Test II rejects $H_0$ if $X = 4$ or 5.

**8.57** Consider testing $H_0: \mu \leq 0$ versus $H_1: \mu > 0$ using 0–1 loss, where $X \sim n(\mu, 1)$. Let $\delta_c$ be the test that rejects $H_0$ if $X > c$. For every test in this problem, there is a $\delta_c$ in the class of tests $\{\delta_c, -\infty \leq c \leq \infty\}$ that has a uniformly smaller (in $\mu$) risk function. Let $\delta$ be the test that rejects $H_0$ if $1 < X < 2$. Find a test $\delta_c$ that is better than $\delta$. (Either prove that the test is better or graph the risk functions for $\delta$ and $\delta_c$ and carefully explain why the proposed test should be better.)

**8.58** Consider the hypothesis testing problem and loss function given in Example 8.3.31, and let $\sigma = n = 1$. Consider tests that reject $H_0$ if $X < -z_\alpha + \theta_0$. Find the value of $\alpha$ that minimizes the maximum value of the risk function, that is, that yields a *minimax* test.

## 8.5 Miscellanea

### 8.5.1 Monotonic Power Function

In this chapter we used the property of MLR quite extensively, particularly in relation to properties of power functions of tests. The concept of *stochastic ordering* can also be used to obtain properties of power functions. (Recall that *stochastic ordering* has already been encountered in previous chapters, for example, in Exercises 1.49, 3.41–3.43, and 5.19. A cdf $F$ is *stochastically greater* than a cdf $G$ if $F(x) \leq G(x)$ for all $x$, with strict inequality for some $x$, which implies that if $X \sim F, Y \sim G$, then $P(X > x) \geq P(Y > x)$ for all $x$, with strict inequality for some $x$. In other words, $F$ gives more probability to greater values.)

In terms of hypothesis testing, it is often the case that the distribution under the alternative is stochastically greater than under the null distribution. For example, if we have a random sample from a $n(\theta, \sigma^2)$ population and are interested in testing $H_0: \theta \leq \theta_0$ versus $H_1: \theta > \theta_0$, it is true that all the distributions in the alternative are stochastically greater than all those in the null. Gilat (1977) uses the property of stochastic ordering, rather than MLR, to prove monotonicity of power functions under general conditions.

### 8.5.2 Likelihood Ratio As Evidence

The *likelihood ratio* $L(\theta_1|\mathbf{x})/L(\theta_0|\mathbf{x}) = f(\mathbf{x}|\theta_1)/f(\mathbf{x}|\theta_0)$ plays an important role in the testing of $H_0: \theta = \theta_0$ versus $H_1: \theta = \theta_1$. This ratio is equal to the LRT statistic $\lambda(\mathbf{x})$ for values of $\mathbf{x}$ that yield small values of $\lambda$. Also, the Neyman–Pearson Lemma says that the UMP level $\alpha$ test of $H_0$ versus $H_1$ can be defined in terms of this ratio. This likelihood ratio also has an important Bayesian interpretation. Suppose $\pi_0$ and $\pi_1$ are our prior probabilities for $\theta_0$ and $\theta_1$. Then, the *posterior odds in favor of* $\theta_1$ are

$$\frac{P(\theta = \theta_1|\mathbf{x})}{P(\theta = \theta_0|\mathbf{x})} = \frac{f(\mathbf{x}|\theta_1)\pi_1/m(\mathbf{x})}{f(\mathbf{x}|\theta_0)\pi_0/m(\mathbf{x})} = \frac{f(\mathbf{x}|\theta_1)}{f(\mathbf{x}|\theta_0)} \cdot \frac{\pi_1}{\pi_0}.$$

$\pi_1/\pi_0$ are the *prior odds in favor of* $\theta_1$. The likelihood ratio is the amount these prior odds should be adjusted, having observed the data $\mathbf{X} = \mathbf{x}$, to obtain the posterior odds. If the likelihood ratio equals 2, then the prior odds are doubled. The likelihood ratio does not depend on the prior probabilities. Thus, it is interpreted as the evidence *in the data* favoring $H_1$ over $H_0$. This kind of interpretation is discussed by Royall (1997).

### 8.5.3 p-Values and Posterior Probabilities

In Section 8.2.2, where Bayes tests were discussed, we saw that the posterior probability that $H_0$ is true is a measure of the evidence the data provide against (or for) the null hypothesis. We also saw, in Section 8.3.4, that p-values provide a measure of data-based evidence against $H_0$. A natural question to ask is whether these two different measures ever agree; that is, can they be *reconciled*? Berger (James, not Roger) and Sellke (1987) contended that, in the two-sided testing problem, these measures could not be reconciled, and the Bayes measure was superior. Casella and Berger (Roger 1987) argued that the two-sided Bayes problem is artificial and that in the more natural one-sided problem, the measures of evidence can be reconciled. This reconciliation makes little difference to Schervish (1996), who argues that, as measures of evidence, p-values have serious logical flaws.

### 8.5.4 Confidence Set p-Values

Berger and Boos (1994) proposed an alternative method for computing p-values. In the common definition of a p-value (Theorem 8.3.27), the "sup" is over the entire null space $\Theta_0$. Berger and Boos proposed taking the sup over a subset of $\Theta_0$ called $C$. This set $C = C(\mathbf{X})$ is determined from the data and has the property that, if $\theta \in \Theta_0$, then $P_\theta(\theta \in C(\mathbf{X})) \geq 1 - \beta$. (See Chapter 9 for a discussion of confidence sets like $C$.) Then the *confidence set p-value* is

$$p_C(\mathbf{x}) = \sup_{\theta \in C(\mathbf{x})} P_\theta\left(W(\mathbf{X}) \geq W(\mathbf{x})\right) + \beta.$$

Berger and Boos showed that $p_C$ is a valid p-value.

There are two potential advantages to $p_C$. The computational advantage is that it may be easier to compute the sup over the smaller set $C$ than over the larger set $\Theta_0$. The statistical advantage is that, having observed $\mathbf{X}$, we have some idea of the value of $\theta$; there is a good chance $\theta \in C$. It seems irrelevant to look at values of $\theta$ that do not appear to be true. The confidence set p-value looks at only those values of $\theta$ in $\Theta_0$ that seem plausible. Berger and Boos (1994) and Silvapulle (1996) give numerous examples of confidence set p-values. Berger (1996) points out that confidence set p-values can produce tests with improved power in the problem of comparing two binomial probabilities.

# Chapter 9

# Interval Estimation

> "I fear," said Holmes, "that if the matter is beyond humanity it is certainly beyond me. Yet we must exhaust all natural explanations before we fall back upon such a theory as this."
>
> **Sherlock Holmes**
> *The Adventure of the Devil's Foot*

## 9.1 Introduction

In Chapter 7 we discussed point estimation of a parameter $\theta$, where the inference is a guess of a single value as the value of $\theta$. In this chapter we discuss interval estimation and, more generally, set estimation. The inference in a set estimation problem is the statement that "$\theta \in C$," where $C \subset \Theta$ and $C = C(\mathbf{x})$ is a set determined by the value of the data $\mathbf{X} = \mathbf{x}$ observed. If $\theta$ is real-valued, then we usually prefer the set estimate $C$ to be an interval. Interval estimators will be the main topic of this chapter.

As in the previous two chapters, this chapter is divided into two parts, the first concerned with finding interval estimators and the second part concerned with evaluating the worth of the estimators. We begin with a formal definition of interval estimator, a definition as vague as the definition of point estimator.

**Definition 9.1.1** An *interval estimate* of a real-valued parameter $\theta$ is any pair of functions, $L(x_1, \ldots, x_n)$ and $U(x_1, \ldots, x_n)$, of a sample that satisfy $L(\mathbf{x}) \leq U(\mathbf{x})$ for all $\mathbf{x} \in \mathcal{X}$. If $\mathbf{X} = \mathbf{x}$ is observed, the inference $L(\mathbf{x}) \leq \theta \leq U(\mathbf{x})$ is made. The random interval $[L(\mathbf{X}), U(\mathbf{X})]$ is called an *interval estimator*.

We will use our previously defined conventions and write $[L(\mathbf{X}), U(\mathbf{X})]$ for an interval estimator of $\theta$ based on the random sample $\mathbf{X} = (X_1, \ldots, X_n)$ and $[L(\mathbf{x}), U(\mathbf{x})]$ for the realized value of the interval. Although in the majority of cases we will work with finite values for $L$ and $U$, there is sometimes interest in *one-sided* interval estimates. For instance, if $L(\mathbf{x}) = -\infty$, then we have the one-sided interval $(-\infty, U(\mathbf{x})]$ and the assertion is that "$\theta \leq U(\mathbf{x})$," with no mention of a lower bound. We could similarly take $U(\mathbf{x}) = \infty$ and have a one-sided interval $[L(\mathbf{x}), \infty)$.

Although the definition mentions a closed interval $[L(\mathbf{x}), U(\mathbf{x})]$, it will sometimes be more natural to use an open interval $(L(\mathbf{x}), U(\mathbf{x}))$ or even a half-open and half-closed interval, as in the previous paragraph. We will use whichever seems most appropriate for the particular problem at hand, although the preference will be for a closed interval.

**Example 9.1.2 (Interval estimator)** For a sample $X_1, X_2, X_3, X_4$ from a n$(\mu, 1)$, an interval estimator of $\mu$ is $[\bar{X} - 1, \bar{X} + 1]$. This means that we will assert that $\mu$ is in this interval. ‖

At this point, it is natural to inquire as to what is gained by using an interval estimator. Previously, we estimated $\mu$ with $\bar{X}$, and now we have the less precise estimator $[\bar{X} - 1, \bar{X} + 1]$. We surely must gain something! By giving up some precision in our estimate (or assertion about $\mu$), we have gained some confidence, or assurance, that our assertion is correct.

**Example 9.1.3 (Continuation of Example 9.1.2)** When we estimate $\mu$ by $\bar{X}$, the probability that we are exactly correct, that is, $P(\bar{X} = \mu)$, is 0. However, with an interval estimator, we have a positive probability of being correct. The probability that $\mu$ is covered by the interval $[\bar{X} - 1, \bar{X} + 1]$ can be calculated as

$$P(\mu \in [\bar{X} - 1, \bar{X} + 1]) = P(\bar{X} - 1 \leq \mu \leq \bar{X} + 1)$$
$$= P(-1 \leq \bar{X} - \mu \leq 1)$$
$$= P\left(-2 \leq \frac{\bar{X} - \mu}{\sqrt{1/4}} \leq 2\right)$$
$$= P(-2 \leq Z \leq 2) \qquad \left(\frac{\bar{X} - \mu}{\sqrt{1/4}} \text{ is standard normal}\right)$$
$$= .9544.$$

Thus we have over a 95% chance of covering the unknown parameter with our interval estimator. Sacrificing some precision in our estimate, in moving from a point to an interval, has resulted in increased confidence that our assertion is correct. ‖

The purpose of using an interval estimator rather than a point estimator is to have some guarantee of capturing the parameter of interest. The certainty of this guarantee is quantified in the following definitions.

**Definition 9.1.4** For an interval estimator $[L(\mathbf{X}), U(\mathbf{X})]$ of a parameter $\theta$, the *coverage probability* of $[L(\mathbf{X}), U(\mathbf{X})]$ is the probability that the random interval $[L(\mathbf{X}), U(\mathbf{X})]$ covers the true parameter, $\theta$. In symbols, it is denoted by either $P_\theta(\theta \in [L(\mathbf{X}), U(\mathbf{X})])$ or $P(\theta \in [L(\mathbf{X}), U(\mathbf{X})]|\theta)$.

**Definition 9.1.5** For an interval estimator $[L(\mathbf{X}), U(\mathbf{X})]$ of a parameter $\theta$, the *confidence coefficient* of $[L(\mathbf{X}), U(\mathbf{X})]$ is the infimum of the coverage probabilities, $\inf_\theta P_\theta(\theta \in [L(\mathbf{X}), U(\mathbf{X})])$.

There are a number of things to be aware of in these definitions. One, it is important to keep in mind that the *interval* is the random quantity, not the parameter. Therefore, when we write probability statements such as $P_\theta(\theta \in [L(\mathbf{X}), U(\mathbf{X})])$, these probability statements refer to $\mathbf{X}$, not $\theta$. In other words, think of $P_\theta(\theta \in [L(\mathbf{X}), U(\mathbf{X})])$, which might look like a statement about a random $\theta$, as the algebraically equivalent $P_\theta(L(\mathbf{X}) \leq \theta, U(\mathbf{X}) \geq \theta)$, a statement about a random $\mathbf{X}$.

Interval estimators, together with a measure of confidence (usually a confidence coefficient), are sometimes known as *confidence intervals*. We will often use this term interchangeably with *interval estimator*. Although we are mainly concerned with confidence *intervals*, we occasionally will work with more general sets. When working in general, and not being quite sure of the exact form of our sets, we will speak of confidence *sets*. A confidence set with confidence coefficient equal to some value, say $1 - \alpha$, is simply called a $1 - \alpha$ *confidence set*.

Another important point is concerned with coverage probabilities and confidence coefficients. Since we do not know the true value of $\theta$, we can only guarantee a coverage probability equal to the infimum, the confidence coefficient. In some cases this does not matter because the coverage probability will be a constant function of $\theta$. In other cases, however, the coverage probability can be a fairly variable function of $\theta$.

**Example 9.1.6 (Scale uniform interval estimator)** Let $X_1, \ldots, X_n$ be a random sample from a uniform$(0, \theta)$ population and let $Y = \max\{X_1, \ldots, X_n\}$. We are interested in an interval estimator of $\theta$. We consider two candidate estimators: $[aY, bY]$, $1 \leq a < b$, and $[Y + c, Y + d]$, $0 \leq c < d$, where $a, b, c,$ and $d$ are specified constants. (Note that $\theta$ is necessarily larger than $y$.) For the first interval we have

$$P_\theta(\theta \in [aY, bY]) = P_\theta(aY \leq \theta \leq bY)$$

$$= P_\theta\left(\frac{1}{b} \leq \frac{Y}{\theta} \leq \frac{1}{a}\right)$$

$$= P_\theta\left(\frac{1}{b} \leq T \leq \frac{1}{a}\right). \qquad (T = Y/\theta)$$

We previously saw (Example 7.3.13) that $f_Y(y) = n y^{n-1}/\theta^n$, $0 \leq y \leq \theta$, so the pdf of $T$ is $f_T(t) = n t^{n-1}$, $0 \leq t \leq 1$. We therefore have

$$P_\theta\left(\frac{1}{b} \leq T \leq \frac{1}{a}\right) = \int_{1/b}^{1/a} n t^{n-1}\, dt = \left(\frac{1}{a}\right)^n - \left(\frac{1}{b}\right)^n.$$

The coverage probability of the first interval is independent of the value of $\theta$, and thus $(\frac{1}{a})^n - (\frac{1}{b})^n$ is the confidence coefficient of the interval.

For the other interval, for $\theta \geq d$ a similar calculation yields

$$P_\theta(\theta \in [Y + c, Y + d]) = P_\theta(Y + c \leq \theta \leq Y + d)$$

$$= P_\theta\left(1 - \frac{d}{\theta} \leq T \leq 1 - \frac{c}{\theta}\right) \qquad (T = Y/\theta)$$

$$= \int_{1-d/\theta}^{1-c/\theta} n t^{n-1}\, dt$$

$$= \left(1 - \frac{c}{\theta}\right)^n - \left(1 - \frac{d}{\theta}\right)^n.$$

In this case, the coverage probability depends on $\theta$. Furthermore, it is straightforward to calculate that for any constants $c$ and $d$,

$$\lim_{\theta \to \infty} \left(1 - \frac{c}{\theta}\right)^n - \left(1 - \frac{d}{\theta}\right)^n = 0,$$

showing that the confidence coefficient of this interval estimator is 0.                                     ∥

## 9.2 Methods of Finding Interval Estimators

We present four subsections of methods of finding estimators. This might seem to indicate that there are four different methods for finding interval estimators. This is really not so; in fact, operationally all of the methods presented in the next four subsections are the same, being based on the strategy of inverting a test statistic. The last subsection, dealing with Bayesian intervals, presents a different construction method.

### 9.2.1 Inverting a Test Statistic

There is a very strong correspondence between hypothesis testing and interval estimation. In fact, we can say in general that every confidence set corresponds to a test and vice versa. Consider the following example.

**Example 9.2.1 (Inverting a normal test)**    Let $X_1, \ldots, X_n$ be iid n$(\mu, \sigma^2)$ and consider testing $H_0$: $\mu = \mu_0$ versus $H_1$: $\mu \neq \mu_0$. For a fixed $\alpha$ level, a reasonable test (in fact, the most powerful unbiased test) has rejection region $\{\mathbf{x}: |\bar{x} - \mu_0| > z_{\alpha/2}\sigma/\sqrt{n}\}$. Note that $H_0$ is accepted for sample points with $|\bar{x} - \mu_0| \leq z_{\alpha/2}\sigma/\sqrt{n}$ or, equivalently,

$$\bar{x} - z_{\alpha/2}\frac{\sigma}{\sqrt{n}} \leq \mu_0 \leq \bar{x} + z_{\alpha/2}\frac{\sigma}{\sqrt{n}}.$$

Since the test has size $\alpha$, this means that $P(H_0 \text{ is rejected}|\mu = \mu_0) = \alpha$ or, stated in another way, $P(H_0 \text{ is accepted}|\mu = \mu_0) = 1 - \alpha$. Combining this with the above characterization of the acceptance region, we can write

$$P\left(\bar{X} - z_{\alpha/2}\frac{\sigma}{\sqrt{n}} \leq \mu_0 \leq \bar{X} + z_{\alpha/2}\frac{\sigma}{\sqrt{n}}\Big| \mu = \mu_0\right) = 1 - \alpha.$$

But this probability statement is true *for every* $\mu_0$. Hence, the statement

$$P_\mu\left(\bar{X} - z_{\alpha/2}\frac{\sigma}{\sqrt{n}} \leq \mu \leq \bar{X} + z_{\alpha/2}\frac{\sigma}{\sqrt{n}}\right) = 1 - \alpha$$

is true. The interval $[\bar{x} - z_{\alpha/2}\sigma/\sqrt{n}, \ \bar{x} + z_{\alpha/2}\sigma/\sqrt{n}]$, obtained by *inverting* the acceptance region of the level $\alpha$ test, is a $1 - \alpha$ confidence interval.                         ∥

We have illustrated the correspondence between confidence sets and tests. The acceptance region of the hypothesis test, the set in the *sample space* for which

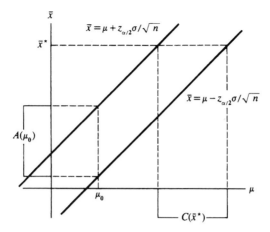

Figure 9.2.1. *Relationship between confidence intervals and acceptance regions for tests. The upper line is $\bar{x} = \mu + z_{\alpha/2}\sigma/\sqrt{n}$ and the lower line is $\bar{x} = \mu - z_{\alpha/2}\sigma/\sqrt{n}$.*

$H_0\colon \mu = \mu_0$ is accepted, is given by

$$A(\mu_0) = \left\{ (x_1, \ldots, x_n) \colon \mu_0 - z_{\alpha/2}\frac{\sigma}{\sqrt{n}} \leq \bar{x} \leq \mu_0 + z_{\alpha/2}\frac{\sigma}{\sqrt{n}} \right\},$$

and the *confidence interval*, the set in the *parameter space* with plausible values of $\mu$, is given by

$$C(x_1, \ldots, x_n) = \left\{ \mu \colon \bar{x} - z_{\alpha/2}\frac{\sigma}{\sqrt{n}} \leq \mu \leq \bar{x} + z_{\alpha/2}\frac{\sigma}{\sqrt{n}} \right\}.$$

These sets are connected to each other by the tautology

$$(x_1, \ldots, x_n) \in A(\mu_0) \Leftrightarrow \mu_0 \in C(x_1, \ldots, x_n).$$

The correspondence between testing and interval estimation for the two-sided normal problem is illustrated in Figure 9.2.1. There it is, perhaps, more easily seen that both tests and intervals ask the same question, but from a slightly different perspective. Both procedures look for consistency between sample statistics and population parameters. The hypothesis test fixes the parameter and asks what sample values (the acceptance region) are consistent with that fixed value. The confidence set fixes the sample value and asks what parameter values (the confidence interval) make this sample value most plausible.

The correspondence between acceptance regions of tests and confidence sets holds in general. The next theorem gives a formal version of this correspondence.

**Theorem 9.2.2**  *For each $\theta_0 \in \Theta$, let $A(\theta_0)$ be the acceptance region of a level $\alpha$ test of $H_0\colon \theta = \theta_0$. For each $\mathbf{x} \in \mathcal{X}$, define a set $C(\mathbf{x})$ in the parameter space by*

(9.2.1)                    $$C(\mathbf{x}) = \{\theta_0 \colon \mathbf{x} \in A(\theta_0)\}.$$

*Then the random set $C(\mathbf{X})$ is a $1 - \alpha$ confidence set. Conversely, let $C(\mathbf{X})$ be a $1 - \alpha$ confidence set. For any $\theta_0 \in \Theta$, define*

$$A(\theta_0) = \{\mathbf{x} \colon \theta_0 \in C(\mathbf{x})\}.$$

*Then $A(\theta_0)$ is the acceptance region of a level $\alpha$ test of $H_0 \colon \theta = \theta_0$.*

**Proof:** For the first part, since $A(\theta_0)$ is the acceptance region of a level $\alpha$ test,

$$P_{\theta_0}(\mathbf{X} \notin A(\theta_0)) \leq \alpha \quad \text{and hence} \quad P_{\theta_0}(\mathbf{X} \in A(\theta_0)) \geq 1 - \alpha.$$

Since $\theta_0$ is arbitrary, write $\theta$ instead of $\theta_0$. The above inequality, together with (9.2.1), shows that the coverage probability of the set $C(\mathbf{X})$ is given by

$$P_\theta(\theta \in C(\mathbf{X})) = P_\theta(\mathbf{X} \in A(\theta)) \geq 1 - \alpha,$$

showing that $C(\mathbf{X})$ is a $1 - \alpha$ confidence set.

For the second part, the Type I Error probability for the test of $H_0 \colon \theta = \theta_0$ with acceptance region $A(\theta_0)$ is

$$P_{\theta_0}(\mathbf{X} \notin A(\theta_0)) = P_{\theta_0}(\theta_0 \notin C(\mathbf{X})) \leq \alpha.$$

So this is a level $\alpha$ test.                                                                 □

Although it is common to talk about inverting a test to obtain a confidence set, Theorem 9.2.2 makes it clear that we really have a family of tests, one for each value of $\theta_0 \in \Theta$, that we invert to obtain one confidence set.

The fact that tests can be inverted to obtain a confidence set and vice versa is theoretically interesting, but the really useful part of Theorem 9.2.2 is the first part. It is a relatively easy task to construct a level $\alpha$ acceptance region. The difficult task is constructing a confidence set. So the method of obtaining a confidence set by inverting an acceptance region is quite useful. All of the techniques we have for obtaining tests can immediately be applied to constructing confidence sets.

In Theorem 9.2.2, we stated only the null hypothesis $H_0 \colon \theta = \theta_0$. All that is required of the acceptance region is

$$P_{\theta_0}(\mathbf{X} \in A(\theta_0)) \geq 1 - \alpha.$$

In practice, when constructing a confidence set by test inversion, we will also have in mind an alternative hypothesis such as $H_1 \colon \theta \neq \theta_0$ or $H_1 \colon \theta > \theta_0$. The alternative will dictate the form of $A(\theta_0)$ that is reasonable, and the form of $A(\theta_0)$ will determine the shape of $C(\mathbf{x})$. Note, however, that we carefully used the word *set* rather than *interval*. This is because there is no guarantee that the confidence set obtained by test inversion will be an interval. In most cases, however, one-sided tests give one-sided intervals, two-sided tests give two-sided intervals, strange-shaped acceptance regions give strange-shaped confidence sets. Later examples will exhibit this.

The properties of the inverted test also carry over (sometimes suitably modified) to the confidence set. For example, unbiased tests, when inverted, will produce unbiased confidence sets. Also, and more important, since we know that we can confine

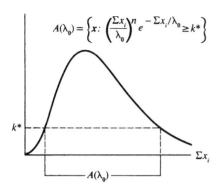
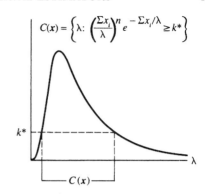

Figure 9.2.2. *Acceptance region and confidence interval for Example 9.2.3. The acceptance region is* $A(\lambda_0) = \left\{ \mathbf{x} : \left( \sum_i x_i/\lambda_0 \right)^n e^{-\sum_i x_i/\lambda_0} \geq k^* \right\}$ *and the confidence region is* $C(\mathbf{x}) = \left\{ \lambda : \left( \sum_i x_i/\lambda \right)^n e^{-\sum_i x_i/\lambda} \geq k^* \right\}$.

attention to sufficient statistics when looking for a good test, it follows that we can confine attention to sufficient statistics when looking for good confidence sets.

The method of test inversion really is most helpful in situations where our intuition deserts us and we have no good idea as to what would constitute a reasonable set. We merely fall back on our all-purpose method for constructing a reasonable test.

**Example 9.2.3 (Inverting an LRT)**  Suppose that we want a confidence interval for the mean, $\lambda$, of an exponential$(\lambda)$ population. We can obtain such an interval by inverting a level $\alpha$ test of $H_0 : \lambda = \lambda_0$ versus $H_1 : \lambda \neq \lambda_0$.

If we take a random sample $X_1, \ldots, X_n$, the LRT statistic is given by

$$\frac{\frac{1}{\lambda_0^n} e^{-\Sigma x_i/\lambda_0}}{\sup_\lambda \frac{1}{\lambda^n} e^{-\Sigma x_i/\lambda}} = \frac{\frac{1}{\lambda_0^n} e^{-\Sigma x_i/\lambda_0}}{\frac{1}{\left(\sum x_i/n\right)^n} e^{-n}} = \left( \frac{\sum x_i}{n\lambda_0} \right)^n e^n e^{-\Sigma x_i/\lambda_0}.$$

For fixed $\lambda_0$, the acceptance region is given by

(9.2.2) $$A(\lambda_0) = \left\{ \mathbf{x} : \left( \frac{\sum x_i}{\lambda_0} \right)^n e^{-\Sigma x_i/\lambda_0} \geq k^* \right\},$$

where $k^*$ is a constant chosen to satisfy $P_{\lambda_0}(\mathbf{X} \in A(\lambda_0)) = 1-\alpha$. (The constant $e^n/n^n$ has been absorbed into $k^*$.) This is a set in the sample space as shown in Figure 9.2.2. Inverting this acceptance region gives the $1 - \alpha$ confidence set

$$C(\mathbf{x}) = \left\{ \lambda : \left( \frac{\sum x_i}{\lambda} \right)^n e^{-\Sigma x_i/\lambda} \geq k^* \right\}.$$

This is an interval in the parameter space as shown in Figure 9.2.2.

The expression defining $C(\mathbf{x})$ depends on $\mathbf{x}$ only through $\sum x_i$. So the confidence interval can be expressed in the form

(9.2.3) $$C(\textstyle\sum x_i) = \{\lambda : L(\textstyle\sum x_i) \leq \lambda \leq U(\textstyle\sum x_i)\},$$

where $L$ and $U$ are functions determined by the constraints that the set (9.2.2) has probability $1 - \alpha$ and

(9.2.4)
$$\left( \frac{\sum x_i}{L(\sum x_i)} \right)^n e^{-\Sigma x_i/L(\Sigma x_i)} = \left( \frac{\sum x_i}{U(\sum x_i)} \right)^n e^{-\Sigma x_i/U(\Sigma x_i)}.$$

If we set

(9.2.5)
$$\frac{\sum x_i}{L(\sum x_i)} = a \quad \text{and} \quad \frac{\sum x_i}{U(\sum x_i)} = b,$$

where $a > b$ are constants, then (9.2.4) becomes

(9.2.6)
$$a^n e^{-a} = b^n e^{-b},$$

which yields easily to numerical solution. To work out some details, let $n = 2$ and note that $\sum X_i \sim \text{gamma}(2, \lambda)$ and $\sum X_i/\lambda \sim \text{gamma}(2, 1)$. Hence, from (9.2.5), the confidence interval becomes $\{\lambda: \frac{1}{a} \sum x_i \leq \lambda \leq \frac{1}{b} \sum x_i\}$, where $a$ and $b$ satisfy

$$P_\lambda \left( \frac{1}{a} \sum X_i \leq \lambda \leq \frac{1}{b} \sum X_i \right) = P\left( b \leq \frac{\sum X_i}{\lambda} \leq a \right) = 1 - \alpha$$

and, from (9.2.6), $a^2 e^{-a} = b^2 e^{-b}$. Then

$$P\left( b \leq \frac{\sum X_i}{\lambda} \leq a \right) = \int_b^a t e^{-t}\, dt$$

$$= e^{-b}(b + 1) - e^{-a}(a + 1). \qquad \binom{\text{integration}}{\text{by parts}}$$

To get, for example, a 90% confidence interval, we must simultaneously satisfy the probability condition and the constraint. To three decimal places, we get $a = 5.480$, $b = .441$, with a confidence coefficient of .90006. Thus,

$$P_\lambda \left( \frac{1}{5.480} \sum X_i \leq \lambda \leq \frac{1}{.441} \sum X_i \right) = .90006. \qquad \|$$

The region obtained by inverting the LRT of $H_0 : \theta = \theta_0$ versus $H_1 : \theta \neq \theta_0$ (Definition 8.2.1) is of the form

$$\text{accept } H_0 \text{ if } \frac{L(\theta_0|\mathbf{x})}{L(\hat{\theta}|\mathbf{x})} \leq k(\theta_0),$$

with the resulting confidence region

(9.2.7)
$$\{\theta : L(\theta|\mathbf{x}) \geq k'(\mathbf{x}, \theta)\},$$

for some function $k'$ that gives $1 - \alpha$ confidence.

In some cases (such as the normal and the gamma distribution) the function $k'$ will not depend on $\theta$. In such cases the likelihood region has a particularly pleasing interpretation, consisting of those values of $\theta$ for which the likelihood is highest. We will also see such intervals arising from optimality considerations in both the frequentist (Theorem 9.3.2) and Bayesian (Corollary 9.3.10) realms.

The test inversion method is completely general in that we can invert any test and obtain a confidence set. In Example 9.2.3 we inverted LRTs, but we could have used a test constructed by any method. Also, note that the inversion of a two-sided test gave a two-sided interval. In the next examples, we invert one-sided tests to get one-sided intervals.

**Example 9.2.4 (Normal one-sided confidence bound)**   Let $X_1, \ldots, X_n$ be a random sample from a $n(\mu, \sigma^2)$ population. Consider constructing a $1 - \alpha$ upper confidence bound for $\mu$. That is, we want a confidence interval of the form $C(\mathbf{x}) = (-\infty, U(\mathbf{x})]$. To obtain such an interval using Theorem 9.2.2, we will invert one-sided tests of $H_0 : \mu = \mu_0$ versus $H_1 : \mu < \mu_0$. (Note that we use the specification of $H_1$ to determine the form of the confidence interval here. $H_1$ specifies "large" values of $\mu_0$, so the confidence set will contain "small" values, values less than a bound. Thus, we will get an upper confidence bound.) The size $\alpha$ LRT of $H_0$ versus $H_1$ rejects $H_0$ if

$$\frac{\bar{X} - \mu_0}{S/\sqrt{n}} < -t_{n-1,\alpha}$$

(similar to Example 8.2.6). Thus the acceptance region for this test is

$$A(\mu_0) = \left\{ \mathbf{x} : \bar{x} \geq \mu_0 - t_{n-1,\alpha} \frac{s}{\sqrt{n}} \right\}$$

and $\mathbf{x} \in A(\mu_0) \Leftrightarrow \bar{x} + t_{n-1,\alpha} s/\sqrt{n} \geq \mu_0$. According to (9.2.1), we define

$$C(\mathbf{x}) = \{\mu_0 : \mathbf{x} \in A(\mu_0)\} = \left\{ \mu_0 : \bar{x} + t_{n-1,\alpha} \frac{s}{\sqrt{n}} \geq \mu_0 \right\}.$$

By Theorem 9.2.2, the random set $C(\mathbf{X}) = (-\infty, \bar{X} + t_{n-1,\alpha} S/\sqrt{n}]$ is a $1-\alpha$ confidence set for $\mu$. We see that, indeed, it is the right form for an upper confidence bound. Inverting the one-sided test gave a one-sided confidence interval.                    ‖

**Example 9.2.5 (Binomial one-sided confidence bound)**   As a more difficult example of a one-sided confidence interval, consider putting a $1 - \alpha$ lower confidence bound on $p$, the success probability from a sequence of Bernoulli trials. That is, we observe $X_1, \ldots, X_n$, where $X_i \sim \text{Bernoulli}(p)$, and we want the interval to be of the form $(L(x_1, \ldots, x_n), 1]$, where $P_p(p \in (L(X_1, \ldots, X_n), 1]) \geq 1 - \alpha$. (The interval we obtain turns out to be open on the left, as will be seen.)

Since we want a one-sided interval that gives a lower confidence bound, we consider inverting the acceptance regions from tests of

$$H_0 : \quad p = p_0 \quad \text{versus} \quad H_1 : \quad p > p_0.$$

To simplify things, we know that we can base our test on $T = \sum_{i=1}^n X_i \sim \text{bino-}$ mial$(n, p)$, since $T$ is sufficient for $p$. (See the Miscellanea section.) Since the binomial distribution has monotone likelihood ratio (see Exercise 8.25), by the Karlin–Rubin Theorem (Theorem 8.3.17) the test that rejects $H_0$ if $T > k(p_0)$ is the UMP test of its size. For each $p_0$, we want to choose the constant $k(p_0)$ (it can be an integer) so that we have a level $\alpha$ test. We cannot get the size of the test to be exactly $\alpha$, except for certain values of $p_0$, because of the discreteness of $T$. But we choose $k(p_0)$ so that the size of the test is as close to $\alpha$ as possible, without being larger. Thus, $k(p_0)$ is defined to be the integer between 0 and $n$ that simultaneously satisfies the inequalities

(9.2.8)
$$\sum_{y=0}^{k(p_0)} \binom{n}{y} p_0^y (1-p_0)^{n-y} \geq 1-\alpha$$

$$\sum_{y=0}^{k(p_0)-1} \binom{n}{y} p_0^y (1-p_0)^{n-y} < 1-\alpha.$$

Because of the MLR property of the binomial, for any $k = 0, \ldots, n$, the quantity

$$f(p_0|k) = \sum_{y=0}^{k} \binom{n}{y} p_0^y (1-p_0)^{n-y}$$

is a decreasing function of $p_0$ (see Exercise 8.26). Of course, $f(0|0) = 1$, so $k(0) = 0$ and $f(p_0|0)$ remains above $1 - \alpha$ for an interval of values. Then, at some point $f(p_0|0) = 1 - \alpha$ and for values of $p_0$ greater than this value, $f(p_0|0) < 1 - \alpha$. So, at this point, $k(p_0)$ increases to 1. This pattern continues. Thus, $k(p_0)$ is an integer-valued *step-function*. It is constant for a range of $p_0$; then it jumps to the next bigger integer. Since $k(p_0)$ is a nondecreasing function of $p_0$, this gives the lower confidence bound. (See Exercise 9.5 for an upper confidence bound.) Solving the inequalities in (9.2.8) for $k(p_0)$ gives both the acceptance region of the test and the confidence set.

For each $p_0$, the acceptance region is given by $A(p_0) = \{t : t \leq k(p_0)\}$, where $k(p_0)$ satisfies (9.2.8). For each value of $t$, the confidence set is $C(t) = \{p_0 : t \leq k(p_0)\}$. This set, in its present form, however, does not do us much practical good. Although it is formally correct and a $1 - \alpha$ confidence set, it is defined implicitly in terms of $p_0$ and we want it to be defined explicitly in terms of $p_0$.

Since $k(p_0)$ is nondecreasing, for a given observation $T = t, k(p_0) < t$ for all $p_0$ less than or equal to some value, call it $k^{-1}(t)$. At $k^{-1}(t)$, $k(p_0)$ jumps up to equal $t$ and $k(p_0) \geq t$ for all $p_0 > k^{-1}(t)$. (Note that at $p_0 = k^{-1}(t)$, $f(p_0|t - 1) = 1 - \alpha$. So (9.2.8) is still satisfied by $k(p_0) = t - 1$. Only for $p_0 > k^{-1}(t)$ is $k(p_0) \geq t$.) Thus, the confidence set is

(9.2.9)      $$C(t) = \{p_0 : t \leq k(p_0)\} = \{p_0 : p_0 > k^{-1}(t)\},$$

and we have constructed a $1-\alpha$ lower confidence bound of the form $C(T) = (k^{-1}(T), 1]$.

The number $k^{-1}(t)$ can be defined as

(9.2.10)      $$k^{-1}(t) = \sup \left\{ p : \sum_{y=0}^{t-1} \binom{n}{y} p^y (1-p)^{n-y} \geq 1-\alpha \right\}.$$

Realize that $k^{-1}(t)$ is not really an inverse of $k(p_0)$ because $k(p_0)$ is not a one-to-one function. However, the expressions in (9.2.8) and (9.2.10) give us well-defined quantities for $k$ and $k^{-1}$.

The problem of binomial confidence bounds was first treated by Clopper and Pearson (1934), who obtained answers similar to these for the two-sided interval (see Exercise 9.21) and started a line of research that is still active today. See Miscellanea 9.5.2.

$\parallel$

### 9.2.2 Pivotal Quantities

The two confidence intervals that we saw in Example 9.1.6 differed in many respects. One important difference was that the coverage probability of the interval $\{aY, bY\}$ did not depend on the value of the parameter $\theta$, while that of $\{Y + c, Y + d\}$ did. This happened because the coverage probability of $\{aY, bY\}$ could be expressed in terms of the quantity $Y/\theta$, a random variable whose distribution does not depend on the parameter, a quantity known as a *pivotal quantity*, or *pivot*.

The use of pivotal quantities for confidence set construction, resulting in what has been called *pivotal inference*, is mainly due to Barnard (1949, 1980) but can be traced as far back as Fisher (1930), who used the term *inverse probability*. Closely related is D. A. S. Fraser's theory of *structural inference* (Fraser 1968, 1979). An interesting discussion of the strengths and weaknesses of these methods is given in Berger and Wolpert (1984).

**Definition 9.2.6**  A random variable $Q(\mathbf{X}, \theta) = Q(X_1, \ldots, X_n, \theta)$ is a *pivotal quantity* (or *pivot*) if the distribution of $Q(\mathbf{X}, \theta)$ is independent of all parameters. That is, if $\mathbf{X} \sim F(\mathbf{x}|\theta)$, then $Q(\mathbf{X}, \theta)$ has the same distribution for all values of $\theta$.

The function $Q(\mathbf{x}, \theta)$ will usually explicitly contain both parameters and statistics, but for any set $\mathcal{A}$, $P_\theta(Q(\mathbf{X}, \theta) \in \mathcal{A})$ cannot depend on $\theta$. The technique of constructing confidence sets from pivots relies on being able to find a pivot and a set $\mathcal{A}$ so that the set $\{\theta: Q(\mathbf{x}, \theta) \in \mathcal{A}\}$ is a set estimate of $\theta$.

**Example 9.2.7 (Location-scale pivots)**  In location and scale cases there are lots of pivotal quantities. We will show a few here; more will be found in Exercise 9.8. Let $X_1, \ldots, X_n$ be a random sample from the indicated pdfs, and let $\bar{X}$ and $S$ be the sample mean and standard deviation. To prove that the quantities in Table 9.2.1 are pivots, we just have to show that their pdfs are independent of parameters (details in Exercise 9.9). Notice that, in particular, if $X_1, \ldots, X_n$ is a random sample from

Table 9.2.1. *Location-scale pivots*

| Form of pdf | Type of pdf | Pivotal quantity |
|---|---|---|
| $f(x - \mu)$ | Location | $\bar{X} - \mu$ |
| $\frac{1}{\sigma}f(\frac{x}{\sigma})$ | Scale | $\frac{\bar{X}}{\sigma}$ |
| $\frac{1}{\sigma}f(\frac{x-\mu}{\sigma})$ | Location–scale | $\frac{\bar{X}-\mu}{S}$ |

a $n(\mu, \sigma^2)$ population, then the $t$ statistic $(\bar{X} - \mu)/(S/\sqrt{n})$ is a pivot because the $t$ distribution does not depend on the parameters $\mu$ and $\sigma^2$. ‖

Of the intervals constructed in Section 9.2.1 using the test inversion method, some turned out to be based on pivots (Examples 9.2.3 and 9.2.4) and some did not (Example 9.2.5). There is no all-purpose strategy for finding pivots. However, we can be a little clever and not rely totally on guesswork. For example, it is a relatively easy task to find pivots for location or scale parameters. In general, *differences* are pivotal for location problems, while *ratios* (or products) are pivotal for scale problems.

**Example 9.2.8 (Gamma pivot)** Suppose that $X_1, \ldots, X_n$ are iid exponential($\lambda$). Then $T = \sum X_i$ is a sufficient statistic for $\lambda$ and $T \sim$ gamma($n, \lambda$). In the gamma pdf $t$ and $\lambda$ appear together as $t/\lambda$ and, in fact the gamma($n, \lambda$) pdf $(\Gamma(n)\lambda^n)^{-1} t^{n-1} e^{-t/\lambda}$ is a scale family. Thus, if $Q(T, \lambda) = 2T/\lambda$, then

$$Q(T, \lambda) \sim \text{gamma}(n, \lambda(2/\lambda)) = \text{gamma}(n, 2),$$

which does not depend on $\lambda$. The quantity $Q(T, \lambda) = 2T/\lambda$ is a pivot with a gamma($n, 2$), or $\chi^2_{2n}$, distribution. ‖

We can sometimes look to the form of the pdf to see if a pivot exists. In the above example, the quantity $t/\lambda$ appeared in the pdf and this turned out to be a pivot. In the normal pdf, the quantity $(\bar{x} - \mu)/\sigma$ appears and this quantity is also a pivot. In general, suppose the pdf of a statistic $T$, $f(t|\theta)$, can be expressed in the form

(9.2.11)
$$f(t|\theta) = g\left(Q(t, \theta)\right) \left| \frac{\partial}{\partial t} Q(t, \theta) \right|$$

for some function $g$ and some monotone function $Q$ (monotone in $t$ for each $\theta$). Then Theorem 2.1.5 can be used to show that $Q(T, \theta)$ is a pivot (see Exercise 9.10).

Once we have a pivot, how do we use it to construct a confidence set? That part is really quite simple. If $Q(\mathbf{X}, \theta)$ is a pivot, then for a specified value of $\alpha$ we can find numbers $a$ and $b$, which do not depend on $\theta$, to satisfy

$$P_\theta(a \leq Q(\mathbf{X}, \theta) \leq b) \geq 1 - \alpha.$$

Then, for each $\theta_0 \in \Theta$,

(9.2.12)
$$A(\theta_0) = \{\mathbf{x} \colon a \leq Q(\mathbf{x}, \theta_0) \leq b\}$$

is the acceptance region for a level $\alpha$ test of $H_0 \colon \theta = \theta_0$. We will use the test inversion method to construct the confidence set, but we are using the pivot to specify the specific form of our acceptance regions. Using Theorem 9.2.2, we invert these tests to obtain

(9.2.13)
$$C(\mathbf{x}) = \{\theta_0 \colon a \leq Q(\mathbf{x}, \theta_0) \leq b\},$$

and $C(\mathbf{X})$ is a $1 - \alpha$ confidence set for $\theta$. If $\theta$ is a real-valued parameter and if, for each $\mathbf{x} \in \mathcal{X}, Q(\mathbf{x}, \theta)$ is a monotone function of $\theta$, then $C(\mathbf{x})$ will be an interval. In fact, if

$Q(\mathbf{x}, \theta)$ is an increasing function of $\theta$, then $C(\mathbf{x})$ has the form $L(\mathbf{x}, a) \leq \theta \leq U(\mathbf{x}, b)$. If $Q(\mathbf{x}, \theta)$ is a decreasing function of $\theta$ (which is typical), then $C(\mathbf{x})$ has the form $L(\mathbf{x}, b) \leq \theta \leq U(\mathbf{x}, a)$.

**Example 9.2.9 (Continuation of Example 9.2.8)** In Example 9.2.3 we obtained a confidence interval for the mean, $\lambda$, of the exponential($\lambda$) pdf by inverting a level $\alpha$ LRT of $H_0: \lambda = \lambda_0$ versus $H_1: \lambda \neq \lambda_0$. Now we also see that if we have a sample $X_1, \ldots, X_n$, we can define $T = \sum X_i$ and $Q(T, \lambda) = 2T/\lambda \sim \chi^2_{2n}$.

If we choose constants $a$ and $b$ to satisfy $P(a \leq \chi^2_{2n} \leq b) = 1 - \alpha$, then

$$P_\lambda\left(a \leq \frac{2T}{\lambda} \leq b\right) = P_\lambda(a \leq Q(T, \lambda) \leq b) = P\left(a \leq \chi^2_{2n} \leq b\right) = 1 - \alpha.$$

Inverting the set $A(\lambda) = \{t: a \leq \frac{2t}{\lambda} \leq b\}$ gives $C(t) = \{\lambda: \frac{2t}{b} \leq \lambda \leq \frac{2t}{a}\}$, which is a $1 - \alpha$ confidence interval. (Notice that the lower endpoint depends on $b$ and the upper endpoint depends on $a$, as mentioned above. $Q(t, \lambda) = 2t/\lambda$ is decreasing in $\lambda$.) For example, if $n = 10$, then consulting a table of $\chi^2$ cutoffs shows that a 95% confidence interval is given by $\{\lambda: \frac{2T}{34.17} \leq \lambda \leq \frac{2T}{9.59}\}$.          ‖

For the location problem, even if the variance is unknown, construction and calculation of pivotal intervals are quite easy. In fact, we have used these ideas already but have not called them by any formal name.

**Example 9.2.10 (Normal pivotal interval)** It follows from Theorem 5.3.1 that if $X_1, \ldots, X_n$ are iid $n(\mu, \sigma^2)$, then $(\bar{X} - \mu)/(\sigma/\sqrt{n})$ is a pivot. If $\sigma^2$ is known, we can use this pivot to calculate a confidence interval for $\mu$. For any constant $a$,

$$P\left(-a \leq \frac{\bar{X} - \mu}{\sigma/\sqrt{n}} \leq a\right) = P(-a \leq Z \leq a), \quad (Z \text{ is standard normal})$$

and (by now) familiar algebraic manipulations give us the confidence interval

$$\left\{\mu: \bar{x} - a\frac{\sigma}{\sqrt{n}} \leq \mu \leq \bar{x} + a\frac{\sigma}{\sqrt{n}}\right\}.$$

If $\sigma^2$ is unknown, we can use the location–scale pivot $\frac{\bar{X} - \mu}{S/\sqrt{n}}$. Since $\frac{\bar{X} - \mu}{S/\sqrt{n}}$ has Student's $t$ distribution,

$$P\left(-a \leq \frac{\bar{X} - \mu}{S/\sqrt{n}} \leq a\right) = P(-a \leq T_{n-1} \leq a).$$

Thus, for any given $\alpha$, if we take $a = t_{n-1, \alpha/2}$, we find that a $1 - \alpha$ confidence interval is given by

(9.2.14)          $$\left\{\mu: \bar{x} - t_{n-1, \alpha/2}\frac{s}{\sqrt{n}} \leq \mu \leq \bar{x} + t_{n-1, \alpha/2}\frac{s}{\sqrt{n}}\right\},$$

which is the classic $1 - \alpha$ confidence interval for $\mu$ based on Student's $t$ distribution.

Continuing with this case, suppose that we also want an interval estimate for $\sigma$. Because $(n-1)S^2/\sigma^2 \sim \chi^2_{n-1}$, $(n-1)S^2/\sigma^2$ is also a pivot. Thus, if we choose $a$ and $b$ to satisfy

$$P\left(a \leq \frac{(n-1)S^2}{\sigma^2} \leq b\right) = P\left(a \leq \chi^2_{n-1} \leq b\right) = 1 - \alpha,$$

we can invert this set to obtain the $1 - \alpha$ confidence interval

$$\left\{\sigma^2 : \frac{(n-1)s^2}{b} \leq \sigma^2 \leq \frac{(n-1)s^2}{a}\right\}$$

or, equivalently,

$$\left\{\sigma : \sqrt{\frac{(n-1)s^2}{b}} \leq \sigma \leq \sqrt{\frac{(n-1)s^2}{a}}\right\}.$$

One choice of $a$ and $b$ that will produce the required interval is $a = \chi^2_{n-1,1-\alpha/2}$ and $b = \chi^2_{n-1,\alpha/2}$. This choice splits the probability equally, putting $\alpha/2$ in each tail of the distribution. The $\chi^2_{n-1}$ distribution, however, is a skewed distribution and it is not immediately clear that an equal probability split is optimal for a skewed distribution. (It is not immediately clear that an equal probability split is optimal for a symmetric distribution, but our intuition makes this latter case more plausible.) In fact, for the chi squared distribution, the equal probability split is not optimal, as will be seen in Section 9.3. (See also Exercise 9.52.)

One final note for this problem. We now have constructed confidence intervals for $\mu$ and $\sigma$ separately. It is entirely plausible that we would be interested in a confidence set for $\mu$ and $\sigma$ *simultaneously*. The Bonferroni Inequality is an easy (and relatively good) method for accomplishing this. (See Exercise 9.14.)                                                        ‖

### 9.2.3 Pivoting the CDF

In previous section we saw that a pivot, $Q$, leads to a confidence set of the form (9.2.13), that is

$$C(\mathbf{x}) = \{\theta_0 : a \leq Q(\mathbf{x}, \theta_0) \leq b\}.$$

If, for every $\mathbf{x}$, the function $Q(\mathbf{x}, \theta)$ is a monotone function of $\theta$, then the confidence set $C(\mathbf{x})$ is guaranteed to be an interval. The pivots that we have seen so far, which were mainly constructed using location and scale transformations, resulted in monotone $Q$ functions and, hence, confidence intervals.

In this section we work with another pivot, one that is totally general and, with minor assumptions, will guarantee an interval.

If in doubt, or in a strange situation, we would recommend constructing a confidence set based on inverting an LRT, if possible. Such a set, although not guaranteed to be optimal, will never be very bad. However, in some cases such a tactic is too difficult, either analytically or computationally; inversion of the acceptance region can sometimes be quite a chore. If the method of this section can be applied, it is rather straightforward to implement and will usually produce a set that is reasonable.

To illustrate the type of trouble that could arise from the test inversion method, without extra conditions on the exact types of acceptance regions used, consider the following example, which illustrates one of the early methods of constructing confidence sets for a binomial success probability.

**Example 9.2.11 (Shortest length binomial set)** Sterne (1954) proposed the following method for constructing binomial confidence sets, a method that produces a set with the shortest length. Given $\alpha$, for each value of $p$ find the size $\alpha$ acceptance region composed of the most probable $x$ values. That is, for each $p$, order the $x = 0, \ldots, n$ values from the most probable to the least probable and put values into the acceptance region $A(p)$ until it has probability $1 - \alpha$. Then use (9.2.1) to invert these acceptance regions to get a $1 - \alpha$ confidence set, which Sterne claimed had length optimality properties.

To see the unexpected problems with this seemingly reasonable construction, consider a small example. Let $X \sim$ binomial$(3, p)$ and use confidence coefficient $1 - \alpha = .442$. Table 9.2.2 gives the acceptance regions obtained by the Sterne construction and the confidence sets derived by inverting this family of tests.

Surprisingly, the confidence set is not a confidence *interval*. This seemingly reasonable construction has led us to an unreasonable procedure. The blame is to be put on the pmf, as it does not behave as we expect. (See Exercise 9.18.)          ‖

We base our confidence interval construction for a parameter $\theta$ on a real-valued statistic $T$ with cdf $F_T(t|\theta)$. (In practice we would usually take $T$ to be a sufficient statistic for $\theta$, but this is not necessary for the following theory to go through.) We will first assume that $T$ is a continuous random variable. The situation where $T$ is discrete is similar but has a few additional technical details to consider. We, therefore, state the discrete case in a separate theorem.

First of all, recall Theorem 2.1.10, the Probability Integral Transformation, which tells us that the random variable $F_T(T|\theta)$ is uniform$(0, 1)$, a pivot. Thus, if $\alpha_1 + \alpha_2 =$

Table 9.2.2. *Acceptance region and confidence set for Sterne's construction,* $X \sim$ binomial$(3, p)$ *and* $1 - \alpha = .442$

| $p$ | Acceptance region = $A(p)$ | $x$ | Confidence set = $C(x)$ |
|---|---|---|---|
| [.000, .238] | {0} | | |
| (.238, .305) | {0, 1} | 0 | [.000, .305) $\cup$ (.362, .366) |
| [.305, .362] | {1} | | |
| (.362, .366) | {0, 1} | 1 | (.238, .634] |
| [.366, .634] | {1, 2} | | |
| (.634, .638) | {2, 3} | 2 | [.366, .762) |
| [.638, .695] | {2} | | |
| (.695, .762) | {2, 3} | 3 | (.634, .638) $\cup$ (.695, 1.00] |
| [.762, 1.00] | {3} | | |

$\alpha$, an $\alpha$-level acceptance region of the hypothesis $H_0 : \theta = \theta_0$ is (see Exercise 9.11)

$$\{t : \alpha_1 \le F_T(t|\theta_0) \le 1 - \alpha_2\},$$

with associated confidence set

$$\{\theta : \alpha_1 \le F_T(t|\theta) \le 1 - \alpha_2\}.$$

Now to guarantee that the confidence set is an interval, we need to have $F_T(t|\theta)$ to be monotone in $\theta$. But we have seen this already, in the definitions of *stochastically increasing* and *stochastically decreasing*. (See the Miscellanea section of Chapter 8 and Exercise 8.26, or Exercises 3.41–3.43.) A family of cdfs $F(t|\theta)$ is *stochastically increasing in $\theta$* (*stochastically decreasing in $\theta$*) if, for each $t \in \mathcal{T}$, the sample space of $T$, $F(t|\theta)$ is a decreasing (increasing) function of $\theta$. In what follows, we need only the fact that $F$ is monotone, either increasing or decreasing. The more statistical concepts of stochastic increasing or decreasing merely serve as interpretational tools.

**Theorem 9.2.12 (Pivoting a continuous cdf)** *Let $T$ be a statistic with continuous cdf $F_T(t|\theta)$. Let $\alpha_1 + \alpha_2 = \alpha$ with $0 < \alpha < 1$ be fixed values. Suppose that for each $t \in \mathcal{T}$, the functions $\theta_L(t)$ and $\theta_U(t)$ can be defined as follows.*

i. *If $F_T(t|\theta)$ is a decreasing function of $\theta$ for each $t$, define $\theta_L(t)$ and $\theta_U(t)$ by*

$$F_T(t|\theta_U(t)) = \alpha_1, \quad F_T(t|\theta_L(t)) = 1 - \alpha_2.$$

ii. *If $F_T(t|\theta)$ is an increasing function of $\theta$ for each $t$, define $\theta_L(t)$ and $\theta_U(t)$ by*

$$F_T(t|\theta_U(t)) = 1 - \alpha_2, \quad F_T(t|\theta_L(t)) = \alpha_1.$$

*Then the random interval $[\theta_L(T), \theta_U(T)]$ is a $1 - \alpha$ confidence interval for $\theta$.*

**Proof:** We will prove only part (i). The proof of part (ii) is similar and is left as Exercise 9.19.

Assume that we have constructed the $1 - \alpha$ acceptance region

$$\{t : \alpha_1 \le F_T(t|\theta_0) \le 1 - \alpha_2\}.$$

Since $F_T(t|\theta)$ is a decreasing function of $\theta$ for each $t$ and $1 - \alpha_2 > \alpha_1$, $\theta_L(t) < \theta_U(t)$, and the values $\theta_L(t)$ and $\theta_U(t)$ are unique. Also,

$$F_T(t|\theta) < \alpha_1 \Leftrightarrow \theta > \theta_U(t),$$

$$F_T(t|\theta) > 1 - \alpha_2 \Leftrightarrow \theta < \theta_L(t),$$

and hence $\{\theta : \alpha_1 \le F_T(t|\theta) \le 1 - \alpha_2\} = \{\theta : \theta_L(T) \le \theta \le \theta_U(T)\}.$ $\qquad\square$

We note that, in the absence of additional information, it is common to choose $\alpha_1 = \alpha_2 = \alpha/2$. Although this may not always be optimal (see Theorem 9.3.2), it is certainly a reasonable strategy in most situations. If a one-sided interval is desired, however, this can easily be achieved by choosing either $\alpha_1$ or $\alpha_2$ equal to 0.

The equations for the stochastically increasing case,

$$(9.2.15) \qquad F_T(t|\theta_U(t)) = \alpha_1, \quad F_T(t|\theta_L(t)) = 1 - \alpha_2,$$

can also be expressed in terms of the pdf of the statistic $T$. The functions $\theta_U(t)$ and $\theta_L(t)$ can be defined to satisfy

$$\int_{-\infty}^t f_T(u|\theta_U(t))\,du = \alpha_1 \quad \text{and} \quad \int_t^\infty f_T(u|\theta_L(t))\,du = \alpha_2.$$

A similar set of equations holds for the stochastically decreasing case.

**Example 9.2.13 (Location exponential interval)** This method can be used to get a confidence interval for the location exponential pdf. (In Exercise 9.25 the answer here is compared to that obtained by likelihood and pivotal methods. See also Exercise 9.41.)

If $X_1, \ldots, X_n$ are iid with pdf $f(x|\mu) = e^{-(x-\mu)}I_{[\mu,\infty)}(x)$, then $Y = \min\{X_1, \ldots, X_n\}$ is sufficient for $\mu$ with pdf

$$f_Y(y|\mu) = ne^{-n(y-\mu)}I_{[\mu,\infty)}(y).$$

Fix $\alpha$ and define $\mu_L(y)$ and $\mu_U(y)$ to satisfy

$$\int_{\mu_U(y)}^y ne^{-n(u-\mu_U(y))}\,du = \frac{\alpha}{2}, \quad \int_y^\infty ne^{-n(u-\mu_L(y))}\,du = \frac{\alpha}{2}.$$

These integrals can be evaluated to give the equations

$$1 - e^{-n(y-\mu_U(y))} = \frac{\alpha}{2}, \quad e^{-n(y-\mu_L(y))} = \frac{\alpha}{2},$$

which give us the solutions

$$\mu_U(y) = y + \frac{1}{n}\log\left(1 - \frac{\alpha}{2}\right), \quad \mu_L(y) = y + \frac{1}{n}\log\left(\frac{\alpha}{2}\right).$$

Hence, the random interval

$$C(Y) = \left\{\mu: Y + \frac{1}{n}\log\left(\frac{\alpha}{2}\right) \le \mu \le Y + \frac{1}{n}\log\left(1 - \frac{\alpha}{2}\right)\right\},$$

a $1 - \alpha$ confidence interval for $\mu$. ∥

Note two things about the use of this method. First, the actual equations (9.2.15) need to be solved only for the value of the statistics actually observed. If $T = t_0$ is observed, then the realized confidence interval on $\theta$ will be $[\theta_L(t_0), \theta_U(t_0)]$. Thus, we need to solve only the two equations

$$\int_{-\infty}^{t_0} f_T(u|\theta_U(t_0))\,du = \alpha_1 \quad \text{and} \quad \int_{t_0}^\infty f_T(u|\theta_L(t_0))\,du = \alpha_2$$

for $\theta_L(t_0)$ and $\theta_U(t_0)$. Second, realize that even if these equations cannot be solved analytically, we really only need to solve them numerically since the proof that we have a $1 - \alpha$ confidence interval did not require an analytic solution.

We now consider the discrete case.

**Theorem 9.2.14 (Pivoting a discrete cdf)** *Let $T$ be a discrete statistic with cdf $F_T(t|\theta) = P(T \le t|\theta)$. Let $\alpha_1 + \alpha_2 = \alpha$ with $0 < \alpha < 1$ be fixed values. Suppose that for each $t \in \mathcal{T}$, $\theta_L(t)$ and $\theta_U(t)$ can be defined as follows.*

i. *If $F_T(t|\theta)$ is a decreasing function of $\theta$ for each $t$, define $\theta_L(t)$ and $\theta_U(t)$ by*

$$P(T \le t|\theta_U(t)) = \alpha_1, \quad P(T \ge t|\theta_L(t)) = \alpha_2.$$

ii. *If $F_T(t|\theta)$ is an increasing function of $\theta$ for each $t$, define $\theta_L(t)$ and $\theta_U(t)$ by*

$$P(T \ge t|\theta_U(t)) = \alpha_1, \quad P(T \le t|\theta_L(t)) = \alpha_2.$$

*Then the random interval $[\theta_L(T), \theta_U(T)]$ is a $1 - \alpha$ confidence interval for $\theta$.*

**Proof:** We will only sketch the proof of part (i). The details, as well as the proof of part (ii), are left to Exercise 9.20.

First recall Exercise 2.10, where it was shown that $F_T(T|\theta)$ is stochastically greater than a uniform random variable, that is, $P_\theta(F_T(T|\theta) \le x) \le x$. Furthermore, this property is shared by $\bar{F}_T(T|\theta) = P(T \ge t|\theta)$, and this implies that the set

$$\{\theta : F_T(T|\theta) \le \alpha_1 \text{ and } \bar{F}_T(T|\theta) \le \alpha_2\}$$

is a $1 - \alpha$ confidence set.

The fact that $F_T(t|\theta)$ is a decreasing function of $\theta$ for each $t$ implies that $\bar{F}(t|\theta)$ is a nondecreasing function of $\theta$ for each $t$. It therefore follows that

$$\theta > \theta_U(t) \Rightarrow F_T(t|\theta) < \tfrac{\alpha}{2},$$

$$\theta < \theta_L(t) \Rightarrow \bar{F}_T(t|\theta) < \tfrac{\alpha}{2},$$

and hence $\{\theta : F_T(T|\theta) \le \alpha_1 \text{ and } \bar{F}_T(T|\theta) \le \alpha_2\} = \{\theta : \theta_L(T) \le \theta \le \theta_U(T)\}$. $\quad\square$

We close this section with an example to illustrate the construction of Theorem 9.2.14. Notice that an alternative interval can be constructed by inverting an LRT (see Exercise 9.23).

**Example 9.2.15 (Poisson interval estimator)**   Let $X_1, \ldots, X_n$ be a random sample from a Poisson population with parameter $\lambda$ and define $Y = \sum X_i$. $Y$ is sufficient for $\lambda$ and $Y \sim \text{Poisson}(n\lambda)$. Applying the above method with $\alpha_1 = \alpha_2 = \alpha/2$, if $Y = y_0$ is observed, we are led to solve for $\lambda$ in the equations

(9.2.16) $$\sum_{k=0}^{y_0} e^{-n\lambda} \frac{(n\lambda)^k}{k!} = \frac{\alpha}{2} \quad \text{and} \quad \sum_{k=y_0}^{\infty} e^{-n\lambda} \frac{(n\lambda)^k}{k!} = \frac{\alpha}{2}.$$

Recall the identity, from Example 3.3.1, linking the Poisson and gamma families. Applying that identity to the sums in (9.2.16), we can write (remembering that $y_0$ is the *observed* value of $Y$)

$$\frac{\alpha}{2} = \sum_{k=0}^{y_0} e^{-n\lambda} \frac{(n\lambda)^k}{k!} = P(Y \le y_0|\lambda) = P\left(\chi^2_{2(y_0+1)} > 2n\lambda\right),$$

where $\chi^2_{2(y_0+1)}$ is a chi squared random variable with $2(y_0+1)$ degrees of freedom. Thus, the solution to the above equation is to take

$$\lambda = \frac{1}{2n}\chi^2_{2(y_0+1),\alpha/2}.$$

Similarly, applying the identity to the other equation in (9.2.16) yields

$$\frac{\alpha}{2} = \sum_{k=y_0}^{\infty} e^{-n\lambda}\frac{(n\lambda)^k}{k!} = P(Y \geq y_0|\lambda) = P\left(\chi^2_{2y_0} < 2n\lambda\right).$$

Doing some algebra, we obtain the $1 - \alpha$ confidence interval for $\lambda$ as

(9.2.17) $$\left\{\lambda: \frac{1}{2n}\chi^2_{2y_0,1-\alpha/2} \leq \lambda \leq \frac{1}{2n}\chi^2_{2(y_0+1),\alpha/2}\right\}.$$

(At $y_0 = 0$ we define $\chi^2_{0,1-\alpha/2} = 0$.)

These intervals were first derived by Garwood (1936). A graph of the coverage probabilities is given in Figure 9.2.5. Notice that the graph is quite jagged. The jumps occur at the endpoints of the different confidence intervals, where terms are added or subtracted from the sum that makes up the coverage probability. (See Exercise 9.24.)

For a numerical example, consider $n = 10$ and observe $y_0 = \sum x_i = 6$. A 90% confidence interval for $\lambda$ is given by

$$\frac{1}{20}\chi^2_{12,.95} \leq \lambda \leq \frac{1}{20}\chi^2_{14,.05},$$

which is

$$.262 \leq \lambda \leq 1.184.$$

Similar derivations, involving the negative binomial and binomial distributions, are given in the exercises.                                                                                    ‖

### 9.2.4 Bayesian Intervals

Thus far, when describing the interactions between the confidence interval and the parameter, we have carefully said that the interval *covers* the parameter, not that the parameter *is inside* the interval. This was done on purpose. We wanted to stress that the random quantity is the interval, not the parameter. Therefore, we tried to make the action verbs apply to the interval and not the parameter.

In Example 9.2.15 we saw that if $y_0 = \sum_{i=1}^{10} x_i = 6$, then a 90% confidence interval for $\lambda$ is $.262 \leq \lambda \leq 1.184$. It is tempting to say (and many experimenters do) that "the probability is 90% that $\lambda$ is in the interval [.262, 1.184]." Within classical statistics, however, such a statement is invalid since the parameter is assumed fixed. Formally, the interval [.262, 1.184] is one of the possible *realized values* of the random interval $[\frac{1}{2n}\chi^2_{2Y,.95}, \frac{1}{2n}\chi^2_{2(Y+1),.05}]$ and, since the parameter $\lambda$ does not move, $\lambda$ is in the *realized interval* [.262, 1.184] with probability either 0 or 1. When we say that the realized

interval [.262, 1.184] has a 90% chance of coverage, we only mean that we know that
90% of the sample points of the random interval cover the true parameter.

In contrast, the Bayesian setup allows us to say that $\lambda$ *is inside* [.262, 1.184] with
some probability, not 0 or 1. This is because, under the Bayesian model, $\lambda$ is a random
variable with a probability distribution. All Bayesian claims of coverage are made with
respect to the posterior distribution of the parameter.

To keep the distinction between Bayesian and classical sets clear, since the sets
make quite different probability assessments, the Bayesian set estimates are referred
to as *credible sets* rather than confidence sets.

Thus, if $\pi(\theta|\mathbf{x})$ is the posterior distribution of $\theta$ given $\mathbf{X} = \mathbf{x}$, then for any set
$A \subset \Theta$, the credible probability of $A$ is

$$(9.2.18) \qquad\qquad P(\theta \in A|\mathbf{x}) = \int_A \pi(\theta|\mathbf{x})\,d\theta,$$

and $A$ is a *credible set* for $\theta$. If $\pi(\theta|\mathbf{x})$ is a pmf, we replace integrals with sums in the
above expressions.

Notice that both the interpretation and construction of the Bayes credible set are
more straightforward than those of a classical confidence set. However, remember that
nothing comes free. The ease of construction and interpretation comes with additional
assumptions. The Bayesian model requires more input than the classical model.

**Example 9.2.16 (Poisson credible set)**  We now construct a credible set for the
problem of Example 9.2.15. Let $X_1, \ldots, X_n$ be iid Poisson($\lambda$) and assume that $\lambda$ has
a gamma prior pdf, $\lambda \sim$ gamma($a, b$). The posterior pdf of $\lambda$ (see Exercise 7.24) is

$$(9.2.19) \qquad\qquad \pi(\lambda|\textstyle\sum X = \sum x) = \text{gamma}(a + \textstyle\sum x, [n + (1/b)]^{-1}).$$

We can form a credible set for $\lambda$ in many different ways, as any set $A$ satisfying
(9.2.18) will do. One simple way is to split the $\alpha$ equally between the upper and lower
endpoints. From (9.2.19) it follows that $\frac{2(nb+1)}{b}\lambda \sim \chi^2_{2(a+\Sigma x_i)}$ (assuming that $a$ is an
integer), and thus a $1 - \alpha$ credible interval is

$$(9.2.20) \qquad \left\{ \lambda \colon \frac{b}{2(nb+1)}\chi^2_{2(\Sigma x+a),1-\alpha/2} \leq \lambda \leq \frac{b}{2(nb+1)}\chi^2_{2(\Sigma x+a),\alpha/2} \right\}.$$

If we take $a = b = 1$, the posterior distribution of $\lambda$ given $\sum X = \sum x$ can then
be expressed as $2(n + 1)\lambda \sim \chi^2_{2(\Sigma x+1)}$. As in Example 9.2.15, assume $n = 10$ and
$\sum x = 6$. Since $\chi^2_{14,.95} = 6.571$ and $\chi^2_{14,.05} = 23.685$, a 90% credible set for $\lambda$ is given
by [.299, 1.077].

The realized 90% credible set is different from the 90% confidence set obtained in
Example 9.2.15, [.262, 1.184]. To better see the differences, look at Figure 9.2.3, which
shows the 90% credible intervals and 90% confidence intervals for a range of $x$ values.
Notice that the credible set has somewhat shorter intervals, and the upper endpoints
are closer to 0. This reflects the prior, which is pulling the intervals toward 0.     ‖

It is important not to confuse credible probability (the Bayes posterior probability)
with coverage probability (the classical probability). The probabilities are very different entities, with different meanings and interpretations. Credible probability comes

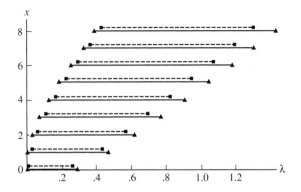

Figure 9.2.3. *The 90% credible intervals (dashed lines) and 90% confidence intervals (solid lines) from Example 9.2.16*

from the posterior distribution, which in turn gets its probability from the prior distribution. Thus, credible probability reflects the experimenter's subjective beliefs, as expressed in the prior distribution and updated with the data to the posterior distribution. A Bayesian assertion of 90% coverage means that the experimenter, upon combining prior knowledge with data, is 90% sure of coverage.

Coverage probability, on the other hand, reflects the uncertainty in the sampling procedure, getting its probability from the objective mechanism of repeated experimental trials. A classical assertion of 90% coverage means that in a long sequence of identical trials, 90% of the realized confidence sets will cover the true parameter.

Statisticians sometimes argue as to which is the better way to do statistics, classical or Bayesian. We do not want to argue or even defend one over another. In fact, we believe that there is no one best way to do statistics; some problems are best solved with classical statistics and some are best solved with Bayesian statistics. The important point to realize is that the solutions may be quite different. A Bayes solution is often not reasonable under classical evaluations and vice versa.

**Example 9.2.17 (Poisson credible and coverage probabilities)** The 90% confidence and credible sets of Example 9.2.16 maintain their respective probability guarantees, but how do they fare under the other criteria? First, lets look at the credible probability of the confidence set (9.2.17), which is given by

$$(9.2.21) \qquad P\left\{ \frac{1}{2n}\chi^2_{2\Sigma x,1-\alpha/2} \leq \lambda \leq \frac{1}{2n}\chi^2_{2(\Sigma x+1),\alpha/2}\right\},$$

where $\lambda$ has the distribution (9.2.19). Figure 9.2.4 shows the credible probability of the set (9.2.20), which is constant at $1-\alpha$, along with the credible probability of the confidence set (9.2.21).

This latter probability seems to be steadily decreasing, and we want to know if it remains above 0 for all values of $\Sigma x_i$ (for each fixed $n$). To do this, we evaluate the probability as $\Sigma x_i \to \infty$. Details are left to Exercise 9.30, but it is the case that, as

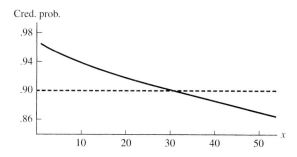

Figure 9.2.4. *Credible probabilities of the 90% credible intervals (dashed line) and 90% confidence intervals (solid line) from Example 9.2.16*

$\Sigma x_i \to \infty$, the probability (9.2.21) $\to 0$ unless $b = 1/n$. Thus, the confidence interval cannot maintain a nonzero credible probability.

The credible set (9.2.20) does not fare much better when evaluated as a confidence set. Figure 9.2.5 suggests that the coverage probability of the credible set is going to 0 as $\lambda \to \infty$. To evaluate the coverage probability, write

$$\lambda = \frac{\lambda}{\chi^2_{2Y}} \chi^2_{2Y},$$

where $\chi^2_{2Y}$ is a chi squared random variable with $2Y$ degrees of freedom, and $Y \sim$ Poisson $(n\lambda)$. Then, as $\lambda \to \infty$, $\lambda/\chi^2_{2Y} \to 1/(2n)$, and the coverage probability of (9.2.20) becomes

$$(9.2.22) \qquad P\left( \frac{nb}{nb+1} \chi^2_{2(Y+a),1-\alpha/2} \leq \chi^2_{2Y} \leq \frac{nb}{nb+1} \chi^2_{2(Y+a),\alpha/2} \right).$$

That this probability goes to 0 as $\lambda \to \infty$ is established in Exercise 9.31.      $\|$

The behavior exhibited in Example 9.2.17 is somewhat typical. Here is an example where the calculations can be done explicitly.

**Example 9.2.18 (Coverage of a normal credible set)**   Let $X_1, \ldots, X_n$ be iid $n(\theta, \sigma^2)$, and let $\theta$ have the prior pdf $n(\mu, \tau^2)$, where $\mu$, $\sigma$, and $\tau$ are all known. In Example 7.2.16 we saw that

$$\pi(\theta|\bar{x}) \sim n(\delta^{\mathrm{B}}(\bar{x}), \mathrm{Var}(\theta|\bar{x})),$$

where

$$\delta^{\mathrm{B}}(\bar{x}) = \frac{\sigma^2}{\sigma^2 + n\tau^2} \mu + \frac{n\tau^2}{\sigma^2 + n\tau^2} \bar{x} \quad \text{and} \quad \mathrm{Var}(\theta|\bar{x}) = \frac{\sigma^2 \tau^2}{\sigma^2 + n\tau^2}.$$

It therefore follows that under the posterior distribution,

$$\frac{\theta - \delta^{\mathrm{B}}(\bar{x})}{\sqrt{\mathrm{Var}(\theta|\bar{x})}} \sim n(0, 1),$$

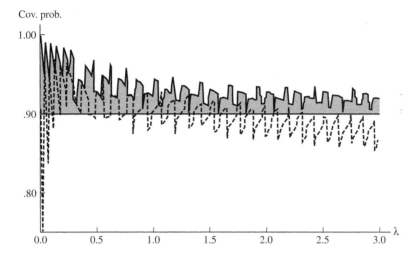

Figure 9.2.5. *Coverage probabilities of the 90% credible intervals (dashed lines) and 90% confidence intervals (solid lines) from Example 9.2.16*

and a $1 - \alpha$ credible set for $\theta$ is given by

$$(9.2.23) \qquad \delta^{\mathrm{B}}(\bar{x}) - z_{\alpha/2}\sqrt{\mathrm{Var}(\theta|\bar{x})} \leq \theta \leq \delta^{\mathrm{B}}(\bar{x}) + z_{\alpha/2}\sqrt{\mathrm{Var}(\theta|\bar{x})}.$$

We now calculate the coverage probability of the Bayesian region (9.2.23). Under the classical model $\bar{X}$ is the random variable, $\theta$ is fixed, and $\bar{X} \sim \mathrm{n}(\theta, \sigma^2/n)$. For ease of notation define $\gamma = \sigma^2/(n\tau^2)$, and from the definitions of $\delta^{\mathrm{B}}(\bar{X})$ and $\mathrm{Var}(\theta|\bar{X})$ and a little algebra, the coverage probability of (9.2.23) is

$$P_\theta\left(|\theta - \delta^{\mathrm{B}}(\bar{X})| \leq z_{\alpha/2}\sqrt{\mathrm{Var}(\theta|\bar{X})}\right)$$

$$= P_\theta\left(\left|\theta - \left(\frac{\gamma}{1+\gamma}\mu + \frac{1}{1+\gamma}\bar{X}\right)\right| \leq z_{\alpha/2}\sqrt{\frac{\sigma^2}{n(1+\gamma)}}\right)$$

$$= P_\theta\left(-\sqrt{1+\gamma}z_{\alpha/2} + \frac{\gamma(\theta - \mu)}{\sigma/\sqrt{n}} \leq Z \leq \sqrt{1+\gamma}z_{\alpha/2} + \frac{\gamma(\theta - \mu)}{\sigma/\sqrt{n}}\right),$$

where the last equality used the fact that $\sqrt{n}(\bar{X} - \theta)/\sigma = Z \sim \mathrm{n}(0, 1)$.

Although we started with a $1 - \alpha$ credible set, we do not have a $1 - \alpha$ confidence set, as can be seen by considering the following parameter configuration. Fix $\theta \neq \mu$ and let $\tau = \sigma/\sqrt{n}$, so that $\gamma = 1$. Also, let $\sigma/\sqrt{n}$ be very small ($\to 0$). Then it is easy to see that the above probability goes to 0, since if $\theta > \mu$ the lower bound goes to infinity, and if $\theta < \mu$ the upper bound goes to $-\infty$. If $\theta = \mu$, however, the coverage probability is bounded away from 0.

On the other hand, the usual $1 - \alpha$ confidence set for $\theta$ is $\{\theta \colon |\theta - \bar{x}| \leq z_{\alpha/2}\sigma/\sqrt{n}\}$. The credible probability of this set (now $\theta \sim \pi(\theta|\bar{x})$) is given by

$$
P_{\bar{x}}\left(|\theta - \bar{x}| \leq z_{\alpha/2}\frac{\sigma}{\sqrt{n}}\right)
$$

$$
= P_{\bar{x}}\left(\left|[\theta - \delta^{\mathrm{B}}(\bar{x})] + [\delta^{\mathrm{B}}(\bar{x}) - \bar{x}]\right| \leq z_{\alpha/2}\frac{\sigma}{\sqrt{n}}\right)
$$

$$
= P_{\bar{x}}\left(-\sqrt{1+\gamma}z_{\alpha/2} + \frac{\gamma(\bar{x} - \mu)}{\sqrt{1+\gamma}\sigma/\sqrt{n}} \leq Z \leq \sqrt{1+\gamma}z_{\alpha/2} + \frac{\gamma(\bar{x} - \mu)}{\sqrt{1+\gamma}\sigma/\sqrt{n}}\right),
$$

where the last equality used the fact that $(\theta - \delta^{\mathrm{B}}(\bar{x}))/\sqrt{\mathrm{Var}(\theta|\bar{x})} = Z \sim \mathrm{n}(0,1)$. Again, it is fairly easy to show that this probability is not bounded away from 0, showing that the confidence set is also not, in general, a credible set. Details are in Exercise 9.32.                                                                                                        ‖

## 9.3 Methods of Evaluating Interval Estimators

We now have seen many methods for deriving confidence sets and, in fact, we can derive different confidence sets for the same problem. In such situations we would, of course, want to choose a best one. Therefore, we now examine some methods and criteria for evaluating set estimators.

In set estimation two quantities vie against each other, size and coverage probability. Naturally, we want our set to have small size and large coverage probability, but such sets are usually difficult to construct. (Clearly, we can have a large coverage probability by increasing the size of our set. The interval $(-\infty, \infty)$ has coverage probability 1!) Before we can optimize a set with respect to size and coverage probability, we must decide how to measure these quantities.

The coverage probability of a confidence set will, except in special cases, be a function of the parameter, so there is not one value to consider but an infinite number of values. For the most part, however, we will measure coverage probability performance by the *confidence coefficient*, the infimum of the coverage probabilities. This is one way, but not the only available way of summarizing the coverage probability information. (For example, we could calculate an average coverage probability.)

When we speak of the *size* of a confidence set we will usually mean the *length* of the confidence set, if the set is an interval. If the set is not an interval, or if we are dealing with a multidimensional set, then length will usually become *volume*. (There are also cases where a size measure other than length is natural, especially if equivariance is a consideration. This topic is treated by Schervish 1995, Chapter 6, and Berger 1985, Chapter 6.)

### 9.3.1 Size and Coverage Probability

We now consider what appears to be a simple, constrained minimization problem. For a given, specified coverage probability find the confidence interval with the shortest length. We first consider an example.

**Example 9.3.1 (Optimizing length)** Let $X_1, \ldots, X_n$ be iid $n(\mu, \sigma^2)$, where $\sigma$ is known. From the method of Section 9.2.2 and the fact that

$$Z = \frac{\bar{X} - \mu}{\sigma/\sqrt{n}}$$

is a pivot with a standard normal distribution, any $a$ and $b$ that satisfy

$$P(a \leq Z \leq b) = 1 - \alpha$$

will give the $1 - \alpha$ confidence interval

$$\left\{ \mu : \bar{x} - b\frac{\sigma}{\sqrt{n}} \leq \mu \leq \bar{x} - a\frac{\sigma}{\sqrt{n}} \right\}.$$

Which choice of $a$ and $b$ is best? More formally, what choice of $a$ and $b$ will minimize the length of the confidence interval while maintaining $1 - \alpha$ coverage? Notice that the length of the confidence interval is equal to $(b - a)\sigma/\sqrt{n}$ but, since the factor $\sigma/\sqrt{n}$ is part of each interval length, it can be ignored and length comparisons can be based on the value of $b - a$. Thus, we want to find a pair of numbers $a$ and $b$ that satisfy $P(a \leq Z \leq b) = 1 - \alpha$ and minimize $b - a$.

In Example 9.2.1 we took $a = -z_{\alpha/2}$ and $b = z_{\alpha/2}$, but no mention was made of optimality. If we take $1 - \alpha = .90$, then any of the following pairs of numbers give 90% intervals:

*Three 90% normal confidence intervals*

| $a$ | $b$ | Probability | | $b - a$ |
|-----|-----|------------|---|---------|
| $-1.34$ | $2.33$ | $P(Z < a) = .09,$ | $P(Z > b) = .01$ | $3.67$ |
| $-1.44$ | $1.96$ | $P(Z < a) = .075,$ | $P(Z > b) = .025$ | $3.40$ |
| $-1.65$ | $1.65$ | $P(Z < a) = .05,$ | $P(Z > b) = .05$ | $3.30$ |

This numerical study suggests that the choice $a = -1.65$ and $b = 1.65$ gives the best interval and, in fact, it does. In this case splitting the probability $\alpha$ equally is an optimal strategy.                                                                          ‖

The strategy of splitting $\alpha$ equally, which is optimal in the above case, is not always optimal. What makes the equal $\alpha$ split optimal in the above case is the fact that the height of the pdf is the same at $-z_{\alpha/2}$ and $z_{\alpha/2}$. We now prove a theorem that will demonstrate this fact, a theorem that is applicable in some generality, needing only the assumption that the pdf is unimodal. Recall the definition of unimodal: A pdf $f(x)$ is *unimodal* if there exists $x^*$ such that $f(x)$ is nondecreasing for $x \leq x^*$ and $f(x)$ is nonincreasing for $x \geq x^*$. (This is a rather weak requirement.)

**Theorem 9.3.2** *Let $f(x)$ be a unimodal pdf. If the interval $[a, b]$ satisfies*

i. $\int_a^b f(x)\,dx = 1 - \alpha,$

ii. $f(a) = f(b) > 0,$ *and*

iii. $a \leq x^* \leq b$, where $x^*$ is a mode of $f(x)$,

*then $[a, b]$ is the shortest among all intervals that satisfy (i).*

**Proof:** Let $[a', b']$ be any interval with $b' - a' < b - a$. We will show that this implies $\int_{a'}^{b'} f(x) \, dx < 1 - \alpha$. The result will be proved only for $a' \leq a$, the proof being similar if $a < a'$. Also, two cases need to be considered, $b' \leq a$ and $b' > a$.

If $b' \leq a$, then $a' \leq b' \leq a \leq x^*$ and

$$\int_{a'}^{b'} f(x) \, dx \leq f(b')(b' - a') \qquad (x \leq b' \leq x^* \Rightarrow f(x) \leq f(b'))$$

$$\leq f(a)(b' - a') \qquad (b' \leq a \leq x^* \Rightarrow f(b') \leq f(a))$$

$$< f(a)(b - a) \qquad (b' - a' < b - a \text{ and } f(a) > 0)$$

$$\leq \int_{a}^{b} f(x) \, dx \qquad \left( \begin{array}{l} \text{(ii), (iii), and unimodality} \\ \Rightarrow f(x) \geq f(a) \text{ for } a \leq x \leq b \end{array} \right)$$

$$= 1 - \alpha, \tag{i}$$

completing the proof in the first case.

If $b' > a$, then $a' \leq a < b' < b$ for, if $b'$ were greater than or equal to $b$, then $b' - a'$ would be greater than or equal to $b - a$. In this case, we can write

$$\int_{a'}^{b'} f(x) \, dx = \int_{a}^{b} f(x) \, dx + \left[ \int_{a'}^{a} f(x) \, dx - \int_{b'}^{b} f(x) \, dx \right]$$

$$= (1 - \alpha) + \left[ \int_{a'}^{a} f(x) \, dx - \int_{b'}^{b} f(x) \, dx \right],$$

and the theorem will be proved if we show that the expression in square brackets is negative. Now, using the unimodality of $f$, the ordering $a' \leq a < b' < b$, and (ii), we have

$$\int_{a'}^{a} f(x) \, dx \leq f(a)(a - a')$$

and

$$\int_{b'}^{b} f(x) \, dx \geq f(b)(b - b').$$

Thus,

$$\int_{a'}^{a} f(x) \, dx - \int_{b'}^{b} f(x) \, dx \leq f(a)(a - a') - f(b)(b - b')$$

$$= f(a) \left[ (a - a') - (b - b') \right] \qquad (f(a) = f(b))$$

$$= f(a) \left[ (b' - a') - (b - a) \right],$$

which is negative if $(b' - a') < (b - a)$ and $f(a) > 0$.  $\square$

If we are willing to put more assumptions on $f$, for instance, that $f$ is continuous, then we can simplify the proof of Theorem 9.3.2. (See Exercise 9.38.)

Recall the discussion after Example 9.2.3 about the form of likelihood regions, which we now see is an optimal construction by Theorem 9.3.2. A similar argument, given in Corollary 9.3.10, shows how this construction yields an optimal Bayesian region. Also, we can see now that the equal $\alpha$ split, which is optimal in Example 9.3.1, will be optimal for any symmetric unimodal pdf (see Exercise 9.39). Theorem 9.3.2 may even apply when the optimality criterion is somewhat different from minimum length.

**Example 9.3.3 (Optimizing expected length)**   For normal intervals based on the pivot $\frac{\bar{X}-\mu}{S/\sqrt{n}}$ we know that the shortest length $1-\alpha$ confidence interval of the form

$$\bar{x} - b\frac{s}{\sqrt{n}} \le \mu \le \bar{x} - a\frac{s}{\sqrt{n}}$$

has $a = -t_{n-1,\alpha/2}$ and $b = t_{n-1,\alpha/2}$. The interval length is a function of $s$, with general form

$$\text{Length}(s) = (b-a)\frac{s}{\sqrt{n}}.$$

It is easy to see that if we had considered the criterion of *expected length* and wanted to find a $1-\alpha$ interval to minimize

$$\text{E}_\sigma(\text{Length}(S)) = (b-a)\frac{\text{E}_\sigma S}{\sqrt{n}} = (b-a)c(n)\frac{\sigma}{\sqrt{n}},$$

then Theorem 9.3.2 applies and the choice $a = -t_{n-1,\alpha/2}$ and $b = t_{n-1,\alpha/2}$ again gives the optimal interval. (The quantity $c(n)$ is a constant dependent only on $n$. See Exercise 7.50.)                                                                                      ‖

In some cases, especially when working outside of the location problem, we must be careful in the application of Theorem 9.3.2. In scale cases in particular, the theorem may not be directly applicable, but a variant may be.

**Example 9.3.4 (Shortest pivotal interval)**   Suppose $X \sim \text{gamma}(k,\beta)$. The quantity $Y = X/\beta$ is a pivot, with $Y \sim \text{gamma}(k,1)$, so we can get a confidence interval by finding constants $a$ and $b$ to satisfy

(9.3.1)                                $P(a \le Y \le b) = 1-\alpha.$

However, blind application of Theorem 9.3.2 will not give the shortest confidence interval. That is, choosing $a$ and $b$ to satisfy (9.3.1) and also $f_Y(a) = f_Y(b)$ is not optimal. This is because, based on (9.3.1), the interval on $\beta$ is of the form

$$\left\{\beta : \frac{x}{b} \le \beta \le \frac{x}{a}\right\},$$

so the length of the interval is $(\frac{1}{a} - \frac{1}{b})x$; that is, it is proportional to $(1/a) - (1/b)$ and not to $b-a$.

Although Theorem 9.3.2 is not directly applicable here, a modified argument can solve this problem. Condition (a) in Theorem 9.3.2 defines $b$ as a function of $a$, say $b(a)$. We must solve the following constrained minimization problem:

$$\text{Minimize, with respect to } a: \quad \frac{1}{a} - \frac{1}{b(a)}$$

$$\text{subject to:} \quad \int_a^{b(a)} f_Y(y)\, dy = 1 - \alpha.$$

Differentiating the first equation with respect to $a$ and setting it equal to 0 yield the identity $db/da = b^2/a^2$. Substituting this in the derivative of the second equation, which must equal 0, gives $f(b)b^2 = f(a)a^2$ (see Exercise 9.42). Equations like these also arise in interval estimation of the variance of a normal distribution; see Example 9.2.10 and Exercise 9.52. Note that the above equations define not the shortest *overall* interval, but the shortest *pivotal* interval, that is, the shortest interval based on the pivot $X/\beta$. For a generalization of this result, involving the Neyman-Pearson Lemma, see Exercise 9.43. ‖

### 9.3.2 Test-Related Optimality

Since there is a one-to-one correspondence between confidence sets and tests of hypotheses (Theorem 9.2.2), there is some correspondence between optimality of tests and optimality of confidence sets. Usually, test-related optimality properties of confidence sets do not directly relate to the size of the set but rather to the probability of the set covering false values.

The probability of covering false values, or the *probability of false coverage*, indirectly measures the size of a confidence set. Intuitively, smaller sets cover fewer values and, hence, are less likely to cover false values. Moreover, we will later see an equation that links size and probability of false coverage.

We first consider the general situation, where $\mathbf{X} \sim f(\mathbf{x}|\theta)$, and we construct a $1-\alpha$ confidence set for $\theta$, $C(\mathbf{x})$, by inverting an acceptance region, $A(\theta)$. The probability of coverage of $C(\mathbf{x})$, that is, the probability of *true coverage*, is the function of $\theta$ given by $P_\theta(\theta \in C(\mathbf{X}))$. The probability of *false coverage* is the function of $\theta$ and $\theta'$ defined by

(9.3.2)
$$
\begin{aligned}
&P_\theta\left(\theta' \in C(\mathbf{X})\right), \theta \neq \theta', &&\text{if } C(\mathbf{X}) = [L(\mathbf{X}), U(\mathbf{X})], \\
&P_\theta\left(\theta' \in C(\mathbf{X})\right), \theta' < \theta, &&\text{if } C(\mathbf{X}) = [L(\mathbf{X}), \infty), \\
&P_\theta\left(\theta' \in C(\mathbf{X})\right), \theta' > \theta, &&\text{if } C(\mathbf{X}) = (-\infty, U(\mathbf{X})],
\end{aligned}
$$

the probability of covering $\theta'$ when $\theta$ is the true parameter.

It makes sense to define the probability of false coverage differently for one-sided and two-sided intervals. For example, if we have a lower confidence bound, we are asserting that $\theta$ is greater than a certain value and false coverage would occur only if we cover values of $\theta$ that are too small. A similar argument leads us to the definitions used for upper confidence bounds and two-sided bounds.

A $1 - \alpha$ confidence set that minimizes the probability of false coverage over a class of $1 - \alpha$ confidence sets is called a *uniformly most accurate* (UMA) confidence set. Thus, for example, we would consider looking for a UMA confidence set among sets of the form $[L(\mathbf{x}), \infty)$. UMA confidence sets are constructed by inverting the acceptance regions of UMP tests, as we will prove below. Unfortunately, although a UMA confidence set is a desirable set, it exists only in rather rare circumstances (as do UMP tests). In particular, since UMP tests are generally one-sided, so are UMA intervals. They make for elegant theory, however. In the next theorem we see that a UMP test of $H_0 \colon \theta = \theta_0$ versus $H_1 \colon \theta > \theta_0$ yields a UMA lower confidence bound.

**Theorem 9.3.5**    Let $\mathbf{X} \sim f(\mathbf{x}|\theta)$, where $\theta$ is a real-valued parameter. For each $\theta_0 \in \Theta$, let $A^*(\theta_0)$ be the UMP level $\alpha$ acceptance region of a test of $H_0 \colon \theta = \theta_0$ versus $H_1 \colon \theta > \theta_0$. Let $C^*(\mathbf{x})$ be the $1 - \alpha$ confidence set formed by inverting the UMP acceptance regions. Then for any other $1 - \alpha$ confidence set $C$,

$$P_\theta(\theta' \in C^*(\mathbf{X})) \le P_\theta(\theta' \in C(\mathbf{X})) \text{ for all } \theta' < \theta.$$

**Proof:** Let $\theta'$ be any value less than $\theta$. Let $A(\theta')$ be the acceptance region of the level $\alpha$ test of $H_0 \colon \theta = \theta'$ obtained by inverting $C$. Since $A^*(\theta')$ is the UMP acceptance region for testing $H_0 \colon \theta = \theta'$ versus $H_1 \colon \theta > \theta'$, and since $\theta > \theta'$, we have

$$P_\theta(\theta' \in C^*(\mathbf{X})) = P_\theta(\mathbf{X} \in A^*(\theta')) \qquad \text{(invert the confidence set)}$$

$$\le P_\theta(\mathbf{X} \in A(\theta')) \qquad \begin{pmatrix} \text{true for any } A \\ \text{since } A^* \text{ is UMP} \end{pmatrix}$$

$$= P_\theta(\theta' \in C(\mathbf{X})). \qquad \begin{pmatrix} \text{invert } A \text{ to} \\ \text{obtain } C \end{pmatrix}$$

Notice that the above inequality is "$\le$" because we are working with probabilities of acceptance regions. This is $1 - \text{power}$, so UMP tests will minimize these acceptance region probabilities. Therefore, we have established that for $\theta' < \theta$, the probability of false coverage is minimized by the interval obtained from inverting the UMP test.                                                                                    □

Recall our discussion in Section 9.2.1. The UMA confidence set in the above theorem is constructed by inverting the family of tests for the hypotheses

$$H_0 \colon \quad \theta = \theta_0 \qquad \text{versus} \qquad H_1 \colon \quad \theta > \theta_0,$$

where the form of the confidence set is governed by the alternative hypothesis. The above alternative hypotheses, which specify that $\theta_0$ is less than a particular value, lead to *lower* confidence bounds; that is, if the sets are intervals, they are of the form $[L(\mathbf{X}), \infty)$.

**Example 9.3.6 (UMA confidence bound)** Let $X_1, \ldots, X_n$ be iid $n(\mu, \sigma^2)$, where $\sigma^2$ is known. The interval

$$C(\bar{x}) = \left\{ \mu : \mu \ge \bar{x} - z_\alpha \frac{\sigma}{\sqrt{n}} \right\}$$

is a $1 - \alpha$ UMA lower confidence bound since it can be obtained by inverting the UMP test of $H_0 \colon \mu = \mu_0$ versus $H_1 \colon \mu > \mu_0$.

The more common two-sided interval,

$$C(\bar{x}) = \left\{ \mu : \bar{x} - z_{\alpha/2} \frac{\sigma}{\sqrt{n}} \le \mu \le \bar{x} + z_{\alpha/2} \frac{\sigma}{\sqrt{n}} \right\},$$

is not UMA, since it is obtained by inverting the two-sided acceptance region from the test of $H_0: \mu = \mu_0$ versus $H_1: \mu \neq \mu_0$, hypotheses for which no UMP test exists.

$\parallel$

In the testing problem, when considering two-sided tests, we found the property of *unbiasedness* to be both compelling and useful. In the confidence interval problem, similar ideas apply. When we deal with two-sided confidence intervals, it is reasonable to restrict consideration to unbiased confidence sets. Remember that an unbiased test is one in which the power in the alternative is always greater than the power in the null. Keep that in mind when reading the following definition.

**Definition 9.3.7**   A $1 - \alpha$ confidence set $C(\mathbf{x})$ is *unbiased* if $P_\theta(\theta' \in C(\mathbf{X})) \le 1 - \alpha$ for all $\theta \neq \theta'$.

Thus, for an unbiased confidence set, the probability of false coverage is never more than the minimum probability of true coverage. Unbiased confidence sets can be obtained by inverting unbiased tests. That is, if $A(\theta_0)$ is an unbiased level $\alpha$ acceptance region of a test of $H_0: \theta = \theta_0$ versus $H_1: \theta \neq \theta_0$ and $C(\mathbf{x})$ is the $1 - \alpha$ confidence set formed by inverting the acceptance regions, then $C(\mathbf{x})$ is an unbiased $1 - \alpha$ confidence set (see Exercise 9.46).

**Example 9.3.8 (Continuation of Example 9.3.6)**   The two-sided normal interval

$$C(\bar{x}) = \left\{ \mu : \bar{x} - z_{\alpha/2} \frac{\sigma}{\sqrt{n}} \le \mu \le \bar{x} + z_{\alpha/2} \frac{\sigma}{\sqrt{n}} \right\}$$

is an unbiased interval. It can be obtained by inverting the unbiased test of $H_0: \mu = \mu_0$ versus $H_1: \mu \neq \mu_0$ given in Example 8.3.20. Similarly, the interval (9.2.14) based on the $t$ distribution is also an unbiased interval, since it also can be obtained by inverting a unbiased test (see Exercise 8.38).

$\parallel$

Sets that minimize the probability of false coverage are also called *Neyman-shortest*. The fact that there is a length connotation to this name is somewhat justified by the following theorem, due to Pratt (1961).

**Theorem 9.3.9 (Pratt)**   *Let $X$ be a real-valued random variable with $X \sim f(x|\theta)$, where $\theta$ is a real-valued parameter. Let $C(x) = [L(x), U(x)]$ be a confidence interval for $\theta$. If $L(x)$ and $U(x)$ are both increasing functions of $x$, then for any value $\theta^*$,*

$$(9.3.3) \qquad \mathrm{E}_{\theta^*}(\mathrm{Length}[C(\mathbf{X})]) = \int_{\theta \neq \theta^*} P_{\theta^*}(\theta \in C(\mathbf{X})) \, d\theta.$$

Theorem 9.3.9 says that the expected length of $C(x)$ is equal to a sum (integral) of the probabilities of false coverage, the integral being taken over all false values of the parameter.

**Proof:** From the definition of expected value we can write

$$\mathrm{E}_{\theta^*}\left(\mathrm{Length}[C(X)]\right) = \int_{\mathcal{X}} \mathrm{Length}[C(x)] f(x|\theta^*)\,dx$$

$$= \int_{\mathcal{X}} [U(x) - L(x)] f(x|\theta^*)\,dx \qquad \text{(definition of length)}$$

$$= \int_{\mathcal{X}} \left[\int_{L(x)}^{U(x)} d\theta\right] f(x|\theta^*)\,dx \qquad \left(\begin{array}{c}\text{using } \theta \text{ as a} \\ \text{dummy variable}\end{array}\right)$$

$$= \int_{\Theta} \left[\int_{U^{-1}(\theta)}^{L^{-1}(\theta)} f(x|\theta^*)\,dx\right] d\theta \qquad \left(\begin{array}{c}\text{invert the order of} \\ \text{integration—see below}\end{array}\right)$$

$$= \int_{\Theta} \left[P_{\theta^*}\left(U^{-1}(\theta) \le X \le L^{-1}(\theta)\right)\right] d\theta \qquad \text{(definition)}$$

$$= \int_{\Theta} \left[P_{\theta^*}\left(\theta \in C(X)\right)\right] d\theta \qquad \left(\begin{array}{c}\text{invert the} \\ \text{acceptance region}\end{array}\right)$$

$$= \int_{\theta \ne \theta^*} \left[P_{\theta^*}\left(\theta \in C(X)\right)\right] d\theta. \qquad \left(\begin{array}{c}\text{one point does} \\ \text{not change value}\end{array}\right)$$

The string of equalities establishes the identity and proves the theorem. The interchange of integrals is formally justified by Fubini's Theorem (Lehmann and Casella 1998, Section 1.2) but is easily seen to be justified as long as all of the integrands are finite. The inversion of the confidence interval is standard, where we use the relationship

$$\theta \in \{\theta: L(x) \le \theta \le U(x)\} \Leftrightarrow x \in \{x: U^{-1}(\theta) \le x \le L^{-1}(\theta)\}\,,$$

which is valid because of the assumption that $L$ and $U$ are increasing. Note that the theorem could be modified to apply to an interval with decreasing endpoints.      □

Theorem 9.3.9 shows that there is a formal relationship between the length of a confidence interval and its probability of false coverage. In the two-sided case, this implies that minimizing the probability of false coverage carries along some guarantee of length optimality. In the one-sided case, however, the analogy does not quite work. In that case, intervals that are set up to minimize the probability of false coverage are concerned with parameters in only a portion of the parameter space and length optimality may not obtain. Madansky (1962) has given an example of a $1 - \alpha$ UMA interval (one-sided) that can be beaten in the sense that another, shorter $1-\alpha$ interval can be constructed. (See Exercise 9.45.) Also, Maatta and Casella (1987) have shown that an interval obtained by inverting a UMP test can be suboptimal when measured against other reasonable criteria.

### 9.3.3 Bayesian Optimality

The goal of obtaining a smallest confidence set with a specified coverage probability can also be attained using Bayesian criteria. If we have a posterior distribution $\pi(\theta|\mathbf{x})$, the posterior distribution of $\theta$ given $\mathbf{X} = \mathbf{x}$, we would like to find the set $C(\mathbf{x})$ that satisfies

$$\text{(i)} \quad \int_{C(\mathbf{x})} \pi(\theta|\mathbf{x})d\mathbf{x} = 1 - \alpha$$

$$\text{(ii)} \quad \text{Size } (C(\mathbf{x})) \le \text{Size } (C'(\mathbf{x}))$$

for any set $C'(\mathbf{x})$ satisfying $\int_{C'(\mathbf{x})} \pi(\theta|\mathbf{x})d\mathbf{x} \ge 1 - \alpha$.

If we take our measure of size to be length, then we can apply Theorem 9.3.2 and obtain the following result.

**Corollary 9.3.10** *If the posterior density $\pi(\theta|\mathbf{x})$ is unimodal, then for a given value of $\alpha$, the shortest credible interval for $\theta$ is given by*

$$\{\theta : \pi(\theta|\mathbf{x}) \ge k\} \quad where \quad \int_{\{\theta : \pi(\theta|\mathbf{x}) \ge k\}} \pi(\theta|\mathbf{x})d\theta = 1 - \alpha.$$

The credible set described in Corollary 9.3.10 is called a *highest posterior density* (HPD) region, as it consists of the values of the parameter for which the posterior density is highest. Notice the similarity in form between the HPD region and the likelihood region.

**Example 9.3.11 (Poisson HPD region)** In Example 9.2.16 we derived a $1 - \alpha$ credible set for a Poisson parameter. We now construct an HPD region. By Corollary 9.3.10, this region is given by $\{\lambda : \pi(\lambda|\sum x) \ge k\}$, where $k$ is chosen so that

$$1 - \alpha = \int_{\{\lambda : \pi(\lambda|\Sigma x) \ge k\}} \pi(\lambda|\textstyle\sum x)\,d\lambda.$$

Recall that the posterior pdf of $\lambda$ is gamma$(a + \sum x, [n + (1/b)]^{-1})$, so we need to find $\lambda_L$ and $\lambda_U$ such that

$$\pi(\lambda_L|\sum x) = \pi(\lambda_U|\sum x) \quad \text{and} \quad \int_{\lambda_L}^{\lambda_U} \pi(\lambda|\sum x)d\lambda = 1 - \alpha.$$

If we take $a = b = 1$ (as in Example 9.2.16), the posterior distribution of $\lambda$ given $\sum X = \sum x$ can be expressed as $2(n+1)\lambda \sim \chi^2_{2(\Sigma x + 1)}$ and, if $n = 10$ and $\sum x = 6$, the 90% HPD credible set for $\lambda$ is given by $[.253, 1.005]$.

In Figure 9.3.1 we show three $1 - \alpha$ intervals for $\lambda$: the $1 - \alpha$ equal-tailed Bayes credible set of Example 9.2.16, the HPD region derived here, and the classical $1 - \alpha$ confidence set of Example 9.2.15.                                                    ‖

The shape of the HPD region is determined by the shape of the posterior distribution. In general, the HPD region is not symmetric about a Bayes point estimator but, like the likelihood region, is rather asymmetric. For the Poisson distribution this is clearly true, as the above example shows. Although it will not always happen, we can usually expect asymmetric HPD regions for scale parameter problems and symmetric HPD regions for location parameter problems.

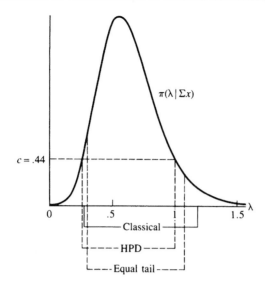

Figure 9.3.1. *Three interval estimators from Example 9.2.16*

**Example 9.3.12 (Normal HPD region)** The equal-tailed credible set derived in
Example 9.2.18 is, in fact, an HPD region. Since the posterior distribution of $\theta$ is
normal with mean $\delta^B$, it follows that $\{\theta : \pi(\theta|\bar{x}) \geq k\} = \{\theta : \theta \in \delta^B \pm k'\}$ for some $k'$
(see Exercise 9.40). So the HPD region is symmetric about the mean $\delta^B(\bar{x})$.        $\|$

### 9.3.4 Loss Function Optimality

In the previous two sections we looked at optimality of interval estimators by first
requiring them to have a minimum coverage probability and then looking for the
shortest interval. However, it is possible to put these requirements together in one
loss function and use decision theory to search for an optimal estimator. In interval
estimation, the action space $\mathcal{A}$ will consist of subsets of the parameter space $\Theta$ and,
more formally, we might talk of "set estimation," since an optimal rule may not
necessarily be an interval. However, practical considerations lead us to mainly consider
set estimators that are intervals and, happily, many optimal procedures turn out to
be intervals.

We use $C$ (for confidence interval) to denote elements of $\mathcal{A}$, with the meaning of
the action $C$ being that the interval estimate "$\theta \in C$" is made. A decision rule $\delta(\mathbf{x})$
simply specifies, for each $\mathbf{x} \in \mathcal{X}$, which set $C \in \mathcal{A}$ will be used as an estimate of $\theta$ if
$\mathbf{X} = \mathbf{x}$ is observed. Thus we will use the notation $C(\mathbf{x})$, as before.

The loss function in an interval estimation problem usually includes two quantities:
a measure of whether the set estimate correctly includes the true value $\theta$ and a
measure of the size of the set estimate. We will, for the most part, consider only sets
$C$ that are intervals, so a natural measure of size is Length$(C)$ = length of $C$. To

express the correctness measure, it is common to use

$$I_C(\theta) = \begin{cases} 1 & \theta \in C \\ 0 & \theta \notin C. \end{cases}$$

That is, $I_C(\theta) = 1$ if the estimate is correct and 0 otherwise. In fact, $I_C(\theta)$ is just the indicator function for the set $C$. But realize that $C$ will be a random set determined by the value of the data $\mathbf{X}$.

The loss function should reflect the fact that a good estimate would have Length($C$) small and $I_C(\theta)$ large. One such loss function is

(9.3.4)                          $$L(\theta, C) = b\,\mathrm{Length}(C) - I_C(\theta),$$

where $b$ is a positive constant that reflects the relative weight that we want to give to the two criteria, a necessary consideration since the two quantities are very different. If there is more concern with correct estimates, then $b$ should be small, while a large $b$ should be used if there is more concern with interval length.

The risk function associated with (9.3.4) is particularly simple, given by

$$\begin{aligned} R(\theta, C) &= b\mathrm{E}_\theta\left[\mathrm{Length}(C(\mathbf{X}))\right] - \mathrm{E}_\theta I_{C(\mathbf{X})}(\theta) \\ &= b\mathrm{E}_\theta\left[\mathrm{Length}(C(\mathbf{X}))\right] - P_\theta(I_{C(\mathbf{X})}(\theta) = 1) \\ &= b\mathrm{E}_\theta\left[\mathrm{Length}(C(\mathbf{X}))\right] - P_\theta(\theta \in C(\mathbf{X})). \end{aligned}$$

The risk has two components, the expected length of the interval and the coverage probability of the interval estimator. The risk reflects the fact that, simultaneously, we want the expected length to be small and the coverage probability to be high, just as in the previous sections. But now, instead of requiring a minimum coverage probability and then minimizing length, the trade-off between these two quantities is specified in the risk function. Perhaps a smaller coverage probability will be acceptable if it results in a greatly decreased length.

By varying the size of $b$ in the loss (9.3.4), we can vary the relative importance of size and coverage probability of interval estimators, something that could not be done previously. As an example of the flexibility of the present setup, consider some limiting cases. If $b = 0$, then size does not matter, only coverage probability, so the interval estimator $C = (-\infty, \infty)$, which has coverage probability 1, is the best decision rule. Similarly, if $b = \infty$, then coverage probability does not matter, so point sets are optimal. Hence, an entire range of decision rules are possible candidates. In the next example, for a specified finite range of $b$, choosing a good rule amounts to using the risk function to decide the confidence coefficient while, if $b$ is outside this range, the optimal decision rule is a point estimator.

**Example 9.3.13 (Normal interval estimator)** Let $X \sim \mathrm{n}(\mu, \sigma^2)$ and assume $\sigma^2$ is known. $X$ would typically be a sample mean and $\sigma^2$ would have the form $\tau^2/n$, where $\tau^2$ is the known population variance and $n$ is the sample size. For each $c \geq 0$, define an interval estimator for $\mu$ by $C(x) = [x - c\sigma, x + c\sigma]$. We will compare these estimators using the loss in (9.3.4). The length of an interval, Length($C(x)$) = $2c\sigma$, does not depend on $x$. Thus, the first term in the risk is $b(2c\sigma)$. The second term in the risk is

$$P_\mu(\mu \in C(X)) = P_\mu(X - c\sigma \le \mu \le X + c\sigma)$$

$$= P_\mu\left(-c \le \frac{X - \mu}{\sigma} \le c\right)$$

$$= 2P(Z \le c) - 1,$$

where $Z \sim n(0, 1)$. Thus, the risk function for an interval estimator in this class is

$$(9.3.5) \qquad R(\mu, C) = b(2c\sigma) - [2P(Z \le c) - 1].$$

The risk function is constant, as it does not depend on $\mu$, and the best interval estimator in this class is the one corresponding to the value $c$ that minimizes (9.3.5). If $b\sigma > 1/\sqrt{2\pi}$, it can be shown that $R(\mu, C)$ is minimized at $c = 0$. That is, the length portion completely overwhelms the coverage probability portion of the loss, and the best *interval* estimator is the *point* estimator $C(x) = [x, x]$. But if $b\sigma \le 1/\sqrt{2\pi}$, the risk is minimized at $c = \sqrt{-2\log(b\sigma\sqrt{2\pi})}$. If we express $c$ as $z_{\alpha/2}$ for some $\alpha$, then the interval estimator that minimizes the risk is just the usual $1 - \alpha$ confidence interval. (See Exercise 9.53 for details.)                                                            ‖

The use of decision theory in interval estimation problems is not as widespread as in point estimation or hypothesis testing problems. One reason for this is the difficulty in choosing $b$ in (9.3.4) (or in Example 9.3.13). We saw in the previous example that a choice that might seem reasonable could lead to unintuitive results, indicating that the loss in (9.3.4) may not be appropriate. Some who would use decision theoretic analysis for other problems still prefer to use only interval estimators with a fixed confidence coefficient $(1 - \alpha)$. They then use the risk function to judge other qualities like the size of the set.

Another difficulty is in the restriction of the shape of the allowable sets in $\mathcal{A}$. Ideally, the loss and risk functions would be used to judge which shapes are best. But one can always add isolated points to an interval estimator and get an improvement in coverage probability with no loss penalty regarding size. In the previous example we could have used the estimator

$$C(x) = [x - c\sigma, x + c\sigma] \cup \{\text{all integer values of } \mu\}.$$

The "length" of these sets is the same as before, but now the coverage probability is 1 for all integer values of $\mu$. Some more sophisticated measure of size must be used to avoid such anomalies. (Joshi 1969 addressed this problem by defining equivalence classes of estimators.)

## 9.4 Exercises _____

**9.1** If $L(x)$ and $U(x)$ satisfy $P_\theta(L(X) \le \theta) = 1 - \alpha_1$ and $P_\theta(U(X) \ge \theta) = 1 - \alpha_2$, and $L(x) \le U(x)$ for all $x$, show that $P_\theta(L(X) \le \theta \le U(X)) = 1 - \alpha_1 - \alpha_2$.

**9.2** Let $X_1, \ldots, X_n$ be iid $n(\theta, 1)$. A 95% confidence interval for $\theta$ is $\bar{x} \pm 1.96/\sqrt{n}$. Let $p$ denote the probability that an additional independent observation, $X_{n+1}$, will fall in this interval. Is $p$ greater than, less than, or equal to .95? Prove your answer.

**9.3** The independent random variables $X_1, \ldots, X_n$ have the common distribution

$$P(X_i \leq x) = \begin{cases} 0 & \text{if } x \leq 0 \\ (x/\beta)^\alpha & \text{if } 0 < x < \beta \\ 1 & \text{if } x \geq \beta. \end{cases}$$

(a) In Exercise 7.10 the MLEs of $\alpha$ and $\beta$ were found. If $\alpha$ is a known constant, $\alpha_0$, find an upper confidence limit for $\beta$ with confidence coefficient .95.

(b) Use the data of Exercise 7.10 to construct an interval estimate for $\beta$. Assume that $\alpha$ is known and equal to its MLE.

**9.4** Let $X_1, \ldots, X_n$ be a random sample from a $n(0, \sigma_X^2)$, and let $Y_1, \ldots, Y_m$ be a random sample from a $n(0, \sigma_Y^2)$, independent of the $X$s. Define $\lambda = \sigma_Y^2/\sigma_X^2$.

(a) Find the level $\alpha$ LRT of $H_0: \lambda = \lambda_0$ versus $H_1: \lambda \neq \lambda_0$.

(b) Express the rejection region of the LRT of part (a) in terms of an $F$ random variable.

(c) Find a $1 - \alpha$ confidence interval for $\lambda$.

**9.5** In Example 9.2.5 a lower confidence bound was put on $p$, the success probability from a sequence of Bernoulli trials. This exercise will derive an upper confidence bound. That is, observing $X_1, \ldots, X_n$, where $X_i \sim \text{Bernoulli}(p)$, we want an interval of the form $[0, U(x_1, \ldots, x_n))$, where $P_p(p \in [0, U(X_1, \ldots, X_n))) \geq 1 - \alpha$.

(a) Show that inversion of the acceptance region of the test

$$H_0: p = p_0 \qquad \text{versus} \qquad H_1: p < p_0$$

will give a confidence interval of the desired confidence level and form.

(b) Find equations, similar to those given in (9.2.8), that can be used to construct the confidence interval.

**9.6** (a) Derive a confidence interval for a binomial $p$ by inverting the LRT of $H_0: p = p_0$ versus $H_1: p \neq p_0$.

(b) Show that the interval is a highest density region from $p^y(1-p)^{n-y}$ and is not equal to the interval in (10.4.4).

**9.7** (a) Find the $1 - \alpha$ confidence set for $a$ that is obtained by inverting the LRT of $H_0: a = a_0$ versus $H_1: a \neq a_0$ based on a sample $X_1, \ldots, X_n$ from a $n(0, a\theta)$ family, where $\theta$ is unknown.

(b) A similar question can be asked about the related family, the $n(0, a\theta^2)$ family. If $X_1, \ldots, X_n$ are iid $n(0, a\theta^2)$, where $\theta$ is unknown, find the $1 - \alpha$ confidence set based on inverting the LRT of $H_0: a = a_0$ versus $H_1: a \neq a_0$.

**9.8** Given a sample $X_1, \ldots, X_n$ from a pdf of the form $\frac{1}{\sigma}f((x - \theta)/\sigma)$, list at least five different pivotal quantities.

**9.9** Show that each of the three quantities listed in Example 9.2.7 is a pivot.

**9.10** (a) Suppose that $T$ is a real-valued statistic. Suppose that $Q(t, \theta)$ is a monotone function of $t$ for each value of $\theta \in \Theta$. Show that if the pdf of $T$, $f(t|\theta)$, can be expressed in the form (9.2.11) for some function $g$, then $Q(T, \theta)$ is a pivot.

(b) Show that (9.2.11) is satisfied by taking $g = 1$ and $Q(t, \theta) = F_\theta(t)$, the cdf of $T$. (This is the Probability Integral Transform.)

**9.11** If $T$ is a continuous random variable with cdf $F_T(t|\theta)$ and $\alpha_1 + \alpha_2 = \alpha$, show that an $\alpha$ level acceptance region of the hypothesis $H_0 : \theta = \theta_0$ is $\{t : \alpha_1 \leq F_T(t|\theta_0) \leq 1 - \alpha_2\}$, with associated confidence $1 - \alpha$ set $\{\theta : \alpha_1 \leq F_T(t|\theta) \leq 1 - \alpha_2\}$.

**9.12** Find a pivotal quantity based on a random sample of size $n$ from a $n(\theta, \theta)$ population, where $\theta > 0$. Use the pivotal quantity to set up a $1 - \alpha$ confidence interval for $\theta$.

**9.13** Let $X$ be a single observation from the beta$(\theta, 1)$ pdf.

(a) Let $Y = -(\log X)^{-1}$. Evaluate the confidence coefficient of the set $[y/2, y]$.

(b) Find a pivotal quantity and use it to set up a confidence interval having the same confidence coefficient as the interval in part (a).

(c) Compare the two confidence intervals.

**9.14** Let $X_1, \ldots, X_n$ be iid $n(\mu, \sigma^2)$, where both parameters are unknown. Simultaneous inference on both $\mu$ and $\sigma$ can be made using the Bonferroni Inequality in a number of ways.

(a) Using the Bonferroni Inequality, combine the two confidence sets

$$\left\{\mu : \bar{x} - \frac{ks}{\sqrt{n}} \leq \mu \leq \bar{x} + \frac{ks}{\sqrt{n}}\right\} \quad \text{and} \quad \left\{\sigma^2 : \frac{(n-1)s^2}{b} \leq \sigma^2 \leq \frac{(n-1)s^2}{a}\right\}$$

into one confidence set for $(\mu, \sigma)$. Show how to choose $a$, $b$, and $k$ to make the simultaneous set a $1 - \alpha$ confidence set.

(b) Using the Bonferroni Inequality, combine the two confidence sets

$$\left\{\mu : \bar{x} - \frac{k\sigma}{\sqrt{n}} \leq \mu \leq \bar{x} + \frac{k\sigma}{\sqrt{n}}\right\} \quad \text{and} \quad \left\{\sigma^2 : \frac{(n-1)s^2}{b} \leq \sigma^2 \leq \frac{(n-1)s^2}{a}\right\}$$

into one confidence set for $(\mu, \sigma)$. Show how to choose $a$, $b$, and $k$ to make the simultaneous set a $1 - \alpha$ confidence set.

(c) Compare the confidence sets in parts (a) and (b).

**9.15** Solve for the roots of the quadratic equation that defines Fieller's confidence set for the ratio of normal means (see Miscellanea 9.5.3). Find conditions on the random variables for which

(a) the parabola opens upward (the confidence set is an interval).

(b) the parabola opens downward (the confidence set is the complement of an interval).

(c) the parabola has no real roots.

In each case, give an interpretation of the meaning of the confidence set. For example, what would you tell an experimenter if, for his data, the parabola had no real roots?

**9.16** Let $X_1, \ldots, X_n$ be iid $n(\theta, \sigma^2)$, where $\sigma^2$ is known. For each of the following hypotheses, write out the acceptance region of a level $\alpha$ test and the $1 - \alpha$ confidence interval that results from inverting the test.

(a) $H_0 : \theta = \theta_0$ versus $H_1 : \theta \neq \theta_0$

(b) $H_0 : \theta \geq \theta_0$ versus $H_1 : \theta < \theta_0$

(c) $H_0 : \theta \leq \theta_0$ versus $H_1 : \theta > \theta_0$

**9.17** Find a $1 - \alpha$ confidence interval for $\theta$, given $X_1, \ldots, X_n$ iid with pdf

(a) $f(x|\theta) = 1, \; \theta - \frac{1}{2} < x < \theta + \frac{1}{2}$.

(b) $f(x|\theta) = 2x/\theta^2, \; 0 < x < \theta, \; \theta > 0$.

**9.18** In this exercise we will investigate some more properties of binomial confidence sets and the Sterne (1954) construction in particular. As in Example 9.2.11, we will again consider the binomial$(3, p)$ distribution.

(a) Draw, as a function of $p$, a graph of the four probability functions $P_p(X = x)$, $x = 0, \ldots, 3$. Identify the maxima of $P_p(X = 1)$ and $P_p(X = 2)$.

(b) Show that for small $\epsilon$, $P_p(X = 0) > P_p(X = 2)$ for $p = \frac{1}{3} + \epsilon$.

(c) Show that the *most probable construction* is to blame for the difficulties with the Sterne sets by showing that the following acceptance regions can be inverted to obtain a $1 - \alpha = .442$ confidence *interval*.

| $p$ | Acceptance region |
|---|---|
| $[.000, .238]$ | $\{0\}$ |
| $(.238, .305)$ | $\{0,\ 1\}$ |
| $[.305, .362]$ | $\{1\}$ |
| $(.362, .634)$ | $\{1,\ 2\}$ |
| $[.634, .695]$ | $\{2\}$ |
| $(.695, .762)$ | $\{2,\ 3\}$ |
| $[.762, 1.00]$ | $\{3\}$ |

(This is essentially Crow's 1956 modification of Sterne's construction; see Miscellanea 9.5.2.)

**9.19** Prove part (b) of Theorem 9.2.12.

**9.20** Some of the details of the proof of Theorem 9.2.14 need to be filled in, and the second part of the theorem needs to be proved.

(a) Show that if $F_T(T|\theta)$ is stochastically greater than or equal to a uniform random variable, then so is $\bar{F}_T(T|\theta)$. That is, if $P_\theta(F_T(T|\theta) \leq x) \leq x$ for every $x, 0 \leq x \leq 1$, then $P_\theta(\bar{F}_T(T|\theta) \leq x) \leq x$ for every $x, 0 \leq x \leq 1$.

(b) Show that for $\alpha_1 + \alpha_2 = \alpha$, the set $\{\theta : F_T(T|\theta) \leq \alpha_1 \text{ and } \bar{F}_T(T|\theta) \leq \alpha_2\}$ is a $1 - \alpha$ confidence set.

(c) If the cdf $F_T(t|\theta)$ is a decreasing function of $\theta$ for each $t$, show that the function $\bar{F}_T(t|\theta)$ defined by $\bar{F}_T(t|\theta) = P(T \geq t|\theta)$ is a nondecreasing function of $\theta$ for each $t$.

(d) Prove part (b) of Theorem 9.2.14.

**9.21** In Example 9.2.15 it was shown that a confidence interval for a Poisson parameter can be expressed in terms of chi squared cutoff points. Use a similar technique to show that if $X \sim$ binomial$(n, p)$, then a $1 - \alpha$ confidence interval for $p$ is

$$\frac{1}{1 + \frac{n-x+1}{x} F_{2(n-x+1), 2x, \alpha/2}} \leq p \leq \frac{\frac{x+1}{n-x} F_{2(x+1), 2(n-x), \alpha/2}}{1 + \frac{x+1}{n-x} F_{2(x+1), 2(n-x), \alpha/2}},$$

where $F_{\nu_1, \nu_2, \alpha}$ is the upper $\alpha$ cutoff from an $F$ distribution with $\nu_1$ and $\nu_2$ degrees of freedom, and we make the endpoint adjustment that the lower endpoint is 0 if $x = 0$ and the upper endpoint is 1 if $x = n$. These are the Clopper and Pearson (1934) intervals.

(*Hint:* Recall the following identity from Exercise 2.40, which can be interpreted in the following way. If $X \sim$ binomial$(n, \theta)$, then $P_\theta(X \geq x) = P(Y \leq \theta)$, where $Y \sim$ beta$(x, n - x + 1)$. Use the properties of the $F$ and beta distributions from Chapter 5.)

**9.22** If $X \sim$ negative binomial$(r, p)$, use the relationship between the binomial and negative binomial to show that a $1 - \alpha$ confidence interval for $p$ is given by

$$\frac{1}{1 + \frac{x+1}{r} F_{2(x+1), 2r, \alpha/2}} \leq p \leq \frac{\frac{r}{x} F_{2r, 2x, \alpha/2}}{1 + \frac{r}{x} F_{2r, 2x, \alpha/2}},$$

with a suitable modification if $x = 0$.

**9.23** (a) Let $X_1, \ldots, X_n$ be a random sample from a Poisson population with parameter $\lambda$ and define $Y = \sum X_i$. In Example 9.2.15 a confidence interval for $\lambda$ was found using the method of Section 9.2.3. Construct another interval for $\lambda$ by inverting an LRT, and compare the intervals.

   (b) The following data, the number of aphids per row in nine rows of a potato field, can be assumed to follow a Poisson distribution:

$$155, \quad 104, \quad 66, \quad 50, \quad 36, \quad 40, \quad 30, \quad 35, \quad 42.$$

   Use these data to construct a 90% LRT confidence interval for the mean number of aphids per row. Also, construct an interval using the method of Example 9.2.15.

**9.24** For $X \sim \text{Poisson}(\lambda)$, show that the coverage probability of the confidence interval $[L(X), U(X)]$ in Example 9.2.15 is given by

$$P_\lambda(\lambda \in [L(X), U(X)]) = \sum_{x=0}^\infty I_{[L(x),U(x)]}(\lambda) \frac{e^{-\lambda}\lambda^x}{x!}$$

and that we can define functions $x_1(\lambda)$ and $x_u(\lambda)$ so that

$$P_\lambda(\lambda \in [L(X), U(X)]) = \sum_{x=x_1(\lambda)}^{x_u(\lambda)} \frac{e^{-\lambda}\lambda^x}{x!}.$$

Hence, explain why the graph of the coverage probability of the Poisson intervals given in Figure 9.2.5 has jumps occurring at the endpoints of the different confidence intervals.

**9.25** If $X_1, \ldots, X_n$ are iid with pdf $f(x|\mu) = e^{-(x-\mu)}I_{[\mu,\infty)}(x)$, then $Y = \min\{X_1, \ldots, X_n\}$ is sufficient for $\mu$ with pdf

$$f_Y(y|\mu) = ne^{-n(y-\mu)}I_{[\mu,\infty)}(y).$$

In Example 9.2.13 a $1 - \alpha$ confidence interval for $\mu$ was found using the method of Section 9.2.3. Compare that interval to $1 - \alpha$ intervals obtained by likelihood and pivotal methods.

**9.26** Let $X_1, \ldots, X_n$ be iid observations from a beta$(\theta, 1)$ pdf and assume that $\theta$ has a gamma$(r, \lambda)$ prior pdf. Find a $1 - \alpha$ Bayes credible set for $\theta$.

**9.27** (a) Let $X_1, \ldots, X_n$ be iid observations from an exponential$(\lambda)$ pdf, where $\lambda$ has the conjugate IG$(a, b)$ prior, an inverted gamma with pdf

$$\pi(\lambda|a, b) = \frac{1}{\Gamma(a)b^a} \left(\frac{1}{\lambda}\right)^{a+1} e^{-1/(b\lambda)}, \quad 0 < \lambda < \infty.$$

   Show how to find a $1 - \alpha$ Bayes HPD credible set for $\lambda$.

   (b) Find a $1 - \alpha$ Bayes HPD credible set for $\sigma^2$, the variance of a normal distribution, based on the sample variance $s^2$ and using a conjugate IG$(a, b)$ prior for $\sigma^2$.

   (c) Starting with the interval from part (b), find the limiting $1 - \alpha$ Bayes HPD credible set for $\sigma^2$ obtained as $a \to 0$ and $b \to \infty$.

**9.28** Let $X_1, \ldots, X_n$ be iid n$(\theta, \sigma^2)$, where both $\theta$ and $\sigma^2$ are unknown, but there is only interest on inference about $\theta$. Consider the prior pdf

$$\pi(\theta, \sigma^2|\mu, \tau^2, a, b) = \frac{1}{\sqrt{2\pi\tau^2\sigma^2}}e^{-(\theta-\mu)^2/(2\tau^2\sigma^2)} \frac{1}{\Gamma(a)b^a}\left(\frac{1}{\sigma^2}\right)^{a+1}e^{-1/(b\sigma^2)},$$

a n$(\mu, \tau^2\sigma^2)$ multiplied by an IG$(a, b)$.

(a) Show that this prior is a conjugate prior for this problem.

(b) Find the posterior distribution of $\theta$ and use it to construct a $1 - \alpha$ credible set for $\theta$.

(c) The classical $1 - \alpha$ confidence set for $\theta$ can be expressed as

$$\left\{\theta : |\theta - \bar{x}|^2 \leq F_{1,n-1,\alpha} \frac{s^2}{n}\right\}.$$

Is there any (limiting) sequence of $\tau^2$, $a$, and $b$ that would allow this set to be approached by a Bayes set from part (b)?

**9.29** Let $X_1, \ldots, X_n$ are a sequence of $n$ Bernoulli($p$) trials.

(a) Calculate a $1 - \alpha$ credible set for $p$ using the conjugate beta($a, b$) prior.

(b) Using the relationship between the beta and $F$ distributions, write the credible set in a form that is comparable to the form of the intervals in Exercise 9.21. Compare the intervals.

**9.30** Complete the credible probability calculation needed in Example 9.2.17.

(a) Assume that $a$ is an integer, and show that $T = \frac{2(nb+1)}{b}\lambda \sim \chi^2_{2(a+\Sigma x)}$.

(b) Show that

$$\frac{\chi^2_\nu - \nu}{\sqrt{2\nu}} \to \mathrm{n}(0, 1)$$

as $\nu \to \infty$. (Use moment generating functions. The limit is difficult to evaluate—take logs and then do a Taylor expansion. Alternatively, see Example A.0.8 in Appendix A.)

(c) Standardize the random variable $T$ of part (a), and write the credible probability (9.2.21) in terms of this variable. Show that the standardized lower cutoff point $\to \infty$ as $\Sigma x_i \to \infty$, and hence the credible probability goes to 0.

**9.31** Complete the coverage probability calculation needed in Example 9.2.17.

(a) If $\chi^2_{2Y}$ is a chi squared random variable with $Y \sim$ Poisson($\lambda$), show that $\mathrm{E}(\chi^2_{2Y}) = 2\lambda$, $\mathrm{Var}(\chi^2_{2Y}) = 8\lambda$, the mgf of $\chi^2_Y$ is given by $\exp\left(-\lambda + \frac{\lambda}{1-2t}\right)$, and

$$\frac{\chi^2_{2Y} - 2\lambda}{\sqrt{8\lambda}} \to \mathrm{n}(0, 1)$$

as $\lambda \to \infty$. (Use moment generating functions.)

(b) Now evaluate (9.2.22) as $\lambda \to \infty$ by first standardizing $\chi^2_{2Y}$. Show that the standardized upper limit goes to $-\infty$ as $\lambda \to \infty$, and hence the coverage probability goes to 0.

**9.32** In this exercise we will calculate the classical coverage probability of the HPD region in (9.2.23), that is, the coverage probability of the Bayes HPD region using the probability model $\bar{X} \sim \mathrm{n}(\theta, \sigma^2/n)$.

(a) Using the definitions given in Example 9.3.12, prove that

$$P_\theta\left(|\theta - \delta^{\mathrm{B}}(\bar{X})| \leq z_{\alpha/2}\sqrt{\mathrm{Var}(\theta|\bar{X})}\right)$$

$$= P_\theta\left[-\sqrt{1+\gamma}z_{\alpha/2} + \frac{\gamma(\theta - \mu)}{\sigma/\sqrt{n}} \leq Z \leq \sqrt{1+\gamma}z_{\alpha/2} + \frac{\gamma(\theta - \mu)}{\sigma/\sqrt{n}}\right].$$

(b) Show that the above set, although a $1 - \alpha$ credible set, is not a $1 - \alpha$ confidence set. (Fix $\theta \neq \mu$, let $\tau = \sigma/\sqrt{n}$, so that $\gamma = 1$. Prove that as $\sigma^2/n \to 0$, the above probability goes to 0.)

(c) If $\theta = \mu$, however, prove that the coverage probability is bounded away from 0. Find the minimum and maximum of this coverage probability.

(d) Now we will look at the other side. The usual $1 - \alpha$ confidence set for $\theta$ is $\{\theta : |\theta - \bar{x}| \leq z_{\alpha/2}\sigma/\sqrt{n}\}$. Show that the credible probability of this set is given by

$$P_{\bar{x}}\left(|\theta - \bar{x}| \leq z_{\alpha/2}\sigma/\sqrt{n}\right)$$

$$= P_{\bar{x}}\left[-\sqrt{1 + \gamma}z_{\alpha/2} + \frac{\gamma(\bar{x} - \mu)}{\sqrt{1 + \gamma}\sigma/\sqrt{n}} \leq Z \leq \sqrt{1 + \gamma}z_{\alpha/2} + \frac{\gamma(\bar{x} - \mu)}{\sqrt{1 + \gamma}\sigma/\sqrt{n}}\right]$$

and that this probability is not bounded away from 0. Hence, the $1 - \alpha$ confidence set is not a $1 - \alpha$ credible set.

**9.33** Let $X \sim n(\mu, 1)$ and consider the confidence interval

$$C_a(x) = \{\mu : \min\{0, (x - a)\} \leq \mu \leq \max\{0, (x + a)\}\}.$$

(a) For $a = 1.645$, prove that the coverage probability of $C_a(x)$ is exactly .95 for all $\mu$, with the exception of $\mu = 0$, where the coverage probability is 1.

(b) Now consider the so-called noninformative prior $\pi(\mu) = 1$. Using this prior and again taking $a = 1.645$, show that the posterior credible probability of $C_a(x)$ is exactly .90 for $-1.645 \leq x \leq 1.645$ and increases to .95 as $|x| \to \infty$.

This type of interval arises in the problem of *bioequivalence*, where the objective is to decide if two treatments (different formulations of a drug, different delivery systems of a treatment) produce the same effect. The formulation of the problem results in "turning around" the roles of the null and alternative hypotheses (see Exercise 8.47), resulting in some interesting statistics. See Berger and Hsu (1996) for a review of bioequivalence and Brown, Casella, and Hwang (1995) for generalizations of the confidence set.

**9.34** Suppose that $X_1, \ldots, X_n$ is a random sample from a $n(\mu, \sigma^2)$ population.

(a) If $\sigma^2$ is known, find a minimum value for $n$ to guarantee that a .95 confidence interval for $\mu$ will have length no more than $\sigma/4$.

(b) If $\sigma^2$ is unknown, find a minimum value for $n$ to guarantee, with probability .90, that a .95 confidence interval for $\mu$ will have length no more than $\sigma/4$.

**9.35** Let $X_1, \ldots, X_n$ be a random sample from a $n(\mu, \sigma^2)$ population. Compare expected lengths of $1 - \alpha$ confidence intervals for $\mu$ that are computed assuming

(a) $\sigma^2$ is known.

(b) $\sigma^2$ is unknown.

**9.36** Let $X_1, \ldots, X_n$ be independent with pdfs $f_{X_i}(x|\theta) = e^{i\theta - x}I_{[i\theta, \infty)}(x)$. Prove that $T = \min_i(X_i/i)$ is a sufficient statistic for $\theta$. Based on $T$, find the $1 - \alpha$ confidence interval for $\theta$ of the form $[T + a, T + b]$ which is of minimum length.

**9.37** Let $X_1, \ldots, X_n$ be iid uniform$(0, \theta)$. Let $Y$ be the largest order statistic. Prove that $Y/\theta$ is a pivotal quantity and show that the interval

$$\left\{\theta : y \leq \theta \leq \frac{y}{\alpha^{1/n}}\right\}$$

is the shortest $1 - \alpha$ pivotal interval.

**9.38** If, in Theorem 9.3.2, we assume that $f$ is continuous, then we can simplify the proof. For fixed $c$, consider the integral $\int_a^{a+c} f(x)dx$.

   (a) Show that $\frac{d}{da} \int_a^{a+c} f(x)\,dx = f(a+c) - f(a)$.

   (b) Prove that the unimodality of $f$ implies that $\int_a^{a+c} f(x)\,dx$ is maximized when $a$ satisfies $f(a+c) - f(a) = 0$.

   (c) Suppose that, given $\alpha$, we choose $c^*$ and $a^*$ to satisfy $\int_{a^*}^{a^*+c^*} f(x)\,dx = 1 - \alpha$ and $f(a^* + c^*) - f(a^*) = 0$. Prove that this is the shortest $1 - \alpha$ interval.

**9.39** Prove a special case of Theorem 9.3.2. Let $X \sim f(x)$, where $f$ is a *symmetric unimodal* pdf. For a fixed value of $1 - \alpha$, of all intervals $[a, b]$ that satisfy $\int_a^b f(x)\,dx = 1 - \alpha$, the shortest is obtained by choosing $a$ and $b$ so that $\int_{-\infty}^a f(x)\,dx = \alpha/2$ and $\int_b^\infty f(x)\,dx = \alpha/2$.

**9.40** Building on Exercise 9.39, show that if $f$ is symmetric, the optimal interval is of the form $m \pm k$, where $m$ is the mode of $f$ and $k$ is a constant. Hence, show that (a) symmetric likelihood functions produce likelihood regions that are symmetric about the MLE if $k'$ does not depend on the parameter (see (9.2.7)), and (b) symmetric posterior densities produce HPD regions that are symmetric about the posterior mean.

**9.41** (a) Prove the following, which is related to Theorem 9.3.2. Let $X \sim f(x)$, where $f$ is a *strictly decreasing* pdf on $[0, \infty)$. For a fixed value of $1 - \alpha$, of all intervals $[a, b]$ that satisfy $\int_a^b f(x)\,dx = 1 - \alpha$, the shortest is obtained by choosing $a = 0$ and $b$ so that $\int_0^b f(x)\,dx = 1 - \alpha$.

   (b) Use the result of part (a) to find the shortest $1 - \alpha$ confidence interval in Example 9.2.13.

**9.42** Referring to Example 9.3.4, to find the shortest pivotal interval for a gamma scale parameter, we had to solve a constrained minimization problem.

   (a) Show that the solution is given by the $a$ and $b$ that satisfy $\int_a^b f_Y(y)\,dy = 1 - \alpha$ and $f(b)b^2 = f(a)a^2$.

   (b) With one observation from a gamma$(k, \beta)$ pdf with known shape parameter $k$, find the shortest $1 - \alpha$ (pivotal) confidence interval of the form $\{\beta : x/b \le \beta \le x/a\}$.

**9.43** Juola (1993) makes the following observation. If we have a pivot $Q(X, \theta)$, a $1 - \alpha$ confidence interval involves finding $a$ and $b$ so that $P(a < Q < b) = 1 - \alpha$. Typically the length of the interval on $\theta$ will be some function of $a$ and $b$ like $b - a$ or $1/b^2 - 1/a^2$. If $Q$ has density $f$ and the length can be expressed as $\int_a^b g(t)\,dt$ the shortest pivotal interval is the solution to

$$\min_{\{a,b\}} \int_a^b g(t)\,dt \text{ subject to } \int_a^b f(t)\,dt = 1 - \alpha$$

or, more generally,

$$\min_C \int_C g(t)\,dt \text{ subject to } \int_C f(t)\,dt \ge 1 - \alpha.$$

   (a) Prove that the solution is $C = \{t : g(t) < \lambda f(t)\}$, where $\lambda$ is chosen so that $\int_C f(t)\,dt = 1 - \alpha$. (*Hint:* You can adapt the proof of Theorem 8.3.12, the Neyman-Pearson Lemma.)

(b) Apply the result in part (a) to get the shortest intervals in Exercises 9.37 and 9.42.

**9.44** (a) Let $X_1, \ldots, X_n$ be iid Poisson($\lambda$). Find a UMA $1 - \alpha$ confidence interval based on inverting the UMP level $\alpha$ test of $H_0: \lambda = \lambda_0$ versus $H_1: \lambda > \lambda_0$.

(b) Let $f(x|\theta)$ be the logistic$(\theta, 1)$ location pdf. Based on one observation, $x$, find the UMA one-sided $1 - \alpha$ confidence interval of the form $\{\theta: \theta \leq U(x)\}$.

**9.45** Let $X_1, \ldots, X_n$ be iid exponential($\lambda$).

(a) Find a UMP size $\alpha$ hypothesis test of $H_0: \lambda = \lambda_0$ versus $H_1: \lambda < \lambda_0$.

(b) Find a UMA $1 - \alpha$ confidence interval based on inverting the test in part (a). Show that the interval can be expressed as

$$C^*(x_1, \ldots, x_n) = \left\{ \lambda: 0 \leq \lambda \leq \frac{2 \sum x_i}{\chi^2_{2n,\alpha}} \right\}.$$

(c) Find the expected length of $C^*(x_1, \ldots, x_n)$.

(d) Madansky (1962) exhibited a $1 - \alpha$ interval whose expected length is shorter than that of the UMA interval. In general, Madansky's interval is difficult to calculate, but in the following situation calculation is relatively simple. Let $1 - \alpha = .3$ and $n = 120$. Madansky's interval is

$$C^M(x_1, \ldots, x_n) = \left\{ \lambda: 0 \leq \lambda \leq -\frac{x_{(1)}}{\log(.99)} \right\},$$

which is a 30% confidence interval. Use the fact that $\chi^2_{240,.7} = 251.046$ to show that the 30% UMA interval satisfies

$$\mathrm{E}\left[\mathrm{Length}\,(C^*(x_1, \ldots, x_n))\right] = .956\lambda > \mathrm{E}\left[\mathrm{Length}\,\big(C^M(x_1, \ldots, x_n)\big)\right] = .829\lambda.$$

**9.46** Show that if $A(\theta_0)$ is an unbiased level $\alpha$ acceptance region of a test of $H_0: \theta = \theta_0$ versus $H_1: \theta \neq \theta_0$ and $C(\mathbf{x})$ is the $1 - \alpha$ confidence set formed by inverting the acceptance regions, then $C(\mathbf{x})$ is an unbiased $1 - \alpha$ confidence set.

**9.47** Let $X_1, \ldots, X_n$ be a random sample from a n$(0, \sigma^2)$ population, where $\sigma^2$ is known. Show that the usual one-sided $1 - \alpha$ upper confidence bound $\{\theta: \theta \leq \bar{x} + z_\alpha \sigma/\sqrt{n}\}$ is unbiased, and so is the corresponding lower confidence bound.

**9.48** Let $X_1, \ldots, X_n$ be a random sample from a n$(0, \sigma^2)$ population, where $\sigma^2$ is unknown.

(a) Show that the interval $\theta \leq \bar{x} + t_{n-1,\alpha} \frac{s}{\sqrt{n}}$ can be derived by inverting the acceptance region of an LRT.

(b) Show that the corresponding two-sided interval in (9.2.14) can also derived by inverting the acceptance region of an LRT.

(c) Show that the intervals in parts (a) and (b) are unbiased intervals.

**9.49** (*Cox's Paradox*)  We are to test

$$H_0: \theta = \theta_0 \qquad \text{versus} \qquad H_1: \theta > \theta_0,$$

where $\theta$ is the mean of one of two normal distributions and $\theta_0$ is a fixed but arbitrary value of $\theta$. We observe the random variable $X$ with distribution

$$X \sim \begin{cases} \mathrm{n}(\theta, 100) & \text{with probability } p \\ \mathrm{n}(\theta, 1) & \text{with probability } 1 - p. \end{cases}$$

(a) Show that the test given by

$$\text{reject } H_0 \text{ if } X > \theta_0 + z_\alpha \sigma,$$

where $\sigma = 1$ or $10$ depending on which population is sampled, is a level $\alpha$ test. Derive a $1 - \alpha$ confidence set by inverting the acceptance region of this test.

(b) Show that a more powerful level $\alpha$ test (for $\alpha > p$) is given by

$$\text{reject } H_0 \text{ if } X > \theta_0 + z_{(\alpha-p)/(1-p)} \text{ and } \sigma = 1; \text{ otherwise } always \text{ reject } H_0.$$

Derive a $1 - \alpha$ confidence set by inverting the acceptance region of this test, and show that it is the empty set with positive probability. (Cox's Paradox states that classic optimal procedures sometimes ignore the information about conditional distributions and provide us with a procedure that, while optimal, is somehow unreasonable; see Cox 1958 or Cornfield 1969.)

**9.50** Let $X \sim f(x|\theta)$, and suppose that the interval $\{\theta : a(X) \leq \theta \leq b(X)\}$ is a UMA confidence set for $\theta$.

(a) Find a UMA confidence set for $1/\theta$. Note that if $a(x) < 0 < b(x)$, this set is $\{1/\theta : 1/b(x) \leq 1/\theta\} \cup \{1/\theta : 1/\theta \leq 1/a(x)\}$. Hence it is possible for the UMA confidence set to be neither an interval nor bounded.

(b) Show that, if $h$ is a strictly increasing function, the set $\{h(\theta) : h(a(X)) \leq h(\theta) \leq h(b(X))\}$ is a UMA confidence set for $h(\theta)$. Can the condition on $h$ be relaxed?

**9.51** If $X_1, \ldots, X_n$ are iid from a location pdf $f(x - \theta)$, show that the confidence set

$$C(x_1, \ldots, x_n) = \{\theta : \bar{x} - k_1 \leq \theta \leq \bar{x} + k_2\},$$

where $k_1$ and $k_2$ are constants, has constant coverage probability. (*Hint:* The pdf of $\bar{X}$ is of the form $f_{\bar{X}}(\bar{x} - \theta)$.)

**9.52** Let $X_1, \ldots, X_n$ be a random sample from a $n(\mu, \sigma^2)$ population, where both $\mu$ and $\sigma^2$ are unknown. Each of the following methods of finding confidence intervals for $\sigma^2$ results in intervals of the form

$$\left\{ \sigma^2 : \frac{(n-1)s^2}{b} \leq \sigma^2 \leq \frac{(n-1)s^2}{a} \right\},$$

but in each case $a$ and $b$ will satisfy different constraints. The intervals given in this exercise are derived by Tate and Klett (1959), who also tabulate some cutoff points.

Define $f_p(t)$ to be the pdf of a $\chi_p^2$ random variable with $p$ degrees of freedom. In order to have a $1 - \alpha$ confidence interval, $a$ and $b$ must satisfy

$$\int_a^b f_{n-1}(t)\, dt = 1 - \alpha,$$

but additional constraints are required to define $a$ and $b$ uniquely. Verify that each of the following constraints can be derived as stated.

(a) *The likelihood ratio interval:* The $1 - \alpha$ confidence interval obtained by inverting the LRT of $H_0 : \sigma = \sigma_0$ versus $H_1 : \sigma \neq \sigma_0$ is of the above form, where $a$ and $b$ also satisfy $f_{n+2}(a) = f_{n+2}(b)$.

(b) *The minimum length interval:* For intervals of the above form, the $1 - \alpha$ confidence interval obtained by minimizing the interval length constrains $a$ and $b$ to satisfy $f_{n+3}(a) = f_{n+3}(b)$.

(c) *The shortest unbiased interval:* For intervals of the above form, the $1 - \alpha$ confidence interval obtained by minimizing the probability of false coverage among all unbiased intervals constrains $a$ and $b$ to satisfy $f_{n+1}(a) = f_{n+1}(b)$. This interval can also be derived by minimizing the ratio of the endpoints.

(d) *The equal-tail interval:* For intervals of the above form, the $1 - \alpha$ confidence interval obtained by requiring that the probability above and below the interval be equal constrains $a$ and $b$ to satisfy

$$\int_0^a f_{n-1}(t) \, dt = \frac{\alpha}{2}, \qquad \int_b^\infty f_{n-1}(t) \, dt = \frac{\alpha}{2}.$$

(This interval, although very common, is clearly nonoptimal no matter what length criterion is used.)

(e) For $\alpha = .1$ and $n = 3$, find the numerical values of $a$ and $b$ for each of the above cases. Compare the length of this intervals.

**9.53** Let $X \sim n(\mu, \sigma^2)$, $\sigma^2$ known. For each $c \geq 0$, define an interval estimator for $\mu$ by $C(x) = [x - c\sigma, x + c\sigma]$ and consider the loss in (9.3.4).

(a) Show that the risk function, $R(\mu, C)$, is given by

$$R(\mu, C) = b(2c\sigma) - P(-c \leq Z \leq c).$$

(b) Using the Fundamental Theorem of Calculus, show that

$$\frac{d}{dc} R(\mu, C) = 2b\sigma - \frac{2}{\sqrt{2\pi}} e^{-c^2/2}$$

and, hence, the derivative is an increasing function of $c$ for $c \geq 0$.

(c) Show that if $b\sigma > 1/\sqrt{2\pi}$, the derivative is positive for all $c \geq 0$ and, hence, $R(\mu, C)$ is minimized at $c = 0$. That is, the best interval estimator is the point estimator $C(x) = [x, x]$.

(d) Show that if $b\sigma \leq 1/\sqrt{2\pi}$, the $c$ that minimizes the risk is $c = \sqrt{-2\log(b\sigma\sqrt{2\pi})}$. Hence, if $b$ is chosen so that $c = z_{\alpha/2}$ for some $\alpha$, then the interval estimator that minimizes the risk is just the usual $1-\alpha$ confidence interval.

**9.54** Let $X \sim n(\mu, \sigma^2)$, but now consider $\sigma^2$ unknown. For each $c \geq 0$, define an interval estimator for $\mu$ by $C(x) = [x - cs, x + cs]$, where $s^2$ is an estimator of $\sigma^2$ independent of $X$, $\nu S^2/\sigma^2 \sim \chi_\nu^2$ (for example, the usual sample variance). Consider a modification of the loss in (9.3.4),

$$L((\mu, \sigma), C) = \frac{b}{\sigma} \text{Length}(C) - I_C(\mu).$$

(a) Show that the risk function, $R((\mu, \sigma), C)$, is given by

$$R((\mu, \sigma), C) = b(2cM) - [2P(T \leq c) - 1],$$

where $T \sim t_\nu$ and $M = ES/\sigma$.

(b) If $b \leq 1/\sqrt{2\pi}$, show that the $c$ that minimizes the risk satisfies

$$b = \frac{1}{\sqrt{2\pi}} \left(\frac{\nu}{\nu + c^2}\right)^{(\nu+1)/2}.$$

(c) Reconcile this problem with the known $\sigma^2$ case. Show that as $\nu \to \infty$, the solution here converges to the solution in the known $\sigma^2$ problem. (Be careful of the rescaling done to the loss function.)

**9.55** The decision theoretic approach to set estimation can be quite useful (see Exercise 9.56) but it can also give some unsettling results, showing the need for thoughtful implementation. Consider again the case of $X \sim n(\mu, \sigma^2)$, $\sigma^2$ unknown, and suppose that we have an interval estimator for $\mu$ by $C(x) = [x - cs, x + cs]$, where $s^2$ is an estimator of $\sigma^2$ independent of $X$, $\nu S^2/\sigma^2 \sim \chi^2_\nu$. This is, of course, the usual $t$ interval, one of the great statistical procedures that has stood the test of time. Consider the loss

$$L((\mu, \sigma), C) = b \text{ Length}(C) - I_C(\mu),$$

similar to that used in Exercise 9.54, but without scaling the length. Construct another procedure $C'$ as

$$C' = \begin{cases} [x - cs, x + cs] & \text{if } s < K \\ \emptyset & \text{if } s \geq K, \end{cases}$$

where $K$ is a positive constant. Notice that $C'$ does *exactly the wrong thing*. When $s^2$ is big and there is a lot of uncertainty, we would want the interval to be wide. But $C'$ is empty! Show that we can find a value of $K$ so that

$$R((\mu, \sigma), C') \leq R((\mu, \sigma), C) \quad \text{for every } (\mu, \sigma)$$

with strict inequality for some $(\mu, \sigma)$.

**9.56** Let $X \sim f(x|\theta)$ and suppose that we want to estimate $\theta$ with an interval estimator $C$ using the loss in (9.3.4). If $\theta$ has the prior pdf $\pi(\theta)$, show that the Bayes rule is given by

$$C^\pi = \{\theta : \pi(\theta|x) \geq b\}.$$

(*Hint*: Write $\text{Length}(C) = \int_C 1 \, d\theta$ and use the Neyman–Pearson Lemma.)

The following two problems relate to Miscellanea 9.5.4.

**9.57** Let $X_1, \ldots, X_n$ be iid $n(\mu, \sigma^2)$, where $\sigma^2$ is known. We know that a $1 - \alpha$ confidence interval for $\mu$ is $\bar{x} \pm z_{\alpha/2} \frac{\sigma}{\sqrt{n}}$.

(a) Show that a $1 - \alpha$ prediction interval for $X_{n+1}$ is $\bar{x} \pm z_{\alpha/2} \sigma \sqrt{1 + \frac{1}{n}}$.

(b) Show that a $1 - \alpha$ tolerance interval for $100p\%$ of the underlying population is given by $\bar{x} \pm \sigma \left( z_{p/2} + \frac{z_{\alpha/2}}{\sqrt{n}} \right)$.

(c) Find a $1 - \alpha$ prediction interval for $X_{n+1}$ if $\sigma^2$ is unknown.

(If $\sigma^2$ is unknown, the $1 - \alpha$ tolerance interval is quite an involved calculation.)

**9.58** Let $X_1, \ldots, X_n$ be iid observations from a population with median $m$. *Distribution-free* intervals can be based on the order statistics $X_{(1)} \leq \cdots \leq X_{(n)}$ in the following way.

(a) Show that the one-sided intervals $(-\infty, x_{(n)}]$ and $[x_{(1)}, \infty)$ are each confidence intervals for $m$ with confidence coefficient $1 - (1/2)^n$, and the confidence coefficient of the interval $[x_{(1)}, x_{(n)}]$ is $1 - 2(1/2)^n$.

(b) Show that the one-sided intervals of part (a) are prediction intervals with coefficient $n/(n+1)$ and the two-sided interval is a prediction interval with coefficient $(n-1)/(n+1)$.

(c) The intervals in part (a) can also be used as tolerance intervals for proportion $p$ of the underlying population. Show that, when considered as tolerance intervals, the one-sided intervals have coefficient $1 - p^n$ and the two-sided interval has coefficient $1 - p^n - n(1 - p)p^{n-1}$. Vardeman (1992) refers to this last calculation as a "nice exercise in order statistics."

## 9.5 Miscellanea

### 9.5.1 Confidence Procedures

Confidence sets and tests can be related formally by defining an entity called a *confidence procedure* (Joshi 1969). If $X \sim f(x|\theta)$, where $x \in \mathcal{X}$ and $\theta \in \Theta$, then a confidence procedure is a set in the space $\mathcal{X} \times \Theta$, the Cartesian product space. It is defined as

$$\{(x, \theta) \colon (x, \theta) \in \mathbf{C}\}$$

for a set $\mathbf{C} \in \mathcal{X} \times \Theta$.

From the confidence procedure we can define two slices, or sections, obtained by holding one of the variables constant. For fixed $x$, we define the $\theta$-*section* or confidence set as

$$C(x) = \{\theta \colon (x, \theta) \in \mathbf{C}\}.$$

For fixed $\theta$, we define the $x$-*section* or acceptance region as

$$A(\theta) = \{x \colon (x, \theta) \in \mathbf{C}\}.$$

Although this development necessitates working with the product space $\mathcal{X} \times \Theta$, which is one reason we do not use it here, it does provide a more straightforward way of seeing the relationship between tests and sets. Figure 9.2.1 illustrates this correspondence in the normal case.

### 9.5.2 Confidence Intervals in Discrete Distributions

The construction of optimal (or at least improved) confidence intervals for parameters from discrete distributions has a long history, as indicated in Example 9.2.11, where we looked at the Sterne (1954) modification to the intervals of Clopper and Pearson (1934). Of course, there are difficulties with the Sterne construction, but the basic idea is sound, and Crow (1956) and Blyth and Still (1983) modified Sterne's construction, with the latter producing the shortest set of exact intervals. Casella (1986) gave an algorithm to find a class of shortest binomial confidence intervals.

The history of Poisson interval research (which often includes other discrete distributions) is similar. The Garwood (1936) construction is exactly the Clopper-Pearson argument applied to the binomial, and Crow and Gardner (1959) improved the intervals. Casella and Robert (1989) found a class of shortest Poisson intervals.

Blyth (1986) produces very accurate approximate intervals for a binomial parameter, Leemis and Trivedi (1996) compare normal and Poisson approximations, and
Agresti and Coull (1998) argue that requiring discrete intervals to maintain coverage above the nominal level may be too stringent. Blaker (2000) constructs improved intervals for binomial, Poisson, and other discrete distributions that have a
nesting property: For $\alpha < \alpha'$, the $1 - \alpha$ intervals contain the corresponding $1 - \alpha'$
intervals.

### 9.5.3 Fieller's Theorem

Fieller's Theorem (Fieller 1954) is a clever argument to get an exact confidence set
on a ratio of normal means.

Given a random sample $(X_1, Y_1), \ldots, (X_n, Y_n)$ from a bivariate normal distribution
with parameters $(\mu_X, \mu_Y, \sigma_X^2, \sigma_Y^2, \rho)$, a confidence set on $\theta = \mu_Y / \mu_X$ can be formed
in the following way. For $i = 1, \ldots, n$, define $Z_{\theta i} = Y_i - \theta X_i$ and $\bar{Z}_\theta = \bar{Y} - \theta \bar{X}$. It
can be shown that $\bar{Z}_\theta$ is normal with mean 0 and variance

$$V_\theta = \frac{1}{n} \left( \sigma_Y^2 - 2\theta \rho \sigma_Y \sigma_X + \theta^2 \sigma_X^2 \right).$$

$V_\theta$ can be estimated with $\hat{V}_\theta$, given by

$$\hat{V}_\theta = \frac{1}{n(n-1)} \sum_{i=1}^{n} (Z_{\theta i} - \bar{Z}_\theta)^2$$

$$= \frac{1}{n-1} \left( S_Y^2 - 2\theta S_{YX} + \theta^2 S_X^2 \right),$$

where

$$S_Y^2 = \frac{1}{n} \sum_{i=1}^{n} (Y_i - \bar{Y})^2, \quad S_X^2 = \frac{1}{n} \sum_{i=1}^{n} (X_i - \bar{X})^2, \quad S_{YX} = \frac{1}{n} \sum_{i=1}^{n} (Y_i - \bar{Y})(X_i - \bar{X}).$$

Furthermore, it also can be shown that $E\hat{V}_\theta = V_\theta$, $\hat{V}_\theta$ is independent of $\bar{Z}_\theta$, and
$(n-1)\hat{V}_\theta / V_\theta \sim \chi_{n-1}^2$. Hence, $\bar{Z}_\theta / \sqrt{\hat{V}_\theta} \sim t_{n-1}$ and the set

$$\left\{ \theta : \frac{\bar{z}_\theta^2}{\hat{v}_\theta} \leq t_{n-1, \alpha/2}^2 \right\}$$

defines a $1 - \alpha$ confidence set for $\theta$, the ratio of the means. This set defines a
parabola in $\theta$, and the roots of the parabola give the endpoints of the confidence
set. Writing the set in terms of the original variables, we get

$$\left\{ \theta : \left( \bar{x}^2 - \frac{t_{n-1, \alpha/2}^2}{n-1} S_x^2 \right) \theta^2 - 2 \left( \bar{x}\bar{y} - \frac{t_{n-1, \alpha/2}^2}{n-1} S_{yx} \right) \theta + \left( \bar{y}^2 - \frac{t_{n-1, \alpha/2}^2}{n-1} S_y^2 \right) \leq 0 \right\}.$$

One interesting feature of this set is that, depending on the roots of the parabola,
it can be an interval, the complement of an interval, or the entire real line (see

Exercise 9.15). Furthermore, to maintain $1 - \alpha$ confidence, this interval must be infinite with positive probability. See Hwang (1995) for an alternative based on bootstrapping, and Tsao and Hwang (1998, 1999) for an alternative confidence approach.

### 9.5.4 What About Other Intervals?

Vardeman (1992) asks the question in the title of this Miscellanea, arguing that mainstream statistics should spend more time on intervals other than two-sided confidence intervals. In particular, he lists (a) one-sided intervals, (b) distribution-free intervals, (c) prediction intervals, and (d) tolerance intervals.

We have seen one-sided intervals, and distribution-free intervals are intervals whose probability guarantee holds with little (or no) assumption on the underlying cdf (see Exercise 9.58). The other two interval definitions, together with the usual confidence interval, provide use with a hierarchy of inferences, each more stringent than the previous.

If $X_1, X_2, \ldots, X_n$ are iid from a population with cdf $F(x|\theta)$, and $C(\mathbf{x}) = [l(\mathbf{x}), u(\mathbf{x})]$ is an interval, for a specified value $1 - \alpha$ it is a

(i) *confidence* interval if $P_\theta[l(\mathbf{X}) \leq \theta \leq u(\mathbf{X})] \geq 1 - \alpha$;
(ii) *prediction* interval if $P_\theta[l(\mathbf{X}) \leq X_{n+1} \leq u(\mathbf{X})] \geq 1 - \alpha$;
(iii) *tolerance* interval if, for a specified value $p$, $P_\theta[F(u(\mathbf{X})|\theta) - F(l(\mathbf{X})|\theta) \geq p] \geq 1 - \alpha$.

So a confidence interval covers a mean, a prediction interval covers a new random variable, and a tolerance interval covers a proportion of the population. Thus, each gives a different inference, with the appropriate one being dictated by the problem at hand.

Chapter 10

# Asymptotic Evaluations

*"I know, my dear Watson, that you share my love of all that is bizarre and outside the conventions and humdrum routine of everyday life."*

**Sherlock Holmes**
*The Red-headed League*

All of the criteria we have considered thus far have been finite-sample criteria. In contrast, we might consider asymptotic properties, properties describing the behavior of a procedure as the sample size becomes infinite. In this section we will look at some of such properties and consider point estimation, hypothesis testing, and interval estimation separately. We will place particular emphasis on the asymptotics of maximum likelihood procedures.

The power of asymptotic evaluations is that, when we let the sample size become infinite, calculations simplify. Evaluations that were impossible in the finite-sample case become routine. This simplification also allows us to examine some other techniques (such as bootstrap and M-estimation) that typically can be evaluated only asymptotically.

Letting the sample size increase without bound (sometimes referred to as "asymptopia") should not be ridiculed as merely a fanciful exercise. Rather, asymptotics uncover the most fundamental properties of a procedure and give us a very powerful and general evaluation tool.

## 10.1 Point Estimation

### 10.1.1 Consistency

The property of consistency seems to be quite a fundamental one, requiring that the estimator converges to the "correct" value as the sample size becomes infinite. It is such a fundamental property that the worth of an inconsistent estimator should be questioned (or at least vigorously investigated).

Consistency (as well as all asymptotic properties) concerns a sequence of estimators rather than a single estimator, although it is common to speak of a "consistent estimator." If we observe $X_1, X_2, \ldots$ according to a distribution $f(x|\theta)$, we can construct a sequence of estimators $W_n = W_n(X_1, \ldots, X_n)$ merely by performing the same estimation procedure for each sample size $n$. For example, $\bar{X}_1 = X_1$, $\bar{X}_2 = (X_1 + X_2)/2$, $\bar{X}_3 = (X_1 + X_2 + X_3)/3$, etc. We can now define a consistent sequence.

**Definition 10.1.1**  A sequence of estimators $W_n = W_n(X_1, \ldots, X_n)$ is a *consistent sequence of estimators* of the parameter $\theta$ if, for every $\epsilon > 0$ and every $\theta \in \Theta$,

(10.1.1)                         $\lim_{n \to \infty} P_\theta(|W_n - \theta| < \epsilon) = 1.$

Informally, (10.1.1) says that as the sample size becomes infinite (and the sample information becomes better and better), the estimator will be arbitrarily close to the parameter with high probability, an eminently desirable property. Or, turning things around, we can say that the probability that a consistent sequence of estimators misses the true parameter is small. An equivalent statement to (10.1.1) is this: For every $\epsilon > 0$ and every $\theta \in \Theta$, a consistent sequence $W_n$ will satisfy

(10.1.2)                         $\lim_{n \to \infty} P_\theta(|W_n - \theta| \geq \epsilon) = 0.$

Definition 10.1.1 should be compared to Definition 5.5.1, the definition of convergence in probability. Definition 10.1.1 says that a consistent sequence of estimators converges in probability to the parameter $\theta$ it is estimating. Whereas Definition 5.5.1 dealt with one sequence of random variables with one probability structure, Definition 10.1.1 deals with an entire family of probability structures, indexed by $\theta$. For each different value of $\theta$, the probability structure associated with the sequence $W_n$ is different. And the definition says that for each value of $\theta$, the probability structure is such that the sequence converges in probability to the true $\theta$. This is the usual difference between a probability definition and a statistics definition. The probability definition deals with one probability structure, but the statistics definition deals with an entire family.

**Example 10.1.2 (Consistency of $\bar{X}$)**  Let $X_1, X_2, \ldots$ be iid n$(\theta, 1)$, and consider the sequence

$$\bar{X}_n = \frac{1}{n} \sum_{i=1}^{n} X_i.$$

Recall that $\bar{X}_n \sim \text{n}(\theta, 1/n)$, so

$$P_\theta(|\bar{X}_n - \theta| < \epsilon) = \int_{\theta - \epsilon}^{\theta + \epsilon} \left(\frac{n}{2\pi}\right)^{\frac{1}{2}} e^{-(n/2)(\bar{x}_n - \theta)^2} d\bar{x}_n \qquad \text{(definition)}$$

$$= \int_{-\epsilon}^{\epsilon} \left(\frac{n}{2\pi}\right)^{\frac{1}{2}} e^{-(n/2)y^2} dy \qquad \text{(substitute } y = \bar{x}_n - \theta)$$

$$= \int_{-\epsilon\sqrt{n}}^{\epsilon\sqrt{n}} \left(\frac{1}{2\pi}\right)^{\frac{1}{2}} e^{-(1/2)t^2} dt \qquad \text{(substitute } t = y\sqrt{n})$$

$$= P(-\epsilon\sqrt{n} < Z < \epsilon\sqrt{n}) \qquad (Z \sim \text{n}(0,1))$$

$$\to 1 \quad \text{as } n \to \infty,$$

and, hence, $\bar{X}_n$ is a consistent sequence of estimators of $\theta$.                                  ‖

In general, a detailed calculation, such as the above, is not necessary to verify consistency. Recall that, for an estimator $W_n$, Chebychev's Inequality states

$$P_\theta(|W_n - \theta| \geq \epsilon) \leq \frac{E_\theta[(W_n - \theta)^2]}{\epsilon^2},$$

so if, for every $\theta \in \Theta$,

$$\lim_{n\to\infty} E_\theta[(W_n - \theta)^2] = 0,$$

then the sequence of estimators is consistent. Furthermore, by (7.3.1),

$$(10.1.3) \qquad\qquad E_\theta[(W_n - \theta)^2] = \text{Var}_\theta\, W_n + [\text{Bias}_\theta W_n]^2.$$

Putting this all together, we can state the following theorem.

**Theorem 10.1.3**  *If $W_n$ is a sequence of estimators of a parameter $\theta$ satisfying*
  i. $\lim_{n\to\infty} \text{Var}_\theta\, W_n = 0$,
  ii. $\lim_{n\to\infty} \text{Bias}_\theta W_n = 0$,
*for every $\theta \in \Theta$, then $W_n$ is a consistent sequence of estimators of $\theta$.*

**Example 10.1.4 (Continuation of Example 10.1.2)**  Since

$$E_\theta \bar{X}_n = \theta \quad \text{and} \quad \text{Var}_\theta\, \bar{X}_n = \frac{1}{n},$$

the conditions of Theorem 10.1.3 are satisfied and the sequence $\bar{X}_n$ is consistent. Furthermore, from Theorem 5.2.6, if there is iid sampling from any population with mean $\theta$, then $\bar{X}_n$ is consistent for $\theta$ as long as the population has a finite variance. ‖

At the beginning of this section we commented that the worth of an inconsistent sequence of estimators should be questioned. Part of the basis for this comment is the fact that there are so many consistent sequences, as the next theorem shows. Its proof is left to Exercise 10.2.

**Theorem 10.1.5**  *Let $W_n$ be a consistent sequence of estimators of a parameter $\theta$. Let $a_1, a_2, \ldots$ and $b_1, b_2, \ldots$ be sequences of constants satisfying*
  i. $\lim_{n\to\infty} a_n = 1$,
  ii. $\lim_{n\to\infty} b_n = 0$.
*Then the sequence $U_n = a_n W_n + b_n$ is a consistent sequence of estimators of $\theta$.*

We close this section with the outline of a more general result concerning the consistency of maximum likelihood estimators. This result shows that MLEs are consistent estimators of their parameters and is the first case we have seen in which a method of finding an estimator guarantees an optimality property.

To have consistency of the MLE, the underlying density (likelihood function) must satisfy certain "regularity conditions" that we will not go into here, but see Miscellanea 10.6.2 for details.

**Theorem 10.1.6 (Consistency of MLEs)** *Let $X_1, X_2, \ldots,$ be iid $f(x|\theta)$, and let $L(\theta|\mathbf{x}) = \prod_{i=1}^{n} f(x_i|\theta)$ be the likelihood function. Let $\hat{\theta}$ denote the MLE of $\theta$. Let $\tau(\theta)$ be a continuous function of $\theta$. Under the regularity conditions in Miscellanea 10.6.2 on $f(x|\theta)$ and, hence, $L(\theta|\mathbf{x})$, for every $\epsilon > 0$ and every $\theta \in \Theta$,*

$$\lim_{n \to \infty} P_\theta(|\tau(\hat{\theta}) - \tau(\theta)| \geq \epsilon) = 0.$$

*That is, $\tau(\hat{\theta})$ is a consistent estimator of $\tau(\theta)$.*

**Proof:** The proof proceeds by showing that $\frac{1}{n} \log L(\hat{\theta}|\mathbf{x})$ converges almost surely to $E_\theta(\log f(X|\theta))$ for every $\theta \in \Theta$. Under some conditions on $f(x|\theta)$, this implies that $\hat{\theta}$ converges to $\theta$ in probability and, hence, $\tau(\hat{\theta})$ converges to $\tau(\theta)$ in probability. For details see Stuart, Ord, and Arnold (1999, Chapter 18).                                      □

### 10.1.2 Efficiency

The property of consistency is concerned with the asymptotic accuracy of an estimator: Does it converge to the parameter that it is estimating? In this section we look at a related property, efficiency, which is concerned with the asymptotic variance of an estimator.

In calculating an asymptotic variance, we are, perhaps, tempted to proceed as follows. Given an estimator $T_n$ based on a sample of size $n$, we calculate the finite-sample variance $\text{Var} \, T_n$, and then evaluate $\lim_{n \to \infty} k_n \text{Var} \, T_n$, where $k_n$ is some normalizing constant. (Note that, in many cases, $\text{Var} \, T_n \to 0$ as $n \to \infty$, so we need a factor $k_n$ to force it to a limit.)

**Definition 10.1.7** For an estimator $T_n$, if $\lim_{n \to \infty} k_n \text{Var} \, T_n = \tau^2 < \infty$, where $\{k_n\}$ is a sequence of constants, then $\tau^2$ is called the *limiting variance* or *limit of the variances*.

**Example 10.1.8 (Limiting variances)** For the mean $\bar{X}_n$ of $n$ iid normal observations with $\text{E} X = \mu$ and $\text{Var} \, X = \sigma^2$, if we take $T_n = \bar{X}_n$, then $\lim n \, \text{Var} \, \bar{X}_n = \sigma^2$ is the limiting variance of $T_n$.

But a troubling thing happens if, for example, we were instead interested in estimating $1/\mu$ using $1/\bar{X}_n$. If we now take $T_n = 1/\bar{X}_n$, we find that the variance is $\text{Var} \, T_n = \infty$, so the limit of the variances is infinity. But recall Example 5.5.23, where we said that the "approximate" mean and variance of $1/\bar{X}_n$ are

$$\text{E}\left(\frac{1}{\bar{X}_n}\right) \approx \frac{1}{\mu},$$

$$\text{Var}\left(\frac{1}{\bar{X}_n}\right) \approx \left(\frac{1}{\mu}\right)^4 \text{Var} \bar{X}_n,$$

and thus by this second calculation the variance is $\text{Var} \, T_n \approx \frac{\sigma^2}{n\mu^4} < \infty$.          ‖

This example points out the problems of using the limit of the variances as a large sample measure. Of course the exact finite sample variance of $1/\bar{X}$ is $\infty$. However, if

$\mu \neq 0$, the region where $1/\bar{X}$ gets very large has probability going to 0. So the second approximation in Example 10.1.8 is more realistic (as well as being much more useful). It is this second approach to calculating large sample variances that we adopt.

**Definition 10.1.9** For an estimator $T_n$, suppose that $k_n(T_n - \tau(\theta)) \to \mathrm{n}(0, \sigma^2)$ in distribution. The parameter $\sigma^2$ is called the *asymptotic variance* or *variance of the limit distribution* of $T_n$.

For calculations of the variances of sample means and other types of averages, the limit variance and the asymptotic variance typically have the same value. But in more complicated cases, the limiting variance will sometimes fail us. It is also interesting to note that it is always the case that the asymptotic variance is smaller than the limiting variance (Lehmann and Casella 1998, Section 6.1). Here is an illustration.

**Example 10.1.10 (Large-sample mixture variances)** The hierarchical model

$$Y_n | W_n = w_n \sim \mathrm{n}\left(0, w_n + (1 - w_n)\sigma_n^2\right),$$

$$W_n \sim \text{Bernoulli}(p_n),$$

can exhibit big discrepancies between the asymptotic and limiting variances. (This is also sometimes described as a mixture model, where we observe $Y_n \sim \mathrm{n}(0, 1)$ with probability $p_n$ and $Y_n \sim \mathrm{n}(0, \sigma_n^2)$ with probability $1 - p_n$.)

First, using Theorem 4.4.7 we have

$$\text{Var}(Y_n) = p_n + (1 - p_n)\sigma_n^2.$$

It then follows that the limiting variance of $Y_n$ is finite only if $\lim_{n\to\infty}(1-p_n)\sigma_n^2 < \infty$.

On the other hand, the asymptotic distribution of $Y_n$ can be directly calculated using

$$P(Y_n < a) = p_n P(Z < a) + (1 - p_n)P(Z < a/\sigma_n).$$

Suppose now we let $p_n \to 1$ and $\sigma_n \to \infty$ in such a way that $(1-p_n)\sigma_n^2 \to \infty$. It then follows that $P(Y_n < a) \to P(Z < a)$, that is, $Y_n \to \mathrm{n}(0, 1)$, and we have

$$\text{limiting variance} = \lim_{n\to\infty} p_n + (1 - p_n)\sigma_n^2 = \infty,$$

$$\text{asymptotic variance} = 1.$$

See Exercise 10.6 for more details. ‖

In the spirit of the Cramér–Rao Lower Bound (Theorem 7.3.9), there is an optimal asymptotic variance.

**Definition 10.1.11** A sequence of estimators $W_n$ is *asymptotically efficient* for a parameter $\tau(\theta)$ if $\sqrt{n}[W_n - \tau(\theta)] \to \mathrm{n}[0, v(\theta)]$ in distribution and

$$v(\theta) = \frac{[\tau'(\theta)]^2}{\mathrm{E}_\theta\left(\left(\frac{\partial}{\partial\theta}\log f(X|\theta)\right)^2\right)};$$

that is, the asymptotic variance of $W_n$ achieves the Cramér–Rao Lower Bound.

Recall that Theorem 10.1.6 stated that, under general conditions, MLEs are consistent. Under somewhat stronger regularity conditions, the same type of theorem holds with respect to asymptotic efficiency so, in general, we can consider MLEs to be consistent and asymptotically efficient. Again, details on the regularity conditions are in Miscellanea 10.6.2.

**Theorem 10.1.12 (Asymptotic efficiency of MLEs)** *Let $X_1, X_2, \ldots$, be iid $f(x|\theta)$, let $\hat{\theta}$ denote the MLE of $\theta$, and let $\tau(\theta)$ be a continuous function of $\theta$. Under the regularity conditions in Miscellanea 10.6.2 on $f(x|\theta)$ and, hence, $L(\theta|\mathbf{x})$,*

$$\sqrt{n}[\tau(\hat{\theta}) - \tau(\theta)] \to \mathrm{n}[0, v(\theta)],$$

*where $v(\theta)$ is the Cramér–Rao Lower Bound. That is, $\tau(\hat{\theta})$ is a consistent and asymptotically efficient estimator of $\tau(\theta)$.*

**Proof:** The proof of this theorem is interesting for its use of Taylor series and its exploiting of the fact that the MLE is defined as the zero of the likelihood function. We will outline the proof showing that $\hat{\theta}$ is asymptotically efficient; the extension to $\tau(\hat{\theta})$ is left to Exercise 10.7.

Recall that $l(\theta|\mathbf{x}) = \sum \log f(x_i|\theta)$ is the log likelihood function. Denote derivatives (with respect to $\theta$) by $l', l'', \ldots$. Now expand the first derivative of the log likelihood around the true value $\theta_0$,

(10.1.4) $$l'(\theta|\mathbf{x}) = l'(\theta_0|\mathbf{x}) + (\theta - \theta_0)l''(\theta_0|\mathbf{x}) + \cdots,$$

where we are going to ignore the higher-order terms (a justifiable maneuver under the regularity conditions).

Now substitute the MLE $\hat{\theta}$ for $\theta$, and realize that the left-hand side of (10.1.4) is 0. Rearranging and multiplying through by $\sqrt{n}$ gives us

(10.1.5) $$\sqrt{n}(\hat{\theta} - \theta_0) = \sqrt{n}\frac{-l'(\theta_0|\mathbf{x})}{l''(\theta_0|\mathbf{x})} = \frac{-\frac{1}{\sqrt{n}}l'(\theta_0|\mathbf{x})}{\frac{1}{n}l''(\theta_0|\mathbf{x})}.$$

If we let $I(\theta_0) = \mathrm{E}[l'(\theta_0|X)]^2 = 1/v(\theta)$ denote the information number for one observation, application of the Central Limit Theorem and the Weak Law of Large Numbers will show (see Exercise 10.8 for details)

(10.1.6)
$$-\frac{1}{\sqrt{n}}l'(\theta_0|\mathbf{X}) \to \mathrm{n}[0, I(\theta_0)], \qquad \text{(in distribution)}$$

$$\frac{1}{n}l''(\theta_0|\mathbf{X}) \to I(\theta_0). \qquad \text{(in probability)}$$

Thus, if we let $W \sim \mathrm{n}[0, I(\theta_0)]$, then $\sqrt{n}(\hat{\theta} - \theta_0)$ converges in distribution to $W/I(\theta_0) \sim \mathrm{n}[0, 1/I(\theta_0)]$, proving the theorem. $\qquad\square$

**Example 10.1.13 (Asymptotic normality and consistency)** The above theorem shows that it is typically the case that MLEs are efficient and consistent. We want to note that this phrase is somewhat redundant, as efficiency is defined only when the estimator is asymptotically normal and, as we will illustrate, asymptotic normality implies consistency. Suppose that

$$\sqrt{n}\frac{W_n - \mu}{\sigma} \to Z \text{ in distribution,}$$

where $Z \sim n(0,1)$. By applying Slutsky's Theorem (Theorem 5.5.17) we conclude

$$W_n - \mu = \left(\frac{\sigma}{\sqrt{n}}\right)\left(\sqrt{n}\frac{W_n - \mu}{\sigma}\right) \to \lim_{n\to\infty}\left(\frac{\sigma}{\sqrt{n}}\right)Z = 0,$$

so $W_n - \mu \to 0$ in distribution. From Theorem 5.5.13 we know that convergence in distribution to a point is equivalent to convergence in probability, so $W_n$ is a consistent estimator of $\mu$.                                    ‖

### 10.1.3 Calculations and Comparisons

The asymptotic formulas developed in the previous sections can provide us with approximate variances for large-sample use. Again, we have to be concerned with regularity conditions (Miscellanea 10.6.2), but these are quite general and almost always satisfied in common circumstances. One condition deserves special mention, however, whose violation can lead to complications, as we have already seen in Example 7.3.13. For the following approximations to be valid, it must be the case that the support of the pdf or pmf, hence likelihood function, must be independent of the parameter.

If an MLE is asymptotically efficient, the asymptotic variance in Theorem 10.1.6 is the Delta Method variance of Theorem 5.5.24 (without the $1/n$ term). Thus, we can use the Cramér–Rao Lower Bound as an approximation to the true variance of the MLE. Suppose that $X_1, \ldots, X_n$ are iid $f(x|\theta)$, $\hat{\theta}$ is the MLE of $\theta$, and $I_n(\theta) = E_\theta\left(\frac{\partial}{\partial\theta}\log L(\theta|\mathbf{X})\right)^2$ is the information number of the sample. From the Delta Method and asymptotic efficiency of MLEs, the variance of $h(\hat{\theta})$ can be approximated by

$$(10.1.7) \qquad \text{Var}(h(\hat{\theta})|\theta) \approx \frac{[h'(\theta)]^2}{I_n(\theta)}$$

$$= \frac{[h'(\theta)]^2}{E_\theta\left(-\frac{\partial^2}{\partial\theta^2}\log L(\theta|\mathbf{X})\right)} \qquad \binom{\text{using the identity}}{\text{of Lemma 7.3.11}}$$

$$\approx \frac{[h'(\theta)]^2|_{\theta=\hat{\theta}}}{-\frac{\partial^2}{\partial\theta^2}\log L(\theta|\mathbf{X})|_{\theta=\hat{\theta}}}. \qquad \binom{\text{the denominator is } \hat{I}_n(\hat{\theta}), \text{ the}}{\text{observed information number}}$$

Furthermore, it has been shown (Efron and Hinkley 1978) that use of the *observed information number* is superior to the *expected information number*, the information number as it appears in the Cramér–Rao Lower Bound.

Notice that the variance estimation process is a two-step procedure, a fact that is somewhat masked by (10.1.7). To estimate $\text{Var}_\theta h(\hat{\theta})$, first we *approximate* $\text{Var}_\theta h(\hat{\theta})$; then we *estimate* the resulting approximation, usually by substituting $\hat{\theta}$ for $\theta$. The resulting estimate can be denoted by $\text{Var}_{\hat{\theta}} h(\hat{\theta})$ or $\widehat{\text{Var}_\theta} h(\hat{\theta})$.

It follows from Theorem 10.1.6 that $-\frac{1}{n}\frac{\partial^2}{\partial\theta^2}\log L\left(\theta|\mathbf{X}\right)|_{\theta=\hat\theta}$ is a consistent estimator of $I(\theta)$, so it follows that $\mathrm{Var}_{\hat\theta}\, h(\hat\theta)$ is a consistent estimator of $\mathrm{Var}_{\theta}\, h(\hat\theta)$.

**Example 10.1.14 (Approximate binomial variance)** In Example 7.2.7 we saw that $\hat p = \sum X_i/n$ is the MLE of $p$, where we have a random sample $X_1, \ldots, X_n$ from a Bernoulli($p$) population. We also know by direct calculation that

$$\mathrm{Var}_p\,\hat p = \frac{p(1-p)}{n},$$

and a reasonable estimate of $\mathrm{Var}_p\,\hat p$ is

(10.1.8)                   $$\widehat{\mathrm{Var}}_p\,\hat p = \frac{\hat p(1-\hat p)}{n}.$$

If we apply the approximation in (10.1.7), with $h(p) = p$, we get as an estimate of $\mathrm{Var}_p\,\hat p$,

$$\widehat{\mathrm{Var}}_p\,\hat p \approx \frac{1}{-\frac{\partial^2}{\partial p^2}\log L(p|\mathbf{x})|_{p=\hat p}}.$$

Recall that

$$\log L(p|\mathbf{x}) = n\hat p \log(p) + n(1-\hat p)\log(1-p),$$

and so

$$\frac{\partial^2}{\partial p^2}\log L(p|\mathbf{x}) = -\frac{n\hat p}{p^2} - \frac{n(1-\hat p)}{(1-p)^2}.$$

Evaluating the second derivative at $p = \hat p$ yields

$$\frac{\partial^2}{\partial p^2}\log L(p|\mathbf{x})\bigg|_{p=\hat p} = -\frac{n\hat p}{\hat p^2} - \frac{n(1-\hat p)}{(1-\hat p)^2} = -\frac{n}{\hat p(1-\hat p)},$$

which gives a variance approximation identical to (10.1.8). We now can apply Theorem 10.1.6 to assert the asymptotic efficiency of $\hat p$ and, in particular, that

$$\sqrt n(\hat p - p) \to \mathrm{n}[0, p(1-p)]$$

in distribution. If we also employ Theorem 5.5.17 (Slutsky's Theorem) we can conclude that

$$\sqrt n \frac{\hat p - p}{\sqrt{\hat p(1-\hat p)}} \to \mathrm{n}[0, 1].$$

Estimating the variance of $\hat p$ is not really that difficult, and it is not necessary to bring in all of the machinery of these approximations. If we move to a slightly more complicated function, however, things can get a bit tricky. Recall that in Exercise 5.5.22 we used the Delta Method to approximate the variance of $\hat p/(1-\hat p)$, an estimate of the odds $p/(1-p)$. Now we see that this estimator is, in fact, the MLE of the odds, and we can estimate its variance by

$$\widehat{\text{Var}}\left(\frac{\hat{p}}{1-\hat{p}}\right) = \frac{\left[\frac{\partial}{\partial p}\left(\frac{p}{1-p}\right)\right]^2\Big|_{p=\hat{p}}}{-\frac{\partial^2}{\partial p^2}\log L(p|\mathbf{x})\Big|_{p=\hat{p}}}$$

$$= \frac{\left[\frac{(1-p)+p}{(1-p)^2}\right]^2\Big|_{p=\hat{p}}}{\frac{n}{p(1-p)}\Big|_{p=\hat{p}}}$$

$$= \frac{\hat{p}}{n(1-\hat{p})^3}.$$

Moreover, we also know that the estimator is asymptotically efficient.     ‖

The MLE variance approximation works well in many cases, but it is not infallible. In particular, we must be careful when the function $h(\theta)$ is not monotone. In such cases, the derivative $h'$ will have a sign change, and that may lead to an underestimated variance approximation. Realize that, since the approximation is based on the Cramér–Rao Lower Bound, it is probably an underestimate. However, nonmonotone functions can make this problem worse.

**Example 10.1.15 (Continuation of Example 10.1.14)** Suppose now that we want to estimate the variance of the Bernoulli distribution, $p(1-p)$. The MLE of this variance is given by $\hat{p}(1-\hat{p})$, and an estimate of the variance of this estimator can be obtained by applying the approximation of (10.1.7). We have

$$\widehat{\text{Var}}\left(\hat{p}(1-\hat{p})\right) = \frac{\left[\frac{\partial}{\partial p}\left(p(1-p)\right)\right]^2\Big|_{p=\hat{p}}}{-\frac{\partial^2}{\partial p^2}\log L(p|\mathbf{x})\Big|_{p=\hat{p}}}$$

$$= \frac{(1-2p)^2\big|_{p=\hat{p}}}{\frac{n}{p(1-p)}\Big|_{p=\hat{p}}}$$

$$= \frac{\hat{p}(1-\hat{p})(1-2\hat{p})^2}{n},$$

which can be 0 if $\hat{p} = \frac{1}{2}$, a clear underestimate of the variance of $\hat{p}(1-\hat{p})$. The fact that the function $p(1-p)$ is not monotone is a cause of this problem.

Using Theorem 10.1.6, we can conclude that our estimator is asymptotically efficient as long as $p \neq 1/2$. If $p = 1/2$ we need to use a second-order approximation as given in Theorem 5.5.26 (see Exercise 10.10).     ‖

The property of asymptotic efficiency gives us a benchmark for what we can hope to attain in asymptotic variance (although see Miscellanea 10.6.1). We also can use the asymptotic variance as a means of comparing estimators, through the idea of *asymptotic relative efficiency*.

**Definition 10.1.16**  If two estimators $W_n$ and $V_n$ satisfy

$$\sqrt{n}[W_n - \tau(\theta)] \to \mathrm{n}[0, \sigma_W^2]$$
$$\sqrt{n}[V_n - \tau(\theta)] \to \mathrm{n}[0, \sigma_V^2]$$

in distribution, the *asymptotic relative efficiency* (ARE) of $V_n$ with respect to $W_n$ is

$$\mathrm{ARE}(V_n, W_n) = \frac{\sigma_W^2}{\sigma_V^2}.$$

**Example 10.1.17 (AREs of Poisson estimators)**  Suppose that $X_1, X_2, \ldots$ are iid Poisson($\lambda$), and we are interested in estimating the 0 probability. For example, the number of customers that come into a bank in a given time period is sometimes modeled as a Poisson random variable, and the 0 probability is the probability that no one will enter the bank in one time period. If $X \sim$ Poisson($\lambda$), then $P(X = 0) = e^{-\lambda}$, and a natural (but somewhat naive) estimator comes from defining $Y_i = I(X_i = 0)$ and using

$$\hat{\tau} = \frac{1}{n} \sum_{i=1}^{n} Y_i.$$

The $Y_i$s are Bernoulli($e^{-\lambda}$), and hence it follows that

$$\mathrm{E}(\hat{\tau}) = e^{-\lambda} \quad \text{and} \quad \mathrm{Var}(\hat{\tau}) = \frac{e^{-\lambda}(1 - e^{-\lambda})}{n}.$$

Alternatively, the MLE of $e^{-\lambda}$ is $e^{-\hat{\lambda}}$, where $\hat{\lambda} = \sum_i X_i/n$ is the MLE of $\lambda$. Using Delta Method approximations, we have that

$$\mathrm{E}(e^{-\hat{\lambda}}) \approx e^{-\lambda} \quad \text{and} \quad \mathrm{Var}(e^{-\hat{\lambda}}) \approx \frac{\lambda e^{-2\lambda}}{n}.$$

Since

$$\sqrt{n}(\hat{\tau} - e^{-\lambda}) \to \mathrm{n}[0, e^{-\lambda}(1 - e^{-\lambda})]$$
$$\sqrt{n}(e^{-\hat{\lambda}} - e^{-\lambda}) \to \mathrm{n}[0, \lambda e^{-2\lambda}]$$

in distribution, the ARE of $\hat{\tau}$ with respect to the MLE $e^{-\hat{\lambda}}$ is

$$\mathrm{ARE}(\hat{\tau}, e^{-\hat{\lambda}}) = \frac{\lambda e^{-2\lambda}}{e^{-\lambda}(1 - e^{-\lambda})} = \frac{\lambda}{e^{\lambda} - 1}.$$

Examination of this function shows that it is strictly decreasing with a maximum of 1 (the best that $\hat{\tau}$ could hope to do) attained at $\lambda = 0$ and tailing off rapidly (being less than 10% when $\lambda = 4$) to asymptote to 0 as $\lambda \to \infty$. (See Exercise 10.9.)　　‖

Since the MLE is typically asymptotically efficient, another estimator cannot hope to beat its asymptotic variance. However, other estimators may have other desirable properties (ease of calculation, robustness to underlying assumptions) that make them desirable. In such situations, the efficiency of the MLE becomes important in calibrating what we are giving up if we use an alternative estimator.

We will look at one last example, contrasting ease of calculation with optimal variance. In the next section the robustness issue will be addressed.

**Example 10.1.18 (Estimating a gamma mean)**   Difficult as it may seem to believe, estimation of the mean of a gamma distribution is not an easy task. Recall that the gamma pdf $f(x|\alpha, \beta)$ is given by

$$f(x|\alpha, \beta) = \frac{1}{\Gamma(\alpha)\beta^\alpha} x^{\alpha-1} e^{-x/\beta}.$$

The mean of this distribution is $\alpha\beta$, and to compute the maximum likelihood estimator we have to deal with the derivative of the gamma function (called the *digamma* function), which is never pleasant. In contrast, the method of moments gives us an easily computable estimate.

To be specific, suppose we have a random sample $X_1, X_2, \ldots, X_n$ from the gamma density above, but reparameterized so the mean, denoted by $\mu = \alpha\beta$, is explicit. This gives

$$f(x|\mu, \beta) = \frac{1}{\Gamma(\mu/\beta)\beta^{\mu/\beta}} x^{\mu/\beta-1} e^{-x/\beta},$$

and the method of moments estimator of $\mu$ is $\bar{X}$, with variance $\beta\mu/n$.

To calculate the MLE, we use the log likelihood

$$l(\mu, \beta|\mathbf{x}) = \sum_{i=1}^{n} \log f(x_i|\mu, \beta).$$

To ease the computations, assume that $\beta$ is known so we solve $\frac{d}{d\mu} l(\mu, \beta|\mathbf{x}) = 0$ to get the MLE $\hat{\mu}$. There is no explicit solution, so we proceed numerically.

By Theorem 10.1.6 we know that $\hat{\mu}$ is asymptotically efficient. The question of interest is how much do we lose by using the easier-to-calculate method of moments estimator. To compare, we calculate the asymptotic relative efficiency,

$$\text{ARE}(\hat{\mu}, \bar{X}) = [\beta\mu] \left[ \text{E} \left( -\frac{d^2}{d\mu^2} l(\mu, \beta|X) \right) \right]$$

and display it in Figure 10.1.1 for a selection of values of $\beta$. Of course, we know that the ARE must be greater than 1, but we see from the figure that for larger values of $\beta$ it pays to do the more complex calculation and use the MLE. (See Exercise 10.11 for an extension, and Example A.0.7 for details on the calculations.)                          ‖

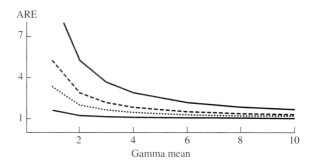

Figure 10.1.1. *Asymptotic relative efficiency of the method of moments estimator versus the MLE of a gamma mean. The four curves correspond to scale parameter values of (1,3,5,10), with the higher curves corresponding to the higher values of the scale parameter.*

### 10.1.4 Bootstrap Standard Errors

The *bootstrap*, which we first saw in Example 1.2.20, provides an alternative means of calculating standard errors. (It can also provide much more; see Miscellanea 10.6.3.)

The bootstrap is based on a simple, yet powerful, idea (whose mathematics can get quite involved).[1] In statistics, we learn about the characteristics of the population by taking samples. As the sample represents the population, analogous characteristics of the sample should give us information about the population characteristics. The bootstrap helps us learn about the sample characteristics by taking *resamples* (that is, we retake samples from the original sample) and use this information to infer to the population. The bootstrap was developed by Efron in the late 1970s, with the original ideas appearing in Efron (1979a, b) and the monograph by Efron (1982). See also Efron (1998) for more recent thoughts and developments.

Let us first look at a simple example where the bootstrap really is not needed.

**Example 10.1.19 (Bootstrapping a variance)** In Example 1.2.20 we calculated all possible averages of four numbers selected from

$$2, 4, 9, 12,$$

where we drew the numbers *with replacement*. This is the simplest form of the bootstrap, sometimes referred to as the *nonparametric bootstrap*. Figure 1.2.2 displays these values in a histogram.

What we have created is a resample of possible values of the sample mean. We saw that there are $\binom{4+4-1}{4} = 35$ distinct possible values, but these values are not equiprobable (and thus cannot be treated like a random sample). The $4^4 = 256$ (nondistinct) resamples are all equally likely, and they can be treated as a random sample. For the $i$th resample, we let $\bar{x}_i^*$ be the mean of that resample. We can then

---

[1] See Lehmann (1999, Section 6.5) for a most readable introduction.

estimate the variance of the sample mean $\bar{X}$ by

$$(10.1.9) \qquad \mathrm{Var}^*(\bar{X}) = \frac{1}{n^n - 1} \sum_{i=1}^{n^n} (\bar{x}_i^* - \bar{\bar{x}}^*)^2,$$

where $\bar{\bar{x}}^* = \frac{1}{n^n} \sum_{i=1}^{n^n} \bar{x}_i^*$, the mean of the resamples. (It is standard to let the notation $*$ denote a bootstrapped, or resampled, value.)

For our example we have that the bootstrap mean and variance are $\bar{\bar{x}}^* = 6.75$ and $\mathrm{Var}^*(\bar{X}) = 3.94$. It turns out that, as far as means and variances are concerned, the bootstrap estimates are almost the same as the usual ones (see Exercise 10.13). ‖

We have now seen how to calculate a bootstrap standard error, but in a problem where it is really not needed. However, the real advantage of the bootstrap is that, like the Delta Method, the variance formula (10.1.9) is applicable to virtually any estimator. Thus, for any estimator $\hat{\theta}(\mathbf{x}) = \hat{\theta}$, we can write

$$(10.1.10) \qquad \mathrm{Var}^*(\hat{\theta}) = \frac{1}{n^n - 1} \sum_{i=1}^{n^n} (\hat{\theta}_i^* - \bar{\hat{\theta}}^*)^2,$$

where $\hat{\theta}_i^*$ is the estimator calculated from the $i$th resample and $\bar{\hat{\theta}}^* = \frac{1}{n^n} \sum_{i=1}^{n^n} \hat{\theta}_i^*$, the mean of the resampled values.

**Example 10.1.20 (Bootstrapping a binomial variance)** In Example 10.1.15, we used the Delta Method to estimate the variance of $\hat{p}(1 - \hat{p})$. Based on a sample of size $n$, we could alternatively estimate this variance by

$$\mathrm{Var}^*(\hat{p}(1 - \hat{p})) = \frac{1}{n^n - 1} \sum_{i=1}^{n^n} (\hat{p}(1 - \hat{p})_i^* - \overline{\hat{p}(1 - \hat{p})^*})^2. \qquad ‖$$

But now a problem pops up. For our Example 10.1.19, with $n = 4$, there were 256 terms in the bootstrap sum. In more typical sample sizes, this number grows so large as to be uncomputable. (Enumerating all the possible resamples when $n > 15$ is virtually impossible, certainly for the authors.) But now we remember that we are statisticians – we take a sample of the resamples.

Thus, for a sample $\mathbf{x} = (x_1, x_2, \ldots, x_n)$ and an estimate $\hat{\theta}(x_1, x_2, \ldots, x_n) = \hat{\theta}$, select $B$ resamples (or *bootstrap samples*) and calculate

$$(10.1.11) \qquad \mathrm{Var}_B^*(\hat{\theta}) = \frac{1}{B - 1} \sum_{i=1}^{B} (\hat{\theta}_i^* - \bar{\hat{\theta}}^*)^2.$$

**Example 10.1.21 (Conclusion of Example 10.1.20)** For a sample of size $n = 24$, we compute the Delta Method variance estimate and the bootstrap variance estimate of $\hat{p}(1 - \hat{p})$ using $B = 1000$. For $\hat{p} \neq 1/2$, we use the first-order Delta Method variance of Example 10.1.15, while for $\hat{p} = 1/2$, we use the second-order variance estimate of Theorem 5.5.26 (see Exercise 10.16). We see in Table 10.1.1 that in all cases the

Table 10.1.1. *Bootstrap and Delta Method variances of $\hat{p}(1-\hat{p})$. The second-order Delta Method (see Theorem 5.5.26) is used when $\hat{p} = 1/2$. The true variance is calculated numerically assuming that $\hat{p} = p$.*

|              | $\hat{p}=1/4$ | $\hat{p}=1/2$ | $\hat{p}=2/3$ |
|--------------|---------------|---------------|---------------|
| Bootstrap    | .00508        | .00555        | .00561        |
| Delta Method | .00195        | .00022        | .00102        |
| True         | .00484        | .00531        | .00519        |

bootstrap variance estimate is closer to the true variance, while the Delta Method variance is an underestimate. (This should not be a surprise, based on (10.1.7), which shows that the Delta Method variance estimate is based on a lower bound.)

The Delta Method is a "first-order" approximation, in that it is based on the first term of a Taylor series expansion. When that term is zeroed out (as when $\hat{p} = 1/2$), we must use the second-order Delta Method. In contrast, the bootstrap can often have "second-order" accuracy, getting more than the first term in an expansion correct (see Miscellanea 10.6.3). So here, the bootstrap automatically corrects for the case $\hat{p} = 1/2$. (Note that $24^{24} \approx 1.33 \times 10^{13}$, an enormous number, so enumerating the bootstrap samples is not feasible.)                                                    ‖

The type of bootstrapping that we have been talking about so far is called the *nonparametric bootstrap*, as we have assumed no functional form for the population pdf or cdf. In contrast, we may also have a *parametric bootstrap*.

Suppose we have a sample $X_1, X_2, \ldots, X_n$ from a distribution with pdf $f(x|\theta)$, where $\theta$ may be a vector of parameters. We can estimate $\theta$ with $\hat{\theta}$, the MLE, and draw samples

$$X_1^*, X_2^*, \ldots, X_n^* \sim f(x|\hat{\theta}).$$

If we take $B$ such samples, we can estimate the variance of $\hat{\theta}$ using (10.1.11). Note that these samples are not resamples of the data, but actual random samples drawn from $f(x|\hat{\theta})$, which is sometimes called the *plug-in distribution*.

**Example 10.1.22 (Parametric bootstrap)** Suppose that we have a sample

$$-1.81, 0.63, 2.22, 2.41, 2.95, 4.16, 4.24, 4.53, 5.09$$

with $\bar{x} = 2.71$ and $s^2 = 4.82$. If we assume that the underlying distribution is normal, then a parametric bootstrap would take samples

$$X_1^*, X_2^*, \ldots, X_n^* \sim n(2.71, 4.82).$$

Based on $B = 1000$ samples, we calculate $\text{Var}_B^*(S^2) = 4.33$. Based on normal theory, the variance of $S^2$ is $2(\sigma^2)^2/8$, which we could estimate with the MLE $2(4.82)^2/8 = 5.81$. The data values were actually generated from a normal distribution with variance 4, so $\text{Var}\, S^2 = 4.00$. The parametric bootstrap is a better estimate here. (In Example 5.6.6 we estimated the distribution of $S^2$ using what we now know is the parametric bootstrap.)                                              ‖

Now that we have an all-purpose method for computing standard errors, how do we know it is a good method? In Example 10.1.21 it seems to do better than the Delta Method, which we know has some good properties. In particular, we know that the Delta Method, which is based on maximum likelihood estimation, will typically produce consistent estimators. Can we say the same for the bootstrap? Although we cannot answer this question in great generality, we say that, in many cases, the bootstrap does provide us with a reasonable estimator that is consistent.

To be a bit more precise, we separate the two distinct pieces in calculating a bootstrap estimator.

a. Establish that (10.1.11) converges to (10.1.10) as $B \to \infty$, that is,

$$\text{Var}_B^*(\hat{\theta}) \overset{B \to \infty}{\to} \text{Var}^*(\hat{\theta}).$$

b. Establish the consistency of the estimator (10.1.10), which uses the entire bootstrap sample, that is,

$$\text{Var}^*(\hat{\theta}) \overset{n \to \infty}{\to} \text{Var}(\hat{\theta}).$$

Part (a) can be established using the Law of Large Numbers (Exercise 10.15). Also notice that all of part (a) takes place in the sample. (Lehmann 1999, Section 6.5, calls $\text{Var}_B^*(\hat{\theta})$ an *approximator* rather than an estimator.)

Establishing part (b) is a bit delicate, and this is where consistency is established. Typically consistency will be obtained in iid sampling, but in more general situations it may not occur. (Lehmann 1999, Section 6.5, gives an example.) For more details on consistency (necessarily at a more advanced level), see Shao and Tu (1995, Section 3.2.2) or Shao (1999, Section 5.5.3).

## 10.2 Robustness

Thus far, we have evaluated the performance of estimators assuming that the underlying model is the correct one. Under this assumption, we have derived estimators that are optimal in some sense. However, if the underlying model is not correct, then we cannot be guaranteed of the optimality of our estimator.

We cannot guard against all possible situations and, moreover, if our model is arrived at through some careful considerations, we shouldn't have to. But we may be concerned about small or medium-sized deviations from our assumed model. This may lead us to the consideration of *robust estimators*. Such estimators will give up optimality at the assumed model in exchange for reasonable performance if the assumed model is not the true model. Thus we have a trade-off, and the more important criterion, optimality or robustness, is probably best decided on a case-by-case basis.

The term "robustness" can have many interpretations, but perhaps it is best summarized by Huber (1981, Section 1.2), who noted:

... any statistical procedure should possess the following desirable features:

(1) It should have a reasonably good (optimal or nearly optimal) efficiency at the assumed model.

(2) It should be robust in the sense that small deviations from the model assumptions should impair the performance only slightly....

(3) Somewhat larger deviations from the model should not cause a catastrophe.

We first look at some simple examples to understand these items better; then we proceed to look at more general robust estimators and measures of robustness.

### 10.2.1 The Mean and the Median

Is the sample mean a robust estimator? It may depend on exactly how we formalize measures of robustness.

**Example 10.2.1 (Robustness of the sample mean)** Let $X_1, X_2, \ldots, X_n$ be iid $n(\mu, \sigma^2)$. We know that $\bar{X}$ has variance $\text{Var}(\bar{X}) = \sigma^2/n$, which is the Cramér–Rao Lower Bound. Hence, $\bar{X}$ satisfies (1) in that it attains the best variance at the assumed model.

To investigate (2), the performance of $\bar{X}$ under small deviations from the model, we first need to decide on what this means. A common interpretation is to use an $\delta$-*contamination* model; that is, for small $\delta$, assume that we observe

$$X_i \sim \begin{cases} n(\mu, \sigma^2) & \text{with probability } 1 - \delta \\ f(x) & \text{with probability } \delta, \end{cases}$$

where $f(x)$ is some other distribution.

Suppose that we take $f(x)$ to be any density with mean $\theta$ and variance $\tau^2$. Then

$$\text{Var}(\bar{X}) = (1 - \delta)\frac{\sigma^2}{n} + \delta\frac{\tau^2}{n} + \frac{\delta(1 - \delta)(\theta - \mu)^2}{n}.$$

This actually looks pretty good for $\bar{X}$, since if $\theta \approx \mu$ and $\sigma \approx \tau$, $\bar{X}$ will be near optimal. We can perturb the model a little more, however, and make things quite bad. Consider what happens if $f(x)$ is a Cauchy pdf. Then it immediately follows that $\text{Var}(\bar{X}) = \infty$. (See Exercises 10.18 for details and 10.19 for another situation.) ‖

Turning to item (3), we ask what happens if there is an usually aberrant observation. Envision a particular set of sample values and then consider the effect of increasing the largest observation. For example, suppose that $X_{(n)} = x$, where $x \to \infty$. The effect of such an observation could be considered "catastrophic." Although none of the distributional properties of $\bar{X}$ are affected, the observed value would be "meaningless." This illustrates the *breakdown value*, an idea attributable to Hampel (1974).

**Definition 10.2.2** Let $X_{(1)} < \cdots < X_{(n)}$ be an ordered sample of size $n$, and let $T_n$ be a statistic based on this sample. $T_n$ has *breakdown value* $b, 0 \le b \le 1$, if, for every $\epsilon > 0$,

$$\lim_{X_{(\{(1-b)n\})} \to \infty} T_n < \infty \quad \text{and} \quad \lim_{X_{(\{(1-(b+\epsilon))n\})} \to \infty} T_n = \infty.$$

(Recall Definition 5.4.2 on percentile notation.)

It is easy to see that the breakdown value of $\bar{X}$ is 0; that is, if any fraction of the sample is driven to infinity, so is the value of $\bar{X}$. In stark contrast, the sample median is unchanged by this change of the sample values. This insensitivity to extreme observations is sometimes considered an asset of the sample median, which has a breakdown value of 50%. (See Exercise 10.20 for more about breakdown values.)

Since the median is improving on the robustness of the mean, we might ask if we are losing anything by switching to a more robust estimator (of course we must!). For example, in the simple normal model of Example 10.2.1, the mean is the best unbiased estimator if the model is true. Therefore it follows that at the normal model (and close to it), the mean is a better estimator. But, the key question is, just how much better is the mean at the normal model? If we can answer this, we can make an informative choice on which estimator to use–and which criterion (optimality or robustness) we consider more important. To answer this question in some generality we call on the criterion of asymptotic relative efficiency.

To compute the ARE of the median with respect to the mean, we must first establish the asymptotic normality of the median and calculate the variance of the asymptotic distribution.

**Example 10.2.3 (Asymptotic normality of the median)** To find the limiting distribution of the median, we resort to an argument similar to that in the proof of Theorems 5.4.3 and 5.4.4, that is, an argument based on the binomial distribution.

Let $X_1, \ldots, X_n$ be a sample from a population with pdf $f$ and cdf $F$ (assumed to be differentiable), with $P(X_i \leq \mu) = 1/2$, so $\mu$ is the population median. Let $M_n$ be the sample median, and consider computing

$$\lim_{n \to \infty} P\left(\sqrt{n}(M_n - \mu) \leq a\right)$$

for some $a$. If we define the random variables $Y_i$ by

$$Y_i = \begin{cases} 1 & \text{if } X_i \leq \mu + a/\sqrt{n} \\ 0 & \text{otherwise,} \end{cases}$$

it follows that $Y_i$ is a Bernoulli random variable with success probability $p_n = F(\mu + a/\sqrt{n})$. To avoid complications, we will assume that $n$ is odd and thus the event $\{M_n \leq \mu + a/\sqrt{n}\}$ is equivalent to the event $\{\sum_i Y_i \geq (n+1)/2\}$.

Some algebra then yields

$$P\left(\sqrt{n}(M_n - \mu) \leq a\right) = P\left(\frac{\sum_i Y_i - np_n}{\sqrt{np_n(1-p_n)}} \geq \frac{(n+1)/2 - np_n}{\sqrt{np_n(1-p_n)}}\right).$$

Now $p_n \to p = F(\mu) = 1/2$, so we expect that an application of the Central Limit Theorem will show that $\frac{\sum_i Y_i - np_n}{\sqrt{np_n(1-p_n)}}$ converges in distribution to $Z$, a standard normal random variable. A straightforward limit calculation will also show that

$$\frac{(n+1)/2 - np_n}{\sqrt{np_n(1-p_n)}} \to -2aF'(\mu) = -2af(\mu).$$

Putting this all together yields that

$$P\left(\sqrt{n}(M_n - \mu) \le a\right) \to P\left(Z \ge -2af(\mu)\right).$$

and thus $\sqrt{n}(M_n - \mu)$ is asymptotically normal with mean 0 and variance $1/[2f(\mu)]^2$. (For details, see Exercise 10.22, and for a rigorous, and more general, development of this result, see Shao 1999, Section 5.3.) ‖

**Example 10.2.4 (AREs of the median to the mean)** As there are simple expressions for the asymptotic variances of the mean and the median, the ARE is easily computed. The following table gives the AREs for three symmetric distributions. We find, as might be expected, that as the tails of the distribution get heavier, the ARE gets bigger. That is, the performance of the median improves in distributions with heavy tails. See Exercise 10.23 for more comparisons.

*Median/mean asymptotic relative efficiencies*

| Normal | Logistic | Double exponential |
|--------|----------|--------------------|
| .64    | .82      | 2                  |

‖

### 10.2.2 M-Estimators

Many of the estimators that we use are the result of minimizing a particular criterion. For example, if $X_1, X_2, \ldots, X_n$ are iid from $f(x|\theta)$, possible estimators are the mean, the minimizer of $\sum(x_i - a)^2$; the median, the minimizer of $\sum|x_i - a|$; and the MLE, the maximizer of $\prod_{i=1}^n f(x_i|\theta)$ (or the minimizer of the negative likelihood). As a systematic way of obtaining a robust estimator, we might attempt to write down a criterion function whose minimum would result in an estimator with desirable robustness properties.

In an attempt at defining a robust criterion, Huber (1964) considered a compromise between the mean and the median. The mean criterion is a square, which gives it sensitivity, but in the "tails" the square gives too much weight to big observations. In contrast, the absolute value criterion of the median does not overweight big or small observations. The compromise is to minimize a criterion function

$$(10.2.1) \qquad \sum_{i=1}^{n} \rho(x_i - a),$$

where $\rho$ is given by

$$(10.2.2) \qquad \rho(x) = \begin{cases} \frac{1}{2}x^2 & \text{if } |x| \le k \\ k|x| - \frac{1}{2}k^2 & \text{if } |x| \ge k. \end{cases}$$

The function $\rho(x)$ acts like $x^2$ for $|x| \le k$ and like $|x|$ for $|x| > k$. Moreover, since $\frac{1}{2}k^2 = k|k| - \frac{1}{2}k^2$, the function is continuous (see Exercise 10.28). In fact $\rho$ is differentiable. The constant $k$, which can also be called a *tuning parameter*, controls the mix, with small values of $k$ yielding a more "median-like" estimator.

Table 10.2.1. *Huber estimators*

| k | 0 | 1 | 2 | 3 | 4 | 5 | 6 | 8 | 10 |
|---|---|---|---|---|---|---|---|---|---|
| Estimate | −.21 | .03 | −.04 | .29 | .41 | .52 | .87 | .97 | 1.33 |

**Example 10.2.5 (Huber estimator)** The estimator defined as the minimizer of (10.2.1) and (10.2.2) is called a *Huber estimator* . To see how the estimator works, and how the choice of $k$ matters, consider the following data set consisting of eight standard normal deviates and three "outliers":

$$\mathbf{x} = -1.28, -.96, -.46, -.44, -.26, -.21, -.063, .39, 3, 6, 9$$

For these data the mean is 1.33 and the median is −.21. As $k$ varies, we get the range of Huber estimates given in Table 10.2.1. We see that as $k$ increases, the Huber estimate varies between the median and the mean, so we interpret increasing $k$ as decreasing robustness to outliers.                                                    ‖

The estimator minimizing (10.2.2) is a special case of the estimators studied by Huber. For a general function $\rho$, we call the estimator minimizing $\sum_i \rho(x_i - \theta)$ an *M-estimator*, a name that is to remind us that these are *maximum-likelihood-type* estimators. Note that if we choose $\rho$ to be the negative log likelihood $-l(\theta|x)$, then the M-estimator is the usual MLE. But with more flexibility in choosing the function to be minimized, estimators with different properties can be derived.

Since minimization of a function is typically done by solving for the zeros of the derivative (when we can take a derivative), defining $\psi = \rho'$, we see that an M-estimator is the solution to

(10.2.3)
$$\sum_{i=1}^{n} \psi(x_i - \theta) = 0.$$

Characterizing an estimator as the root of an equation is particularly useful for getting properties of the estimator, for arguments like those used for likelihood estimators can be extended. In particular, look at Section 10.1.2, especially the proof of Theorem 10.1.12. We assume that the function $\rho(x)$ is symmetric, and its derivative $\psi(x)$ is monotone increasing (which ensures that the root of (10.2.3) is the unique minimum). Then, as in the proof of Theorem 10.1.12, we write a Taylor expansion for $\psi$ as

$$\sum_{i=1}^{n} \psi(x_i - \theta) = \sum_{i=1}^{n} \psi(x_i - \theta_0) + (\theta - \theta_0) \sum_{i=1}^{n} \psi'(x_i - \theta_0) + \cdots,$$

where $\theta_0$ is the true value, and we ignore the higher-order terms. Let $\hat{\theta}_M$ be the solution to (10.2.3) and substitute this for $\theta$ to obtain

$$0 = \sum_{i=1}^{n} \psi(x_i - \theta_0) + (\hat{\theta}_M - \theta_0) \sum_{i=1}^{n} \psi'(x_i - \theta_0) + \cdots,$$

where the left-hand side is 0 because $\hat{\theta}_M$ is the solution. Now, again analogous to the proof of Theorem 10.1.12, we rearrange terms, divide through by $\sqrt{n}$, and ignore the remainder terms to get

$$\sqrt{n}(\hat{\theta}_M - \theta_0) = \frac{\frac{-1}{\sqrt{n}}\sum_{i=1}^{n}\psi(x_i - \theta_0)}{\frac{1}{n}\sum_{i=1}^{n}\psi'(x_i - \theta_0)}.$$

Now we assume that $\theta_0$ satisfies $E_{\theta_0}\psi(X - \theta_0) = 0$ (which is usually taken as the definition of $\theta_0$). It follows that

$$(10.2.4) \quad \frac{-1}{\sqrt{n}}\sum_{i=1}^{n}\psi(X_i - \theta_0) = \sqrt{n}\left[\frac{-1}{n}\sum_{i=1}^{n}\psi(X_i - \theta_0)\right] \rightarrow n\left(0, E_{\theta_0}\psi(X - \theta_0)^2\right)$$

in distribution, and the Law of Large Numbers yields

$$(10.2.5) \quad \frac{1}{n}\sum_{i=1}^{n}\psi'(x_i - \theta_0) \rightarrow E_{\theta_0}\psi'(X - \theta_0)$$

in probability. Putting this all together we have

$$(10.2.6) \quad \sqrt{n}(\hat{\theta}_M - \theta_0) \rightarrow n\left(0, \frac{E_{\theta_0}\psi(X - \theta_0)^2}{[E_{\theta_0}\psi'(X - \theta_0)]^2}\right).$$

**Example 10.2.6 (Limit distribution of the Huber estimator)** If $X_1, \ldots, X_n$ are iid from a pdf $f(x - \theta)$, where $f$ is symmetric around 0, then for $\rho$ given by (10.2.2) we have

$$(10.2.7) \quad \psi(x) = \begin{cases} x & \text{if } |x| \leq k \\ k & \text{if } x > k \\ -k & \text{if } x < -k \end{cases}$$

and thus

$$E_\theta\psi(X - \theta) = \int_{\theta-k}^{\theta+k}(x - \theta)f(x - \theta)\,dx$$

$$(10.2.8) \quad \qquad - k\int_{-\infty}^{\theta-k}f(x - \theta)\,dx + k\int_{\theta+k}^{\infty}f(x - \theta)\,dx$$

$$= \int_{-k}^{k}yf(y)\,dy - k\int_{-\infty}^{-k}f(y)\,dy + k\int_{k}^{\infty}f(y)\,dy = 0,$$

where we substitute $y = x - \theta$. The integrals add to 0 by the symmetry of $f$. Thus, the Huber estimator has the correct mean (see Exercise 10.25).

To calculate the variance we need the expected value of $\psi'$. While $\psi$ is not differentiable, beyond the points of nondifferentiability ($x = \pm k$) $\psi'$ will be 0. Thus, we only need deal with the expectation for $|x| \leq k$, and we have

$$E_\theta \psi'(X - \theta) = \int_{\theta-k}^{\theta+k} f(x - \theta)\, dx = P_0(|X| \leq k),$$

$$E_\theta \psi(X - \theta)^2 = \int_{\theta-k}^{\theta+k} (x - \theta)^2 f(x - \theta)\, dx + k^2 \int_{\theta+k}^{\infty} f(x - \theta)\, dx + k^2 \int_{-\infty}^{\theta-k} f(x - \theta)\, dx$$

$$= \int_{-k}^{k} x^2 f(x)\, dx + 2k^2 \int_{k}^{\infty} f(x)\, dx.$$

Thus we can conclude that the Huber estimator is asymptotically normal with mean $\theta$ and asymptotic variance

$$\frac{\int_{-k}^{k} x^2 f(x)\, dx + 2k^2 P_0(|X| > k)}{[P_0(|X| \leq k)]^2}. \qquad \|$$

As we did in Example 10.2.4, we now examine the ARE of the Huber estimator for a variety of distributions.

**Example 10.2.7 (ARE of the Huber estimator)** As the Huber estimator is, in a sense, a mean/median compromise, we'll look at its relative efficiency with respect to both of these estimators.

*Huber estimator asymptotic relative efficiencies, $k = 1.5$*

|            | Normal | Logistic | Double exponential |
|------------|--------|----------|--------------------|
| vs. mean   | .96    | 1.08     | 1.37               |
| vs. median | 1.51   | 1.31     | .68                |

The Huber estimator behaves similarly to the mean for the normal and logistic distributions and is an improvement on the median. For the double exponential it is an improvement over the mean but not as good as the median. Recall that the mean is the MLE for the normal, and the median is the MLE for the double exponential (so AREs $< 1$ are expected). The Huber estimator has performance similar to the MLEs for these distributions but also seems to maintain reasonable performance in other cases. $\qquad \|$

We see that an M-estimator is a compromise between robustness and efficiency. We now look a bit more closely at what we may be giving up, in terms of efficiency, to gain robustness.

Let us look more closely at the asymptotic variance in (10.2.6). The denominator of the variance contains the term $E_{\theta_0} \psi'(X - \theta_0)$, which we can write as

$$E_\theta \psi'(X - \theta) = \int \psi'(x - \theta) f(x - \theta)\, dx = -\int \left[ \frac{\partial}{\partial \theta} \psi(x - \theta) \right] f(x - \theta)\, dx.$$

Now we use the differentiation product rule to get

$$\frac{d}{d\theta} \int \psi(x-\theta) f(x-\theta)\, dx = \int \left[ \frac{d}{d\theta} \psi(x - \theta) \right] f(x-\theta)\, dx + \int \psi(x-\theta) \left[ \frac{d}{d\theta} f(x - \theta) \right] dx.$$

The left hand side is 0 because $E_\theta \psi(X - \theta) = 0$, so we have

$$- \int \left[ \frac{d}{d\theta} \psi(x - \theta) \right] f(x - \theta) \, dx = \int \psi(x - \theta) \left[ \frac{d}{d\theta} f(x - \theta) \right] dx$$

$$= \int \psi(x - \theta) \left[ \frac{d}{d\theta} \log f(x - \theta) \right] f(x - \theta) \, dx,$$

where we use the fact that $\frac{d}{dy} g(y)/g(y) = \frac{d}{dy} \log g(y)$. This last expression can be written $E_\theta[\psi(X - \theta)l'(\theta|X)]$, where $l(\theta|X)$ is the log likelihood, yielding the identity

$$E_\theta \psi'(X - \theta) = -E_\theta \left[ \frac{d}{d\theta} \psi(X - \theta) \right] = E_\theta[\psi(X - \theta)l'(\theta|X)]$$

(which, when we choose $\psi = l'$, yields the (we hope) familiar equation $-E_\theta[l''(\theta|X)] = E_\theta l'(\theta|X)^2$; see Lemma 7.3.11).

It is now a simple matter to compare the asymptotic variance of an M-estimator to that of the MLE. Recall that the asymptotic variance of the MLE, $\hat{\theta}$, is given by $1/E_\theta l'(\theta|X)^2$, so we have

$$(10.2.9) \qquad \text{ARE}(\hat{\theta}_M, \hat{\theta}) = \frac{[E_\theta \psi(X - \theta_0)l'(\theta|X)]^2}{E_\theta \psi(X - \theta)^2 E_\theta l'(\theta|X)^2} \leq 1$$

by virtue of the Cauchy-Swartz Inequality. Thus, an M-estimator is always less efficient than the MLE, and matches its efficiency only if $\psi$ is proportional to $l'$ (see Exercise 10.29).

In this section we did not try to classify all types of robust estimators, but rather we were content with some examples. There are many good books that treat robustness in detail; the interested reader might try Staudte and Sheather (1990) or Hettmansperger and McKean (1998).

## 10.3 Hypothesis Testing

As in Section 10.1, this section describes a few methods for deriving *some* tests in complicated problems. We are thinking of problems in which no optimal test, as defined in earlier sections, exists (for example, no UMP unbiased test exists) or is known. In such situations, the derivation of any reasonable test might be of use. In two subsections, we will discuss large-sample properties of likelihood ratio tests and other approximate large-sample tests.

### 10.3.1 Asymptotic Distribution of LRTs

One of the most useful methods for complicated models is the likelihood ratio method of test construction because it gives an explicit definition of the test statistic,

$$\lambda(\mathbf{x}) = \frac{\sup\limits_{\Theta_0} L(\theta|\mathbf{x})}{\sup\limits_{\Theta} L(\theta|\mathbf{x})},$$

and an explicit form for the rejection region, $\{\mathbf{x} : \lambda(\mathbf{x}) \leq c\}$. After the data $\mathbf{X} = \mathbf{x}$ are observed, the likelihood function, $L(\theta|\mathbf{x})$, is a completely defined function of the variable $\theta$. Even if the two suprema of $L(\theta|\mathbf{x})$, over the sets $\Theta_0$ and $\Theta$, cannot be analytically obtained, they can usually be computed numerically. Thus, the test statistic $\lambda(\mathbf{x})$ can be obtained for the observed data point even if no convenient formula defining $\lambda(\mathbf{x})$ is available.

To define a level $\alpha$ test, the constant $c$ must be chosen so that

$$(10.3.1) \qquad\qquad \sup_{\theta \in \Theta_0} P_\theta\left(\lambda(\mathbf{X}) \leq c\right) \leq \alpha.$$

If we cannot derive a simple formula for $\lambda(\mathbf{x})$, it might seem that it is hopeless to derive the sampling distribution of $\lambda(\mathbf{X})$ and thus know how to pick $c$ to ensure (10.3.1). However, if we appeal to asymptotics, we can get an approximate answer.

Analogous to Theorem 10.1.12, we have the following result.

**Theorem 10.3.1 (Asymptotic distribution of the LRT—simple $H_0$)** *For testing $H_0 : \theta = \theta_0$ versus $H_1 : \theta \neq \theta_0$, suppose $X_1, \ldots, X_n$ are iid $f(x|\theta)$, $\hat{\theta}$ is the MLE of $\theta$, and $f(x|\theta)$ satisfies the regularity conditions in Miscellanea 10.6.2. Then under $H_0$, as $n \to \infty$,*

$$-2 \log \lambda(\mathbf{X}) \to \chi_1^2 \ \textit{in distribution},$$

*where $\chi_1^2$ is a $\chi^2$ random variable with 1 degree of freedom.*

**Proof:** First expand $\log L(\theta|\mathbf{x}) = l(\theta|\mathbf{x})$ in a Taylor series around $\hat{\theta}$, giving

$$l(\theta|\mathbf{x}) = l(\hat{\theta}|\mathbf{x}) + l'(\hat{\theta}|\mathbf{x})(\theta - \hat{\theta}) + l''(\hat{\theta}|\mathbf{x})\frac{(\theta - \hat{\theta})^2}{2!} + \cdots.$$

Now substitute the expansion for $l(\theta_0|\mathbf{x})$ in $-2 \log \lambda(\mathbf{x}) = -2l(\theta_0|\mathbf{x}) + 2l(\hat{\theta}|\mathbf{x})$, and get

$$-2 \log \lambda(\mathbf{x}) \approx -l''(\hat{\theta}|\mathbf{x})(\theta_0 - \hat{\theta})^2,$$

where we use the fact that $l'(\hat{\theta}|\mathbf{x}) = 0$. Since $-l''(\hat{\theta}|\mathbf{x})$ is the observed information $\hat{I}_n(\hat{\theta})$ and $\frac{1}{n}\hat{I}_n(\hat{\theta}) \to I(\theta_0)$ it follows from Theorem 10.1.12 and Slutsky's Theorem (Theorem 5.5.17) that $-2 \log \lambda(\mathbf{X}) \to \chi_1^2$.                    $\square$

**Example 10.3.2 (Poisson LRT)** For testing $H_0 : \lambda = \lambda_0$ versus $H_1 : \lambda \neq \lambda_0$ based on observing $X_1, \ldots, X_n$ iid Poisson($\lambda$), we have

$$-2 \log \lambda(\mathbf{x}) = -2 \log \left(\frac{e^{-n\lambda_0}\lambda_0^{\Sigma x_i}}{e^{-n\hat{\lambda}}\hat{\lambda}^{\Sigma x_i}}\right) = 2n\left[(\lambda_0 - \hat{\lambda}) - \hat{\lambda}\log(\lambda_0/\hat{\lambda})\right],$$

where $\hat{\lambda} = \Sigma x_i/n$ is the MLE of $\lambda$. Applying Theorem 10.3.1, we would reject $H_0$ at level $\alpha$ if $-2 \log \lambda(\mathbf{x}) > \chi_{1,\alpha}^2$.

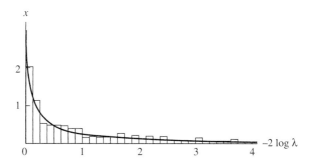

Figure 10.3.1. *Histogram of* 10,000 *values of* $-2 \log \lambda(\mathbf{x})$ *along with the pdf of a* $\chi_1^2$, $\lambda_0 = 5$ *and* $n = 25$

To get some idea of the accuracy of the asymptotics, here is a small simulation of the test statistic. For $\lambda_0 = 5$ and $n = 25$, Figure 10.3.1 shows a histogram of 10,000 values of $-2 \log \lambda(\mathbf{x})$ along with the pdf of a $\chi_1^2$. The match seems to be reasonable. Moreover, a comparison of the simulated (exact) and $\chi_1^2$ (approximate) cutoff points in the following table shows that the cutoffs are remarkably similar.

*Simulated (exact) and approximate percentiles of the Poisson LRT statistic*

| Percentile | .80 | .90 | .95 | .99 |
|---|---|---|---|---|
| Simulated | 1.630 | 2.726 | 3.744 | 6.304 |
| $\chi^2$ | 1.642 | 2.706 | 3.841 | 6.635 |

‖

Theorem 10.3.1 can be extended to the cases where the null hypothesis concerns a vector of parameters. The following generalization, which we state without proof, allows us to ensure (10.3.1) is true, at least for large samples. A complete discussion of this topic may be found in Stuart, Ord, and Arnold (1999, Chapter 22).

**Theorem 10.3.3** *Let* $X_1, \ldots, X_n$ *be a random sample from a pdf or pmf* $f(x|\theta)$. *Under the regularity conditions in Miscellanea 10.6.2, if* $\theta \in \Theta_0$, *then the distribution of the statistic* $-2 \log \lambda(\mathbf{X})$ *converges to a chi squared distribution as the sample size* $n \to \infty$. *The degrees of freedom of the limiting distribution is the difference between the number of free parameters specified by* $\theta \in \Theta_0$ *and the number of free parameters specified by* $\theta \in \Theta$.

Rejection of $H_0 : \theta \in \Theta_0$ for small values of $\lambda(\mathbf{X})$ is equivalent to rejection for large values of $-2 \log \lambda(\mathbf{X})$. Thus,

$$H_0 \text{ is rejected if and only if } -2 \log \lambda(\mathbf{X}) \geq \chi_{\nu,\alpha}^2,$$

where $\nu$ is the degrees of freedom specified in Theorem 10.3.3. The Type I Error probability will be approximately $\alpha$ if $\theta \in \Theta_0$ and the sample size is large. In this way, (10.3.1) will be approximately satisfied for large sample sizes and an *asymptotic size* $\alpha$ *test* has been defined. Note that the theorem will actually imply only that

$$\lim_{n \to \infty} P_\theta(\text{reject } H_0) = \alpha \quad \text{for each } \theta \in \Theta_0,$$

not that the $\sup_{\theta \in \Theta_0} P_\theta(\text{reject} H_0)$ converges to $\alpha$. This is usually the case for asymptotic size $\alpha$ tests.

The computation of the degrees of freedom for the test statistic is usually straightforward. Most often, $\Theta$ can be represented as a subset of $q$-dimensional Euclidian space that contains an open subset in $\Re^q$, and $\Theta_0$ can be represented as a subset of $p$-dimensional Euclidian space that contains an open subset in $\Re^p$, where $p < q$. Then $q - p = \nu$ is the degrees of freedom for the test statistic.

**Example 10.3.4 (Multinomial LRT)** Let $\theta = (p_1, p_2, p_3, p_4, p_5)$, where the $p_j$s are nonnegative and sum to 1. Suppose $X_1, \ldots, X_n$ are iid discrete random variables and $P_\theta(X_i = j) = p_j, j = 1, \ldots, 5$. Thus the pmf of $X_i$ is $f(j|\theta) = p_j$ and the likelihood function is

$$L(\theta|\mathbf{x}) = \prod_{i=1}^n f(x_i|\theta) = p_1^{y_1} p_2^{y_2} p_3^{y_3} p_4^{y_4} p_5^{y_5},$$

where $y_j$ = number of $x_1, \ldots, x_n$ equal to $j$. Consider testing

$$H_0: p_1 = p_2 = p_3 \text{ and } p_4 = p_5 \quad \text{versus} \quad H_1: H_0 \text{ is not true.}$$

The full parameter space, $\Theta$, is really a four-dimensional set. Since $p_5 = 1 - p_1 - p_2 - p_3 - p_4$, there are only four free parameters. The parameter set is defined by

$$\sum_{j=1}^4 p_j \leq 1 \quad \text{and} \quad p_j \geq 0, \quad j = 1, \ldots, 4,$$

a subset of $\Re^4$ containing an open subset of $\Re^4$. Thus $q = 4$. There is only one free parameter in the set specified by $H_0$ because, once $p_1, 0 \leq p_1 \leq \frac{1}{3}$, is fixed, $p_2 = p_3$ must equal $p_1$ and $p_4 = p_5$ must equal $\frac{1-3p_1}{2}$. Thus $p = 1$, and the degrees of freedom is $\nu = 4 - 1 = 3$.

To calculate $\lambda(\mathbf{x})$, the MLE of $\theta$ under both $\Theta_0$ and $\Theta$ must be determined. By setting

$$\frac{\partial}{\partial p_j} \log L(\theta|\mathbf{x}) = 0 \quad \text{for each of } j = 1, \ldots, 4,$$

and using the facts that $p_5 = 1 - p_1 - p_2 - p_3 - p_4$ and $y_5 = n - y_1 - y_2 - y_3 - y_4$, we can verify that the MLE of $p_j$ under $\Theta$ is $\hat{p}_j = y_j/n$. Under $H_0$, the likelihood function reduces to

$$L(\theta|\mathbf{x}) = p_1^{y_1+y_2+y_3} \left(\frac{1 - 3p_1}{2}\right)^{y_4+y_5}.$$

Again, the usual method of setting the derivative equal to 0 shows that the MLE of $p_1$ under $H_0$ is $\hat{p}_{10} = (y_1 + y_2 + y_3)/(3n)$. Then $\hat{p}_{10} = \hat{p}_{20} = \hat{p}_{30}$ and $\hat{p}_{40} = \hat{p}_{50} = (1 - 3\hat{p}_{10})/2$. Substituting these values and the $\hat{p}_j$ values into $L(\theta|\mathbf{x})$ and combining terms with the same exponent yield

$$\lambda(\mathbf{x}) =$$

$$\left(\frac{y_1 + y_2 + y_3}{3y_1}\right)^{y_1} \left(\frac{y_1 + y_2 + y_3}{3y_2}\right)^{y_2} \left(\frac{y_1 + y_2 + y_3}{3y_3}\right)^{y_3} \left(\frac{y_4 + y_5}{2y_4}\right)^{y_4} \left(\frac{y_4 + y_5}{2y_5}\right)^{y_5}.$$

Thus the test statistic is

$$(10.3.2) \qquad -2 \log \lambda(\mathbf{x}) = 2 \sum_{i=1}^{5} y_i \log \left( \frac{y_i}{m_i} \right),$$

where $m_1 = m_2 = m_3 = (y_1 + y_2 + y_3)/3$ and $m_4 = m_5 = (y_4 + y_5)/2$. The asymptotic size $\alpha$ test rejects $H_0$ if $-2 \log \lambda(\mathbf{x}) \geq \chi^2_{3,\alpha}$. This example is one of a large class of testing problems for which the asymptotic theory of the likelihood ratio test is extensively used. ‖

### 10.3.2 Other Large-Sample Tests

Another common method of constructing a large-sample test statistic is based on an estimator that has an asymptotic normal distribution. Suppose we wish to test a hypothesis about a real-valued parameter $\theta$, and $W_n = W(X_1, \ldots, X_n)$ is a point estimator of $\theta$, based on a sample of size $n$, that has been derived by some method. For example, $W_n$ might be the MLE of $\theta$. An approximate test, based on a normal approximation, can be justified in the following way. If $\sigma_n^2$ denotes the variance of $W_n$ and if we can use some form of the Central Limit Theorem to show that, as $n \to \infty, (W_n - \theta)/\sigma_n$ converges in distribution to a standard normal random variable, then $(W_n - \theta)/\sigma_n$ can be compared to a $n(0, 1)$ distribution. We therefore have the basis for an approximate test.

There are, of course, many details to be verified in the argument of the previous paragraph, but this idea does have application in many situations. For example, if $W_n$ is an MLE, Theorem 10.1.12 can be used to validate the above arguments. Note that the distribution of $W_n$ and, perhaps, the value of $\sigma_n$ depend on the value of $\theta$. The convergence, therefore, more formally says that for each fixed value of $\theta \in \Theta$, if we use the corresponding distribution for $W_n$ and the corresponding value for $\sigma_n$, $(W_n - \theta)/\sigma_n$ converges to a standard normal. If, for each $n$, $\sigma_n$ is a calculable constant (which may depend on $\theta$ but not any other unknown parameters), then a test based on $(W_n - \theta)/\sigma_n$ might be derived.

In some instances, $\sigma_n$ also depends on unknown parameters. In such a case, we look for an estimate $S_n$ of $\sigma_n$ with the property that $\sigma_n/S_n$ converges in probability to 1. Then, using Slutsky's Theorem (as in Example 5.5.18) we can deduce that $(W_n - \theta)/S_n$ also converges in distribution to a standard normal distribution. A large-sample test may be based on this fact.

Suppose we wish to test the two-sided hypothesis $H_0 : \theta = \theta_0$ versus $H_1 : \theta \neq \theta_0$. An approximate test can be based on the statistic $Z_n = (W_n - \theta_0)/S_n$ and would reject $H_0$ if and only if $Z_n < -z_{\alpha/2}$ or $Z_n > z_{\alpha/2}$. If $H_0$ is true, then $\theta = \theta_0$ and $Z_n$ converges in distribution to $Z \sim n(0, 1)$. Thus, the Type I Error probability,

$$P_{\theta_0}(Z_n < -z_{\alpha/2} \text{ or } Z_n > z_{\alpha/2}) \to P(Z < -z_{\alpha/2} \text{ or } Z > z_{\alpha/2}) = \alpha,$$

and this is an asymptotically size $\alpha$ test.

Now consider an alternative parameter value $\theta \neq \theta_0$. We can write

$$(10.3.3) \qquad Z_n = \frac{W_n - \theta_0}{S_n} = \frac{W_n - \theta}{S_n} + \frac{\theta - \theta_0}{S_n}.$$

No matter what the value of $\theta$, the term $(W_n - \theta)/S_n \to n(0, 1)$. Typically, it is also the case that $\sigma_n \to 0$ as $n \to \infty$. (Recall, $\sigma_n = \text{Var}\, W_n$, and estimators typically become more precise as $n \to \infty$.) Thus, $S_n$ will converge in probability to 0 and the term $(\theta - \theta_0)/S_n$ will converge to $+\infty$ or $-\infty$ in probability, depending on whether $(\theta - \theta_0)$ is positive or negative. Thus, $Z_n$ will converge to $+\infty$ or $-\infty$ in probability and

$$P_\theta(\text{reject } H_0) = P_\theta(Z_n < -z_{\alpha/2} \text{ or } Z_n > z_{\alpha/2}) \to 1 \quad \text{as } n \to \infty.$$

In this way, a test with asymptotic size $\alpha$ and asymptotic power 1 can be constructed.

If we wish to test the one-sided hypothesis $H_0 : \theta \le \theta_0$ versus $H_1 : \theta > \theta_0$, a similar test might be constructed. Again, the test statistic $Z_n = (W_n - \theta_0)/S_n$ would be used and the test would reject $H_0$ if and only if $Z_n > z_\alpha$. Using reasoning similar to the above, we could conclude that the power function of this test converges to 0, $\alpha$, or 1 according as $\theta < \theta_0, \theta = \theta_0$, or $\theta > \theta_0$. Thus this test too has reasonable asymptotic power properties.

In general, a *Wald test* is a test based on a statistic of the form

$$Z_n = \frac{W_n - \theta_0}{S_n},$$

where $\theta_0$ is a hypothesized value of the parameter $\theta$, $W_n$ is an estimator of $\theta$, and $S_n$ is a standard error for $W_n$, an estimate of the standard deviation of $W_n$. If $W_n$ is the MLE of $\theta$, then, as discussed in Section 10.1.3, $1/\sqrt{I_n(W_n)}$ is a reasonable standard error for $W_n$. Alternatively, $1/\sqrt{\hat{I}_n(W_n)}$, where

$$\hat{I}_n(W_n) = -\frac{\partial^2}{\partial \theta^2} \log L(\theta|\mathbf{X})\Big|_{\theta = W_n}$$

is the observed information number, is often used (see (10.1.7)).

**Example 10.3.5 (Large-sample binomial tests)** Let $X_1, \ldots, X_n$ be a random sample from a Bernoulli($p$) population. Consider testing $H_0 : p \le p_0$ versus $H_1 : p > p_0$, where $0 < p_0 < 1$ is a specified value. The MLE of $p$, based on a sample of size $n$, is $\hat{p}_n = \sum_{i=1}^n X_i/n$. Since $\hat{p}_n$ is just a sample mean, the Central Limit Theorem applies and states that for any $p$, $0 < p < 1$, $(\hat{p}_n - p)/\sigma_n$ converges to a standard normal random variable. Here $\sigma_n = \sqrt{p(1-p)/n}$, a value that depends on the unknown parameter $p$. A reasonable estimate of $\sigma_n$ is $S_n = \sqrt{\hat{p}_n(1-\hat{p}_n)/n}$, and it can be shown (see Exercise 5.32) that $\sigma_n/S_n$ converges in probability to 1. Thus, for any $p$, $0 < p < 1$,

$$\frac{\hat{p}_n - p}{\sqrt{\dfrac{\hat{p}_n(1-\hat{p}_n)}{n}}} \to n(0, 1).$$

The Wald test statistic $Z_n$ is defined by replacing $p$ by $p_0$, and the large-sample Wald test rejects $H_0$ if $Z_n > z_\alpha$. As an alternative estimate of $\sigma_n$, it is easily checked that

$1/I_n(\hat{p}_n) = \hat{p}_n(1 - \hat{p}_n)/n$. So, the same statistic $Z_n$ obtains if we use the information number to derive a standard error for $\hat{p}_n$.

If there was interest in testing the two-sided hypothesis $H_0 : p = p_0$ versus $H_1 : p \neq p_0$, where $0 < p_0 < 1$ is a specified value, the above strategy is again applicable. However, in this case, there is an alternative approximate test. By the Central Limit Theorem, for any $p$, $0 < p < 1$,

$$\frac{\hat{p}_n - p}{\sqrt{p(1-p)/n}} \to n(0, 1).$$

Therefore, if the null hypothesis is true, the statistic

(10.3.4)              $$Z'_n = \frac{\hat{p}_n - p_0}{\sqrt{p_0(1-p_0)/n}} \sim n(0, 1) \qquad \text{(approximately)}.$$

The approximate level $\alpha$ test rejects $H_0$ if $|Z'_n| > z_{\alpha/2}$.

In cases where both tests are applicable, for example, when testing $H_0: p = p_0$, it is not clear which test is to be preferred. The power functions (actual, not approximate) cross one another, so each test is more powerful in a certain portion of the parameter space. (Ghosh 1979) gives some insights into this problem. A related binomial controversy, that of the two-sample problem, is discussed by Robbins 1977 and Eberhardt and Fligner 1977. Two different test statistics for this problem are given in Exercise 10.31.)

Of course, any comparison of power functions is confounded by the fact that these are *approximate* tests and do not necessarily maintain level $\alpha$. The use of a continuity correction (see Example 3.3.2) can help in this problem. In many cases, approximate procedures that use the continuity correction turn out to be *conservative*; that is, they maintain their nominal $\alpha$ level (see Example 10.4.6).                    ‖

Equation (10.3.4) is a special case of another useful large-sample test, the *score test*. The *score statistic* is defined to be

$$S(\theta) = \frac{\partial}{\partial \theta} \log f(\mathbf{X}|\theta) = \frac{\partial}{\partial \theta} \log L(\theta|\mathbf{X}).$$

From (7.3.8) we know that, for all $\theta$, $E_\theta S(\theta) = 0$. In particular, if we are testing $H_0: \theta = \theta_0$ and if $H_0$ is true, then $S(\theta_0)$ has mean 0. Furthermore, from (7.3.10),

$$\text{Var}_\theta S(\theta) = E_\theta\left(\left(\frac{\partial}{\partial \theta} \log L(\theta|\mathbf{X})\right)^2\right) = -E_\theta\left(\frac{\partial^2}{\partial \theta^2} \log L(\theta|\mathbf{X})\right) = I_n(\theta);$$

the information number is the variance of the score statistic. The test statistic for the score test is

$$Z_S = S(\theta_0)/\sqrt{I_n(\theta_0)}.$$

If $H_0$ is true, $Z_S$ has mean 0 and variance 1. From Theorem 10.1.12 it follows that $Z_S$ converges to a standard normal random variable if $H_0$ is true. Thus, the approximate level $\alpha$ score test rejects $H_0$ if $|Z_S| > z_{\alpha/2}$. If $H_0$ is composite, then $\hat{\theta}_0$, an estimate of $\theta$ assuming $H_0$ is true, replaces $\theta_0$ in $Z_S$. If $\hat{\theta}_0$ is the restricted MLE, the restricted maximization might be accomplished using Lagrange multipliers. Thus, the score test is sometimes called the *Lagrange multiplier test*.

**Example 10.3.6 (Binomial score test)** Consider again the Bernoulli model from Example 10.3.5, and consider testing $H_0 : p = p_0$ versus $H_1 : p \neq p_0$. Straightforward calculations yield

$$S(p) = \frac{\hat{p}_n - p}{p(1-p)/n} \quad \text{and} \quad I_n(p) = \frac{n}{p(1-p)}.$$

Hence, the score statistic is

$$Z_S = \frac{S(p_0)}{\sqrt{I_n(p_0)}} = \frac{\hat{p}_n - p_0}{\sqrt{p_0(1-p_0)/n}},$$

the same as (10.3.4).                                                        ‖

One last class of approximate tests to be considered are robust tests (see Miscellanea 10.6.6). From Section 10.2, we saw that if $X_1, \ldots, X_n$ are iid from a location family and $\hat{\theta}_M$ is an M-estimator, then

$$(10.3.5) \qquad\qquad \sqrt{n}(\hat{\theta}_M - \theta_0) \to \mathrm{n}\left(0, \mathrm{Var}_{\theta_0}(\hat{\theta}_M)\right),$$

where $\mathrm{Var}_{\theta_0}(\hat{\theta}_M) = \frac{\mathrm{E}_{\theta_0}\psi(X-\theta_0)^2}{[\mathrm{E}_{\theta_0}\psi'(X-\theta_0)]^2}$ is the asymptotic variance. Thus, we can construct a "generalized" score statistic,

$$Z_{GS} = \sqrt{n}\frac{\hat{\theta}_M - \theta_0}{\sqrt{\mathrm{Var}_{\theta_0}(\hat{\theta}_M)}},$$

or a generalized Wald statistic,

$$Z_{GW} = \sqrt{n}\frac{\hat{\theta}_M - \theta_0}{\sqrt{\widehat{\mathrm{Var}}_{\theta_0}(\hat{\theta}_M)}},$$

where $\widehat{\mathrm{Var}}_{\theta_0}(\hat{\theta}_M)$ can be any consistent estimator. For example, we could use a bootstrap estimate of standard error, or simply substitute an estimator into (10.2.6) and use

$$(10.3.6) \qquad\qquad \widehat{\mathrm{Var}}_1(\hat{\theta}_M) = \frac{\frac{1}{n}\sum_{i=1}^{n}[\psi(x_i - \hat{\theta}_M)]^2}{\left[\frac{1}{n}\sum_{i=1}^{n}\psi'(x_i - \hat{\theta}_M)\right]^2}.$$

The choice of variance estimate can be important; see Boos (1992) or Carroll, Ruppert, and Stefanski (1995, Appendix A.3) for guidance.

**Example 10.3.7 (Tests based on the Huber estimator)** If $X_1, \ldots, X_n$ are iid from a pdf $f(x - \theta)$, where $f$ is symmetric around 0, then for the Huber M-estimator using the $\rho$ function in (10.2.2) and the $\psi$ function (10.2.7), we have an asymptotic variance

(10.3.7)
$$\frac{\int_{-k}^{k} x^2 f(x) dx + k^2 P_0(|X| > k)}{[P_0(|X| \leq k)]^2}.$$

Therefore, based on the asymptotic normality of the M-estimator, we can (for example) test $H_0 : \theta = \theta_0$ vs. $H_1 : \theta \neq \theta_0$ at level $\alpha$ by rejecting $H_0$ if $|Z_{GS}| > z_{\alpha/2}$. To be a bit more practical, we will look at the approximate tests that use an estimated standard error. We will use the statistic $Z_{GW}$, but we will base our variance estimate on (10.3.7), that is

$$\widehat{\text{Var}}_2(\hat{\theta}_M) =$$

(10.3.8)
$$\frac{\frac{1}{n} \sum_{i=1}^{n} (x_i - \hat{\theta}_M)^2 I(|x_i - \hat{\theta}_M| < k) + k^2 \left( \frac{1}{n} \sum_{i=1}^{n} I(|x_i - \hat{\theta}_M| > k) \right)}{\left( 1 - \frac{1}{n} \sum_{i=1}^{n} I(|x_i - \hat{\theta}_M| < k) \right)^2}.$$

Also, we added a "naive" test, $Z_N$, that uses a simple variance estimate

(10.3.9)
$$\widehat{\text{Var}}_3(\hat{\theta}_M) = \frac{1}{n} \sum_{i=1}^{n} (x_i - \hat{\theta}_M)^2.$$

How do these tests fare? Analytical evaluation is difficult, but the small simulation in Table 10.3.1 shows that the $z_{\alpha/2}$ cutoffs are generally too small (neglecting to account for variation in the variance estimates), as the actual size is typically greater than the nominal size. However, there is consistency across a range of distributions, with the double exponential being the best case. (This last occurrence is not totally surprising, as the Huber estimator enjoys an optimality property against distributions with exponential tails; see Huber 1981, Chapter 4.)                    ||

## 10.4 Interval Estimation

As we have done in the previous two sections, we now explore some approximate and asymptotic versions of confidence sets. Our purpose is, as before, to illustrate some methods that will be of use in more complicated situations, methods that will get *some* answer. The answers obtained here are almost certainly not the best but are certainly not the worst. In many cases, however, they are the best that we can do.

We start, as previously, with approximations based on MLEs.

### 10.4.1 Approximate Maximum Likelihood Intervals

From the discussion in Section 10.1.2, and using Theorem 10.1.12, we have a general method to get an asymptotic distribution for a MLE. Hence, we have a general method to construct a confidence interval.

Table 10.3.1. *Power, at specified parameter values, of nominal* $\alpha = .1$ *tests based on* $Z_{GW}$ *and* $Z_N$, *for a sample of size* $n = 15$ *(10,000 simulations)*

| | Underlying pdf | | | | | | | |
|---|---|---|---|---|---|---|---|---|
| | Normal | | $t_5$ | | Logistic | | Double exponential | |
| | $Z_{GW}$ | $Z_N$ | $Z_{GW}$ | $Z_N$ | $Z_{GW}$ | $Z_N$ | $Z_{GW}$ | $Z_N$ |
| $\theta_0$ | .16 | .16 | .14 | .13 | .15 | .15 | .11 | .09 |
| $\theta_0 + .25\sigma$ | .27 | .29 | .29 | .27 | .27 | .27 | .31 | .26 |
| $\theta_0 + .5\sigma$ | .58 | .60 | .65 | .63 | .59 | .60 | .70 | .64 |
| $\theta_0 + .75\sigma$ | .85 | .87 | .89 | .89 | .85 | .87 | .92 | .90 |
| $\theta_0 + 1\sigma$ | .96 | .97 | .98 | .97 | .96 | .97 | .98 | .98 |
| $\theta_0 + 2\sigma$ | 1. | 1. | 1. | 1. | 1. | 1. | 1. | 1. |

If $X_1, \ldots, X_n$ are iid $f(x|\theta)$ and $\hat{\theta}$ is the MLE of $\theta$, then from (10.1.7) the variance of a function $h(\hat{\theta})$ can be approximated by

$$\widehat{\mathrm{Var}}(h(\hat{\theta})|\theta) \approx \frac{[h'(\theta)]^2|_{\theta=\hat{\theta}}}{-\frac{\partial^2}{\partial\theta^2}\log L(\theta|\mathbf{x})|_{\theta=\hat{\theta}}}.$$

Now, for a fixed but arbitrary value of $\theta$, we are interested in the asymptotic distribution of

$$\frac{h(\hat{\theta}) - h(\theta)}{\sqrt{\widehat{\mathrm{Var}}(h(\hat{\theta})|\theta)}}.$$

It follows from Theorem 10.1.12 and Slutsky's Theorem (Theorem 5.5.17) (see Exercise 10.33) that

$$\frac{h(\hat{\theta}) - h(\theta)}{\sqrt{\widehat{\mathrm{Var}}(h(\hat{\theta})|\theta)}} \to \mathrm{n}(0, 1),$$

giving the approximate confidence interval

$$h(\hat{\theta}) - z_{\alpha/2}\sqrt{\widehat{\mathrm{Var}}(h(\hat{\theta})|\theta)} \le h(\theta) \le h(\hat{\theta}) + z_{\alpha/2}\sqrt{\widehat{\mathrm{Var}}(h(\hat{\theta})|\theta)}.$$

**Example 10.4.1 (Continuation of Example 10.1.14)** We have a random sample $X_1, \ldots, X_n$ from a Bernoulli($p$) population. We saw that we could estimate the odds ratio $p/(1-p)$ by its MLE $\hat{p}/(1-\hat{p})$ and that this estimate has approximate variance

$$\widehat{\mathrm{Var}}\left(\frac{\hat{p}}{1-\hat{p}}\right) \approx \frac{\hat{p}}{n(1-\hat{p})^3}.$$

We therefore can construct the approximate confidence interval

$$\frac{\hat{p}}{1-\hat{p}} - z_{\alpha/2}\sqrt{\widehat{\mathrm{Var}}\left(\frac{\hat{p}}{1-\hat{p}}\right)} \le \frac{p}{1-p} \le \frac{\hat{p}}{1-\hat{p}} + z_{\alpha/2}\sqrt{\widehat{\mathrm{Var}}\left(\frac{\hat{p}}{1-\hat{p}}\right)}. \qquad \|$$

A more restrictive form of the likelihood approximation, but one that, when applicable, gives better intervals, is based on the score statistic (see Section 10.3.2). The random quantity

$$(10.4.1) \qquad Q(\mathbf{X}|\theta) = \frac{\frac{\partial}{\partial\theta}\log L(\theta|\mathbf{X})}{\sqrt{-\mathrm{E}_\theta\left(\frac{\partial^2}{\partial\theta^2}\log L(\theta|\mathbf{X})\right)}}$$

has a $n(0,1)$ distribution asymptotically as $n \to \infty$. Thus, the set

$$(10.4.2) \qquad \left\{\theta\colon |Q(\mathbf{x}|\theta)| \le z_{\alpha/2}\right\}$$

is an approximate $1 - \alpha$ confidence set. Notice that, applying results from Section 7.3.2, we have

$$\mathrm{E}_\theta(Q(\mathbf{X}|\theta)) = \frac{\mathrm{E}_\theta\left(\frac{\partial}{\partial\theta}\log L(\theta|\mathbf{X})\right)}{\sqrt{-\mathrm{E}_\theta\left(\frac{\partial^2}{\partial\theta^2}\log L(\theta|\mathbf{X})\right)}} = 0$$

and

$$(10.4.3) \qquad \mathrm{Var}_\theta(Q(\mathbf{X}|\theta)) = \frac{\mathrm{Var}_\theta\left(\frac{\partial}{\partial\theta}\log L(\theta|\mathbf{X})\right)}{-\mathrm{E}_\theta\left(\frac{\partial^2}{\partial\theta^2}\log L(\theta|\mathbf{X})\right)} = 1,$$

and so this approximation exactly matches the first two moments of a $n(0,1)$ random variable. Wilks (1938) proved that these intervals have an asymptotic optimality property; they are, asymptotically, the shortest in a certain class of intervals.

Of course, these intervals are not totally general and may not always be applicable to a function $h(\theta)$. We must be able to express (10.4.2) as a function of $h(\theta)$.

**Example 10.4.2 (Binomial score interval)** Again using a binomial example, if $Y = \sum_{i=1}^n X_i$, where each $X_i$ is an independent Bernoulli$(p)$ random variable, we have

$$Q(Y|p) = \frac{\frac{\partial}{\partial p}\log L(p|Y)}{\sqrt{-\mathrm{E}_p\left(\frac{\partial^2}{\partial p^2}\log L(p|Y)\right)}}$$

$$= \frac{\frac{y}{p} - \frac{n-y}{1-p}}{\sqrt{\frac{n}{p(1-p)}}}$$

$$= \frac{\hat{p} - p}{\sqrt{p(1-p)/n}},$$

where $\hat{p} = y/n$. From (10.4.2), an approximate $1 - \alpha$ confidence interval is given by

$$(10.4.4) \qquad \left\{p\colon \left|\frac{\hat{p} - p}{\sqrt{p(1-p)/n}}\right| \le z_{\alpha/2}\right\}.$$

This is the interval that results from inverting the score statistic (see Example 10.3.6). To calculate this interval we need to solve a quadratic in $p$: see Example 10.4.6 for details. ‖

In Section 10.3 we derived another likelihood test based on the fact that $-2 \log \lambda(\mathbf{X})$ has an asymptotic chi squared distribution. This suggests that if $X_1, \ldots, X_n$ are iid $f(x|\theta)$ and $\hat{\theta}$ is the MLE of $\theta$, then the set

$$(10.4.5) \qquad \left\{ \theta : -2 \log \left( \frac{L(\theta|\mathbf{x})}{L(\hat{\theta}|\mathbf{x})} \right) \leq \chi^2_{1,\alpha} \right\}$$

is an approximate $1 - \alpha$ confidence interval. This is indeed the case and gives us yet another approximate likelihood interval.

Of course, (10.4.5) is just the highest likelihood region (9.2.7) that we originally derived by inverting the LRT statistic. However, we now have an automatic way of attaching an approximate confidence level.

**Example 10.4.3 (Binomial LRT interval)** For $Y = \sum_{i=1}^{n} X_i$, where each $X_i$ is an independent Bernoulli($p$) random variable, we have the approximate $1 - \alpha$ confidence set

$$\left\{ p : -2 \log \left( \frac{p^y (1-p)^{n-y}}{\hat{p}^y (1-\hat{p})^{n-y}} \right) \leq \chi^2_{1,\alpha} \right\}.$$

This confidence set, along with the intervals based on the score and Wald tests, are compared in Example 10.4.7. ‖

### 10.4.2 Other Large-Sample Intervals

Most approximate confidence intervals are based on either finding approximate (or asymptotic) pivots or inverting approximate level $\alpha$ test statistics. If we have any statistics $W$ and $V$ and a parameter $\theta$ such that, as $n \to \infty$,

$$\frac{W - \theta}{V} \to n(0, 1),$$

then we can form the approximate confidence interval for $\theta$ given by

$$W - z_{\alpha/2} V \leq \theta \leq W + z_{\alpha/2} V.$$

which is essentially a Wald-type interval. Direct application of the Central Limit Theorem, together with Slutsky's Theorem, will usually give an approximate confidence interval. (Note that the approximate maximum likelihood intervals of the previous section all reflect this strategy.)

**Example 10.4.4 (Approximate interval)** If $X_1, \ldots, X_n$ are iid with mean $\mu$ and variance $\sigma^2$, then, from the Central Limit Theorem,

$$\frac{\bar{X} - \mu}{\sigma/\sqrt{n}} \to n(0, 1).$$

Table 10.4.1. *Confidence coefficient for the pivotal interval (10.4.6),* $n = 15$, *based on* 10,000 *simulations*

| Nominal level | Underlying pdf | | | |
|---|---|---|---|---|
| | Normal | $t_5$ | Logistic | Double Exponential |
| $1 - \alpha = .90$ | .879 | .864 | .880 | .876 |
| $1 - \alpha = .95$ | .931 | .924 | .931 | .933 |

Moreover, from Slutsky's Theorem, if $S^2 \to \sigma^2$ in probability, then

$$\frac{\bar{X} - \mu}{S/\sqrt{n}} \to n(0, 1),$$

giving the approximate $1 - \alpha$ confidence interval

(10.4.6) $$\bar{x} - z_{\alpha/2} s/\sqrt{n} \le \mu \le \bar{x} + z_{\alpha/2} s/\sqrt{n}.$$

To see how good the approximation is, we present a small simulation to calculate the exact coverage probability of the approximate interval for a variety of pdfs. Note that, since the interval is pivotal, the coverage probability does not depend on the parameter value; it is constant and hence is the confidence coefficient. We see from Table 10.4.1 that even for a sample size as small as $n = 15$, the pivotal confidence interval does a reasonable job, but clearly does not achieve the nominal confidence coefficient. This is, no doubt, due to the optimism of using the $z_{\alpha/2}$ cutoff, which does not account for the variability in $S$. As the sample size increases, the approximation will improve. ‖

In the above example, we could get an approximate confidence interval without specifying the form of the sampling distribution. We should be able to do better when we do specify the form.

**Example 10.4.5 (Approximate Poisson interval)** If $X_1, \ldots, X_n$ are iid Poisson($\lambda$), then we know that

$$\frac{\bar{X} - \lambda}{S/\sqrt{n}} \to n(0, 1).$$

However, this is true even if we did not sample from a Poisson population. Using the Poisson assumption, we know that $\text{Var}(X) = \lambda = E\bar{X}$ and $\bar{X}$ is a good estimator of $\lambda$ (see Example 7.3.12). Thus, using the Poisson assumption, we could also get an approximate confidence interval from the fact that

$$\frac{\bar{X} - \lambda}{\sqrt{\bar{X}/n}} \to n(0, 1),$$

which is the interval that results from inverting the Wald test. We can use the Poisson assumption in another way. Since $\text{Var}(X) = \lambda$, it follows that

$$\frac{\bar{X} - \lambda}{\sqrt{\lambda/n}} \to n(0, 1),$$

resulting in the interval corresponding to the score test, which is also the likelihood interval of (10.4.2) and is best according to Wilks (1938) (see Exercise 10.40).     ‖

Generally speaking, a reasonable rule of thumb is to use as few estimates and as many parameters as possible in an approximation. This is sensible for a very simple reason. Parameters are fixed and do not introduce any added variability into an approximation, while each statistic brings more variability along with it.

**Example 10.4.6 (More on the binomial score interval)** For a random sample $X_1, \ldots, X_n$ from a Bernoulli$(p)$ population, we saw in Example 10.3.5 that, as $n \to \infty$, both

$$\frac{\hat{p} - p}{\sqrt{\hat{p}(1 - \hat{p})/n}} \quad \text{and} \quad \frac{\hat{p} - p}{\sqrt{p(1 - p)/n}}$$

converge in distribution to a standard normal random variable, where $\hat{p} = \sum x_i/n$. In Example 10.3.5 we saw that we could base tests on either approximation, with the former being the Wald test and the latter the score test. We also know that we can use either approximation to form a confidence interval for $p$. However, the score test approximation (with fewer statistics and more parameter values) will give the interval (10.4.4) from Example 10.4.2, which is the asymptotically optimal one; that is,

$$\left\{ p: \left| \frac{\hat{p} - p}{\sqrt{p(1 - p)/n}} \right| \leq z_{\alpha/2} \right\}$$

is the better approximate interval.

It is not immediately clear what this interval looks like, but we can explicitly solve for the set of values. If we square both sides and rearrange terms, we are looking for the set of values of $p$ that satisfy

$$\left\{ p: (\hat{p} - p)^2 \leq z_{\alpha/2}^2 \frac{p(1 - p)}{n} \right\}.$$

This inequality is a quadratic in $p$, which can be put in a more familiar form through some further rearrangement:

$$\left\{ p: \left( 1 + \frac{z_{\alpha/2}^2}{n} \right) p^2 - \left( 2\hat{p} + \frac{z_{\alpha/2}^2}{n} \right) p + \hat{p}^2 \leq 0 \right\}.$$

Since the coefficient of $p^2$ in the quadratic is positive, the quadratic opens upward and, thus, the inequality is satisfied if $p$ lies between the two roots of the quadratic. These two roots are

(10.4.7)
$$\frac{2\hat{p} + z_{\alpha/2}^2/n \pm \sqrt{(2\hat{p} + z_{\alpha/2}^2/n)^2 - 4\hat{p}^2(1 + z_{\alpha/2}^2/n)}}{2(1 + z_{\alpha/2}^2/n)},$$

and the roots define the endpoints of the confidence interval for $p$. Although the expressions for the roots are somewhat nasty, the interval is, in fact, a very good

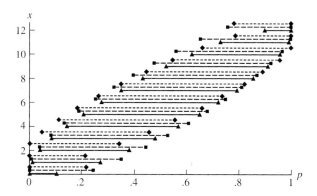

Figure 10.4.1. *Intervals for a binomial proportion from the LRT procedure (solid lines), the score procedure (long dashes), and the modified Wald procedure (short dashes)*

interval for $p$. The interval can be further improved, however, by using a continuity correction (see Example 3.3.2). To do this, we would solve two separate quadratics (see Exercise 10.45),

$$\left| \frac{\hat{p} + \frac{1}{2n} - p}{\sqrt{p(1-p)/n}} \right| \le z_{\alpha/2}, \quad \text{(larger root = upper interval endpoint)}$$

$$\left| \frac{\hat{p} - \frac{1}{2n} - p}{\sqrt{p(1-p)/n}} \right| \le z_{\alpha/2}. \quad \text{(smaller root = lower interval endpoint)}$$

At the endpoints there are obvious modifications. If $\sum x_i = 0$, then the lower interval endpoint is taken to be 0, while, if $\sum x_i = n$, then the upper interval endpoint is taken to be 1. See Blyth (1986) for some good approximations.                     ‖

We now have seen three intervals for a binomial proportion: those based on the Wald and score statistics and the LRT interval of Example 10.4.3. Typically the Wald interval is least preferred, but it would be interesting to compare all three.

**Example 10.4.7 (Comparison of binomial intervals)**   For $Y = \sum_{i=1}^{n} X_i$, $X_1$, ..., $X_n$ iid from a Bernoulli($p$) population, the Wald interval is

$$(10.4.8) \qquad \hat{p} - z_{\alpha/2}\sqrt{\frac{\hat{p}(1-\hat{p})}{n}} \le p \le \hat{p} + z_{\alpha/2}\sqrt{\frac{\hat{p}(1-\hat{p})}{n}},$$

the score interval (with continuity correction) is described in Example 10.4.6, and the approximate LRT is given in Example 10.4.3. To compare them, we look at an example.

For $n = 12$, Figure 10.4.1 shows the realized intervals for the three procedures. The LRT procedure produces the shortest intervals, and the score procedure the longest. For this picture we have made two modifications to the Wald interval. First, at $y = 0$ the unmodified interval is $(0, 0)$, so we have changed the upper endpoint to

Cov. prob.

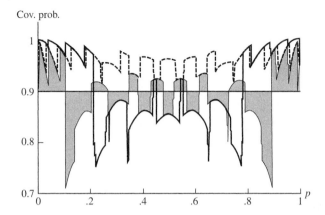

Figure 10.4.2. *Coverage probabilities for nominal .9 confidence procedures for a binomial proportion: the LRT procedure (thin solid lines, shaded grey), the score procedure (dashes), and the modified Wald procedure (thick solid lines)*

$1 - (\alpha/2)^{1/n}$, with a similar modification to the lower interval at $y = n$. Also, there are some instances where the endpoints of the Wald interval go outside $[0, 1]$; these have been truncated.

In Figure 10.4.2 we see that the longer length of the score interval is reflected in its higher coverage probability. Indeed, the score interval is the only one (of the three) that maintains a coverage probability above .9, and hence is the only interval with confidence coefficient .9. The LRT and Wald intervals appear to be too short, and their coverage probabilities are too far below .9 for them to be acceptable. Of course, their performance will improve with increasing $n$.

So it appears that the continuity corrected score interval, although longer, is the interval of choice for small $n$ (but see Exercise 10.44 for another option). The LRT and Wald procedures produce intervals that are just too short for small $n$, with the Wald interval also suffering from endpoint maladies.                    ‖

As we did in Section 10.3.2, we briefly look at intervals based on robust estimators.

**Example 10.4.8 (Intervals based on the Huber estimator)** In a development similar to Example 10.3.7, we can form asymptotic confidence intervals based on the Huber M-estimator. If $X_1, \ldots, X_n$ are iid from a pdf $f(x - \theta)$, where $f$ is symmetric around 0, we have the approximate interval for $\theta$,

$$\hat{\theta}_M \pm z_{\alpha/2}\sqrt{\frac{\text{Var}(\hat{\theta}_M)}{n}},$$

where $\text{Var}(\hat{\theta}_M)$ is given by (10.3.7). Now we replace $\text{Var}(\hat{\theta}_M)$ by the estimates (10.3.8) and (10.3.9) to get Wald-type intervals. To evaluate these intervals, we produce a table similar to Table 10.4.1. It is interesting that, with the exception of the double exponential distribution, the intervals in Table 10.4.2 fare worse than those in Table

Table 10.4.2. *Confidence coefficients for nominal* $1 - \alpha = .9$ *intervals based on Huber's M-estimator,* $n = 15$, *based on 10,000 simulations*

|  | Underlying pdf | | | |
|---|---|---|---|---|
| Nominal level | Normal | $t_5$ | Logistic | Double exponential |
| Variance estimate (10.3.8) | .844 | .856 | .855 | .889 |
| Variance estimate (10.3.9) | .837 | .867 | .855 | .910 |

10.4.1, which are based on the usual mean and variance. We do not have a good explanation for this, except to once again blame it on the overoptimism of the $z_{\alpha/2}$ cutoff. ‖

Thus far, all of the approximations mentioned have been based on letting $n \to \infty$. However, there are other situations where we might use approximate intervals. In Example 9.2.17 we needed approximations as the parameter went to infinity. In another situation, in Example 2.3.13 we saw that for certain parameter configurations, the Poisson distribution can be used to approximate the binomial. This suggests that, if such a parameter configuration is believed to be likely, then an approximate binomial interval can be based on the Poisson distribution. In that spirit we illustrate the following somewhat unusual case.

**Example 10.4.9 (Negative binomial interval)** Let $X_1, \ldots, X_n$ be iid negative binomial$(r, p)$. We assume that $r$ is known and we are interested in a confidence interval for $p$. Using the fact that $Y = \sum X_i \sim$ negative binomial$(nr, p)$, we can form intervals in a number of ways. Using a variation of the binomial–$F$ distribution relationship, we can form an exact confidence interval (see Exercise 9.22) or we can use a normal approximation (see Exercise 10.41). There is another approximation that does not rely on large $n$, but rather small $p$.

In Exercise 2.38 it is established that, as $p \to 0$,

$$2pY \to \chi^2_{2nr} \quad \text{in distribution.}$$

So, for small $p$, $2pY$ is a pivot! Using this fact, we can construct a pivotal $1 - \alpha$ confidence interval, valid for small $p$:

$$\left\{ p: \frac{\chi^2_{2nr, 1-\alpha/2}}{2y} \le p \le \frac{\chi^2_{2nr, \alpha/2}}{2y} \right\}.$$

Details are in Exercise 10.47. ‖

## 10.5 Exercises

**10.1** A random sample $X_1, \ldots, X_n$ is drawn from a population with pdf

$$f(x|\theta) = \frac{1}{2}(1 + \theta x), \quad -1 < x < 1, \quad -1 < \theta < 1.$$

Find a consistent estimator of $\theta$ and show that it is consistent.

**10.2** Prove Theorem 10.1.5.

**10.3** A random sample $X_1, \ldots, X_n$ is drawn from a population that is $n(\theta, \theta)$, where $\theta > 0$.

(a) Show that the MLE of $\theta$, $\hat{\theta}$, is a root of the quadratic equation $\theta^2 + \theta - W = 0$, where $W = (1/n) \sum_{i=1}^{n} X_i^2$, and determine which root equals the MLE.

(b) Find the approximate variance of $\hat{\theta}$ using the techniques of Section 10.1.3.

**10.4** A variation of the model in Exercise 7.19 is to let the random variables $Y_1, \ldots, Y_n$ satisfy

$$Y_i = \beta X_i + \epsilon_i, \quad i = 1, \ldots, n,$$

where $X_1, \ldots, X_n$ are independent $n(\mu, \tau^2)$ random variables, $\epsilon_1, \ldots, \epsilon_n$ are iid $n(0, \sigma^2)$, and the $X$s and $\epsilon$s are independent. Exact variance calculations become quite difficult, so we might resort to approximations. In terms of $\mu$, $\tau^2$, and $\sigma^2$, find approximate means and variances for

(a) $\sum X_i Y_i / \sum X_i^2$.

(b) $\sum Y_i / \sum X_i$.

(c) $\sum (Y_i / X_i) / n$.

**10.5** For the situation of Example 10.1.8 show that for $T_n = \sqrt{n}/\bar{X}_n$:

(a) $\text{Var}(T_n) = \infty$.

(b) If $\mu \neq 0$ and we delete the interval $(-\delta, \delta)$ from the sample space, then $\text{Var}(T_n) < \infty$.

(c) If $\mu \neq 0$, the probability content of the interval $(-\delta, \delta)$ approaches 0 as $n \to \infty$.

**10.6** For the situation of Example 10.1.10 show that

(a) $EY_n = 0$ and $\text{Var}(Y_n) = p_n + (1 - p_n)\sigma_n^2$.

(b) $P(Y_n < a) \to P(Z < a)$, and hence $Y_n \to n(0, 1)$ (recall that $p_n \to 1$, $\sigma_n \to \infty$, and $(1 - p_n)\sigma_n^2 \to \infty$).

**10.7** In the proof of Theorem 10.1.12 it was shown that the MLE $\hat{\theta}$ is an asymptotically efficient estimator of $\theta$. Show that if $\tau(\theta)$ is a continuous and differentiable function of $\theta$, then $\tau(\hat{\theta})$ is a consistent and asymptotically efficient estimator of $\tau(\theta)$.

**10.8** Finish the proof of Theorem 10.1.6 by establishing the two convergence results in (10.1.6).

(a) Show that

$$\frac{1}{\sqrt{n}} l'(\theta_0 | \mathbf{X}) = \sqrt{n} \left[ \frac{1}{n} \sum_i W_i \right],$$

where $W_i = \frac{\frac{d}{d\theta} f(X_i | \theta)}{f(X_i | \theta)}$ has mean 0 and variance $I(\theta_0)$. Now use the Central Limit Theorem to establish the convergence to $n[0, I(\theta_0)]$.

(b) Show that

$$-\frac{1}{n} l''(\theta_0 | \mathbf{X}) = \frac{1}{n} \sum_i W_i^2 - \frac{1}{n} \sum_i \frac{\frac{d^2}{d\theta^2} f(X_i | \theta)}{f(X_i | \theta)}$$

and that the mean of the first piece is $I(\theta_0)$ and the mean of the second piece is 0. Apply the WLLN.

**10.9** Suppose that $X_1, \ldots, X_n$ are iid Poisson($\lambda$). Find the best unbiased estimator of

(a) $e^{-\lambda}$, the probability that $X = 0$.

(b) $\lambda e^{-\lambda}$, the probability that $X = 1$.

(c) For the best unbiased estimators of parts (a) and (b), calculate the asymptotic relative efficiency with respect to the MLE. Which estimators do you prefer? Why?

(d) A preliminary test of a possible carcinogenic compound can be performed by measuring the mutation rate of microorganisms exposed to the compound. An experimenter places the compound in 15 petri dishes and records the following number of mutant colonies:

$$10, \ 7, \ 8, \ 13, \ 8, \ 9, \ 5, \ 7, \ 6, \ 8, \ 3, \ 6, \ 6, \ 3, \ 5.$$

Estimate $e^{-\lambda}$, the probability that no mutant colonies emerge, and $\lambda e^{-\lambda}$, the probability that one mutant colony will emerge. Calculate both the best unbiased estimator and the MLE.

**10.10** Continue the calculations of Example 10.1.15, where the properties of the estimator of $p(1 - p)$ were examined.

(a) Show that, if $p \neq 1/2$, the MLE $\hat{p}(1 - \hat{p})$ is asymptotically efficient.

(b) If $p = 1/2$, use Theorem 5.5.26 to find a limiting distribution of $\hat{p}(1 - \hat{p})$.

(c) Calculate the exact expression for $\text{Var}[\hat{p}(1 - \hat{p})]$. Is the reason for the failure of the approximations any clearer?

**10.11** This problem will look at some details and extensions of the calculation in Example 10.1.18.

(a) Reproduce Figure 10.1.1, calculating the ARE for known $\beta$. (You can follow the calculations in Example A.0.7, or do your own programming.)

(b) Verify that the ARE($\bar{X}, \hat{\mu}$) comparison is the same whether $\beta$ is known or unknown.

(c) For estimation of $\beta$ with known $\mu$, show that the method of moment estimate and MLEs are the same. (It may be easier to use the $(\alpha, \beta)$ parameterization.)

(d) For estimation of $\beta$ with unknown $\mu$, the method of moment estimate and MLEs are not the same. Compare these estimates using asymptotic relative efficiency, and produce a figure like Figure 10.1.1, where the different curves correspond to different values of $\mu$.

**10.12** Verify that the superefficient estimator $d_n$ of Miscellanea 10.6.1 is asymptotically normal with variance $v(\theta) = 1$ when $\theta \neq 0$ and $v(\theta) = a^2$ when $\theta = 0$. (See Lehmann and Casella 1998, Section 6.2, for more on superefficient estimators.)

**10.13** Refer to Example 10.1.19.

(a) Verify that the bootstrap mean and variance of the sample $2, 4, 9, 12$ are 6.75 and 3.94, respectively.

(b) Verify that 6.75 is the mean of the original sample.

(c) Verify that, when we divide by $n$ instead of $n - 1$, the bootstrap variance of the mean, and the usual estimate of the variance of the mean are the same.

(d) Show how to calculate the bootstrap mean and standard error using the $\binom{4+4-1}{4} = 35$ distinct possible resamples.

(e) Establish parts (b) and (c) for a general sample $X_1, X_2, \ldots, X_n$.

**10.14** In each of the following situations we will look at the parametric and nonparametric bootstrap. Compare the estimates, and discuss advantages and disadvantages of the methods.

(a) Referring to Example 10.1.22, estimate the variance of $S^2$ using a nonparametric bootstrap.

(b) In Example 5.6.6 we essentially did a parametric bootstrap of the distribution of $S^2$ from a Poisson sample. Use the nonparametric bootstrap to provide an alternative histogram of the distribution.

(c) In Example 10.1.18 we looked at the problem of estimating a gamma mean. Suppose that we have a random sample

$$0.28, 0.98, 1.36, 1.38, 2.4, 7.42$$

from a gamma$(\alpha, \beta)$ distribution. Estimate the mean and variance of the distribution using maximum likelihood and bootstrapping.

**10.15** (a) Show that $\text{Var}_B^*(\hat{\theta})$ of (10.1.11) converges to $\text{Var}^*(\hat{\theta})$ of (10.1.10) as $B \to \infty$.

(b) For fixed $B$ and $i = 1, 2, \ldots$, calculate the bootstrap variance $\text{Var}_{B_i}^*(\hat{\theta})$. Use the Law of Large Numbers to show $(1/m)\sum_{i=1}^{m} \text{Var}_{B_i}^*(\hat{\theta}) \to \text{Var}^*(\hat{\theta})$ as $m \to \infty$.

**10.16** For the situation of Example 10.1.21, if we observed that $\hat{p} = 1/2$, we might use a variance estimate from Theorem 5.5.26. Show that this variance estimate would be equal to $2[\text{Var}(\hat{p})]^2$.

(a) If we observe $\hat{p} = 11/24$, verify that this variance estimate is .00007.

(b) Using simulation, calculate the "exact variance" of $\hat{p}(1 - \hat{p})$ when $n = 24$ and $p = 11/24$. Verify that it is equal to .00529.

(c) Why do you think the Delta Method is so bad in this case? Might the second-order Delta Method do any better? What about the bootstrap estimate?

**10.17** Efron (1982) analyzes data on law school admission, with the object being to examine the correlation between the LSAT (Law School Admission Test) score and the first-year GPA (grade point average). For each of 15 law schools, we have the pair of data points (average LSAT, average GPA):

| | | | | |
|---|---|---|---|---|
| (576, 3.39) | (635, 3.30) | (558, 2.81) | (578, 3.03) | (666, 3.44) |
| (580, 3.07) | (555, 3.00) | (661, 3.43) | (651, 3.36) | (605, 3.13) |
| (653, 3.12) | (575, 2.74) | (545, 2.76) | (572, 2.88) | (594, 2.96) |

(a) Calculate the correlation coefficient between LSAT score and GPA.

(b) Use the nonparametric bootstrap to estimate the standard deviation of the correlation coefficient. Use $B = 1000$ resamples, and also plot them in a histogram.

(c) Use the parametric bootstrap to estimate the standard deviation of the correlation coefficient. Assume that (LSAT, GRE) has a bivariate normal distribution, and estimate the five parameters. Then generate 1000 samples of 15 pairs from this bivariate normal distribution.

(d) If $(X, Y)$ are bivariate normal with correlation coefficient $\rho$ and sample correlation $r$, then the Delta Method can be used to show that

$$\sqrt{n}(r - \rho) \to \text{n}(0, (1 - \rho^2)^2).$$

Use this fact to estimate the standard deviation of $r$. How does it compare to the bootstrap estimates? Draw an approximate pdf of $r$.

(e) Fisher's z-transformation is a *variance-stabilizing* transformation for the correlation coefficient (see Exercise 11.4). If $(X, Y)$ are bivariate normal with correlation coefficient $\varrho$ and sample correlation $r$, then

$$\frac{1}{2}\left[\log\left(\frac{1 + r}{1 - r}\right) - \log\left(\frac{1 + \rho}{1 - \rho}\right)\right]$$

is approximately normal. Use this fact to draw an approximate pdf of $r$.
(Establishing the normality result in part (d) involves some tedious matrix calculations; see Lehmann and Casella 1998, Example 6.5). The z-transformation
of part (e) yields faster convergence to normality than the Delta Method of
part (d). Diaconis and Holmes 1994 do an exhaustive bootstrap for this problem,
enumerating all $77,558,760$ correlation coefficients.)

**10.18** For the situation of Exercise 10.2.1, that is, if $X_1, X_2, \ldots, X_n$ are iid, where $X_i \sim$
$n(\mu, \sigma^2)$ with probability $1 - \delta$ and $X_i \sim f(x)$ with probability $\delta$, where $f(x)$ is any
density with mean $\theta$ and variance $\tau^2$, show that

$$\text{Var}(\bar{X}) = (1 - \delta)\frac{\sigma^2}{n} + \delta\frac{\tau^2}{n} + \frac{\delta(1 - \delta)(\theta - \mu)^2}{n}.$$

Also deduce that contamination with a Cauchy pdf will always result in an infinite
variance. (*Hint*: Write this mixture model as a hierarchical model. Let $Y = 0$ with
probability $1 - \delta$ and $Y = 1$ with probability $\delta$. Then $\text{Var}(X_i) = \text{E}[\text{Var}(X_i)|Y] +$
$\text{Var}(\text{E}[X_i|Y])$.)

**10.19** Another way in which underlying assumptions can be violated is if there is correlation
in the sampling, which can seriously affect the properties of the sample mean. Suppose
we introduce correlation in the case discussed in Exercise 10.2.1; that is, we observe
$X_1, \ldots, X_n$, where $X_i \sim n(\theta, \sigma^2)$, but the $X_i$s are no longer independent.

(a) For the equicorrelated case, that is, $\text{Corr}(X_i, X_j) = \rho$ for $i \neq j$, show that

$$\text{Var}(\bar{X}) = \frac{\sigma^2}{n} + \frac{n-1}{n}\rho\sigma^2,$$

so $\text{Var}(\bar{X}) \not\to 0$ as $n \to \infty$.

(b) If the $X_i$s are observed through time (or distance), it is sometimes assumed that
the correlation decreases with time (or distance), with one specific model being
$\text{Corr}(X_i, X_j) = \rho^{|i-j|}$. Show that in this case

$$\text{Var}(\bar{X}) = \frac{\sigma^2}{n} + \frac{2\sigma^2}{n^2}\frac{\rho}{1-\rho}\left(n - \frac{1-\rho^n}{1-\rho}\right),$$

so $\text{Var}(\bar{X}) \to 0$ as $n \to \infty$. (See Miscellanea 5.8.2 for another effect of correlation.)

(c) The correlation structure in part (b) arises in an *autoregressive AR(1) model*,
where we assume that $X_{i+1} = \rho X_i + \delta_i$, with $\delta_i$ iid $n(0,1)$. If $|\rho| < 1$ and we
define $\sigma^2 = 1/(1 - \rho^2)$, show that $\text{Corr}(X_1, X_i) = \rho^{i-1}$.

**10.20** Refer to Definition 10.2.2 about breakdown values.

(a) If $T_n = \bar{X}_n$, the sample mean, show that $b = 0$.

(b) If $T_n = M_n$, the sample median, show that $b = .5$.
An estimator that "splits the difference" between the mean and the median in
terms of sensitivity is the $\alpha$-*trimmed* mean, $0 < \alpha < \frac{1}{2}$, defined as follows. $\bar{X}_n^\alpha$,
the $\alpha$-trimmed mean, is computed by deleting the $\alpha n$ smallest observations and
the $\alpha n$ largest observations, and taking the arithmetic mean of the remaining
observations.

(c) If $T_n = \bar{X}_n^\alpha$, the $\alpha$-trimmed mean of the sample, $0 < \alpha < \frac{1}{2}$, show that $0 < b < \frac{1}{2}$.

**10.21** The breakdown performance of the mean and the median continues with their scale
estimate counterparts. For a sample $X_1, \ldots, X_n$:

(a) Show that the breakdown value of the sample variance $S^2 = \sum(X_i - \bar{X})^2/(n-1)$ is 0.

(b) A robust alternative is the *median absolute deviation*, or MAD, the median of $|X_1 - \mathrm{M}|, |X_2 - \mathrm{M}|, \ldots, |X_n - \mathrm{M}|$, where M is the sample median. Show that this estimator has a breakdown value of 50%.

**10.22** This exercise will look at some of the details of Example 10.2.3.

(a) Verify that, if $n$ is odd, then

$$P\left(\sqrt{n}(M_n - \mu) \le a\right) = P\left(\frac{\sum_i Y_i - np_n}{\sqrt{np_n(1 - p_n)}} \ge \frac{(n+1)/2 - np_n}{\sqrt{np_n(1 - p_n)}}\right).$$

(b) Verify that $p_n \to p = F(\mu) = 1/2$ and

$$\frac{(n+1)/2 - np_n}{\sqrt{np_n(1 - p_n)}} \to -2aF'(\mu) = -2af(\mu).$$

(*Hint*: Establish that $\frac{(n+1)/2 - np_n}{\sqrt{n}}$ is the limit form of a derivative.)

(c) Explain how to go from the statement that

$$P\left(\sqrt{n}(M_n - \mu) \le a\right) \to P\left(Z \ge -2af(\mu)\right)$$

to the conclusion that $\sqrt{n}(M_n - \mu)$ is asymptotically normal with mean 0 and variance $1/[2f(\mu)]^2$.

(Note that the CLT would directly apply only if $p_n$ did not depend on $n$. As it does, more work needs to be done to rigorously conclude limiting normality. When the work is done, the result is as expected.)

**10.23** In this exercise we will further explore the ARE of the median to the mean, $\mathrm{ARE}(M_n, \bar{X})$.

(a) Verify the three AREs given in Example 10.2.4.

(b) Show that $\mathrm{ARE}(M_n, \bar{X})$ is unaffected by scale changes. That is, it doesn't matter whether the underlying pdf is $f(x)$ or $(1/\sigma)f(x/\sigma)$.

(c) Calculate $\mathrm{ARE}(M_n, \bar{X})$ when the underlying distribution is Student's $t$ with $\nu$ degrees of freedom, for $\nu = 3, 5, 10, 25, 50, \infty$. What can you conclude about the ARE and the tails of the distribution?

(d) Calculate $\mathrm{ARE}(M_n, \bar{X})$ when the underlying pdf is the *Tukey model*

$$X \sim \begin{cases} n(0, 1) & \text{with probability } 1 - \delta \\ n(0, \sigma^2) & \text{with probability } \delta. \end{cases}$$

Calculate the ARE for a range of $\delta$ and $\sigma$. What can you conclude about the relative performance of the mean and the median?

**10.24** Assuming that $\theta_0$ satisfies $\mathrm{E}_{\theta_0}\psi(X - \theta_0) = 0$, show that (10.2.4) and (10.2.5) imply (10.2.6).

**10.25** If $f(x)$ is a pdf symmetric around 0 and $\rho$ is a symmetric function, show that $\int \psi(x - \theta)f(x - \theta)\,dx = 0$, where $\psi = \rho'$ is an odd function. Show that this then implies that if $X_1, \ldots, X_n$ are iid from $f(x - \theta)$ and $\hat{\theta}_M$ is the minimizer of $\sum_i \rho(x_i - \theta)$, then $\hat{\theta}_M$ is asymptotically normal with mean equal to the true value of $\theta$.

**10.26** Here we look at some details in the calculations in Example 10.2.6.

(a) Verify the expressions for $E_\theta \psi'(X - \theta)$ and $E_\theta[\psi(X - \theta)]^2$, and hence verify the formula for the variance of $\hat{\theta}_M$.

(b) When calculating the expected value of $\psi'$, we noted that $\psi$ was not differentiable, but we could work with the differentiable portion. Another approach is to realize that the expected value of $\psi$ is differentiable, and that in (10.2.5) we could write

$$\frac{1}{n} \sum_{i=1}^n \psi'(x_i - \theta_0) \rightarrow \frac{d}{d\theta} E_{\theta_0} \psi(X - \theta) \Big|_{\theta=\theta_0}.$$

Show that this is the same limit as in (10.2.5).

**10.27** Consider the situation of Example 10.6.2.

(a) Verify that $IF(\bar{X}, x) = x - \mu$.

(b) For the median we have $T(F) = m$ if $P(X \le m) = 1/2$ or $m = F^{-1}(1/2)$. If $X \sim F_\delta$, show that

$$P(X \le a) = \begin{cases} (1-\delta)F(a) & \text{if } x > a \\ (1-\delta)F(a) + \delta & \text{otherwise} \end{cases}$$

and thus

$$T(F_\delta) = \begin{cases} F^{-1}\left(\frac{1}{2(1-\delta)}\right) & \text{if } x > F^{-1}\left(\frac{1}{2(1-\delta)}\right) \\ F^{-1}\left(\frac{1/2-\delta}{1-\delta}\right) & \text{otherwise.} \end{cases}$$

(c) Show that

$$\frac{1}{\delta}\left[F^{-1}\left(\frac{1}{2(1-\delta)}\right) - F^{-1}\left(\frac{1}{2}\right)\right] \rightarrow \frac{1}{2f(m)},$$

and complete the argument to calculate $IF(M, x)$.

(*Hint*: Write $a_\delta = F^{-1}\left(\frac{1}{2(1-\delta)}\right)$, and argue that the limit is $a'_\delta|_{\delta=0}$. This latter quantity can be calculated using implicit differentiation and the fact that $(1-\delta)^{-1} = 2F(a_\delta)$.)

**10.28** Show that if $\rho$ is defined by (10.2.2), then both $\rho$ and $\rho'$ are continuous.

**10.29** From (10.2.9) we know that an M-estimator can never be more efficient than a maximum likelihood estimator. However, we also know when it can be as efficient.

(a) Show that (10.2.9) is an equality if we choose $\psi(x - \theta) = cl'(\theta|x)$, where $l$ is the log likelihood and $c$ is a constant.

(b) For each of the following distributions, verify that the corresponding $\psi$ functions give asymptotically efficient M-estimators.

(i) Normal: $f(x) = e^{-x^2/2}/(\sqrt{2\pi})$, $\psi(x) = x$

(ii) Logistic: $f(x) = e^{-x}/(1 + e^{-x})^2$, $\psi(x) = \tanh(x)$, where $\tanh(x)$ is the *hyperbolic tangent*

(iii) Cauchy: $f(x) = [\pi(1 + x^2)]^{-1}$, $\psi(x) = 2x/(1 + x^2)$

(iv) Least informative distribution:

$$f(x) = \begin{cases} Ce^{-x^2/2} & |x| \le c \\ Ce^{-c|x|+c^2/2} & |x| > c \end{cases}$$

with $\psi(x) = \max\{-c, \min(c, x)\}$ and $C$ and $c$ are constants.

(See Huber 1981, Section 3.5, for more details.)

**10.30** For M-estimators there is a connection between the $\psi$ function and the breakdown value. The details are rather involved (Huber 1981, Section 3.2) but they can be summarized as follows: If $\psi$ is a bounded function, then the breakdown value of the associated M-estimator is given by

$$b^* = \frac{\eta}{1+\eta}, \text{ where } \eta = \min\left\{-\frac{\psi(-\infty)}{\psi(\infty)}, -\frac{\psi(\infty)}{\psi(-\infty)}\right\}.$$

(a) Calculate the breakdown value of the efficient M-estimators of Exercise 10.29. Which ones are both efficient and robust?

(b) Calculate the breakdown value of these other M-estimators
   (i) The Huber estimator given by (10.2.1)
   (ii) Tukey's biweight: $\psi(x) = x(c^2 - x^2)$ for $|x| \le c$ and 0 otherwise, where $c$ is a constant
   (iii) Andrew's sine wave: $\psi(x) = c\sin(x/c)$ for $|x| \le c\pi$ and 0 otherwise

(c) Evaluate the AREs of the estimators in part (b) with respect to the MLE when the underlying distribution is (i) normal and (ii) double exponential.

**10.31** Binomial data gathered from more than one population are often presented in a *contingency table*. For the case of two populations, the table might look like this:

|  | Population 1 | 2 | Total |
|---|---|---|---|
| Successes | $S_1$ | $S_2$ | $S = S_1 + S_2$ |
| Failures | $F_1$ | $F_2$ | $F = F_1 + F_2$ |
| Total | $n_1$ | $n_2$ | $n = n_1 + n_2$ |

where Population 1 is binomial$(n_1, p_1)$, with $S_1$ successes and $F_1$ failures, Population 2 is binomial$(n_2, p_2)$, with $S_2$ successes and $F_2$ failures, and $S_1$ and $S_2$ are independent. A hypothesis that is usually of interest is

$$H_0: p_1 = p_2 \quad \text{versus} \quad H_1: p_1 \ne p_2.$$

(a) Show that a test can be based on the statistic

$$T = \frac{(\hat{p}_1 - \hat{p}_2)^2}{\left(\frac{1}{n_1} + \frac{1}{n_2}\right)(\hat{p}(1-\hat{p}))},$$

where $\hat{p}_1 = S_1/n_1$, $\hat{p}_2 = S_2/n_2$, and $\hat{p} = (S_1 + S_2)/(n_1 + n_2)$. Also, show that as $n_1, n_2 \to \infty$, the distribution of $T$ approaches $\chi_1^2$. (This is a special case of a test known as a *chi squared test of independence*.)

(b) Another way of measuring departure from $H_0$ is by calculating an *expected frequency table*. This table is constructed by conditioning on the marginal totals and filling in the table according to $H_0: p_1 = p_2$, that is,

Expected frequencies

|  | 1 | 2 | Total |
|---|---|---|---|
| Successes | $\dfrac{n_1 S}{n_1 + n_2}$ | $\dfrac{n_2 S}{n_1 + n_2}$ | $S = S_1 + S_2$ |
| Failures | $\dfrac{n_1 F}{n_1 + n_2}$ | $\dfrac{n_2 F}{n_1 + n_2}$ | $F = F_1 + F_2$ |
| Total | $n_1$ | $n_2$ | $n = n_1 + n_2$ |

Using the expected frequency table, a statistic $T^*$ is computed by going through the cells of the tables and computing

$$T^* = \sum \frac{(\text{observed} - \text{expected})^2}{\text{expected}}$$

$$= \frac{\left(S_1 - \frac{n_1 S}{n_1 + n_2}\right)^2}{\frac{n_1 S}{n_1 + n_2}} + \cdots + \frac{\left(F_2 - \frac{n_2 F}{n_1 + n_2}\right)^2}{\frac{n_2 F}{n_1 + n_2}}.$$

Show, algebraically, that $T^* = T$ and hence that $T^*$ is asymptotically chi squared.

(c) Another statistic that could be used to test equality of $p_1$ and $p_2$ is

$$T^{**} = \frac{\hat{p}_1 - \hat{p}_2}{\sqrt{\frac{\hat{p}_1(1 - \hat{p}_1)}{n_1} + \frac{\hat{p}_2(1 - \hat{p}_2)}{n_2}}}.$$

Show that, under $H_0$, $T^{**}$ is asymptotically $n(0, 1)$, and hence its square is asymptotically $\chi_1^2$. Furthermore, show that $(T^{**})^2 \neq T^*$.

(d) Under what circumstances is one statistic preferable to the other?

(e) A famous medical experiment was conducted by Joseph Lister in the late 1800s. Mortality associated with surgery was quite high, and Lister conjectured that the use of a disinfectant, carbolic acid, would help. Over a period of several years Lister performed 75 amputations with and without using carbolic acid. The data are given here:

|  |  | Carbolic acid used? | |
|---|---|:---:|:---:|
|  |  | Yes | No |
| Patient lived? | Yes | 34 | 19 |
|  | No | 6 | 16 |

Use these data to test whether the use of carbolic acid is associated with patient mortality.

**10.32** (a) Let $(X_1, \ldots, X_n) \sim \text{multinomial}(m, p_1, \ldots, p_n)$. Consider testing $H_0 : p_1 = p_2$ versus $H_1 : p_1 \neq p_2$. A test that is often used, called *McNemar's Test*, rejects $H_0$ if

$$\frac{(X_1 - X_2)^2}{X_1 + X_2} > \chi_{1,\alpha}^2.$$

Show that this test statistic has the form (as in Exercise 10.31)

$$\sum_1^n \frac{(\text{observed} - \text{expected})^2}{\text{expected}},$$

where the $X_i$s are the observed cell frequencies and the expected cell frequencies are the MLEs of $mp_i$, under the assumption that $p_1 = p_2$.

(b) McNemar's Test is often used in the following type of problem. Subjects are asked if they agree or disagree with a statement. Then they read some information

about the statement and are asked again if they agree or disagree. The numbers
of responses in each category are summarized in a $2 \times 2$ table like this:

|  |  | Before | |
|---|---|---|---|
|  |  | Agree | Disagree |
| After | Agree | $X_3$ | $X_2$ |
|  | Disagree | $X_1$ | $X_4$ |

The hypothesis $H_0 : p_1 = p_2$ states that the proportion of people who change
from agree to disagree is the same as the proportion of people who change from
disagree to agree. Another hypothesis that might be tested is that the proportion
of those who initially agree and then change is the same as the proportion of
those who initially disagree and then change. Express this hypothesis in terms of
conditional probabilities and show that it is different from the above $H_0$. (This
hypothesis can be tested with a $\chi^2$ test like those in Exercise 10.31.)

**10.33** Fill in the gap in Theorem 10.3.1. Use Theorem 10.1.12 and Slutsky's Theorem (The-
orem 5.5.17) to show that $\sqrt{-l''(\hat{\theta}|\mathbf{x})}\,(\hat{\theta}-\theta_0) \to n(0,1)$, and therefore $-2\log\lambda(\mathbf{X}) \to \chi_1^2$.

**10.34** For testing $H_0 : p = p_0$ versus $H_1 : p \neq p_0$, suppose we observe $X_1, \ldots, X_n$ iid
Bernoulli($p$).

(a) Derive an expression for $-2\log\lambda(\mathbf{x})$, where $\lambda(\mathbf{x})$ is the LRT statistic.

(b) As in Example 10.3.2, simulate the distribution of $-2\log\lambda(\mathbf{x})$ and compare it to
the $\chi^2$ approximation.

**10.35** Let $X_1, \ldots, X_n$ be a random sample from a $n(\mu, \sigma^2)$ population.

(a) If $\mu$ is unknown and $\sigma^2$ is known, show that $Z = \sqrt{n}(\bar{X} - \mu_0)/\sigma$ is a Wald
statistic for testing $H_0 : \mu = \mu_0$.

(b) If $\sigma^2$ is unknown and $\mu$ is known, find a Wald statistic for testing $H_0 : \sigma = \sigma_0$.

**10.36** Let $X_1, \ldots, X_n$ be a random sample from a gamma($\alpha, \beta$) population. Assume $\alpha$ is
known and $\beta$ is unknown. Consider testing $H_0 : \beta = \beta_0$.

(a) What is the MLE of $\beta$?

(b) Derive a Wald statistic for testing $H_0$, using the MLE in both the numerator and
denominator of the statistic.

(c) Repeat part (b) but using the sample standard deviation in the standard error.

**10.37** Let $X_1, \ldots, X_n$ be a random sample from a $n(\mu, \sigma^2)$ population.

(a) If $\mu$ is unknown and $\sigma^2$ is known, show that $Z = \sqrt{n}(\bar{X} - \mu_0)/\sigma$ is a score statistic
for testing $H_0 : \mu = \mu_0$.

(b) If $\sigma^2$ is unknown and $\mu$ is known, find a score statistic for testing $H_0 : \sigma = \sigma_0$.

**10.38** Let $X_1, \ldots, X_n$ be a random sample from a gamma($\alpha, \beta$) population. Assume $\alpha$ is
known and $\beta$ is unknown. Consider testing $H_0 : \beta = \beta_0$. Derive a score statistic for
testing $H_0$.

**10.39** Expand the comparisons made in Example 10.3.7.

(a) Another test based on Huber's M-estimator would be one that used a variance
estimate, based on (10.3.6). Examine the performance of such a test statistic,
and comment on its desirability (or lack of) as an alternative to either (10.3.8)
or (10.3.9).

(b) Another test based on Huber's M-estimator would be one that used a variance
from a bootstrap calculation. Examine the performance of such a test statistic.

(c) A robust competitor to $\hat{\theta}_M$ is the median. Examine the performance of tests of a location parameter based on the median.

**10.40** In Example 10.4.5 we saw that the Poisson assumption, together with the Central Limit Theorem, could be used to form an approximate interval based on the fact that

$$\frac{\bar{X} - \lambda}{\sqrt{\lambda/n}} \rightarrow n(0, 1).$$

Show that this approximation is optimal according to Wilks (1938). That is, show that

$$\frac{\bar{X} - \lambda}{\sqrt{\lambda/n}} = \frac{\frac{\partial}{\partial \lambda} \log L(\lambda|\mathbf{X})}{\sqrt{-E_\lambda \left(\frac{\partial^2}{\partial \lambda^2} \log L(\lambda|\mathbf{X})\right)}}.$$

**10.41** Let $X_1, \ldots, X_n$ be iid negative binomial$(r, p)$. We want to construct some approximate confidence intervals for the negative binomial parameters.

(a) Calculate Wilks' approximation (10.4.1) and show how to form confidence intervals with this expression.

(b) Find an approximate $1 - \alpha$ confidence interval for the *mean* of the negative binomial distribution. Show how to incorporate the continuity correction into your interval.

(c) The aphid data of Exercise 9.23 can also be modeled using the negative binomial distribution. Construct an approximate 90% confidence interval for the aphid data using the results of part (b). Compare the interval to the Poisson-based intervals of Exercise 9.23.

**10.42** Show that (10.4.5) is equivalent to the highest likelihood region (9.2.7) in that for any fixed $\alpha$ level, they will produce the same confidence set.

**10.43** In Example 10.4.7, two modifications were made to the Wald interval.

(a) At $y = 0$ the upper interval endpoint was changed to $1 - (\alpha/2)^{1/n}$, and at $y = n$ the lower interval endpoint was changed to $(\alpha/2)^{1/n}$. Justify the choice of these endpoints. (*Hint*: see Section 9.2.3.)

(b) The second modification was to truncate all intervals to be within $[0, 1]$. Show that this change, together with the one in part (a), results in an improvement over the original Wald interval.

**10.44** Agresti and Coull (1998) "strongly recommend" the score interval for a binomial parameter but are concerned that a formula such as (10.4.7) might be a bit formidable for an elementary course in statistics. To produce a reasonable binomial interval with an easier formula, they suggest the following modification to the Wald interval: Add 2 successes and 2 failures; then use the original Wald formula (10.4.8). That is, use $\hat{p} = (y + 2)/(n + 4)$ instead of $\hat{p} = y/n$. Using both length and coverage probability, compare this interval to the binomial score interval. Do you agree that it is a reasonable alternative to the score interval?

(Samuels and Lu 1992 suggest another modification to the Wald interval based on sample sizes. Agresti and Caffo 2000 extend these improved approximate intervals to the two sample problem.)

**10.45** Solve for the endpoints of the approximate binomial confidence interval, with continuity correction, given in Example 10.4.6. Show that this interval is wider than the corresponding interval without continuity correction, and that the continuity corrected interval has a uniformly higher coverage probability. (In fact, the coverage probability of the uncorrected interval does not maintain $1 - \alpha$; it dips below this level for some parameter values. The corrected interval does maintain a coverage probability greater than $1 - \alpha$ for all parameter values.)

**10.46** Expand the comparisons made in Example 10.4.8.

(a) Produce a table similar to Table 10.4.2 that examines the robustness of intervals for a location parameter based on the median. (Intervals based on the mean are done in Table 10.4.1.)

(b) Another interval based on Huber's M-estimator would be one that used a variance from a bootstrap calculation. Examine the robustness of such an interval.

**10.47** Let $X_1, \ldots, X_n$ be iid negative binomial$(r, p)$.

(a) Complete the details of Example 10.4.9; that is, show that for small $p$, the interval

$$\left\{ p: \frac{\chi^2_{2nr,1-\alpha/2}}{2\sum x} \leq p \leq \frac{\chi^2_{2nr,\alpha/2}}{2\sum x} \right\}$$

is an approximate $1 - \alpha$ confidence interval.

(b) Show how to choose the endpoints in order to obtain a minimum length $1 - \alpha$ interval.

**10.48** For the case of Fieller's confidence set (see Miscellanea 9.5.3), that is, given a random sample $(X_1, Y_1), \ldots, (X_n, Y_n)$ from a bivariate normal distribution with parameters $(\mu_X, \mu_Y, \sigma_X^2, \sigma_Y^2, \rho)$, find an approximate confidence interval for $\theta = \mu_Y / \mu_X$. Use the approximate moment calculations in Example 5.5.27 and apply the Central Limit Theorem.

# 10.6 Miscellanea

## 10.6.1 Superefficiency

Although the Cramér–Rao Lower Bound of Theorem 7.3.9 is a bona fide lower bound on the variance, the lower bound of Definition 10.1.11 and Theorem 10.1.6, which refers to the asymptotic variance, can be violated. An example of an estimator that beats the bound of Definition 10.1.11 was given by Hodges (see LeCam 1953).

If $X_1, \ldots, X_n$ are iid n$(\theta, 1)$, the Cramér–Rao Lower Bound for unbiased estimators of $\theta$ is $v(\theta) = 1/n$. The estimator

$$d_n = \begin{cases} \bar{X} & \text{if } |\bar{X}| \geq 1/n^{1/4} \\ a\bar{X} & \text{if } |\bar{X}| < 1/n^{1/4} \end{cases}$$

satisfies

$$\sqrt{n}(d_n - \theta) \to n[0, v(\theta)],$$

in distribution, where $v(\theta) = 1$ when $\theta \neq 0$ and $v(\theta) = a^2$ when $\theta = 0$. If $a < 1$, inequality (7.2.5) is therefore violated at $\theta = 0$.

Although estimators such as $d_n$, called *superefficient*, can be constructed in some generality, they are more of a theoretical oddity than a practical concern. This is because the values of $\theta$ for which the variance goes below the bound are a set of Lebesgue measure 0. However, the existence of superefficient estimators serves to remind us to always be careful in our examination of assumptions for establishing properties of estimators (and to be careful in general!).

### 10.6.2 Suitable Regularity Conditions

The phrase "under suitable regularity conditions" is a somewhat abused phrase, as with enough assumptions we can probably prove whatever we want. However, "regularity conditions" are typically very technical, rather boring, and usually satisfied in most reasonable problems. But they are a necessary evil, so we should deal with them. To be complete, we present a set of regularity conditions that suffice to rigorously establish Theorems 10.1.6 and 10.1.12. These are not the most general conditions but are sufficiently general for many applications (with a notable exception being if the MLE is on the boundary of the parameter space). Be forewarned, the following is not for the fainthearted and can be skipped without sacrificing much in the way of understanding.

These conditions mainly relate to differentiability of the density and the ability to interchange differentiation and integration (as in the conditions for Theorem 7.3.9). For more details and generality, see Stuart, Ord, and Arnold (1999, Chapter 18), Ferguson (1996, Part 4), or Lehmann and Casella (1998, Section 6.3).

The following four assumptions are sufficient to prove Theorem 10.1.6, consistency of MLEs:

$(A1)$ We observe $X_1, \ldots, X_n$, where $X_i \sim f(x|\theta)$ are iid.

$(A2)$ The parameter is *identifiable*; that is, if $\theta \neq \theta'$, then $f(x|\theta) \neq f(x|\theta')$.

$(A3)$ The densities $f(x|\theta)$ have common support, and $f(x|\theta)$ is differentiable in $\theta$.

$(A4)$ The parameter space $\Omega$ contains an open set $\omega$ of which the true parameter value $\theta_0$ is an interior point.

The next two assumptions, together with $(A1)$–$(A4)$ are sufficient to prove Theorem 10.1.12, asymptotic normality and efficiency of MLEs.

$(A5)$ For every $x \in \mathcal{X}$, the density $f(x|\theta)$ is three times differentiable with respect to $\theta$, the third derivative is continuous in $\theta$, and $\int f(x|\theta)\, dx$ can be differentiated three times under the integral sign.

$(A6)$ For any $\theta_0 \in \Omega$, there exists a positive number $c$ and a function $M(x)$ (both of which may depend on $\theta_0$) such that

$$\left| \frac{\partial^3}{\partial \theta^3} \log f(x|\theta) \right| \leq M(x) \quad \text{for all } x \in \mathcal{X}, \quad \theta_0 - c < \theta < \theta_0 + c,$$

with $\mathrm{E}_{\theta_0}[M(X)] < \infty$ .

*10.6.3 More on the Bootstrap*

   *Theory*

The theory behind the bootstrap is quite sophisticated, being based on *Edge-worth expansions.* These are expansions (in the spirit of Taylor series expansions) of distribution functions around a normal distribution. As an example, for $X_1, \ldots, X_n$ iid with density $f$ with mean and variance $\mu$ and $\sigma^2$, an Edgeworth expansion of the cdf of $\frac{\sqrt{n}(\bar{X}-\mu)}{\sigma}$ is (Hall 1992, Equation 2.17)

$$P\left(\frac{\sqrt{n}(\bar{X}-\mu)}{\sigma} \le w\right) = \Phi(w) + \phi(w)\left[\frac{-1}{6\sqrt{n}}\kappa(w^2-1) + Rn\right]$$

where $nRn$ is bounded, $\Phi$ and $\phi$ are, respectively, the distribution and density function of a standard normal and $\kappa = E(X_1 - \mu)^3$ is the skewness. The first term in the expansion is the "usual" normal approximation, and as we add more terms, the expansion becomes more accurate.

The amazing thing about the bootstrap is that in some cases it automatically gets the second term in the expansion correct (hence achieving "second-order" accuracy). This does not happen in all cases, but one case in which it does occur is in bootstrapping a pivotal quantity. The Edgeworth theory of bootstrap is given a thorough treatment by Hall (1992); see also Shao and Tu (1995).

   *Practice*

We have used the bootstrap only to calculate standard errors, but it has many other uses, with perhaps the most popular being the construction of confidence intervals. There are also many variations of the bootstrap developed for different situations. In particular, dealing with dependent data is somewhat delicate. For an introduction to the many uses of the bootstrap and much more, see Efron and Tibshirani (1993).

   *Limitations*

Although the bootstrap is perhaps the single most important development in statistical methodology in recent times, it is not without its limitations and detractors. Outside of the cases of iid sampling and pivotal quantities, the bootstrap is less automatic but still can be extremely useful. For an interesting treatment of these issues, see LePage and Billard (1992) or Young (1994).

*10.6.4 Influence Functions*

A measure of catastrophic occurrences that does consider distributional properties is the *influence function,* which also measures the effect of an aberrant observation. The influence function has an interpretation as a derivative, which also turns out to have some interesting consequences.

The influence function of a statistic is actually calculated using its population counterpart. For example, the influence function of the sample mean is calculated using the population mean, as it seeks to measure the influence of perturbing the population. Similarly, the influence function of the sample median is calculated using the population median. To treat this idea in a consistent manner, it makes sense to think of an estimator as a function that operates on the cdf $F$ or its sample counterpart, the empirical cdf (Definition 1.5.1) $F_n$. Such functions, that actually have other functions as arguments are known as *functionals*.

Note that for a sample $X_1, X_2, \ldots, X_n$, knowledge of the sample is equivalent to knowledge of the empirical cdf $F_n$, as $F_n$ has a jump of size $1/n$ at each $X_i$. Thus, a statistic $T = T(X_1, X_2, \ldots, X_n)$ can equivalently be written $T(F_n)$. In doing so, we can then denote its population counterpart as $T(F)$.

**Definition 10.6.1** For a sample $X_1, X_2, \ldots, X_n$ from a population with cdf $F$, the *influence function* of a statistic $T = T(F_n)$ at a point $x$ is

$$IF(T, x) = \lim_{\delta \to 0} \frac{1}{\delta} [T(F_\delta) - T(F)],$$

where $X \sim F_\delta$ if

$$X \sim \begin{cases} F & \text{with probability } 1 - \delta \\ x & \text{with probability } \delta, \end{cases}$$

that is, $F_\delta$ is a mixture of $F$ and a point $x$.

**Example 10.6.2 (Influence functions of the mean and median)** Suppose that we have a population with continuous cdf $F$ and pdf $f$. Let $\mu$ denote the population mean and $\bar{X}$ the sample mean, and let $T(\cdot)$ be the functional that calculates the mean of a population. Thus $T(F_n) = \bar{X}$, $T(F) = \mu$, and

$$T(F_\delta) = (1 - \delta)\mu + \delta x,$$

so $IF(\bar{X}, x) = x - \mu$, and as $x$ gets larger, its influence on $\bar{X}$ becomes increasingly large.

For the median $M$, we have (see Exercise 10.27)

$$IF(M, x) = \begin{cases} \frac{1}{2f(m)} & \text{if } x > m \\ -\frac{1}{2f(m)} & \text{otherwise.} \end{cases}$$

So, in contrast to the mean, the median has a bounded influence function.    ‖

Why is a bounded influence function important? To answer that, we look at the influence function of an M-estimator, of which the mean and median are special cases.

Let $\hat{\theta}_M$ be the M-estimator that is the solution to $\sum_i \psi(x_i - \theta) = 0$, where $X_1, \ldots, X_n$ are iid with cdf $F$. In Section 10.2.2 we saw that $\hat{\theta}_M$ will be a consistent estimator of the value $\theta_0$ that satisfies $E_{\theta_0} \psi(X - \theta_0) = 0$. The influence function of $\hat{\theta}_M$ is

$$IF(\hat{\theta}_M, x) = \frac{\psi(x - \theta_0)}{-\int \psi'(t - \theta_0) f(t)\, dt} = \frac{\psi(x - \theta_0)}{-E_0(\psi'(X - \theta_0))}.$$

Now if we recall (10.2.6), we see that the expected square of the influence function gives the asymptotic variance of $\hat{\theta}_M$, that is,

$$\sqrt{n}(\hat{\theta}_M - \theta_0) \to \text{n}\left(0, \text{E}_{\theta_0}[IF(\hat{\theta}_M, X)]^2\right)$$

in distribution. Thus, the influence function is directly related to the asymptotic variance.

### 10.6.5 Bootstrap Intervals

In Section 10.1.4 we saw the bootstrap to be a simple, general technique for obtaining a standard error of any statistic. In calculating these standard errors, we actually construct a distribution of a statistic, the *bootstrap* distribution. Then, a natural question arises. Is there a simple, general method of using the bootstrap distribution to make a confidence statement? The bootstrap can indeed be used to construct very good confidence intervals but, alas, the simplicity of application that it enjoys in calculating standard errors does not carry over into confidence intervals.

Methods based on using percentiles of the bootstrap distribution, or on bootstrapping a *t*-statistic (pivot), would seem to have potential for being generally applicable. However, Efron and Tibshirani (1993, Section 13.4) note that "neither of these intervals works well in general." Hall (1992, Chapter 3) prefers the *t*-statistic method and points out that bootstrapping a pivot is a superior technique in general.

Percentile and percentile-*t* intervals are only the tip of a vast development of bootstrap confidence intervals, many of which are excellent performers. However, we cannot summarize these procedures in one simple recipe; different problems will require different techniques.

### 10.6.6 Robust Intervals

Although we went into some detail about robustness of point estimators in Section 10.2, aside from Examples 10.3.7 and 10.4.8, we did not give much detail about robust tests and confidence intervals. This is not a comment on the importance of the subject but has more to do with space.

When we examined point estimators for robustness properties, the main concerns had to do with performance under deviations (both small and large) from the underlying assumptions. The same concerns are carried over to tests and intervals, with the expectation that robust point estimators will lead to robust tests and intervals. In particular, we would want robust tests to maintain power and robust intervals to maintain coverage over a range of deviations from the underlying model. That this is the case is indicated by the fact (see Staudte and Sheather 1990, Section 5.3.3) that the power function of a test can be related to the influence function of the point estimate on which it is based. Of course, this immediately implies that coverage properties of a related interval estimate can also be related to the influence function.

A nice introduction to robust tests, through estimating equations and score tests, is given by Boos (1992). The books by Staudte and Sheather (1990) and Hettmansperger and McKean (1998) are also excellent sources, as is the now-classic book by Huber (1981).

Chapter 11

# Analysis of Variance and Regression

*"I've wasted time enough," said Lestrade rising. "I believe in hard work and not in sitting by the fire spinning fine theories."*

**Inspector Lestrade**
*The Adventure of the Noble Bachelor*

## 11.1 Introduction

Up until now, we have modeled a random variable with a pdf or pmf that depended on parameters to be estimated. In many situations, some of which follow, a random variable can be modeled not only with unknown parameters but also with known (and sometimes controllable) covariates. This chapter describes the methodologies of analysis of variance (ANOVA) and regression analysis. They are based on an underlying assumption of a linear relationship and form a large core of the statistical methods that are used in practice.

The analysis of variance (commonly referred to as the ANOVA) is one of the most widely used statistical techniques. A basic idea of the ANOVA, that of partitioning variation, is a fundamental idea of experimental statistics. The ANOVA belies its name in that it is not concerned with analyzing variances but rather with analyzing *variation in means*.

We will study a common type of ANOVA, the oneway ANOVA. For a thorough treatment of the different facets of ANOVA designs, there is the classic text by Cochran and Cox (1957) or the more modern, but still somewhat classic, treatments by Dean and Voss (1999) and Kuehl (2000). The text by Neter, Wasserman, and Whitmore (1993) provides a guide to overall strategies in experimental statistics.

The technique of regression, in particular linear regression, probably wins the prize as the most popular statistical tool. There are all forms of regression: linear, nonlinear, simple, multiple, parametric, nonparametric, etc. In this chapter we will look at the simplest case, linear regression with one predictor variable. (This is usually called *simple* linear regression, as opposed to *multiple* linear regression, which deals with many predictor variables.)

A major purpose of regression is to explore the dependence of one variable on others. In simple linear regression, the mean of a random variable, $Y$, is modeled as a function of another observable variable, $x$, by the relationship $EY = \alpha + \beta x$. In general, the function that gives $EY$ as a function of $x$ is called the *population regression function*.

Good overall references for regression models are Christensen (1996) and Draper and Smith (1998). A more theoretical treatment is given in Stuart, Ord, and Arnold (1999, Chapter 27).

## 11.2 Oneway Analysis of Variance

In its simplest form, the ANOVA is a method of estimating the means of several populations, populations often assumed to be normally distributed. The heart of the ANOVA, however, lies in the topic of statistical design. How can we get the most information on the most populations with the fewest observations? The ANOVA design question is not our major concern, however; we will be concerned with inference, that is, with estimation and testing, in the ANOVA.

Classic ANOVA had testing as its main goal—in particular, testing what is known as "the ANOVA null hypothesis." But more recently, especially in the light of greater computing power, experimenters have realized that testing one hypothesis (a somewhat ludicrous one at that, as we shall see) does not make for good experimental inference. Thus, although we will derive the test of the ANOVA null, it is far from the most important part of an analysis of variance. More important is estimation, both point and interval. In particular, inference based on *contrasts* (to be defined) is of major importance.

In the oneway analysis of variance (also known as the oneway classification) we assume that data, $Y_{ij}$, are observed according to a model

$$(11.2.1) \qquad Y_{ij} = \theta_i + \epsilon_{ij}, \quad i = 1, \ldots, k, \quad j = 1, \ldots, n_i,$$

where the $\theta_i$ are unknown parameters and the $\epsilon_{ij}$ are error random variables.

**Example 11.2.1 (Oneway ANOVA)** Schematically, the data, $y_{ij}$, from a oneway ANOVA will look like this:

| | Treatments | | | | |
|---|---|---|---|---|---|
| 1 | 2 | 3 | ... | k |
| $y_{11}$ | $y_{21}$ | $y_{31}$ | $\cdots$ | $y_{k1}$ |
| $y_{12}$ | $y_{22}$ | $y_{32}$ | $\cdots$ | $y_{k2}$ |
| $\vdots$ | $\vdots$ | $\vdots$ | $\cdots$ | $y_{k3}$ |
| | | $y_{3n_3}$ | | $\vdots$ |
| $y_{1n_1}$ | | | | |
| | $y_{2n_2}$ | | | $y_{kn_k}$ |

Note that we do not assume that there are equal numbers of observations in each treatment group.

As an example, consider the following experiment performed to assess the relative effects of three toxins and a control on the liver of a certain species of trout. The data are the amounts of deterioration (in standard units) of the liver in each sacrificed fish.

| Toxin 1 | Toxin 2 | Toxin 3 | Control |
|---------|---------|---------|---------|
| 28 | 33 | 18 | 11 |
| 23 | 36 | 21 | 14 |
| 14 | 34 | 20 | 11 |
| 27 | 29 | 22 | 16 |
|    | 31 | 24 |    |
|    | 34 |    |    |

‖

Without loss of generality we can assume that $E\epsilon_{ij} = 0$, since if not, we can rescale the $\epsilon_{ij}$ and absorb the leftover mean into $\theta_i$. Thus it follows that

$$EY_{ij} = \theta_i, \quad j = 1, \ldots, n_i,$$

so the $\theta_i$s are the means of the $Y_{ij}$s. The $\theta_i$s are usually referred to as *treatment means*, since the index often corresponds to different treatments or to *levels* of a particular treatment, such as dosage levels of a particular drug.

There is an alternative model to (11.2.1), sometimes called the *overparameterized model*, which can be written as

(11.2.2) $$Y_{ij} = \mu + \tau_i + \epsilon_{ij}, \quad i = 1, \ldots, k, \quad j = 1, \ldots, n_i,$$

where, again, $E\epsilon_{ij} = 0$. It follows from this model that

$$EY_{ij} = \mu + \tau_i.$$

In this formulation we think of $\mu$ as a grand mean, that is, the common mean level of the treatments. The parameters $\tau_i$ then denote the unique effect due to treatment $i$, the deviation from the mean level that is caused by the treatment. However, we cannot estimate both $\tau_i$ and $\mu$ separately, because there are problems with *identifiability*.

**Definition 11.2.2**   A parameter $\theta$ for a family of distributions $\{f(x|\theta) : \theta \in \Theta\}$ is *identifiable* if distinct values of $\theta$ correspond to distinct pdfs or pmfs. That is, if $\theta \neq \theta'$, then $f(x|\theta)$ is not the same function of $x$ as $f(x|\theta')$.

Identifiability is a property of the model, not of an estimator or estimation procedure. However, if the model is not identifiable, then there is difficulty in doing inference. For example, if $f(x|\theta) = f(x|\theta')$, then observations from both distributions look exactly the same and we would have no way of knowing whether the true value of the parameter was $\theta$ or $\theta'$. In particular, both $\theta$ and $\theta'$ would give the likelihood function the same value.

Realize that problems with identifiability can usually be solved by redefining the model. One reason that we have not encountered identifiability problems before is that our models have not only made intuitive sense but also were identifiable (for example, modeling a normal population in terms of its mean and variance). Here, however, we have a model, (11.2.2), that makes intuitive sense but is not identifiable. In Chapter 12 we will see a parameterization of the bivariate normal distribution that models a situation well but is not identifiable.

In the parameterization of (11.2.2), there are $k + 1$ parameters, $(\mu, \tau_1, \ldots, \tau_k)$, but only $k$ means, $EY_{ij}, i = 1, \ldots, k$. Without any further restriction on the parameters, more than one set of values for $(\mu, \tau_1, \ldots, \tau_k)$ will lead to the same distribution. It is common in this model to add the restriction that $\sum_{i=1}^{k} \tau_i = 0$, which effectively reduces the number of parameters to $k$ and makes the model identifiable. The restriction also has the effect of giving the $\tau_i$s an interpretation as deviations from an overall mean level. (See Exercise 11.5.)

For the oneway ANOVA the model (11.2.1), the *cell means model*, which has a more straightforward interpretation, is the one that we prefer to use. In more complicated ANOVAs, however, there is sometimes an interpretive advantage in model (11.2.2).

### 11.2.1 Model and Distribution Assumptions

Under model (11.2.1), a minimum assumption that is needed before any estimation can be done is that $E\epsilon_{ij} = 0$ and $\text{Var}\,\epsilon_{ij} < \infty$ for all $i, j$. Under these assumptions, we can do some estimation of the $\theta_i$s (as in Exercise 7.41). However, to do any confidence interval estimation or testing, we need distributional assumptions. Here are the classic ANOVA assumptions.

*Oneway ANOVA assumptions*

Random variables $Y_{ij}$ are observed according to the model

$$Y_{ij} = \theta_i + \epsilon_{ij}, \quad i = 1, \ldots, k, \quad j = 1, \ldots, n_i,$$

where

(i) $E\epsilon_{ij} = 0, \text{Var}\,\epsilon_{ij} = \sigma_i^2 < \infty$, for all $i, j$. $\text{Cov}(\epsilon_{ij}, \epsilon_{i'j'}) = 0$ for all $i$, $i'$, $j$, and $j'$ unless $i = i'$ and $j = j'$.

(ii) The $\epsilon_{ij}$ are independent and normally distributed (normal errors).

(iii) $\sigma_i^2 = \sigma^2$ for all $i$ (equality of variance, also known as *homoscedasticity*).

Without assumption (ii) we could do only point estimation and possibly look for estimators that minimize variance within a class, but we could not do interval estimation or testing. If we assume some distribution other than normal, intervals and tests can be quite difficult (but still possible) to derive. Of course, with reasonable sample sizes and populations that are not too asymmetric, we have the Central Limit Theorem (CLT) to rely on.

The equality of variance assumption is also quite important. Interestingly, its importance is linked to the normality assumption. In general, if it is suspected that the data badly violate the ANOVA assumptions, a first course of attack is usually to try to transform the data nonlinearly. This is done as an attempt to more closely satisfy the ANOVA assumptions, a generally easier alternative than finding another model for the untransformed data. A number of common transformations can be found in Snedecor and Cochran (1989); also see Exercises 11.1 and 11.2. (Other research on transformations has been concerned with the Box–Cox family of power transformations. See Exercise 11.3.)

The classic paper of Box (1954) shows that the robustness of the ANOVA to the assumption of normality depends on how equal the variances are (equal being better). The problem of estimating means when variances are unequal, known as the Behrens–Fisher problem, has a rich statistical history which can be traced back to Fisher (1935, 1939). A full account of the Behrens–Fisher problem can be found in Stuart, Ord, and Arnold (1999).

For the remainder of this chapter we will do what is done in most of the experimental situations and we will assume that the three classic assumptions hold. If the data are such that transformations and the CLT are needed, we assume that such measures have been taken.

### 11.2.2 The Classic ANOVA Hypothesis

The classic ANOVA test is a test of the null hypothesis

$$H_0: \quad \theta_1 = \theta_2 = \cdots = \theta_k,$$

a hypothesis that, in many cases, is silly, uninteresting, and not true. An experimenter would not usually believe that the different treatments have *exactly* the same mean. More reasonably, an experiment is done to find out which treatments are better (for example, have a higher mean), and the real interest in the ANOVA is not in testing but in estimation. (There are some specialized situations where there is interest in the ANOVA null in its own right.) Most situations are like the following.

**Example 11.2.3 (The ANOVA hypothesis)** The ANOVA evolved as a method of analyzing agricultural experiments. For example, in a study of the effect of various fertilizers on the zinc content of spinach plants $(y_{ij})$, five treatments are investigated. Each treatment consists of a mixture of fertilizer material (magnesium, potassium, and zinc) and the data look like the layout of Example 11.2.1. The five treatments, in pounds per acre, are

| Treatment | Magnesium | Potassium | Zinc |
|:---:|:---:|:---:|:---:|
| 1 | 0 | 0 | 0 |
| 2 | 0 | 200 | 0 |
| 3 | 50 | 200 | 0 |
| 4 | 200 | 200 | 0 |
| 5 | 0 | 200 | 15 |

The classic ANOVA null hypothesis is really of no interest since the experimenter is sure that the different fertilizer mixtures have some different effects. The interest is in quantifying these effects.     ‖

We will spend some time with the ANOVA null but mostly use it as a means to an end. Recall the connection between testing and interval estimation established in Chapter 9. By using this connection, we can derive confidence regions by deriving, then inverting, appropriate tests (an easier route here).

The alternative to the ANOVA null is simply that the means are not all equal; that is, we test

(11.2.3)        $H_0$:   $\theta_1 = \theta_2 = \cdots = \theta_k$        versus        $H_1$:   $\theta_i \neq \theta_j$, for some $i, j$.

Equivalently, we can specify $H_1$ as $H_1$: not $H_0$. Realize that if $H_0$ is rejected, we can conclude only that there is *some* difference in the $\theta_i$s, but we can make no inference as to where this difference might be. (Note that if $H_1$ is accepted, we are *not* saying that all of the $\theta_i$s are different, merely that at least two are.)

One problem with the ANOVA hypotheses, a problem shared by many multivariate hypotheses, is that the interpretation of the hypotheses is not easy. What would be more useful, rather than concluding just that some $\theta_i$s are different, is a statistical description of the $\theta_i$s. Such a description can be obtained by breaking down the ANOVA hypotheses into smaller, more easily describable pieces.

We have already encountered methods for breaking down complicated hypotheses into smaller, more easily understood pieces—the union–intersection and intersection–union methods of Chapter 8. For the ANOVA, the union–intersection method is best suited, as the ANOVA null is the intersection of more easily understood univariate hypotheses, hypotheses expressed in terms of *contrasts*. Furthermore, in the cases we will consider, the resulting tests based on the union–intersection method are identical to LRTs (see Exercise 11.13). Hence, they enjoy all the properties of likelihood tests.

**Definition 11.2.4**   Let $\mathbf{t} = (t_1, \ldots, t_k)$ be a set of variables, either parameters or statistics, and let $\mathbf{a} = (a_1, \ldots, a_k)$ be known constants. The function

(11.2.4)                                $$\sum_{i=1}^{k} a_i t_i$$

is called a *linear combination* of the $t_i$s. If, furthermore, $\sum a_i = 0$, it is called a *contrast*.

Contrasts are important because they can be used to compare treatment means. For example, if we have means $\theta_1, \ldots, \theta_k$ and constants $\mathbf{a} = (1, -1, 0, \ldots, 0)$, then

$$\sum_{i=1}^{k} a_i \theta_i = \theta_1 - \theta_2$$

is a contrast that compares $\theta_1$ to $\theta_2$. (See Exercise 11.10 for more about contrasts.)

The power of the union–intersection approach is increased understanding. The individual null hypotheses, of which the ANOVA null hypothesis is the intersection, are quite easy to visualize.

**Theorem 11.2.5**   *Let $\theta = (\theta_1, \ldots, \theta_k)$ be arbitrary parameters. Then*

$$\theta_1 = \theta_2 = \cdots = \theta_k \Leftrightarrow \sum_{i=1}^{k} a_i \theta_i = 0 \quad \text{for all } \mathbf{a} \in \mathcal{A},$$

*where $\mathcal{A}$ is the set of constants satisfying $\mathcal{A} = \{\mathbf{a} = (a_1, \ldots, a_k) : \sum a_i = 0\}$; that is, all contrasts must satisfy $\sum a_i \theta_i = 0$.*

**Proof:** If $\theta_1 = \cdots = \theta_k = \theta$, then

$$\sum_{i=1}^{k} a_i \theta_i = \sum_{i=1}^{k} a_i \theta = \theta \sum_{i=1}^{k} a_i = 0, \quad \text{(because } \mathbf{a} \text{ satisfies } \sum a_i = 0\text{)}$$

proving one implication ($\Rightarrow$). To prove the other implication, consider the set of $a_i \in \mathcal{A}$ given by

$$\mathbf{a}_1 = (1, -1, 0, \ldots, 0), \quad \mathbf{a}_2 = (0, 1, -1, 0, \ldots, 0), \quad \ldots, \quad \mathbf{a}_{k-1} = (0, \ldots, 0, 1, -1).$$

(The set $(\mathbf{a}_1, \mathbf{a}_2, \ldots, \mathbf{a}_{k-1})$ *spans* the elements of $\mathcal{A}$. That is, any $\mathbf{a} \in \mathcal{A}$ can be written as a linear combination of $(\mathbf{a}_1, \mathbf{a}_2, \ldots, \mathbf{a}_{k-1})$.) Forming contrasts with these $\mathbf{a}_i$s, we get that

$$\mathbf{a}_1 \Rightarrow \theta_1 = \theta_2, \quad \mathbf{a}_2 \Rightarrow \theta_2 = \theta_3, \quad \ldots, \quad \mathbf{a}_{k-1} \Rightarrow \theta_{k-1} = \theta_k,$$

which, taken together, imply that $\theta_1 = \cdots = \theta_k$, proving the theorem. $\qquad \square$

It immediately follows from Theorem 11.2.5 that the ANOVA null can be expressed as a hypothesis about contrasts. That is, the null hypothesis is true if and only if the hypothesis

$$H_0: \quad \sum_{i=1}^{k} a_i \theta_i = 0 \quad \text{for all } (a_1, \ldots, a_k) \text{ such that } \sum_{i=1}^{k} a_i = 0$$

is true. Moreover, if $H_0$ is false, we now know that there must be at least one nonzero contrast. That is, the ANOVA alternative, $H_1$: not all $\theta_i$s equal, is equivalent to the alternative

$$H_1: \quad \sum_{i=1}^{k} a_i \theta_i \neq 0 \quad \text{for some } (a_1, \ldots, a_k) \text{ such that } \sum_{i=1}^{k} a_i = 0.$$

Thus, we have gained in that the use of contrasts leaves us with hypotheses that are a little easier to understand and perhaps are a little easier to interpret. The real gain, however, is that the use of contrasts now allows us to think and operate in a univariate manner.

### 11.2.3 Inferences Regarding Linear Combinations of Means

Linear combinations, in particular contrasts, play an extremely important role in the analysis of variance. Through understanding and analyzing the contrasts, we can make meaningful inferences about the $\theta_i$s. In the previous section we showed that the ANOVA null is really a statement about contrasts. In fact, most interesting inferences in an ANOVA can be expressed as contrasts or sets of contrasts. We start simply with inference about a single linear combination.

Working under the oneway ANOVA assumptions, we have that

$$Y_{ij} \sim n(\theta_i, \sigma^2), \quad i = 1, \ldots, k, \quad j = 1, \ldots, n_i.$$

Therefore,

$$\bar{Y}_{i\cdot} = \frac{1}{n_i} \sum_{j=1}^{n_i} Y_{ij} \sim \text{n}(\theta_i, \sigma^2/n_i), \quad i = 1, \ldots, k.$$

*A note on notation*: It is a common convention that if a subscript is replaced by a · (dot), it means that subscript has been summed over. Thus, $Y_{i\cdot} = \sum_{j=1}^{n_i} Y_{ij}$ and $Y_{\cdot j} = \sum_{i=1}^{k} Y_{ij}$. The addition of a "bar" indicates that a mean is taken, as in $\bar{Y}_{i\cdot}$ above. If both subscripts are summed over and the overall mean (called the *grand mean*) is calculated, we will break this rule to keep notation a little simpler and write $\bar{Y} = (1/N) \sum_{i=1}^{k} \sum_{j=1}^{n_i} Y_{ij}$, where $N = \sum_{i=1}^{k} n_i$.

For any constants $\mathbf{a} = (a_1, \ldots, a_k)$, $\sum_{i=1}^{k} a_i \bar{Y}_{i\cdot}$ is also normal (see Exercise 11.8) with

$$\text{E}\left( \sum_{i=1}^{k} a_i \bar{Y}_{i\cdot} \right) = \sum_{i=1}^{k} a_i \theta_i \quad \text{and} \quad \text{Var}\left( \sum_{i=1}^{k} a_i \bar{Y}_{i\cdot} \right) = \sigma^2 \sum_{i=1}^{k} \frac{a_i^2}{n_i},$$

and furthermore

$$\frac{\sum_{i=1}^{k} a_i \bar{Y}_{i\cdot} - \sum_{i=1}^{k} a_i \theta_i}{\sqrt{\sigma^2 \sum_{i=1}^{k} a_i^2/n_i}} \sim \text{n}(0, 1).$$

Although this is nice, we are usually in the situation of wanting to make inferences about the $\theta_i$s without knowledge of $\sigma$. Therefore, we want to replace $\sigma$ with an estimate. In each population, if we denote the sample variance by $S_i^2$, that is,

$$S_i^2 = \frac{1}{n_i - 1} \sum_{j=1}^{n_i} (Y_{ij} - \bar{Y}_{i\cdot})^2, \quad i = 1, \ldots, k,$$

then $S_i^2$ is an estimate of $\sigma^2$ and $(n_i - 1)S_i^2/\sigma^2 \sim \chi_{n_i-1}^2$. Furthermore, under the ANOVA assumptions, since each $S_i^2$ estimates the same $\sigma^2$, we can improve the estimators by combining them. We thus use the pooled estimator of $\sigma^2$, $S_p^2$, given by

$$(11.2.5) \qquad S_p^2 = \frac{1}{N-k} \sum_{i=1}^{k} (n_i - 1)S_i^2 = \frac{1}{N-k} \sum_{i=1}^{k} \sum_{j=1}^{n_i} (Y_{ij} - \bar{Y}_{i\cdot})^2.$$

Note that $N - k = \sum(n_i - 1)$. Since the $S_i^2$s are independent, Lemma 5.3.2 shows that $(N - k)S_p^2/\sigma^2 \sim \chi_{N-k}^2$. Also, $S_p^2$ is independent of each $\bar{Y}_{i\cdot}$ (see Exercise 11.6) and thus

$$(11.2.6) \qquad \frac{\sum_{i=1}^{k} a_i \bar{Y}_{i\cdot} - \sum_{i=1}^{k} a_i \theta_i}{\sqrt{S_p^2 \sum_{i=1}^{k} a_i^2/n_i}} \sim t_{N-k},$$

Student's $t$ with $N - k$ degrees of freedom.

To test

$$H_0: \sum_{i=1}^{k} a_i\theta_i = 0 \qquad \text{versus} \qquad H_1: \sum_{i=1}^{k} a_i\theta_i \neq 0$$

at level $\alpha$, we would reject $H_0$ if

(11.2.7)
$$\left| \frac{\sum_{i=1}^{k} a_i\bar{Y}_{i\cdot}}{\sqrt{S_p^2 \sum_{i=1}^{k} a_i^2/n_i}} \right| > t_{N-k,\alpha/2}.$$

(Exercise 11.9 shows some other tests involving linear combinations.) Furthermore, (11.2.6) defines a pivot that can be inverted to give an interval estimator of $\sum a_i\theta_i$. With probability $1 - \alpha$,

$$\sum_{i=1}^{k} a_i\bar{Y}_{i\cdot} - t_{N-k,\alpha/2}\sqrt{S_p^2 \sum_{i=1}^{k} \frac{a_i^2}{n_i}} \leq \sum_{i=1}^{k} a_i\theta_i$$

(11.2.8)
$$\leq \sum_{i=1}^{k} a_i\bar{Y}_{i\cdot} + t_{N-k,\alpha/2}\sqrt{S_p^2 \sum_{i=1}^{k} \frac{a_i^2}{n_i}}.$$

**Example 11.2.6 (ANOVA contrasts)**   Special values of $\mathbf{a}$ will give particular tests or confidence intervals. For example, to compare treatments 1 and 2, take $\mathbf{a} = (1, -1, 0, \ldots, 0)$. Then, using (11.2.6), to test $H_0: \theta_1 = \theta_2$ versus $H_1: \theta_1 \neq \theta_2$, we would reject $H_0$ if

$$\left| \frac{\bar{Y}_{1\cdot} - \bar{Y}_{2\cdot}}{\sqrt{S_p^2\left(\frac{1}{n_1} + \frac{1}{n_2}\right)}} \right| > t_{N-k,\alpha/2}.$$

Note, the difference between this test and the two-sample $t$ test (see Exercise 8.41) is that here information from treatments $3, \ldots, k$, as well as treatments 1 and 2, is used to estimate $\sigma^2$.

Alternatively, to compare treatment 1 to the average of treatments 2 and 3 (for example, treatment 1 might be a control, 2 and 3 might be experimental treatments, and we are looking for some overall effect), we would take $\mathbf{a} = (1, -\frac{1}{2}, -\frac{1}{2}, 0, \ldots, 0)$ and reject $H_0: \theta_1 = \frac{1}{2}(\theta_2 + \theta_3)$ if

$$\left| \frac{\bar{Y}_{1\cdot} - \frac{1}{2}\bar{Y}_{2\cdot} - \frac{1}{2}\bar{Y}_{3\cdot}}{\sqrt{S_p^2\left(\frac{1}{n_1} + \frac{1}{4n_2} + \frac{1}{4n_3}\right)}} \right| > t_{N-k,\alpha/2}.$$

Using either (11.2.6) or (11.2.8), we have a way of testing or estimating any linear combination in the ANOVA. By judiciously choosing our linear combination we can learn much about the treatment means. For example, if we look at the contrasts $\theta_1 - \theta_2, \theta_2 - \theta_3$, and $\theta_1 - \theta_3$, we can learn something about the ordering of the $\theta_i$s. (Of course, we have to be careful of the overall $\alpha$ level when doing a number of tests or intervals, but we can use the Bonferroni Inequality. See Example 11.2.9.)

We also must use some care in drawing formal conclusions from combinations of contrasts. Consider the hypotheses

$$H_0: \quad \theta_1 = \frac{1}{2}(\theta_2 + \theta_3) \qquad \text{versus} \qquad H_1: \quad \theta_1 < \frac{1}{2}(\theta_2 + \theta_3)$$

and

$$H_0: \quad \theta_2 = \theta_3 \qquad \text{versus} \qquad H_1: \quad \theta_2 < \theta_3.$$

If we reject both null hypotheses, we can conclude that $\theta_3$ is greater than both $\theta_1$ and $\theta_2$, although we can draw no formal conclusion about the ordering of $\theta_2$ and $\theta_1$ from these two tests. (See Exercise 11.10.)                                                  ‖

Now we will use these univariate results about linear combinations and the relationship between the ANOVA null hypothesis and contrasts given in Theorem 11.2.5 to derive a test of the ANOVA null hypothesis.

### 11.2.4 The ANOVA F Test

In the previous section we saw how to deal with single linear combinations and, in particular, contrasts in the ANOVA. Also, in Section 11.2, we saw that the ANOVA null hypothesis is equivalent to a hypothesis about contrasts. In this section we will use this equivalence, together with the union–intersection methodology of Chapter 8, to derive a test of the ANOVA hypothesis.

From Theorem 11.2.5, the ANOVA hypothesis test can be written

$$H_0: \quad \sum_{i=1}^{k} a_i\theta_i = 0 \text{ for all } \mathbf{a} \in \mathcal{A} \qquad \text{versus} \qquad H_1: \quad \sum_{i=1}^{k} a_i\theta_i \neq 0 \text{ for some } \mathbf{a} \in \mathcal{A},$$

where $\mathcal{A} = \{\mathbf{a} = (a_1, \ldots, a_k) : \sum_{i=1}^{k} a_i = 0\}$. To see this more clearly as a union–intersection test, define, for each $\mathbf{a}$, the set

$$\Theta_{\mathbf{a}} = \{\theta = (\theta_1, \ldots, \theta_k) : \sum_{i=1}^{k} a_i\theta_i = 0\}.$$

Then we have

$$\theta \in \{\theta: \theta_1 = \theta_2 = \cdots = \theta_k\} \Leftrightarrow \theta \in \Theta_{\mathbf{a}} \qquad \text{for all } \mathbf{a} \in \mathcal{A} \Leftrightarrow \theta \in \bigcap_{\mathbf{a} \in \mathcal{A}} \Theta_{\mathbf{a}},$$

showing that the ANOVA null can be written as an intersection.

Now, recalling the union–intersection methodology from Section 8.2.3, we would reject $H_0: \theta \in \cap_{\mathbf{a} \in \mathcal{A}} \Theta_{\mathbf{a}}$ (and, hence, the ANOVA null) if we can reject

$$H_{0_{\mathbf{a}}}: \quad \theta \in \Theta_{\mathbf{a}} \qquad \text{versus} \qquad H_{1_{\mathbf{a}}}: \quad \theta \notin \Theta_{\mathbf{a}}$$

for any **a**. We test $H_{0_a}$ with the $t$ statistic of (11.2.6),

$$(11.2.9) \qquad T_{\mathbf{a}} = \left| \frac{\sum_{i=1}^{k} a_i \bar{Y}_{i\cdot} - \sum_{i=1}^{k} a_i \theta_i}{\sqrt{S_p^2 \sum_{i=1}^{k} a_i^2 / n_i}} \right|.$$

We then reject $H_{0_a}$ if $T_{\mathbf{a}} > k$ for some constant $k$. From the union–intersection methodology, it follows that if we could reject for any **a**, we could reject for the **a** that maximizes $T_{\mathbf{a}}$. Thus, the union–intersection test of the ANOVA null is to reject $H_0$ if $\sup_{\mathbf{a}} T_{\mathbf{a}} > k$, where $k$ is chosen so that $P_{H_0}(\sup_{\mathbf{a}} T_{\mathbf{a}} > k) = \alpha$.

Calculation of $\sup_{\mathbf{a}} T_{\mathbf{a}}$ is not straightforward, although with a little care it is not difficult. The calculation is that of a constrained maximum, similar to problems previously encountered (see, for example, Exercise 7.41, where a constrained minimum is calculated). We will attack the problem in a manner similar to what we have done previously and use the Cauchy–Schwarz Inequality. (Alternatively, a method such as Lagrange multipliers could be used, but then we would have to use second-order conditions to verify that a maximum has been found.)

Most of the technical maximization arguments will be given in the following lemma and the lemma will then be applied to obtain the supremum of $T_{\mathbf{a}}$. The lemma is just a statement about constrained maxima of quadratic functions. The proof of the lemma may be skipped by the fainthearted.

**Lemma 11.2.7**　*Let $(v_1, \ldots, v_k)$ be constants and let $(c_1, \ldots, c_k)$ be positive constants. Then, for $\mathcal{A} = \{\mathbf{a} = (a_1, \ldots, a_k) : \sum a_i = 0\}$,*

$$(11.2.10) \qquad \max_{\mathbf{a} \in \mathcal{A}} \left\{ \frac{\left( \sum_{i=1}^{k} a_i v_i \right)^2}{\sum_{i=1}^{k} a_i^2 / c_i} \right\} = \sum_{i=1}^{k} c_i (v_i - \bar{v}_c)^2,$$

*where $\bar{v}_c = \sum c_i v_i / \sum c_i$. The maximum is attained at any **a** of the form $a_i = K c_i (v_i - \bar{v}_c)$, where $K$ is a nonzero constant.*

**Proof:** Define $\mathcal{B} = \{\mathbf{b} = (b_1, \ldots, b_k) : \sum b_i = 0 \text{ and } \sum b_i^2 / c_i = 1\}$. For any $\mathbf{a} \in \mathcal{A}$, define $\mathbf{b} = (b_1, \ldots, b_k)$ by

$$b_i = \frac{a_i}{\sqrt{\sum_{i=1}^{k} a_i^2 / c_i}}$$

and note that $\mathbf{b} \in \mathcal{B}$. For any $\mathbf{a} \in \mathcal{A}$,

$$\frac{\left( \sum_{i=1}^{k} a_i v_i \right)^2}{\sum_{i=1}^{k} a_i^2 / c_i} = \left( \sum_{i=1}^{k} b_i v_i \right)^2.$$

We will find an upper bound on $(\sum b_i v_i)^2$ for $\mathbf{b} \in \mathcal{B}$, and then we will show that the maximizing **a** given in the lemma achieves the upper bound.

Since we are dealing with the sum of products, the Cauchy–Schwarz Inequality (see Section 4.7) is a natural thing to try, but we have to be careful to build in the constraints involving the $c_i$s. We can do this in the following way. Define $C = \sum c_i$ and write

$$\frac{1}{C^2}\left(\sum_{i=1}^{k} b_i v_i\right)^2 = \left\{\sum_{i=1}^{k}\left(\frac{b_i}{c_i}\right)(v_i)\left(\frac{c_i}{C}\right)\right\}^2.$$

This is the square of a *covariance* for a probability measure defined by the ratios $c_i/C$. Formally, if we define random variables $B$ and $V$ by

$$P\left(B = \frac{b_i}{c_i}, V = v_i\right) = \frac{c_i}{C}, \quad i = 1,\ldots,k,$$

then $EB = \sum(b_i/c_i)(c_i/C) = \sum b_i/C = 0$. Thus,

$$\left\{\sum_{i=1}^{k}\left(\frac{b_i}{c_i}\right)(v_i)\left(\frac{c_i}{C}\right)\right\}^2 = (\text{E}\,BV)^2$$

$$= (\text{Cov}(B,V))^2 \qquad\qquad (EB = 0)$$

$$\leq (\text{Var }B)(\text{Var }V) \qquad\qquad \text{(Cauchy–Schwarz Inequality)}$$

$$= \left(\sum_{i=1}^{k}\left(\frac{b_i}{c_i}\right)^2\left(\frac{c_i}{C}\right)\right)\left(\sum_{i=1}^{k}(v_i - \bar{v}_c)^2\left(\frac{c_i}{C}\right)\right). \quad \left(\bar{v}_c = \frac{\sum c_i v_i}{\sum c_i}\right)$$

Using the fact that $\sum b_i^2/c_i = 1$ and canceling common terms, we obtain

(11.2.11) $$\left(\sum_{i=1}^{k} b_i v_i\right)^2 \leq \sum_{i=1}^{k} c_i(v_i - \bar{v}_c)^2 \quad \text{for any } \mathbf{b} \in \mathcal{B}.$$

Finally, we see that if $a_i = Kc_i(v_i - \bar{v}_c)$ for any nonzero constant $K$, then $\mathbf{a} \in \mathcal{A}$ and

$$b_i = \frac{Kc_i(v_i - \bar{v}_c)}{\sqrt{\sum_{i=1}^{k}(Kc_i(v_i - \bar{v}_c))^2/c_i}} = \frac{c_i(v_i - \bar{v}_c)}{\sqrt{\sum_{i=1}^{k} c_i(v_i - \bar{v}_c)^2}}.$$

Since $\sum c_i(v_i - \bar{v}_c) = 0$,

$$\sum_{i=1}^{k} b_i v_i = \frac{\sum_{i=1}^{k} c_i(v_i - \bar{v}_c)v_i}{\sqrt{\sum_{i=1}^{k} c_i(v_i - \bar{v}_c)^2}}$$

$$= \frac{\sum_{i=1}^{k} c_i(v_i - \bar{v}_c)^2}{\sqrt{\sum_{i=1}^{k} c_i(v_i - \bar{v}_c)^2}} = \sqrt{\sum_{i=1}^{k} c_i(v_i - \bar{v}_c)^2},$$

and the inequality in (11.2.11) is an equality. Thus, the upper bound is attained and the function is maximized at such an $\mathbf{a}$. $\qquad\square$

Returning to $T_{\mathbf{a}}$ of (11.2.9), we see that maximizing $T_{\mathbf{a}}$ is equivalent to maximizing $T_{\mathbf{a}}^2$. We have

$$T_{\mathbf{a}}^2 = \frac{\left(\sum_{i=1}^k a_i \bar{Y}_{i\cdot} - \sum_{i=1}^k a_i \theta_i\right)^2}{S_p^2 \sum_{i=1}^k a_i^2/n_i} = \frac{\left(\sum_{i=1}^k a_i \bar{U}_i\right)^2}{S_p^2 \sum_{i=1}^k a_i^2/n_i}. \qquad (\bar{U}_i = \bar{Y}_{i\cdot} - \theta_i)$$

Noting that $S_p^2$ has no effect on the maximization, we can apply Lemma 11.2.7 to the above expression to get the following theorem.

**Theorem 11.2.8** *For $T_{\mathbf{a}}$ defined in expression (11.2.9),*

$$(11.2.12) \qquad \sup_{\mathbf{a}:\sum a_i = 0} T_{\mathbf{a}}^2 = \frac{\sum_{i=1}^k n_i \left((\bar{Y}_{i\cdot} - \bar{\bar{Y}}) - (\theta_i - \bar{\theta})\right)^2}{S_p^2},$$

*where $\bar{\bar{Y}} = \sum n_i \bar{Y}_{i\cdot}/\sum n_i$ and $\bar{\theta} = \sum n_i \theta_i/\sum n_i$. Furthermore, under the ANOVA assumptions,*

$$(11.2.13) \qquad \sup_{\mathbf{a}:\sum a_i = 0} T_{\mathbf{a}}^2 \sim (k-1) F_{k-1, N-k},$$

*that is, $\sup_{\mathbf{a}:\sum a_i = 0} T_{\mathbf{a}}^2/(k-1)$ has an F distribution with $k-1$ and $N-k$ degrees of freedom. (Recall that $N = \sum n_i$.)*

**Proof:** To prove (11.2.12), use Lemma 11.2.7 and identify $v_i$ with $\bar{U}_i$ and $c_i$ with $n_i$. The result is immediate.

To prove (11.2.13), we must show that the numerator and denominator of (11.2.12) are independent chi squared random variables, each divided by its degrees of freedom. From the ANOVA assumptions two things follow. The numerator and denominator are independent and $S_p^2 \sim \sigma^2 \chi_{N-k}^2/(N-k)$. A little work must be done to show that

$$\frac{1}{\sigma^2} \sum_{i=1}^k n_i \left((\bar{Y}_{i\cdot} - \bar{\bar{Y}}) - (\theta_i - \bar{\theta})\right)^2 \sim \chi_{k-1}^2.$$

This can be done, however, and is left as an exercise. (See Exercise 11.7.)  $\square$

If $H_0: \theta_1 = \theta_2 = \cdots = \theta_k$ is true, $\theta_i = \bar{\theta}$ for all $i = 1, \ldots, k$ and the $\theta_i - \bar{\theta}$ terms drop out of (11.2.12). Thus, for an $\alpha$ level test of the ANOVA hypotheses

$$H_0: \quad \theta_1 = \theta_2 = \cdots = \theta_k \qquad \text{versus} \qquad H_1: \quad \theta_i \neq \theta_j \text{ for some } i, j,$$

we reject $H_0$ if

$$(11.2.14) \qquad \frac{\sum_{i=1}^k n_i \left((\bar{Y}_{i\cdot} - \bar{\bar{Y}})\right)^2}{S_p^2} > (k-1) F_{k-1, N-k, \alpha}.$$

This rejection region is usually written as

$$\text{reject } H_0 \text{ if } F = \frac{\sum_{i=1}^{k} n_i \left( (\bar{Y}_{i\cdot} - \bar{\bar{Y}}) \right)^2 / (k-1)}{S_p^2} > F_{k-1, N-k, \alpha}.$$

and the test statistic $F$ is called the *ANOVA F statistic*.

### 11.2.5 Simultaneous Estimation of Contrasts

We have already seen how to estimate and test a single contrast in the ANOVA; the $t$ statistic and interval are given in (11.2.6) and (11.2.8). However, in the ANOVA we are often in the position of wanting to make more than one inference and we know that the simultaneous inference from many $\alpha$ level tests is not necessarily at level $\alpha$. In the context of the ANOVA this problem has already been mentioned.

**Example 11.2.9 (Pairwise differences)** Many times there is interest in pairwise differences of means. Thus, if an ANOVA has means $\theta_1, \ldots, \theta_k$, there may be interest in interval estimates of $\theta_1 - \theta_2$, $\theta_2 - \theta_3$, $\theta_3 - \theta_4$, etc. With the Bonferroni Inequality, we can build a simultaneous inference statement. Define

$$C_{ij} = \left\{ \theta_i - \theta_j : \theta_i - \theta_j \in \bar{Y}_{i\cdot} - \bar{Y}_{j\cdot} \pm t_{N-k, \alpha/2} \sqrt{S_p^2 \left( \frac{1}{n_i} + \frac{1}{n_j} \right)} \right\}.$$

Then $P(C_{ij}) = 1 - \alpha$ for *each* $C_{ij}$, but, for example, $P(C_{12} \text{ and } C_{23}) < 1 - \alpha$. However, this last inference is the kind that we want to make in the ANOVA.

Recall the Bonferroni Inequality, given in expression (1.2.10), which states that for any sets $A_1, \ldots, A_n$,

$$P \left( \bigcap_{i=1}^{n} A_i \right) \geq \sum_{i=1}^{n} P(A_i) - (n-1).$$

In this case we want to bound $P(\cap_{i,j} C_{ij})$, the probability that all of the pairwise intervals cover their respective differences.

If we want to make a simultaneous $1 - \alpha$ statement about the coverage of $m$ confidence sets, then, from the Bonferroni Inequality, we can construct each confidence set to be of level $\gamma$, where $\gamma$ satisfies

$$1 - \alpha = \sum_{i=1}^{m} \gamma - (m-1),$$

or, equivalently,

$$\gamma = 1 - \frac{\alpha}{m}.$$

A slight generalization is also possible in that it is not necessary to require each individual inference at the same level. We can construct each confidence set to be of level $\gamma_i$, where $\gamma_i$ satisfies

$$1 - \alpha = \sum_{i=1}^{m} \gamma_i - (m-1).$$

In an ANOVA with $k$ treatments, simultaneous inference on all $k(k-1)/2$ pairwise differences can be made with confidence $1 - \alpha$ if each $t$ interval has confidence $1 - 2\alpha/[k(k-1)]$.                                                                                    ‖

An alternative and quite elegant approach to simultaneous inference is given by Scheffé (1959). Scheffé's procedure, sometimes called the $S$ method, allows for simultaneous confidence intervals (or tests) on *all* contrasts. (Exercise 11.14 shows that Scheffé's method can also be used to set up simultaneous intervals for any linear combination, not just for contrasts.) The procedure allows us to set a confidence coefficient that will be valid for *all contrast intervals simultaneously*, not just a specified group. The Scheffé procedure would be preferred if a large number of contrasts are to be examined. If the number of contrasts is small, the Bonferroni bound will almost certainly be smaller. (See the Miscellanea section for a discussion of other types of multiple comparison procedures.)

The proof that the Scheffé procedure has simultaneous $1 - \alpha$ coverage on all contrasts follows easily from the union–intersection nature of the ANOVA test.

**Theorem 11.2.10**  *Under the ANOVA assumptions, if* $\mathbf{M} = \sqrt{(k-1)F_{k-1,N-k,\alpha}}$, *then the probability is* $1 - \alpha$ *that*

$$\sum_{i=1}^{k} a_i \bar{Y}_{i\cdot} - \mathbf{M}\sqrt{S_p^2 \sum_{i=1}^{k} \frac{a_i^2}{n_i}} \;\leq\; \sum_{i=1}^{k} a_i \theta_i \;\leq\; \sum_{i=1}^{k} a_i \bar{Y}_{i\cdot} + \mathbf{M}\sqrt{S_p^2 \sum_{i=1}^{k} \frac{a_i^2}{n_i}}$$

*simultaneously for all* $\mathbf{a} \in \mathcal{A} = \{\mathbf{a} = (a_1, \ldots, a_k) \colon \sum a_i = 0\}$.

**Proof:** The simultaneous probability statement requires $\mathbf{M}$ to satisfy

$$P\left( \left| \sum_{i=1}^{k} a_i \bar{Y}_{i\cdot} - \sum_{i=1}^{k} a_i \theta_i \right| \leq \mathbf{M}\sqrt{S_p^2 \sum_{i=1}^{k} \frac{a_i^2}{n_i}} \text{ for all } \mathbf{a} \in \mathcal{A} \right) = 1 - \alpha$$

or, equivalently,

$$P(T_{\mathbf{a}}^2 \leq \mathbf{M}^2 \text{ for all } \mathbf{a} \in \mathcal{A}) = 1 - \alpha,$$

where $T_{\mathbf{a}}$ is defined in (11.2.9). However, since

$$P(T_{\mathbf{a}}^2 \leq \mathbf{M}^2 \text{ for all } \mathbf{a} \in \mathcal{A}) = P\left( \sup_{\mathbf{a} \colon \sum a_i = 0} T_{\mathbf{a}}^2 \leq \mathbf{M}^2 \right),$$

Theorem 11.2.8 shows that choosing $\mathbf{M}^2 = (k-1)F_{k-1,N-k,\alpha}$ satisfies the probability requirement.                                                                                          □

One of the real strengths of the Scheffé procedure is that it allows legitimate "data snooping." That is, in classic statistics it is taboo to test hypotheses that have been suggested by the data, since this can bias the results and, hence, invalidate the inference. (We normally would not test $H_0 : \theta_1 = \theta_2$ just because we noticed that $\bar{Y}_1.$ was different from $\bar{Y}_2.$. See Exercise 11.18.) However, with Scheffé's procedure such a strategy is legitimate. The intervals or tests are valid for *all* contrasts. Whether they have been suggested by the data makes no difference. They already have been taken care of by the Scheffé procedure.

Of course, we must pay for all of the inferential power offered by the Scheffé procedure. The payment is in the form of the lengths of the intervals. In order to guarantee the simultaneous confidence level, the intervals may be quite long. For example, it can be shown (see Exercise 11.15) that if we compare the $t$ and $F$ distributions, for any $\nu$, $\alpha$, and $k$, the cutoff points satisfy

$$t_{\nu,\alpha/2} \leq \sqrt{(k-1)F_{k-1,\nu,\alpha}},$$

and so the Scheffé intervals are always wider, sometimes much wider, than the single-contrast intervals (another argument in favor of the doctrine that nothing substitutes for careful planning and preparation in experimentation). The interval length phenomenon carries over to testing. It also follows from the above inequality that Scheffé tests are less powerful than $t$ tests.

### 11.2.6 Partitioning Sums of Squares

The ANOVA provides a useful way of thinking about the way in which different treatments affect a measured variable—the idea of allocating variation to different sources. The basic idea of allocating variation can be summarized in the following identity.

**Theorem 11.2.11**  *For any numbers $y_{ij}, i = 1, \ldots, k$, and $j = 1, \ldots, n_i$,*

$$(11.2.15) \qquad \sum_{i=1}^{k}\sum_{j=1}^{n_i}(y_{ij} - \bar{\bar{y}})^2 = \sum_{i=1}^{k} n_i(\bar{y}_{i.} - \bar{\bar{y}})^2 + \sum_{i=1}^{k}\sum_{j=1}^{n_i}(y_{ij} - \bar{y}_{i.})^2,$$

*where $\bar{y}_{i.} = \frac{1}{n_i}\sum_j y_{ij}$ and $\bar{\bar{y}} = \sum_i n_i \bar{y}_{i.}/\sum_i n_i$.*

**Proof:** The proof is quite simple and relies only on the fact that, when we are dealing with means, the cross-term often disappears. Write

$$\sum_{i=1}^{k}\sum_{j=1}^{n_i}(y_{ij} - \bar{\bar{y}})^2 = \sum_{i=1}^{k}\sum_{j=1}^{n_i}\left((y_{ij} - \bar{y}_{i.}) + (\bar{y}_{i.} - \bar{\bar{y}})\right)^2,$$

expand the right-hand side, and regroup terms. (See Exercise 11.21.)          □

The sums in (11.2.15) are called *sums of squares* and are thought of as measuring variation in the data ascribable to different sources. (They are sometimes called *corrected sums of squares*, where the word *corrected* refers to the fact that a mean has been subtracted.) In particular, the terms in the oneway ANOVA model,

$$Y_{ij} = \theta_i + \epsilon_{ij},$$

are in one-to-one correspondence with the terms in (11.2.15). Equation (11.2.15) shows how to allocate variation to the treatments (variation *between* treatments) and to random error (variation *within* treatments). The left-hand side of (11.2.15) measures variation without regard to categorization by treatments, while the two terms on the right-hand side measure variation due only to treatments and variation due only to random error, respectively. The fact that these sources of variation satisfy the above identity shows that the variation in the data, measured by sums of squares, is additive in the same way as the ANOVA model.

One reason it is easier to deal with sums of squares is that, under normality, corrected sums of squares are chi squared random variables and we have already seen that independent chi squareds can be added to get new chi squareds.

Under the ANOVA assumptions, in particular if $Y_{ij} \sim n(\theta_i, \sigma^2)$, it is easy to show that

$$(11.2.16) \qquad \frac{1}{\sigma^2} \sum_{i=1}^{k} \sum_{j=1}^{n_i} (Y_{ij} - \bar{Y}_{i\cdot})^2 \sim \chi^2_{N-k},$$

because for each $i = 1, \ldots, k$, $\frac{1}{\sigma^2}\sum_{j=1}^{n_i}(Y_{ij} - \bar{Y}_{i\cdot})^2 \sim \chi^2_{n_i-1}$, all independent, and, for independent chi squared random variables, $\sum_{i=1}^{k}\chi^2_{n_i-1} \sim \chi^2_{N-k}$. Furthermore, if $\theta_i = \theta_j$ for every $i, j$, then

$$(11.2.17) \qquad \frac{1}{\sigma^2}\sum_{i=1}^{k} n_i(\bar{Y}_{i\cdot} - \bar{\bar{Y}})^2 \sim \chi^2_{k-1} \quad \text{and} \quad \frac{1}{\sigma^2}\sum_{i=1}^{k}\sum_{j=1}^{n_i}(Y_{ij} - \bar{\bar{Y}})^2 \sim \chi^2_{N-1}.$$

Thus, under $H_0: \theta_1 = \cdots = \theta_k$, the sum of squares partitioning of (11.2.15) is a partitioning of chi squared random variables. When scaled, the left-hand side is distributed as a $\chi^2_{N-1}$, and the right-hand side is the sum of two independent random variables distributed, respectively, as $\chi^2_{k-1}$ and $\chi^2_{N-k}$. Note that the $\chi^2$ partitioning is true only if the terms on the right-hand side of (11.2.15) are independent, which follows in this case from the normality in the ANOVA assumptions. The partitioning of $\chi^2$s does hold in a slightly more general context, and a characterization of this is sometimes referred to as Cochran's Theorem. (See Searle 1971 and also the Miscellanea section.)

In general, it is possible to partition a sum of squares into sums of squares of uncorrelated contrasts, each with 1 degree of freedom. If the sum of squares has $\nu$ degrees of freedom and is $\chi^2_\nu$, it is possible to partition it into $\nu$ independent terms, each of which is $\chi^2_1$.

The quantity $(\sum a_i \bar{Y}_{i\cdot})^2 / (\sum a_i^2/n_i)$ is called the *contrast sum of squares* for a treatment contrast $\sum a_i \bar{Y}_{i\cdot}$. In a oneway ANOVA it is always possible to find sets of constants $\mathbf{a}^{(l)} = (a_1^{(l)}, \ldots, a_k^{(l)})$, $l = 1, \ldots, k-1$, to satisfy

$$\sum_{i=1}^{k} n_i(\bar{Y}_{i\cdot} - \bar{\bar{Y}})^2 = \frac{\sum_{i=1}^{k} a_i^{(1)}\bar{Y}_{i\cdot}^2}{\sum_{i=1}^{k}(a_i^{(1)})^2/n_i} + \frac{\sum_{i=1}^{k} a_i^{(2)}\bar{Y}_{i\cdot}^2}{\sum_{i=1}^{k}(a_i^{(2)})^2/n_i} + \cdots + \frac{\sum_{i=1}^{k} a_i^{(k-1)}\bar{Y}_{i\cdot}^2}{\sum_{i=1}^{k}(a_i^{(k-1)})^2/n_i}$$

Table 11.2.1. *ANOVA table for oneway classification*

| Source of variation | Degrees of freedom | Sum of squares | Mean square | $F$ statistic |
|---|---|---|---|---|
| Between treatment groups | $k - 1$ | $\text{SSB} = \sum n_i(\bar{y}_i - \bar{\bar{y}})^2$ | $\text{MSB} = \text{SSB}/(k-1)$ | $F = \dfrac{\text{MSB}}{\text{MSW}}$ |
| Within treatment groups | $N - k$ | $\text{SSW} = \sum\sum(y_{ij} - \bar{y}_{i\cdot})^2$ | $\text{MSW} = \text{SSW}/(N-k)$ | |
| Total | $N - 1$ | $\text{SST} = \sum\sum(y_{ij} - \bar{\bar{y}})^2$ | | |

and

(11.2.18) $$\sum_{i=1}^{k} \frac{a_i^{(l)} a_i^{(l')}}{n_i} = 0 \quad \text{for all } l \neq l'.$$

Thus, the individual contrast sums of squares are all uncorrelated and hence independent under normality (Lemma 5.3.3). When suitably normalized, the left-hand side of (11.2.18) is distributed as a $\chi^2_{k-1}$ and the right-hand side is $k-1$ $\chi^2_1$s. (Such contrasts are called *orthogonal contrasts*. See Exercises 11.10 and 11.11.)

It is common to summarize the results of an ANOVA $F$ test in a standard form, called an ANOVA table, shown in Table 11.2.1. The table also gives a number of useful, intermediate statistics. The headings should be self-explanatory.

**Example 11.2.12 (Continuation of Example 11.2.1)** The ANOVA table for the fish toxin data is

| Source of variation | Degrees of freedom | Sum of squares | Mean square | $F$ statistic |
|---|---|---|---|---|
| Treatments | 3 | 995.90 | 331.97 | 26.09 |
| Within | 15 | 190.83 | 12.72 | |
| Total | 18 | 1,186.73 | | |

The $F$ statistic of 26.09 is highly significant, showing that there is strong evidence the toxins produce different effects.      ‖

It follows from equation (11.2.15) that the sum of squares column "adds"—that is, $\text{SSB} + \text{SSW} = \text{SST}$. Similarly, the degrees of freedom column adds. The mean square column, however, does not, as these are means rather than sums.

The ANOVA table contains no new statistics; it merely gives an orderly form for calculation and presentation. The $F$ statistic is exactly the same as derived before and, moreover, MSW is the usual pooled, unbiased estimator of $\sigma^2$, $S_p^2$ of (11.2.5) (see Exercise 11.22).

## 11.3 Simple Linear Regression

In the analysis of variance we looked at how one factor (variable) influenced the means of a response variable. We now turn to simple linear regression, where we try to better understand the functional dependence of one variable on another. In particular, in simple linear regression we have a relationship of the form

$$(11.3.1) \qquad\qquad Y_i = \alpha + \beta x_i + \epsilon_i,$$

where $Y_i$ is a random variable and $x_i$ is another observable variable. The quantities $\alpha$ and $\beta$, the *intercept* and *slope* of the regression, are assumed to be fixed and unknown parameters and $\epsilon_i$ is, necessarily, a random variable. It is also common to suppose that $\mathrm{E}\epsilon_i = 0$ (otherwise we could just rescale the excess into $\alpha$), so that, from (11.3.1), we have

$$(11.3.2) \qquad\qquad \mathrm{E}Y_i = \alpha + \beta x_i.$$

In general, the function that gives $\mathrm{E}Y$ as a function of $x$ is called the *population regression function*. Equation (11.3.2) defines the population regression function for simple linear regression.

One main purpose of regression is to predict $Y_i$ from knowledge of $x_i$ using a relationship like (11.3.2). In common usage this is often interpreted as saying that $Y_i$ *depends* on $x_i$. It is common to refer to $Y_i$ as the *dependent* variable and to refer to $x_i$ as the *independent* variable. This terminology is confusing, however, since this use of the word *independent* is different from our previous usage. (The $x_i$s are not necessarily random variables, so they cannot be statistically "independent" according to our usual meaning.) We will not use this confusing terminology but will use alternative, more descriptive terminology, referring to $Y_i$ as the *response* variable and to $x_i$ as the *predictor* variable.

Actually, to keep straight the fact that our inferences about the relationship between $Y_i$ and $x_i$ assume knowledge of $x_i$, we could write (11.3.2) as

$$(11.3.3) \qquad\qquad \mathrm{E}(Y_i \mid x_i) = \alpha + \beta x_i.$$

We will tend to use (11.3.3) to reinforce the conditional aspect of any inferences.

Recall that in Chapter 4 we encountered the word *regression* in connection with conditional expectations (see Exercise 4.13). There, the regression of $Y$ on $X$ was defined as $\mathrm{E}(Y|x)$, the conditional expectation of $Y$ given $X = x$. More generally, the word *regression* is used in statistics to signify a relationship between variables. When we refer to *regression that is linear*, we can mean that the conditional expectation of $Y$ given $X = x$ is a linear function of $x$. Note that, in equation (11.3.3), it does not matter whether $x_i$ is fixed and known or it is a realization of the observable random variable $X_i$. In either case, equation (11.3.3) has the same interpretation. This will not be the case in Section 11.3.4, however, when we will be concerned with inference using the joint distribution of $X_i$ and $Y_i$.

The term *linear regression* refers to a specification that is *linear in the parameters*. Thus, the specifications $E(Y_i|x_i) = \alpha + \beta x_i^2$ and $E(\log Y_i|x_i) = \alpha + \beta(1/x_i)$ both specify linear regressions. The first specifies a linear relationship between $Y_i$ and $x_i^2$, and the second between $\log Y_i$ and $1/x_i$. In contrast, the specification $E(Y_i|x_i) = \alpha + \beta^2 x_i$ does not specify a linear regression.

The term *regression* has an interesting history, dating back to the work of Sir Francis Galton in the 1800s. (See Freedman *et al.* 1991 for more details or Stigler 1986 for an in-depth historical treatment.) Galton investigated the relationship between heights of fathers and heights of sons. He found, not surprisingly, that tall fathers tend to have tall sons and short fathers tend to have short sons. However, he also found that very tall fathers tend to have shorter sons and very short fathers tend to have taller sons. (Think about it—it makes sense.) Galton called this phenomenon *regression toward the mean* (employing the usual meaning of *regression*, "to go back"), and from this usage we get the present use of the word *regression*.

**Example 11.3.1 (Predicting grape crops)** A more modern use of regression is to predict crop yields of grapes. In July, the grape vines produce clusters of berries, and a count of these clusters can be used to predict the final crop yield at harvest time. Typical data are like the following, which give the cluster counts and yields (tons/acre) for a number of years.

| Year | Yield ($Y$) | Cluster count ($x$) |
|------|-------------|---------------------|
| 1971 | 5.6 | 116.37 |
| 1973 | 3.2 | 82.77 |
| 1974 | 4.5 | 110.68 |
| 1975 | 4.2 | 97.50 |
| 1976 | 5.2 | 115.88 |
| 1977 | 2.7 | 80.19 |
| 1978 | 4.8 | 125.24 |
| 1979 | 4.9 | 116.15 |
| 1980 | 4.7 | 117.36 |
| 1981 | 4.1 | 93.31 |
| 1982 | 4.4 | 107.46 |
| 1983 | 5.4 | 122.30 |

The data from 1972 are missing because the crop was destroyed by a hurricane. A plot of these data would show that there is a strong linear relationship. ‖

When we write an equation like (11.3.3) we are implicitly making the assumption that the regression of $Y$ on $X$ *is* linear. That is, the conditional expectation of $Y$, given that $X = x$, is a linear function of $x$. This assumption may not be justified, because there may be no underlying theory to support a linear relationship. However, since a linear relationship is so convenient to work with, we might want to assume that the regression of $Y$ on $X$ can be adequately approximated by a linear function. Thus, we really do not expect (11.3.3) to hold, but instead we hope that

$$(11.3.4) \qquad\qquad \mathrm{E}(Y_i|x_i) \approx \alpha + \beta x_i$$

is a reasonable approximation. If we start from the (rather strong) assumption that the pair $(X_i, Y_i)$ has a bivariate normal distribution, it immediately follows that the regression of $Y$ on $X$ is linear. In this case, the conditional expectation $\mathrm{E}(Y|x)$ is linear in the parameters (see Definition 4.5.10 and the subsequent discussion).

There is one final distinction to be made. When we do a regression analysis, that is, when we investigate the relationship between a predictor and a response variable, there are two steps to the analysis. The first step is a totally data-oriented one, in which we attempt only to summarize the observed data. (This step is always done, since we almost always calculate sample means and variances or some other summary statistic. However, this part of the analysis now tends to get more complicated.) It is important to keep in mind that this "data fitting" step is not a matter of statistical inference. Since we are interested only in the data at hand, we do not have to make any assumptions about parameters.

The second step in the regression analysis is the statistical one, in which we attempt to infer conclusions about the relationship in the population, that is, about the population regression function. To do this, we need to make assumptions about the population. In particular, if we want to make inferences about the slope and intercept of a population linear relationship, we need to assume that there are parameters that correspond to these quantities.

In a simple linear regression problem, we observe data consisting of $n$ pairs of observations, $(x_1, y_1), \ldots, (x_n, y_n)$. In this section, we will consider a number of different models for these data. The different models will entail different assumptions about whether $x$ or $y$ or both are observed values of random variables $X$ or $Y$.

In each model we will be interested in investigating a linear relationship between $x$ and $y$. The $n$ data points will not fall exactly on a straight line, but we will be interested in *summarizing* the sample information by *fitting a line* to the observed data points. We will find that many different approaches lead us to the same line.

Based on the data $(x_1, y_1), \ldots, (x_n, y_n)$, define the following quantities. The *sample means* are

$$(11.3.5) \qquad\qquad \bar{x} = \frac{1}{n}\sum_{i=1}^{n} x_i \quad \text{and} \quad \bar{y} = \frac{1}{n}\sum_{i=1}^{n} y_i.$$

The *sums of squares* are

$$(11.3.6) \qquad\qquad S_{xx} = \sum_{i=1}^{n}(x_i - \bar{x})^2 \quad \text{and} \quad S_{yy} = \sum_{i=1}^{n}(y_i - \bar{y})^2,$$

and the *sum of cross-products* is

$$(11.3.7) \qquad\qquad S_{xy} = \sum_{i=1}^{n}(x_i - \bar{x})(y_i - \bar{y}).$$

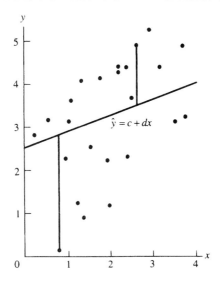

Figure 11.3.1. *Data from Table 11.3.1: Vertical distances that are measured by RSS*

Then the most common estimates of $\alpha$ and $\beta$ in (11.3.4), which we will subsequently justify under various models, are denoted by $a$ and $b$, respectively, and are given by

$$(11.3.8) \qquad\qquad b = \frac{S_{xy}}{S_{xx}} \quad \text{and} \quad a = \bar{y} - b\bar{x}.$$

### 11.3.1 Least Squares: A Mathematical Solution

Our first derivation of estimates for $\alpha$ and $\beta$ makes no statistical assumptions about the observations $(x_i, y_i)$. Simply consider $(x_1, y_1), \ldots, (x_n, y_n)$ as $n$ pairs of numbers plotted in a scatterplot as in Figure 11.3.1. (The 24 data points pictured in Figure 11.3.1 are listed in Table 11.3.1.) Think of drawing through this cloud of points a straight line that comes "as close as possible" to all the points.

Table 11.3.1. *Data pictured in Figure 11.3.1*

| $x$ | $y$ | $x$ | $y$ | $x$ | $y$ | $x$ | $y$ |
|------|------|------|------|------|------|------|------|
| 3.74 | 3.22 | 0.20 | 2.81 | 1.22 | 1.23 | 1.76 | 4.12 |
| 3.66 | 4.87 | 2.50 | 3.71 | 1.00 | 3.13 | 0.51 | 3.16 |
| 0.78 | 0.12 | 3.50 | 3.11 | 1.29 | 4.05 | 2.17 | 4.40 |
| 2.40 | 2.31 | 1.35 | 0.90 | 0.95 | 2.28 | 1.99 | 1.18 |
| 2.18 | 4.25 | 2.36 | 4.39 | 1.05 | 3.60 | 1.53 | 2.54 |
| 1.93 | 2.24 | 3.13 | 4.36 | 2.92 | 5.39 | 2.60 | 4.89 |
| $\bar{x} = 1.95$ | $\bar{y} = 3.18$ | $S_{xx} = 22.82$ | | $S_{yy} = 43.62$ | | $S_{xy} = 15.48$ | |

For any line $y = c + dx$, the *residual sum of squares* (RSS) is defined to be

$$\text{RSS} = \sum_{i=1}^{n}(y_i - (c + dx_i))^2.$$

The RSS measures the *vertical* distance from each data point to the line $c + dx$ and then sums the squares of these distances. (Two such distances are shown in Figure 11.3.1.) The *least squares estimates* of $\alpha$ and $\beta$ are defined to be those values $a$ and $b$ such that the line $a + bx$ minimizes RSS. That is, the least squares estimates, $a$ and $b$, satisfy

$$\min_{c,d} \sum_{i=1}^{n}(y_i - (c + dx_i))^2 = \sum_{i=1}^{n}(y_i - (a + bx_i))^2.$$

This function of two variables, $c$ and $d$, can be minimized in the following way. For any fixed value of $d$, the value of $c$ that gives the minimum value can be found by writing

$$\sum_{i=1}^{n}(y_i - (c + dx_i))^2 = \sum_{i=1}^{n}((y_i - dx_i) - c)^2.$$

From Theorem 5.2.4, the minimizing value of $c$ is

(11.3.9) $$c = \frac{1}{n}\sum_{i=1}^{n}(y_i - dx_i) = \bar{y} - d\bar{x}.$$

Thus, for a given value of $d$, the minimum value of RSS is

$$\sum_{i=1}^{n}((y_i - dx_i) - (\bar{y} - d\bar{x}))^2 = \sum_{i=1}^{n}((y_i - \bar{y}) - d(x_i - \bar{x}))^2 = S_{yy} - 2dS_{xy} + d^2 S_{xx}.$$

The value of $d$ that gives the overall minimum value of RSS is obtained by setting the derivative of this quadratic function of $d$ equal to 0. The minimizing value is

(11.3.10) $$d = \frac{S_{xy}}{S_{xx}}.$$

This value is, indeed, a minimum since the coefficient of $d^2$ is positive. Thus, by (11.3.9) and (11.3.10), $a$ and $b$ from (11.3.8) are the values of $c$ and $d$ that minimize the residual sum of squares.

The RSS is only one of many reasonable ways of measuring the distance from the line $c + dx$ to the data points. For example, rather than using vertical distances we could use horizontal distances. This is equivalent to graphing the $y$ variable on the horizontal axis and the $x$ variable on the vertical axis and using vertical distances as we did above. Using the above results (interchanging the roles of $x$ and $y$), we find the least squares line is $\hat{x} = a' + b'y$, where

$$b' = \frac{S_{xy}}{S_{yy}} \quad \text{and} \quad a' = \bar{x} - b'\bar{y}.$$

Reexpressing the line so that $y$ is a function of $x$, we obtain $\hat{y} = -(a'/b') + (1/b')x$.

Usually the line obtained by considering horizontal distances is different from the line obtained by considering vertical distances. From the values in Table 11.3.1, the *regression of y on x* (vertical distances) is $\hat{y} = 1.86 + .68x$. The *regression of x on y* (horizontal distances) is $\hat{y} = -2.31 + 2.82x$. In Figure 12.2.2, these two lines are shown (along with a third line discussed in Section 12.2). If these two lines were the same, then the slopes would be the same and $b/(1/b')$ would equal 1. But, in fact, $b/(1/b') \le 1$ with equality only in special cases. Note that

$$\frac{b}{1/b'} = bb' = \frac{(S_{xy})^2}{S_{xx}S_{yy}}.$$

Using the version of Hölder's Inequality in (4.7.9) with $p = q = 2, a_i = x_i - \bar{x}$, and $b_i = y_i - \bar{y}$, we see that $(S_{xy})^2 \le S_{xx}S_{yy}$ and, hence, the ratio is less than 1.

If $x$ is the predictor variable, $y$ is the response variable, and we think of predicting $y$ from $x$, then the vertical distance measured in RSS is reasonable. It measures the distance from $y_i$ to the predicted value of $y_i, \hat{y}_i = c + dx_i$. But if we do not make this distinction between $x$ and $y$, then it is unsettling that another reasonable criterion, horizontal distance, gives a different line.

The least squares method should be considered only as a method of "fitting a line" to a set of data, not as a method of statistical inference. We have no basis for constructing confidence intervals or testing hypotheses because, in this section, we have not used any statistical model for the data. When we think of $a$ and $b$ in the context of this section, it might be better to call them least squares *solutions* rather than least squares *estimates* because they are the solutions of the mathematical problem of minimizing the RSS rather than estimates derived from a statistical model. But, as we shall see, these least squares solutions have optimality properties in certain statistical models.

### 11.3.2 Best Linear Unbiased Estimators: A Statistical Solution

In this section we show that the estimates $a$ and $b$ from (11.3.8) are optimal in the class of linear unbiased estimates under a fairly general statistical model. The model is described as follows. Assume that the values $x_1, \ldots, x_n$ are known, fixed values. (Think of them as values the experimenter has chosen and set in a laboratory experiment.) The values $y_1, \ldots, y_n$ are observed values of uncorrelated random variables $Y_1, \ldots, Y_n$. The linear relationship assumed between the $x$s and the $y$s is

(11.3.11)                          $EY_i = \alpha + \beta x_i, \quad i = 1, \ldots, n,$

where we also assume that

(11.3.12)                          $\text{Var } Y_i = \sigma^2.$

There is no subscript in $\sigma^2$ because we are assuming that all the $Y_i$s have the same (unknown) variance. These assumptions about the first two moments of the $Y_i$s are the only assumptions we need to make to proceed with the derivation in this subsection. For example, we do not need to specify a probability distribution for the $Y_1, \ldots, Y_n$.

The model in (11.3.11) and (11.3.12) can also be expressed in this way. We assume that

(11.3.13) $$Y_i = \alpha + \beta x_i + \epsilon_i, \quad i = 1, \ldots, n,$$

where $\epsilon_1, \ldots, \epsilon_n$ are uncorrelated random variables with

(11.3.14) $$E\,\epsilon_i = 0 \quad \text{and} \quad \text{Var}\,\epsilon_i = \sigma^2.$$

The $\epsilon_1, \ldots, \epsilon_n$ are called the *random errors*. Since $Y_i$ depends only on $\epsilon_i$ and the $\epsilon_i$s are uncorrelated, the $Y_i$s are uncorrelated. Also, from (11.3.13) and (11.3.14), the expressions for $EY_i$ and $\text{Var}\,Y_i$ in (11.3.11) and (11.3.12) are easily verified.

To derive estimators for the parameters $\alpha$ and $\beta$, we restrict attention to the class of *linear estimators*. An estimator is a linear estimator if it is of the form

(11.3.15) $$\sum_{i=1}^{n} d_i Y_i,$$

where $d_1, \ldots, d_n$ are known, fixed constants. (Exercise 7.39 concerns linear estimators of a population mean.) Among the class of linear estimators, we further restrict attention to unbiased estimators. This restricts the values of $d_1, \ldots, d_n$ that can be used.

An unbiased estimator of the slope $\beta$ must satisfy

$$E \sum_{i=1}^{n} d_i Y_i = \beta,$$

regardless of the true value of the parameters $\alpha$ and $\beta$. This implies that

$$\beta = E \sum_{i=1}^{n} d_i Y_i \;=\; \sum_{i=1}^{n} d_i EY_i \;=\; \sum_{i=1}^{n} d_i(\alpha + \beta x_i)$$

$$= \alpha \left( \sum_{i=1}^{n} d_i \right) + \beta \left( \sum_{i=1}^{n} d_i x_i \right).$$

This equality is true for *all* $\alpha$ and $\beta$ if and only if

(11.3.16) $$\sum_{i=1}^{n} d_i = 0 \quad \text{and} \quad \sum_{i=1}^{n} d_i x_i = 1.$$

Thus, $d_1, \ldots, d_n$ must satisfy (11.3.16) in order for the estimator to be an unbiased estimator of $\beta$.

In Chapter 7 we called an unbiased estimator "best" if it had the smallest variance among all unbiased estimators. Similarly, an estimator is the *best linear unbiased estimator* (*BLUE*) if it is the linear unbiased estimator with the smallest variance. We will now show that the choice of $d_i = (x_i - \bar{x})/S_{xx}$ that defines the estimator $b = S_{xY}/S_{xx}$ is the best choice in that it results in the linear unbiased estimator of $\beta$

with the smallest variance. (The $d_i$s must be known, fixed constants but the $x_i$s are known, fixed constants, so this choice of $d_i$s is legitimate.)

*A note on notation:* The notation $S_{xY}$ stresses the fact that $S_{xY}$ is a random variable that is a function of the random variables $Y_1, \ldots, Y_n$. $S_{xY}$ also depends on the nonrandom quantities $x_1, \ldots, x_n$.

Because $Y_1, \ldots, Y_n$ are uncorrelated with equal variance $\sigma^2$, the variance of *any* linear estimator is given by

$$\text{Var} \sum_{i=1}^{n} d_i Y_i = \sum_{i=1}^{n} d_i^2 \text{Var } Y_i = \sum_{i=1}^{n} d_i^2 \sigma^2 = \sigma^2 \sum_{i=1}^{n} d_i^2.$$

The BLUE of $\beta$ is, therefore, defined by constants $d_1, \ldots, d_n$ that satisfy (11.3.16) and have the minimum value of $\sum_{i=1}^{n} d_i^2$. (The presence of $\sigma^2$ has no effect on the minimization over linear estimators since it appears as a multiple of the variance of every linear estimator.)

The minimizing values of the constants $d_1, \ldots, d_n$ can now be found by using Lemma 11.2.7. To apply the lemma to our minimization problem, make the following correspondences, where the left-hand sides are notation from Lemma 11.2.7 and the right-hand sides are our current notation. Let

$$k = n, \quad v_i = x_i, \quad c_i = 1, \quad \text{and} \quad a_i = d_i,$$

which implies $\bar{v}_c = \bar{x}$. If $d_i$ is of the form

(11.3.17) $$d_i = Kc_i(v_i - \bar{v}_c) = K(x_i - \bar{x}), \quad i = 1, \ldots, n,$$

then, by Lemma 11.2.7, $d_1, \ldots, d_n$ maximize

(11.3.18) $$\frac{\left( \sum_{i=1}^{n} d_i x_i \right)^2}{\sum_{i=1}^{n} d_i^2}$$

among all $d_1, \ldots, d_n$ that satisfy $\sum d_i = 0$. Furthermore, since

$$\left\{ (d_1, \ldots, d_n) : \sum d_i = 0, \sum d_i x_i = 1 \right\} \subset \left\{ (d_1, \ldots, d_n) : \sum d_i = 0 \right\},$$

if $d_i$s of the form (11.3.17) also satisfy (11.3.16), they certainly maximize (11.3.18) among all $d_1, \ldots, d_n$ that satisfy (11.3.16). (Since the set over which the maximum is taken is smaller, the maximum cannot be larger.) Now, using (11.3.17), we have

$$\sum_{i=1}^{n} d_i x_i = \sum_{i=1}^{n} K(x_i - \bar{x}) x_i = K S_{xx}.$$

The second constraint in (11.3.16) is satisfied if $K = \frac{1}{S_{xx}}$. Therefore, with $d_1, \ldots, d_n$ defined by

(11.3.19) $$d_i = \frac{(x_i - \bar{x})}{S_{xx}}, \quad i = 1, \ldots, n,$$

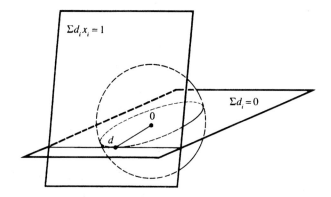

Figure 11.3.2. *Geometric description of the BLUE*

both constraints of (11.3.16) are satisfied and this set of $d_i$s produces the maximum. Finally, note that for all $d_1, \ldots, d_n$ that satisfy (11.3.16),

$$\frac{\left(\sum_{i=1}^n d_i x_i\right)^2}{\sum_{i=1}^n d_i^2} = \frac{1}{\sum_{i=1}^n d_i^2}.$$

Thus, for $d_1, \ldots, d_n$ that satisfy (11.3.16), maximization of (11.3.18) is equivalent to minimization of $\sum d_i^2$. Hence, we can conclude that the $d_i$s defined in (11.3.19) give the minimum value of $\sum d_i^2$ among all $d_i$s that satisfy (11.3.16), and the linear unbiased estimator defined by these $d_i$s, namely,

$$b = \sum_{i=1}^n \frac{(x_i - \bar{x})}{S_{xx}} y_i = \frac{S_{xy}}{S_{xx}},$$

is the BLUE of $\beta$.

A geometric description of this construction of the BLUE of $\beta$ is given in Figure 11.3.2, where we take $n = 3$. The figure shows three-dimensional space with coordinates $d_1, d_2$, and $d_3$. The two planes represent the vectors $(d_1, d_2, d_3)$ that satisfy the two linear constraints in (11.3.16), and the line where the two planes intersect consists of the vectors $(d_1, d_2, d_3)$ that satisfy both equalities. For any point on the line, $\sum_{i=1}^n d_i^2$ is the square of the distance from the point to the origin **0**. The vector $(d_1, d_2, d_3)$ that defines the BLUE is the point on the line that is closest to **0**. The sphere in the figure is the smallest sphere that intersects the line, and the point of intersection is the point $(d_1, d_2, d_3)$ that defines the BLUE of $\beta$. This, we have shown, is the point with $d_i = (x_i - \bar{x})/S_{xx}$.

The variance of $b$ is

$$(11.3.20) \qquad \operatorname{Var} b = \sigma^2 \sum_{i=1}^n d_i^2 = \frac{\sigma^2}{S_{xx}} = \frac{\sigma^2}{\sum_{i=1}^n (x_i - \bar{x})^2}.$$

Since $x_1, \ldots, x_n$ are values chosen by the experimenter, they can be chosen to make $S_{xx}$ large and the variance of the estimator small. That is, the experimenter can *design the experiment* to make the estimator more precise. Suppose that all the $x_1, \ldots, x_n$ must be chosen in an interval $[e, f]$. Then, if $n$ is even, the choice of $x_1, \ldots, x_n$ that makes $S_{xx}$ as large as possible is to take half of the $x_i$s equal to $e$ and half equal to $f$ (see Exercise 11.26). This would be the best design in that it would give the most precise estimate of the slope $\beta$ if the experimenter were certain that the model described by (11.3.11) and (11.3.12) was correct. In practice, however, this design is seldom used because an experimenter is hardly ever certain of the model. This *two-point design* gives information about the value of $E(Y|x)$ at only two values, $x = e$ and $x = f$. If the population regression function $E(Y|x)$, which gives the mean of $Y$ as a function of $x$, is nonlinear, it could never be detected from data obtained using the "optimal" two-point design.

We have shown that $b$ is the BLUE of $\beta$. A similar analysis will show that $a$ is the BLUE of the intercept $\alpha$. The constants $d_1, \ldots, d_n$ that define a linear estimator of $\alpha$ must satisfy

$$(11.3.21) \qquad \sum_{i=1}^{n} d_i = 1 \quad \text{and} \quad \sum_{i=1}^{n} d_i x_i = 0.$$

The details of this derivation are left as Exercise 11.27. The fact that least squares estimators are BLUEs holds in other linear models also. This general result is called the *Gauss–Markov Theorem* (see Christensen 1996; Lehmann and Casella 1998, Section 3.4, or the more general treatment in Harville 1981).

### 11.3.3 Models and Distribution Assumptions

In this section, we will introduce two more models for paired data $(x_1, y_1), \ldots, (x_n, y_n)$ that are called simple linear regression models.

To obtain the least squares estimates in Section 11.3.1, we used no statistical model. We simply solved a mathematical minimization problem. Thus, we could not derive any statistical properties about the estimators obtained by this method because there were no probability models to work with. There are not really any parameters for which we could construct hypothesis tests or confidence intervals.

In Section 11.3.2 we made some statistical assumptions about the data. Specifically, we made assumptions about the first two moments, the mean, variance, and covariance of the data. These are all statistical assumptions related to probability models for the data, and we derived statistical properties for the estimators. The properties of unbiasedness and minimum variance, which we proved for the estimators $a$ and $b$ of the parameters $\alpha$ and $\beta$, are statistical properties.

To obtain these properties we did not have to specify a complete probability model for the data, only assumptions about the first two moments. We were able to obtain a general optimality property under these minimal assumptions, but the optimality was only in a restricted class of estimators—linear unbiased estimators. We were not able to derive exact tests and confidence intervals under this model because the model does not specify enough about the probability distribution of the data. We now present two statistical models that completely specify the probabilistic structure of the data.

*Conditional normal model*

The *conditional normal model* is the most common simple linear regression model and the most straightforward to analyze. The observed data are the $n$ pairs, $(x_1, y_1), \ldots, (x_n, y_n)$. The values of the predictor variable, $x_1, \ldots, x_n$, are considered to be known, fixed constants. As in Section 11.3.2, think of them as being chosen and set by the experimenter. The values of the response variable, $y_1, \ldots, y_n$, are observed values of random variables, $Y_1, \ldots, Y_n$. The random variables $Y_1, \ldots, Y_n$ are assumed to be independent. Furthermore, the distribution of the $Y_i$s is normal, specifically,

$$(11.3.22) \qquad Y_i \sim \mathrm{n}(\alpha + \beta x_i, \sigma^2), \quad i = 1, \ldots, n.$$

Thus the population regression function is a linear function of $x$, that is, $\mathrm{E}(Y|x) = \alpha + \beta x$, and all the $Y_i$s have the same variance, $\sigma^2$. The conditional normal model can be expressed similar to (11.3.13) and (11.3.14), namely,

$$(11.3.23) \qquad Y_i = \alpha + \beta x_i + \epsilon_i, \quad i = 1, \ldots, n,$$

where $\epsilon_1, \ldots, \epsilon_n$ are iid $\mathrm{n}(0, \sigma^2)$ random variables.

The conditional normal model is a special case of the model considered in Section 11.3.2. The population regression function, $\mathrm{E}(Y|x) = \alpha + \beta x$, and the variance, $\mathrm{Var}\, Y = \sigma^2$, are as in that model. The uncorrelatedness of $Y_1, \ldots, Y_n$ (or, equivalently, $\epsilon_1, \ldots, \epsilon_n$) has been strengthened to independence. And, of course, rather than just the first two moments of the distribution of $Y_1, \ldots, Y_n$, the exact form of the probability distribution is now specified.

The joint pdf of $Y_1, \ldots, Y_n$ is the product of the marginal pdfs because of the independence. It is given by

$$f(\mathbf{y}|\alpha, \beta, \sigma^2) = f(y_1, \ldots, y_n|\alpha, \beta, \sigma^2)$$

$$= \prod_{i=1}^{n} f(y_i|\alpha, \beta, \sigma^2)$$

$$(11.3.24) \qquad = \prod_{i=1}^{n} \frac{1}{\sqrt{2\pi}\sigma} \exp\left[-(y_i - (\alpha + \beta x_i))^2/(2\sigma^2)\right]$$

$$= \frac{1}{(2\pi)^{n/2}\sigma^n} \exp\left[-\left(\sum_{i=1}^{n}(y_i - \alpha - \beta x_i)^2\right)/(2\sigma^2)\right].$$

It is this joint probability distribution that will be used to develop the statistical procedures in Sections 11.3.4 and 11.3.5. For example, the expression in (11.3.24) will be used to find MLEs of $\alpha, \beta$, and $\sigma^2$.

*Bivariate normal model*

In all the previous models we have discussed, the values of the predictor variable, $x_1, \ldots, x_n$, have been fixed, known constants. But sometimes these values are actually observed values of random variables, $X_1, \ldots, X_n$. In Galton's example in Section 11.3,

$x_1, \ldots, x_n$ were observed heights of fathers. But the experimenter certainly did not choose these heights before collecting the data. Thus it is necessary to consider models in which the predictor variable, as well as the response variable, is random. One such model that is fairly simple is the *bivariate normal model*. A more complex model is discussed in Section 12.2.

In the bivariate normal model the data $(x_1, y_1), \ldots, (x_n, y_n)$ are observed values of the bivariate random vectors $(X_1, Y_1), \ldots, (X_n, Y_n)$. The random vectors are independent and the joint distribution of $(X_i, Y_i)$ is assumed to be bivariate normal. Specifically, it is assumed that

$$(X_i, Y_i) \sim \text{bivariate normal}(\mu_X, \mu_Y, \sigma_X^2, \sigma_Y^2, \rho).$$

The joint pdf and various properties of a bivariate normal distribution are given in Definition 4.5.10 and the subsequent discussion. The joint pdf of all the data $(X_1, Y_1), \ldots, (X_n, Y_n)$ is the product of these bivariate pdfs.

In a simple linear regression analysis, we are still thinking of $x$ as the predictor variable and $y$ as the response variable. That is, we are most interested in predicting the value of $y$ having observed the value of $x$. This naturally leads to basing inference on the conditional distribution of $Y$ given $X = x$. For a bivariate normal model, the conditional distribution of $Y$ given $X = x$ is normal. The population regression function is now a true conditional expectation, as the notation suggests, and is

$$(11.3.25) \quad \text{E}(Y|x) = \mu_Y + \rho\frac{\sigma_Y}{\sigma_X}(x - \mu_X) = \left[\mu_Y - \rho\frac{\sigma_Y}{\sigma_X}\mu_X\right] + \left[\rho\frac{\sigma_Y}{\sigma_X}\right]x.$$

The bivariate normal model *implies* that the population regression is a linear function of $x$. We need not assume this as in the previous models. Here $\text{E}(Y|x) = \alpha + \beta x$, where $\beta = \rho\frac{\sigma_Y}{\sigma_X}$ and $\alpha = \mu_Y - \rho\frac{\sigma_Y}{\sigma_X}\mu_X$. Also, as in the conditional normal model, the conditional variance of the response variable $Y$ does not depend on $x$,

$$(11.3.26) \quad\quad\quad\quad \text{Var}(Y|x) = \sigma_Y^2(1 - \rho^2).$$

For the bivariate normal model, the linear regression analysis is almost always carried out using the conditional distribution of $(Y_1, \ldots, Y_n)$ given $X_1 = x_1, \ldots, X_n = x_n$, rather than the unconditional distribution of $(X_1, Y_1), \ldots, (X_n, Y_n)$. But then we are in the same situation as the conditional normal model described above. The fact that $x_1, \ldots, x_n$ are observed values of random variables is immaterial if we condition on these values and, in general, in simple linear regression we do not use the fact of bivariate normality except to define the conditional distribution. (Indeed, for the most part, the marginal distribution of $X$ is of no consequence whatsoever. In linear regression it is the conditional distribution that matters.) Inference based on point estimators, intervals, or tests is the same for the two models. See Brown (1990b) for an alternative view.

### 11.3.4 Estimation and Testing with Normal Errors

In this and the next subsections we develop inference procedures under the conditional normal model, the regression model defined by (11.3.22) or (11.3.23).

First, we find the maximum likelihood estimates of the three parameters, $\alpha, \beta$, and $\sigma^2$. Using the joint pdf in (11.3.24), we see that the log likelihood function is

$$\log L(\alpha, \beta, \sigma^2 | \mathbf{x}, \mathbf{y}) = -\frac{n}{2} \log(2\pi) - \frac{n}{2} \log \sigma^2 - \frac{\sum_{i=1}^n (y_i - \alpha - \beta x_i)^2}{2\sigma^2}.$$

For any fixed value of $\sigma^2$, $\log L$ is maximized as a function of $\alpha$ and $\beta$ by those values, $\hat\alpha$ and $\hat\beta$, that minimize

$$\sum_{i=1}^n (y_i - \alpha - \beta x_i)^2.$$

But this function is just the RSS from Section 11.3.1! There we found that the minimizing values are

$$\hat\beta = b = \frac{S_{xy}}{S_{xx}} \quad \text{and} \quad \hat\alpha = a = \bar{y} - b\bar{x} = \bar{y} - \hat\beta \bar{x}.$$

Thus, the least squares estimators of $\alpha$ and $\beta$ are also the MLEs of $\alpha$ and $\beta$. The values $\hat\alpha$ and $\hat\beta$ are the maximizing values for any fixed value of $\sigma^2$. Now, substituting in the log likelihood, to find the MLE of $\sigma^2$ we need to maximize

$$-\frac{n}{2} \log(2\pi) - \frac{n}{2} \log \sigma^2 - \frac{\sum_{i=1}^n (y_i - \hat\alpha - \hat\beta x_i)^2}{2\sigma^2}.$$

This maximization is similar to finding the MLE of $\sigma^2$ in ordinary normal sampling (see Example 7.2.11), and we leave the details to Exercise 11.28. The MLE of $\sigma^2$, under the conditional normal model, is

$$\hat\sigma^2 = \frac{1}{n} \sum_{i=1}^n (y_i - \hat\alpha - \hat\beta x_i)^2,$$

the RSS, evaluated at the least squares line, divided by the sample size. Henceforth, when we refer to RSS we mean the RSS evaluated at the least squares line.

In Section 11.3.2, we showed that $\hat\alpha$ and $\hat\beta$ were linear unbiased estimators of $\alpha$ and $\beta$. However, $\hat\sigma^2$ is not an unbiased estimator of $\sigma^2$. For the calculation of $E\hat\sigma^2$ and in many subsequent calculations, the following lemma will be useful.

**Lemma 11.3.2**   *Let $Y_1, \ldots, Y_n$ be uncorrelated random variables with $\operatorname{Var} Y_i = \sigma^2$ for all $i = 1, \ldots, n$. Let $c_1, \ldots, c_n$ and $d_1, \ldots, d_n$ be two sets of constants. Then*

$$\operatorname{Cov}\left(\sum_{i=1}^n c_i Y_i, \sum_{i=1}^n d_i Y_i\right) = \left(\sum_{i=1}^n c_i d_i\right) \sigma^2.$$

**Proof:** This type of result has been encountered before. It is similar to Lemma 5.3.3 and Exercise 11.11. However, here we do not need either normality or independence of $Y_1, \ldots, Y_n$.   □

We next find the bias in $\sigma^2$. From (11.3.23) we have

$$\epsilon_i = Y_i - \alpha - \beta x_i.$$

We define the *residuals from the regression* to be

(11.3.27) $$\hat\epsilon_i = Y_i - \hat\alpha - \hat\beta x_i,$$

and thus

$$\hat\sigma^2 = \frac{1}{n}\sum_{i=1}^n \hat\epsilon_i^2 = \frac{1}{n}\text{RSS}.$$

It can be calculated (see Exercise 11.29) that

$$\text{E}\hat\epsilon_i = 0,$$

and a lengthy calculation (also in Exercise 11.29) gives

(11.3.28)

$$\text{Var}\,\hat\epsilon_i = \text{E}\hat\epsilon_i^2 = \left(\frac{n-2}{n} + \frac{1}{S_{xx}}\left(\frac{1}{n}\sum_{j=1}^n x_j^2 + x_i^2 - 2(x_i - \bar x)^2 - 2x_i\bar x\right)\right)\sigma^2.$$

Thus,

$$\text{E}\hat\sigma^2 = \frac{1}{n}\sum_{i=1}^n \text{E}\hat\epsilon_i^2$$

$$= \frac{1}{n}\sum_{i=1}^n\left[\frac{n-2}{n} + \frac{1}{S_{xx}}\left(\frac{1}{n}\sum_{j=1}^n x_j^2 + x_i^2 - 2(x_i - \bar x)^2 - 2x_i\bar x\right)\right]\sigma^2$$

$$= \left[\frac{n-2}{n} + \frac{1}{nS_{xx}}\left\{\sum_{j=1}^n x_j^2 + \sum_{i=1}^n x_i^2 - 2S_{xx} - 2\frac{1}{n}\left(\sum_{i=1}^n x_i\right)^2\right\}\right]\sigma^2$$

$$\qquad\qquad\qquad\qquad\qquad\qquad\qquad\qquad \left(\sum x_i\bar x = \tfrac{1}{n}(\sum x_i)^2\right)$$

$$= \left(\frac{n-2}{n} + 0\right)\sigma^2 \qquad\qquad\qquad\qquad \left(\sum x_i^2 - \tfrac{1}{n}(\sum x_i)^2 = S_{xx}\right)$$

$$= \frac{n-2}{n}\sigma^2.$$

The MLE $\hat\sigma^2$ is a biased estimator of $\sigma^2$. The more commonly used estimator of $\sigma^2$, which is unbiased, is

(11.3.29) $$S^2 = \frac{n}{n-2}\hat\sigma^2 = \frac{1}{n-2}\sum_{i=1}^n (y_i - \hat\alpha - \hat\beta x_i)^2 = \frac{1}{n-2}\sum_{i=1}^n \hat\epsilon_i^2.$$

To develop estimation and testing procedures, based on these estimators, we need to know their sampling distributions. These are summarized in the following theorem.

**Theorem 11.3.3** *Under the conditional normal regression model (11.3.22), the sampling distributions of the estimators $\hat{\alpha}$, $\hat{\beta}$, and $S^2$ are*

$$\hat{\alpha} \sim \text{n}\left(\alpha, \frac{\sigma^2}{nS_{xx}}\sum_{i=1}^{n} x_i^2\right), \qquad \hat{\beta} \sim \text{n}\left(\beta, \frac{\sigma^2}{S_{xx}}\right),$$

*with*

$$\text{Cov}(\hat{\alpha}, \hat{\beta}) = \frac{-\sigma^2 \bar{x}}{S_{xx}}.$$

*Furthermore, $(\hat{\alpha}, \hat{\beta})$ and $S^2$ are independent and*

$$\frac{(n-2)S^2}{\sigma^2} \sim \chi^2_{n-2}.$$

**Proof:** We first show that $\hat{\alpha}$ and $\hat{\beta}$ have the indicated normal distributions. The estimators $\hat{\alpha}$ and $\hat{\beta}$ are both linear functions of the independent normal random variables $Y_1, \ldots, Y_n$. Thus, by Corollary 4.6.10, they both have normal distributions. Specifically, in Section 11.3.2, we showed that $\hat{\beta} = \sum_{i=1}^{n} d_i Y_i$, where the $d_i$ are given in (11.3.19), and we also showed that

$$\text{E}\hat{\beta} = \beta \quad \text{and} \quad \text{Var}\,\hat{\beta} = \frac{\sigma^2}{S_{xx}}.$$

The estimator $\hat{\alpha} = \bar{Y} - \hat{\beta}\bar{x}$ can be expressed as $\hat{\alpha} = \sum_{i=1}^{n} c_i Y_i$, where

$$c_i = \frac{1}{n} - \frac{(x_i - \bar{x})\bar{x}}{S_{xx}},$$

and thus it is straightforward to verify that

$$\text{E}\hat{\alpha} = \sum_{i=1}^{n} c_i \text{E}Y_i = \sum_{i=1}^{n} \left(\frac{1}{n} - \frac{(x_i - \bar{x})\bar{x}}{S_{xx}}\right)(\alpha + \beta x_i) = \alpha,$$

$$\text{Var}\,\hat{\alpha} = \sigma^2 \sum_{i=1}^{n} c_i^2 = \sigma^2\left[\frac{1}{nS_{xx}}\sum_{i=1}^{n} x_i^2\right],$$

showing that $\hat{\alpha}$ and $\hat{\beta}$ have the specified distributions. Also, $\text{Cov}(\hat{\alpha}, \hat{\beta})$ is easily calculated using Lemma 11.3.2. Details are left to Exercise 11.30.

We next show that $\hat{\alpha}$ and $\hat{\beta}$ are independent of $S^2$, a fact that will follow from Lemma 11.3.2 and Lemma 5.3.3. From the definition of $\hat{\epsilon}_i$ in (11.3.27), we can write

(11.3.30)
$$\hat{\epsilon}_i = \sum_{j=1}^{n} [\delta_{ij} - (c_j + d_j x_i)]\, Y_j,$$

where

$$\delta_{ij} = \begin{cases} 1 & \text{if } i = j \\ 0 & \text{if } i \neq j \end{cases}, \quad c_j = \frac{1}{n} - \frac{(x_j - \bar{x})\bar{x}}{S_{xx}}, \quad \text{and} \quad d_j = \frac{(x_j - \bar{x})}{S_{xx}}.$$

Since $\hat{\alpha} = \sum c_i Y_i$ and $\hat{\beta} = \sum d_i Y_i$, application of Lemma 11.3.2 together with some algebra will show that

$$\text{Cov}(\hat{\epsilon}_i, \hat{\alpha}) = \text{Cov}(\hat{\epsilon}_i, \hat{\beta}) = 0, \quad i = 1, \ldots, n.$$

Details are left to Exercise 11.31. Thus, it follows from Lemma 5.3.3 that, under normal sampling, $S^2 = \sum \hat{\epsilon}_i^2/(n-2)$ is independent of $\hat{\alpha}$ and $\hat{\beta}$.

To prove that $(n-2)S^2/\sigma^2 \sim \chi^2_{n-2}$, we write $(n-2)S^2$ as the sum of $n-2$ independent random variables, each of which has a $\chi^2_1$ distribution. That is, we find constants $a_{ij}, i = 1, \ldots, n$ and $j = 1, \ldots, n-2$, that satisfy

$$(11.3.31) \qquad \sum_{i=1}^{n} \hat{\epsilon}_i^2 = \sum_{j=1}^{n-2} \left( \sum_{i=1}^{n} a_{ij} Y_i \right)^2,$$

where

$$\sum_{i=1}^{n} a_{ij} = 0, \quad j = 1, \ldots, n-2, \qquad \text{and} \qquad \sum_{i=1}^{n} a_{ij} a_{ij'} = 0, \quad j \neq j'.$$

The details are somewhat involved because of the general nature of the $x_i$s. We omit details. $\qquad\qquad\qquad\qquad\qquad\qquad\qquad\qquad\qquad\qquad\qquad\qquad\qquad\qquad\qquad\qquad\square$

The RSS from the *linear* regression contains information about the worth of a polynomial fit of a higher order, over and above a linear fit. Since, in this model, we assume that the population regression is linear, the variation in this higher-order fit is just random variation. Robson (1959) gives a general recursion formula for finding coefficients for such higher-order polynomial fits, a formula that can be adapted to explicitly find the $a_{ij}$s of (11.3.31). Alternatively, Cochran's Theorem (see Miscellanea 11.5.1) can be used to establish that $\sum \hat{\epsilon}_i^2/\sigma^2 \sim \chi^2_{n-2}$.

Inferences regarding the two parameters $\alpha$ and $\beta$ are usually based on the following two Student's $t$ distributions. Their derivations follow immediately from the normal and $\chi^2$ distributions and the independence in Theorem 11.3.3. We have

$$(11.3.32) \qquad \frac{\hat{\alpha} - \alpha}{S\sqrt{(\sum_{i=1}^{n} x_i^2)/(nS_{xx})}} \sim t_{n-2}$$

and

$$(11.3.33) \qquad \frac{\hat{\beta} - \beta}{S/\sqrt{S_{xx}}} \sim t_{n-2}.$$

The joint distribution of these two $t$ statistics is called a *bivariate Student's t distribution*. This distribution is derived in a manner analogous to the univariate case. We use the fact that the joint distribution of $\hat{\alpha}$ and $\hat{\beta}$ is bivariate normal and the same variance estimate $S$ is used in both univariate $t$ statistics. This joint distribution would be used if we wanted to do simultaneous inference regarding $\alpha$ and $\beta$. However, we shall deal only with the inferences regarding one parameter at a time.

Usually there is more interest in $\beta$ than in $\alpha$. The parameter $\alpha$ is the expected value of $Y$ at $x = 0, \text{E}(Y|x = 0)$. Depending on the problem, this may or may not be an interesting quantity. In particular, the value $x = 0$ may not be a reasonable value for the predictor variable. However, $\beta$ is the rate of change of $E(Y|x)$ as a function of $x$. That is, $\beta$ is the amount that $E(Y|x)$ changes if $x$ is changed by one unit. Thus, this parameter relates to the entire range of $x$ values and contains the information about whatever linear relationship exists between $Y$ and $x$. (See Exercise 11.33.) Furthermore, the value $\beta = 0$ is of particular interest.

If $\beta = 0$, then $E(Y|x) = \alpha + \beta x = \alpha$ and $Y \sim n(\alpha, \sigma^2)$, which does not depend on $x$. In a well-thought-out experiment leading to a regression analysis we do not expect this to be the case, but we would be interested in knowing this if it were true.

The test that $\beta = 0$ is quite similar to the ANOVA test that all treatments are equal. In the ANOVA the null hypothesis states that the treatments are unrelated to the response *in any way*, while in linear regression the null hypothesis $\beta = 0$ states that the treatments ($x$) are unrelated to the response in a linear way.

To test

$$(11.3.34) \qquad\qquad H_0: \quad \beta = 0 \qquad \text{versus} \qquad H_1: \quad \beta \neq 0$$

using (11.3.33), we reject $H_0$ at level $\alpha$ if

$$\left| \frac{\hat{\beta} - 0}{S/\sqrt{S_{xx}}} \right| > t_{n-2, \alpha/2}$$

or, equivalently, if

$$(11.3.35) \qquad\qquad \frac{\hat{\beta}^2}{S^2/S_{xx}} > F_{1, n-2, \alpha}.$$

Recalling the formula for $\hat{\beta}$ and that RSS$= \sum \hat{\epsilon}_i^2$, we have

$$\frac{\hat{\beta}^2}{S^2/S_{xx}} = \frac{S_{xy}^2/S_{xx}}{\text{RSS}/(n-2)} = \frac{\text{Regression sum of squares}}{\text{Residual sum of squares/df}}.$$

This last formula is summarized in the *regression ANOVA table*, which is like the ANOVA tables encountered in Section 11.2. For simple linear regression, the table, resulting in the test given in (11.3.35), is given in Table 11.3.2. Note that the table involves only a hypothesis about $\beta$. The parameter $\alpha$ and the estimate $\hat{\alpha}$ play the same role here as the grand mean did in Section 11.2. They merely serve to locate the overall level of the data and are "corrected" for in the sums of squares.

**Example 11.3.4 (Continuation of Example 11.3.1)**    The regression ANOVA for the grape crop yield data follows.

*ANOVA table for grape data*

| Source of variation | Degrees of freedom | Sum of squares | Mean square | F statistic |
|---|---|---|---|---|
| Regression | 1 | 6.66 | 6.66 | 50.23 |
| Residual | 10 | 1.33 | .133 | |
| Total | 11 | 7.99 | | |

This shows a highly significant slope of the regression line.        ‖

Table 11.3.2. *ANOVA table for simple linear regression*

| Source of variation | Degrees of freedom | Sum of squares | Mean square | $F$ statistic |
|---|---|---|---|---|
| Regression (slope) | 1 | Reg. SS $= S_{xy}^2/S_{xx}$ | MS(Reg) $=$ Reg. SS | $F = \dfrac{\text{MS(Reg)}}{\text{MS(Resid)}}$ |
| Residual | $n-2$ | RSS $= \sum \hat{\epsilon}_i^2$ | MS(Resid) $=$ RSS$/(n-2)$ | |
| Total | $n-1$ | SST $=$ $\sum (y_i - \bar{y})^2$ | | |

We draw one final parallel with the analysis of variance. It may not be obvious from Table 11.3.2, but the partitioning of the sum of squares of the ANOVA has an analogue in regression. We have

Total sum of squares = Regression sum of squares + Residual sum of squares

$$(11.3.36) \qquad \sum_{i=1}^{n}(y_i - \bar{y})^2 = \sum_{i=1}^{n}(\hat{y}_i - \bar{y})^2 + \sum_{i=1}^{n}(y_i - \hat{y}_i)^2,$$

where $\hat{y}_i = \hat{\alpha} + \hat{\beta}x_i$. Notice the similarity of these sums of squares to those in ANOVA. The total sum of squares is, of course, the same. The RSS measures deviation of the fitted line from the observed values, and the regression sum of squares, analogous to the ANOVA treatment sum of squares, measures the deviation of predicted values ("treatment means") from the grand mean. Also, as in the ANOVA, the sum of squares identity is valid because of the disappearance of the cross-term (see Exercise 11.34). The total and residual sums of squares in (11.3.36) are clearly the same as in Table 11.3.2. But the regression sum of squares looks different. However, they are equal (see Exercise 11.34); that is,

$$\sum_{i=1}^{n}(\hat{y}_i - \bar{y})^2 = \frac{S_{xy}^2}{S_{xx}}.$$

The expression $S_{xy}^2/S_{xx}$ is easier to use for computing and provides the link with the $t$ test. But $\sum_{i=1}^{n}(\hat{y}_i - \bar{y})^2$ is the more easily interpreted expression.

A statistic that is used to quantify how well the fitted line describes the data is the *coefficient of determination*. It is defined as the ratio of the regression sum of squares to the total sum of squares. It is usually referred to as $r^2$ and can be written in the various forms

$$r^2 = \frac{\text{Regression sum of squares}}{\text{Total sum of squares}} = \frac{\sum_{i=1}^{n}(\hat{y}_i - \bar{y})^2}{\sum_{i=1}^{n}(y_i - \bar{y})^2} = \frac{S_{xy}^2}{S_{xx}S_{yy}}.$$

The coefficient of determination measures the proportion of the total variation in $y_1, \ldots, y_n$ (measured by $S_{yy}$) that is explained by the fitted line (measured by the regression sum of squares). From (11.3.36), $0 \le r^2 \le 1$. If $y_1, \ldots, y_n$ all fall exactly on the fitted line, then $y_i = \hat{y}_i$ for all $i$ and $r^2 = 1$. If $y_1, \ldots, y_n$ are not close to the fitted line, then the residual sum of squares will be large and $r^2$ will be near 0. The coefficient of determination can also be (perhaps more straightforwardly) derived as the square of the sample correlation coefficient of the $n$ pairs $(y_1, x_1), \ldots, (y_n, x_n)$ or of the $n$ pairs $(y_1, \hat{y}_1), \ldots, (y_n, \hat{y}_n)$.

Expression (11.3.33) can be used to construct a $100(1-\alpha)\%$ confidence interval for $\beta$ given by

$$(11.3.37) \qquad \hat{\beta} - t_{n-2,\alpha/2} \frac{S}{\sqrt{S_{xx}}} < \beta < \hat{\beta} + t_{n-2,\alpha/2} \frac{S}{\sqrt{S_{xx}}}.$$

Also, a level $\alpha$ test of $H_0 : \beta = \beta_0$ versus $H_1 : \beta \neq \beta_0$ rejects $H_0$ if

$$(11.3.38) \qquad \left| \frac{\hat{\beta} - \beta_0}{S/\sqrt{S_{xx}}} \right| > t_{n-2,\alpha/2}.$$

As mentioned above, it is common to test $H_0 : \beta = 0$ versus $H_1 : \beta \neq 0$ to determine if there is some linear relationship between the predictor and response variables. However, the above test is more general, since any value of $\beta_0$ can be specified. The regression ANOVA, which is locked into a "recipe," can test only $H_0 : \beta = 0$.

### 11.3.5 Estimation and Prediction at a Specified $x = x_0$

Associated with a specified value of the predictor variable, say $x = x_0$, there is a population of $Y$ values. In fact, according to the conditional normal model, a random observation from this population is $Y \sim n(\alpha + \beta x_0, \sigma^2)$. After observing the regression data $(x_1, y_1), \ldots, (x_n, y_n)$ and estimating the parameters $\alpha, \beta$, and $\sigma^2$, perhaps the experimenter is going to set $x = x_0$ and obtain a new observation, call it $Y_0$. There might be interest in estimating the mean of the population from which this observation will be drawn, or even predicting what this observation will be. We will now discuss these types of inferences.

We assume that $(x_1, Y_1), \ldots, (x_n, Y_n)$ satisfy the conditional normal regression model, and based on these $n$ observations we have the estimates $\hat{\alpha}, \hat{\beta}$, and $S^2$. Let $x_0$ be a specified value of the predictor variable. First, consider estimating the mean of the $Y$ population associated with $x_0$, that is, $E(Y|x_0) = \alpha + \beta x_0$. The obvious choice for our point estimator is $\hat{\alpha} + \hat{\beta} x_0$. This is an unbiased estimator since $E(\hat{\alpha} + \hat{\beta} x_0) = E\hat{\alpha} + (E\hat{\beta}) x_0 = \alpha + \beta x_0$. Using the moments given in Theorem 11.3.3, we can also calculate

$$\text{Var}\,(\hat{\alpha} + \hat{\beta} x_0) = \text{Var}\,\hat{\alpha} + (\text{Var}\,\hat{\beta}) x_0^2 + 2x_0 \,\text{Cov}(\hat{\alpha}, \hat{\beta})$$

$$= \frac{\sigma^2}{nS_{xx}} \sum_{i=1}^{n} x_i^2 + \frac{\sigma^2 x_0^2}{S_{xx}} - \frac{2\sigma^2 x_0 \bar{x}}{S_{xx}}$$

$$= \frac{\sigma^2}{S_{xx}} \left( \frac{1}{n} \sum_{i=1}^{n} x_i^2 - \bar{x}^2 + \bar{x}^2 - 2x_0 \bar{x} + x_0^2 \right) \qquad (\pm \bar{x})$$

$$= \frac{\sigma^2}{S_{xx}}\left(\frac{1}{n}\left[\sum_{i=1}^{n}x_i^2 - \frac{1}{n}\left(\sum_{i=1}^{n}x_i\right)^2\right] + (x_0 - \bar{x})^2\right) \quad \begin{pmatrix}\text{recombine}\\ \text{terms}\end{pmatrix}$$

$$= \sigma^2\left(\frac{1}{n} + \frac{(x_0 - \bar{x})^2}{S_{xx}}\right). \qquad \left(\sum x_i^2 - \frac{1}{n}\left(\sum x_i\right)^2 = S_{xx}\right)$$

Finally, since $\hat{\alpha}$ and $\hat{\beta}$ are both linear functions of $Y_1, \ldots, Y_n$, so is $\hat{\alpha} + \hat{\beta}x_0$. Thus $\hat{\alpha} + \hat{\beta}x_0$ has a normal distribution, specifically,

$$(11.3.39) \qquad \hat{\alpha} + \hat{\beta}x_0 \sim \text{n}\left(\alpha + \beta x_0, \sigma^2\left(\frac{1}{n} + \frac{(x_0 - \bar{x})^2}{S_{xx}}\right)\right).$$

By Theorem 11.3.3, $(\hat{\alpha}, \hat{\beta})$ and $S^2$ are independent. Thus $S^2$ is also independent of $\hat{\alpha} + \hat{\beta}x_0$ (Theorem 4.6.12) and

$$\frac{\hat{\alpha} + \hat{\beta}x_0 - (\alpha + \beta x_0)}{S\sqrt{\frac{1}{n} + \frac{(x_0 - \bar{x})^2}{S_{xx}}}} \sim t_{n-2}.$$

This pivot can be inverted to give the $100(1 - \alpha)\%$ confidence interval for $\alpha + \beta x_0$,

$$\hat{\alpha} + \hat{\beta}x_0 - t_{n-2,\alpha/2}S\sqrt{\frac{1}{n} + \frac{(x_0 - \bar{x})^2}{S_{xx}}}$$

$$(11.3.40) \qquad \leq \quad \alpha + \beta x_0 \quad \leq \quad \hat{\alpha} + \hat{\beta}x_0 + t_{n-2,\alpha/2}S\sqrt{\frac{1}{n} + \frac{(x_0 - \bar{x})^2}{S_{xx}}}.$$

The length of the confidence interval for $\alpha + \beta x_0$ depends on the values of $x_1, \ldots, x_n$ through the value of $(x_0 - \bar{x})^2/S_{xx}$. It is clear that the length of the interval is shorter if $x_0$ is near $\bar{x}$ and minimized at $x_0 = \bar{x}$. Thus, in designing the experiment, the experimenter should choose the values $x_1, \ldots, x_n$ so that the value $x_0$, at which the mean is to be estimated, is at or near $\bar{x}$. It is only reasonable that we can estimate more precisely near the center of the data we observed.

A type of inference we have not discussed until now is *prediction* of an, as yet, unobserved random variable $Y$, a type of inference that is of interest in a regression setting. For example, suppose that $x$ is a college applicant's measure of high school performance. A college admissions officer might want to use $x$ to predict $Y$, the student's grade point average after one year of college. Clearly, $Y$ has not been observed yet since the student has not even been admitted! The college has data on former students, $(x_1, y_1), \ldots, (x_n, y_n)$, giving their high school performances and one-year GPAs. These data might be used to predict the new student's GPA.

**Definition 11.3.5** A $100(1 - \alpha)\%$ *prediction interval* for an unobserved random variable $Y$ based on the observed data $\mathbf{X}$ is a random interval $[L(\mathbf{X}), U(\mathbf{X})]$ with the property that

$$P_\theta(L(\mathbf{X}) \leq Y \leq U(\mathbf{X})) \geq 1 - \alpha$$

for all values of the parameter $\theta$.

Note the similarity in the definitions of a prediction interval and a confidence interval. The difference is that a prediction interval is an interval on a random variable, rather than a parameter. Intuitively, since a random variable is more variable than a parameter (which is constant), we expect a prediction interval to be wider than a confidence interval of the same level. In the special case of linear regression, we see that this is the case.

We assume that the new observation $Y_0$ to be taken at $x = x_0$ has a $n(\alpha + \beta x_0, \sigma^2)$ distribution, independent of the previous data, $(x_1, Y_1), \ldots, (x_n, Y_n)$. The estimators $\hat{\alpha}, \hat{\beta},$ and $S^2$ are calculated from the previous data and, thus, $Y_0$ is independent of $\hat{\alpha}, \hat{\beta},$ and $S^2$. Using (11.3.39), we find that $Y_0 - (\hat{\alpha} + \hat{\beta} x_0)$ has a normal distribution with mean $E(Y_0 - (\hat{\alpha} + \hat{\beta} x_0)) = \alpha + \beta x_0 - (\alpha + \beta x_0) = 0$ and variance

$$\text{Var}\,(Y_0 - (\hat{\alpha} + \hat{\beta} x_0)) \;=\; \text{Var}\, Y_0 + \text{Var}\,(\hat{\alpha} + \hat{\beta} x_0) \;=\; \sigma^2 + \sigma^2\left(\frac{1}{n} + \frac{(x_0 - \bar{x})^2}{S_{xx}}\right).$$

Using the independence of $S^2$ and $Y_0 - (\hat{\alpha} + \hat{\beta} x_0)$, we see that

$$T = \frac{Y_0 - (\hat{\alpha} + \hat{\beta} x_0)}{S\sqrt{1 + \frac{1}{n} + \frac{(x_0 - \bar{x})^2}{S_{xx}}}} \sim t_{n-2},$$

which can be rearranged in the usual way to obtain the $100(1 - \alpha)\%$ prediction interval,

$$\hat{\alpha} + \hat{\beta} x_0 - t_{n-2, \alpha/2} S\sqrt{1 + \frac{1}{n} + \frac{(x_0 - \bar{x})^2}{S_{xx}}}$$

(11.3.41) $$< \; Y_0 \; < \; \hat{\alpha} + \hat{\beta} x_0 + t_{n-2, \alpha/2} S\sqrt{1 + \frac{1}{n} + \frac{(x_0 - \bar{x})^2}{S_{xx}}}.$$

Since the endpoints of this interval depend only on the observed data, (11.3.41) defines a prediction interval for the new observation $Y_0$.

### 11.3.6 Simultaneous Estimation and Confidence Bands

In the previous section we looked at prediction at a single value $x_0$. In some circumstances, however, there may be interest in prediction at many $x_0$s. For example, in the previously mentioned grade point average prediction problem, an admissions officer probably has interest in predicting the grade point average of many applicants, which naturally leads to prediction at many $x_0$s.

The problem encountered is the (by now) familiar problem of simultaneous inference. That is, how do we control the overall confidence level for the simultaneous inference? In the previous section, we saw that a $1 - \alpha$ confidence interval for the mean of the $Y$ population associated with $x_0$, that is, $E(Y|x_0) = \alpha + \beta x_0$, is given by

$$\hat{\alpha} + \hat{\beta}x_0 - t_{n-2,\alpha/2}S\sqrt{\frac{1}{n} + \frac{(x_0 - \bar{x})^2}{S_{xx}}}$$

$$< \quad \alpha + \beta x_0 \quad < \quad \hat{\alpha} + \hat{\beta}x_0 + t_{n-2,\alpha/2}S\sqrt{\frac{1}{n} + \frac{(x_0 - \bar{x})^2}{S_{xx}}}.$$

Now suppose that we want to make an inference about the $Y$ population mean at a number of $x_0$ values. For example, we might want intervals for $\mathrm{E}(Y|x_{0i}), i = 1, \ldots, m$. We know that if we set up $m$ intervals as above, each at level $1 - \alpha$, the overall inference will not be at the $1 - \alpha$ level.

A simple and reasonably good solution is to use the Bonferroni Inequality, as used in Example 11.2.9. Using the inequality, we can state that the probability is at least $1 - \alpha$ that

$$\hat{\alpha} + \hat{\beta}x_{0i} - t_{n-2,\alpha/(2m)}S\sqrt{\frac{1}{n} + \frac{(x_{0i} - \bar{x})^2}{S_{xx}}}$$

(11.3.42)    $$< \quad \alpha + \beta x_{0i} \quad < \quad \hat{\alpha} + \hat{\beta}x_{0i} + t_{n-2,\alpha/(2m)}S\sqrt{\frac{1}{n} + \frac{(x_{0i} - \bar{x})^2}{S_{xx}}}$$

simultaneously for $i = 1, \ldots, m$. (See Exercise 11.39.)

We can take simultaneous inference in regression one step further. Realize that our assumption about the population regression line implies that the equation $\mathrm{E}(Y|x) = \alpha + \beta x$ holds *for all* $x$; hence, we should be able to make inferences at all $x$. Thus, we want to make a statement like (11.3.42), but we want it to hold for all $x$. As might be expected, as he did for the ANOVA, Scheffé derived a solution for this problem. We summarize the result for the case of simple linear regression in the following theorem.

**Theorem 11.3.6** *Under the conditional normal regression model (11.3.22), the probability is at least $1 - \alpha$ that*

$$\hat{\alpha} + \hat{\beta}x - M_\alpha S\sqrt{\frac{1}{n} + \frac{(x - \bar{x})^2}{S_{xx}}}$$

(11.3.43)    $$< \quad \alpha + \beta x \quad < \quad \hat{\alpha} + \hat{\beta}x + M_\alpha S\sqrt{\frac{1}{n} + \frac{(x - \bar{x})^2}{S_{xx}}}$$

*simultaneously for all $x$, where $M_\alpha = \sqrt{2F_{2,n-2,\alpha}}$.*

**Proof:** If we rearrange terms, it should be clear that the conclusion of the theorem is true if we can find a constant $M_\alpha$ that satisfies

$$P\left(\frac{\left((\hat{\alpha} + \hat{\beta}x) - (\alpha + \beta x)\right)^2}{S^2\left[\frac{1}{n} + \frac{(x-\bar{x})^2}{S_{xx}}\right]} \leq M_\alpha^2 \text{ for all } x\right) = 1 - \alpha$$

or, equivalently,

$$P\left(\max_x \frac{\left((\hat{\alpha} + \hat{\beta}x) - (\alpha + \beta x)\right)^2}{S^2\left[\frac{1}{n} + \frac{(x-\bar{x})^2}{S_{xx}}\right]} \le M_\alpha^2\right) = 1 - \alpha.$$

The parameterization given in Exercise 11.32, which results in independent estimators for $\alpha$ and $\beta$, makes the above maximization easier. Write

$$\hat{\alpha} + \hat{\beta}x = \bar{Y} + \hat{\beta}(x - \bar{x}),$$

$$\alpha + \beta x = \mu_{\bar{Y}} + \beta(x - \bar{x}), \qquad (\mu_{\bar{Y}} = \mathrm{E}\bar{Y} = \alpha + \beta\bar{x})$$

and, for notational convenience, define $t = x - \bar{x}$. We then have

$$\frac{\left((\hat{\alpha} + \hat{\beta}x) - (\alpha + \beta x)\right)^2}{S^2\left[\frac{1}{n} + \frac{(x-\bar{x})^2}{S_{xx}}\right]} = \frac{\left((\bar{Y} - \mu_{\bar{Y}}) + (\hat{\beta} - \beta)t\right)^2}{S^2\left[\frac{1}{n} + \frac{t^2}{S_{xx}}\right]},$$

and we want to find $M_\alpha$ to satisfy

$$P\left(\max_t \frac{\left((\bar{Y} - \mu_{\bar{Y}}) + (\hat{\beta} - \beta)t\right)^2}{S^2\left[\frac{1}{n} + \frac{t^2}{S_{xx}}\right]} \le M_\alpha^2\right) = 1 - \alpha.$$

Note that $S^2$ plays no role in the maximization, merely being a constant. Applying the result of Exercise 11.40, a direct application of calculus, we obtain

$$\max_t \frac{\left((\bar{Y} - \mu_{\bar{Y}}) + (\hat{\beta} - \beta)t\right)^2}{S^2\left[\frac{1}{n} + \frac{t^2}{S_{xx}}\right]} = \frac{n(\bar{Y} - \mu_{\bar{Y}})^2 + S_{xx}(\hat{\beta} - \beta)^2}{S^2}$$

$$(11.3.44) \qquad\qquad = \frac{\frac{(\bar{Y} - \mu_{\bar{Y}})^2}{\sigma^2/n} + \frac{(\hat{\beta} - \beta)^2}{\sigma^2/S_{xx}}}{S^2/\sigma^2}. \qquad \text{(multiply by } \sigma^2/\sigma^2)$$

From Theorem 11.3.3 and Exercise 11.32, we see that this last expression is the quotient of independent chi squared random variables, the denominator being divided by its degrees of freedom. The numerator is the sum of two independent random variables, each of which has a $\chi_1^2$ distribution. Thus the numerator is distributed as $\chi_2^2$, the distribution of the quotient is

$$\frac{\frac{(\bar{Y} - \mu_{\bar{Y}})^2}{\sigma^2/n} + \frac{(\hat{\beta} - \beta)^2}{\sigma^2/S_{xx}}}{S^2/\sigma^2} \sim 2F_{2,n-2},$$

and

$$P\left(\max_t \frac{\left((\bar{Y} - \mu_{\bar{Y}}) + (\hat{\beta} - \beta)t\right)^2}{S^2\left[\frac{1}{n} + \frac{t^2}{S_{xx}}\right]} \le M_\alpha^2\right) = 1 - \alpha$$

if $M_\alpha = \sqrt{2F_{2,n-2}}$, proving the theorem. $\qquad\qquad\qquad\qquad\qquad\qquad \square$

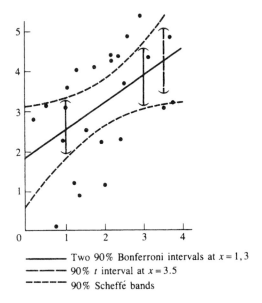

Figure 11.3.3. *Scheffé bands, t interval (at $x = 3.5$), and Bonferroni intervals (at $x = 1$ and $x = 3$) for data in Table 11.3.1*

Since (11.3.43) is true for all $x$, it actually gives a *confidence band* on the entire population regression line. That is, as a confidence interval covers a single-valued parameter, a confidence band covers an entire line with a band. An example of the Scheffé band is given in Figure 11.3.3, along with two Bonferroni intervals and a single $t$ interval. Notice that, although it is not the case in Figure 11.3.3, it is possible for the Bonferroni intervals to be *wider* than the Scheffé bands, even though the Bonferroni inference (necessarily) pertains to fewer intervals. This will be the case whenever

$$t_{n-2,\alpha/(2m)} > 2F_{2,n-2,\alpha},$$

where $m$ is defined as in (11.3.42). The inequality will always be satisfied for large enough $m$, so there will always be a point where it pays to switch from Bonferroni to Scheffé, even if there is interest in only a finite number of $x$s. This "phenomenon," that we seem to get something for nothing, occurs because the Bonferroni Inequality is an all-purpose bound while the Scheffé band is an exact solution for the problem at hand. (The actual coverage probability for the Bonferroni intervals is higher than $1 - \alpha$.) There are many variations on the Scheffé band. Some variations have different shapes and some guarantee coverage for only a particular interval of $x$ values. See the Miscellanea section for a discussion of these alternative bands.

In theory, the proof of Theorem 11.3.6, with suitable modifications, can result in simultaneous prediction intervals. (In fact, the maximization of the function in Exercise 11.40 gives the result almost immediately.) The problem, however, is that the resulting statistic does not have a particularly nice distribution.

Finally, we note a problem about using procedures like the Scheffé band to make inferences at $x$ values that are outside the range of the observed $x$s. Such procedures are based on the assumption that we *know* the population regression function is linear for all $x$. Although it may be reasonable to assume the regression function is linear over the range of $x$s observed, *extrapolation* to $x$s outside the observed range is usually unwise. (Since there are no data outside the observed range, we cannot check whether the regression becomes nonlinear.) This caveat also applies to the procedures in Section 11.3.5.

## 11.4 Exercises

**11.1** An ANOVA variance-stabilizing transformation stabilizes variances in the following approximate way. Let $Y$ have mean $\theta$ and variance $v(\theta)$.

(a) Use arguments as in Section 10.1.3 to show that a one-term Taylor series approximation of the variance of $g(y)$ is given by $\text{Var}\,(g(Y)) = [\frac{d}{d\theta}g(\theta)]^2 v(\theta)$.

(b) Show that the approximate variance of $g^*(Y)$ is independent of $\theta$, where $g^*(y) = \int[1/\sqrt{v(y)}]dy$.

**11.2** Verify that the following transformations are approximately variance-stabilizing in the sense of Exercise 11.1.

(a) $Y \sim$ Poisson, $g^*(y) = \sqrt{y}$

(b) $Y \sim$ binomial$(n, p), g^*(y) = \sin^{-1}(\sqrt{y/n})$

(c) $Y$ has variance $v(\theta) = K\theta^2$ for some constant $K, g^*(y) = \log(y)$.

(Conditions for the existence of variance-stabilizing transformations go back at least to Curtiss 1943, with refinements given by Bar-Lev and Enis 1988, 1990.)

**11.3** The Box–Cox family of power transformations (Box and Cox 1964) is defined by

$$g^*_\lambda(y) = \begin{cases} (y^\lambda - 1)/\lambda & \text{if } \lambda \neq 0 \\ \log y & \text{if } \lambda = 0, \end{cases}$$

where $\lambda$ is a free parameter.

(a) Show that, for each $y, g^*_\lambda(y)$ is continuous in $\lambda$. In particular, show that

$$\lim_{\lambda \to 0}(y^\lambda - 1)/\lambda = \log y.$$

(b) Find the function $v(\theta)$, the approximate variance of $Y$, that $g^*_\lambda(y)$ stabilizes. (Note that $v(\theta)$ will most likely also depend on $\lambda$.)

Analysis of transformed data in general and the Box–Cox power transformation in particular has been the topic of some controversy in the statistical literature. See Bickel and Doksum (1981), Box and Cox (1982), and Hinkley and Runger (1984).

**11.4** A most famous (and useful) variance-stabilizing transformation is Fisher's z-transformation, which we have already encountered in Exercise 10.17. Here we will look at a few more details. Suppose that $(X, Y)$ are bivariate normal with correlation coefficient $\varrho$ and sample correlation $r$.

(a) Starting from Exercise 10.17, part (d), use the Delta Method to show that

$$\frac{1}{2}\left[\log\left(\frac{1+r}{1-r}\right) - \log\left(\frac{1+\varrho}{1-\varrho}\right)\right]$$

is approximately normal with mean 0 and variance $1/n$.

(b) Fisher actually used a somewhat more accurate expansion (Stuart and Ord 1987, Section 16.33) and established that the quantity in part (a) is approximately normal with

$$\text{mean} = \frac{\varrho}{2(n-1)} \quad \text{and} \quad \text{variance} = \frac{1}{n-1} + \frac{4-\varrho^2}{2(n-1)^2}.$$

Show that for small $\varrho$ and moderate $n$, we can approximate this mean and variance by 0 and $1/(n-3)$, which is the most popular form of Fisher's z-transformation.

**11.5** Suppose that random variables $Y_{ij}$ are observed according to the overparameterized oneway ANOVA model in (11.2.2). Show that, without some restriction on the parameters, this model is not identifiable by exhibiting two distinct collections of parameters that lead to exactly the same distribution of the $Y_{ij}$s.

**11.6** Under the oneway ANOVA assumptions:

(a) Show that the set of statistics $(\bar{Y}_1, \bar{Y}_2, \ldots, \bar{Y}_k, S_p^2)$ is sufficient for $(\theta_1, \theta_2, \ldots, \theta_k, \sigma^2)$.

(b) Show that $S_p^2 = \frac{1}{N-k}\sum_{i=1}^{k}(n_i - 1)S_i^2$ is independent of each $\bar{Y}_i, i = 1, \ldots, k$. (See Lemma 5.3.3.)

(c) If $\sigma^2$ is known, explain how the ANOVA data are equivalent to their canonical version in Miscellanea 11.5.6.

**11.7** Complete the proof of Theorem 11.2.8 by showing that

$$\frac{1}{\sigma^2}\sum_{i=1}^{k} n_i \left((\bar{Y}_{i\cdot} - \bar{\bar{Y}}) - (\theta_i - \bar{\theta})\right)^2 \sim \chi_{k-1}^2.$$

(*Hint:* Define $\bar{U}_i = \bar{Y}_{i\cdot} - \theta_i, i = 1, \ldots, k$. Show that $\bar{U}_i$ are independent n$(0, \sigma^2/n_i)$. Then adapt the induction argument of Lemma 5.3.2 to show that $\sum n_i(\bar{U}_i - \bar{\bar{U}})^2/\sigma^2 \sim \chi_{k-1}^2$, where $\bar{\bar{U}} = \sum n_i \bar{U}_i / \sum n_i$.)

**11.8** Show that under the oneway ANOVA assumptions, for any set of constants $\mathbf{a} = (a_1, \ldots, a_k)$, the quantity $\sum a_i \bar{Y}_{i\cdot}$ is normally distributed with mean $\sum a_i \theta_i$ and variance $\sigma^2 \sum a_i^2/n_i$. (See Corollary 4.6.10.)

**11.9** Using an argument similar to that which led to the $t$ test in (11.2.7), show how to construct a $t$ test for

(a) $H_0: \sum a_i \theta_i = \delta$ versus $H_1: \sum a_i \theta_i \neq \delta$.

(b) $H_0: \sum a_i \theta_i \leq \delta$ versus $H_1: \sum a_i \theta_i > \delta$, where $\delta$ is a specified constant.

**11.10** Suppose we have a oneway ANOVA with five treatments. Denote the treatment means by $\theta_1, \ldots, \theta_5$, where $\theta_1$ is a control and $\theta_2, \ldots, \theta_5$ are alternative new treatments, and assume that an equal number of observations per treatment is taken. Consider the four contrasts $\sum a_i \theta_i$ defined by

$$\mathbf{a}_1 = \left(1, -\frac{1}{4}, -\frac{1}{4}, -\frac{1}{4}, -\frac{1}{4}\right),$$

$$\mathbf{a}_2 = \left(0, 1, -\frac{1}{3}, -\frac{1}{3}, -\frac{1}{3}\right),$$

$$\mathbf{a}_3 = \left(0, 0, 1, -\frac{1}{2}, -\frac{1}{2}\right),$$

$$\mathbf{a}_4 = (0, 0, 0, 1, -1).$$

(a) Argue that the results of the four $t$ tests using these contrasts can lead to conclusions about the ordering of $\theta_1, \ldots, \theta_5$. What conclusions might be made?

(b) Show that any two contrasts $\sum a_i \bar{Y}_i$. formed from the four $\mathbf{a}_i$s in part (a) are uncorrelated. (Recall that these are called orthogonal contrasts.)

(c) For the fertilizer experiment of Example 11.2.3, the following contrasts were planned:

$$\mathbf{a}_1 = (-1, 1, 0, 0, 0),$$

$$\mathbf{a}_2 = \left(0, -1, \frac{1}{2}, \frac{1}{2}, 0\right),$$

$$\mathbf{a}_3 = (0, 0, 1, -1, 0),$$

$$\mathbf{a}_4 = (0, -1, 0, 0, 1, ).$$

Show that these contrasts are not orthogonal. Interpret these contrasts in the context of the fertilizer experiment, and argue that they are a sensible set of contrasts.

**11.11** For any sets of constants $\mathbf{a} = (a_1, \ldots, a_k)$ and $\mathbf{b} = (b_1, \ldots, b_k)$, show that under the oneway ANOVA assumptions,

$$\text{Cov}\left(\sum a_i \bar{Y}_i., \sum b_i \bar{Y}_i.\right) = \sigma^2 \sum \frac{a_i b_i}{n_i}.$$

Hence, in the oneway ANOVA, contrasts are uncorrelated (orthogonal) if $\sum a_i b_i / n_i = 0$.

**11.12** Suppose that we have a oneway ANOVA with equal numbers of observations on each treatment, that is, $n_i = n, i = 1, \ldots, k$. In this case the $F$ test can be considered an average $t$ test.

(a) Show that a $t$ test of $H_0: \theta_i = \theta_{i'}$ versus $H_1: \theta_i \neq \theta_{i'}$ can be based on the statistic

$$t_{ii'}^2 = \frac{(\bar{Y}_i. - \bar{Y}_{i'}.)^2}{S_p^2(2/n)}.$$

(b) Show that

$$\frac{1}{k(k-1)} \sum_{i, i'} t_{ii'}^2 = F,$$

where $F$ is the usual ANOVA $F$ statistic. (*Hint:* See Exercise 5.8(a).) (*Communicated by George McCabe, who learned it from John Tukey.*)

**11.13** Under the oneway ANOVA assumptions, show that the likelihood ratio test of $H_0: \theta_1 = \theta_2 = \cdots = \theta_k$ is given by the $F$ test of (11.2.14).

**11.14** The Scheffé simultaneous interval procedure actually works for all linear combinations, not just contrasts. Show that under the oneway ANOVA assumptions, if $\mathbf{M} = \sqrt{k F_{k, N-k, \alpha}}$ (note the change in the numerator degrees of freedom), then the probability is $1 - \alpha$ that

$$\sum_{i=1}^{k} a_i \bar{Y}_i. - \mathbf{M}\sqrt{S_p^2 \sum_{i=1}^{k} \frac{a_i^2}{n_i}} \leq \sum_{i=1}^{k} a_i \theta_i \leq \sum_{i=1}^{k} a_i \bar{Y}_i. + \mathbf{M}\sqrt{S_p^2 \sum_{i=1}^{k} \frac{a_i^2}{n_i}}$$

simultaneously for all $\mathbf{a} = (a_1, \ldots, a_k)$. It is probably easiest to proceed by first establishing, in the spirit of Lemma 11.2.7, that if $v_1, \ldots, v_k$ are constants and $c_1, \ldots, c_k$ are positive constants, then

$$\max_{\mathbf{a}} \left\{ \frac{\left( \sum_{i=1}^k a_i v_i \right)^2}{\sum_{i=1}^k a_i^2 / c_i} \right\} = \sum_{i=1}^k c_i v_i^2.$$

The proof of Theorem 11.2.10 can then be adapted to establish the result.

**11.15** (a) Show that for the $t$ and $F$ distributions, for any $\nu$, $\alpha$, and $k$,

$$t_{\nu, \alpha/2} \leq \sqrt{(k-1) F_{k-1, \nu, \alpha}}.$$

(Recall the relationship between the $t$ and the $F$. This inequality is a consequence of the fact that the distributions $k F_{k, \nu}$ are stochastically increasing in $k$ for fixed $\nu$ but is actually a weaker statement. See Exercise 5.19.)

(b) Explain how the above inequality shows that the simultaneous Scheffé intervals are always wider than the single-contrast intervals.

(c) Show that it also follows from the above inequality that Scheffé tests are less powerful than $t$ tests.

**11.16** In Theorem 11.2.5 we saw that the ANOVA null is equivalent to all contrasts being 0. We can also write the ANOVA null as the intersection over another set of hypotheses.

(a) Show that the hypotheses

$$H_0: \quad \theta_1 = \theta_2 = \cdots = \theta_k \qquad \text{versus} \qquad H_1: \quad \theta_i \neq \theta_j \text{ for some } i, j$$

and the hypotheses

$$H_0: \quad \theta_i - \theta_j = 0 \text{ for all } i, j \qquad \text{versus} \qquad H_1: \quad \theta_i - \theta_j \neq 0 \text{ for some } i, j$$

are equivalent.

(b) Express $H_0$ and $H_1$ of the ANOVA test as unions and intersections of the sets

$$\Theta_{ij} = \{\theta = (\theta_1, \ldots, \theta_k): \theta_i - \theta_j = 0\}.$$

Describe how these expressions can be used to construct another (different) union–intersection test of the ANOVA null hypothesis. (See Miscellanea 11.5.2.)

**11.17** A multiple comparison procedure called the *Protected LSD* (Protected Least Significant Difference) is performed as follows. If the ANOVA $F$ test rejects $H_0$ at level $\alpha$, then for each pair of means $\theta_i$ and $\theta_{i'}$, declare the means different if

$$\frac{|\bar{Y}_{i\cdot} - \bar{Y}_{i'\cdot}|}{\sqrt{S_p^2 \left( \frac{1}{n_i} + \frac{1}{n_{i'}} \right)}} > t_{\alpha/2, N-k}.$$

Note that each $t$ test is done at the same $\alpha$ level as the ANOVA $F$ test. Here we are using an *experimentwise* $\alpha$ level, where

$$\text{experimentwise } \alpha = P \left( \begin{array}{c} \text{at least one false} \\ \text{assertion of difference} \end{array} \middle| \begin{array}{c} \text{all the means} \\ \text{are equal} \end{array} \right).$$

(a) Prove that no matter how many means are in the experiment, simultaneous inference from the Protected LSD is made at level $\alpha$.

(b) The *ordinary* (or *unprotected*) LSD simply does the individual $t$ tests, at level $\alpha$, no matter what the outcome of the ANOVA $F$ test. Show that the ordinary LSD can have an experimentwise error rate greater than $\alpha$. (The unprotected LSD does maintain a *comparisonwise* error rate of $\alpha$.)

(c) Perform the LSD procedure on the fish toxin data of Example 11.2.1. What are the conclusions?

**11.18** Demonstrate that "data snooping," that is, testing hypotheses that are suggested by the data, is generally not a good practice.

(a) Show that, for any random variable $Y$ and constants $a$ and $b$ with $a > b$ and $P(Y > b) < 1$, $P(Y > a|Y > b) > P(Y > a)$.

(b) Apply the inequality in part (a) to the size of a data-suggested hypothesis test by letting $Y$ be a test statistic and $a$ be a cutoff point.

**11.19** Let $X_i \sim \text{gamma}(\lambda_i, 1)$ independently for $i = 1, \ldots, n$. Define $Y_i = X_{i+1}/\left(\sum_{j=1}^{i} X_j\right)$, $i = 1, \ldots, n-1$, and $Y_n = \sum_{i=1}^{n} X_i$.

(a) Find the joint and marginal distributions of $Y_i, i = 1, \ldots, n$.

(b) Connect your results to any distributions that are commonly employed in the ANOVA.

**11.20** Assume the oneway ANOVA null hypothesis is true.

(a) Show that $\sum n_i(\bar{Y}_{i\cdot} - \bar{\bar{Y}})^2/(k-1)$ gives an unbiased estimate of $\sigma^2$.

(b) Show how to use the method of Example 5.3.5 to derive the ANOVA $F$ test.

**11.21** (a) Illustrate the partitioning of the sums of squares in the ANOVA by calculating the complete ANOVA table for the following data. To determine diet quality, male weanling rats were fed diets with various protein levels. Each of 15 rats was randomly assigned to one of three diets, and their weight gain in grams was recorded.

| Low | Medium | High |
|-----|--------|------|
| 3.89 | 8.54 | 20.39 |
| 3.87 | 9.32 | 24.22 |
| 3.26 | 8.76 | 30.91 |
| 2.70 | 9.30 | 22.78 |
| 3.82 | 10.45 | 26.33 |

Diet protein level

(b) Analytically verify the partitioning of the ANOVA sums of squares by completing the proof of Theorem 11.2.11.

(c) Illustrate the relationship between the $t$ and $F$ statistics, given in Exercise 11.12(b), using the data of part (a).

**11.22** Calculate the expected values of MSB and MSW given in the oneway ANOVA table. (Such expectations are formally known as *expected mean squares* and can be used to help identify $F$ tests in complicated ANOVAs. An algorithm exists for calculating expected mean squares. See, for example, Kirk 1982 for details about the algorithm.)

**11.23** Use the model in Miscellanea 11.5.3.

(a) Show that the mean and variance of $Y_{ij}$ are $EY_{ij} = \mu + \tau_i$ and $\mathrm{Var}\, Y_{ij} = \sigma_B^2 + \sigma^2$.

(b) If $\sum a_i = 0$, show that the unconditional variance of $\sum a_i \bar{Y}_i$ is $\mathrm{Var}\left(\sum a_i \bar{Y}_i.\right) = \frac{1}{r}(\sigma^2 + \sigma_B^2)(1 - \rho) \sum a_i^2$, where $\rho = $ intraclass correlation.

**11.24** The form of the Stein estimator of Miscellanea 11.5.6 can be justified somewhat by an *empirical Bayes* argument given in Efron and Morris (1972), which can be quite useful in data analysis. Such an argument may have been known by Stein (1956), although he makes no mention of it. Let $X_i \sim n(\theta_i, 1), i = 1, \ldots, p$, and $\theta_i$ be iid $n(0, \tau^2)$.

(a) Show that the $X_i$s, marginally, are iid $n(0, \tau^2 + 1)$, and, hence, $\sum X_i^2/(\tau^2 + 1) \sim \chi_p^2$.

(b) Using the marginal distribution, show that $E(1 - ((p-2)/\sum_{j=1}^{p} X_j^2)) = \tau^2/(\tau^2 + 1)$ if $p \geq 3$. Thus, the Stein estimator of Miscellanea 11.5.6 is an empirical Bayes version of the Bayes estimator $\delta_i^\tau(\mathbf{X}) = [\tau^2/(\tau^2 + 1)]X_i$.

(c) Show that the argument fails if $p < 3$ by showing that $E(1/Y) = \infty$ if $Y \sim \chi_p^2$ with $p < 3$.

**11.25** In Section 11.3.1, we found the least squares estimators of $\alpha$ and $\beta$ by a two-stage minimization. This minimization can also be done using partial derivatives.

(a) Compute $\frac{\partial \mathrm{RSS}}{\partial c}$ and $\frac{\partial \mathrm{RSS}}{\partial d}$ and set them equal to 0. Show that the resulting two equations can be written as

$$nc + \left(\sum_{i=1}^{n} x_i\right) d = \sum_{i=1}^{n} y_i \quad \text{and} \quad \left(\sum_{i=1}^{n} x_i\right) c + \left(\sum_{i=1}^{n} x_i^2\right) d = \sum_{i=1}^{n} x_i y_i.$$

(These equations are called the *normal equations* for this minimization problem.)

(b) Show that $c = a$ and $d = b$ are the solutions to the normal equations.

(c) Check the second partial derivative condition to verify that the point $c = a$ and $d = b$ is indeed the minimum of RSS.

**11.26** Suppose $n$ is an even number. The values of the predictor variable, $x_1, \ldots, x_n$, all must be chosen to be in the interval $[e, f]$. Show that the choice that maximizes $S_{xx}$ is for half of the $x_i$ equal to $e$ and the other half equal to $f$. (This was the choice mentioned in Section 11.3.2 that minimizes $\mathrm{Var}\, b$.)

**11.27** Observations $(x_i, Y_i)$, $i = 1, \ldots, n$, follow the model $Y_i = \alpha + \beta x_i + \epsilon_i$, where $E\,\epsilon_i = 0$, $\mathrm{Var}\,\epsilon_i = \sigma^2$, and $\mathrm{Cov}(\epsilon_i, \epsilon_j) = 0$ if $i \neq j$. Find the best linear unbiased estimator of $\alpha$.

**11.28** Show that in the conditional normal model for simple linear regression, the MLE of $\sigma^2$ is given by

$$\hat{\sigma}^2 = \frac{1}{n} \sum_{i=1}^{n} (y_i - \hat{\alpha} - \hat{\beta} x_i)^2.$$

**11.29** Consider the residuals $\hat{\epsilon}_1, \ldots, \hat{\epsilon}_n$ defined in Section 11.3.4 by $\hat{\epsilon}_i = Y_i - \hat{\alpha} - \hat{\beta} x_i$.

(a) Show that $E\hat{\epsilon}_i = 0$.

(b) Verify that

$$\mathrm{Var}\,\hat{\epsilon}_i = \mathrm{Var}\, Y_i + \mathrm{Var}\,\hat{\alpha} + x_i^2 \mathrm{Var}\,\hat{\beta} - 2\mathrm{Cov}(Y_i, \hat{\alpha}) - 2x_i \mathrm{Cov}(Y_i, \hat{\beta}) + 2x_i \mathrm{Cov}(\hat{\alpha}, \hat{\beta}).$$

(c) Use Lemma 11.3.2 to show that

$$\text{Cov}(Y_i, \hat{\alpha}) = \sigma^2 \left( \frac{1}{n} + \frac{(x_i - \bar{x})\bar{x}}{S_{xx}} \right) \quad \text{and} \quad \text{Cov}(Y_i, \hat{\beta}) = \sigma^2 \frac{x_i - \bar{x}}{S_{xx}},$$

and use these to verify (11.3.28).

**11.30** Fill in the details about the distribution of $\hat{\alpha}$ left out of the proof of Theorem 11.3.3.

(a) Show that the estimator $\hat{\alpha} = \bar{y} - \hat{\beta}\bar{x}$ can be expressed as $\hat{\alpha} = \sum_{i=1}^{n} c_i Y_i$, where

$$c_i = \frac{1}{n} - \frac{(x_i - \bar{x})\bar{x}}{S_{xx}}.$$

(b) Verify that

$$\text{E}\hat{\alpha} = \alpha \quad \text{and} \quad \text{Var}\,\hat{\alpha} = \sigma^2 \left[ \frac{1}{nS_{xx}} \sum_{i=1}^{n} x_i^2 \right].$$

(c) Verify that

$$\text{Cov}(\hat{\alpha}, \hat{\beta}) = -\frac{\sigma^2 \bar{x}}{S_{xx}}.$$

**11.31** Verify the claim in Theorem 11.3.3, that $\hat{\epsilon}_i$ is uncorrelated with $\hat{\alpha}$ and $\hat{\beta}$. (Show that $\hat{\epsilon}_i = \sum e_j Y_j$, where the $e_j$s are given by (11.3.30). Then, using the facts that we can write $\hat{\alpha} = \sum c_j Y_j$ and $\hat{\beta} = \sum d_j Y_j$, verify that $\sum e_j c_j = \sum e_j d_j = 0$ and apply Lemma 11.3.2.)

**11.32** Observations $(x_i, Y_i), i = 1, \ldots, n$, are made according to the model

$$Y_i = \alpha + \beta x_i + \epsilon_i,$$

where $x_1, \ldots, x_n$ are fixed constants and $\epsilon_1, \ldots, \epsilon_n$ are iid $n(0, \sigma^2)$. The model is then reparameterized as

$$Y_i = \alpha' + \beta'(x_i - \bar{x}) + \epsilon_i.$$

Let $\hat{\alpha}$ and $\hat{\beta}$ denote the MLEs of $\alpha$ and $\beta$, respectively, and $\hat{\alpha}'$ and $\hat{\beta}'$ denote the MLEs of $\alpha'$ and $\beta'$, respectively.

(a) Show that $\hat{\beta}' = \hat{\beta}$.

(b) Show that $\hat{\alpha}' \neq \hat{\alpha}$. In fact, show that $\hat{\alpha}' = \bar{Y}$. Find the distribution of $\hat{\alpha}'$.

(c) Show that $\hat{\alpha}'$ and $\hat{\beta}'$ are uncorrelated and, hence, independent under normality.

**11.33** Observations $(X_i, Y_i), i = 1, \ldots, n$, are made from a bivariate normal population with parameters $(\mu_X, \mu_Y, \sigma_X^2, \sigma_Y^2, \rho)$, and the model $Y_i = \alpha + \beta x_i + \epsilon_i$ is going to be fit.

(a) Argue that the hypothesis $H_0 : \beta = 0$ is true if and only if the hypothesis $H_0 : \rho = 0$ is true. (See (11.3.25).)

(b) Show algebraically that

$$\frac{\hat{\beta}}{S/\sqrt{S_{xx}}} = \sqrt{n-2}\, \frac{r}{\sqrt{1-r^2}},$$

where $r$ is the sample correlation coefficient, the MLE of $\rho$.

(c) Show how to test $H_0 : \rho = 0$, given only $r^2$ and $n$, using Student's $t$ with $n - 2$ degrees of freedom (see (11.3.33)). (Fisher derived an approximate confidence interval for $\rho$, using a variance-stabilizing transformation. See Exercise 11.4.)

**11.34** (a) Illustrate the partitioning of the sum of squares for simple linear regression by calculating the regression ANOVA table for the following data. Parents are often interested in predicting the eventual heights of their children. The following is a portion of the data taken from a study that might have been suggested by Galton's analysis.

| Height (inches) at age 2 $(x)$ | 39 | 30 | 32 | 34 | 35 | 36 | 36 | 30 |
|---|---|---|---|---|---|---|---|---|
| Height (inches) as an adult $(y)$ | 71 | 63 | 63 | 67 | 68 | 68 | 70 | 64 |

(b) Analytically establish the partitioning of the sum of squares for simple linear regression by verifying (11.3.36).

(c) Prove that the two expressions for the regression sum of squares are, in fact, equal; that is, show that

$$\sum_{i=1}^{n}(\hat{y}_i - \bar{y})^2 = \frac{S_{xy}^2}{S_{xx}}.$$

(d) Show that the *coefficient of determination*, $r^2$, given by

$$r^2 = \frac{\sum_{i=1}^{n}(\hat{y}_i - \bar{y})^2}{\sum_{i=1}^{n}(y_i - \bar{y})^2}$$

can be derived as the square of the sample correlation coefficient either of the $n$ pairs $(y_1, x_1), \ldots, (y_n, x_n)$ or of the $n$ pairs $(y_1, \hat{y}_1), \ldots, (y_n, \hat{y}_n)$.

**11.35** Observations $Y_1, \ldots, Y_n$ are described by the relationship $Y_i = \theta x_i^2 + \epsilon_i$, where $x_1, \ldots, x_n$ are fixed constants and $\epsilon_1, \ldots, \epsilon_n$ are iid $n(0, \sigma^2)$.

(a) Find the least squares estimator of $\theta$.
(b) Find the MLE of $\theta$.
(c) Find the best unbiased estimator of $\theta$.

**11.36** Observations $Y_1, \ldots, Y_n$ are made according to the model $Y_i = \alpha + \beta x_i + \epsilon_i$, where $x_1, \ldots, x_n$ are fixed constants and $\epsilon_1, \ldots, \epsilon_n$ are iid $n(0, \sigma^2)$. Let $\hat{\alpha}$ and $\hat{\beta}$ denote MLEs of $\alpha$ and $\beta$.

(a) Assume that $x_1, \ldots, x_n$ are observed values of iid random variables $X_1, \ldots, X_n$ with distribution $n(\mu_X, \sigma_X^2)$. Prove that when we take expectations over the joint distribution of $X$ and $Y$, we still get $E\hat{\alpha} = \alpha$ and $E\hat{\beta} = \beta$.

(b) The phenomenon of part (a) does not carry over to the covariance. Calculate the unconditional covariance of $\hat{\alpha}$ and $\hat{\beta}$ (using the joint distribution of $X$ and $Y$).

**11.37** We observe random variables $Y_1, \ldots, Y_n$ that are mutually independent, each with a normal distribution with variance $\sigma^2$. Furthermore, $EY_i = \beta x_i$, where $\beta$ is an unknown parameter and $x_1, \ldots, x_n$ are fixed constants not all equal to 0.

(a) Find the MLE of $\beta$. Compute its mean and variance.
(b) Compute the Cramér–Rao Lower Bound for the variance of an unbiased estimator of $\beta$.
(c) Find a best unbiased estimator of $\beta$.

(d) If you could place the values $x_1, \ldots, x_n$ anywhere within a given nondegenerate closed interval $[A, B]$, where would you place these values? Justify your answer.

(e) For a given positive value $r$, the *maximum probability estimator of $\beta$ with respect to $r$* is the value of $D$ that maximizes the integral

$$\int_{D-r}^{D+r} f(y_1, \ldots, y_n | \beta) d\beta,$$

where $f(y_1, \ldots, y_n | \beta)$ is the joint pdf of $Y_1, \ldots, Y_n$. Find this estimator.

**11.38** An ecologist takes data $(x_i, Y_i)$, $i = 1, \ldots, n$, where $x_i$ is the size of an area and $Y_i$ is the number of moss plants in the area. We model the data by $Y_i \sim \text{Poisson}(\theta x_i)$, $Y_i$s independent.

(a) Show that the least squares estimator of $\theta$ is $\sum x_i Y_i / \sum x_i^2$. Show that this estimator has variance $\theta \sum x_i^3 / (\sum x_i^2)^2$. Also, compute its bias.

(b) Show that the MLE of $\theta$ is $\sum Y_i / \sum x_i$ and has variance $\theta / \sum x_i$. Compute its bias.

(c) Find a best unbiased estimator of $\theta$ and show that its variance attains the Cramér–Rao Lower Bound.

**11.39** Verify that the simultaneous confidence intervals in (11.3.42) have the claimed coverage probability.

**11.40** (a) Prove that if $a$, $b$, $c$, and $d$ are constants, with $c > 0$ and $d > 0$, then

$$\max_t \frac{(a + bt)^2}{c + dt^2} = \frac{a^2}{c} + \frac{b^2}{d}.$$

(b) Use part (a) to verify equation (11.3.44) and hence fill in the gap in Theorem 11.3.6.

(c) Use part (a) to find a Scheffé-type simultaneous band using the prediction intervals of (11.3.41). That is, rewriting the prediction intervals as was done in Theorem 11.3.6, show that

$$\max_t \frac{\left( (\bar{Y} - \mu_{\bar{Y}}) + (\hat{\beta} - \beta)t \right)^2}{S^2 \left[ 1 + \frac{1}{n} + \frac{t^2}{S_{xx}} \right]} = \frac{\frac{n}{n+1}(\bar{Y} - \mu_{\bar{Y}})^2 + S_{xx}(\hat{\beta} - \beta)^2}{S^2}.$$

(d) The distribution of the maximum is not easy to write down, but we could approximate it. Approximate the statistic by using moment matching, as done in Example 7.2.3.

**11.41** In the discussion in Example 12.4.2, note that there was one observation from the potoroo data that had a missing value. Suppose that on the 24th animal it was observed that $O_2 = 16.3$.

(a) Write down the observed data and expected complete data log likelihood functions.

(b) Describe the E step and the M step of an EM algorithm to find the MLEs.

(c) Find the MLEs using all 24 observations.

(d) Actually, the $O_2$ reading on the 24th animal was not observed, but rather the $CO_2$ was observed to be 4.2 (and the $O_2$ was missing). Set up the EM algorithm in this case and find the MLEs. (This is a much harder problem, as you now have to take expectations over the $x$s. This means you have to formulate the regression problem using the bivariate normal distribution.)

## 11.5 Miscellanea _____

### 11.5.1 Cochran's Theorem

Sums of squares of normal random variables, when properly scaled and centered, are distributed as chi squared random variables. This type of result is first due to Cochran (1934). Cochran's Theorem gives necessary and sufficient conditions on the scaling required for squared and summed iid normal random variables to be distributed as a chi squared random variable. The conditions are not difficult, but they are best stated in terms of properties of matrices and will not be treated here. It is an immediate consequence of Cochran's Theorem that in the oneway ANOVA, the $\chi^2$ random variables partition as discussed in Section 11.2.6. Furthermore, another consequence is that in the Randomized Complete Blocks ANOVA (see Miscellanea 11.5.3), the mean squares all have chi squared distributions.

Cochran's Theorem has been generalized to the extent that necessary and sufficient conditions are known for the distribution of squared normals (not necessarily iid) to be chi squared. See Stuart and Ord (1987, Chapter 15) for details.

### 11.5.2 Multiple Comparisons

We have seen two ways of doing simultaneous inference in this chapter: the Scheffé procedure and use of the Bonferroni Inequality. There is a plethora of other simultaneous inference procedures. Most are concerned with inference on pairwise comparisons, that is, differences between means. These procedures can be applied to estimate treatment means in the oneway ANOVA.

A method due to Tukey (see Miller 1981), sometimes known as the $Q$ method, applies a Scheffé-type maximization argument but over only pairwise differences, not all contrasts. The $Q$ distribution is the distribution of

$$Q = \max_{i,j} \left| \frac{(\bar{Y}_{i\cdot} - \bar{Y}_{j\cdot}) - (\theta_i - \theta_j)}{\sqrt{S_p^2 \left( \frac{1}{n} + \frac{1}{n} \right)}} \right|,$$

where $n_i = n$ for all $i$. (Hayter 1984 has shown that if $n_i \neq n_j$ and the $n$ above is replaced by the harmonic mean $n_h$, where $1/n_h = \frac{1}{2}((1/n_i) + (1/n_j))$, the resulting procedure is conservative.) The $Q$ method is an improvement over Scheffé's $S$ method in that if there is interest only in pairwise differences, the $Q$ method is more powerful (shorter intervals). This is easy to see because, by definition, the $Q$ maximization will produce a smaller maximum than the $S$ method.

Other types of multiple comparison procedures that deal with pairwise differences are more powerful than the $S$ method. Some procedures are the LSD (Least Significant Difference) Procedure, Protected LSD, Duncan's Procedure, and Student–Neumann–Keuls' Procedure. These last two are *multiple range* procedures. The cutoff point to which comparisons are made changes between comparisons.

One difficulty in fully understanding multiple comparison procedures is that the definition of Type I Error is not inviolate. Some of these procedures have changed the definition of Type I Error for multiple comparisons, so exactly what is meant by "$\alpha$ level" is not always clear. Some of the types of error rates considered are called *experimentwise error rate, comparisonwise error rate,* and *familywise error rate*. Miller (1981) and Hsu (1996) are good references for this topic. A humorous but illuminating treatment of this subject is given in Carmer and Walker (1982).

### 11.5.3 Randomized Complete Block Designs

Section 11.2 was concerned with a *oneway* classification of the data; that is, there was only one categorization (treatment) in the experiment. In general, the ANOVA allows for many types of categorization, with one of the most commonly used ANOVAs being the Randomized Complete Block (RCB) ANOVA.

A *block* (or *blocking factor*) is categorization that is in an experiment for the express purpose of removing variation. In contrast to a treatment, there is usually no interest in finding block differences. The practice of blocking originated in agriculture, where experimenters took advantage of similar growing conditions to control experimental variances. To model this, the actual blocks in the experiment were considered to be a random sample from a large population of blocks (which makes them a *random* factor).

### RCB ANOVA assumptions

Random variables $Y_{ij}$ are observed according to the model

$$Y_{ij}|\mathbf{b} = \mu + \tau_i + b_j + \epsilon_{ij}, \quad i = 1, \ldots, k, \quad j = 1, \ldots, r,$$

where:

(i) The random variables $\epsilon_{ij} \sim$ iid $n(0, \sigma^2)$ for $i = 1, \ldots, k$ and $j = 1, \ldots, r$ (normal errors with equal variances).

(ii) The random variables $B_1, \ldots, B_r$, whose realized (but unobserved) values are the blocks $b_1, \ldots, b_r$, are iid $n(0, \sigma_B^2)$ and are independent of $\epsilon_{ij}$ for all $i, j$.

The mean and variance of $Y_{ij}$ are

$$\mathrm{E}\,Y_{ij} = \mu + \tau_i \quad \text{and} \quad \mathrm{Var}\,Y_{ij} = \sigma_B^2 + \sigma^2.$$

Moreover, although the $Y_{ij}$s are uncorrelated conditionally, there is correlation in the blocks unconditionally. The correlation between $Y_{ij}$ and $Y_{i'j}$ in block $j$, with $i \neq i'$, is

$$\frac{\mathrm{Cov}(Y_{ij}, Y_{i'j})}{\sqrt{(\mathrm{Var}\,Y_{ij})(\mathrm{Var}\,Y_{i'j})}} = \frac{\sigma_B^2}{\sigma_B^2 + \sigma^2},$$

a quantity called the *intraclass correlation*. Thus, the model implies not only that there is correlation in the blocks but also that there is positive correlation. This is a consequence of the additive model and the assumption that the $\epsilon$s and $B$s are independent (see Exercise 11.23). Even though the $Y_{ij}$s are not independent, the intraclass correlation structure still results in an analysis of variance where ratios of mean squares have the $F$ distribution (see Miscellanea 11.5.1).

### 11.5.4 Other Types of Analyses of Variance

The two types of ANOVAs that we have considered, oneway ANOVAs and RCB ANOVAs, are the simplest types. For example, an extension of a complete block design is an *incomplete* block design. Sometimes there are physical constraints that prohibit putting all treatments in each block and an incomplete block design is needed. Deciding how to arrange the treatments in such a design is both difficult and critical. Of course, as the design gets more complicated, so does the analysis.

Study of the subject of statistical design, which is concerned with getting the most information from the fewest observations, leads to more complicated and more efficient ANOVAs in many situations. ANOVAs based on designs such as *fractional factorials, Latin squares,* and *balanced incomplete blocks* can be efficient methods of gathering much information about a phenomenon. Good overall references for this subject are Cochran and Cox (1957), Dean and Voss (1999), and Kuehl (2000).

### 11.5.5 Shapes of Confidence Bands

Confidence bands come in many shapes, not just the *hyperbolic* shape defined by the Scheffé band. For example, Gafarian (1964) showed how to construct a *straight-line* band over a finite interval. Gafarian-type bands allow statements of the form

$$P(\hat{\alpha} + \hat{\beta}x - d_\alpha \leq \alpha + \beta x \leq \hat{\alpha} + \hat{\beta}x + d_\alpha \text{ for all } x \in [a, b]) = 1 - \alpha.$$

Gafarian gave tables of $d_\alpha$. A finite-width band must, necessarily, apply only to a finite range of $x$. Any band of level $1 - \alpha$ must have infinite length as $|x| \to \infty$.

Casella and Strawderman (1980), among others, showed how to construct Scheffé-type bands over finite intervals, thereby reducing width while maintaining the same confidence as the infinite Scheffé band. Naiman (1983) compared performance of straight-line and Scheffé bands over finite intervals. Under his criterion, one of average width, the Scheffé band is superior. In some cases, an experimenter might be more comfortable with the interpretation of a straight-line band, however.

Shapes other than straight-line and hyperbolic are possible. Piegorsch (1985) investigated and characterized the shapes that are admissible in the sense that their probability statements cannot be improved upon. He obtained "growth conditions" that must be satisfied by an admissible band. Naiman (1983, 1984, 1987) and Naiman and Wynn (1992, 1997) have developed this theory to a very high level, establishing useful inequalities and geometric identities to further improve inferences.

### 11.5.6 Stein's Paradox

One part of the analysis of variance is concerned with the simultaneous estimation of a collection of normal means. Developments in this particular problem, starting with Stein (1956), have had a profound effect on both the theory and applications of point estimation.

A canonical version of the analysis of variance is to observe $\mathbf{X} = (X_1, \ldots, X_p)$, independent normal random variables with $X_i \sim n(\theta_i, 1)$, $i = 1, \ldots, p$, with the objective being the estimation of $\boldsymbol{\theta} = (\theta_1, \ldots, \theta_p)$. Our usual estimate of $\theta_i$ would be $X_i$, but Stein (1956) established the surprising result that, if $p \geq 3$, the estimator of $\theta_i$ given by

$$\delta_i^S(\mathbf{X}) = \left(1 - \frac{p-2}{\sum_{i=1}^{p} X_i^2}\right) X_i$$

is a better estimator of $\theta_i$ in the sense that

$$\sum_{i=1}^{p} E_\theta \left(X_i - \theta_i\right)^2 \geq \sum_{i=1}^{p} E_\theta \left(\delta_i^S(\mathbf{X}) - \theta_i\right)^2.$$

That is, the summed mean squared of Stein's estimator is always smaller, and usually strictly smaller, than that of $\mathbf{X}$.

Notice that the estimators are being compared using the sum of the component-wise mean squared errors, and each $\delta_i^S$ can be a function of the entire vector $(X_1, \ldots, X_p)$. Thus, all of the data can be used in estimating each mean. Since the $X_i$s are independent, we might think that restricting $\delta_i^S$ to be just a function of $X_i$ would be enough. However, by summing the mean squared errors, we tie the components together.

In the oneway ANOVA we observe

$$\bar{Y}_{i\cdot} \sim n\left(\theta_i, \frac{\sigma^2}{n_i}\right), \quad i = 1, \ldots, k, \quad \text{independent},$$

where the $\bar{Y}_{i\cdot}$s are the cell means. The Stein estimator takes the form

$$\delta_i^S(\bar{Y}_{1\cdot}, \ldots, \bar{Y}_{k\cdot}) = \left(1 - \frac{(k-2)\sigma^2}{\sum n_j \bar{Y}_{j\cdot}^2}\right)^+ \bar{Y}_{i\cdot}, \quad i = 1, \ldots, k.$$

This Stein-type estimator can further be improved by choosing a meaningful place toward which to shrink (the above estimator shrinks toward 0). One such estimator, due to Lindley (1962), shrinks toward the grand mean of the observations. It is given by

$$\delta_i^L(\bar{Y}_{1\cdot}, \ldots, \bar{Y}_{k\cdot}) = \bar{\bar{Y}} + \left(1 - \frac{(k-3)\sigma^2}{\sum n_j (\bar{Y}_{j\cdot} - \bar{\bar{Y}})^2}\right)^+ (\bar{Y}_{i\cdot} - \bar{\bar{Y}}), \quad i = 1, \ldots, k.$$

Other choices of a shrinkage target might be even more appropriate. Discussion of this, including methods for improving on confidence statements, such as the Scheffé $S$ method, is given in Casella and Hwang (1987). Morris (1983) also discusses applications of these types of estimators.

There have been many theoretical developments using Stein-type estimators, not only in point estimation but also in confidence set estimation, where it has been shown that recentering at a Stein estimator can result in increased coverage probability and reduced size. There is also a strong connection between Stein estimators and empirical Bayes estimators (see Miscellanea 7.5.6), first uncovered in a series of papers by Efron and Morris (1972, 1973, 1975), where the components of $\boldsymbol{\theta}$ are tied together using a common prior distribution. An introduction to the theory and some applications of Stein estimators is given in Lehmann and Casella (1998, Chapter 5).

Chapter 12

# Regression Models

*"So startling would his results appear to the uninitiated that until they learned
the processes by which he had arrived at them they might well consider him as
a necromancer."*

**Dr. Watson, speaking about Sherlock Holmes**
*A Study in Scarlet*

## 12.1 Introduction

Chapter 11 was concerned with what could be called "classic linear models." Both
the ANOVA and simple linear regression are based on an underlying linear model
with normal errors. In this chapter we look at some extensions of this model that
have proven to be useful in practical problems.

In Section 12.2, the linear model is extended to models with errors in the predictor,
which is called regression with errors in variables (EIV). In this model the predictor
variable $X$ now becomes a random variable like the response variable $Y$. Estimation
in this model encounters many unforeseen difficulties and can be very different from
the simple linear regression model.

The linear model is further generalized in Section 12.3, where we look at logistic
regression. Here, the response variable is discrete, a Bernoulli variable. The Bernoulli
mean is a bounded function, and a linear model on a bounded function can run into
problems (especially at the boundaries). Because of this we transform the mean to
an unbounded parameter (using the logit transformation) and model the transformed
parameter as a linear function of the predictor variable. When a linear model is put
on a function of a response mean, it becomes a *generalized linear model*.

Lastly, in Section 12.4, we look at robustness in the setting of linear regression. In
contrast to the other sections in this chapter where we change the model, now we
change the fitting criterion. The development parallels that of Section 10.2.2, where
we looked at robust point estimates. That is, we replace the least squares criterion
with one based on a $\rho$-function that results in estimates that are less sensitive to
underlying observations (but retain some efficiency).

## 12.2 Regression with Errors in Variables

Regression with *errors in variables* (EIV), also known as the *measurement error
model*, is so fundamentally different from the simple linear regression of Section 11.3

that it is probably best thought of as a completely different topic. It is presented as a generalization of the usual regression model mainly for traditional reasons. However, the problems that arise with this model are very different.

The models of this section are generalizations of simple linear regression in that we will work with models of the form

$$(12.2.1) \qquad\qquad Y_i = \alpha + \beta x_i + \epsilon_i,$$

but now we do not assume that the $x$s are known. Instead, we can measure a random variable whose mean is $x_i$. (In keeping with our notational conventions, we will speak of measuring a random variable $X_i$ whose mean is not $x_i$ but $\xi_i$.)

The intention here is to illustrate different approaches to the EIV model, showing some of the standard solutions and the (sometimes) unexpected difficulties that arise. For a more thorough introduction to this problem, there are the review article by Gleser (1991); books by Fuller (1987) and Carroll, Ruppert, and Stefanski (1995); and a volume edited by Brown and Fuller (1991). Kendall and Stuart (1979, Chapter 29) also treat this topic in some detail.

In the general EIV model we assume that we observe pairs $(x_i, y_i)$ sampled from random variables $(X_i, Y_i)$ whose means satisfy the linear relationship

$$(12.2.2) \qquad\qquad \mathrm{E}Y_i = \alpha + \beta(\mathrm{E}X_i).$$

If we define

$$\mathrm{E}Y_i = \eta_i \quad \text{and} \quad \mathrm{E}X_i = \xi_i,$$

then the relationship (12.2.2) becomes

$$(12.2.3) \qquad\qquad \eta_i = \alpha + \beta\xi_i,$$

a linear relationship between the means of the random variables.

The variables $\xi_i$ and $\eta_i$ are sometimes called *latent variables*, a term that refers to quantities that cannot be directly measured. Latent variables may be not only impossible to measure directly but impossible to measure *at all*. For example, the IQ of a person is impossible to measure. We can measure a score on an IQ test, but we can never measure the variable IQ. Relationships between IQ and other variables, however, are often hypothesized.

The model specified in (12.2.2) really makes no distinction between $X$ and $Y$. If we are interested in a regression, however, there should be a reason for choosing $Y$ as the response and $X$ as the predictor. Keeping this specification in mind, of regressing $Y$ on $X$, we define the *errors in variables model* or *measurement error model* as this. Observe independent pairs $(X_i, Y_i)$, $i = 1, \ldots, n$, according to

$$Y_i = \alpha + \beta\xi_i + \epsilon_i, \quad \epsilon_i \sim \mathrm{n}(0, \sigma_\epsilon^2),$$

$$(12.2.4) \qquad X_i = \xi_i + \delta_i, \qquad\quad \delta_i \sim \mathrm{n}(0, \sigma_\delta^2).$$

Note that the assumption of normality, although common, is not necessary. Other distributions can be used. In fact, some of the problems encountered with this model are caused by the normality assumption. (See, for example, Solari 1969.)

**Example 12.2.1 (Estimating atmospheric pressure)** The EIV regression model arises fairly naturally in situations where the $x$ variable is observed along with the $y$ variable (rather than being controlled). For example, in the 1800s the Scottish physicist J. D. Forbes tried to use measurements on the boiling temperature of water to estimate altitude above sea level. To do this, he simultaneously measured boiling temperature and atmospheric pressure (from which altitude can be obtained). Since barometers were quite fragile in the 1800s, it would be useful to estimate pressure, or more precisely log(pressure), from temperature. The data observed at nine locales are

| Boiling point (°F) | log(pressure) (log(Hg)) |
|:---:|:---:|
| 194.5 | 1.3179 |
| 197.9 | 1.3502 |
| 199.4 | 1.3646 |
| 200.9 | 1.3782 |
| 201.4 | 1.3806 |
| 203.6 | 1.4004 |
| 209.5 | 1.4547 |
| 210.7 | 1.4630 |
| 212.2 | 1.4780 |

and an EIV model is reasonable for this situation.                                                 ||

A number of special cases of the model (12.2.4) have already been seen. If $\delta_i = 0$, then the model becomes simple linear regression (since there is now no measurement error, we can directly observe the $\xi_i$s). If $\alpha = 0$, then we have

$$Y_i \sim n(\eta_i, \sigma_\epsilon^2), \quad i = 1, \ldots, n,$$
$$X_i \sim n(\xi_i, \sigma_\delta^2), \quad i = 1, \ldots, n,$$

where, possibly, $\sigma_\delta^2 \neq \sigma_\epsilon^2$, a version of the Behrens–Fisher problem.

*12.2.1 Functional and Structural Relationships*

There are two different types of relationship that can be specified in the EIV model: one that specifies a *functional* linear relationship and one describing a *structural* linear relationship. The different relationship specifications can lead to different estimators with different properties. As said by Moran (1971), "This is not very happy terminology, but we will stick to it because the distinction is essential. . . ." Some interpretations of this terminology are given in the Miscellanea section. For now we merely present the two models.

*Linear functional relationship model*

This is the model as presented in (12.2.4) where we have random variables $X_i$ and $Y_i$, with $\mathrm{E}X_i = \xi_i$, $\mathrm{E}Y_i = \eta_i$, and we assume the *functional relationship*

$$\eta_i = \alpha + \beta \xi_i.$$

We observe pairs $(X_i, Y_i), i = 1, \ldots, n$, according to

$$
\begin{aligned}
Y_i &= \alpha + \beta \xi_i + \epsilon_i, \quad & \epsilon_i &\sim \mathrm{n}(0, \sigma_\epsilon^2), \\
X_i &= \xi_i + \delta_i, \quad & \delta_i &\sim \mathrm{n}(0, \sigma_\delta^2),
\end{aligned}
$$

(12.2.5)

where the $\xi_i$s are fixed, unknown parameters and the $\epsilon_i$s and $\delta_i$s are independent. The parameters of main interest are $\alpha$ and $\beta$, and inference on these parameters is made using the joint distribution of $((X_1, Y_1), \ldots, (X_n, Y_n))$, *conditional on* $\xi_1, \ldots, \xi_n$.

*Linear structural relationship model*

This model can be thought of as an extension of the functional relationship model, extended through the following hierarchy. As in the functional relationship model, we have random variables $X_i$ and $Y_i$, with $\mathrm{E}X_i = \xi_i$, $\mathrm{E}Y_i = \eta_i$, and we assume the *functional relationship* $\eta_i = \alpha + \beta \xi_i$. But now we assume that the parameters $\xi_1, \ldots, \xi_n$ are themselves a random sample from a common population. Thus, conditional on $\xi_1, \ldots, \xi_n$, we observe pairs $(X_i, Y_i), i = 1, \ldots, n$, according to

$$
\begin{aligned}
Y_i &= \alpha + \beta \xi_i + \epsilon_i, \quad & \epsilon_i &\sim \mathrm{n}(0, \sigma_\epsilon^2), \\
X_i &= \xi_i + \delta_i, \quad & \delta_i &\sim \mathrm{n}(0, \sigma_\delta^2),
\end{aligned}
$$

(12.2.6)

and also

$$\xi_i \sim \text{iid } \mathrm{n}(\xi, \sigma_\xi^2).$$

As before, the $\epsilon_i$s and $\delta_i$s are independent and they are also independent of the $\xi_i$s. As in the functional relationship model, the parameters of main interest are $\alpha$ and $\beta$. Here, however, the inference on these parameters is made using the joint distribution of $((X_1, Y_1), \ldots, (X_n, Y_n))$, *unconditional on* $\xi_1, \ldots, \xi_n$. (That is, $\xi_1, \ldots, \xi_n$ are integrated out according to the distribution in (12.2.6).)

The two models are quite similar in that statistical properties of estimators in one model (for example, consistency) often carry over into the other model. More precisely, estimators that are consistent in the functional model are also consistent in the structural model (Nussbaum 1976 or Gleser 1983). This makes sense, as the functional model is a "conditional version" of the structural model. Estimators that are consistent in the functional model must be so for all values of the $\xi_i$s so are necessarily consistent in the structural model, which averages over the $\xi_i$s. The converse implication is false. However, there is a useful implication that goes from the structural to the functional relationship model. If a parameter is not *identifiable* in the structural model, it is also not identifiable in the functional model. (See Definition 11.2.2.)

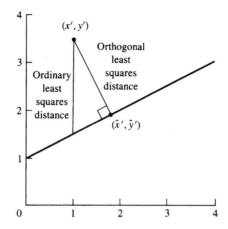

Figure 12.2.1. *Distance minimized by orthogonal least squares*

As we shall see, the models share similar problems and, in certain situations, similar likelihood solutions. It is probably easier to do statistical theory in the structural model, while the functional model often seems to be the more reasonable model for many situations. Thus, the underlying similarities come in handy.

As already mentioned, one of the major differences in the models is in the inferences about $\alpha$ and $\beta$, the parameters that describe the regression relationship. This difference is of utmost importance and cannot be stressed too often. In the functional relationship model, this inference is made conditional on $\xi_1, \ldots, \xi_n$, using the joint distribution of $X$ and $Y$ conditional on $\xi_1, \ldots, \xi_n$. On the other hand, in the structural relationship model, this inference is made unconditional on $\xi_1, \ldots, \xi_n$, using the marginal distribution of $X$ and $Y$ with $\xi_1, \ldots, \xi_n$ integrated out.

### 12.2.2 A Least Squares Solution

As in Section 11.3.1, we forget statistics for a while and try to find the "best" line through the observed points $(x_i, y_i)$, $i = 1, \ldots, n$. Previously, when it was assumed that the $x$s were measured without error, it made sense to consider minimization of vertical distances. This distance measure implicitly assumes that the $x$ value is correct and results in *ordinary* least squares. Here, however, there is no reason to consider vertical distances, since the $x$s now have error associated with them. In fact, statistically speaking, ordinary least squares has some problems in EIV models (see the Miscellanea section).

One way to take account of the fact that the $x$s also have error in their measurement is to perform *orthogonal least squares*, that is, to find the line that minimizes orthogonal (perpendicular to the line) distances rather than vertical distances (see Figure 12.2.1). This distance measure does not favor the $x$ variable, as does ordinary least squares, but rather treats both variables equitably. It is also known as the method of *total least squares*. From Figure 12.2.1, for a particular data point $(x', y')$, the point on a line $y = a + bx$ that is closest when we measure distance orthogonally is given by (see Exercise 12.1)

$$(12.2.7) \quad \hat{x}' = \frac{by' + x' - ab}{1 + b^2}, \quad \text{and} \quad \hat{y}' = a + \frac{b}{1 + b^2}(by' + x' - ab).$$

Now assume that we have data $(x_i, y_i), i = 1, \ldots, n$. The squared distance between an observed point $(x_i, y_i)$ and the closest point on the line $y = a + bx$ is $(x_i - \hat{x}_i)^2 + (y_i - \hat{y}_i)^2$, where $\hat{x}_i$ and $\hat{y}_i$ are defined by (12.2.7). The *total least squares problem* is to minimize, over all $a$ and $b$, the quantity

$$\sum_{i=1}^{n} \left( (x_i - \hat{x}_i)^2 + (y_i - \hat{y}_i)^2 \right).$$

It is straightforward to establish that we have

$$\sum_{i=1}^{n} \left( (x_i - \hat{x}_i)^2 + (y_i - \hat{y}_i)^2 \right)$$

$$= \sum_{i=1}^{n} \left( \frac{b^2}{(1 + b^2)^2} [y_i - (a + bx_i)]^2 + \frac{1}{(1 + b^2)^2} [y_i - (a + bx_i)]^2 \right)$$

$$(12.2.8) \qquad = \frac{1}{1 + b^2} \sum_{i=1}^{n} (y_i - (a + bx_i))^2.$$

For fixed $b$, the term in front of the sum is a constant. Thus, the minimizing choice of $a$ in the sum is $a = \bar{y} - b\bar{x}$, just as in (11.3.9). If we substitute back into (12.2.8), the total least squares solution is the one that minimizes, over all $b$,

$$(12.2.9) \qquad\qquad \frac{1}{1 + b^2} \sum_{i=1}^{n} ((y_i - \bar{y}) - b(x_i - \bar{x}))^2.$$

As in (11.3.6) and (11.3.6), we define the sums of squares and cross-products by

$$(12.2.10) \quad S_{xx} = \sum_{i=1}^{n}(x_i - \bar{x})^2, \quad S_{yy} = \sum_{i=1}^{n}(y_i - \bar{y})^2, \quad S_{xy} = \sum_{i=1}^{n}(x_i - \bar{x})(y_i - \bar{y}).$$

Expanding the square and summing show that (12.2.9) becomes

$$\frac{1}{1 + b^2} \left[ S_{yy} - 2bS_{xy} + b^2 S_{xx} \right].$$

Standard calculus methods will give the minimum (see Exercise 12.2), and we find the orthogonal least squares line given by $y = a + bx$, with

$$(12.2.11) \qquad a = \bar{y} - b\bar{x} \quad \text{and} \quad b = \frac{-(S_{xx} - S_{yy}) + \sqrt{(S_{xx} - S_{yy})^2 + 4S_{xy}^2}}{2S_{xy}}.$$

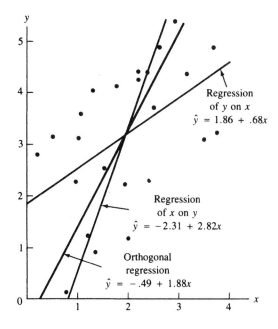

Figure 12.2.2. *Three regression lines for data in Table 11.3.1*

As might be expected, this line is different from the least squares line. In fact, as we shall see, this line always lies between the ordinary regression of $y$ on $x$ and the ordinary regression of $x$ on $y$. This is illustrated in Figure 12.2.2, where the data in Table 11.3.1 were used to calculate the orthogonal least squares line $\hat{y} = -.49 + 1.88x$.

In simple linear regression we saw that, under normality, the ordinary least squares solutions for $\alpha$ and $\beta$ were the same as the MLEs. Here, the orthogonal least squares solution is the MLE only in a special case, when we make certain assumptions about the parameters.

The difficulties to be encountered with likelihood estimation once again illustrate the differences between a mathematical solution and a statistical solution. We obtained a mathematical least squares solution to the line fitting problem without much difficulty. This will not happen with the likelihood solution.

### 12.2.3 Maximum Likelihood Estimation

We first consider the maximum likelihood solution of the functional linear relationship model, the situation for the structural relationship model being similar and, in some respects, easier. With the normality assumption, the functional relationship model can be expressed as

$$Y_i \sim n(\alpha + \beta\xi_i, \sigma_\epsilon^2) \quad \text{and} \quad X_i \sim n(\xi_i, \sigma_\delta^2), \quad i = 1, \ldots, n,$$

where the $X_i$s and $Y_i$s are independent. Given observations $(\mathbf{x}, \mathbf{y}) = ((x_1, y_1), \ldots,$ $(x_n, y_n))$, the likelihood function is

(12.2.12)

$$L\left(\alpha, \beta, \xi_1, \ldots, \xi_n, \sigma_\delta^2, \sigma_\epsilon^2 | (\mathbf{x}, \mathbf{y})\right) =$$

$$\frac{1}{(2\pi)^n} \frac{1}{(\sigma_\delta^2 \sigma_\epsilon^2)^{n/2}} \exp\left[-\sum_{i=1}^n \frac{(x_i - \xi_i)^2}{2\sigma_\delta^2}\right] \exp\left[-\sum_{i=1}^n \frac{(y_i - (\alpha + \beta\xi_i))^2}{2\sigma_\epsilon^2}\right].$$

The problem with this likelihood function is that it does not have a finite maximum. To see this, take the parameter configuration $\xi_i = x_i$ and then let $\sigma_\delta^2 \to 0$. The value of the function goes to infinity, showing that there is no maximum likelihood solution. In fact, Solari (1969) has shown that if the equations defining the first derivative of $L$ are set equal to 0 and solved, the result is a saddle point, not a maximum. Notice that as long as we have total control over the parameters, we can always force the likelihood function to infinity. In particular, we can always take a variance to 0, while keeping the exponential term bounded.

We will make the common assumption, which not only is reasonable but also alleviates many problems, that $\sigma_\delta^2 = \lambda\sigma_\epsilon^2$, where $\lambda > 0$ is fixed and known. (See Kendall and Stuart 1979, Chapter 29, for a discussion of other assumptions on the variances.) This assumption is one of the least restrictive, saying that we know only the ratio of the variances, not the individual values. Moreover, the resulting model is relatively well behaved.

Under this assumption, we can write the likelihood function as

(12.2.13)    $L\left(\alpha, \beta, \xi_1, \ldots, \xi_n, \sigma_\delta^2 | (\mathbf{x}, \mathbf{y})\right)$

$$= \frac{1}{(2\pi)^n} \frac{\lambda^{n/2}}{(\sigma_\delta^2)^n} \exp\left[-\sum_{i=1}^n \frac{(x_i - \xi_i)^2 + \lambda(y_i - (\alpha + \beta\xi_i))^2}{2\sigma_\delta^2}\right],$$

which we can now maximize. We will perform the maximization in stages, making sure that, at each step, we have a maximum before proceeding to the next step. By examining the function (12.2.13), we can determine a reasonable order of maximization.

First, for each value of $\alpha$, $\beta$, and $\sigma_\delta^2$, to maximize $L$ with respect to $\xi_1, \ldots, \xi_n$ we minimize $\sum_{i=1}^n \left((x_i - \xi_i)^2 + \lambda(y_i - (\alpha + \beta\xi_i))^2\right)$. (See Exercise 12.3 for details.) For each $i$, we have a quadratic in $\xi_i$ and the minimum is attained at

$$\xi_i^* = \frac{x_i + \lambda\beta(y_i - \alpha)}{1 + \lambda\beta^2}.$$

On substituting back we get

$$\sum_{i=1}^n \left((x_i - \xi_i^*)^2 + \lambda(y_i - (\alpha + \beta\xi_i^*))^2\right) = \frac{\lambda}{1 + \lambda\beta^2} \sum_{i=1}^n (y_i - (\alpha + \beta x_i))^2.$$

The likelihood function now becomes

$$\max_{\xi_1,\ldots,\xi_n} L(\alpha,\beta,\xi_1,\ldots,\xi_n,\sigma_\delta^2|(\mathbf{x},\mathbf{y}))$$

$$(12.2.14) \qquad = \frac{1}{(2\pi)^n} \frac{\lambda^{n/2}}{(\sigma_\delta^2)^n} \exp\left\{-\frac{1}{2\sigma_\delta^2}\left[\frac{\lambda}{1+\lambda\beta^2}\sum_{i=1}^n (y_i - (\alpha+\beta x_i))^2\right]\right\}.$$

Now, we can maximize with respect to $\alpha$ and $\beta$, but a little work will show that we have already done this in the orthogonal least squares solution! Yes, there is somewhat of a correspondence between orthogonal least squares and maximum likelihood in the EIV model and we are about to exploit it. Define

$$(12.2.15) \qquad \alpha^* = \sqrt{\lambda}\alpha, \quad \beta^* = \sqrt{\lambda}\beta, \quad y_i^* = \sqrt{\lambda}y_i, \quad i = 1,\ldots,n.$$

The exponent of (12.2.14) becomes

$$\frac{\lambda}{1+\lambda\beta^2}\sum_{i=1}^n (y_i - (\alpha+\beta x_i))^2 = \frac{1}{1+\beta^{*2}}\sum_{i=1}^n (y_i^* - (\alpha^*+\beta^* x_i))^2,$$

which is identical to the expression in the orthogonal least squares problem. From (12.2.11) we know the minimizing values of $\alpha^*$ and $\beta^*$, and using (12.2.15) we obtain our MLEs for the slope and intercept:

$$(12.2.16) \quad \hat{\alpha} = \bar{y} - \hat{\beta}\bar{x} \quad \text{and} \quad \hat{\beta} = \frac{-(S_{xx} - \lambda S_{yy}) + \sqrt{(S_{xx} - \lambda S_{yy})^2 + 4\lambda S_{xy}^2}}{2\lambda S_{xy}}.$$

It is clear from the formula that, at $\lambda = 1$, the MLEs agree with the orthogonal least squares solutions. This makes sense. The orthogonal least squares solution treated $x$ and $y$ as having the same magnitude of error and this translates into a variance ratio of 1. Carrying this argument further, we can relate this solution to ordinary least squares or maximum likelihood when the $x$s are assumed to be fixed. If the $x$s are fixed, their variance is 0 and hence $\lambda = 0$. The maximum likelihood solution for general $\lambda$ does reduce to ordinary least squares in this case. This relationship, among others, is explored in Exercise 12.4.

Putting (12.2.16) together with (12.2.14), we now have almost completely maximized the likelihood. We have

$$(12.2.17) \qquad \max_{\alpha,\beta,\xi_1,\ldots,\xi_n} L(\alpha,\beta,\xi_1,\ldots,\xi_n,\sigma_\delta^2|(\mathbf{x},\mathbf{y}))$$

$$= \frac{1}{(2\pi)^n}\frac{\lambda^{n/2}}{(\sigma_\delta^2)^n}\exp\left[-\frac{1}{2\sigma_\delta^2}\frac{\lambda}{1+\lambda\beta^2}\sum_{i=1}^n \left(y_i - (\hat{\alpha}+\hat{\beta}x_i)\right)^2\right].$$

Now maximizing $L$ with respect to $\sigma_\delta^2$ is very similar to finding the MLE of $\sigma^2$ in ordinary normal sampling (see Example 7.2.11), the major difference being the exponent of $n$, rather than $n/2$, on $\sigma_\delta^2$. The details are left to Exercise 12.5. The resulting MLE for $\sigma_\delta^2$ is

(12.2.18) $$\hat\sigma_\delta^2 = \frac{1}{2n}\frac{\lambda}{1+\lambda\hat\beta^2}\sum_{i=1}^n\left(y_i-(\hat\alpha+\hat\beta x_i)\right)^2.$$

From the properties of MLEs, it follows that the MLE of $\sigma_\epsilon^2$ is given by $\hat\sigma_\epsilon^2 = \hat\sigma_\delta^2/\lambda$ and $\hat\xi_i = \hat\alpha + \hat\beta x_i$. Although the $\hat\xi_i$s are not usually of interest, they can sometimes be useful if prediction is desired. Also, the $\hat\xi_i$s are useful in examining the adequacy of the fit (see Fuller 1987).

It is interesting to note that although $\hat\alpha$ and $\hat\beta$ are consistent estimators, $\sigma_\delta^2$ is not. More precisely, as $n \to \infty$,

$$\hat\alpha \to \alpha \text{ in probability,}$$

$$\hat\beta \to \beta \text{ in probability,}$$

but

$$\hat\sigma_\delta^2 \to \frac{1}{2}\sigma_\delta^2 \text{ in probability.}$$

General results on consistency in EIV functional relationship models have been obtained by Gleser (1981).

We now turn to the linear structural relationship model. Recall that here we assume that we observe pairs $(X_i, Y_i), i = 1, \ldots, n$, according to

$$Y_i \sim \text{n}(\alpha + \beta\xi_i, \sigma_\epsilon^2),$$

$$X_i \sim \text{n}(\xi_i, \sigma_\delta^2),$$

$$\xi_i \sim \text{n}(\xi, \sigma_\xi^2),$$

where the $\xi_i$s are independent and, given the $\xi_i$s, the $X_i$s and $Y_i$s are independent. As mentioned before, inference about $\alpha$ and $\beta$ will be made from the marginal distribution of $X_i$ and $Y_i$, that is, the distribution obtained by integrating out $\xi_i$. If we integrate out $\xi_i$, we obtain the marginal distribution of $(X_i, Y_i)$ (see Exercise 12.6):

(12.2.19)  $(X_i, Y_i) \sim$  bivariate normal$(\xi, \alpha + \beta\xi, \sigma_\delta^2 + \sigma_\xi^2, \sigma_\epsilon^2 + \beta^2\sigma_\xi^2, \beta\sigma_\xi^2).$

Notice the similarity of the correlation structure to that of the RCB ANOVA (see Miscellanea 11.5.3). There, conditional on blocks, the observations were uncorrelated, but unconditionally, there was correlation (the intraclass correlation). Here, the functional relationship model, which is conditional on the $\xi_i$s, has uncorrelated observations, but the structural relationship model, where we infer unconditional on the $\xi_i$s, has correlated observations. The $\xi_i$s are playing a role similar to blocks and the correlation that appears here is similar to the intraclass correlation. (In fact, it is identical to the intraclass correlation if $\beta = 1$ and $\sigma_\delta^2 = \sigma_\epsilon^2$.)

To proceed with likelihood estimation in this case, given observations $(\mathbf{x}, \mathbf{y}) = ((x_1, y_1), \ldots, (x_n, y_n))$, the likelihood function is that of a bivariate normal, as was encountered in Exercise 7.18. There, it was seen that the likelihood estimators in the bivariate normal could be found by equating sample quantities to population quantities. Hence, to find the MLEs of $\alpha$, $\beta$, $\xi$, $\sigma_\epsilon^2$, $\sigma_\delta^2$, and $\sigma_\xi^2$, we solve

$$\bar{y} = \hat{\alpha} + \hat{\beta}\hat{\xi},$$

$$\bar{x} = \hat{\xi},$$

(12.2.20)
$$\frac{1}{n}S_{yy} = \hat{\sigma}_\epsilon^2 + \hat{\beta}^2\hat{\sigma}_\xi^2,$$

$$\frac{1}{n}S_{xx} = \hat{\sigma}_\delta^2 + \hat{\sigma}_\xi^2,$$

$$\frac{1}{n}S_{xy} = \hat{\beta}\hat{\sigma}_\xi^2.$$

Note that we have five equations, but there are six unknowns, so the system is indeterminate. That is, the system of equations does not have a unique solution and there is no unique value of the parameter vector $(\alpha, \beta, \xi, \sigma_\epsilon^2, \sigma_\delta^2, \sigma_\xi^2)$ that maximizes the likelihood.

Before we go on, realize that the variances of $X_i$ and $Y_i$ here are different from the variances in the functional relationship model. There we were working conditional on $\xi_1, \ldots, \xi_n$, and here we are working marginally with respect to the $\xi_i$s. So, for example, in the functional relationship model we write $\text{Var} X_i = \sigma_\delta^2$ (where it is understood that this variance is conditional on $\xi_1, \ldots, \xi_n$), while in the structural model we write $\text{Var} X_i = \sigma_\delta^2 + \sigma_\xi^2$ (where it is understood that this variance is unconditional on $\xi_1, \ldots, \xi_n$). This should not be a source of confusion.

A solution to the equations in (12.2.20) implies a restriction on $\hat{\beta}$, a restriction that we have already encountered in the functional relationship case (see Exercise 12.4). From the above equations involving the variances and covariance, it is straightforward to deduce that

$$\hat{\sigma}_\delta^2 \geq 0 \quad \text{only if } S_{xx} \geq \tfrac{1}{\beta}S_{xy},$$

$$\hat{\sigma}_\epsilon^2 \geq 0 \quad \text{only if } S_{yy} \geq \hat{\beta}S_{xy},$$

which together imply that

$$\frac{|S_{xy}|}{S_{xx}} \leq |\hat{\beta}| \leq \frac{S_{yy}}{|S_{xy}|}.$$

(The bounds on $\hat{\beta}$ are established in Exercise 12.9.)

We now address the identifiability problem in the structural relationship case, a problem that can be expected since, in (12.2.19) we have more parameters than are needed to specify the distribution. To make the structural linear relationship model identifiable, we must make an assumption that reduces the number of parameters to five. It fortunately happens that the assumption about variances made for the functional relationship solves the identifiability problem here. Thus, we assume that $\sigma_\delta^2 = \lambda\sigma_\epsilon^2$, where $\lambda$ is known. This reduces the number of unknown parameters to five and makes the model identifiable. (See Exercise 12.8.) More restrictive assumptions, such as assuming that $\sigma_\delta^2$ is known, may lead to MLEs of variances that have the value 0. Kendall and Stuart (1979, Chapter 29) have a full discussion of this.

Once we assume that $\sigma_\delta^2 = \lambda\sigma_\epsilon^2$, the maximum likelihood estimates for $\hat{\alpha}$ and $\hat{\beta}$ in this model are the same as in the functional relationship model and are given by (12.2.16). The variance estimates are different, however, and are given by

$$\hat{\sigma}_\delta^2 = \frac{1}{n}\left(S_{xx} - \frac{S_{xy}}{\hat{\beta}}\right),$$

(12.2.21)
$$\hat{\sigma}_\epsilon^2 = \frac{\hat{\sigma}_\delta^2}{\lambda} = \frac{1}{n}(S_{yy} - \hat{\beta}S_{xy}),$$

$$\hat{\sigma}_\xi^2 = \frac{1}{n}\frac{S_{xy}}{\hat{\beta}}.$$

(Exercise 12.10 shows this and also explores the relationship between variance estimates here and in the functional model.) Note that, in contrast to what happened in the functional relationship model, these estimators are all consistent in the linear structural relationship model (when $\sigma_\delta^2 = \lambda\sigma_\epsilon^2$).

### 12.2.4 Confidence Sets

As might be expected, the construction of confidence sets in the EIV model is a difficult task. A complete treatment of the subject needs machinery that we have not developed. In particular, we will concentrate here only on confidence sets for the slope, $\beta$.

As a first attack, we could use the approximate likelihood method of Section 10.4.1 to construct approximate confidence intervals. In practice this is probably what is most often done and is not totally unreasonable. However, these approximate intervals cannot maintain a nominal $1-\alpha$ confidence level. In fact, results of Gleser and Hwang (1987) yield the rather unsettling result that any interval estimator of the slope whose length is *always finite* will have confidence coefficient equal to 0!

For definiteness, in the remainder of this section we will assume that we are in the structural relationship case of the EIV model. The confidence set results presented are valid in both the structural and functional cases and, in particular, the formulas remain the same. We continue to assume that $\sigma_\delta^2 = \lambda\sigma_\epsilon^2$, where $\lambda$ is known.

Gleser and Hwang (1987) identify the parameter

$$\tau^2 = \frac{\sigma_\xi^2}{\sigma_\delta^2}$$

as determining the amount of information potentially available in the data to determine the slope $\beta$. They show that, as $\tau^2 \to 0$, the coverage probability of any finite-length confidence interval on $\beta$ must also go to 0. To see why this is plausible, note that $\tau^2 = 0$ implies that the $\xi_i$s do not vary and it would be impossible to fit a unique straight line.

An approximate confidence interval for $\beta$ can be constructed by using the fact that the estimator

$$\hat{\sigma}_\beta^2 = \frac{(1 + \lambda\hat{\beta}^2)^2(S_{xx}S_{yy} - S_{xy}^2)}{(S_{xx} - \lambda S_{yy})^2 + 4\lambda S_{xy}^2}$$

is a consistent estimator of $\sigma_\beta^2$, the true variance of $\hat{\beta}$. Hence, using the CLT together with Slutsky's Theorem (see Section 5.5), we can show that the interval

$$\hat{\beta} - \frac{z_{\alpha/2}\hat{\sigma}_\beta}{\sqrt{n}} \leq \beta \leq \hat{\beta} + \frac{z_{\alpha/2}\hat{\sigma}_\beta}{\sqrt{n}}$$

is an approximate $1 - \alpha$ confidence interval for $\beta$. However, since it has finite length, it cannot maintain $1 - \alpha$ coverage for all parameter values.

Gleser (1987) considers a modification of this interval and reports the infimum of its coverage probabilities as a function of $\tau^2$. Gleser's modification, $C_G(\hat{\beta})$, is

$$(12.2.22) \qquad \hat{\beta} - \frac{t_{n-2,\alpha/2}\hat{\sigma}_\beta}{\sqrt{n-2}} \leq \beta \leq \hat{\beta} + \frac{t_{n-2,\alpha/2}\hat{\sigma}_\beta}{\sqrt{n-2}}.$$

Again using the CLT together with Slutsky's Theorem, we can show that this is an approximate $1 - \alpha$ confidence interval for $\beta$. Since this interval also has finite length, it also cannot maintain $1 - \alpha$ coverage for all parameter values. Gleser does some finite-sample numerical calculations and gives bounds on the infima of the coverage probabilities as a function of $\tau^2$. For reasonable values of $n$ ($\geq 10$), the coverage probability of a nominal 90% interval will be at least 80% if $\tau^2 \geq .25$. As $\tau^2$ or $n$ increases, this performance improves.

In contrast to $C_G(\hat{\beta})$ of (12.2.22), which has finite length but no guaranteed coverage probability, we now look at an exact confidence set that, as it must, has infinite length. The set, known as the Creasy–Williams confidence set, is due to Creasy (1956) and Williams (1959) and is based on the fact (see Exercise 12.11) that if $\sigma_\delta^2 = \lambda\sigma_\epsilon^2$, then

$$\text{Cov}(\beta\lambda Y_i + X_i, Y_i - \beta X_i) = 0.$$

Define $r_\lambda(\beta)$ to be the sample correlation coefficient between $\beta\lambda Y_i + X_i$ and $Y_i - \beta X_i$, that is,

$$r_\lambda(\beta) = \frac{\sum_{i=1}^n ((\beta\lambda y_i + x_i) - (\beta\lambda\bar{y} + \bar{x}))((y_i - \beta x_i) - (\bar{y} - \beta\bar{x}))}{\sqrt{\sum_{i=1}^n ((\beta\lambda y_i + x_i) - (\beta\lambda\bar{y} + \bar{x}))^2 \sum_{i=1}^n ((y_i - \beta x_i) - (\bar{y} - \beta\bar{x}))^2}}$$

$$(12.2.23) \qquad = \frac{\beta\lambda S_{yy} + (1 - \beta^2\lambda)S_{xy} - \beta S_{xx}}{\sqrt{(\beta^2\lambda^2 S_{yy} + 2\beta\lambda S_{xy} + S_{xx})(S_{yy} - 2\beta S_{xy} + \beta^2 S_{xx})}}.$$

Since $\beta\lambda Y_i + X_i$ and $Y_i - \beta X_i$ are bivariate normal with correlation 0, it follows (see Exercise 11.33) that

$$\frac{\sqrt{n-2}\, r_\lambda(\beta)}{\sqrt{1 - r_\lambda^2(\beta)}} \sim t_{n-2}$$

for any value of $\beta$. Thus, we have identified a pivotal quantity and we conclude that

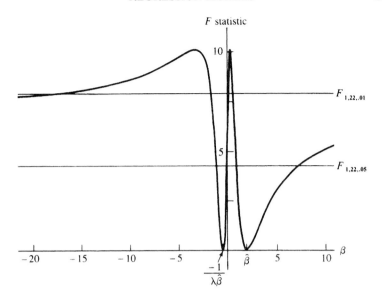

Figure 12.2.3. *F statistic defining Creasy–Williams confidence set, $\lambda = 1$*

the set

(12.2.24)
$$\left\{ \beta : \frac{(n-2)r_\lambda^2(\beta)}{1 - r_\lambda^2(\beta)} \leq F_{1, n-2, \alpha} \right\}$$

is a $1 - \alpha$ confidence set for $\beta$ (see Exercise 12.11).

Although this confidence set is a $1 - \alpha$ set, it suffers from defects similar to those of Fieller's intervals. The function describing the set (12.2.24) has two minima, where the function is 0. The confidence set can consist of two finite disjoint intervals, one finite and two infinite disjoint intervals, or the entire real line. For example, the graph of the $F$ statistic function for the data in Table 11.3.1 with $\lambda = 1$ is in Figure 12.2.3. The confidence set is all the $\beta$s where the function is less than or equal to $F_{1,22,\alpha}$. For $\alpha = .05$ and $F_{1,22,.05} = 4.30$, the confidence set is $[-1.13, -.14] \cup [.89, 7.38]$. For $\alpha = .01$ and $F_{1,22,.01} = 7.95$, the confidence set is $(-\infty, -18.18] \cup [-1.68, .06] \cup [.60, \infty)$.

Furthermore, for every value of $\beta$, $-r_\lambda(\beta) = r_\lambda(-1/(\lambda\beta))$ (see Exercise 12.12) so that if $\beta$ is in the confidence set, so is $-1/(\lambda\beta)$. Using this confidence set, we cannot distinguish $\beta$ from $-1/(\lambda\beta)$ and this confidence set always contains both positive and negative values. We can never determine the sign of the slope from this confidence set!

The confidence set given in (12.2.24) is not exactly the one discussed by Creasy (1956) but rather a modification. She was actually interested in estimating $\phi$, the angle that $\beta$ makes with the $x$-axis, that is, $\beta = \tan(\phi)$, and confidence sets there have fewer problems. Estimation of $\phi$ is perhaps more natural in EIV models (see, for example, Anderson 1976), but we seem to be more inclined to estimate $\alpha$ and $\beta$.

Most of the other standard statistical analyses that can be done in the ordinary linear regression case have analogues in EIV models. For example, we can test hypotheses about $\beta$ or estimate values of $\mathrm{E}\,Y_i$. More about these topics can be found in Fuller (1987) or Kendall and Stuart (1979, Chapter 29).

## 12.3 Logistic Regression

The conditional normal model of Section 11.3.3 is an example of a *generalized linear model* (GLM). A GLM describes a relationship between the mean of a response variable $Y$ and an independent variable $x$. But the relationship may be more complicated than the $\mathrm{E}\,Y_i = \alpha + \beta x_i$ of (11.3.2). Many different models can be expressed as GLMs. In this section, we will concentrate on a specific GLM, the logistic regression model.

### 12.3.1 The Model

A GLM consists of three components: the random component, the systematic component, and the link function.

(1) The response variables $Y_1, \ldots, Y_n$ are the *random component*. They are assumed to be independent random variables, each with a distribution from a specified exponential family. The $Y_i$s are not identically distributed, but they each have a distribution from the same family: binomial, Poisson, normal, etc.

(2) The *systematic component* is the model. It is the function of the predictor variable $x_i$, linear in the parameters, that is related to the mean of $Y_i$. So the systematic component could be $\alpha + \beta x_i$ or $\alpha + \beta/x_i$, for example. We will consider only $\alpha + \beta x_i$ here.

(3) Finally, the *link function* $g(\mu)$ links the two components by asserting that $g(\mu_i) = \alpha + \beta x_i$, where $\mu_i = \mathrm{E}\,Y_i$.

The conditional normal regression model of Section 11.3.3 is an example of a GLM. In that model, the responses $Y_i$s all have normal distributions. Of course, the normal family is an exponential family, which is the random component. The form of the regression function is $\alpha + \beta x_i$ in this model, which is the systematic component. Finally, the relationship $\mu_i = \mathrm{E}\,Y_i = \alpha + \beta x_i$ is assumed. This means the link function is $g(\mu) = \mu$. This simple link function is called the *identity link*.

Another very useful GLM is the *logistic regression model*. In this model, the responses $Y_1, \ldots, Y_n$ are independent and $Y_i \sim \text{Bernoulli}(\pi_i)$. (The Bernoulli family is an exponential family.) Recall, $\mathrm{E}\,Y_i = \pi_i = P(Y_i = 1)$. In this model, $\pi_i$ is assumed to be related to $x_i$ by

$$(12.3.1) \qquad\qquad \log\left(\frac{\pi_i}{1 - \pi_i}\right) = \alpha + \beta x_i.$$

The left-hand side is the log of the odds of success for $Y_i$. The model assumes this log-odds (or *logit*) is a linear function of the predictor $x$. The Bernoulli pmf can be written in exponential family form as

$$\pi^y (1 - \pi)^{1-y} = (1 - \pi) \exp\left\{ y \log\left(\frac{\pi_i}{1 - \pi_i}\right) \right\}.$$

The term $\log(\pi/(1-\pi))$ is the natural parameter of this exponential family, and in (12.3.1) the link function $g(\pi) = \log(\pi/(1-\pi))$ is used. When the natural parameter is used in this way, it is called the *canonical link*.

Equation (12.3.1) can be rewritten as

$$\pi_i = \frac{e^{\alpha + \beta x_i}}{1 + e^{\alpha + \beta x_i}}$$

or, more generally,

$$(12.3.2) \qquad\qquad \pi(x) = \frac{e^{\alpha + \beta x}}{1 + e^{\alpha + \beta x}}.$$

We see that $0 < \pi(x) < 1$, which seems appropriate because $\pi(x)$ is a probability. But, if it is possible that $\pi(x) = 0$ or $1$ for some $x$, then this model is not appropriate. If we examine $\pi(x)$ more closely, its derivative can be written

$$(12.3.3) \qquad\qquad \frac{d\,\pi(x)}{dx} = \beta\pi(x)(1 - \pi(x)).$$

As the term $\pi(x)(1 - \pi(x))$ is always positive, the derivative of $\pi(x)$ is positive, $0$, or negative according as $\beta$ is positive, $0$, or negative. If $\beta$ is positive, $\pi(x)$ is a strictly increasing function of $x$; if $\beta$ is negative, $\pi(x)$ is a strictly decreasing function of $x$; if $\beta = 0$, $\pi(x) = e^{\alpha}/(1+e^{\alpha})$ for all $x$. As in simple linear regression, if $\beta = 0$, there is no relationship between $\pi$ and $x$. Also, in a logistic regression model, $\pi(-\alpha/\beta) = 1/2$. A logistic regression function exhibits this kind of symmetry; for any $c$, $\pi((-\alpha/\beta)+c) = 1 - \pi((-\alpha/\beta) - c)$.

The parameters $\alpha$ and $\beta$ have meanings similar to those in simple linear regression. Setting $x = 0$ in (12.3.1) yields that $\alpha$ is the log-odds of success at $x = 0$. Evaluating (12.3.1) at $x$ and $x + 1$ yields, for any $x$,

$$\log\left(\frac{\pi(x+1)}{1 - \pi(x+1)}\right) - \log\left(\frac{\pi(x)}{1 - \pi(x)}\right) = \alpha + \beta(x+1) - \alpha - \beta(x) = \beta.$$

Thus, $\beta$ is the change in the log-odds of success corresponding to a one-unit increase in $x$. In simple linear regression, $\beta$ is the change in the mean of $Y$ corresponding to a one-unit increase in $x$. Exponentiating both sides of this equality yields

$$(12.3.4) \qquad\qquad e^{\beta} = \frac{\pi(x+1)/(1 - \pi(x+1))}{\pi(x)/(1 - \pi(x))}.$$

The right-hand side is the *odds ratio* comparing the odds of success at $x+1$ to the odds of success at $x$. (Recall that in Examples 5.5.19 and 5.5.22 we looked at estimating odds.) In a logistic regression model this ratio is constant as a function of $x$. Finally,

$$(12.3.5) \qquad\qquad \frac{\pi(x+1)}{1 - \pi(x+1)} = e^{\beta}\frac{\pi(x)}{1 - \pi(x)};$$

that is, $e^{\beta}$ is the multiplicative change in the odds of success corresponding to a one-unit increase in $x$.

Equation (12.3.2) suggests other ways of modeling the Bernoulli success probability $\pi(x)$ as a function of the predictor variable $x$. Recall that $F(w) = e^w/(1 + e^w)$ is the cdf of a logistic$(0, 1)$ distribution. In (12.3.2) we have assumed $\pi(x) = F(\alpha + \beta x)$. We can define other models for $\pi(x)$ by using other continuous cdfs. If $F(w)$ is the standard normal cdf, the model is called probit regression (see Exercise 12.17). If a Gumbel cdf is used, the link function is called the log-log link.

### 12.3.2 Estimation

In linear regression, where we use a model such as $Y_i = \alpha + \beta x_i + \varepsilon_i$, the technique of least squares was an option for calculating estimates of $\alpha$ and $\beta$. This is no longer the case here. In the model (12.3.1) with $Y_i \sim$ Bernoulli$(\pi_i)$, we no longer have a direct connection between $Y_i$ and $\alpha + \beta x_i$ (which is why we need a link function). Thus, least squares is no longer an option.

The estimation method that is most commonly used is maximum likelihood. In the general model we have $Y_i \sim$ Bernoulli$(\pi_i)$, where $\pi(x) = F(\alpha + \beta x)$. If we let $F_i = F(\alpha + \beta x_i)$, then the likelihood function is

$$L(\alpha, \beta | \mathbf{y}) = \prod_{i=1}^{n} \pi(x_i)^{y_i} (1 - \pi(x_i))^{1-y_i} = \prod_{i=1}^{n} F_i^{y_i} (1 - F_i)^{1-y_i}$$

with log likelihood

$$\log L(\alpha, \beta | \mathbf{y}) = \sum_{i=1}^{n} \left\{ \log(1 - F_i) + y_i \log \left( \frac{F_i}{1 - F_i} \right) \right\}.$$

We obtain the likelihood equations by differentiating the log likelihood with respect to $\alpha$ and $\beta$. Let $dF(w)/dw = f(w)$, the pdf corresponding to $F(w)$, and let $f_i = f(\alpha + \beta x_i)$. Then

$$\frac{\partial \log(1 - F_i)}{\partial \alpha} = -\frac{f_i}{1 - F_i} = -\frac{F_i f_i}{F_i(1 - F_i)}$$

and

(12.3.6) $$\frac{\partial}{\partial \alpha} \log \left( \frac{F_i}{1 - F_i} \right) = \frac{f_i}{F_i(1 - F_i)}.$$

Hence,

(12.3.7) $$\frac{\partial}{\partial \alpha} \log L(\alpha, \beta | \mathbf{y}) = \sum_{i=1}^{n} (y_i - F_i) \frac{f_i}{F_i(1 - F_i)}.$$

A similar calculation yields

(12.3.8) $$\frac{\partial}{\partial \beta} \log L(\alpha, \beta | \mathbf{y}) = \sum_{i=1}^{n} (y_i - F_i) \frac{f_i}{F_i(1 - F_i)} x_i.$$

For logistic regression, with $F(w) = e^w/(1 + e^w)$, $f_i/[F_i(1 - F_i)] = 1$, and (12.3.7) and (12.3.8) are somewhat simpler.

The MLEs are obtained by setting (12.3.7) and (12.3.8) equal to 0 and solving for $\alpha$ and $\beta$. These are nonlinear equations in $\alpha$ and $\beta$ and must be solved numerically. (This will be discussed later.) For logistic and probit regression, the log likelihood is strictly concave. Hence, if the likelihood equations have a solution, it is unique and it is the MLE. However, for some extreme data the likelihood equations do not have a solution. The maximum of the likelihood occurs in some limit as the parameters go to $\pm\infty$. See Exercise 12.16 for an example. This is because the logistic model assumes $0 < \pi(x) < 1$, but, for certain data sets, the maximum of the logistic likelihood occurs in a limit with $\pi(x) = 0$ or 1. The probability of obtaining such data converges to 0 if the logistic model is true.

**Example 12.3.1 (Challenger data)** A by now infamous data set is that of space shuttle O-ring failures, which have been linked to temperature. The data in Table 12.3.1 give the temperatures at takeoff and whether or not an O-ring failed.

Solving (12.3.6) and (12.3.7) using $F(\alpha + \beta x_i) = e^{\alpha+\beta x_i}/(1 + e^{\alpha+\beta x_i})$ yields MLEs $\hat{\alpha} = 15.043$ and $\hat{\beta} = -.232$. Figure 12.3.1 shows the fitted curve along with the data.

The space shuttle Challenger exploded during takeoff, killing the seven astronauts aboard. The explosion was the result of an O-ring failure, believed to be caused by the unusually cold weather ($31°$ F) at the time of launch. The MLE of the probability of O-ring failure at $31°$ is .9996. (See Dalal, Fowlkes, and Hoadley 1989 for the full story.)

Table 12.3.1. *Temperature at flight time (°F) and failure of O-rings (1 = failure, 0 = success)*

| Flight no. | 14 | 9 | 23 | 10 | 1 | 5 | 13 | 15 | 4 | 3 | 8 | 17 |
|---|---|---|---|---|---|---|---|---|---|---|---|---|
| Failure | 1 | 1 | 1 | 1 | 0 | 0 | 0 | 0 | 0 | 0 | 0 | 0 |
| Temp. | 53 | 57 | 58 | 63 | 66 | 67 | 67 | 67 | 68 | 69 | 70 | 70 |
| Flight no. | 2 | 11 | 6 | 7 | 16 | 21 | 19 | 22 | 12 | 20 | 18 | |
| Failure | 1 | 1 | 0 | 0 | 0 | 1 | 0 | 0 | 0 | 0 | 0 | |
| Temp. | 70 | 70 | 72 | 73 | 75 | 75 | 76 | 76 | 78 | 79 | 81 | |

$\parallel$

We have, thus far, assumed that at each value of $x_i$, we observe the result of only one Bernoulli trial. Although this often the case, there are many situations in which there are multiple Bernoulli observations at each value of $x$. We now revisit the likelihood solution in this more general case.

Suppose there are $J$ different values of the predictor $x$ in the data set $x_1, \ldots, x_J$. Let $n_j$ denote the number of Bernoulli observations at $x_j$, and let $Y_j^*$ denote the number of successes in these $n_j$ observations. Thus, $Y_j^* \sim$ binomial$(n_j, \pi(x_j))$. Then the likelihood is

$$L(\alpha, \beta|\mathbf{y}^*) = \prod_{j=1}^{J} \pi(x_j)^{y_j^*}(1 - \pi(x_j))^{n_j - y_j^*} = \prod_{j=1}^{J} F_j^{y_j^*}(1 - F_j)^{n_j - y_j^*},$$

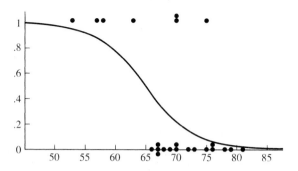

Figure 12.3.1. *The data of Table 12.3.1 along with the fitted logistic curve*

and the likelihood equations are

$$0 = \sum_{j=1}^{J} (y_j^* - n_j F_j) \frac{f_j}{F_j (1 - F_j)}$$

$$0 = \sum_{j=1}^{J} (y_j^* - n_j F_j) \frac{f_j}{F_j (1 - F_j)} x_j.$$

As we have estimated the parameters of the logistic regression using maximum likelihood, we can next get approximate variances using MLE asymptotics. However, we have to proceed in a more general way. In Section 10.1.3 we saw how to approximate the variance of the MLE using the information number. We use the same strategy here, but as there are two parameters, there is an *information matrix* given by the $2 \times 2$ matrix

$$(12.3.9) \quad I(\theta_1, \theta_2) = \begin{pmatrix} -\frac{\partial^2}{\partial \theta_1^2} \log L(\theta_1, \theta_2 | \mathbf{y}) & -\frac{\partial^2}{\partial \theta_1 \partial \theta_2} \log L(\theta_1, \theta_2 | \mathbf{y}) \\ -\frac{\partial^2}{\partial \theta_1 \partial \theta_2} \log L(\theta_1, \theta_2 | \mathbf{y}) & -\frac{\partial^2}{\partial \theta_2^2} \log L(\theta_1, \theta_2 | \mathbf{y}) \end{pmatrix}$$

For logistic regression, the information matrix is given by

$$(12.3.10) \quad I(\alpha, \beta) = \begin{pmatrix} \sum_{j=1}^{J} n_j F_j (1 - F_j) & \sum_{j=1}^{J} x_j n_j F_j (1 - F_j) \\ \sum_{j=1}^{J} x_j n_j F_j (1 - F_j) & \sum_{j=1}^{J} x_j^2 n_j F_j (1 - F_j) \end{pmatrix},$$

and the variances of the MLEs $\hat{\alpha}$ and $\hat{\beta}$ are usually approximated using this matrix. Note that the elements of $I(\alpha, \beta)$ do not depend on $Y_1^*, \ldots, Y_J^*$. Thus, the observed information is the same as the information in this case.

In Section 10.1.3, we used the approximation (10.1.7), namely, $\mathrm{Var}(h(\hat{\theta}) | \theta) \approx [h'(\hat{\theta})]^2 / I(\hat{\theta})$, where $I(\cdot)$ was the information number. Here, we cannot do exactly the same with the information matrix, but rather we need to get the inverse of the matrix and use the inverse elements to approximate the variance. Recall that the inverse of a $2 \times 2$ matrix is given by

$$\begin{pmatrix} a & b \\ c & d \end{pmatrix}^{-1} = \frac{1}{ad - bc} \begin{pmatrix} d & -b \\ -c & a \end{pmatrix}.$$

To obtain the approximate variances, the MLEs are used to estimate the parameters in the matrix (12.3.10), and the estimates of the variances, $[\text{se}(\hat{\alpha})]^2$ and $[\text{se}(\hat{\beta})]^2$, are the diagonal elements of the inverse of $I(\hat{\alpha}, \hat{\beta})$. (The notation $\text{se}(\hat{\alpha})$ stands for *standard error of* $(\hat{\alpha})$.)

**Example 12.3.2 (Challenger data continued)** The estimated information matrix of the estimates from the Challenger data is given by

$$I(\hat{\alpha}, \hat{\beta}) = \begin{pmatrix} \sum_{j=1}^{J} \hat{F}_j(1 - \hat{F}_j) & \sum_{j=1}^{J} x_j \hat{F}_j(1 - \hat{F}_j) \\ \sum_{j=1}^{J} x_j \hat{F}_j(1 - \hat{F}_j) & \sum_{j=1}^{J} x_j^2 \hat{F}_j(1 - \hat{F}_j j) \end{pmatrix} = \begin{pmatrix} 3.15 & 214.75 \\ 214.75 & 14728.5 \end{pmatrix},$$

where $\hat{F}_j = e^{\hat{\alpha} + \hat{\beta} x_j}/(1 + e^{\hat{\alpha} + \hat{\beta} x_j})$ and has inverse

$$I(\hat{\alpha}, \hat{\beta})^{-1} = \begin{pmatrix} 54.44 & -.80 \\ -.80 & .012 \end{pmatrix}.$$

The likelihood asymptotics tell use that, for example, $\hat{\beta} \pm z_{\alpha/2}\text{se}(\hat{\beta})$ is, for large samples, an approximate $100(1 - \alpha)\%$ confidence interval for $\beta$. So for the Challenger data we have a 95% confidence interval of

$$\beta \in -.232 \pm 1.96 \times \sqrt{.012} \quad \Rightarrow \quad -.447 \le \beta \le -.017,$$

supporting the conclusion that $\beta < 0$.                                                    ‖

It is, perhaps, most common in this model to test the hypothesis $H_0 : \beta = 0$, because, as in simple linear regression, this hypothesis states there is no relationship between the predictor and response variables. The Wald test statistic, $Z = \hat{\beta}/\text{se}(\hat{\beta})$, has approximately a standard normal distribution if $H_0$ is true and the sample size is large. Thus, $H_0$ can be rejected if $|Z| \ge z_{\alpha/2}$. Alternatively, $H_0$ can be tested with the log LRT statistic

$$-2 \log \lambda(\mathbf{y}^*) = 2[\log L(\hat{\alpha}, \hat{\beta}|\mathbf{y}^*) - L(\hat{\alpha}_0, 0|\mathbf{y}^*)],$$

where $\hat{\alpha}_0$ is the MLE of $\alpha$ assuming $\beta = 0$. With standard binomial arguments (see Exercise 12.20), it can be shown that $\hat{\alpha}_0 = \sum_{i=1}^{n} y_i/n = \sum_{j=1}^{J} y_j^*/\sum_{j=1}^{J} n_j$. Therefore, under $H_0$, $-2 \log \lambda$ has an approximate $\chi_1^2$ distribution, and we can reject $H_0$ at level $\alpha$ if $-2 \log \lambda \ge \chi_{1,\alpha}^2$.

We have introduced only the simplest logistic regression and generalized linear models. Much more can be found in standard texts such as Agresti (1990).

## 12.4 Robust Regression

As in Section 10.2, we now want to take a look at the performance of our procedures if the underlying model is not the correct one, and we will take a look at some robust alternatives to least squares estimation, starting with a comparison analogous to the mean/median comparison.

Recall that, when observing $x_1, x_2, \ldots, x_n$, we can define the mean and the median as minimizers of the following quantities:

$$\text{mean: } \min_m \left\{ \sum_{i=1}^n (x_i - m)^2 \right\}, \qquad \text{median: } \min_m \left\{ \sum_{i=1}^n |x_i - m| \right\}.$$

For simple linear regression, observing $(y_1, x_1), (y_2, x_2), \ldots, (y_n, x_n)$, we know that least squares regression estimates satisfy

$$\text{least squares: } \min_{a,b} \left\{ \sum_{i=1}^n [y_i - (a + bx_i)]^2 \right\},$$

and we analogously define *least absolute deviation (LAD)* regression estimates by

$$\text{least absolute deviation: } \min_{a,b} \left\{ \sum_{i=1}^n |y_i - (a + bx_i)| \right\}.$$

(The LAD estimates may not be unique. See Exercise 12.25.)

Thus, we see that the least squares estimators are the regression analogues of the sample mean. This should make us wonder about their robustness performance (as listed in items (1)–(3) of Section 10.2).

**Example 12.4.1 (Robustness of least squares estimates)** If we observe $(y_1, x_1)$, $(y_2, x_2), \ldots, (y_n, x_n)$, where

$$Y_i = \alpha + \beta x_i + \varepsilon_i,$$

and the $\varepsilon_i$ are uncorrelated with $\mathrm{E}\,\varepsilon_i = 0$ and $\mathrm{Var}\,\varepsilon_i = \sigma^2$, we know that the least squares estimator $b$ with variance $\sigma^2 / \sum (x_i - \bar{x})^2$ is the BLUE of $\beta$, satisfying (1) of Section 10.2.

To investigate how $b$ performs for small deviations, we assume that

$$\mathrm{Var}(\varepsilon_i) = \begin{cases} \sigma^2 & \text{with probability } 1 - \delta \\ \tau^2 & \text{with probability } \delta. \end{cases}$$

Writing $b = \sum d_i Y_i$, where $d_i = (x_i - \bar{x}) / \sum (x_i - \bar{x})^2$, we now have

$$\mathrm{Var}(b) = \sum_{i=1}^n d_i^2 \, \mathrm{Var}(\varepsilon_i) = \frac{(1 - \delta)\sigma^2 + \delta\tau^2}{\sum_{i=1}^n (x_i - \bar{x})^2}.$$

This shows that, as with the sample mean, for small perturbations $b$ performs pretty well. (But we could, of course, blow things up by contaminating with a Cauchy pdf,

Table 12.4.1. *Values of $CO_2$ and $O_2$ in the pouches of 23 potoroos (McPherson 1990)*

| Animal | 1 | 2 | 3 | 4 | 5 | 6 | 7 | 8 |
|--------|-----|------|------|------|------|-----|-----|------|
| % $O_2$ | 20 | 19.6 | 19.6 | 19.4 | 18.4 | 19 | 19 | 18.3 |
| % $CO_2$ | 1 | 1.2 | 1.1 | 1.4 | 2.3 | 1.7 | 1.7 | 2.4 |

| Animal | 9 | 10 | 11 | 12 | 13 | 14 | 15 | 16 |
|--------|------|------|------|------|------|------|-----|------|
| % $O_2$ | 18.2 | 18.6 | 19.2 | 18.2 | 18.7 | 18.5 | 18 | 17.4 |
| % $CO_2$ | 2.1 | 2.1 | 1.2 | 2.3 | 1.9 | 2.4 | 2.6 | 2.9 |

| Animal | 17 | 18 | 19 | 20 | 21 | 22 | 23 |
|--------|------|------|------|------|------|------|------|
| % $O_2$ | 16.5 | 17.2 | 17.3 | 17.8 | 17.3 | 18.4 | 16.9 |
| % $CO_2$ | 4.0 | 3.3 | 3.0 | 3.4 | 2.9 | 1.9 | 3.9 |

for example.) The behavior of the least squares intercept $a$ is similar (see Exercise 12.22). See also Exercise 12.23 to see how bias contamination affects things. ‖

We next look at the effect of a "catastrophic" observation and compare least squares to its median-analogous alternative, least absolute deviation regression.

**Example 12.4.2 (Catastrophic observations)** McPherson (1990) describes an experiment in which the levels of carbon dioxide ($CO_2$) and oxygen ($O_2$) were measured in the pouches of 24 *potoroos* (a marsupial). Interest is in the regression of $CO_2$ on $O_2$, where the experimenter expects a slope of $-1$. The data for 23 animals (one had missing values) are given in Table 12.4.1. For the original data, the least squares and LAD lines are quite close:

$$\text{least squares: } y = 18.67 - .89x,$$

$$\text{least absolute deviation: } y = 18.59 - .89x.$$

However, an aberrant observation can upset least squares. When entering the data the $O_2$ value of 18 on Animal 15 was incorrectly entered as 10 (we really did this). For this new (incorrect) data set we have

$$\text{least squares: } y = 6.41 - .23x,$$

$$\text{least absolute deviation: } y = 15.95 - .75x,$$

showing that the aberrant observation had much less of an effect on LAD. See Figure 12.4.1 for a display of the regression lines.

These calculations illustrate the resistance of LAD, as opposed to least squares. Since we have the mean/median analogy, we can surmise that this behavior is reflected in breakdown values, which are 0% for least squares and 50% for LAD. ‖

However, the mean/median analogy continues. Although the LAD estimator is robust to catastrophic observations, it loses much in efficiency with respect to the least squares estimator (see also Exercise 12.25).

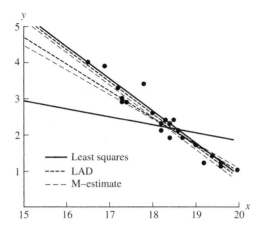

Figure 12.4.1. *Least squares, LAD, and M-estimate fits for the data of Table 12.4.1, for both the original data and the data with* $(18, 2.6)$ *mistyped as* $(10, 2.6)$*. The LAD and M-estimate lines are quite similar, while the least squares line reacts to the changed data.*

**Example 12.4.3 (Asymptotic normality of the LAD estimator)** We adapt the argument leading to (10.2.6) to derive the asymptotic distribution of the LAD estimator. Also, to simplify things, we consider only the model

$$Y_i = \beta x_i + \varepsilon_i,$$

that is, we set $\alpha = 0$. (This avoids having to deal with a bivariate limiting distribution.)

In the terminology of M-estimators, the LAD estimator is obtained by minimizing

$$\sum_{i=1}^n \rho(y_i - \beta x_i) = \sum_{i=1}^n |y_i - \beta x_i|$$

(12.4.1)
$$= \sum_{i=1}^n (y_i - \beta x_i) I(y_i > \beta x_i) - (y_i - \beta x_i) I(y_i < \beta x_i).$$

We then calculate $\psi = \rho'$ and solve $\sum_i \psi(y_i - \beta x_i) = 0$ for $\beta$, where

$$\psi(y_i - \beta x_i) = x_i I(y_i > \beta x_i) - x_i I(y_i < \beta x_i).$$

If $\hat{\beta}_L$ is the solution, expand $\psi$ in a Taylor series around $\beta$:

$$\sum_{i=1}^n \psi(y_i - \hat{\beta}_L x_i) = \sum_{i=1}^n \psi(y_i - \beta x_i) + (\hat{\beta}_L - \beta) \frac{d}{d\hat{\beta}_L} \sum_{i=1}^n \psi(y_i - \hat{\beta}_L x_i) \bigg|_{\hat{\beta}_L = \beta} + \cdots.$$

Although the left-hand side of the equation is not equal to 0, we assume that it approaches 0 as $n \to \infty$ (see Exercise 12.27). Then rearranging we obtain

(12.4.2)
$$\sqrt{n}(\hat{\beta}_L - \beta) = \frac{\frac{-1}{\sqrt{n}} \sum_{i=1}^{n} \psi(y_i - \beta x_i)}{\frac{1}{n} \frac{d}{d\hat{\beta}_L} \sum_{i=1}^{n} \psi(y_i - \hat{\beta}_L x_i)\big|_{\hat{\beta}_L = \beta}}.$$

First look at the numerator. As $E_\beta \psi(Y_i - \hat{\beta}_L x_i) = 0$ and $\mathrm{Var}\, \psi(Y_i - \hat{\beta}_L x_i) = x_i^2$, it follows that

(12.4.3)
$$\frac{-1}{\sqrt{n}} \sum_{i=1}^{n} \psi(Y_i - \hat{\beta} x_i) = \sqrt{n} \left[ \frac{-1}{n} \sum_{i=1}^{n} \psi(Y_i - \hat{\beta} x_i) \right] \to \mathrm{n}\left(0, \sigma_x^2\right),$$

where $\sigma_x^2 = \lim_{n \to \infty} \frac{1}{n} \sum_{i=1}^{n} x_i^2$. Turning to the denominator, we must be a bit careful as $\psi$ has points of nondifferentiability. We therefore first apply the Law of Large Numbers before differentiating, and use the approximation

(12.4.4)
$$\frac{1}{n} \frac{d}{d\beta_0} \sum_{i=1}^{n} \psi(y_i - \beta_0 x_i) \approx \frac{1}{n} \sum_{i=1}^{n} \frac{d}{d\beta_0} E_\beta[\psi(Y_i - \beta_0 x_i)]$$
$$= \frac{1}{n} \sum_{i=1}^{n} \frac{d}{d\beta_0} [x_i P_\beta(Y_i > \beta_0 x_i) - x_i P_\beta(Y_i < \beta_0 x_i)]$$
$$= \frac{1}{n} \sum_{i=1}^{n} x_i^2 f(\beta_0 x_i - \beta x_i) + x_i^2 f(\beta_0 x_i - \beta x_i).$$

If we now evaluate the derivative at $\beta_0 = \beta$, we have

$$\frac{1}{n} \frac{d}{d\beta_0} \sum_{i=1}^{n} \psi(y_i - \beta_0 x_i)\bigg|_{\beta_0 = \beta} \approx 2f(0) \frac{1}{n} \sum_{i=1}^{n} x_i^2,$$

and putting this together with (12.4.2) and (12.4.3), we have

(12.4.5)
$$\sqrt{n}(\hat{\beta}_L - \beta) \to \mathrm{n}\left(0, \frac{1}{4f(0)^2 \sigma_x^2}\right).$$

Finally, for the case of $\alpha = 0$, the least squares estimator is $\hat{\beta} = \sum_{i=1}^{n} x_i y_i / \sum_{i=1}^{n} x_i^2$ and satisfies

$$\sqrt{n}(\hat{\beta} - \beta) \to \mathrm{n}\left(0, \frac{1}{\sigma_x^2},\right)$$

so that the asymptotic relative efficiency of $\hat{\beta}_L$ to $\hat{\beta}$ is

$$\mathrm{ARE}(\hat{\beta}_L, \hat{\beta}) = \frac{1/\sigma_x^2}{1/(4f(0)^2 \sigma_x^2)} = 4f(0)^2,$$

whose values are the same as those given in Table 10.2.1, comparing the median to the mean. Thus, for normal errors the ARE of the LAD estimator to least squares is only 64%, showing that the LAD estimator gives up a good bit of efficiency with respect to least squares.                                                                ‖

So we are in the same situation that we encountered in Section 10.2.1. The LAD alternative to least squares seems to lose too much in efficiency if the errors are truly normal. The compromise, once again, is an M-estimator. We can construct one by minimizing a function analogous to (10.2.2), which would be $\sum_i \rho_i(\alpha, \beta)$, where

$$(12.4.6) \qquad \rho_i(\alpha, \beta) = \begin{cases} \frac{1}{2}(y_i - \alpha - \beta x_i)^2 & \text{if } |y_i - \alpha - \beta x_i| \leq k \\ k|y_i - \alpha - \beta x_i| - \frac{1}{2}k^2 & \text{if } |y_i - \alpha - \beta x_i| > k, \end{cases}$$

where $k$ is a tuning parameter.

**Example 12.4.4 (Regression M-estimator)**   Using the function (12.4.6) with $k = 1.5\sigma$, we fit the M-estimators of $\alpha$ and $\beta$ for the data of Table 12.4.1. The results were

$$\text{M-estimate for original data: } y = 18.5 - .89x$$

$$\text{M-estimate for mistyped data: } y = 14.67 - .68x,$$

where we estimated $\sigma$ by .23, the standard deviation of the residuals from the least squares fit.

Thus we see that the M-estimate is somewhat more resistant than the least squares line, behaving more like the LAD fit when there are outliers.                    ‖

As in Section 10.2, we expect the ARE of the M-estimator to be better than that of the LAD. This is the case, however, the calculations become very involved (even more so than for the LAD) so we will not give the details here. Huber (1981, Chapter 7) gives a detailed treatment of M-estimator asymptotics; see also Portnoy (1987). We content ourselves with an evaluation of the M-estimator through a small simulation study, reproducing a table like Table 10.2.3.

**Example 12.4.5 (Simulation of regression AREs)**   For the model $Y_i = \alpha + \beta x_i + \varepsilon_i$, $i = 1, 2, \ldots, 5$, we take the $x_i$s to be $(-2, -1, 0, 1, 2)$, $\alpha = 0$, and $\beta = 1$. We generate $\varepsilon_i$ from normals, double exponentials, and logistic distributions and calculate the variance of the least squares, LAD, and M-estimator. These are presented in the following table.

*Regression M-estimator AREs, $k = 1.5$ (based on 10,000 simulations)*

|                    | Normal | Logistic | Double exponential |
|--------------------|--------|----------|--------------------|
| vs. least squares  | 0.98   | 1.03     | 1.07               |
| vs. LAD            | 1.39   | 1.27     | 1.14               |

The M-estimator variance is similar to that of least squares for all three distributions and is a uniform improvement over LAD. The dominance of the M-estimator over LAD is more striking than that of the Huber estimator over the median (as given in Table 10.2.3).                    ‖

## 12.5 Exercises _____

**12.1** Verify the expressions in (12.2.7). (*Hint:* Use the Pythagorean Theorem.)

**12.2** Show that the extrema of

$$f(b) = \frac{1}{1 + b^2} \left[ S_{yy} - 2b S_{xy} + b^2 S_{xx} \right]$$

are given by

$$b = \frac{-(S_{xx} - S_{yy}) \pm \sqrt{(S_{xx} - S_{yy})^2 + 4S_{xy}^2}}{2S_{xy}}.$$

Show that the "+" solution gives the minimum of $f(b)$.

**12.3** In maximizing the likelihood (12.2.13), we first minimized, for each value of $\alpha$, $\beta$, and $\sigma_\delta^2$, the function

$$f(\xi_1, \ldots, \xi_n) = \sum_{i=1}^{n} \left( (x_i - \xi_i)^2 + \lambda(y_i - (\alpha + \beta\xi_i))^2 \right)$$

with respect to $\xi_1, \ldots, \xi_n$.

(a) Prove that this function is minimized at

$$\xi_i^* = \frac{x_i + \lambda\beta(y_i - \alpha)}{1 + \lambda\beta^2}.$$

(b) Show that the function

$$D_\lambda \left( (x, y), (\xi, \alpha + \beta\xi) \right) = (x - \xi)^2 + \lambda(y - (\alpha + \beta\xi))^2$$

defines a *metric* between the points $(x, y)$ and $(\xi, \alpha + \beta\xi)$. A *metric* is a distance measure, a function $D$ that measures the distance between two points $A$ and $B$. A metric satisfies the following four properties:
  i. $D(A, A) = 0$.
  ii. $D(A, B) > 0$ if $A \neq B$.
  iii. $D(A, B) = D(B, A)$ (reflexive).
  iv. $D(A, B) \leq D(A, C) + D(C, B)$ (triangle inequality).

**12.4** Consider the MLE of the slope in the EIV model

$$\hat{\beta}(\lambda) = \frac{-(S_{xx} - \lambda S_{yy}) + \sqrt{(S_{xx} - \lambda S_{yy})^2 + 4\lambda S_{xy}^2}}{2\lambda S_{xy}},$$

where $\lambda = \sigma_\delta^2 / \sigma_\epsilon^2$ is assumed known.

(a) Show that $\lim_{\lambda \to 0} \hat{\beta}(\lambda) = S_{xy}/S_{xx}$, the slope of the ordinary regression of $y$ on $x$.

(b) Show that $\lim_{\lambda \to \infty} \hat{\beta}(\lambda) = S_{yy}/S_{xy}$, the reciprocal of the slope of the ordinary regression of $x$ on $y$.

(c) Show that $\hat{\beta}(\lambda)$ is, in fact, monotone in $\lambda$ and is increasing if $S_{xy} > 0$ and decreasing if $S_{xy} < 0$.

(d) Show that the orthogonal least squares line ($\lambda = 1$) is always between the lines given by the ordinary regressions of $y$ on $x$ and of $x$ on $y$.

(e) The following data were collected in a study to examine the relationship between brain weight and body weight in a number of animal species.

| Species | Body weight (kg) $(x)$ | Brain weight (g) $(y)$ |
|---|---|---|
| Arctic fox | 3.385 | 44.50 |
| Owl monkey | .480 | 15.50 |
| Mountain beaver | 1.350 | 8.10 |
| Guinea pig | 1.040 | 5.50 |
| Chinchilla | .425 | 6.40 |
| Ground squirrel | .101 | 4.00 |
| Tree hyrax | 2.000 | 12.30 |
| Big brown bat | .023 | .30 |

Calculate the MLE of the slope assuming the EIV model. Also, calculate the least squares slopes of the regressions of $y$ on $x$ and of $x$ on $y$, and show how these quantities bound the MLE.

**12.5** In the EIV functional relationship model, where $\lambda = \sigma_\delta^2/\sigma_\epsilon^2$ is assumed known, show that the MLE of $\sigma_\delta^2$ is given by (12.2.18).

**12.6** Show that in the linear structural relationship model (12.2.6), if we integrate out $\xi_i$, the marginal distribution of $(X_i, Y_i)$ is given by (12.2.19).

**12.7** Consider a linear structural relationship model where we assume that $\xi_i$ has an improper distribution, $\xi_i \sim \text{uniform}(-\infty, \infty)$.

(a) Show that for each $i$,

$$\int_{-\infty}^{\infty} \frac{1}{(2\pi)} \frac{1}{\sigma_\delta \sigma_\epsilon} \exp\left[-\left(\frac{(x_i - \xi_i)^2}{2\sigma_\delta^2}\right)\right] \exp\left[-\left(\frac{(y_i - (\alpha + \beta\xi_i))^2}{2\sigma_\epsilon^2}\right)\right] d\xi_i$$

$$= \frac{1}{\sqrt{2\pi}} \frac{1}{\sqrt{\beta^2\sigma_\delta^2 + \sigma_\epsilon^2}} \exp\left[-\frac{1}{2}\frac{(y_i - (\alpha + \beta x_i))^2}{\beta^2\sigma_\delta^2 + \sigma_\epsilon^2}\right].$$

(Completing the square in the exponential makes the integration easy.)

(b) The result of the integration in part (a) looks like a pdf, and if we consider it a pdf of $Y$ conditional on $X$, then we seem to have a linear relationship between $X$ and $Y$. Thus, it is sometimes said that this "limiting case" of the structural relationship leads to simple linear regression and ordinary least squares. Explain why this interpretation of the above function is wrong.

**12.8** Verify the nonidentifiability problems in the structural relationship model in the following ways.

(a) Produce two different sets of parameters that give the same marginal distribution to $(X_i, Y_i)$.

(b) Show that there are at least two distinct parameter vectors that yield the same solution to the equations given in (12.2.20).

**12.9** In the structural relationship model, the solution to the equations in (12.2.20) implies a restriction on $\hat{\beta}$, the same restriction seen in the functional relationship case (see Exercise 12.4).

(a) Show that in (12.2.20), the MLE of $\sigma_\delta^2$ is nonnegative only if $S_{xx} \geq (1/\hat{\beta})S_{xy}$. Also, the MLE of $\sigma_\epsilon^2$ is nonnegative only if $S_{yy} \geq \hat{\beta}S_{xy}$.

(b) Show that the restrictions in part (a), together with the rest of the equations in (12.2.20), imply that

$$\frac{|S_{xy}|}{S_{xx}} \leq |\hat{\beta}| \leq \frac{S_{yy}}{|S_{xy}|}.$$

**12.10** (a) Derive the MLEs for $(\alpha, \beta, \sigma_\epsilon^2, \sigma_\delta^2, \sigma_\xi^2)$ in the structural relationship model by solving the equations (12.2.20) under the assumption that $\sigma_\delta^2 = \lambda \sigma_\epsilon^2$.

(b) Calculate the MLEs for $(\alpha, \beta, \sigma_\epsilon^2, \sigma_\delta^2, \sigma_\xi^2)$ for the data of Exercise 12.4 by assuming the structural relationship model holds and that $\sigma_\delta^2 = \lambda \sigma_\epsilon^2$.

(c) Verify the relationship between variance estimates in the functional and structural relationship models. In particular, show that

$$\widehat{\text{Var}}\, X_i(\text{structural}) = 2\widehat{\text{Var}}\, X_i(\text{functional}).$$

That is, verify

$$\left(S_{xx} - \frac{S_{xy}}{\hat{\beta}}\right) = \frac{\lambda}{1 + \lambda \hat{\beta}^2} \sum_{i=1}^n \left(y_i - (\hat{\alpha} + \hat{\beta} x_i)\right)^2.$$

(d) Verify the following equality, which is implicit in the MLE variance estimates given in (12.2.21). Show that

$$S_{xx} - \frac{S_{xy}}{\hat{\beta}} = \lambda(S_{yy} - \hat{\beta} S_{xy}).$$

**12.11** (a) Show that for random variables $X$ and $Y$ and constants $a$, $b$, $c$, $d$,

$$\text{Cov}(aY + bX, cY + dX) = ac\text{Var}\, Y + (bc + ad)\text{Cov}(X, Y) + bd\text{Var}\, X.$$

(b) Use the result in part (a) to verify that in the structural relationship model with $\sigma_\delta^2 = \lambda \sigma_\epsilon^2$,

$$\text{Cov}(\beta \lambda Y_i + X_i, Y_i - \beta X_i) = 0,$$

the identity on which the Creasy–Williams confidence set is based.

(c) Use the results of part (b) to show that

$$\frac{\sqrt{(n-2)}\, r_\lambda(\beta)}{\sqrt{1 - r_\lambda^2(\beta)}} \sim t_{n-2}$$

for any value of $\beta$, where $r_\lambda(\beta)$ is given in (12.2.23). Also, show that the confidence set defined in (12.2.24) has constant coverage probability equal to $1 - \alpha$.

**12.12** Verify the following facts about $\hat{\beta}$ (the MLE of $\beta$ when we assume $\sigma_\delta^2 = \lambda \sigma_\epsilon^2$), $r_\lambda(\beta)$ of (12.2.23), and $C_\lambda(\hat{\beta})$, the Creasy–Williams confidence set of (12.2.24).

(a) $\hat{\beta}$ and $-1/(\lambda \hat{\beta})$ are the two roots of the quadratic equation defining the zeros of the first derivative of the likelihood function (12.2.14).

(b) $r_\lambda(\beta) = -r_\lambda(-1/(\lambda \beta))$ for every $\beta$.

(c) If $\beta \in C_\lambda(\hat{\beta})$, then $-1/(\lambda \beta) \in C_\lambda(\hat{\beta})$.

**12.13** There is an interesting connection between the Creasy–Williams confidence set of (12.2.24) and the interval $C_G(\hat{\beta})$ of (12.2.22).

(a) Show that

$$C_G(\hat{\beta}) = \left\{ \beta : \frac{(\beta - \hat{\beta})^2}{\hat{\sigma}_{\hat{\beta}}^2/(n-2)} \leq F_{1,n-2,\alpha} \right\},$$

where $\hat{\beta}$ is the MLE of $\beta$ and $\hat{\sigma}_{\hat{\beta}}^2$ is the previously defined consistent estimator of $\sigma_{\hat{\beta}}^2$.

(b) Show that the Creasy–Williams set can be written in the form

$$\left\{ \beta : \frac{(\beta - \hat{\beta})^2}{\hat{\sigma}_{\hat{\beta}}^2/(n-2)} \left[ \frac{(1 + \lambda\beta\hat{\beta})^2}{(1 + \lambda\beta^2)^2} \right] \leq F_{1,n-2,\alpha} \right\}.$$

Hence $C_G(\hat{\beta})$ can be derived by replacing the term in square brackets with 1, its probability limit. (In deriving this representation, the fact that $\hat{\beta}$ and $-1/(\lambda\hat{\beta})$ are roots of the numerator of $r_\lambda(\beta)$ is of great help. In particular, the fact that

$$\frac{r_\lambda^2(\beta)}{1 - r_\lambda^2(\beta)} = \frac{\lambda^2 S_{xy}^2 (\beta - \hat{\beta})^2 (\beta + (1/\lambda\hat{\beta}))^2}{(1 + \lambda\beta^2)^2 (S_{xx}S_{yy} - S_{xy}^2)}$$

is straightforward to establish.)

**12.14** Graph the logistic regression function $\pi(x)$ from (12.3.2) for these three cases: $\alpha = \beta = 1$, $\alpha = \beta = 2$, and $\alpha = \beta = 3$.

**12.15** For the logistic regression function in (12.3.2), verify these relationships.

(a) $\pi(-\alpha/\beta) = 1/2$
(b) $\pi((-\alpha/\beta) + c) = 1 - \pi((-\alpha/\beta) - c)$ for any $c$
(c) (12.3.3) for $d\,\pi(x)/dx$
(d) (12.3.4) about the odds ratio
(e) (12.3.5) about the multiplicative change in odds
(f) (12.3.6) and (12.3.8) regarding the likelihood equations for a Bernoulli GLM
(g) For logistic regression, $f_i/(F_i(1 - F_i)) = \beta$ in (12.3.7) and (12.3.8)

**12.16** Consider this logistic regression data. Only two values, $x = 0$ and 1, are observed. For $x = 0$ there are 10 successes in 10 trials. For $x = 1$ there are 5 successes in 10 trials. Show that the logistic regression MLEs $\hat{\alpha}$ and $\hat{\beta}$ do not exist for these data by verifying the following.

(a) The MLEs for $\pi(0)$ and $\pi(1)$, not restricted by (12.3.2), are given by $\hat{\pi}(0) = 1$ and $\hat{\pi}(1) = .5$.
(b) The overall maximum of the likelihood function given by the estimates in part (a) can not be achieved at any finite values of the logistic regression parameters $\alpha$ and $\beta$, but can be achieved in the limit as $\beta \to -\infty$ and $\alpha = -\beta$.

**12.17** In *probit regression*, the link function is the standard normal cdf $\Phi(x) = P(Z \leq x)$, where $Z \sim n(0, 1)$. Thus, in this model we observe $(Y_1, x_1), (Y_2, x_2), \ldots, (Y_n, x_n)$, where $Y_i \sim$ Bernoulli$(\pi_i)$ and $\pi_i = \Phi(\alpha + \beta x_i)$.

(a) Write out the likelihood function and show how to solve for the MLEs of $\alpha$ and $\beta$.
(b) Fit the probit model to the data of Table 12.3.1. Comment on any differences from the logistic fit.

**12.18** Brown and Rothery (1993, Chapter 4) discuss a generalization of the linear logistic model to the quadratic model

$$\log\left(\frac{\pi_i}{1-\pi_i}\right) = \alpha + \beta x_i + \gamma x_i^2.$$

(a) Write out the likelihood function and show how to solve for the MLEs of $\alpha$, $\beta$, and $\gamma$.

(b) Using the log LRT, show how to test the hypothesis $H_0 : \gamma = 0$, that is, that the model is really linear logistic.

(c) Fit the quadratic logistic model to the data in the table on survival of sparrowhawks of different ages.

| Age | 1 | 2 | 3 | 4 | 5 | 6 | 7 | 8 | 9 |
|---|---|---|---|---|---|---|---|---|---|
| No. of birds | 77 | 149 | 182 | 118 | 78 | 46 | 27 | 10 | 4 |
| No. surviving | 35 | 89 | 130 | 79 | 52 | 28 | 14 | 3 | 1 |

(d) Decide whether the better model for the sparrowhawks is linear or quadratic, that is, test $H_0 : \gamma = 0$.

**12.19** For the logistic regression model:

(a) Show that $\left(\sum_{j=1}^{J} Y_j^*, \sum_{j=1}^{J} Y_j^* x_j\right)$ is a sufficient statistic for $(\alpha, \beta)$.

(b) Verify the formula for the logistic regression information matrix in (12.3.10).

**12.20** Consider a logistic regression model and assume $\beta = 0$.

(a) If $0 < \sum_{i=1}^{n} y_i < n$, show that the MLE of $\pi(x)$ (which does not depend on $x$ in this case) is $\hat{\pi} = \sum_{i=1}^{n} y_i/n$.

(b) If $0 < \sum_{i=1}^{n} y_i < n$, show that the MLE of $\alpha$ is $\hat{\alpha}_0 = \log\left((\sum_{i=1}^{n} y_i)/(n - \sum_{i=1}^{n} y_i)\right)$.

(c) Show that if $\sum_{i=1}^{n} y_i = 0$ or $n$, $\hat{\alpha}_0$ does not exist, but the LRT statistic for testing $H_0 : \beta = 0$ is still well defined.

**12.21** Let $Y \sim$ binomial$(n, \pi)$, and let $\hat{\pi} = Y/n$ denote the MLE of $\pi$. Let $W = \log\left(\hat{\pi}/(1-\hat{\pi})\right)$ denote the sample logit, the MLE of $\log\left(\pi/(1-\pi)\right)$. Use the Delta Method to show that $1/(n\hat{\pi}(1-\hat{\pi}))$ is a reasonable estimate of Var $W$.

**12.22** In Example 12.4.1 we examined how small perturbations affected the least squares estimate of slope. Perform the analogous calculation and assess the robustness (to small perturbations) of the least squares estimate of intercept.

**12.23** In Example 12.4.1, in contrast to Example 10.2.1, when we introduced the contaminated distribution for $\varepsilon_i$, we did not introduce a bias. Show that if we had, it would not have mattered. That is, if we assume

$$(\mathrm{E}\,\varepsilon_i, \mathrm{Var}\,\varepsilon_i) = \begin{cases} (0, \sigma^2) & \text{with probability } 1 - \delta \\ (\mu, \tau^2) & \text{with probability } \delta, \end{cases}$$

then:

(a) the least squares estimator $b$ would still be an unbiased estimator of $\beta$.

(b) the least squares estimator $a$ has expectation $\alpha + \delta\mu$, so the model may just as well be assumed to be $Y_i = \alpha + \delta\mu + \beta x_i + \varepsilon_i$.

**12.24** For the model $Y_i = \beta x_i + \varepsilon_i$, show that the LAD estimator is given by $t_{(k^*+1)}$, where $t_i = y_i/x_i$, $t_{(1)} \leq \cdots \leq t_{(n)}$ and, if $x_{(i)}$ is the $x$ value paired with $t_{(i)}$, $k^*$ satisfies $\sum_{i=1}^{k^*} |x_{(i)}| \leq \sum_{i=k^*+1}^{n} |x_{(i)}|$ and $\sum_{i=1}^{k^*+1} |x_{(i)}| > \sum_{i=k^*+2}^{n} |x_{(i)}|$.

**12.25** A problem with the LAD regression line is that it is not always uniquely defined.

    (a) Show that, for a data set with three observations, $(x_1, y_1), (x_1, y_2)$, and $(x_3, y_3)$ (note the first two $x$s are the same), any line that goes through $(x_3, y_3)$ and lies between $(x_1, y_1)$ and $(x_1, y_2)$ is a least absolute deviation line.

    (b) For three individuals, measurements are taken on heart rate ($x$, in beats per minute) and oxygen consumption ($y$, in ml/kg). The $(x, y)$ pairs are (127, 14.4), (127, 11.9), and (136, 17.9). Calculate the slope and intercept of the least squares line and the range of the least absolute deviation lines.

There seems to be some disagreement over the value of the least absolute deviation line. It is certainly more robust than least squares but can be very difficult to compute (but see Portnoy and Koenker 1997 for an efficient computing algorithm). It also seems that Ellis (1998) questions its robustness, and in a discussion Portnoy and Mizera (1998) question Ellis.

*Exercises 12.26–12.28 will look at some of the details of Example 12.4.3.*

**12.26** (a) Throughout Example 12.4.3 we assumed that $\frac{1}{n}\sum_{i=1}^{n} x_i^2 \to \sigma_x^2 < \infty$. Show that this condition is satisfied by (i) $x_i = 1$ (the case of the ordinary median) and (ii) $|x_i| \leq 1$ (the case of bounded $x_i$).

    (b) Show that, under the conditions on $x_i$ in part (a), $\frac{1}{n}\sum_{i=1}^{n} \psi(y_i - \hat{\beta}_L x_i) \to 0$ in probability.

**12.27** (a) Verify that $\frac{-1}{\sqrt{n}}\sum_{i=1}^{n} \psi(Y_i - \hat{\beta}x_i) \to n\left(0, \sigma_x^2\right)$.

    (b) Verify that $\frac{1}{n}\sum_{i=1}^{n} \frac{d}{d\beta_0} E_\beta[\psi(Y_i - \beta_0 x_i)]\Big|_{\beta_0 = \beta} = 2f(0)\frac{1}{n}\sum_{i=1}^{n} x_i^2$, and, with part (a), conclude that $\sqrt{n}(\hat{\beta}_L - \beta) \to n\left(0, \frac{1}{4f(0)^2\sigma_x^2}\right)$.

**12.28** Show that the least squares estimator is given by $\hat{\beta} = \sum_{i=1}^{n} x_i y_i / \sum_{i=1}^{n} x_i^2$, and $\sqrt{n}(\hat{\beta} - \beta) \to n(0, 1/\sigma_x^2)$.

**12.29** Using a Taylor series argument as in Example 12.4.3, derive the asymptotic distribution of the median in iid sampling.

**12.30** For the data of Table 12.4.1, use the parametric bootstrap to assess the standard error from the LAD and M-estimator fit. In particular:

    (a) Fit the line $y = \alpha + \beta x$ to get estimates $\tilde{\alpha}$ and $\tilde{\beta}$.

    (b) Calculate the residual mean squared error $\hat{\sigma}^2 = \frac{1}{n-2}\sum_{i=1}^{n}[y_i - (\tilde{\alpha} + \tilde{\beta}x_i)]^2$.

    (c) Generate new residuals from $n(0, \hat{\sigma}^2)$ and re-estimate $\alpha$ and $\beta$.

    (d) Do part (c) $B$ times and calculate the standard deviation of $\tilde{\alpha}$ and $\tilde{\beta}$.

    (e) Repeat parts (a)–(d) using both the double exponential and Laplace distributions for the errors. Compare your answers to the normal.

**12.31** For the data of Table 12.4.1, we could also use the nonparametric bootstrap to assess the standard error from the LAD and M-estimator fit.

    (a) Fit the line $y = \alpha + \beta x$ to get estimates $\tilde{\alpha}$ and $\tilde{\beta}$.

    (b) Generate new residuals by resampling from the fitted residuals and re-estimate $\alpha$ and $\beta$.

    (c) Do part (c) $B$ times and calculate the standard deviation of $\tilde{\alpha}$ and $\tilde{\beta}$.

# 12.6 Miscellanea _____

### 12.6.1 The Meaning of Functional and Structural

The names *functional* and *structural* are, in themselves, a prime source of confusion in the EIV model. Kendall and Stuart (1979, Chapter 29) give a detailed discussion of these concepts, distinguishing among relationships between *mathematical* (nonrandom) variables and relationships between random variables. One way to see the relationship is to write the models in a hierarchy in which the structural relationship model is obtained by putting a distribution on the parameters of the functional model:

$$
\begin{matrix}
\text{Functional} \\
\text{relationship} \\
\text{model}
\end{matrix}
\left\{
\begin{matrix}
\mathrm{E}(Y_i|\xi_i) = \alpha + \beta\xi_i + \epsilon_i & \epsilon_i \sim \mathrm{n}(0, \sigma_\epsilon^2) \\
\mathrm{E}(X_i|\xi_i) = \xi_i + \delta_i & \delta_i \sim \mathrm{n}(0, \sigma_\delta^2) \\
& \xi_i \sim \mathrm{n}(\xi, \sigma_\xi^2)
\end{matrix}
\right\}
\begin{matrix}
\text{Structural} \\
\text{relationship} \\
\text{model}
\end{matrix}
$$

The difference in the words may be understood through the following distinction, not a universally accepted one. In the subject of calculus, for example, we often see the equation $y = f(x)$, an equation that describes a *functional* relationship, that is, a relationship that is *assumed to exist between variables*. Thus, from the idea that a functional relationship is an assumed relationship between two variables, the equation $\eta_i = \alpha + \beta\xi_i$, where $\eta_i = \mathrm{E}(Y_i|\xi_i)$, is a functional (hypothesized) relationship in either the functional or structural relationship model.

On the other hand, a *structural* relationship is a relationship that *arises from the hypothesized structure of the problem*. Thus, in the structural relationship model, the relationship $\eta = \mathrm{E}Y_i = \alpha + \beta\xi = \alpha + \beta\mathrm{E}X_i$ can be deduced from the structure of the model; hence it is a structural relationship.

To make these ideas clearer, consider the case of simple linear regression where we assume that there is no error in the $x$s. The equation $\mathrm{E}(Y_i|x_i) = \alpha + \beta x_i$ is a functional relationship, a relationship that is hypothesized to exist between $\mathrm{E}(Y_i|x_i)$ and $x_i$. We can, however, also do simple linear regression under the assumption that the pair $(X_i, Y_i)$ has a bivariate normal distribution and we operate conditional on the $x_i$s. In this case, the relationship $\mathrm{E}(Y_i|x_i) = \alpha + \beta x_i$ follows from the structure of the hypothesized model and hence is a structural relationship.

Notice that, with these meanings, the distinction in terminology becomes a matter of taste. In any model we can deduce structural relationships from functional relationships, and vice versa. The important distinction is whether the nuisance parameters, the $\xi_i$s, are integrated out before inference is done.

### 12.6.2 Consistency of Ordinary Least Squares in EIV Models

In general it is not a good idea to use the ordinary least squares estimator to estimate the slope in EIV regression. This is because the estimator is inconsistent.

Suppose that we assume a linear structural relationship (12.2.6). We have

$$\hat\beta = \frac{\sum_{i=1}^n (X_i - \bar X)(Y_i - \bar Y)}{\displaystyle\sum_{i=1}^n (X_i - \bar X)^2}$$

$$= \frac{\frac{1}{n}\sum_{i=1}^n (X_i - \bar X)(Y_i - \bar Y)}{\frac{1}{n}\displaystyle\sum_{i=1}^n (X_i - \bar X)^2}$$

$$\longrightarrow \frac{\mathrm{Cov}(X,Y)}{\mathrm{Var}\,X} \qquad (\text{as } n \to \infty, \text{ using the WLLN})$$

$$= \frac{\beta\sigma_\xi^2}{\sigma_\delta^2 + \sigma_\xi^2}, \qquad\qquad (\text{from } (12.2.19))$$

showing that $\hat\beta$ cannot be consistent. The same type of result can be obtained in the functional relationship case.

The behavior of $\hat\beta$ in EIV models is treated in Cochran (1968). Carroll, Gallo, and Gleser (1985) and Gleser, Carroll, and Gallo (1987) investigated conditions under which functions of the ordinary least squares estimator are consistent.

### 12.6.3 Instrumental Variables in EIV Models

The concept of instrumental variables goes back at least to Wald (1940), who constructed a consistent estimator of the slope with their help. To see what an instrumental variable is, write the EIV model in the form

$$Y_i = \alpha + \beta\xi_i + \epsilon_i,$$
$$X_i = \xi_i + \delta_i,$$

and do some algebra to get

$$Y_i = \alpha + \beta X_i + [\epsilon_i - \beta\delta_i].$$

An *instrumental variable*, $Z_i$, is a random variable that predicts $X_i$ well but is uncorrelated with $\nu_i = \epsilon_i - \beta\delta_i$. If such a variable can be identified, it can be used to improve predictions. In particular, it can be used to construct a consistent estimator of $\beta$.

Wald (1940) showed that, under fairly general conditions, the estimator

$$\hat\beta_{\mathrm W} = \frac{\bar Y_{(1)} - \bar Y_{(2)}}{\bar X_{(1)} - \bar X_{(2)}}$$

is a consistent estimator of $\beta$ in identifiable models, where the subscripts refer to two groupings of the data. A variable $Z_i$, which takes on only two values to define the grouping, is an instrumental variable. See Moran (1971) for a discussion of Wald's estimator.

Although instrumental variables can be of great help, there can be some problems associated with their use. For example, Feldstein (1974) showed instances where the use of instrumental variables can be detrimental. Moran (1971) discussed the difficulty of verifying the conditions needed to ensure consistency of a simple estimator like $\hat{\beta}_W$. Fuller (1987) provided an in-depth discussion of instrumental variables. A model proposed by Berkson (1950) exploited a correlation structure similar to that used with instrumental variables.

### 12.6.4 Logistic Likelihood Equations

In a logistic regression model, the likelihood equations are nonlinear in the parameters, and they must be solved numerically. The most commonly used method for solving these equations is the *Newton–Raphson method*. This method begins with an initial guess $(\hat{\alpha}^{(1)}, \hat{\beta}^{(1)})$ for the value of the MLEs. Then the log likelihood is approximated with a quadratic function, its second-order Taylor series about the point $(\hat{\alpha}^{(1)}, \hat{\beta}^{(1)})$. The next guess for the values of the MLEs, $(\hat{\alpha}^{(2)}, \hat{\beta}^{(2)})$, is the maximum of this quadratic function. Now another quadratic approximation is used, this one centered at $(\hat{\alpha}^{(2)}, \hat{\beta}^{(2)})$, and its maximum is the next guess for the values of the MLEs. The Taylor series approximations involve the first and second derivatives of the log likelihood. These are evaluated at the current guess $(\hat{\alpha}^{(t)}, \hat{\beta}^{(t)})$. These are the same second derivatives that appear in the information matrix in (12.3.10). Thus, a byproduct of this method of solving the likelihood equations is estimates of the variances and covariance of $\hat{\alpha}$ and $\hat{\beta}$. The convergence of the guesses $(\hat{\alpha}^{(t)}, \hat{\beta}^{(t)})$ to the MLEs $(\hat{\alpha}, \hat{\beta})$ is usually rapid for logistic regression models. It often takes only a few iterations to obtain satisfactory approximations.

The Newton–Raphson method is also called *iteratively reweighted least squares*. At each stage, the next guess for $(\hat{\alpha}, \hat{\beta})$ can be expressed as the solution of a least squares problem. But, this is a least squares problem in which the different terms in the sum of squares function are given different weights. In this case the weights are $n_j F_j^{(t)}(1 - F_j^{(t)})$, where $F_j^{(t)} = F(\hat{\alpha}^{(t)} + \hat{\beta}^{(t)} x_j)$ and $F$ is the logistic cdf. This is the inverse of an approximation to the variance of the $j$th sample logit (see Exercise 12.21). The weights are recalculated at each stage, because the current guesses for the MLEs are used. That leads to the name "iteratively reweighted." Thus, the Newton–Raphson method is approximately the result of using the sample logits as the data and performing a weighted least squares to estimate the parameters.

### 12.6.5 More on Robust Regression

Robust alternatives to least squares have been an object of study for many years, and there is a vast body of literature addressing a variety of problems. In Section 12.4 we saw only a brief introduction to robust regression, but some of the advantages and difficulties should be apparent. There are many good books that treat robust regression in detail, including Hettmansperger and McKean (1996), Staudte and Sheather (1990), and Huber (1981). Some other topics that have received a lot of attention are discussed below.

*Trimming and transforming*

The work of Carroll, Ruppert, and co-authors has addressed many facets of robust regression. Ruppert and Carroll (1979) is a careful treatment of the asymptotics of *trimming* (the trimmed mean is discussed in Exercise 10.20), while Carroll and Ruppert (1985) examine alternatives to least squares when the errors are not identical. Later work looked at the advantages of transformations in regression (Carroll and Ruppert 1985, 1988).

*Other robust alternatives*

We looked at only the LAD estimator and one M-estimator. There are, of course, many other choices of robust estimators. One popular alternative is the least median of squares (LMS) estimator of Rousseeuw (1984); see also Rousseeuw and Leroy 1987). There are also *R-estimators*, rank-based regression estimators (see the review paper by Draper 1988). More recently, there has been work on *data depth* (Liu 1990, Liu and Singh 1992) with applications to regression in finding the *deepest line* (Rousseeuw and Hubert 1999).

*Computing*

From a practical view, computation of robust estimates can be quite challenging, as we are often faced with a difficult minimization problem. The review paper by Portnoy and Koenker (1997) is concerned with computation of LAD estimates. Hawkins (1993, 1994, 1995) has a number of algorithms for computing LMS and related estimates.

# Computer Algebra

Computer algebra systems allow for symbolic manipulation of expressions. They can be particularly helpful when we are faced with tedious, rote calculations (such as taking the second derivative of a ratio). In this appendix we illustrate the use of such a system in various problems. Although use of a computer algebra system is in no way necessary to understand and use statistics, it not only can relieve some of the tedium but also can lead us to new insights and more applicable answers.

Realize that there are many computer algebra systems, and there are numerous calculations with which they can be helpful (sums, integrals, simulations, etc.). The purpose of this appendix is not to teach the use of these systems or to display all their possible uses, but rather to illustrate some of the possibilities.

We illustrate our calculations using the package Mathematica®. There are other packages, such as Maple®, that could also be used for these calculations.

*Chapter 1*

**Example A.0.1 (Unordered sampling)**  We illustrate Mathematica code for enumerating the unordered outcomes from sampling with replacement from $\{2, 4, 9, 12\}$, as described in Example 1.2.20. After enumerating the outcomes and calculating the multinomial weights, the outcomes and weights are sorted. Note that to produce the histogram of Figure 1.2.2 requires a bit more work. For example, there are two distinct outcomes that have average value 8, so to produce a picture like Figure 1.2.2, the outcomes $\{8, \frac{3}{128}\}$ and $\{8, \frac{3}{64}\}$ need to be combined into $\{8, \frac{9}{128}\}$.

Enumeration such as this gets very time consuming if the set has more than 7 numbers, which results in $\binom{13}{7} = 27{,}132$ unordered outcomes.

(1) *The "DiscreteMath" package contains functions for counting permutations and combinations.*

```
In[1]:= Needs["DiscreteMath`Combinatorica`"]
```

(2) *We let $x = $ collection of numbers. The number of distinct samples is* Numberof-Compositions[n,m] $= \binom{n+m-1}{n}$.

```
In[2]:= x ={2,4,9,12};
n=Length[x];
ncomp=NumberOfCompositions[n,n]
```

**Out[4]**= 35

(3) *We enumerate the samples* (w), *calculate the average of each of the samples* (avg), *and calculate the weights for each value* (wt). *The weight is the multinomial co-efficient that corresponds to the configuration.*

```
In[5]:= w = Compositions[n, n];
wt = n!/(Apply[Times, Factorial /@ w, 1]*n^n);
avg = w . x/n;
Sort[Transpose[{avg, wt}]]
```

**Out[8]** $= \{2, 1/256\}, \{5/2, 1/64\}, \{3, 3/128\}, \{7/2, 1/64\}, \{15/4, 1/64\},$
$\{4, 1/256\}, \{17/4, 3/64\}, \{9/2, 1/64\}, \{19/4, 3/64\}, \{5, 3/64\},$
$\{21/4, 1/64\}, \{11/2, 3/128\}, \{11/2, 3/64\}, \{6, 1/64\}, \{6, 3/64\},$
$\{25/4, 3/64\}, \{13/2, 3/128\}, \{27/4, 3/32\}, \{7, 3/128\}, \{29/4, 1/64\},$
$\{29/4, 3/64\}, \{15/2, 3/64\}, \{31/4, 1/64\}, \{8, 3/128\}, \{8, 3/64\},$
$\{17/2, 3/64\}, \{35/4, 3/64\}, \{9, 1/256\}, \{37/4, 3/64\}, \{19/2, 1/64\},$
$\{39/4, 1/64\}, \{10, 1/64\}, \{21/2, 3/128\}, \{45/4, 1/64\}, \{12, 1/256\}$     ‖

*Chapter 2*

**Example A.0.2 (Univariate transformation)**   Exercise 2.1(a) is a standard uni-variate change of variable. Such calculations are usually easy for a computer algebra program.

(1) *Enter* $f(x)$ *and solve for the transformed variable.*

```
In[1]:=f[x_]  := 42*(x^5)*(1 - x)
sol = Solve[y == x^3, x]
```

**Out[2]** $= \{\{x- > y^{1/3}\}, \{x- > -(-1)^{1/3}y^{1/3}\}, \{x- > -(-1)^{2/3}y^{2/3}\}\}$

(2) *Calculate the Jacobean.*

```
In[3]:= D[x/.sol[[1]],y]
```

**Out[3]** $= \dfrac{1}{3y^{2/3}}$

(3) *Calculate the density of the transformed variable.*

```
In[4]:= f[x/.sol[[1]]]* D[x/.sol[[1]],y]
```

**Out[4]** $= 14(1 - y^{1/3})y$     ‖

*Chapter 4*

**Example A.0.3 (Bivariate transformations)**   We illustrate some bivariate trans-formations. It is also possible, with similar code, to do multivariate transformations. In the first calculation we illustrate Example 4.3.4 to obtain the distribution of the sum of normal variables. Then we do Example 4.3.3, which does a bivariate transfor-mation of a product of beta densities and then marginalizes.

(1) *Sum of normal variables. Out[4] is the joint distribution, and Out[5] is the marginal.*

```
In[1]:=f[x_,y_]:=(1/(2*Pi))*E^(-x^2/2)*E^(-y^2/2)
So:=Solve[{u==x+y,v==x-y},{x,y}]
g:=f[x/.So,y/.So]*Abs[Det[Outer[D,First[{x,y}/.So],{u,v}]]]
Simplify[g]
```

$$\mathbf{Out[4]} = \left\{ \frac{e^{-\frac{u^2}{4} - \frac{v^2}{4}}}{4\pi} \right\}$$

```
In[5]:= Integrate[g,{v,0,Infinity}]
```

$$\mathbf{Out[5]} = \left\{ \frac{e^{-\frac{u^2}{4}}}{4\sqrt{\pi}} \right\}$$

(2) *Product of beta variables. (The package "ContinuousDistributions" contains pdfs and cdfs of many standard distributions.) Out[10] is the joint density of the product of the beta variables, and Out[11] is the density of u. The* If *statement is read* If(test, true, false), *so if the test is true, the middle value is taken. In most situations the test will be true, and the marginal density is the given beta density.*

```
In[6]:= Needs["Statistics`ContinuousDistributions`"]

    Clear[f, g, u, v, x, y, a, b, c]

    f[x_, y_] := PDF[BetaDistribution[a, b],x]
                *PDF[BetaDistribution[a + b, c], y]

    So := Solve[{u == x*y, v == x}, {x, y}]

    g[u_, v_] = f[x /. So, y /. So]
                *Abs[Det[Outer[D, First[{x, y} /. So], {u, v}]]]

    Integrate[g[u, v] ,{v,0,1}]
```

$$\mathbf{Out[10]} = \left\{ \frac{(1 - \frac{u}{v})^{-1+c}(1 - v)^{-1+b}(\frac{u}{v})^{-1+a+b}v^{-1+a}}{\mathrm{Abs}[v]\mathrm{Beta}[a, b]\mathrm{Beta}[a + b, c]} \right\}$$

$$\mathbf{Out[11]} = \Big\{ \mathrm{If}\big(\mathrm{Re}[b] > 0\,\&\&\,\mathrm{Re}[b + c] < 1\,\&\&\,\mathrm{Im}[u] == 0\,\&\&\,u > 0\,\&\&\,u < 1\big),$$

$$\frac{1}{\mathrm{Gamma}[1 - c]} \left( (1 - u)^{-1+b+c} \left( \frac{-1}{u} \right)^{b+c} (-u)^c u^{-1+a+b}\mathrm{Gamma}[b]\mathrm{Gamma}[1 - b - c] \right),$$

$$\int_0^1 \frac{(1 - \frac{u}{v})^{-1+c}(1 - v)^{-1+b}(\frac{u}{v})^{-1+a+b}v^{-1+a}}{\mathrm{Abs}[v]\mathrm{Beta}[a, b]\mathrm{Beta}[a + b, c]}\, dv \Big\}$$

‖

**Example A.0.4 (Normal probability)** The calculation asked for in Exercise 4.14(a) can be easily handled. We first do a direct calculation, and Mathematica easily does the integral numerically but doesn't find the closed form expression. Note that the answer is given in terms of the *erf* function, defined by

$$\mathrm{erf}(z) = \frac{2}{\sqrt{\pi}} \int_0^z e^{-t^2} dt.$$

If we re-express the probability in terms of a chi squared random variable, then the closed-form expression is found.

(1) *To evaluate the integral we set up the integrand and the limits of integration.*

```
In[1]:= Needs["Statistics`ContinuousDistributions`"]

    Clear[f, g, x, y]

    f[x_, y_] = PDF[NormalDistribution[0, 1],x]
            *PDF[NormalDistribution[0, 1],y]

    g[x_] = Sqrt[1 - x^2]
```

$$\mathbf{Out[3]} = \frac{e^{-\frac{x^2}{2} - \frac{y^2}{2}}}{2\pi}$$

(2) *We now evaluate the double integral and get the* Erf *function. The command $N[\%]$ numerically evaluates the previous line.*

```
In[5]:= Integrate[f[x, y], {x, -1, 1}, {y, -g[x], g[x]}]

    N[%]
```

$$\mathbf{Out[5]} = \frac{\int_{-1}^{1} e^{-\frac{x^2}{2}} \mathrm{Erf}\left(\frac{\sqrt{1-x^2}}{\sqrt{2}}\right) dx}{\sqrt{2\pi}}$$

$\mathbf{Out[6]} = 0.393469$

(3) *We of course know that $X^2 + Y^2$ is a chi squared random variable with 2 degrees of freedom. If we use that fact we get a closed-form answer.*

```
In[7]: = Clear[f, t]
f[t_] = PDF[ChiSquareDistribution[2], t];
Integrate[f[t], {t, 0, 1}]
N[%]
```

$$\mathbf{Out[10]} = \frac{1}{2}\left(2 - \frac{2}{\sqrt{e}}\right)$$

$\mathbf{Out[11]} = 0.393469$ ‖

*Chapter 5*

**Example A.0.5 (Density of a sum)** The calculation done in Example 5.2.10, which illustrates Theorem 5.2.9, is quite an involved one. We illustrate such a calculation in three cases: normal, Cauchy, and Student's *t*.

There are two points to note.

(1) To correctly interpret the answers, some knowledge of complex analysis may be needed. For the normal case, the answer is reported conditional on the value of the real part of the (possibly complex-valued) variable $z$. In the Cauchy example, it is important to know that $I^2 = -1$, so we have

$$\frac{2}{\pi(-2I + z)(2I + z)} = \frac{2}{\pi(4 + z^2)}.$$

(2) When we add Student's $t$ variables, it seems that if the sum of the degrees of freedom is even, then a closed-form expression exists. If not, then the integration must be done numerically. (This is an empirical observation that we discovered by playing around with computer algebra systems.)

We also note that later versions of computer algebra programs may avoid the complex numbers here. However, they will pop up in other calculations, so it is best to be prepared to deal with them.

(1) *The density of the sum of two normals*

```
In[1]:= Clear[f,  x,  y,  z]

      f[x_] = Exp[(-x^2/2]/(2 Pi);
      Integrate[f[y]* f[z - y], {y,-Infinity,Infinity}]
```

$$\mathbf{Out[3]} = \text{If}\left[\text{Re}[z] < 0, \frac{e^{-\frac{z^2}{4}}}{2\pi}, \int_{-\infty}^{\infty} \frac{E^{-\frac{y^2}{2} - \frac{1}{2}(-y+z)^2}}{2\pi} dy\right]$$

(2) *The density of the sum of two Cauchys*

```
In[4]:= Clear[f, x, y, z]
f[x_] = 1/(Pi*(1+x^2));
Integrate[f[y]* f[z - y], {y,-Infinity,Infinity}]
```

$$\mathbf{Out[6]} = \frac{2}{\pi(-2I + z)(2I + z)}$$

(3) *The density of the sum of two t's with 5 degrees of freedom*

```
In[7]:= Needs["Statistics`ContinuousDistributions`"]
      Clear[f, x, y, z]
      f[x_] = PDF[StudentTDistribution[5, x]
      Integrate[f[y]* f[z - y], {y,-Infinity,Infinity}]
```

$$\mathbf{Out[10]} = \frac{400\sqrt{5}(8400 + 120z^2 + z^4)}{3\pi(20 + z^2)^5}$$

‖

**Example A.0.6 (Fourth moment of sum of uniforms)** Exercise 5.51 asks for the fourth moment of a sum of 12 uniform random variables. Deriving the density is somewhat painful because of its piecewise nature (but also can be done using computer algebra). However, using mgfs simplifies things.

(1) *First calculate the mgf of $X_1$, a uniform random variable, and then of $\sum_{i=1}^{12} X_i$, where the $X_i$s are independent.*

```
In[1]:= M[t_] = Integrate[Exp[t*x], {x, 0, 1}]
In[2]:= Msum[t_] = M[t]^12
```

$$\textbf{Out[1]} = \frac{-1 + e^t}{t}$$

$$\textbf{Out[2]} = \frac{(-1 + e^t)^{12}}{t^{12}}$$

(2) *Calculate the fourth derivative of the mgf of $\sum_{i=1}^{12} X_i - 6$. It is too big to print out.*

```
In[3]:= g[t_]=D[Exp[-6*t]*Msum[t],{t,4}];
```

(3) *The value $g[0]$ is the fourth moment; however, just substituting $0$ results in division by $0$, so we have to do this calculation as a limit.*

```
In[4]:= g[0]
```

Power:infy: Infinite expression $\frac{1}{0^{16}}$ encountered.

```
In[5]:= Limit[g[t],t->0]
```

$$\textbf{Out[5]} = \frac{29}{10} \qquad\qquad \|$$

*Chapter 7*

**Example A.0.7 (ARE for a gamma mean)**   The calculations done in Example 10.1.18, which led to Figure 10.1.1, were done in Mathematica. The following code will produce one of the ARE graphs.

(1) *The second derivative of the log likelihood is taken symbolically.*

```
In[1]:= Needs["Statistics`ContinuousDistributions`"]
        Clear[m, b, x]
        f[x_, m_, b_] = PDF[GammaDistribution[m/b, b], x];
        loglike2[m_, b_, x_] = D[Log[f[x, m, b]], {m, 2}];
```

(2) *The asymptotic variance is calculated by integrating this second derivative with respect to the density.*

```
In[5]:= var[m_, b_] := 1/(-Integrate[loglike2[m, b, x]*f[x, m, b],
{x, 0, Infinity}])
```

(3) *The following code sets up the plot.*

```
In[6]:=
mu = {1, 2, 3, 4, 6, 8, 10};
beta = 5;
mlevar = Table[var[mu[[i]], beta], {i, 1, 7}];
momvar = Table[mu[[i]]*beta, {i, 1, 7}];
ARE = momvar/mlevar
```

**Out[10]** = {5.25348, 2.91014, 2.18173, 1.83958, 1.52085, 1.37349, 1.28987}

```
In[11]:=ListPlot[Transpose[{mu, ARE}], PlotJoined -> True,
    PlotRange -> {{0, mu[[7]]}, {0, 8}}, AxesLabel -> {"Gamma mean",
    "ARE"}]
```
||

*Chapter 9*

**Example A.0.8 (Limit of chi squared mgfs)**   Calculation of the limiting distribution in Exercise 9.30(b) is delicate but is rather straightforward in Mathematica.

(1) *First calculate the mgf of a chi squared random variable with n degrees of freedom. (Of course, this step is really not necessary.)*

```
In[1]:= Needs["Statistics'ContinuousDistributions'"]
f[x_] = PDF[ChiSquareDistribution[n], x];
Integrate[Exp[t*x]*f[x], {x, 0, Infinity}]
```

**Out[3]** =

$$\frac{2^{n/2}\text{If}\left[\text{Re}[n] > 0\&\&\text{Re}[t] < \tfrac{1}{2}, \left(\tfrac{1}{2} - t\right)^{n/2} \text{Gamma}[\tfrac{n}{2}], \int_0^\infty e^{-\frac{x}{2}+tx}x^{-1+\frac{n}{2}}dx\right]}{\text{Gamma}[\tfrac{n}{2}]}$$

(2) *As the test condition is satisfied, the mgf of $\chi_n^2$ is the middle term. Now take the limit of the mgf of $\frac{\chi_n^2 - n}{\sqrt{2n}}$.*

```
In[4]:M[t_] = (1 - 2*t)^(-n/2);
Limit[Exp[-n*t/Sqrt[2*n]]*M[t/Sqrt[2*n]], n -> Infinity]
```

**Out[5]** = $e^{\frac{t^2}{2}}$
||

# Table of Common Distributions

**Bernoulli**$(p)$

| | |
|---|---|
| *pmf* | $P(X = x\|p) = p^x(1-p)^{1-x}; \quad x = 0,1; \quad 0 \leq p \leq 1$ |
| *mean and variance* | $EX = p, \quad \text{Var } X = p(1-p)$ |
| *mgf* | $M_X(t) = (1-p) + pe^t$ |

**Binomial**$(n, p)$

| | |
|---|---|
| *pmf* | $P(X = x\|n, p) = \binom{n}{x}p^x(1-p)^{n-x}; \quad x = 0,1,2,\ldots,n; \quad 0 \leq p \leq 1$ |
| *mean and variance* | $EX = np, \quad \text{Var } X = np(1-p)$ |
| *mgf* | $M_X(t) = [pe^t + (1-p)]^n$ |
| *notes* | Related to Binomial Theorem (Theorem 3.2.2). The *multinomial* distribution (Definition 4.6.2) is a multivariate version of the binomial distribution. |

**Discrete uniform**

| | |
|---|---|
| *pmf* | $P(X = x\|N) = \frac{1}{N}; \quad x = 1,2,\ldots,N; \quad N = 1,2,\ldots$ |
| *mean and variance* | $EX = \frac{N+1}{2}, \quad \text{Var } X = \frac{(N+1)(N-1)}{12}$ |
| *mgf* | $M_X(t) = \frac{1}{N}\sum_{i=1}^{N} e^{it}$ |

**Geometric**$(p)$

| | |
|---|---|
| *pmf* | $P(X = x\|p) = p(1-p)^{x-1}; \quad x = 1,2,\ldots; \quad 0 \leq p \leq 1$ |
| *mean and variance* | $EX = \frac{1}{p}, \quad \text{Var } X = \frac{1-p}{p^2}$ |

mgf        $M_X(t) = \frac{pe^t}{1-(1-p)e^t}, \quad t < -\log(1-p)$

notes      $Y = X - 1$ is negative binomial$(1, p)$. The distribution is *memoryless*:
           $P(X > s | X > t) = P(X > s - t)$.

---

### *Hypergeometric*

pmf        $P(X = x | N, M, K) = \frac{\binom{M}{x}\binom{N-M}{K-x}}{\binom{N}{K}}; \quad x = 0, 1, 2, \ldots, K;$

           $M - (N - K) \le x \le M; \quad N, M, K \ge 0$

mean and   $EX = \frac{KM}{N}, \quad \text{Var}\, X = \frac{KM}{N}\frac{(N-M)(N-K)}{N(N-1)}$
variance

notes      If $K \ll M$ and $N$, the range $x = 0, 1, 2, \ldots, K$ will be appropriate.

---

### *Negative binomial*$(r, p)$

pmf        $P(X = x | r, p) = \binom{r+x-1}{x} p^r (1-p)^x; \quad x = 0, 1, \ldots; \quad 0 \le p \le 1$

mean and   $EX = \frac{r(1-p)}{p}, \quad \text{Var}\, X = \frac{r(1-p)}{p^2}$
variance

mgf        $M_X(t) = \left(\frac{p}{1-(1-p)e^t}\right)^r, \quad t < -\log(1-p)$

notes      An alternate form of the pmf is given by $P(Y = y | r, p) = \binom{y-1}{r-1} p^r (1 - p)^{y-r}$, $y = r, r + 1, \ldots$. The random variable $Y = X + r$. The negative binomial can be derived as a gamma mixture of Poissons. (See Exercise 4.32.)

---

### *Poisson*$(\lambda)$

pmf        $P(X = x | \lambda) = \frac{e^{-\lambda}\lambda^x}{x!}; \quad x = 0, 1, \ldots; \quad 0 \le \lambda < \infty$

mean and   $EX = \lambda, \quad \text{Var}\, X = \lambda$
variance

mgf        $M_X(t) = e^{\lambda(e^t - 1)}$

---

## Continuous Distributions

---

### $Beta(\alpha, \beta)$

*pdf*  $f(x|\alpha, \beta) = \frac{1}{B(\alpha,\beta)} x^{\alpha-1}(1-x)^{\beta-1}, \quad 0 \le x \le 1, \quad \alpha > 0, \quad \beta > 0$

*mean and variance*  $EX = \frac{\alpha}{\alpha+\beta}, \quad \operatorname{Var} X = \frac{\alpha\beta}{(\alpha+\beta)^2(\alpha+\beta+1)}$

*mgf*  $M_X(t) = 1 + \sum_{k=1}^{\infty} \left( \prod_{r=0}^{k-1} \frac{\alpha+r}{\alpha+\beta+r} \right) \frac{t^k}{k!}$

*notes*  The constant in the beta pdf can be defined in terms of gamma functions, $B(\alpha, \beta) = \frac{\Gamma(\alpha)\Gamma(\beta)}{\Gamma(\alpha+\beta)}$. Equation (3.2.18) gives a general expression for the moments.

---

### $Cauchy(\theta, \sigma)$

*pdf*  $f(x|\theta, \sigma) = \frac{1}{\pi\sigma} \frac{1}{1+\left(\frac{x-\theta}{\sigma}\right)^2}, \quad -\infty < x < \infty; \quad -\infty < \theta < \infty, \quad \sigma > 0$

*mean and variance*  do not exist

*mgf*  does not exist

*notes*  Special case of Student's $t$, when degrees of freedom $= 1$. Also, if $X$ and $Y$ are independent $n(0,1)$, $X/Y$ is Cauchy.

---

### $Chi\ squared(p)$

*pdf*  $f(x|p) = \frac{1}{\Gamma(p/2)2^{p/2}} x^{(p/2)-1} e^{-x/2}; \quad 0 \le x < \infty; \quad p = 1, 2, \ldots$

*mean and variance*  $EX = p, \quad \operatorname{Var} X = 2p$

*mgf*  $M_X(t) = \left(\frac{1}{1-2t}\right)^{p/2}, \quad t < \frac{1}{2}$

*notes*  Special case of the gamma distribution.

---

### $Double\ exponential(\mu, \sigma)$

*pdf*  $f(x|\mu, \sigma) = \frac{1}{2\sigma} e^{-|x-\mu|/\sigma}, \quad -\infty < x < \infty, \quad -\infty < \mu < \infty, \quad \sigma > 0$

*mean and variance*  $EX = \mu, \quad \operatorname{Var} X = 2\sigma^2$

*mgf*  $M_X(t) = \frac{e^{\mu t}}{1-(\sigma t)^2}, \quad |t| < \frac{1}{\sigma}$

*notes*  Also known as the *Laplace* distribution.

---

## Exponential($\beta$)

pdf
$$f(x|\beta) = \frac{1}{\beta}e^{-x/\beta}, \quad 0 \le x < \infty, \quad \beta > 0$$

mean and variance
$$EX = \beta, \quad \operatorname{Var} X = \beta^2$$

mgf
$$M_X(t) = \frac{1}{1-\beta t}, \quad t < \frac{1}{\beta}$$

notes
Special case of the gamma distribution. Has the *memoryless* property. Has many special cases: $Y = X^{1/\gamma}$ is *Weibull*, $Y = \sqrt{2X/\beta}$ is *Rayleigh*, $Y = \alpha - \gamma \log(X/\beta)$ is *Gumbel*.

---

## F

pdf
$$f(x|\nu_1, \nu_2) = \frac{\Gamma\left(\frac{\nu_1+\nu_2}{2}\right)}{\Gamma\left(\frac{\nu_1}{2}\right)\Gamma\left(\frac{\nu_2}{2}\right)} \left(\frac{\nu_1}{\nu_2}\right)^{\nu_1/2} \frac{x^{(\nu_1-2)/2}}{\left(1+\left(\frac{\nu_1}{\nu_2}\right)x\right)^{(\nu_1+\nu_2)/2}};$$
$$0 \le x < \infty; \quad \nu_1, \nu_2 = 1, \ldots$$

mean and variance
$$EX = \frac{\nu_2}{\nu_2-2}, \quad \nu_2 > 2,$$

$$\operatorname{Var} X = 2\left(\frac{\nu_2}{\nu_2-2}\right)^2 \frac{(\nu_1+\nu_2-2)}{\nu_1(\nu_2-4)}, \quad \nu_2 > 4$$

moments
(mgf does not exist)
$$EX^n = \frac{\Gamma\left(\frac{\nu_1+2n}{2}\right)\Gamma\left(\frac{\nu_2-2n}{2}\right)}{\Gamma\left(\frac{\nu_1}{2}\right)\Gamma\left(\frac{\nu_2}{2}\right)} \left(\frac{\nu_2}{\nu_1}\right)^n, \quad n < \frac{\nu_2}{2}$$

notes
Related to chi squared ($F_{\nu_1, \nu_2} = \left(\frac{\chi^2_{\nu_1}}{\nu_1}\right) / \left(\frac{\chi^2_{\nu_2}}{\nu_2}\right)$), where the $\chi^2$s are independent) and $t$ ($F_{1,\nu} = t^2_\nu$).

---

## Gamma($\alpha, \beta$)

pdf
$$f(x|\alpha, \beta) = \frac{1}{\Gamma(\alpha)\beta^\alpha} x^{\alpha-1} e^{-x/\beta}, \quad 0 \le x < \infty, \quad \alpha, \beta > 0$$

mean and variance
$$EX = \alpha\beta, \quad \operatorname{Var} X = \alpha\beta^2$$

mgf
$$M_X(t) = \left(\frac{1}{1-\beta t}\right)^\alpha, \quad t < \frac{1}{\beta}$$

notes
Some special cases are exponential ($\alpha = 1$) and chi squared ($\alpha = p/2$, $\beta = 2$). If $\alpha = \frac{3}{2}$, $Y = \sqrt{X/\beta}$ is *Maxwell*. $Y = 1/X$ has the *inverted gamma distribution*. Can also be related to the Poisson (Example 3.2.1).

---

## Logistic($\mu, \beta$)

pdf
$$f(x|\mu, \beta) = \frac{1}{\beta} \frac{e^{-(x-\mu)/\beta}}{[1+e^{-(x-\mu)/\beta}]^2}, \quad -\infty < x < \infty, \quad -\infty < \mu < \infty, \quad \beta > 0$$

mean and variance
$$EX = \mu, \quad \operatorname{Var} X = \frac{\pi^2\beta^2}{3}$$

*mgf*        $M_X(t) = e^{\mu t} \Gamma(1 - \beta t) \Gamma(1 + \beta t), \quad |t| < \frac{1}{\beta}$

*notes*      The cdf is given by $F(x|\mu, \beta) = \frac{1}{1 + e^{-(x-\mu)/\beta}}$.

---

**Lognormal**$(\mu, \sigma^2)$

*pdf*        $f(x|\mu, \sigma^2) = \frac{1}{\sqrt{2\pi}\sigma} \frac{e^{-(\log x - \mu)^2/(2\sigma^2)}}{x}, \quad 0 \le x < \infty, \quad -\infty < \mu < \infty,$
             $\sigma > 0$

*mean and*   $EX = e^{\mu + (\sigma^2/2)}, \quad \operatorname{Var} X = e^{2(\mu + \sigma^2)} - e^{2\mu + \sigma^2}$
*variance*

*moments*    $EX^n = e^{n\mu + n^2\sigma^2/2}$
*(mgf does not exist)*

*notes*      Example 2.3.5 gives another distribution with the same moments.

---

**Normal**$(\mu, \sigma^2)$

*pdf*        $f(x|\mu, \sigma^2) = \frac{1}{\sqrt{2\pi}\sigma} e^{-(x-\mu)^2/(2\sigma^2)}, \quad -\infty < x < \infty, \quad -\infty < \mu < \infty,$
             $\sigma > 0$

*mean and*   $EX = \mu, \quad \operatorname{Var} X = \sigma^2$
*variance*

*mgf*        $M_X(t) = e^{\mu t + \sigma^2 t^2/2}$

*notes*      Sometimes called the *Gaussian* distribution.

---

**Pareto**$(\alpha, \beta)$

*pdf*        $f(x|\alpha, \beta) = \frac{\beta \alpha^\beta}{x^{\beta+1}}, \quad a < x < \infty, \quad \alpha > 0, \quad \beta > 0$

*mean and*   $EX = \frac{\beta \alpha}{\beta - 1}, \quad \beta > 1, \quad \operatorname{Var} X = \frac{\beta \alpha^2}{(\beta-1)^2(\beta-2)}, \quad \beta > 2$
*variance*

*mgf*        does not exist

---

**t**

*pdf*        $f(x|\nu) = \frac{\Gamma\left(\frac{\nu+1}{2}\right)}{\Gamma\left(\frac{\nu}{2}\right)} \frac{1}{\sqrt{\nu\pi}} \frac{1}{\left(1 + \left(\frac{x^2}{\nu}\right)\right)^{(\nu+1)/2}}, \quad -\infty < x < \infty, \quad \nu = 1, \ldots$

*mean and*   $EX = 0, \quad \nu > 1, \quad \operatorname{Var} X = \frac{\nu}{\nu-2}, \quad \nu > 2$
*variance*

*moments*    $EX^n = \frac{\Gamma\left(\frac{n+1}{2}\right)\Gamma\left(\frac{\nu-n}{2}\right)}{\sqrt{\pi}\Gamma\left(\frac{\nu}{2}\right)} \nu^{n/2}$ if $n < \nu$ and even,
*(mgf does not exist)*
             $EX^n = 0$ if $n < \nu$ and odd.

*notes*      Related to $F$ $(F_{1,\nu} = t_\nu^2)$.

---

### $\textbf{\textit{Uniform}}(a, b)$

| | |
|---|---|
| *pdf* | $f(x\|a, b) = \frac{1}{b-a}, \quad a \leq x \leq b$ |
| *mean and variance* | $\mathrm{E}X = \frac{b+a}{2}, \quad \mathrm{Var}\,X = \frac{(b-a)^2}{12}$ |
| *mgf* | $M_X(t) = \frac{e^{bt} - e^{at}}{(b-a)t}$ |
| *notes* | If $a = 0$ and $b = 1$, this is a special case of the beta ($\alpha = \beta = 1$). |

### $\textbf{\textit{Weibull}}(\gamma, \beta)$

| | |
|---|---|
| *pdf* | $f(x\|\gamma, \beta) = \frac{\gamma}{\beta} x^{\gamma - 1} e^{-x^\gamma / \beta}, \quad 0 \leq x < \infty, \quad \gamma > 0, \quad \beta > 0$ |
| *mean and variance* | $\mathrm{E}X = \beta^{1/\gamma} \Gamma\left(1 + \frac{1}{\gamma}\right), \quad \mathrm{Var}\,X = \beta^{2/\gamma} \left[\Gamma\left(1 + \frac{2}{\gamma}\right) - \Gamma^2\left(1 + \frac{1}{\gamma}\right)\right]$ |
| *moments* | $\mathrm{E}X^n = \beta^{n/\gamma} \Gamma\left(1 + \frac{n}{\gamma}\right)$ |
| *notes* | The mgf exists only for $\gamma \geq 1$. Its form is not very useful. A special case is exponential ($\gamma = 1$). |

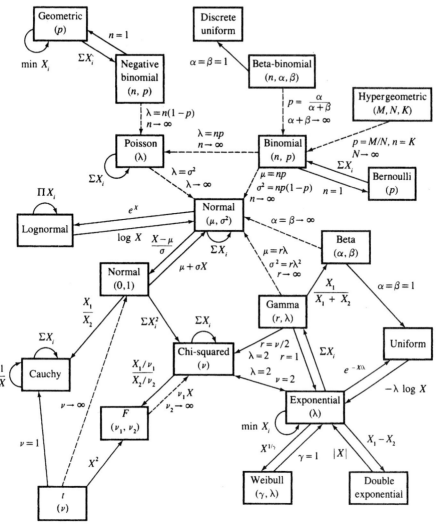

**Relationships among common distributions.** Solid lines represent transformations and special cases, dashed lines represent limits. Adapted from Leemis (1986).

# References

1. Agresti, A. (1990). *Categorical Data Analysis*. New York: Wiley.

2. Agresti, A., and Caffo, B. (2000). Simple and Effective Confidence Intervals for Proportions and Differences of Proportions Result from Adding Two Successes and Two Failures. *Amer. Statist.* **54** 280–288.

3. Agresti, A., and Coull, B. A. (1998). Approximate Is Better Than "Exact" for Interval Estimation of Binomial Proportions. *Amer. Statist.* **52** 119–126.

4. Anderson, T. W. (1976). Estimation of Linear Functional Relationships: Approximate Distributions and Connection with Simultaneous Equations in Econometrics (with discussion). *J. Roy. Statist. Soc. Ser. B* **38** 1–36.

5. Anderson, T. W. (1984). *An Introduction to Multivariate Statistical Analysis*, 2nd edition. New York: Wiley.

6. Balanda, K. P., and MacGillivray, H. L. (1988). Kurtosis: A Critical Review. *Amer. Statist.* **42** 111–119.

7. Bar–Lev, S. K., and Enis, P. (1988). On the Classical Choice of Variance Stabilizing Transformations and an Application for a Poisson Variate. *Biometrika* **75** 803–804.

8. Bar–Lev, S. K., and Enis, P. (1990). On the Construction of Classes of Variance Stabilizing Transformations. *Statist. Prob. Let.* **10** 95–100.

9. Barlow, R., and Proschan, F. (1975). *Statistical Theory of Life Testing*. New York: Holt, Rinehart and Winston.

10. Barnard, G. A. (1949). Statistical Inference (with discussion). *J. Roy. Statist. Soc. Ser. B* **11** 115–139.

11. Barnard, G. A. (1980). Pivotal Inference and the Bayesian Controversy (with discussion). *Bayesian Statistics* (J. M. Bernardo, M. H. DeGroot, D. V. Lindley, and A. F. M. Smith, eds.). Valencia: University Press.

12. Barr, D. R., and Zehna, P. W. (1983). *Probability: Modeling Uncertainty*. Reading, MA: Addison–Wesley.

13. Basu, D. (1959). The Family of Ancillary Statistics. *Sankhyā, Ser. A* **21** 247–256.

14. Bechhofer, R. E. (1954). A Single–Sample Multiple Decision Procedure for Ranking Means of Normal Populations with Known Variances. *Ann. Math. Statist.* **25** 16–39.

15. Behboodian, J. (1990). Examples of Uncorrelated Dependent Random Variables Using a Bivariate Mixture. *Amer. Statist.* **44** 218.

16. Berg, C. (1988). The Cube of a Normal Distribution Is Indeterminate. *Ann. Prob.* **16** 910–913.

17. Berger, J. O. (1984). The Robust Bayesian Viewpoint (with discussion). *Robustness of Bayesian Analysis* (J. Kadane, ed.), 63–144. Amsterdam: North–Holland.

18. Berger, J. O. (1985). *Statistical Decision Theory and Bayesian Analysis*, 2nd edition. New York: Springer-Verlag.

19. Berger, J. O. (1990). Robust Bayesian Analysis: Sensitivity to the Prior. *J. Statist. Plan. Inf.* **25** 303–328.

20. Berger, J. O. (1994). An Overview of Robust Bayesian Analysis (with discussion). *Test* **3** 5–124.

21. Berger, J. O., and Sellke, T. (1987). Testing a Point Null Hypothesis: The Irreconcilability of P Values and Evidence (with discussion). *J. Amer. Statist. Assoc.* **82** 112–122.

22. Berger, J. O., and Wolpert, R. W. (1984). *The Likelihood Principle*. Institute of Mathematical Statistics Lecture Notes—Monograph Series. Hayward, CA: IMS.

23. Berger, R. L. (1982). Multiparameter Hypothesis Testing and Acceptance Sampling. *Technometrics* **24** 295–300.

24. Berger, R. L. (1996). More Powerful Tests from Confidence Interval P Values. *Amer. Statist.* **50** 314–318.

25. Berger, R. L., and Boos, D. D. (1994). P Values Maximized over a Confidence Set for the Nuisance Parameter. *J. Amer. Statist. Assoc.* **89** 1012–1016.

26. Berger, R. L., and Casella, G. (1992). Deriving Generalized Means as Least Squares and Maximum Likelihood Estimates. *Amer. Statist.* **46** 279–282.

27. Berger, R. L., and Hsu, J. C. (1996) Bioequivalence Trials, Intersection–Union Tests and Equivalence Confidence Sets (with discussion). *Statist. Sci.* **11** 283–319.

28. Berkson, J. (1950). Are There Two Regressions? *J. Amer. Statist. Assoc.* **45** 164–180.

29. Bernardo, J. M., and Smith, A. F. M. (1994). *Bayesian Theory*. New York: Wiley.

30. Betteley, I. G. (1977). The Addition Law for Expectations. *Amer. Statist.* **31** 33–35.

31. Bickel, P. J., and Doksum, K. A. (1981). An Analysis of Transformations Revisited. *J. Amer. Statist. Assoc.* **76** 296–311.

32. Bickel, P. J., and Mallows, C. L. (1988). A Note on Unbiased Bayes Estimates. *Amer. Statist.* **42** 132–134.

33. Billingsley, P. (1995). *Probability and Measure*, 3rd edition. New York: Wiley.

34. Binder, D. A. (1993). Letter to the Editor about the "Exchange Paradox." *Amer. Statist.* **47** 160.

35. Birnbaum, A. (1962). On the Foundations of Statistical Inference (with discussion). *J. Amer. Statist. Assoc.* **57** 269–306.

36. Blachman, N. M. (1996). Letter to the Editor about the "Exchange Paradox." *Amer. Statist.* **50** 98–99.

37. Blaker, H. (2000). Confidence Curves and Improved Exact Confidence Intervals for Discrete Distributions. *Can. J. Statist.* **28**.

38. Bloch, D. A., and Moses, L. E. (1988). Nonoptimally Weighted Least Squares. *Amer. Statist.* **42** 50–53.

39. Blyth, C. R. (1986). Approximate Binomial Confidence Limits. *J. Amer. Statist. Assoc.* **81** 843–855; correction **84** 636.

40. Blyth, C. R., and Still, H. A. (1983). Binomial Confidence Intervals. *J. Amer. Statist. Assoc.* **78** 108–116.

41. Boos, D. D. (1992). On Generalized Score Tests (Com: 93V47 P311–312). *Amer. Statist.* **46** 327–333.

42. Boos, D. D., and Hughes–Oliver, J. M. (1998). Applications of Basu's Theorem. *Amer. Statist.* **52** 218–221.

43. Box, G. E. P. (1954). Some Theorems on Quadratic Forms Applied in the Study of Analysis of Variances Problems I: Effect of Inequality of Variance in the One–Way Classification. *Ann. Math. Statist.* **25** 290–302.

44. Box, G. E. P., and Cox, D. R. (1964). An Analysis of Transformations (with discussion). *J. Roy. Statist. Soc. Ser. B* **26** 211–252.

45. Box, G. E. P., and Cox, D. R. (1982). An Analysis of Transformations Revisited, Rebutted. *J. Amer. Statist. Assoc.* **77** 209–210.

46. Box, G. E. P., and Muller, M. (1958). A Note on the Generation of Random Normal Variates. *Ann. Math. Statist.* **29** 610–611.

47. Boyles, R. A. (1983). On the Convergence of the EM Algorithm. *J. Roy. Statist. Soc. Ser. B* **45** 47–50.

48. Brewster, J. F., and Zidek, J. V. (1974). Improving on Equivariant Estimators. *Ann. Statist.* **2** 21–38.

49. Brown, D., and Rothery, P. (1993). *Models in Biology: Mathematics, Statistics and Computing.* New York: Wiley.

50. Brown, L. D. (1986). *Fundamentals of Statistical Exponential Families with Applications in Statistical Decision Theory.* Institute of Mathematical Statistics Lecture Notes—Monograph Series. Hayward, CA: IMS.

51. Brown, L. D. (1990a). Comment on the Paper by Maatta and Casella. *Statist. Sci.* **5** 103–106.

52. Brown, L. D. (1990b). An Ancillarity Paradox Which Occurs in Multiple Linear Regression (with discussion). *Ann. Statist.* **18** 471–492.

53. Brown, L. D., Casella, G., and Hwang, J. T. G. (1995). Optimal Confidence Sets, Bioequivalence, and the Limaçon of Pascal. *J. Amer. Statist. Assoc.* **90** 880–889.

54. Brown, L. D., and Purves, R. (1973). Measurable Selections of Extrema. *Ann. Statist.* **1** 902–912.

55. Brown, P. J., and Fuller, W. A. (1991). *Statistical Analysis of Measurement Error Models and Applications.* Providence, R.I.: American Mathematical Society.

56. Buehler, R. J. (1982). Some Ancillary Statistics and Their Properties (with discussion). *J. Amer. Statist. Assoc.* **77** 581–594.

57. Carlin, B. P., and Louis, T. A. (1996). *Bayes and Empirical Bayes Methods for Data Analysis.* London: Chapman and Hall.

58. Carmer, S. G., and Walker, W. M. (1982). Baby Bear's Dilemma: A Statistical Tale. *Agronomy Journal* **74** 122–124.

59. Carroll, R. J., Gallo, P., and Gleser, L. J. (1985). Comparison of Least Squares and Errors–in–Variables Regression, with Special Reference to Randomized Analysis of Covariance. *J. Amer. Statist. Assoc.* **80** 929–932.

60. Carroll, R. J., and Ruppert, D. (1985). Transformations: A Robust Analysis. *Technometrics* **27** 1–12.

61. Carroll, R. J., and Ruppert, D. (1988). *Transformation and Weighting in Regression*. London: Chapman and Hall.

62. Carroll, R. J., Ruppert, D., and Stefanski, L. A. (1995). *Measurement Error in Nonlinear Models*. London: Chapman and Hall.

63. Casella, G. (1985). An Introduction to Empirical Bayes Data Analysis. *Amer. Statist.* **39** 83–87.

64. Casella, G. (1986). Refining Binomial Confidence Intervals. *Can. J. Statist.* **14** 113–129.

65. Casella, G. (1992). Illustrating Empirical Bayes Methods. *Chemolab* **16** 107–125.

66. Casella, G., and Berger, R. L. (1987). Reconciling Bayesian and Frequentist Evidence in the One–Sided Testing Problem (with discussion). *J. Amer. Statist. Assoc.* **82** 106–111.

67. Casella, G., and George, E. I. (1992). Explaining the Gibbs Sampler. *Amer. Statist.* **46** 167–174.

68. Casella, G., and Hwang, J. T. (1987). Employing Vague Prior Information in the Construction of Confidence Sets. *J. Mult. Analysis* **21** 79–104.

69. Casella, G., and Robert, C. (1989). Refining Poisson Confidence Intervals. *Can. J. Statist.* **17** 45–57.

70. Casella, G., and Robert, C. P. (1996). Rao–Blackwellisation of Sampling Schemes. *Biometrika* **83** 81–94.

71. Casella, G., and Strawderman, W. E. (1980). Confidence Bands for Linear Regression with Restricted Predictor Variables. *J. Amer. Statist. Assoc.* **75** 862–868.

72. Chapman, D. G., and Robbins, H. (1951). Minimum Variance Estimation Without Regularity Assumptions. *Ann. Math. Statist.* **22** 581–586.

73. Chib, S., and Greenberg, E. (1995). Understanding the Metropolis–Hastings Algorithm. *Ann. Math. Statist.* **49** 327–335.

74. Chikkara, R. S., and Folks, J. L. (1989). *The Inverse Gaussian Distribution: Theory, Methodology, and Applications*. New York: Marcel Dekker.

75. Christensen, R. (1996). *Plane Answers to Complex Questions. The Theory of Linear Models*, 2nd edition. New York: Springer-Verlag.

76. Christensen, R., and Utts, J. (1992). Bayesian Resolution of the "Exchange Paradox." *Amer. Statist.* **46** 274–278.

77. Chun, Y. H. (1999). On the Information Economics Approach to the Generalized Game Show Problem. *Amer. Statist.* **53** 43–51.

78. Chung, K. L. (1974). *A Course in Probability Theory*. New York: Academic Press.

79. Clopper, C. J., and Pearson, E. S. (1934). The Use of Confidence or Fiducial Limits Illustrated in the Case of the Binomial. *Biometrika* **26** 404–413. (Also in *The Selected Papers of E. S. Pearson*, New York: Cambridge University Press, 1966.)

80. Cochran, W. G. (1934). The Distribution of Quadratic Forms in a Normal System with Applications to the Analysis of Variance. *Proceedings of the Cambridge Philosophical Society* **30** 178–191.

81. Cochran, W. G. (1968). Errors of Measurement in Statistics. *Technometrics* **10** 637–666.

82. Cochran, W. G., and Cox, G. M. (1957). *Experimental Designs*, 2nd edition. New York: Wiley.

83. Cornfield, J. (1969). The Bayesian Outlook and Its Application (with discussion). *Biometrika* **25** 617–657.

84. Cox, D. R. (1958). Some Problems Connected with Statistical Inference. *Ann. Math. Statist.* **29** 357–372.

85. Cox, D. R. (1971). The Choice Between Ancillary Statistics. *J. Roy. Statist. Soc. Ser. B* **33** 251–255.

86. Cox, D. R., and Hinkley, D. V. (1974). *Theoretical Statistics*. London: Chapman and Hall.

87. Creasy, M. A. (1956). Confidence Limits for the Gradient in the Linear Functional Relationship. *J. Roy. Statist. Soc. Ser. B* **18** 65–69.

88. Crow, E. L. (1956). Confidence Intervals for a Proportion. *Biometrika* **43** 423–425.

89. Crow, E. L., and Gardner, R. S. (1959). Confidence Intervals for the Expectation of a Poisson Variable. *Biometrika* **46** 441–453.

90. Curtiss, J. H. (1943). On Transformations Used in the Analysis of Variance. *Ann. Math. Statist.* **14** 107–122.

91. Dalal, S. R., Fowlkes, E. B., and Hoadley, B. (1989). Risk Analysis of the Space Shuttle: Pre–Challenger Prediction of Failure. *J. Amer. Statist. Assoc.* **84** 945–957.

92. David, H. A. (1985). Bias of $S^2$ Under Dependence. *Amer. Statist.* **39** 201.

93. Davidson, R. R., and Solomon, D. L. (1974). Moment–Type Estimation in the Exponential Family. *Communications in Statistics* **3** 1101–1108.

94. Dean, A., and Voss, D. (1999). *Design and Analysis of Experiments*. New York: Springer-Verlag.

95. deFinetti, B. (1972). *Probability, Induction, and Statistics*. New York: Wiley.

96. DeGroot, M. H. (1986). *Probability and Statistics*, 2nd edition. New York: Addison–Wesley.

97. Dempster, A. P., Laird, N. M., and Rubin, D. B. (1977). Maximum Likelihood from Incomplete Data via the EM Algorithm. *J. Roy. Statist. Soc. Ser. B* **39** 1–22.

98. Devroye, L. (1985). *Non-Uniform Random Variate Generation*. New York: Springer-Verlag.

99. Diaconis, P., and Holmes, S. (1994). Gray Codes for Randomization Procedures. *Statistics and Computing* **4** 287–302.

100. Diaconis, P., and Mosteller, F. (1989). Methods for Studying Coincidences. *J. Amer. Statist. Assoc.* **84** 853–861.

101. Draper, D. (1988). Rank–Based Robust Analysis of Linear Models. I. Exposition and Review (with discussion). *Statist. Sci.* **3** 239–271.

102. Draper, N. R., and Smith, H. (1998). *Applied Regression Analysis*, 3rd edition. New York: Wiley.

103. Durbin, J. (1970). On Birnbaum's Theorem and the Relation Between Sufficiency, Conditionality, and Likelihood. *J. Amer. Statist. Assoc.* **65** 395–398.

104. Dynkin, E. B. (1951). Necessary and Sufficient Statistics for a Family of Probability Distributions. English translation in *Selected Translations in Mathematical Statistics and Probability* **1** (1961), 23–41.

105. Eberhardt, K. R., and Fligner, M. A. (1977). A Comparison of Two Tests for Equality of Two Proportions. *Amer. Statist.* **31** 151–155.

106. Efron, B. F. (1979a). Bootstrap Methods: Another Look at the Jackknife. *Ann. Statist.* **7** 1–26.

107. Efron, B. F. (1979b). Computers and the Theory of Statistics: Thinking the Unthinkable. *SIAM Review* **21** 460–480.

108. Efron, B. F. (1982). *The Jackknife, the Bootstrap, and Other Resampling Plans.* Philadelphia: Society for Industrial and Applied Mathematics.

109. Efron, B. F. (1998). R. A. Fisher in the 21st Century (with discussion). *Statist. Sci.* **13** 95–122.

110. Efron, B. F., and Hinkley, D. V. (1978). Assessing the Accuracy of the Maximum Likelihood Estimator: Observed Versus Expected Fisher Information. *Biometrika* **65** 457–487.

111. Efron, B. F., and Morris, C. N. (1972). Limiting the Risk of Bayes and Empirical Bayes Estimators Part II: The Empirical Bayes Case. *J. Amer. Statist. Assoc.* **67** 130–139.

112. Efron, B. F., and Morris, C. N. (1973). Stein's Estimation Rule and Its Competitors—An Empirical Bayes Approach. *J. Amer. Statist. Assoc.* **68** 117–130.

113. Efron, B. F., and Morris, C. N. (1975). Data Analysis Using Stein's Estimator and Its Generalizations. *J. Amer. Statist. Assoc.* **70** 311–319.

114. Efron, B., and Tibshirani, R. J. (1993). *An Introduction to the Bootstrap.* London: Chapman and Hall.

115. Ellis, S. (1998). Instability of Least Squares, Least Absolute Deviation and Least Median of Squares Linear Regression (with discussion). *Statist. Sci.* **13** 337–350.

116. Feldman, D., and Fox, M. (1968). Estimation of the Parameter $n$ in the Binomial Distribution. *J. Amer. Statist. Assoc.* **63** 150–158.

117. Feldstein, M. (1974). Errors in Variables: A Consistent Estimator with Smaller MSE in Finite Samples. *J. Amer. Statist. Assoc.* **69** 990–996.

118. Feller, W. (1968). *An Introduction to Probability Theory and Its Applications, Volume I.* New York: Wiley.

119. Feller, W. (1971). *An Introduction to Probability Theory and Its Applications, Volume II.* New York: Wiley.

120. Ferguson, T. S. (1996). *A Course in Large Sample Theory.* London: Chapman and Hall.

121. Fieller, E. C. (1954). Some Problems in Interval Estimation. *J. Roy. Statist. Soc. Ser. B* **16** 175–185.

122. Finch, S. J., Mendell, N. R., and Thode, H. C. (1989). Probabilistic Measures of Adequacy of a Numerical Search for a Global Maximum. *J. Amer. Statist. Assoc.* **84** 1020–1023.

123. Fisher, R. A. (1925). Theory of Statistical Estimation. *Proceedings of the Cambridge Philosophical Society* **22** 700–725.

124. Fisher, R. A. (1930). Inverse Probability. *Proceedings of the Cambridge Philosophical Society* **26** 528–535.

125. Fisher, R. A. (1935). The Fiducial Argument in Statistical Inference. *Annals of Eugenics* **6** 391–398. (Also in R. A. Fisher, *Contributions to Mathematical Statistics*, New York: Wiley, 1950.)

126. Fisher, R. A. (1939). The Comparison of Samples with Possibly Unequal Variances. *Annals of Eugenics* **9** 174–180. (Also in R. A. Fisher, *Contributions to Mathematical Statistics*, New York: Wiley, 1950.)

127. Fraser, D. A. S. (1968). *The Structure of Inference*. New York: Wiley.

128. Fraser, D. A. S. (1979). *Inference and Linear Models*. New York: McGraw–Hill.

129. Freedman, D., Pisani, R., Purves, R., and Adhikari, A. (1991). *Statistics*, 2nd edition. New York: Norton.

130. Fuller, W. A. (1987). *Measurement Error Models*. New York: Wiley.

131. Gafarian, A. V. (1964). Confidence Bands in Straight Line Regression. *J. Amer. Statist. Assoc.* **59** 182–213.

132. Gardner, M. (1961). *The Second Scientific American Book of Mathematical Puzzles and Diversions*. New York: Simon & Schuster.

133. Garwood, F. (1936). Fiducial Limits for the Poisson Distribution. *Biometrika* **28** 437–442.

134. Gelfand, A. E., and Smith, A. F. M. (1990). Sampling–Based Approaches to Calculating Marginal Densities. *J. Amer. Statist. Assoc.* **85** 398–409.

135. Gelman, A., and Meng, X.–L. (1991). A Note on Bivariate Distributions That Are Conditionally Normal. *Amer. Statist.* **45** 125–126.

136. Gelman, A., and Rubin, D. B. (1992). Inference from Iterative Simulation Using Multiple Sequences (with discussion). *Statist. Sci.* **7** 457–511.

137. Geman, S., and Geman, D. (1984). Stochastic Relaxation, Gibbs Distributions and the Bayesian Restoration of Images. *IEEE Trans. Pattern Anal. Mach. Intell.* **6** 721–741.

138. Geyer, C. J., and Thompson, E. A. (1992). Constrained Monte Carlo Maximum Likelihood for Dependent Data (with discussion). *J. Roy. Statist. Soc. Ser. B* **54** 657–699.

139. Ghosh, B. K. (1979). A Comparison of Some Approximate Confidence Intervals for the Binomial Parameter. *J. Amer. Statist. Assoc.* **74** 894–900.

140. Ghosh, M., and Meeden, G. (1977). On the Non–Attainability of Chebychev Bounds. *Amer. Statist.* **31** 35–36.

141. Gianola, D., and Fernando, R. L. (1986). Bayesian Methods in Animal Breeding Theory. *Journal of Animal Science* **63** 217–244.

142. Gilat, D. (1977). Monotonicity of a Power Function: An Elementary Probabilistic Proof. *Amer. Statist.* **31** 91–93.

143. Gleser, L. J. (1981). Estimation in a Multivariate Errors–in–Variables Regression Model: Large–Sample Results. *Ann. Statist.* **9** 24–44.

144. Gleser, L. J. (1983). Functional, Structural, and Ultrastructural Errors–in–Variables Models. *Proceedings of the Business and Economic Statistics Section*, 57–66. Alexandria, VA: American Statistical Association.

145. Gleser, L. J. (1987). Confidence Intervals for the Slope in a Linear Errors–in–Variables Regression Model. *Advances in Multivariate Statistical Analysis*, (K. Gupta, ed.), 85–109. Dordrecht: D. Reidel.

146. Gleser, L. J. (1989). The Gamma Distribution As a Mixture of Exponential Distributions. *Amer. Statist.* **43** 115–117.

147. Gleser, L. J. (1991). Measurement Error Models (with discussion). *Chem. Int. Lab. Sys.* **10** 45–67.

148. Gleser, L. J., Carroll, R. J., and Gallo, P. (1987). The Limiting Distribution of Least Squares in an Errors–in–Variables Regression Model. *Ann. Statist.* **15** 220–233.

149. Gleser, L. J., and Healy, J. D. (1976). Estimating the Mean of a Normal Distribution with Known Coefficient of Variation. *J. Amer. Statist. Assoc.* **71** 977–981.

150. Gleser, L. J., and Hwang, J. T. (1987). The Nonexistence of $100(1 - \alpha)\%$ Confidence Sets of Finite Expected Diameter in Errors–in–Variables and Related Models. *Ann. Statist.* **15** 1351–1362.

151. Gnedenko, B. V. (1978). *The Theory of Probability.* Moscow: MIR Publishers.

152. Groeneveld, R. A. (1991). An Influence Function Approach to Describing the Skewness of a Distribution. *Amer. Statist.* **45** 97–102.

153. Guenther, W. C. (1978). Some Easily Found Minimum Variance Unbiased Estimators. *Amer. Statist.* **32** 29–33.

154. Hall, P. (1992). *The Bootstrap and Edgeworth Expansion.* New York: Springer-Verlag.

155. Halmos, P. R., and Savage, L. J. (1949). Applications of the Radon–Nikodym Theorem to the Theory of Sufficient Statistics. *Ann. Math. Statist.* **20** 225–241.

156. Hampel, F. R. (1974). The Influence Curve and Its Role in Robust Estimation. *J. Amer. Statist. Assoc.* **69** 383–393.

157. Hanley, J. A. (1992). Jumping to Coincidences: Defying Odds in the Realm of the Preposterous. *Amer. Statist.* **46** 197–202.

158. Hardy, G. H., Littlewood, J. E., and Polya, G. (1952). *Inequalities*, 2nd edition. London: Cambridge University Press.

159. Hartley, H. O. (1958). Maximum Likelihood Estimation from Incomplete Data. *Biometrics* **14** 174–194.

160. Harville, D. A. (1981). Unbiased and Minimum–Variance Unbiased Estimation of Estimable Functions for Fixed Linear Models with Arbitrary Covariance Structure. *Ann. Statist.* **9** 633–637.

161. Hawkins, D. M. (1993). The Feasible Set Algorithm for Least Median of Squares Regression. *Computational Statistics and Data Analysis* **16** 81–101.

162. Hawkins, D. M. (1994). The Feasible Solution Algorithm for Least Trimmed Squares Regression. *Computational Statistics and Data Analysis* **17** 186–196.

163. Hawkins, D. M. (1995). Convergence of the Feasible Solution Algorithm for Least Median of Squares Regression. *Computational Statistics and Data Analysis* **19** 519–538.

164. Hayter, A. J. (1984). A Proof of the Conjecture That the Tukey–Kramer Multiple Comparison Procedure Is Conservative. *Ann. Statist.* **12** 61–75.

165. Hettmansperger, T. P., and McKean, J. W. (1998). *Robust Nonparametric Statistical Methods.* London: Kendall's Library of Statistics, 5. Edward Arnold; New York: Wiley.

166. Hinkley, D. V. (1980). Likelihood. *Can. J. Statist.* **8** 151–163.

167. Hinkley, D. V., Reid, N., and Snell, L. (1991). *Statistical Theory and Modelling. In Honor of Sir David Cox.* London: Chapman and Hall.

168. Hinkley, D. V., and Runger, G. (1984). The Analysis of Transformed Data (with discussion). *J. Amer. Statist. Assoc.* **79** 302–320.

169. Hsu, J. C. (1996). *Multiple Comparisons: Theory and Methods*. London: Chapman and Hall.

170. Huber, P. J. (1964). Robust Estimation of a Location Parameter. *Ann. Math. Statist.* **35** 73–101.

171. Huber, P. J. (1981). *Robust Statistics*. New York: Wiley.

172. Hudson, H. M. (1978). A Natural Identity for Exponential Families with Applications in Multiparameter Estimation. *Ann. Statist.* **6** 473–484.

173. Huzurbazar, V. S. (1949). On a Property of Distributions Admitting Sufficient Statistics. *Biometrika* **36** 71–74.

174. Hwang, J. T. (1982). Improving on Standard Estimators in Discrete Exponential Families with Applications to Poisson and Negative Binomial Cases. *Ann. Statist.* **10** 857–867.

175. Hwang, J. T. (1995). Fieller's Problems and Resampling Techniques. *Statistica Sinica* **5** 161–171.

176. James, W., and Stein, C. (1961). Estimation with Quadratic Loss. *Proceedings of the Fourth Berkeley Symposium on Mathematical Statistics and Probability* **1** 361–380. Berkeley: University of California Press.

177. Johnson, N. L., and Kotz, S. (1969–1972). *Distributions in Statistics* (4 vols.). New York: Wiley.

178. Johnson, N. L., Kotz, S., and Balakrishnan, N. (1994). *Continuous Univariate Distributions, Volume 1*, 2nd edition. New York: Wiley.

179. Johnson, N. L., Kotz, S., and Balakrishnan, N. (1995). *Continuous Univariate Distributions, Volume 2*, 2nd edition. New York: Wiley.

180. Johnson, N. L., Kotz, S., and Kemp, A. W. (1992). *Univariate Discrete Distributions*, 2nd edition. New York: Wiley.

181. Jones, M. C. (1999). Distributional Relationships Arising from Simple Trigonometric Formulas. *Amer. Statist.* **53** 99–102.

182. Joshi, S. M., and Nabar, S. P. (1989). Linear Estimators for the Parameter in the Problem of the Nile. *Amer. Statist.* **43** 40–41.

183. Joshi, V. M. (1969). Admissibility of the Usual Confidence Sets for the Mean of a Univariate or Bivariate Normal Population. *Ann. Math. Statist.* **40** 1042–1067.

184. Juola, R. C. (1993). More on Shortest Confidence Intervals. *Amer. Statist.* **47** 117–119.

185. Kalbfleisch, J. D. (1975). Sufficiency and Conditionality. *Biometrika* **62** 251–268.

186. Kalbfleisch, J. D., and Prentice, R. L. (1980). *The Statistical Analysis of Failure Time Data*. New York: Wiley.

187. Karlin, S., and Ost, F. (1988). Maximal Length of Common Words Among Random Letter Sequences. *Ann. Prob.* **16** 535–563.

188. Kelker, D. (1970). Distribution Theory of Spherical Distributions and a Location–Scale Parameter Generalization. *Sankhyā, Ser. A* **32** 419–430.

189. Kendall, M., and Stuart, A. (1979). *The Advanced Theory of Statistics, Volume II: Inference and Relationship*, 4th edition. New York: Macmillan.

190. Kirk, R. E. (1982). *Experimental Design: Procedures for the Behavorial Sciences*, 2nd edition. Pacific Grove, CA: Brooks/Cole.

191. Koopmans, L. H. (1993). A Note on Using the Moment Generating Function to Teach the Laws of Large Numbers. *Amer. Statist.* **47** 199–202.

192. Kuehl, R. O. (2000). *Design of Experiments: Statistical Principles of Research Design and Analysis*, 2nd edition. Pacific Grove, CA: Duxbury.

193. Lange, N., Billard, L., Conquest, L., Ryan, L., Brillinger, D., and Greenhouse, J. (eds.). (1994). *Case Studies in Biometry.* New York: Wiley–Interscience.

194. Le Cam, L. (1953). On Some Asymptotic Properties of Maximum Likelihood Estimates and Related Bayes' Estimates. *Univ. of Calif. Publ. in Statist.* **1** 277–330.

195. Leemis, L. M. (1986). Relationships Among Common Univariate Distributions. *Amer. Statist.* **40** 143–146.

196. Leemis, L. M., and Trivedi, K. S. (1996). A Comparison of Approximate Interval Estimators for the Bernoulli Parameter. *Amer. Statist.* **50** 63–68.

197. Lehmann, E. L. (1981). An Interpretation of Completeness and Basu's Theorem. *J. Amer. Statist. Assoc.* **76** 335–340.

198. Lehmann, E. L. (1986). *Testing Statistical Hypotheses*, 2nd edition. New York: Wiley.

199. Lehmann, E. L. (1999). *Introduction to Large-Sample Theory.* New York: Springer-Verlag.

200. Lehmann, E. L., and Casella, G. (1998). *Theory of Point Estimation*, 2nd edition. New York: Springer-Verlag.

201. Lehmann, E. L., and Scheffé, H. (1950, 1955, 1956). Completeness, Similar Regions, and Unbiased Estimation. *Sankhyā, Ser. A* **10** 305–340; **15** 219–236; correction **17** 250.

202. Lehmann, E. L., and Scholz, F. W. (1992). Ancillarity. *Current Issues in Statistical Inference: Essays in Honor of D. Basu* (M. Ghosh and P. K. Pathak, eds.). Hayward, CA: IMS Monograph Series, 32–51.

203. LePage, R., and Billard, L. (eds.). (1992). *Exploring the Limits of Bootstrap.* New York: Wiley.

204. Lindley, D. V. (1957). A Statistical Paradox. *Biometrika* **44** 187–192.

205. Lindley, D. V. (1962). Discussion of the Article by Stein. *J. Roy. Statist. Soc. Ser. B* **24** 265–296.

206. Lindley, D. V., and Phillips, L. D. (1976). Inference for a Bernoulli Process (a Bayesian View). *Amer. Statist.* **30** 112–119.

207. Lindley, D. V., and Smith, A. F. M. (1972). Bayes Estimates for the Linear Model. *J. Roy. Statist. Soc. Ser. B* **34** 1–41.

208. Little, R. J. A., and Rubin, D. B. (1987). *Statistical Analysis with Missing Data.* New York: Wiley.

209. Liu, R. Y. (1990). On a Notion of Data Depth Based on Random Simplices. *Ann. Statist.* **18** 405–414.

210. Liu, R. Y., and Singh, K. (1992). Ordering Directional Data: Concepts of Data Depth on Circles and Spheres. *Ann. Statist.* **20** 1468–1484.

211. Luceño, A. (1997). Further Evidence Supporting the Numerical Usefulness of Characteristic Functions. *Amer. Statist.* **51** 233–234.

212. Maatta, J. M., and Casella, G. (1987). Conditional Properties of Interval Estimators of the Normal Variance. *Ann. Statist.* **15** 1372–1388.

213. Madansky, A. (1962). More on Length of Confidence Intervals. *J. Amer. Statist. Assoc.* **57** 586–589.

214. Marshall, A. W., and Olkin, I. (1979). *Inequalities: Theory of Majorization and Its Applications.* New York: Academic Press.

215. McCullagh, P. (1994). Does the Moment Generating Function Characterize a Distribution? *Amer. Statist.* **48** 208.

216. McLachlan, G., and Krishnan, T. (1997). *The EM Algotithm and Extensions.* New York: Wiley.

217. McPherson, G. (1990). *Statistics in Scientific Inverstigation.* New York: Springer-Verlag.

218. Mengersen, K. L., and Tweedie, R. L. (1996). Rates of Convergence of the Hastings and Metropolis Algorithms. *Ann. Statist.* **24** 101–121.

219. Metropolis, N., Rosenbluth, A. W., Rosenbluth, M. N., Teller, A. H., and Teller, E. (1953). Equations of State Calculations by Fast Computing Machines. *J. Chem. Phys.* **21** 1087–1092.

220. Miller, R. G. (1974). The Jackknife—A Review. *Biometrika* **61** 1–15.

221. Miller, R. G. (1981). *Simultaneous Statistical Inference*, 2nd edition. New York: Springer-Verlag.

222. Moran, P. A. P. (1971). Estimating Structural and Functional Relationships. *J. Mult. Analysis.* **1** 232–255.

223. Morgan, J. P., Chaganty, N. R., Dahiya, R. C., and Doviak, M. J. (1991). Let's Make a Deal: The Player's Dilemma (with discussion). *Amer. Statist.* **45** 284–289.

224. Morris, C. N. (1982). Natural Exponential Families with Quadratic Variance Functions. *Ann. Statist.* **10** 65–80.

225. Morris, C. N. (1983). Parametric Empirical Bayes Inference: Theory and Applications (with discussion). *J. Amer. Statist. Assoc.* **78** 47–65.

226. Morrison, D. G. (1978). A Probability Model for Forced Binary Choices. *Amer. Statist.* **32** 23–25.

227. Naiman, D. Q. (1983). Comparing Scheffé–Type to Constant Width Confidence Bounds in Regression. *J. Amer. Statist. Assoc.* **78** 906–912.

228. Naiman, D. Q. (1984). Optimal Simultaneous Confidence Bounds. *Ann. Statist.* **12** 702–715.

229. Naiman, D. Q. (1987). Simultaneous Confidence Bounds in Multiple Regression Using Predictor Variable Constraints. *J. Amer. Statist. Assoc.* **82** 214–219.

230. Naiman, D. Q., and Wynn, H. P. (1992). Inclusion–Exclusion–Bonferroni Identities and Inequalities for Discrete Tube–Like Problems via Euler Characteristics. *Ann. Statist.* **20** 43–76.

231. Naiman, D. Q., and Wynn, H. P. (1997). Abstract Tubes, Improved Inclusion–Exclusion Identities and Inequalities and Importance Sampling. *Ann. Statist.* **25** 1954–1983.

232. Neter, J., Wasserman, W., and Whitmore, G. A. (1993). *Applied Statistics.* Boston: Allyn & Bacon.

233. Neyman, J. (1935). Su un Teorema Concernente le Cosiddette Statistiche Sufficienti. *Inst. Ital. Atti. Giorn.* **6** 320–334.

234. Noorbaloochi, S., and Meeden, G. (1983). Unbiasedness As the Dual of Being Bayes. *J. Amer. Statist. Assoc.* **78** 619–623.

235. Norton, R. M. (1984). The Double Exponential Distribution: Using Calculus to Find an MLE. *Amer. Statist.* **38** 135–136.

236. Nussbaum, M. (1976). Maximum Likelihood and Least Squares Estimation of Linear Functional Relationships. *Mathematische Operationsforschung und Statistik, Ser. Statistik* **7** 23–49.

237. Olkin, I., Petkau, A. J., and Zidek, J. V. (1981). A Comparison of $n$ Estimators for the Binomial Distribution. *J. Amer. Statist. Assoc.* **76** 637–642.

238. Pal, N., and Berry, J. (1992). On Invariance and Maximum Likelihood Estimation. *Amer. Statist.* **46** 209–212.

239. Park, C. G., Park, T., and Shin, D. W. (1996). A Simple Method for Generating Correlated Binary Variates. *Amer. Statist.* **50** 306-310.

240. Pena, E. A., and Rohatgi, V. (1994). Some Comments About Sufficiency and Unbiased Estimation. *Amer. Statist.* **48** 242–243.

241. Piegorsch, W. W. (1985). Admissible and Optimal Confidence Bands in Simple Linear Regression. *Ann. Statist.* **13** 801–810.

242. Pitman, E. J. G. (1939). The Estimation of the Location and Scale Parameters of a Continuous Population of Any Given Form. *Biometrika* **30** 200–215.

243. Portnoy, S. (1987). A Central Limit Theorem Applicable to Robust Regression Estimators. *J. Mult. Analysis.* **22** 24–50.

244. Portnoy, S., and Koenker, R. (1997). The Gaussian Hare and the Laplacian Tortoise: Computability of Squared–Error Versus Absolute–Error Estimators. *Statist. Sci.* **12** 279–300.

245. Portnoy, S., and Mizera, I. (1998). Discussion of the Paper by Ellis. *Statist. Sci.* **13** 344–347.

246. Pratt, J. W. (1961). Length of Confidence Intervals. *J. Amer. Statist. Assoc.* **56** 549–567.

247. Proschan, M. A., and Presnell, B. (1998). Expect the Unexpected from Conditional Expectation. *Amer. Statist.* **52** 248–252.

248. Pukelsheim, F. (1994). The Three Sigma Rule. *Amer. Statist.* **48** 88–91.

249. Quenouille, M. H. (1956). Notes on Bias in Estimation. *Biometrika* **43** 353–360.

250. Reid, N. (1995). The Role of Conditioning in Inference (with discussion). *Statist. Sci.* **10** 138–166.

251. Resnick, S. I. (1999). *A Probability Path*. Basel: Birkhauser.

252. Ridgeway, T. (1993). Letter to the Editor about the "Exchange Paradox." *Amer. Statist.* **47** 311.

253. Ripley, B. D. (1987). *Stochastic Simulation*. New York: Wiley.

254. Robbins, H. (1977). A Fundamental Question of Practical Statistics (letter to the editor). *Amer. Statist.* **31** 97.

255. Robert, C. P. (1994). *The Bayesian Choice*. New York: Springer-Verlag.

256. Robert, C. P., and Casella, G. (1999). *Monte Carlo Statistical Methods*. New York: Springer-Verlag.

257. Robson, D. S. (1959). A Simple Method for Constructing Orthogonal Polynomials When the Independent Variable Is Unequally Spaced. *Biometrics* **15** 187–191.

258. Romano, J. P., and Siegel, A. F. (1986). *Counterexamples in Probability and Statistics*. Pacific Grove, CA: Wadsworth and Brooks/Cole.

259. Ross, S. M. (1988). *A First Course in Probability Theory*, 3rd edition. New York: Macmillan.

260. Ross, S. M. (1994). Letter to the Editor about the "Exchange Paradox." *Amer. Statist.* **48** 267.

261. Ross, S. M. (1996). Bayesians Should Not Resample a Prior Sample to Learn About the Posterior. *Amer. Statist.* **50** 116.

262. Rousseeuw, P. J. (1984). Least Median of Squares Regression. *J. Amer. Statist. Assoc.* **79** 871–880.

263. Rousseeuw, P. J., and Hubert, M. (1999). Regression Depth (with discussion). *J. Amer. Statist. Assoc.* **94** 388–433.

264. Rousseeuw, P. J., and Leroy, A. M. (1987). *Robust Regression and Outlier Detection*. New York: Wiley.

265. Royall, R. M. (1997). *Statistical Evidence: A Likelihood Paradigm*. London: Chapman and Hall.

266. Rubin, D. B. (1988). Using the SIR Algorithm to Simulate Posterior Distributions. *Bayesian Statistics 3* (J. M. Bernardo, M. H. DeGroot, D. V. Lindley, and A. F. M. Smith, eds.), 395–402. Cambridge, MA: Oxford University Press.

267. Rudin, W. (1976). *Principles of Real Analysis*. New York: McGraw–Hill.

268. Ruppert, D. (1987). What Is Kurtosis? *Amer. Statist.* **41** 1–5.

269. Ruppert, D., and Carroll, R. J. (1979). Trimmed Least Squares Estimation in the Linear Model. *J. Amer. Statist. Assoc.* **75** 828–838.

270. Russell, K. G. (1991). Estimating the Value of $e$ by Simulation. *Amer. Statist.* **45** 66–68.

271. Samuels, M. L., and Lu, T.–F. C. (1992). Sample Size Requirements for the Back–of–the–Envelope Binomial Confidence Interval. *Amer. Statist.* **46** 228–231.

272. Satterthwaite, F. E. (1946). An Approximate Distribution of Estimates of Variance Components. *Biometrics Bulletin* (now called *Biometrics*) **2** 110–114.

273. Saw, J. G., Yang, M. C. K., and Mo, T. C. (1984). Chebychev's Inequality with Estimated Mean and Variance. *Amer. Statist.* **38** 130–132.

274. Schafer, J. L. (1997). *Analysis of Incomplete Multivariate Data*. London: Chapman and Hall.

275. Scheffé, H. (1959). *The Analysis of Variance*. New York: Wiley.

276. Schervish, M. J. (1995). *Theory of Statistics*. New York: Springer-Verlag.

277. Schervish, M. J. (1996). P Values: What They Are and What They Are Not. *Amer. Statist.* **50** 203–206.

278. Schuirmann, D. J. (1987). A Comparison of the Two One–Sided Tests Procedure and the Power Approach for Assessing the Equivalence of Average Bioavailability. *J. Pharmacokinetics and Biopharmaceutics* **15** 657–680.

279. Schwager, S. J. (1984). Bonferroni Sometimes Loses. *Amer. Statist.* **38** 192–197.

280. Schwager, S. J. (1985). Reply to Worsley. *Amer. Statist.* **39** 236.

281. Schwarz, C. J., and Samanta, M. (1991). An Inductive Proof of the Sampling Distributions for the MLEs of the Parameters in an Inverse Gaussian Distribution. *Amer. Statist.* **45** 223–225.

282. Searle, S. R. (1971). *Linear Models.* New York: Wiley.

283. Searle, S. R. (1982). *Matrix Algebra Useful for Statistics.* New York: Wiley.

284. Searls, D. T., and Intarapanich, P. (1990). A Note on an Estimator for the Variance That Utilizes the Kurtosis. *Amer. Statist.* **44** 295–296.

285. Selvin, S. (1975). A Problem in Probability (letter to the editor). *Amer. Statist.* **29** 67.

286. Seshadri, V. (1993). *The Inverse Gaussian Distribution. A Case Study in Exponential Families.* New York: Clarendon Press.

287. Shao, J. (1999). *Mathematical Statistics.* New York: Springer-Verlag.

288. Shao, J., and Tu, D. (1995). *The Jackknife and the Bootstrap.* New York: Springer-Verlag.

289. Shier, D. P. (1988). The Monotonicity of Power Means Using Entropy. *Amer. Statist.* **42** 203–204.

290. Shuster, J. J. (1991). The Statistician in a Reverse Cocaine Sting. *Amer. Statist.* **45** 123–124.

291. Silvapulle, M. J. (1996). A Test in the Presence of Nuisance Parameters. *J. Amer. Statist. Assoc.* **91** 1690–1693.

292. Smith, A. F. M., and Gelfand, A. E. (1992). Bayesian Statistics Without Tears: A Sampling–Resampling Perspective. *Amer. Statist.* **46** 84–88.

293. Smith, A. F. M., and Roberts, G. O. (1993). Bayesian Computation via the Gibbs Sampler and Related Markov Chain Monte Carlo Methods (with discussion). *J. Roy. Statist. Soc. Ser. B* **55** 3–24.

294. Snedecor, G. W., and Cochran, W. G. (1989). *Statistical Methods*, 8th edition. Ames: Iowa State University Press.

295. Solari, M. E. (1969). The "Maximum Likelihood Solution" of the Problem of Estimating a Linear Functional Relationship. *J. Roy. Statist. Soc. Ser. B* **31** 372–375.

296. Solomon, D. L. (1975). A Note on the Non–Equivalence of the Neyman–Pearson and Generalized Likelihood Ratio Tests for Testing a Simple Null Versus a Simple Alternative Hypothesis. *Amer. Statist.* **29** 101–102.

297. Solomon, D. L. (1983). The Spatial Distribution of Cabbage Butterfly Eggs. *Life Science Models, Volume 4* (H. Marcus–Roberts and M. Thompson, eds.), 350–366. New York: Springer-Verlag.

298. Sprott, D. A., and Farewell, V. T. (1993). The Difference Between Two Normal Means. *Amer. Statist.* **47** 126–128.

299. Staudte, R. G., and Sheather, S. J. (1990). *Robust Estimation and Testing.* New York: Wiley.

300. Stefanski, L. A. (1996). A Note on the Arithmetic–Geometric–Harmonic Mean Inequalities. *Amer. Statist.* **50** 246–247.

301. Stein, C. (1956). Inadmissibility of the Usual Estimator for the Mean of a Multivariate Normal Distribution. *Proceedings of the Third Berkeley Symposium on Mathematical Statistics and Probability* **1** 197–206. Berkeley: University of California Press.

302. Stein, C. (1964). Inadmissibility of the Usual Estimator for the Variance of a Normal Distribution with Unknown Mean. *Ann. Inst. Statist. Math.* **16** 155–160.

303. Stein, C. (1973). Estimation of the Mean of a Multivariate Distribution. *Proceedings of the Prague Symposium on Asymptotic Statistics.* Prague: Charles Univ. 345–381.

304. Stein, C. (1981). Estimation of the Mean of a Multivariate Normal Distribution. *Ann. Statist.* **9** 1135–1151.

305. Sterne, T. E. (1954). Some Remarks on Confidence or Fiducial Limits. *Biometrika* **41** 275–278.

306. Stigler, S. M. (1983). Who Discovered Bayes' Theorem? *Amer. Statist.* **37** 290–296.

307. Stigler, S. M. (1984). Kruskal's Proof of the Joint Distribution of $\bar{X}$ and $S^2$. *Amer. Statist.* **38** 134–135.

308. Stigler, S. M. (1986). *The History of Statistics: The Measurement of Uncertainty Before 1900.* Cambridge, MA: Harvard University Press.

309. Stuart, A., and Ord, J. K. (1987). *Kendall's Advanced Theory of Statistics, Volume I: Distribution Theory*, 5th edition. New York: Oxford University Press.

310. Stuart, A., and Ord, J. K. (1991). *Kendall's Advanced Theory of Statistics, Volume II*, 5th edition. New York: Oxford University Press.

311. Stuart, A., Ord, J. K., and Arnold, S. (1999). *Advanced Theory of Statistics, Volume 2A: Classical Inference and the Linear Model*, 6th edition. London: Oxford University Press.

312. Tanner, M. A. (1996). *Tools for Statistical Inference: Observed Data and Data Augmentation Methods*, 3rd edition. New York: Springer-Verlag.

313. Tate, R. F., and Klett, G. W. (1959). Optimal Confidence Intervals for the Variance of a Normal Distribution. *J. Amer. Statist. Assoc.* **54** 674–682.

314. Tierney, J. (1991). Behind Monte Hall's Door: Puzzle, Debate and Answer? *The New York Times,* July 21, 1991.

315. Tierney, L. (1994). Markov Chains for Exploring Posterior Distributions (with discussion). *Ann. Statist.* **22** 1701–1786.

316. Tsao, C. A., and Hwang, J. T. (1998). Improved Confidence Estimators for Fieller's Confidence Sets. *Can. J. Statist.* **26** 299–310.

317. Tsao, C. A., and Hwang, J. T. (1999). Generalized Bayes Confidence Estimators for Fieller's Confidence Sets. *Statistica Sinica* **9** 795–810.

318. Tukey, J. W. (1977). *Exploratory Data Analysis.* Reading, MA: Addison–Wesley.

319. Tweedie, M. C. K. (1957). Statistical Properties of Inverse Gaussian Distributions I. *Ann. Math. Statist.* **28** 362–377.

320. Vardeman, S. B. (1987). Discussion of the Articles by Casella and Berger and Berger and Sellke. *J. Amer. Statist. Assoc.* **82** 130–131.

321. Vardeman, S. B. (1992). What About Other Intervals? *Amer. Statist.* **46** 193–197; correction **47** 238.

322. vos Savant, M. (1990). Ask Marilyn. *Parade Magazine*, September 9, 15.

323. vos Savant, M. (1991). Letter to the Editor. *Amer. Statist.* **45** 347.

324. Wald, A. (1940). The Fitting of Straight Lines When Both Variables Are Subject to Error. *Ann. Math. Statist.* **11** 284–300.

325. Waller, L. A. (1995). Does the Characteristic Function Numerically Distinguish Distributions? *Amer. Statist.* **49** 150–151.

326. Waller, L. A., Turnbull, B. W., and Hardin, J. M. (1995). Obtaining Distribution Functions by Numerical Inversion of Characteristic Functions with Applications. *Amer. Statist.* **49** 346–350.

327. Wassermann, L. (1992). Recent Methodological Advances in Robust Bayesian Inference (with discussion). *Bayesian Statistics 4. Proceedings of the Fourth Valencia International Meeting*, 483–502. Oxford: Clarendon Press.

328. Westlake, W. J. (1981). Bioequivalence Testing – A Need to Rethink. *Biometrics* **37** 591–593.

329. Widder, D. V. (1946). *The Laplace Transform.* Princeton, NJ: Princeton University Press.

330. Wilks, S. S. (1938). Shortest Average Confidence Intervals from Large Samples. *Ann. Math. Statist.* **9** 166–175.

331. Williams, E. J. (1959). *Regression Analysis.* New York: Wiley.

332. Worsley, K. J. (1982). An Improved Bonferroni Inequality and Applications. *Biometrika* **69** 297–302.

333. Worsley, K. J. (1985). Bonferroni (Improved) Wins Again. *Amer. Statist.* **39** 235.

334. Wright, T. (1992). Lagrange's Identity Reveals Correlation Coefficient and Straight–Line Connection. *Amer. Statist.* **46** 106–107.

335. Wu, C. F. J. (1983). On the Convergence of the EM Algorithm. *Ann. Statist.* **11** 95–103.

336. Young, G. A. (1994). Bootstrap: More Than a Stab in the Dark? (with discussion). *Statist. Sci.* **9** 382–415.

337. Zehna, P. W. (1966). Invariance of Maximum Likelihood Estimators. *Ann. Math. Statist.* **37** 744.

338. Zellner, A. (1986). Bayesian Estimation and Prediction Using Asymmetric Loss Functions. *J. Amer. Statist. Assoc.* **81** 446–451.

# Author Index

# Subject Index